Games, Scales, and Suslin Cardinals: The Cabal Seminar, Volume I

The proceedings of the Los Angeles Caltech–UCLA "Cabal Seminar" were originally published in the 1970s and 1980s. *Games, Scales, and Suslin Cardinals* is the first of a series of four books collecting the seminal papers from the original volumes together with extensive unpublished material, new papers on related topics, and discussion of research developments since the publication of the original volumes.

Focusing on the subjects of "Games and Scales" (Part I) and "Suslin Cardinals, Partition Properties, Homogeneity" (Part II), each of the two sections is preceded by an introductory survey putting the papers into present context. This volume will be an invaluable reference for anyone interested in higher set theory.

Alexander S. Kechris is Professor of Mathematics at the California Institute of Technology. He is the recipient of numerous honors, including the J. S. Guggenheim Memorial Foundation Fellowship and the Carol Karp Prize of the Association for Symbolic Logic, and is a member of the Scientific Research Board of the American Institute of Mathematics.

Benedikt Löwe is Universitair Docent in Logic and Scientific Director of the Graduate Programme in Logic at the Institute for Logic, Language and Computation of the Universiteit van Amsterdam. He is an editor of the *Journal of Logic, Language and Information* and managing editor of *Tbilisi Mathematical Journal*. He is a board member of the DVMLG and the EACSL.

John R. Steel is Professor of Mathematics at the University of California, Berkeley. Prior to that, he was a professor in the mathematics department at UCLA. He is a recipient of the Carol Karp Prize of the Association for Symbolic Logic and of a Humboldt Prize. Steel is a former Fellow at the Wissenschaftskolleg zu Berlin and the Sloan Foundation.

LECTURE NOTES IN LOGIC

A Publication of
The Association for Symbolic Logic

This series serves researchers, teachers, and students in the field of symbolic logic, broadly interpreted. The aim of the series is to bring publications to the logic community with the least possible delay and to provide rapid dissemination of the latest research. Scientific quality is the overriding criterion by which submissions are evaluated.

See end of book for a list of the books in the series. More information can be found at http://www.aslonline.org/books-lnl.html.

LECTURE NOTES IN LOGIC 31

Games, Scales, and Suslin Cardinals: The Cabal Seminar, Volume I

Edited by

ALEXANDER S. KECHRIS
California Institute of Technology
BENEDIKT LÖWE
Universiteit van Amsterdam
JOHN R. STEEL
University of California, Berkeley

ASL
ASSOCIATION FOR SYMBOLIC LOGIC

CAMBRIDGE
UNIVERSITY PRESS

CAMBRIDGE
UNIVERSITY PRESS

University Printing House, Cambridge CB2 8BS, United Kingdom

One Liberty Plaza, 20th Floor, New York, NY 10006, USA

477 Williamstown Road, Port Melbourne, VIC 3207, Australia

314-321, 3rd Floor, Plot 3, Splendor Forum, Jasola District Centre, New Delhi - 110025, India

79 Anson Road, #06-04/06, Singapore 079906

Cambridge University Press is part of the University of Cambridge.

It furthers the University's mission by disseminating knowledge in the pursuit of education, learning and research at the highest international levels of excellence.

www.cambridge.org
Information on this title: www.cambridge.org/9780521899512

Association for Symbolic Logic
David Marker, Publisher
Department of Mathematics, Statistics, and Computer Science (M/C249)
University of Illinois at Chicago
851 S. Morgan St.
Chicago, IL 60607, USA
www.aslonline.org

First published 2008

A catalogue record for this publication is available from the British Library

Library of Congress Cataloging in Publication data

Games, scales, and suslin cardinals / edited by Alexander S. Kechris, Benedikt Löwe, John R. Steel.
p. cm. – (The Cabal seminar ; v. 1) (Lecture notes in logic ; 31)
Includes bibliographical references.
ISBN 978-0-521-89951-2 (hardback)
1. Game theory. 2. Set theory. 3. Measure theory. I. Kechris, A. S., 1946– II. Löwe, Benedikt. III. Steel, J. R. (John R.), 1948– IV. Title. V. Series.
QA269.G376 2008
519.3–dc22 2008019576

ISBN 978-0-521-89951-2 Hardback

CONTENTS

Preface ix

PART I: GAMES AND SCALES

John R. Steel
Games and scales. Introduction to Part I 3

Alexander S. Kechris and Yiannis N. Moschovakis
Notes on the theory of scales 28

Itay Neeman
Propagation of the scale property using games 75

John R. Steel
Scales on Σ_1^1-sets 90

Yiannis N. Moschovakis
Inductive scales on inductive sets 94

Yiannis N. Moschovakis
Scales on coinductive sets 102

Donald A. Martin and John R. Steel
The extent of scales in $\mathbf{L}(\mathbb{R})$ 110

Donald A. Martin
The largest countable this, that, and the other 121

John R. Steel
Scales in $\mathbf{L}(\mathbb{R})$ 130

John R. Steel
Scales in $\mathbf{K}(\mathbb{R})$ 176

Donald A. Martin
The real game quantifier propagates scales 209

John R. Steel
Long games 223

John R. Steel
The length-ω_1 open game quantifier propagates scales 260

PART II: SUSLIN CARDINALS, PARTITION PROPERTIES, HOMOGENEITY

Steve Jackson
Suslin cardinals, partition properties, homogeneity. Introduction to Part II 273

Alexander S. Kechris
Suslin cardinals, κ-Suslin sets, and the scale property in the hyperprojective hierarchy 314

Alexander S. Kechris, Eugene M. Kleinberg, Yiannis N. Moschovakis, and W. Hugh Woodin
The axiom of determinacy, strong partition properties, and nonsingular measures 333

Alexander S. Kechris and W. Hugh Woodin
The equivalence of partition properties and determinacy 355

Alexander S. Kechris and W. Hugh Woodin
Generic codes for uncountable ordinals, partition properties, and elementary embeddings 379

Alexander S. Kechris
A coding theorem for measures 398

Donald A. Martin and John R. Steel
The tree of a Moschovakis scale is homogeneous 404

Donald A. Martin and W. Hugh Woodin
Weakly homogeneous trees 421

Bibliography 439

PREFACE

This is the first of four volumes containing reprints of the papers in the original Cabal Seminar volumes of the Springer Lecture Notes in Mathematics series [CABAL i, CABAL ii, CABAL iii, CABAL iv], unpublished material, and new papers. We have grouped the papers of the original Cabal Seminar volumes according to their topics. This volume contains the papers on "Games and Scales" (Part I) and "Suslin Cardinals, Partition Properties, Homogeneity" (Part II). Each of the parts contains an introductory survey (written by John Steel and Steve Jackson, respectively) putting the papers into a present-day context. Table 1 gives an overview of the papers in this volume with their original references.

This volume must not be understood as a historical edition of old papers. In the 1980s, there were a number of results obtained by the researchers associated with the Cabal Seminar, some of which were intended for a fifth Cabal volume that was never published. We include some of these papers in this volume, together with papers reporting on new developments related to the research of the Cabal Seminar. These papers are Steel's "Scales in $\mathbf{K}(\mathbb{R})$" and "The Length-ω_1 Open Game Quantifier Propagates Scales" in Part Iand "The Equivalence of Partition Properties and Determinacy" and "Generic Codes for Uncountable Ordinals, Partition Properties, and Elementary Embeddings" by Kechris and Woodin, "The Tree of a Moschovakis Scale is Homogeneous" by Martin and Steel, and "Weakly Homogeneous Trees" by Martin and Woodin in Part II. We also added a new expository paper, "Propagation of the Scale Property Using Games" by Neeman, and modernized and made uniform the notation and layout, and we have given the authors the opportunity to make corrections to their original papers. The new LaTeX layout has resulted in changes in the numbering of sections and theorems.

The typing and design were partially funded by NSF Grant DMS-0100745; Johan van Benthem's Spinoza project Logic in Action; the Institute for Logic, Language and Computation; the Association for Symbolic Logic; and the DFG-NWO collaborative project "Determinacy and Combinatorics" (DFG: KO 1353/3-1; NWO: DN 61-532). A lot of people were involved in typing,

	PART I	
Steel	*Games and scales* *Introduction to Part I*	NEW
Kechris, Moschovakis	*Notes on the theory of scales*	[CABAL i, pp. 1–53]
Neeman	*Propagation of the scale property using games*	NEW
Steel	*Scales on Σ^1_1-sets*	[CABAL iii, pp. 72–76]
Moschovakis	*Inductive scales on inductive sets*	[CABAL i, pp. 185–192]
Moschovakis	*Scales on coinductive sets*	[CABAL iii, pp. 77–85]
Martin, Steel	*The extent of scales in* $\mathbf{L}(\mathbb{R})$	[CABAL iii, pp. 86–96]
Martin	*The largest countable this, that, and the other*	[CABAL iii, pp. 97–106]
Steel	*Scales in* $\mathbf{L}(\mathbb{R})$	[CABAL iii, pp. 107–156]
Steel	*Scales in* $\mathbf{K}(\mathbb{R})$	NEW
Martin	*The real game quantifier propagates scales*	[CABAL iii, pp. 157–171]
Steel	*Long games*	[CABAL iv, pp. 56–97]
Steel	*The length-ω_1 open game quantifier propagates scales*	NEW
	PART II	
Jackson	*Suslin cardinals, partition properties, homogeneity* *Introduction to Part II*	NEW
Kechris	*Suslin cardinals, κ-Suslin sets, and the scale property in the hyperprojective hierarchy*	[CABAL ii, pp. 127–146]
Kechris, Kleinberg, Moschovakis, Woodin	*The axiom of determinacy, strong partition properties, and nonsingular measures*	[CABAL ii, pp. 75–100]
Kechris, Woodin	*The equivalence of partition properties and determinacy*	NEW
Kechris, Woodin	*Generic codes for uncountable ordinals, partition properties, and elementary embeddings*	NEW
Kechris	*A coding theorem for measures*	[CABAL iv, pp. 103–109]
Martin, Steel	*The tree of a Moschovakis scale is homogeneous*	NEW
Martin, Woodin	*Weakly homogeneous trees*	NEW

TABLE 1.

laying out, and proofreading the papers. We should like to thank (in alphabetic order) Edgar Andrade, Stefan Bold, Samson de Jager, Leona Kershaw, Tomasz Polacik, Dorothée Reuther, and Philipp Rohde for their important contribution. Very special thanks are due to Samson de Jager, who coordinated the typesetting effort in the final two years of the project.

REFERENCES

Alexander S. Kechris, Donald A. Martin, and Yiannis N. Moschovakis
[Cabal ii] *Cabal seminar* 77–79, Lecture Notes in Mathematics, no. 839, Berlin, Springer, 1981.
[Cabal iii] *Cabal seminar* 79–81, Lecture Notes in Mathematics, no. 1019, Berlin, Springer, 1983.

Alexander S. Kechris, Donald A. Martin, and John R. Steel
[Cabal iv] *Cabal seminar* 81–85, Lecture Notes in Mathematics, no. 1333, Berlin, Springer, 1988.

Alexander S. Kechris and Yiannis N. Moschovakis
[Cabal i] *Cabal seminar* 76–77, Lecture Notes in Mathematics, no. 689, Berlin, Springer, 1978.

The Editors
Alexander S. Kechris, *Pasadena, CA*
Benedikt Löwe, *Amsterdam*
John R. Steel, *Berkeley, CA*

PART I: GAMES AND SCALES

GAMES AND SCALES
INTRODUCTION TO PART I

JOHN R. STEEL

The construction and use of *Suslin representations* for sets of reals lies at the heart of descriptive set theory. Indeed, virtually every paper in descriptive set theory in the Cabal Seminar volumes deals with such representations in one way or another. Most of the papers in the section to follow focus on the construction of optimally definable Suslin representations via game-theoretic methods. In this introduction, we shall attempt to put those papers in a broader historical and mathematical context. We shall also give a short synopsis of the papers themselves, and describe some of the work done later to which they are related.

§1. **Some definitions and history.** A **tree** on a set X is a subset of $X^{<\omega}$ closed under initial segments. If T is a tree on $X \times Y$, then we regard T as a set of pairs (s, t) of sequences with $\mathrm{dom}(s) = \mathrm{dom}(t)$. If T is a tree, we use $[T]$ for the set of infinite branches of T, and if T is on $X \times Y$, we write

$$\mathrm{p}[T] = \{x \in {}^{\omega}X : \exists y \in {}^{\omega}Y \forall n < \omega((x \restriction n, y \restriction n) \in T)\}.$$

We call $\mathrm{p}[T]$ the **projection** of T, and say that T is a **Suslin representation** of $\mathrm{p}[T]$, or that $\mathrm{p}[T]$ is Y**-Suslin** via T. For $s \in X^{<\omega}$, let $T_s = \{u : (s, u) \in T\}$, and put $T_x = \bigcup_n T_{x \restriction n}$. Then $x \in \mathrm{p}[T]$ iff $[T_x] \neq \varnothing$ iff T_x is illfounded.

Any set $A \subseteq {}^{\omega}X$ is trivially A-Suslin. For the most part, useful Suslin representations come from trees on some $X \times Y$ such that Y is wellordered. Assuming (as we do) the Axiom of Choice (AC), this is no restriction on Y, but we can parlay it into an important and useful restriction by requiring in addition that T be *definable* in some way or other. A variant of this approach is to require that T belong to a model of AD. If T is definable, and X and Y are definably wellordered, and $\mathrm{p}[T]$ is nonempty, then the leftmost branch (x, f) of T gives us a definable element x of $\mathrm{p}[T]$. (Here "leftmost" can be determined by the lexicographic order on $X \times Y$.)

The simplest nontrivial X to consider are the countable ones. This is by far the most well-studied case in the Cabal volumes. In this case, one may regard $\mathrm{p}[T]$ as a subset of the Baire space ${}^{\omega}\omega$, that is, as a set of "logician's reals".

The Cabal Seminar. Volume I: Games, Scales and Suslin Cardinals
Edited by A. S. Kechris, B. Löwe, J. R. Steel
Lecture Notes in Logic, 31

Thus if A is a nonempty set of reals, κ is an ordinal, and A is κ-Suslin via a definable tree, then A has a definable element.

Suslin representations were first discovered in 1917 by Suslin [Sus17], who isolated the class of ω-Suslin sets of reals, showed that it properly includes the Borel sets, and showed that sets in this class have various regularity properties. (For example, they are all Baire and Lebesgue measurable, and have the perfect set property.) Suslin also found a beautiful characterization of the Borel sets of reals as those which are both ω-Suslin and have ω-Suslin complements. (The ω-Suslin sets of reals are precisely the $\undertilde{\Sigma}^1_1$ sets of reals, almost by definition.)

Definable Suslin representations yield definable elements, and in the "boldface" setting of classical descriptive set theory, this comes out as a uniformization result. Here we say that a function f **uniformizes** a relation R iff $\mathrm{dom}(f) = \{x : \exists y R(x, y)\}$, and $\forall x \in \mathrm{dom}(f) R(x, f(x))$. If R is a $\undertilde{\Sigma}^1_1$ relation, say $R = \mathrm{p}[T]$ where T is a tree on $(\omega \times \omega) \times \omega$, then we can use leftmost branches to uniformize R: let $f(x) = y$, where (y, h) is the leftmost branch of T_x. One can calculate that for any open set U, $f^{-1}(U)$ is in the σ-algebra generated by the $\undertilde{\Sigma}^1_1$ sets, and is therefore Lebesgue and Baire measurable. This classical uniformization result was proved by Jankov and von Neumann around 1940 [vN49]. The "lightface", effective refinement of a uniformization theorem is a *basis theorem*, where we say a pointclass Λ is a basis for a pointclass Γ just in case every nonempty set of reals in Γ has a member which is in Λ. Kleene [Kle55] proved the lightface version of the Jankov-von Neumann result. He observed that if $A \subseteq \omega^\omega$ is lightface Σ^1_1, then $A = \mathrm{p}[T]$ for some recursive tree T, and that the leftmost branch of T is recursive in the set W of all Gödel numbers of wellfounded trees on ω. Thus $\{x : x \leq_T W\}$ is a basis for Σ^1_1.

In 1935–38, toward the end of the classical period, Novikoff and Kondô constructed definable, ω_1-Suslin representations for arbitrary $\undertilde{\Sigma}^1_2$ sets, and used them to show every $\undertilde{\Sigma}^1_2$ relation has a $\undertilde{\Sigma}^1_2$ uniformization. (See [LN35, Kon38].) The effective refinement of this landmark theorem is due to Addison, who showed that the ω_1-Suslin reprentations of nonempty lightface Σ^1_2 sets constructed by Novikoff and Kondô yield, via leftmost branches, lightface Δ^1_2 elements for such sets.

Logicians often meet Suslin representations through the Shoenfield Absoluteness theorem. Shoenfield [Sch61] showed that a certain tree T on $\omega \times \omega_1$ which comes from the Novikoff-Kondô construction is in $\mathbf{L}$. Because wellfoundedness is absolute to transitive models of ZF, he was able to conclude that the leftmost branch of T is in $\mathbf{L}$, and thus, that every nonempty Σ^1_2 set of reals has an element in $\mathbf{L}$. From this it follows easily that $\mathbf{L}$ is Σ^1_2 correct. This method of using definable Suslin representations to obtain correctness and absoluteness results for models of set theory is very important.

In addition to definability, there is a second very useful property a Suslin representation might have. We call a tree T on $X \times Y$ **homogeneous** just in case there is a family $\langle \mu_s : s \in X^{<\omega} \rangle$ such that

(1) for all s, μ_s is a countably complete 2-valued measure (i.e. ultrafilter) on $\{u : (s,u) \in T\}$,
(2) if $s \subseteq t$, and $\mu_s(A) = 1$, then $\mu_t(\{u : u \upharpoonright \mathrm{dom}(s) \in A\}) = 1$, and
(3) for any $x \in \mathrm{p}[T]$ and any $\langle A_i : i < \omega\rangle$ such that $\mu_{x\upharpoonright i}(A_i) = 1$ for all i, there is a $f \in Y^\omega$ such that $f \upharpoonright i \in A_i$ for all i.

We say T is **κ-homogeneous** if the measures μ_s can be taken to be κ-additive. If T is κ-homogeneous, then we also call $\mathrm{p}[T]$ a κ-homogeneously Suslin set. We write Hom_κ for the pointclass of κ-homogeneous sets, and Hom_∞ for the pointclass $\bigcap_\kappa \mathrm{Hom}_\kappa$.

The concept of homogeneity is implicit in Martin's 1968 proof [Mar70A] of Π^1_1 determinacy, and was first explicitly isolated by Martin and Kechris. Martin showed that if κ is a measurable cardinal, then every Π^1_1 set of reals is κ homogeneous, via a Shoenfield tree on $\omega \times \kappa$. He also showed that every homogeneously Suslin set of reals is determined. Martin's proof became the template for all later proofs of definable determinacy from large cardinal hypotheses. Indeed, the standard characterization of descriptive set theory, as the study of the good behavior of definable sets of reals, would perhaps be more accurate if one replaced "definable" by "∞-homogeneously Suslin".

There are two natural weakenings of homogeneity. First, a tree T on $X \times (\omega \times Y)$ is **κ-weakly homogeneous** just in case it is κ- homogeneous when viewed as a tree on $(X \times \omega) \times Y$. Thus the weakly homogeneous subsets of ${}^\omega X$ are just the existential real quantifications of a homogeneous subsets of ${}^\omega X \times {}^\omega\omega$, and Martin's [Mar70A] shows in effect that whenever κ is measurable, all $\underset{\sim}{\Sigma}^1_2$ sets of reals are κ-weakly homogeneous. Second, a pair of trees S and T, on $X \times Y$ and $X \times Z$ respectively, are **κ-absolute complements** iff

$$\mathbf{V}[G] \models \mathrm{p}[S] = {}^\omega X \setminus \mathrm{p}[T]$$

whenever G is $\mathbf{V}$-generic for a poset of cardinality $< \kappa$. The fundamental Martin-Solovay construction, also from 1968 (see [MS69]), shows that every κ-weakly homogeneous tree has a κ-absolute complement. The projection of a κ-absolutely complemented tree is said to be **κ-universally Baire**. This concept was first explicitly isolated and studied by Feng, Magidor, and Woodin in [FMW92]. Any universally Baire set has the Baire property and is Lebesgue measurable, but one cannot show in ZFC alone that such sets must be determined. (See [FMW92].) On the other hand, if there are arbitrarily large Woodin cardinals, then for any set of reals A, A is κ-homogeneous for all κ iff A is κ-weakly homogeneous for all κ iff A is κ-universally Baire for all κ. (This is work of Martin, Solovay, Steel, and Woodin; see [Lar04, Theorem 3.3.13] for one exposition, and [SteA] for another.)

Although our discussion of homogeneity has focussed on its use in situations where the Axiom of Choice and the existence of large cardinals is assumed, the concept is also quite important in contexts in which full AD is assumed. AD gives us not just measures, but homogeneity measures; indeed, assuming AD, a set of reals is homogeneously Suslin iff both it and its complement are Θ-Suslin. (This result of Martin from the 80's can be found in [MS89].) The analysis of homogeneity measures is a central theme in the work of Kunen, Martin, and Jackson [Sol78A, Jac88, Jac99] which located the projective ordinals among the alephs. The reader should see Jackson's surveys [Jac07A] and [Jac07B] for more on homogeneity and the projective ordinals in the AD context.

§2. Construction methods. One could group the methods for producing useful Suslin representations as follows:

(1) the Martin-Solovay construction,
(2) trees to produce an elementary submodel, and
(3) scale constructions using comparison games.

We discuss these methods briefly:

2.1. The Martin-Solovay construction. The Martin-Solovay construction makes use of homogeneity. If T on $X \times Y$ is is κ- weakly homogeneous via the system of measures $\vec{\mu}$, and $|X| < \kappa$, then the construction produces a tree $\mathrm{ms}(T, \vec{\mu})$ which is a κ-absolute complement for T. The construction of $\mathrm{ms}(T, \vec{\mu})$ is effective, and its basic properties can be proved to hold in $\mathsf{ZF} + \mathsf{DC}$. Martin and Solovay [MS69] applied it with T the Shoenfield tree for Σ^1_2 and $\vec{\mu}$ its weak homogeneity measures implicit in Martin's [Mar70A]. They showed thereby that if κ is measurable, then for any Σ^1_3 formula φ, there is a tree U such that $p[U] = \{x \in \omega^\omega : \varphi(x)\}$ is true in every generic extension of $\mathbf{V}$ by a poset of size $< \kappa$.

The Martin-Solovay tree $\mathrm{ms}(T, \vec{\mu})$ is definable from T and $\vec{\mu}$. Now suppose T be on $\omega \times Y$. There is a simple variant of $\mathrm{ms}(T, \vec{\mu})$ which is definable from T and the restrictions of the measures in $\vec{\mu}$ to $\bigcup\{\mathbf{L}[T, x] : x \in {}^\omega\omega\}$. Let us call this variant $\mathrm{ms}^*(T, \vec{\mu})$. If T is the Shoenfield tree, so that $T \in \mathbf{L}$, then one can define these restricted weak homogeneity measures, and hence $\mathrm{ms}^*(T, \vec{\mu})$ itself, from the sharp function on the reals. Martin and Solovay showed this way that Δ^1_4 is a basis for Π^1_2, and Mansfield later improved their result by showing the class of Π^1_3 singletons is a basis for Π^1_2. (See [Man71].) These results are not optimal, however. We do not know whether one can get the optimal basis and uniformization results in the projective hierarchy using the Martin-Solovay construction.

Under appropriate large cardinal hypotheses, the Martin-Solovay tree is itself homogeneous. (See [MS07] for a precise statement.) Thus under the appropriate large cardinal hypotheses, one can show via the Martin-Solovay construction that the pointclass Hom_∞ is closed under complements and real quantification.

2.2. The tree to produce an elementary submodel. If a set A of reals admits a definition with certain condensation and generic absoluteness properties, then A is universally Baire. More precisely, let κ be a cardinal, and $\varphi(v_0, v_1)$ a formula in the language of set theory, and t any set. Let $\tau > \kappa$, $X \prec \mathbf{V}_\tau$ be countable, and let M be the transitive collapse of X, with $\bar{\kappa}$ and $\bar{t}$ the images of κ and t under collapse. We say X is **generically $\langle\varphi, A\rangle$-correct** iff whenever g is M-generic for a poset of size $< \bar{\kappa}$ in M, then for all reals $y \in M[g]$,

$$y \in A \Leftrightarrow M[g] \models \varphi[y, \bar{t}].$$

If the set of generically $\langle\varphi, A\rangle$ correct X is club in $\wp_{\omega_1}(\mathbf{V}_\tau)$, then A admits a κ-absolutely complemented Suslin representation T. The construction of T is relatively straightforward: if $(y, f) \in [T]$, then f will have built an X in our club of generically correct hulls, together with a proof that $M[g] \models \varphi[y, \bar{t}]$, for some g generic over the collapse M of X. (We are not certain as to the origin of this construction. Woodin made early use of it. See [FMW92] or [SteA].)

One can use either stationary tower forcing (cf. the *Tree Production Lemma*, [Lar04] or [SteA]) or iterations to make reals generic [Ste07B, § 7] to obtain, for various interesting $\langle\varphi, A\rangle$, a club of generically $\langle\varphi, A\rangle$-correct X.

If one replaces $\mathbf{V}_\tau$ by an appropriate direct limit of mice, then the tree to produce an elementary submodel becomes definable, at a level corresponding to the definability of the iteration strategies for the mice in question. See the concluding paragraphs of [Ste95A], and [Ste07B, § 8]. One can use this to get optimal basis and uniformization results for various pointclasses, for example $(\Sigma^2_1)^{\mathrm{L}(\mathbb{R})}$. It is difficult to obtain the optimal basis and uniformization results for Π^1_3 by these methods, but, building on work of Hugh Woodin, Itay Neeman has succeeded in doing so. (This work is unpublished.)

2.3. Propagation of scales using comparison games. The simplest method for obtaining optimally definable Suslin representations makes direct use of the determinacy of certain infinite games. It was discovered in 1971 by Moschovakis, who used it to extend the Novikoff-Kondô-Addison theorems to the higher levels of the projective hierarchy. (The original paper is [Mos71A]; see also [KM78B] and [Mos80, Chapter 6].) As part of this work, Moschovakis introduced the basic notion of a *scale*, which we now describe.

Let T be a tree on $\omega \times \lambda$, and $A = \mathrm{p}[T]$. One can get a "small" subtree of T which still projects to A by considering only ordinals $< \lambda$ which appear in some leftmost branch. The *scale of T* does this, then records the resulting subtree as a sequence of **norms**, i.e. ordinal-valued functions, on A. More precisely, for $x \in A$ and $n < \omega$, put

$$\varphi_n(x) = |\langle l_x(0), ..., l_x(n)\rangle|_{\mathrm{lex}},$$

where for $u \in \lambda^{n+1}$, $|u|_{\mathrm{lex}}$ is the ordinal rank of u in the lexicographic order on λ^n. Then

$$\vec{\varphi} = \langle\varphi_n : n < \omega\rangle$$

is the scale of T. It has the properties:

(a) Suppose that $x_i \in A$ for all $i < \omega$, and $x_i \to x$ as $i \to \infty$, and for all n, $\varphi_n(x_i)$ is eventually constant as $n \to \infty$, then
   (i) (limit property) $x \in A$, and
   (ii) (lower semi-continuity) for all n, $\varphi_n(x) \leq$ the eventual value of $\varphi_n(x_i)$ as $i \to \infty$.

(b) (refinement property) if $x, y \in A$ and $\varphi_n(x) < \varphi_n(y)$, then $\varphi_m(x) < \varphi_m(y)$ for all $m > n$.

A sequence of norms on A with property (a) is called a **scale on** A. Any scale on A can be easily transformed into a scale on A with the refinement property. If $\vec{\varphi}$ is a scale on A, then we define the tree of $\vec{\varphi}$ to be

$$T_{\vec{\varphi}} = \{(\langle x(0), ..., x(n-1)\rangle, \langle \varphi_0(x), ..., \varphi_{n-1}(x)\rangle) : n < \omega \text{ and } x \in A\}.$$

It is not hard to see that $\mathrm{p}[T_{\vec{\varphi}}] = A$. If $\vec{\varphi}$ has the refinement property, and $\vec{\psi}$ is the scale of $T_{\vec{\varphi}}$, then $\vec{\psi}$ is equivalent to $\vec{\varphi}$, in the sense that for all n, x and y, $\psi_n(x) \leq \psi_n(y)$ iff $\varphi_n(x) \leq \varphi_n(y)$. The reader should see [KM78B, 6B] and [Jac07B, § 2] for more on the relationship between scales and Suslin representations.

There are least two benefits to considering the scale of a tree: first, it becomes easier to state and prove optimal definability results, and second, the construction of Suslin representations using comparison games becomes clearer. Concerning definability, we have

Definition 2.1. Let Γ be a pointclass, and $\vec{\varphi}$ a scale on A, where $A \in \Gamma$; then we call $\vec{\varphi}$ a **Γ-scale on** A just in case the relations

$$R(n, x, y) \Leftrightarrow x \in A \wedge (y \notin A \vee \varphi_n(x) \leq \varphi_n(y)),$$

and

$$S(n, x, y) \Leftrightarrow x \in A \wedge (y \notin A \vee \varphi_n(x) < \varphi(y))$$

are each in Γ. We say Γ **has the scale property** just in case every set in Γ admits a Γ-scale, and write Scale(Γ) in this case.

Moschovakis showed that if Γ is a pointclass which is closed under universal real quantification, has other mild closure properties, and has the scale property, then every Γ relation has a Γ uniformization, and the Γ singletons are a basis for Γ. [KM78B, 3A-1]. He also showed that assuming $\underset{\sim}{\Delta}^1_{2n}$ determinacy, both Π^1_{2n+1} and Σ^1_{2n+2} have the scale property [KM78B, 3B, 3C]. From this, one gets the natural generalization of Novikoff-Kondô-Addison to the higher levels of the projective hierachy.

Moschovakis' construction of scales goes by *propagating* them from a set A to a set B obtained from A via certain logical operations. One starts with the fact that Σ^0_1 has the scale property, and uses these propagation theorems to obtain definable scales on more complicated sets. The propagation works at the level of the individual norms in the scales.

For example, if φ is a norm of A, where $A \subseteq X \times {}^{\omega}Y$, and

$$B(y) \Leftrightarrow \exists x A(x, y),$$

then we obtain the "inf" norm on B by setting

$$\psi(y) = \inf\{\varphi(x, y) : A(x, y)\}.$$

If either X is an ordinal, or $X = {}^{\omega}\omega$, then inf norms can be used to transform a scale on A into a scale on B. (See [KM78B, 3B-2].) This transformation has a simple meaning in terms of the tree of the scale; if $X = {}^{\omega}\omega$, it corresponds to regarding a tree on $(Y \times \omega) \times \kappa$ as a tree on $Y \times (\omega \times \kappa)$.

Definable scales do not propagate under negation or universal quantification over ordinals. (Otherwise, it would be possible to assign to each countable ordinal α a scale on the set of wellorders of ω of order type α, in a definable way. This would then yield a definable function picking a codes for the countable ordinals.) Moschovakis' main advance in [Mos71A] was to show that universal quantification over the reals propagates definable scales. Here it is definitely important to work with scales, rather than their associated trees. As before, the propagation takes place at the level of individual norms. Let φ be a norm on A, where $A \subseteq \mathbb{R} \times Y$, and let

$$B(y) \Leftrightarrow \forall x A(x, y).$$

To each $y \in B$, we associate $f_y \colon \mathbb{R} \to \mathrm{OR}$, where

$$f_y(x) = \varphi(x, y).$$

Our norm on B records an ordinal measure of the growth rate of f_y. Namely, given $f, g \colon \mathbb{R} \to \mathrm{OR}$, we let $G(f, g)$ be the game on ω: I plays out x_0, II plays out x_1, the players alternating moves as usual. Player II wins iff $f(x_0) \leq g(x_1)$. (Thus a winning strategy for II witnesses that g grows at least as fast as f, in an effective way.) Now put

$$f \leq^* g \Leftrightarrow \text{II has a winning strategy in } G(f, g).$$

Granted full AD, one can show $\leq^*$ is a prewellorder of all the ordinal-valued functions on $\mathbb{R}$, and granted only determinacy for sets simply definable from φ, one can show that $\leq^*$ prewellorders the f_y for $y \in B$. Our norm on B is then given by

$$\psi(y) = \text{ ordinal rank of } f_y \text{ in } \leq^* \restriction \{f_z : B(z)\}.$$

(See [KM78B, 2C-1].) The norm ψ is generally called the "fake sup" norm obtained from φ; the ordinal $\psi(y)$ measures how difficult it is to verify $A(x, y)$ at arbitrary x.

The fake-sup construction was first used in [AM68], to propagate the prewellordering property, which involves only one norm. Granted enough determinacy, the construction can be used to transform a scale on A into a scale on B, where $B(y) \Leftrightarrow \forall x A(x, y)$. The key additional idea is to record, for each basic neighborhood N_s, the ordinal rank of $f_y \restriction N_s$ in $\leq^* \restriction \{f_z \restriction N_s : B(z)\}$. See [KM78B, 3C-1].

Using more sophisticated comparison games, one can combine the techniques for propagating scales under universal and existential real quantification, as well as existential ordinal quantification. This leads to the propagation of scales under various *game quantifiers*. We shall discuss these results in more detail in the next section.

Although the fake-sup method of propagating scales was invented in order to obtain optimally definable scales, one can show that under AD, the tree of the scale it produces is very often homogeneous. (The tree of any scale is the surjective image of $\mathbb{R}$, so it is too small to be homogeneous in $\mathbf{V}$.) See [MS07], where it is also shown that the tree very often has the "generic codes" property of [KW07].

§3. **Individual papers.** We pass to an extended table of contents for the papers in the block to follow, together with pointers to some related results and literature. We also include a number of proof sketches. Some of these sketches will only make sense to readers with significant background knowledge. We have included references to fuller explanations in the literature when possible.

Notes on the theory of scales [KM78B].

This is a survey paper, written in 1971. It is still an excellent starting point for anyone seeking basic information regarding the construction and use of scales under determinacy hypotheses. It is truly remarkable how much of the descriptive set theory that is founded on large cardinals and determinacy emerged in the early years of the subject.

The paper begins in §2 – §4 with the inf and fake-sup constructions, and their corollaries regarding the scale property and uniformization in the projective hierarchy.

THEOREM 3.1 (Moschovakis 1970). Assume all $\underset{\sim}{\mathbf{\Delta}}^1_{2n}$ games are determined; then

(1) Π^1_{2n+1} and Σ^1_{2n+2} have the scale property, and hence
(2) every Π^1_{2n+1} (respectively Σ^1_{2n+2}) relation on $\mathbb{R}$ can be uniformized by a Π^1_{2n+1} (respectively Σ^1_{2n+2}) function.

In §6, the **projective ordinals**

$$\underset{\sim}{\delta}^1_n := \sup\{\alpha : \alpha \text{ is the order type of a } \underset{\sim}{\mathbf{\Delta}}^1_n \text{ prewellorder of } \mathbb{R}\}$$

are introduced. One can show that, assuming PD, any Π^1_{2n+1}-norm on a complete Π^1_{2n+1} set has length $\underset{\sim}{\delta}^1_{2n+1}$; see [Mos80, 4C.14]. From the scale property for Π^1_{2n+1} one then gets that all Π^1_{2n+1} sets are $\underset{\sim}{\delta}^1_{2n+1}$-Suslin, and thence that all Σ^1_{2n+2} sets are $\underset{\sim}{\delta}^1_{2n+1}$-Suslin. (For $n = 0$, this reduces to the classical

Novikoff-Kondô result that all Σ^1_2 sets are ω_1-Suslin.) The size of the projective ordinals, both in inner models of AD, and in the full universe $\mathbf{V}$, is therefore a very important topic. It is a classical result that $\underset{\sim}{\delta}^1_1 = \omega_1$, while the size of the larger projective ordinals has been the subject of much later work, some of which will be collected in a block of papers in a later volume in this series.

§7 proves the Kunen-Martin theorem:

THEOREM 3.2 (Kunen, Martin). Every κ-Suslin wellfounded relation on $\mathbb{R}$ has rank $< \kappa^+$.

This basic result has important corollaries concerning the sizes of the projective ordinals. For example, because all Σ^1_{2n+2} sets are $\underset{\sim}{\delta}^1_{2n+1}$-Suslin, we have that $\underset{\sim}{\delta}^1_{2n+2} \leq (\underset{\sim}{\delta}^1_{2n+1})^+$, and in particular, $\underset{\sim}{\delta}^1_2 \leq \omega_2$.

§8 investigates the way in which Suslin representations yield ∞-Borel representations. It is shown that κ-Suslin sets are κ^{++}-Borel (i.e. can be built up from open sets using complementation and wellordered unions of length $< \kappa^{++}$). Of course, if CH holds, then *every* set of reals is a union of ω_1 singletons; the true content of the result of §8 lies in the fact that the κ^{++}-Borel representation is definable from the κ-Suslin representation. §8 also shows that, assuming PD, every $\underset{\sim}{\Delta}^1_{2n+1}$ set is $\underset{\sim}{\delta}^1_{2n+1}$-Borel. If $n = 0$, this is just Suslin's original theorem. In order to obtain a converse when $n > 0$, we must impose a definability restriction on our $\underset{\sim}{\delta}^1_{2n+1}$-Borel representation, since again, it could be that every set of reals is $\omega_1 + 1$-Borel. One way to do that is to assume full AD, and Martin showed that indeed, assuming AD, every $\underset{\sim}{\Delta}^1_{2n+1}$ set is $\underset{\sim}{\delta}^1_{2n+1}$-Borel. So we have

THEOREM 3.3 (Martin, Moschovakis). Assume AD; then the $\underset{\sim}{\Delta}^1_{2n+1}$ sets of reals are precisely the $\underset{\sim}{\delta}^1_{2n+1}$-Borel sets.

See [Mos80, 7D.9]. This fully generalizes Suslin's 1917 theorem to the higher levels of the projective hierarchy.

§5 and §9 introduce inner models, obtained from Suslin representations, which have certain degrees of correctness. In §5, it is shown that for $n \geq 2$, there is a unique, minimal Σ^1_n-correct inner model M^*_n containing all the ordinals; the model is obtained by closing under constructibility and an optimally definable Skolem function for Σ^1_n. (Kechris and Moschovakis call this model M_n—not to be confused with $\mathbf{M}_n$; see below.) §9 considers the model $\mathbf{L}[T]$, where T is the tree of a Π^1_{2n+1} scale on a complete Π^1_{2n+1} set. These models have proved more important in later work than the M^*_n. It is shown that if $n = 0$, then $\mathbf{L}[T] = \mathbf{L}$; in particular, $\mathbf{L}[T]$ is independent of the Π^1_{2n+1} scale and complete set chosen. Moschovakis conjectured that $\mathbf{L}[T]$ is independent of these choices if $n > 0$ as well, and more vaguely, that it is a "correct higher level analog of $\mathbf{L}$".

Becker's paper [Bec78] contains an excellent summary of what was known in 1977 about the models of §5 and §9. The independence conjecture, which inspired a great deal of work, became the third Victoria Delfino problem. Harrington and Kechris [HK81] made a significant advance by showing that the reals of $\mathbf{L}[T]$, where T is the tree of any Π^1_{2n+1} scale on a complete Π^1_{2n+1} set, are the largest countable Σ^1_{2n+2} set of reals, and hence independent of the choice of T. Building on this work, Moschovakis made a step forward in the late 70's with the introduction of the model H_Γ, for Γ a pointclass which *resembles* Π^1_1 in a certain technical sense, and has the scale property. (See [Mos80, 8G.17 ff.].) Assuming $\underset{\sim}{\mathbf{\Delta}}^1_{2n}$-determinacy, the pointclass Π^1_{2n+1} is an example of such a Γ, but there are many more examples. The model H_Γ is of the form $\mathbf{L}[U]$, where U is a universal $\exists^{\mathbb{R}}\Gamma$ (in the codes) subset of the prewellordering ordinal of Γ, and one can think of it as a fragment of **HOD** corresponding to Γ-definability. Using the Harrington-Kechris work, Moschovakis showed that H_Γ is independent of the universal set and Γ-norm used to define U, that it includes $\mathbf{L}[T]$, for the tree T of a Γ scale on a complete Γ set, and that $\mathbb{R} \cap H_\Gamma$ is the largest countable $\exists^{\mathbb{R}}\Gamma$ set of reals. (See [Mos80, 8G.17 ff.].) Moschovakis' results require a bit more than $\underset{\sim}{\Gamma}$-determinacy.

The independence of $\mathbf{L}[T]$ was finally proved by Becker and Kechris [BK84], who showed

THEOREM 3.4 (Becker, Kechris 1984). Let Γ be a pointclass which resembles Π^1_1 and has the scale property, and suppose AD holds in $\mathbf{L}(\underset{\sim}{\Gamma}, \mathbb{R})$. Let T be the tree of any Γ-scale on a complete Γ set; then $\mathbf{L}[T] = H_\Gamma$.

The Becker-Kechris proof makes heavy use of a class of games introduced by Martin in order to obtain an approximation to Theorem 3.4.

Not long after the last of the Cabal Seminar volumes appeared, our understanding of the large cardinal side of the "equivalence" between large cardinals and determinacy caught up with our understanding of the determinacy side. This equivalence is mediated by the canonical inner models for large cardinal hypotheses, which are sometimes called *extender models*. We can now identify each of the models of §5 and §9 as an extender model, and thereby understand it much more deeply than we could using only pure descriptive set theory. For example, most nontrivial facts in the first order theory of $\mathbf{L}[T]$ (e.g., that the GCH, and Jensen's diamond and square principles, hold in $\mathbf{L}[T]$) seem to require its identification as an extender model for proof. The identifications are as follows: Here and in the rest of the paper, for $0 \le n \le \omega$, we let $\mathbf{M}_n$ be the minimal iterable proper class extender model with n Woodin cardinals. If $n \ge 2$ is even, then M_n^* is $\mathbf{L}[\mathbf{M}_{n-2}|\gamma]$, where γ is least such that $\gamma = \omega_1^{\mathbf{L}[\mathbf{M}_{n-2}|\gamma]}$ and $\mathbf{L}[\mathbf{M}_{n-2}|\gamma]$ is Σ^1_n-correct. (For $n > 2$, we have that $\gamma < \omega_1^{\mathbf{M}_{n-2}}$.) If n is odd, then M_n^* is the minimal proper class extender model Q such that if S is an initial segment of Q projecting to ω, then $\mathbf{M}_{n-2}(S)^\#$ is an initial segment of Q. These identifications are implicit in [Ste95B]. Finally, if $n \ge 3$ is odd, and T is the tree of a Π^1_n scale on a complete Π^1_n set, then there is an iterate Q

of $\mathbf{M}_{n-1}$ such that $\mathbf{L}[T] = \mathbf{L}[Q|\underset{\sim}{\delta}^1_n]$. This identification is implicit in [Ste95A], where the parallel fact with the pointclass Π_n replaced by $\Sigma_1^{\mathbf{L}(\mathbb{R})}$, and $\mathbf{M}_{n-1}$ replaced by M_ω, is proved. So we have

THEOREM 3.5 (Steel 1994). Assume there are ω Woodin cardinals with a measurable above them all, and let $\Gamma = \Pi^1_{2n+1}$ or $\Gamma = \Sigma_1^{\mathbf{L}(\mathbb{R})}$; then H_Γ is an iterable extender model.

In a similar vein, the prewellordering and scale theorems of §2 - §4 can now be proved using extender models. In the prewellordering case, the proof is due to Woodin, and in the scale case, to Neeman; in neither case is the proof published, but see [Ste95B]. These proofs require significantly more theory than the comparsion game approach, but in some ways they give deeper insight into the meaning of the norms being constructed.

Finally, Suslin and ∞-Borel representations are related to Lebesgue measurability, the Baire property, and the perfect set property in §10 and §11. Solovay's breakthrough results from 1966 on the regularity of Σ^1_2 sets under large cardinal hypotheses [Sol66] are thereby extended to other pointclasses. A basic result on the existence of largest countable sets is proved (in effect):

THEOREM 3.6 (Kechris, Moschovakis). Suppose Γ is adequate, ω-parametrized, has the scale property, and is closed under $\exists^{\mathbb{R}}$, and suppose all $\underset{\sim}{\Gamma}$ games are determined; then there is a largest countable Γ set of reals.

When it exists, the largest countable Γ set is called C_Γ. The theorem is implicit in the proof of Theorem 11B-2, which proves the existence of C_Γ for $\Gamma = \Sigma^1_{2n}$. Kechris' paper [Kec75] contains further basic information in this area. The sets C_Γ are quite important, partly because many of them show up naturally as the set of reals in some canoncal inner model. For example, Solovay showed that $C_{\Sigma^1_2} = \mathbb{R} \cap \mathbf{L}$ [KM78B, 11B-1], and we now know that for any n, $C_{\Sigma^1_{2n+2}} = \mathbb{R} \cap \mathbf{M}_{2n}$. (See [Ste95B]. Note that $M_0 = \mathbf{L}$.) In general, under the hypotheses of Theorem 3.4, we have $C_{\exists^{\mathbb{R}}\Gamma} = \mathbb{R} \cap H_\Gamma$. (See [Mos80, 8G.29].)

Kechris [Kec75] shows that assuming $\underset{\sim}{\Pi}^1_{2n+1}$- determinacy, there is a largest countable Π^1_{2n+1} set of reals $C_{\Pi^1_{2n+1}}$. This result is due to Guaspari, Kechris, and Sacks for $n = 0$, in which case $C_{\Pi^1_{2n+1}}$ has an inner-model-theoretic meaning as the set of reals Δ^1_{2n+1}-equivalent to the first order theory of some level of $\mathbf{M}_{2n}$ projecting to ω. It is open whether this characterization of $C_{\Pi^1_{2n+1}}$ holds also for $n > 0$.

Propagation of the scale property using games [Nee07].
Scales on $\underset{\sim}{\Sigma}^1_1$ sets [Ste83B].

Moschovakis unified his results on scale propagation under the real quantifiers into a single theorem on the propagation of scales under the *game quantifier* on ω. Letting $A \subseteq \mathbb{R} \times \mathbb{R}$, we put

$$\Game y A(x, y) \Leftrightarrow \exists n_0 \forall n_1 \exists n_2 \forall n_3 ... A(x, \langle n_i \ : \ i < \omega \rangle),$$

where we interpret the right hand side as meaning its quantifier string has a Skolem function, that is, that player I wins the game on ω with payoff $A_x = \{y \ : \ A(x, y)\}$. We write $\Game A$ for $\{x \ : \ \Game y A(x, y)\}$, and if Γ is a pointclass, we set $\Game\Gamma = \{\Game A \ : \ A \in \Gamma\}$. The following is often called the *third periodicity* theorem. It dates from approximately 1973; see [Mos73] or [Mos80, 6E].

THEOREM 3.7 (Moschovakis). Let Γ be an adequate, ω-parameterized pointclass closed under quantification over ω, and suppose $\Gamma(x)$-determinacy holds for all reals x. Suppose Γ has the scale property; then

(a) $\Game\Gamma$ has the scale property, and
(b) if G is a game on ω with payoff set in Γ, and the player whose payoff is Γ has a winning strategy in G, then that player has a $\Game\Gamma$ winning strategy.

The proof involves a more sophisticated comparison game: given a norm φ on A, one gets a norm on $\Game A$ using comparison games in which the two players play out the games with payoff A_{x_1} and A_{x_2} simultaneously, in different roles on the two boards, each trying to win in his role as player I with lower φ-norm than the other. The first paper in the present pair gives a thorough exposition of the proof of this theorem. (See also [Mos80, 6E].)

It is easy to see that $\Game\Pi^1_n = \Sigma^1_{n+1}$, and assuming $\underset{\sim}{\Sigma}^1_n$-determinacy, that $\Game\Sigma^1_n = \Pi^1_{n+1}$. Setting $\Sigma^1_0 = \Sigma^0_1$, this is true for $n = 0$ as well. Thus Theorem 3.7 subsumes Theorem 3.1. Part (b) of Theorem 3.7, on the existence of canonical winning strategies, is very useful. In the special case of projective sets, we get

COROLLARY 3.8 (Moschovakis). Assume $\underset{\sim}{\Delta}^1_{2n}$-determinacy, and let G be a game with Σ^1_{2n} payoff, and suppose the player with Σ^1_{2n} payoff has a winning strategy; then he has a Δ^1_{2n+1} winning strategy.

Moschovakis' proof used $\underset{\sim}{\Sigma}^1_{2n}$-determinacy, but Martin later showed this follows from $\underset{\sim}{\Delta}^1_{2n}$-determinacy, so we have stated the theorem in its sharper form. Of course, we also get Δ^1_{2n+2} strategies for games won by a player with Π^1_{2n+1} payoff from Corollary 3.8, but this already follows easily from the basis theorem for Π^1_{2n+1}. It is easy to see that these definability bounds on winning strategies are optimal.

It is natural to ask what are the optimally definable scales and winning strategies for the projective pointclasses which zig when they should have zagged, that is, for Σ^1_{2n+1} and Π^1_{2n+2}. The second paper in this pair gives part of the answer. Let α-Π^1_1 be the αth level of the difference hierarchy over Π^1_1

(see [Ste83B]). and let

$$\Lambda_0 = \bigcup_{k<\omega} \omega k\text{-}\Pi^1_1.$$

Steel gives a simple proof in [Ste83B] that every Σ^1_1 set admits a very good scale whose associated prewellorders are each in Λ_0, and in fact, each set in Λ_0 admits a very good scale whose associated prewellorders are all in Λ_0. (We are *not* demanding that the sequence of prewellorders be in Λ_0.) Now let $⅁^{(n)} = ⅁....⅁$ be the n-fold composition of the game quantifier on ω; then the proof of third periodicity theorem easily gives

THEOREM 3.9 (Steel 1980). Let $n \geq 1$, and suppose all $⅁^{(n-1)}\Lambda_0(x)$ games are determined, for all reals x; then

(a) every $⅁^{(n)}\Lambda_0$ set admits a very good scale, all of whose norms are $⅁^{(n)}\Lambda_0$, and

(b) if G is a game with payoff in $⅁^{(n-1)}\Lambda_0$, then there is a winning strategy σ for G such that for any k, $\sigma \restriction \{p : \mathrm{lh}(p) \leq k\}$ is in $⅁^{(n)}\Lambda_0$.

It is easy to see that for $n \geq 1$,

$$(\Sigma^1_n \cup \Pi^1_n) \subseteq ⅁^{(n-1)} \bigcup_{k<\omega} \omega k\text{-}\Pi^1_1 \subseteq \Delta^1_{n+1}.$$

The best bounds on the definability of very good scales and winning strategies for Σ^1_n sets (n odd) and Π^1_n sets (n even) are just those given by Theorem 3.9 and this inclusion. That the bounds cannot be improved follows from Martin [Mar83A]; see below.

(We should note here that Busch [Bus76] showed that every Σ^1_1 set admits a scale all of whose prewellorders are $(\omega + 3)$-Π^1_1. However, the Busch scale is not very good, and transforming it to a very good scale involves taking intersections, which drives us up to Λ_0. The third periodicity propagation technique requires, in effect, that the input scale be very good.)

The progress of inner model theory has shed some light on these results. Neeman [Nee95] gives an inner-model-theoretic proof that every Σ^1_{2n} game won by the player with Σ^1_{2n} payoff has as Δ^1_{2n+1} winning strategy, as a byproduct of his proof of Σ^1_{2n} determinacy from the existence and iterability of $\mathbf{M}_{2n-1}$. Neeman's work also gives an insight into the pointclasses $⅁^{(n)}\Lambda_0$. For $n \geq 0$, let T^n_k be the theory in $\mathbf{M}_n$ of its first k indiscernibles. Thus the reals in $\mathbf{M}_n$ are just those reals which are recursive in some T^n_k. One can show that every $⅁^{n+1}\omega k$-Π^1_1 real is recursive in T^n_k, and that T^n_k itself is $⅁^{(n+1)}\omega(k+1)$-Π^1_1. The proof is an induction on n, with the base case $n = 0$ being due to Martin, as part of his proof of ωk-Π^1_1-determinacy from the existence of the sharp of $\mathbf{M}_0 = \mathbf{L}$. (Here is a proof sketch of the $n > 0$ case for experts: To reduce a $⅁^{n+1}\omega k$-Π^1_1 real to T^n_k, we ask questions about what is forced in collapse of the bottom Woodin of $\mathbf{M}_n$ about its first k indiscernibles. The answer we get will reflect $\Game^{n+1}\omega k$-Π^1_1 truth because every real, and in particular a winning strategy witnessing or refuting the outer $\Game$ quantifier, is generic over an iterate of $\mathbf{M}_n$ for this collapse. To show that T^n_k itself is $\Game^{(n+1)}\omega(k+1)$-Π^1_1, we use a game in which the players play a putative $\mathbf{M}^\#_n$'s, say P and Q respectively, and then the two are coiterated inside $\mathbf{M}_{n-1}(\langle P, Q\rangle)$.)

It follows that $\mathbf{M}^\#_n$ is Turing equivalent to the set of true $\Game^{(n+1)}\Lambda_0$ sentences. From this we see that any game with $\Game^{(n)}\Lambda_0$ payoff has a winning strategy which is recursive in $\mathbf{M}^\#_n$. (By Theorem 3.14(b) below, no better definability bound is possible.) In particular, every nonempty Σ^1_{2n+1} set has a member recursive in $\mathbf{M}^\#_{2n}$, using the trivial game in which I must play a member of the set and a witness to the Π^1_{2n} matrix, and II does nothing. This gives us an inner-model-theoretic proof of Martin and Solovay's generalization of the Kleene Basis Theorem for Σ^1_1 [KMS83, 5.6].

Inductive scales on inductive sets [Mos78].
Scales on coinductive sets [Mos83].
The extent of scales in $\mathbf{L}(\mathbb{R})$ [MS83].

It is natural to try to extend the civilizing influence of definable scales to more complicated sets. The remaining papers in this block use the comparison game construction to do that, while showing that, most of the time, the scales produced are definable in the simplest possible logical form.

The papers in this group, which represent work done in late 1979, exploit the uniformities in the comparison game method of propagating scales. Let us use $\exists^{\mathrm{Ord}}$, $\exists^{\mathbb{R}}$, and $\forall^{\mathbb{R}}$ to stand for existential quantification over the ordinals, over the reals, and universal quantification over the reals, respectively. Because the propagation of scales under these operations is uniform in the scales, one gets inductive scales on inductive sets; this is done in [Mos78]. (A set is inductive iff it is $\Sigma_1^{\mathbf{J}_{\kappa_{\mathbb{R}}}(\mathbb{R})}$, where $\kappa_{\mathbb{R}}$ is least κ such that $\mathbf{J}_\kappa(\mathbb{R}) \models \mathsf{KP}$.) Since AD implies that the pointclass of inductive sets is closed under real quantification and wellordered unions, it seemed at first that one needed a radically new idea to go further. (One cannot hope to show that the class of scaled sets is closed under complement!) The existence of definable scales for coinductive sets became the second Victoria Delfino problem. However, it turned out that what was missing was more in the nature of a subtle observation: the comparison game propagation of scales under $\exists^{\mathrm{or}}$, $\exists^{\mathbb{R}}$, and $\forall^{\mathbb{R}}$ acts at the level of individual norms–it corresponds, in each case, to a *continuous* operation on the input scale. Moschovakis realized this, and realized that it could be used to define scales on any set A definable in the form

$$A(x) \Leftrightarrow \exists x_0 \exists \alpha_0 \forall x_1 \exists x_2 \exists \alpha_1 \forall x_3 ... \forall n R(\langle x_0|n, ..., x_n|n\rangle, \langle \alpha_0, ..., \alpha_n\rangle),$$

where R is definable, the α's are ordinals, and the x's are reals. (This is done by simultaneously defining scales on each of the ω-many sets defined by the formula on the right with some initial segment of its quantifiers removed. The scale on any such set is obtained from the scale on the set corresponding to removing one more quantifier by the continuous operation corresponding to that quantifier.)

The expression displayed on the right hand side above gives what is called a **closed game representation** of A: it asserts that player I wins the infinite game in which he plays the even x's and the α's, while II plays the odd x's, and the payoff indicated by the matrix is closed in the product of the Baire topology on $\mathbb{R}$ and the discrete topology on the ordinals. What [Mos83] shows, in effect, is that granted sufficient determinacy, any set with a closed game representation admits a definable scale. (The converse is trivial.) In the special case that the game involves no ordinal moves, one gets

THEOREM 3.10 (Moschovakis 1979). *Suppose all games in $\mathbf{J}_{\kappa_{\mathbb{R}}+1}(\mathbb{R})$ are determined, and let A be coinductive; then A admits a scale whose associated prewellorders are each in $\mathbf{J}_{\kappa_{\mathbb{R}}+1}(\mathbb{R})$.*

Martin and Steel showed in [MS83] that in fact, every $\Sigma_1^{\mathbf{L}(\mathbb{R})}$ set admits a closed game representation in $\mathbf{L}(\mathbb{R})$, which together with Moschovakis' work and some simple definability calculations implies

THEOREM 3.11 (Martin, Steel 1979). *Assume $\mathsf{AD}^{\mathbf{L}(\mathbb{R})}$; then the pointclass $\Sigma_1^{\mathbf{L}(\mathbb{R})}$ has the scale property.*

Kechris and Solovay had observed earlier that, assuming AD, the relation "$x, y \in \mathbb{R}$ and y is not ordinal definable from x" is ordinal definable, but admits no uniformization, and hence no scale, which is ordinal definable from a real. (Let f be a uniformizing function, and suppose f is ordinal definable from x; then $f(x)$ is ordinal definable from x, a contradiction.) If $\mathbf{V}=\mathbf{L}(\mathbb{R})$, then this relation is Π_1, while every set whatsoever is ordinal definable from a real, so we have a Π_1 set which admits no scale at all. A simple Wadge argument then shows that assuming $\mathsf{AD}^{\mathbf{L}(\mathbb{R})}$, the sets admitting scales in $\mathbf{L}(\mathbb{R})$ are precisely the $\Sigma_1^{\mathbf{L}(\mathbb{R})}$ sets.

Under suitable large cardinal assumptions, one can construct natural models of AD properly larger than $\mathbf{L}(\mathbb{R})$. These models, and $\mathbf{L}(\mathbb{R})$ itself, all satisfy a certain strengthening of AD called AD^+. The theory AD^+ was isolated by Woodin, and part of his work is the following far-reaching generalization of Theorem 3.11:

THEOREM 3.12 (Woodin, mid 90's). *Assume AD^+; then the pointclass Σ_1^2 has the scale property.*

Note here that $\Sigma_1^{\mathbf{L}(\mathbb{R})} = (\Sigma_1^2)^{\mathbf{L}(\mathbb{R})}$, so that Woodin's theorem reduces to that of Martin and Steel if our model of AD^+ is $\mathbf{L}(\mathbb{R})$. No proof of Theorem 3.12 has been published as yet, but the theory AD^+ is described in [Woo99, Section

9.1], where Theorem 3.12 and related results are stated as Theorem 9.7. A proof that Σ_1^2 has the scale property in those models of AD^+ obtained from models with large cardinals via the standard means, i.e. the derived model construction, is exposited in [SteA, §7].

The largest countable this, that, and the other [Mar83A].

Moschovakis [Mos83] shows that the norms of the scale on a coinductive set it constructs are each first order definable over $\mathbf{J}_{\kappa_{\mathbb{R}}}(\mathbb{R})$. It is natural to ask whether one can do better: does every coinductive set admit a scale such that for some fixed $n < \omega$, all the norms are $\utilde{\Sigma}_n^{\mathbf{J}_{\kappa_{\mathbb{R}}}(\mathbb{R})}$? In [Mar83A], Martin proves an important reflection result which implies that the answer is "no". Let us write $y \in \mathbf{OD}^\alpha(x)$ to mean that y is ordinal definable from x over $\mathbf{J}_\alpha(\mathbb{R})$.

THEOREM 3.13 (Martin 1980). Assume inductive determinacy, and suppose $x, y \in \mathbb{R}$ and $y \in \mathbf{OD}^{\kappa_{\mathbb{R}}}(x)$; then $y \in \mathbf{OD}^\alpha(x)$ for some $\alpha < \kappa_{\mathbb{R}}$.

(Though not literally stated in [Mar83A], this is a fairly direct consequence of [Mar83A, Lemma 4.1].) Now the relation "$x, y \in \mathbb{R}$ and $\forall \alpha < \kappa_{\mathbb{R}}(y \notin \mathbf{OD}^\alpha(x))$" is coinductive, and by the Kechris-Solovay argument, it cannot be uniformized by a function in $\mathbf{J}_{\kappa_{\mathbb{R}}+1}(\mathbb{R})$, and hence it admits no scale whose sequence of associated prewellorders is in $\mathbf{J}_{\kappa_{\mathbb{R}}}(\mathbb{R})$. Thus one cannot improve Moschovakis' definability bound.

Martin's reflection result is part of a characterization of C_Γ, of the largest countable Γ set of reals, for various pointclasses Γ. Letting $\mathrm{IND} = \Sigma_1^{\mathbf{J}_{\kappa_{\mathbb{R}}}}(\mathbb{R})$ be the pointclass of (lightface) inductive sets, it is easy to see that assuming inductive determinacy, $C_{\mathrm{IND}} = \{y : y \in \mathbf{OD}^{\kappa_{\mathbb{R}}}(\varnothing)\}$. (See [Ste83A, 2.11].) This characterizes the members of C_{IND} in terms of definability from ordinals, in a way which parallels Kechris' characterization of $C_{\Sigma^1_{2n}}$ as $\{y : y$ is Δ^1_{2n} in a countable ordinal$\}$. Martin found characterizations of C_{IND} and $C_{\Sigma^1_{2n}}$ in terms of definability wihout parameters:

THEOREM 3.14 (Martin 1980). Assume all games in $\mathbf{J}_{\kappa_{\mathbb{R}}+1}(\mathbb{R})$ are determined; then for any real y

(a) $y \in C_{\mathrm{IND}}$ iff y is definable over $\mathbf{J}_{\kappa_{\mathbb{R}}}(\mathbb{R})$ from no parameters, and
(b) $y \in C_{\Sigma^1_{2n}}$ iff y is $\Game^{(2n-1)} \bigcup_{k<\omega} \omega k\text{-}\Pi^1_1$.

The left-to-right directions make use of the existence of scales on coinductive sets in case (a), and on Π^1_{2n} sets in case (b), which are definable in the appropriate way: each norm being definable over $\mathbf{J}_{\kappa_{\mathbb{R}}}(\mathbb{R})$ in case (a), each norm being $\Game^{(2n-1)} \bigcup_{k<\omega} \omega k\text{-}\Pi^1_1$ in case (b). The right-to-left directions are reflection arguments, and they show that the definability bounds for scales used in the other direction are best possible.

Soon after Martin proved Theorem 3.14, Kechris and Woodin used his technique, among other ideas, to prove a vaguely similar reflection result: if there is a non-determined game in $\mathbf{J}_{\kappa_{\mathbb{R}}+1}(\mathbb{R})$, then there is a non-determined game in $\mathbf{J}_{\kappa_{\mathbb{R}}}(\mathbb{R})$. It follows that the hypothesis of Moschovakis' scale existence result Theorem 3.10 can be reduced to $\mathbf{J}_{\kappa_{\mathbb{R}}}(\mathbb{R})$-determinacy. See [KW83], which is reprinted in Part II of this volume. Somewhat more general results along the same lines were a key ingredient in the proof of

THEOREM 3.15 (Kechris, Woodin 1980). *If there is a non-determined game in* $\mathbf{L}(\mathbb{R})$, *then there is a non-determined game whose payoff is Suslin in* $\mathbf{L}(\mathbb{R})$.

The structure of this proof has played an important role in later proofs of $\mathsf{AD}^{\mathbf{L}(\mathbb{R})}$ under various hypotheses. See below.

Kechris-Woodin [KW83] also proves something along the lines of Theorem 3.14(b): $\underset{\sim}{\Delta}^1_{2n}$-determinacy implies $\Game^{2n-1}\bigcup_{k<\omega}\omega k$-$\Pi^1_1$-determinacy. (Martin had proved $\underset{\sim}{\Delta}^1_{2n}$-determinacy implies $\underset{\sim}{\Sigma}^1_{2n}$-determinacy earlier, by a different method. See [KS85].)

Once again, the progress of inner model theory has given us deeper insight into these results of Martin and Kechris-Woodin. Itay Neeman found an inner-model-theoretic proof of Theorem 3.14(b). Let $\mathbf{M}_n$ be the minimal iterable proper class extender model with n Woodin cardinals, and T^n_k be the theory in $\mathbf{M}_n$ of its first k indiscernibles. By [Ste95B], the reals in $C_{\Sigma^1_{2n}}$ are precisely the reals in $\mathbf{M}_{2n}$, and hence are just those reals which are recursive in some T^{2n}_k. But by [Nee95], every $\Game^{2n-1}\omega k$-Π^1_1 real is recursive in T_k, and that T_k itself is $\Game^{(2n-1)}\omega(k+1)$-$\Pi^1_1$. (We sketched this proof above.) Theorem 3.14(b) now follows easily. Mitch Rudominer found an inner-model-theoretic proof of Theorem 3.14(a); in this case, the role of $\mathbf{M}_{2n}$ is played by the minimal iterable extender model P having ω Woodin cardinals, which by minimality are cofinal in the ordinals of P, and the role of T^{2n}_k is played by the Σ_0 theory in P of its first k Woodin cardinals. See [Rud99].

It is worth noting that the set of reals in $\mathbf{M}_{2n+1}$ is also well known from descriptive set theory; it is the set Q_{2n+1}. See [KMS83] for the many characterizations of this set. In general, the reals of $\mathbf{M}_n$, for any n, can be characterized in terms of ordinal definability as those reals which are Δ^1_{n+2} in a countable ordinal.

Neeman and Woodin have proved the Kechris-Woodin theorem within the projective hierarchy by the methods of inner model theory, and at the same time generalized it to the odd levels as well. Woodin (unpublished) showed that for any $n \geq 1$, $\underset{\sim}{\Pi}^1_n$-determinacy implies that for all reals x, $\mathbf{M}_n(x)^\#$ exists, and Neeman [Nee95] showed that the existence of these mice implies $\Game^{n-1}\bigcup_{k<\omega}\omega k$-$\Pi^1_1$-determinacy. For $n = 1$, these are results of Harrington [Har78] Martin (unpublished) respectively. It is known from work of Kechris and Solovay [KS85] that these "transfer results" for determinacy in the projective hierarchy cannot be improved.

No one has as yet proved that $\mathbf{J}_{\kappa_{\mathbb{R}}}(\mathbb{R})$-determinacy implies $\mathbf{J}_{\kappa_{\mathbb{R}}+1}(\mathbb{R})$-determinacy by purely inner-model-theoretic methods.

Scales in $\mathbf{L}(\mathbb{R})$ [Ste83A].
Scales in $\mathbf{K}(\mathbb{R})$ [Ste07C].

Given that a set of reals admits a scale in $\mathbf{L}(\mathbb{R})$, it is natural to ask what is the least level of the Levy hierarchy for $\mathbf{L}(\mathbb{R})$ at which such a scale appears. The papers in the last group answered this question in some important cases. The paper [Ste83A], work of Steel from 1980, knits the arguments of those papers together into a complete answer.

It turns out that the (lexicographically) least $\langle\gamma, n\rangle$ such that A admits a $\underset{\sim}{\Sigma}_n^{\mathbf{J}_\gamma(\mathbb{R})}$ scale is the lexicographically least $\langle\gamma, n\rangle$ such that $A \in \underset{\sim}{\Sigma}_n^{\mathbf{J}_\gamma(\mathbb{R})}$ and $\underset{\sim}{\Sigma}_n^{\mathbf{J}_\gamma(\mathbb{R})}$ has the scale property. [Ste83A] characterizes those pointclasses $\underset{\sim}{\Sigma}_n^{\mathbf{J}_\gamma(\mathbb{R})}$ which have the scale property in terms of reflection properties. The key concept is that of a Σ_1-*gap*. Let us say $\mathcal{M} \prec_1^{\mathbb{R}} \mathcal{N}$ iff $\mathcal{M}$ is an elementary submodel of $\mathcal{N}$ with respect to Σ_1 formulae about parameters from $\mathbb{R} \cup \{\mathbb{R}\}$.

Definition 3.16. The interval $[\alpha, \beta]$ is a Σ_1**-gap** iff
1. $\mathbf{J}_\alpha(\mathbb{R}) \prec_1^{\mathbb{R}} \mathbf{J}_\beta(\mathbb{R})$,
2. $\forall\gamma < \alpha(\mathbf{J}_\gamma(\mathbb{R}) \not\prec_1^{\mathbb{R}} \mathbf{J}_\alpha(\mathbb{R}))$, and
3. $\mathbf{J}_\beta(\mathbb{R}) \not\prec_1^{\mathbb{R}} \mathbf{J}_{\beta+1}(\mathbb{R})$.

With the convention that $[(\underset{\sim}{\delta}_1^2)^{\mathbf{L}(\mathbb{R})}, \infty]$ is also a Σ_1 gap, we have that Σ_1 gaps partition OR. In order to determine whether $\underset{\sim}{\Sigma}_n^{\mathbf{J}_\gamma(\mathbb{R})}$ has the scale property, we consider the unique Σ_1 gap $[\alpha, \beta]$ to which γ belongs. [Ste83A] shows that, assuming enough determinacy,

(1) if $\alpha < \gamma < \beta$ then $\underset{\sim}{\Sigma}_n^{\mathbf{J}_\gamma(\mathbb{R})}$ does not have the scale property (Kechris, Solovay),
(2) if $\gamma = \alpha$ and $n = 1$, then $\underset{\sim}{\Sigma}_n^{\mathbf{J}_\gamma(\mathbb{R})}$ has the scale property,
(3) if $\gamma = \alpha$, $n > 1$, and α is admissible, or if $\gamma = \beta$ and $[\alpha, \beta]$ is a strong gap, then $\underset{\sim}{\Sigma}_n^{\mathbf{J}_\gamma(\mathbb{R})}$ does not have the scale property (Martin),
(4) if $\gamma = \alpha$, $n > 1$, and α is inadmissible, or if $\gamma = \beta$ and $[\alpha, \beta]$ is a weak gap, and $\rho_n(\mathbf{J}_\beta(\mathbb{R})) = \mathbb{R}$, then $\underset{\sim}{\Sigma}_n^{\mathbf{J}_\gamma(\mathbb{R})}$ or its dual class has the scale property, according to the "zig-zag" pattern.

The weak/strong distinction for gaps is motivated by Martin's proof of Theorem 3.14; strong gaps have a reflection property which is used in this argument.

The most important applications of this analysis occur in inductive proofs of $\mathsf{AD}^{\mathbf{L}(\mathbb{R})}$. The first of these, which set the pattern, is the Kechris-Woodin proof of Theorem 3.15, that Suslin determinacy implies determinacy in $\mathbf{L}(\mathbb{R})$. Their argument goes roughly as follows: let β be least such that there is a non-determined game in $\mathbf{J}_{\beta+1}(\mathbb{R})$. The failure of determinacy is a Σ_1 fact, so β ends a gap of the form $[\alpha, \beta]$. Setting $\gamma = \beta$, and letting n be least such that there is a non-determined $\Sigma_n^{\mathbf{J}_\beta(\mathbb{R})}$ game, we have enough determinacy to prove (1)-(4) above. (This comes from inspecting the proof of [Ste83A], and using an observation of Kechris.) The Kechris-Woodin result that $\mathbf{J}_{\kappa_\mathbb{R}}(\mathbb{R})$-determinacy implies $\mathbf{J}_{\kappa_\mathbb{R}+1}(\mathbb{R})$-determinacy generalizes routinely so as to show that we cannot be in case (3). In all other cases, we have enough determinacy to show that either $\Sigma_n^{\mathbf{J}_\beta(\mathbb{R})}$ or $\Sigma_{n+1}^{\mathbf{J}_\beta(\mathbb{R})}$ has the scale property. Thus the payoff of our non-determined game is Suslin in $\mathbf{L}(\mathbb{R})$, a contradiction.

Kechris and Woodin used Theorem 3.15 to show that if $\mathbf{V}=\mathbf{L}(\mathbb{R})$ and there are arbitrarily large strong partition cardinals below Θ, then AD holds. With the proofs of determinacy from Woodin cardinals in the mid-80's [MS89], and the advances in techniques for constructing correct inner models with Woodin cardinals in the late 80's and 1990 [MS94, MS94, Ste96], it became possible to use this pattern of argument to prove $\mathsf{AD}^{\mathbf{L}(\mathbb{R})}$ under many different hypotheses. Woodin pioneered this *core model induction* method, in his 1990 proof that the Proper Forcing Axiom together with the existence of an inaccessible cardinal implies $\mathsf{AD}^{\mathbf{L}(\mathbb{R})}$. In such an argument, one proves not just determinacy, but the existence of correct mice "certifying" the determinacy in question, by an induction. The scale analysis is used to get a definable scale on a set coding truth at the next level of correctness, and then core model theory is used to construct mice correct at that level. The method has been used a number of times since 1990, by Woodin and others. (See [Ket00, Zob00].) Indeed, if the large cardinal strength of a theory T is not close to the surface, it is highly likely that any proof that T implies $\mathsf{AD}^{\mathbf{L}(\mathbb{R})}$ will use the core model induction method. Only very recently has a paper describing and using the method been published; see [Ste05].

[Ste07C] extends the scale analysis of [Ste83A] to $\mathbf{K}(\mathbb{R})$. This is useful in constructing mice with more than infinitely many Woodin cardinals by the core model induction method.

The real game quantifier propagates scales [Mar83B].
Long games [Ste88].
The length-ω_1 open game quantifier propagates scales [Ste07A].

These papers push the comparison game construction of scales as far as it has been pushed to date.

It is natural to ask whether there is a propagation theorem behind Moschovakis' Theorem 3.10. Martin obtains a positive answer in [Mar83B], showing that the game quantifier corresponding to games of length ω on the reals propagates definable scales. In fact, he shows that the game quantifier corresponding to games of any fixed countable length propagates definable scales. (Note here that a game of length α on $\mathbb{R}$ can be simulated by a game of length $\omega\alpha$ on ω.) He also proves the other half of third periodicity, that there are definable winning strategies for these games. Both theorems require the determinacy of the games in question as a hypothesis.

Steel extends these results to the game quantifiers corresponding to clopen games of length ω_1 in [Ste88], and then finally to the game quantifier corresponding to open games of length ω_1 in [Ste07A]. These latter results represent the limit of what has been done in this direction, so let us state them more precisely.

Let $A \subseteq \mathbb{R} \times \omega^{<\omega_1}$. For $x \in \mathbb{R}$, consider the game of length ω_1 in which I and II play natural numbers, alternating as usual, with I going first at limit ordinals. We fix some natural way of representing runs of such games by $p \in {}^{\omega_1}\omega$, and then declare that p is a winning run for I in G_{A_x} iff $\exists\alpha < \omega_1 A(x, p \restriction \alpha)$. Thus G_{A_x} is open for I in the topology on ${}^{\omega_1}\omega$ whose basis consists of sets of the form $\{p \in {}^{\omega_1}\omega : q \subset p\}$, where $q \in \omega^{<\omega_1}$. We then put

$$x \in \Game^{\omega_1} A \Leftrightarrow \text{I has a winning strategy in } G_{A_x}.$$

In order to calculate definability, we code countable sequences q of natural numbers by reals q^* in some simple way. Putting $A^*(x, y) \Leftrightarrow \exists q(y = q^* \wedge A(x, q))$, we then define, for any pointclass Γ,

$$\Game^{\omega_1} \text{ open-}\Gamma = \{\Game^{\omega_1} A : A^* \in \Gamma\}.$$

The main result of [Ste07A] is

Theorem 3.17 (Steel). Let Γ be an adequate pointclass with the scale property, and suppose that all (boldface) length ω_1 open-$\underset{\sim}{\Gamma}$ games are determined; then

(a) $\Game^{\omega_1}$ open-Γ has the scale property, and
(b) if the player with open payoff has a winning strategy in a length ω_1 open-Γ game, then he has a $\Game^{\omega_1}$ open Γ winning strategy.

Thus the determinacy of length ω_1 open-$\underset{\sim}{\Delta}^1_2$ games implies that $\Game^{\omega_1}$ open $-$ Δ^1_2 has the scale property. It is not known how to construct definable scales on the complements of $\Game^{\omega_1}$ open-Δ^1_2 sets, or show that if the closed player wins such a game, he has a definable winning strategy, under any definable determinacy or large cardinal assumption. We should note here that assuming the determinacy of the long games involved, the complements of $\Game^{\omega_1}$ open-Δ^1_2 sets are just the $\Game^{\omega_1}$ closed-Δ^1_2 sets. in the natural meaning of this term. All $\Game^{\omega_1}$ closed-Δ^1_2 sets are Σ^2_1, and assuming CH, all Σ^2_1 sets are $\Game^{\omega_1}$ closed-Δ^1_2. Woodin's Σ^2_1 absoluteness theorem [Lar04, Theorem 3.2.1] can be formulated without referring to CH as follows: if there are arbitrarily large measurable Woodin cardinals, then $\Game^{\omega_1}$ closed-Δ^1_2 sentences are absolute between set-generic extensions of $\mathbf{V}$.

[Ste88, §3] uses a weaker form of this theorem to show that the determinacy of games ending at the first Σ_2 admissible relative to the play, with

Δ^1_2-in-the-codes payoff, implies that there is an inner model of $\mathsf{AD}_{\mathbb{R}}$ containing all reals and ordinals. This was the first proof that some form of definable determinacy implies that there is an inner model of $\mathsf{AD}_{\mathbb{R}}$. In the early 90's, Woodin showed how to obtain such a model directly, under the optimal large cardinal hypothesis. This work is unpublished, but is exposited in [SteA].

[Ste88, §4] also proves the determinacy of games ending at the first Σ_1 admissible relative to the play in the theory $\mathsf{ZF} + \mathsf{AD} + \mathsf{DC}+$ "every set of reals admits a scale" + "ω_1 is $\wp(\mathbb{R})$-supercompact". The latter had been proved to have a model containing all reals and ordinals assuming ZFC+ "there is a supercompact cardinal" by Woodin (unpublished, but see [SteA]), so [Ste88, §4] completed the proof of a small fragment of clopen determinacy beyond fixed countable length from large cardinals. Neeman [Nee06] has since proved this fragment of clopen determinacy directly, from essentially the optimal large cardinal hypotheses.

Although it is not known how to construct definable scales on $\Game^{\omega_1}$closed-Δ^1_2 sets under any definable determinacy or large cardinal assumption, one can do so under the assumption that there are sufficiently strong countable iterable structures. What one needs is essentially a countable, $\omega_1 + 1$-iterable model $\mathsf{ZF}^- +$ "there is a measurable Woodin cardnal". (A slightly weaker theory suffices.) Woodin, and later independently Steel, showed that $\Game^{\omega_1}$closed-Δ^1_2 truth can be reduced to truth in any such model, by asking what is forced in the extender algebra at its measurable Woodin cardinal. This yields a recursive function t such that if φ is a $\Game^{\omega_1}$closed-Δ^1_2 formula, and x is a real, then

$$\varphi(x) \Leftrightarrow M \models t(\varphi)[x],$$

where M is any iterable model as above such that $x \in M$. (Sadly, this work is still unpublished. The arguments of [Ste07B, §7] are a prototype lower down.) We can now use the tree-to-produce-an-elementary-submodel method to get a definable scale. For let M_0 be the minimal extender model satisfying $\mathsf{ZF}^- +$ "there is a measurable Woodin cardinal". M_0 exists and is iterable by our assumption. Because it is minimal, M_0 is pointwise definable, and hence by [Ste07B, 4.10] it has a *unique* $\omega_1 + 1$ iteration strategy with the Dodd-Jensen property. We can then define the direct limit M_∞ of all countable iterates of M_0 under the iteration maps given by this strategy, as in [Ste95A]. Our Suslin representation T of the universal $\Game^{\omega_1}$closed-Δ^1_2 set is definable from M_∞, and hence definable. Roughly speaking, T builds along each branch a putative $\Game^{\omega_1}$closed-Δ^1_2 truth $\varphi(x)$, a model N and an embedding of N into M_∞, and a proof that x is N-generic for the extender algebra at the bottom Woodin cardinal of N, and that $N[x] \models t(\varphi)[x]$.

It is much harder to construct definable winning strategies for $\Game^{\omega_1}$closed-Δ^1_2 games granted the existence of an iterable model of $\mathsf{ZF}^- +$ "there is a measurable Woodin cardnal". One must show that there are any strategies at all, in the first place! Itay Neeman [Nee04, Nee] has made great progress on the questions of the existence and definability of winning strategies in length ω_1 open-Hom_∞ games. He shows that these strategies can be obtained by a logically simple transformation from iteration strategies for appropriate mice. In the case of length ω_1 closed-Δ^1_2, the appropriate mouse is essentially the minimal iterable mouse with a measurable Woodin cardinal. Letting M_0 be as above, Neeman's work shows that any length ω_1 closed-Δ^1_2 game has a winning strategy which is easily definable from the unique iteration strategy for M_0, which by its uniqueness is itself definable.

The problem of constructing iteration strategies for mice at the level of M_0 and beyond has been one of the fundamental open problems in pure set theory since the mid 1980's. *Carthago delenda est*!

REFERENCES

JOHN W. ADDISON AND YIANNIS N. MOSCHOVAKIS

[AM68] *Some consequences of the axiom of definable determinateness*, ***Proceedings of the National Academy of Sciences of the United States of America***, no. 59, 1968, pp. 708–712.

JAMES BAUMGARTNER, DONALD A. MARTIN, AND SAHARON SHELAH

[BMS84] ***Axiomatic set theory***. *Proceedings of the AMS-IMS-SIAM joint summer research conference held in Boulder, Colo., June* 19–25, 1983, Contemporary Mathematics, vol. 31, Amer. Math. Soc., Providence, RI, 1984.

HOWARD S. BECKER

[Bec78] *Partially playful universes*, In Kechris and Moschovakis [CABAL i], pp. 55–90.

HOWARD S. BECKER AND ALEXANDER S. KECHRIS

[BK84] *Sets of ordinals constructible from trees and the third Victoria Delfino problem*, In Baumgartner et al. [BMS84], pp. 13–29.

DOUGLAS R. BUSCH

[Bus76] *λ-Scales, κ-Souslin sets and a new definition of analytic sets*, ***The Journal of Symbolic Logic***, vol. 41 (1976), p. 373.

QI FENG, MENACHEM MAGIDOR, AND W. HUGH WOODIN

[FMW92] *Universally Baire sets of reals*, In Judah et al. [JJW92], pp. 203–242.

LEO A. HARRINGTON

[Har78] *Analytic determinacy and* $0^{\#}$, ***The Journal of Symbolic Logic***, vol. 43 (1978), pp. 685–693.

LEO A. HARRINGTON AND ALEXANDER S. KECHRIS

[HK81] *On the determinacy of games on ordinals*, ***Annals of Mathematical Logic***, vol. 20 (1981), pp. 109–154.

STEPHEN JACKSON

[Jac88] AD *and the projective ordinals*, In Kechris et al. [CABAL iv], pp. 117–220.

[Jac99] ***A computation of*** $\underset{\sim}{\delta}^1_5$, vol. 140, Memoirs of the AMS, no. 670, American Mathematical Society, July 1999.

[Jac07A] *Structural consequences of* AD, In Kanamori and Foreman [KF07].

[Jac07B] *Suslin cardinals, partition properties, homogeneity. Introduction to Part II*, this volume, 2007.

H. JUDAH, W. JUST, AND W. HUGH WOODIN

[JJW92] ***Set theory of the continuum***, MSRI publications, vol. 26, Springer-Verlag, 1992.

AKIHIRO KANAMORI AND MATTHEW FOREMAN

[KF07] ***Handbook of set theory into the 21st century***, Springer, 2007.

ALEXANDER S. KECHRIS

[Kec75] *The theory of countable analytical sets*, ***Transactions of the American Mathematical Society***, vol. 202 (1975), pp. 259–297.

ALEXANDER S. KECHRIS, DONALD A. MARTIN, AND YIANNIS N. MOSCHOVAKIS

[CABAL iii] *Cabal seminar* 79–81, Lecture Notes in Mathematics, no. 1019, Berlin, Springer, 1983.

ALEXANDER S. KECHRIS, DONALD A. MARTIN, AND ROBERT M. SOLOVAY

[KMS83] *Introduction to Q-theory*, In Kechris et al. [CABAL iii], pp. 199–282.

ALEXANDER S. KECHRIS, DONALD A. MARTIN, AND JOHN R. STEEL

[CABAL iv] *Cabal seminar* 81–85, Lecture Notes in Mathematics, no. 1333, Berlin, Springer, 1988.

ALEXANDER S. KECHRIS AND YIANNIS N. MOSCHOVAKIS

[CABAL i] *Cabal seminar* 76–77, Lecture Notes in Mathematics, no. 689, Berlin, Springer, 1978.

[KM78B] *Notes on the theory of scales*, this volume, originally published in *Cabal Seminar* 76–77 [CABAL i], pp. 1–53.

ALEXANDER S. KECHRIS AND ROBERT M. SOLOVAY

[KS85] *On the relative consistency strength of determinacy hypotheses*, ***Transactions of the American Mathematical Society***, vol. 290 (1985), no. 1, pp. 179–211.

ALEXANDER S. KECHRIS AND W. HUGH WOODIN

[KW83] *Equivalence of determinacy and partition properties*, ***Proceedings of the National Academy of Sciences of the United States of America***, vol. 80 (1983), no. 6 i., pp. 1783–1786.

[KW07] *Generic codes for uncountable ordinals*, this volume, originally circulated manuscript, 2007.

RICHARD O. KETCHERSID

[Ket00] ***Toward*** $\mathsf{AD}_{\mathbb{R}}$ ***from the continuum hypothesis and an*** ω_1***-dense ideal***, Ph.D. thesis, Berkeley, 2000.

STEPHEN C. KLEENE

[Kle55] *Arithmetical predicates and function quantifiers*, ***Transactions of the American Mathematical Society***, vol. 79 (1955), pp. 312–340.

MOTOKITI KONDÔ

[Kon38] *Sur l'uniformization des complementaires analytiques et les ensembles projectifs de la seconde classe*, ***Japanese Journal of Mathematics***, vol. 15 (1938), pp. 197–230.

PAUL B. LARSON

[Lar04] ***The stationary tower: Notes on a course by W. Hugh Woodin***, University Lecture Series (AMS), vol. 32, Providence, RI, 2004.

N. N. LUZIN AND P. S. NOVIKOV

[LN35] *Choix effectif d'un point dans un complemetaire analytique arbitraire, donne par un crible*, ***Fundamenta Mathematicae***, vol. 25 (1935), pp. 559–560.

RICHARD MANSFIELD

[Man71] *A Souslin operation on* Π^1_2, ***Israel Journal of Mathematics***, vol. 9 (1971), no. 3, pp. 367–379.

DONALD A. MARTIN

[Mar70A] *Measurable cardinals and analytic games*, ***Fundamenta Mathematicae***, (1970), no. LXVI, pp. 287–291.

[Mar83A] *The largest countable this, that, and the other*, this volume, originally published in Kechris et al. [CABAL iii], pp. 97–106.

[Mar83B] *The real game quantifier propagates scales*, this volume, originally published in Kechris et al. [CABAL iii], pp. 157–171.

DONALD A. MARTIN AND ROBERT M. SOLOVAY

[MS69] *A basis theorem for Σ^1_3 sets of reals*, ***Annals of Mathematics***, vol. 89 (1969), pp. 138–160.

DONALD A. MARTIN AND JOHN R. STEEL

[MS83] *The extent of scales in* $\mathbf{L}(\mathbb{R})$, this volume, originally published in Kechris et al. [CABAL iii], pp. 86–96.

[MS89] *A proof of projective determinacy*, ***Journal of the American Mathematical Society***, vol. 2 (1989), pp. 71–125.

[MS94] *Iteration trees*, ***Journal of the American Mathematical Society***, vol. 7 (1994), pp. 1–73.

[MS07] *The tree of a Moschovakis scale is homogeneous*, this volume, 2007.

WILLIAM J. MITCHELL AND JOHN R. STEEL

[MS94] ***Fine structure and iteration trees***, Lecture Notes in Logic, vol. 3, Springer-Verlag, Berlin, 1994.

YIANNIS N. MOSCHOVAKIS

[Mos71A] *Uniformization in a playful universe*, ***Bulletin of the American Mathematical Society***, vol. 77 (1971), pp. 731–736.

[Mos73] *Analytical definability in a playful universe*, ***Logic, methodology, and philosophy of science IV*** (Patrick Suppes, Leon Henkin, Athanase Joja, and Gr. C. Moisil, editors), North-Holland, 1973, pp. 77–83.

[Mos78] *Inductive scales on inductive sets*, this volume, originally published in Kechris and Moschovakis [CABAL i], pp. 185–192.

[Mos80] ***Descriptive set theory***, Studies in Logic and the Foundations of Mathematics, no. 100, North-Holland, Amsterdam, 1980.

[Mos83] *Scales on coinductive sets*, this volume, originally published in Kechris et al. [CABAL iii], pp. 77–85.

ITAY NEEMAN

[Nee] *Games of length* ω_1, ***Journal of Mathematical Logic***, to appear.

[Nee95] *Optimal proofs of determinacy*, ***The Bulletin of Symbolic Logic***, vol. 1 (1995), pp. 327–339.

[Nee04] ***The determinacy of long games***, de Gruyter Series in Logic and its Applications, vol. 7, Walter de Gruyter, Berlin, 2004.

[Nee06] *Determinacy for games ending at the first admissible relative to the play*, ***The Journal of Symbolic Logic***, vol. 71 (2006), no. 2, pp. 425–459.

[Nee07] *Propagation of the scale property using games*, this volume, 2007.

MITCHELL RUDOMINER

[Rud99] *The largest countable inductive set is a mouse set*, ***The Journal of Symbolic Logic***, vol. 64 (1999), pp. 443–459.

JOSEPH R. SCHOENFIELD

[Sch61] *The problem of predicativity*, ***Essays on the foundations of mathematics*** (Y. Bar-Hillel et al., editors), Magnes Press, Jerusalem, 1961, pp. 132–139.

ROBERT M. SOLOVAY

[Sol66] *On the cardinality of Σ^1_2 set of reals*, ***Foundations of Mathematics: Symposium papers commemorating the 60th birthday of Kurt Gödel*** (Jack J. Bulloff, Thomas C. Holyoke, and S. W. Hahn, editors), Springer-Verlag, 1966, pp. 58–73.

[Sol78A] *A $\underset{\sim}{\Delta}^1_3$ coding of the subsets of $^{\omega}\omega$*, In Kechris and Moschovakis [CABAL i], pp. 133–150.

JOHN R. STEEL

[SteA] *The derived model theorem*, Available at `http://www.math.berkeley.edu/~steel`.

[Ste83A] *Scales in* $\mathbf{L}(\mathbb{R})$, this volume, originally published in Kechris et al. [CABAL iii], pp. 107–156.

[Ste83B] *Scales on Σ^1_1-sets*, this volume, originally published in Kechris et al. [CABAL iii], pp. 72–76.

[Ste88] *Long games*, this volume, originally published in Kechris et al. [CABAL iv], pp. 56–97.

[Ste95A] $\mathbf{HOD}^{\mathbf{L}(\mathbb{R})}$ *is a core model below* Θ, ***The Bulletin of Symbolic Logic***, vol. 1 (1995), pp. 75–84.

[Ste95B] *Projectively wellordered inner models*, ***Annals of Pure and Applied Logic***, vol. 74 (1995), pp. 77–104.

[Ste96] ***The core model iterability problem***, Lecture Notes in Logic, no. 8, Springer-Verlag, Berlin, 1996.

[Ste05] PFA *implies* $\mathsf{AD}^{\mathbf{L}(\mathbb{R})}$, ***The Journal of Symbolic Logic***, vol. 70 (2005), no. 4, pp. 1255–1296.

[Ste07A] *The length-ω_1 open game quantifier propagates scales*, this volume, 2007.

[Ste07B] *An outline of inner model theory*, In Kanamori and Foreman [KF07].

[Ste07C] *Scales in* $\mathrm{K}(\mathbb{R})$, this volume, 2007.

M. Y. SUSLIN

[Sus17] *Sur une définition des ensembles mesurables* B *sans nombres transfinis*, ***Comptes Rendus Hebdomadaires des Séances de l'Académie des Sciences***, vol. 164 (1917), pp. 88–91.

JOHN VON NEUMANN

[vN49] *On rings of operators, reduction theory*, ***Annals of Mathematics***, vol. 50 (1949), pp. 448–451.

W. HUGH WOODIN

[Woo99] ***The axiom of determinacy, forcing axioms, and the nonstationary ideal***, De Gruyter Series in Logic and its Applications, Walter de Gruyter, Berlin, 1999.

A. STUART ZOBLE

[Zob00] ***Stationary reflection and the determinacy of inductive games***, Ph.D. thesis, U.C. Berkeley, 2000.

DEPARTMENT OF MATHEMATICS
UNIVERSITY OF CALIFORNIA
BERKELEY, CA 94720, USA
E-mail: steel@math.berkeley.edu

NOTES ON THE THEORY OF SCALES

ALEXANDER S. KECHRIS AND YIANNIS N. MOSCHOVAKIS

Editorial Note. These informal notes were written in the summer of 1971 and were distributed fairly widely, but with the intention of publishing the results separately. Despite this intention, they were included in [CABAL i] because "it seemed like a good idea to include them ... particularly since there are many references to them in the literature." The notes are reproduced here in their [CABAL i] form (including the increasingly inaccurate remark "This paper is not meant for publication"), which added a few comments regarding progress since 1971.

It is the purpose of these notes to give an informal exposition of several recent results in Descriptive Set Theory, all centering around the notion of a **scale**. This was first isolated explicitly in the generalization of the Uniformization Theorem on the hypothesis of projective determinacy [Mos71A], but is surely implicit in some of the classical proofs. It has turned out that scales have many applications beyond the Uniformization Theorem, both in producing new results and in providing more elegant proofs of known results. Among the new results the Kunen-Martin theorem on the length of wellfounded relations is perhaps the most important. As for new proofs, one can now establish the beautiful results of Solovay on the regularity of $\underset{\sim}{\Sigma}^1_2$ sets (Lebesgue measurability, property of Baire, *etc.*) without any use of Cohen's forcing—in fact the new arguments are very simple and much in the spirit of classical Descriptive Set Theory. (Some of these new arguments are due to Solovay again.)

This paper is not meant for publication; Moschovakis wants to keep his results for his forthcoming book [Mos80], Kechris is holding his for his Ph.D. thesis and the theorems which belong to neither of us will be presumably written up by their authors. In many ways this can be considered a first draft of part of [Mos80]. The point of making it available now is that sometimes books remain "forthcoming" for a long time, despite the best of authors' intentions. We believe that these results are interesting enough to deserve early—if incomplete—dissemination.

This draft should be comprehensible to one with some knowledge of classical Descriptive Set Theory, recursive functions with real arguments and at

During the preparation of this manuscript, both authors were partially supported by NSF Grant #GP-27964, and Moschovakis was the recipient of a Sloan Foundation Fellowship.

The Cabal Seminar. Volume I: Games, Scales and Suslin Cardinals
Edited by A. S. Kechris, B. Löwe, J. R. Steel
Lecture Notes in Logic, 31

least the basic definitions of games and determinacy. Some parts depend on a knowledge of the theory of indiscernibles for **L**, but they are independent of the main results. Except for a few inessential changes (explained in §1), we shall follow the notation and terminology of [Mos70A, §1, §2]; it will be convenient and space-saving to assume that the reader is familiar with this material, although the rest of [Mos70A] is not relevant to this work, except for one result which will be identified when used.

§1. Preliminaries. As usual, $\omega = \{0, 1, 2, \dots\}$ and $\mathbb{R} = {}^{\omega}\omega$ = the set of **reals**. We study subsets of the product spaces

$$\mathfrak{X} = X_1 \times \cdots \times X_k \ (X_i = \omega \text{ or } X_i = \mathbb{R})$$

which we call **pointsets**. Sometimes we think of these as **relations** and write interchangeably,

$$x \in A \iff A(x).$$

A **pointclass** is a class of pointsets, not necessarily all in the same product space. Thus Σ^1_1 consists of all relations expressible in the form

$$(\exists\alpha)(\forall n)R(\bar{\alpha}(n), x)$$

(R recursive) and similarly for Π^1_1, *etc.*

If $A \subseteq \mathfrak{X} \times \omega$, put

$$\exists^{\omega} A = \{x : \exists n A(x, n)\},$$
$$\forall^{\omega} A = \{x : \forall n A(x, n)\}$$

and if $A \subseteq \mathfrak{X} \times \mathbb{R}$, put

$$\exists^{\mathbb{R}} A = \{x : \exists\alpha A(x, \alpha)\},$$
$$\forall^{\mathbb{R}} A = \{x : \forall\alpha A(x, \alpha)\}.$$

If Γ is a pointclass, define

$$\check{\Gamma} = \{\mathfrak{X} \setminus A : A \subseteq \mathfrak{X}, A \in \Gamma\}$$

and call it the **dual class** of Γ, define

$$\underset{\sim}{\Gamma} = \{A : \text{ for some } B \subseteq \mathbb{R} \times \mathfrak{X} \text{ and some } \alpha_0 \in \mathbb{R}, x \in A \iff \langle \alpha_0, x\rangle \in B\},$$

and for any operation Φ on pointsets, define

$$\Phi\Gamma = \{\Phi A : A \in \Gamma\}.$$

A pointclass Γ is **adequate** if it contains all recursive sets and is closed under disjunction, conjunction, bounded number quantification of both kinds and substitution of recursive functions. All the usual arithmetical, analytical and projective classes are adequate.

If κ is an ordinal, then κ^+ is the least cardinal greater than κ. By "κ^λ" we always mean ordinal exponentiation.

We work here entirely within ZF+DC, Zermelo-Fraenkel set theory with **Dependent Choices**,

$$(\forall u \in x)(\exists v)\langle u, v\rangle \in r \Longrightarrow (\exists f)(\forall n)\langle f(n), f(n+1)\rangle \in r. \tag{DC}$$

We always state all additional hypotheses, including the full **Axiom of Choice**, AC, when we need them.

§2. Norms and the Prewellordering property. The **Prewellordering property** on a pointclass Γ was formulated in order to extend elegantly to Σ^1_2 some of the basic results about Π^1_1. It was later shown that if **Projective Determinacy** (PD) holds, then the Prewellordering property can be established for all Π^1_n (for odd n) and Σ^1_k (for even k), so that the same results could be extended to all analytical classes of the right kind and index. Our main purpose in this section is to establish the elementary facts about the Prewellordering property and prove this theorem.

2.1. Definition and elementary properties. A **norm** on a pointset A is any function $\varphi\colon A \twoheadrightarrow \kappa$ from A onto some ordinal κ, the **length** of φ. Each norm φ determines uniquely a **prewellordering** (reflexive, transitive, connected, wellfounded relation) $\leq^\varphi$ on A given by

$$x \leq^\varphi y \Longleftrightarrow \varphi(x) \leq \varphi(y);$$

conversely each prewellordering $\preccurlyeq$ on A determines a unique norm φ such that $\preccurlyeq \,=\, \leq^\varphi$.

There are, of course, many trivial norms on a pointset, for example the constant 0 function. The concept becomes nontrivial when we place definability conditions on a norm in the following way:

Let Γ be a pointclass, $\varphi\colon A \twoheadrightarrow \kappa$ a norm on some pointset. We call φ a **Γ-norm** if there exist relations $\leq^\varphi_\Gamma, \leq^\varphi_{\check\Gamma}$ in Γ and $\check\Gamma$ respectively such that for every y,

$$y \in A \Longrightarrow \forall x\{(x \in A \wedge \varphi(x) \leq \varphi(y)) \Longleftrightarrow x \leq^\varphi_\Gamma y \Longleftrightarrow x \leq^\varphi_{\check\Gamma} y\}. \tag{1}$$

Notice that if Γ is adequate and φ is a Γ-norm on A we can also define relations $<^\varphi_\Gamma, <^\varphi_{\check\Gamma}$ in $\Gamma, \check\Gamma$ respectively such that for every y,

$$y \in A \Longrightarrow \forall x\{(x \in A \wedge \varphi(x) < \varphi(y)) \Longleftrightarrow x <^\varphi_\Gamma y \Longleftrightarrow x <^\varphi_{\check\Gamma} y\}. \tag{2}$$

In fact we put

$$x <^\varphi_\Gamma y \Longleftrightarrow x \leq^\varphi_\Gamma y \wedge \neg(x \leq^\varphi_{\check\Gamma} y)$$
$$x <^\varphi_{\check\Gamma} y \Longleftrightarrow x \leq^\varphi_{\check\Gamma} y \wedge \neg(x \leq^\varphi_\Gamma y).$$

It is quite important for the applications that the definition of a Γ-norm be precisely that given by (1). Notice that for Γ adequate and $A \in \Gamma$, this is stronger than simply requiring that $\leq^\varphi \in \Gamma$, but weaker than insisting that $\leq^\varphi \in \Gamma \cap \check\Gamma$ (which implies $A \in \Gamma \cap \check\Gamma$).

Finally put:

$$\mathrm{PWO}(\Gamma) \iff \text{Every pointset } A \text{ in } \Gamma \text{ admits a } \Gamma\text{-norm.}$$

This notion is not interesting unless Γ is at least adequate. For adequate, parametrized Γ it is equivalent to that defined in [Mos70A, §2], where there is also a discussion of some properties that it implies (see also [AM68]). We only give here a short proof of the Reduction property from this form of the Prewellordering property.

THEOREM 2.1. Assume Γ is adequate and $\mathrm{PWO}(\Gamma)$; then $\mathrm{Reduction}(\Gamma)$, i.e., if $A, B \in \Gamma$, $A \subseteq \mathfrak{X}$, $B \subseteq \mathfrak{X}$, then there exists $A_1 \subseteq A$, $B_1 \subseteq B$, $A_1 \in \Gamma$, $B_1 \in \Gamma$ such that $A_1 \cap B_1 = \varnothing$, $A_1 \cup B_1 = A \cup B$.

PROOF. Let $A, B \subseteq \mathfrak{X}$, $A, B \in \Gamma$ be given. Define $C = (A \times \{0\}) \cup (B \times \{1\})$ (notice that $C \subseteq \mathfrak{X} \times \omega$). Since Γ is adequate, $C \in \Gamma$. Let φ be a Γ-norm on C, and put

$$x \in A_1 \iff x \in A \wedge \neg(\langle x, 1\rangle \leq^{\varphi}_{\Gamma} \langle x, 0\rangle)$$
$$x \in B_1 \iff x \in B \wedge \neg(\langle x, 0\rangle <^{\varphi}_{\Gamma} \langle x, 1\rangle).$$

An easy checking shows that A_1, B_1 have all the required properties. ⊣

The following trivial observation will save us having to deal separately, with the "lightface" and "boldface" cases:

PROPOSITION 2.2. If Γ is adequate and $\mathrm{PWO}(\Gamma)$, then

$$\mathrm{PWO}(\utilde{\Gamma}).$$

PROOF. Let $A \in \utilde{\Gamma}$, $A \subseteq \mathfrak{X}$. Then for some α_0 and some $B \subseteq \mathbb{R} \times \mathfrak{X}$, $B \in \Gamma$ we have $A = \{\alpha : \langle \alpha_0, \alpha\rangle \in B\}$. Find a Γ-norm $\bar{\varphi}$ for B and define the following ordinal map on A

$$\psi(\alpha) = \bar{\varphi}(\alpha_0, \alpha).$$

Then let $\alpha \preccurlyeq \beta \iff \psi(\alpha) \leq \psi(\beta)$ and let φ be the norm on A such that $\preccurlyeq \,=\, \leq^{\varphi}$. (We define the norm in this roundabout way because ψ need not be *onto* an ordinal.) Clearly φ is a $\utilde{\Gamma}$-norm on A. ⊣

2.2. Establishing the Prewellordering property. We first prove

THEOREM 2.3. $\mathrm{PWO}(\Pi^1_1)$.

PROOF. Let $A \in \Pi^1_1$, $A \subseteq \mathcal{X}$. For $\alpha \in {}^{\omega}\omega$, let

$$\leq_{\alpha} := \{\langle m, n\rangle : \alpha(\ulcorner m, n\urcorner) = 0\}.$$

Now put

$$\mathrm{LOR} = \{\alpha : \leq_{\alpha} \text{ is a linear ordering}\}$$
$$\mathrm{WO} = \{\alpha : \leq_{\alpha} \text{ is a wellordering}\}$$

Then for some recursive function $f : \mathfrak{X} \to \mathbb{R}$, $f(x) \in \text{LOR}$, for all $x \in \mathfrak{X}$ and $x \in A \iff f(x) \in \text{WO}$. Define on A

$$\psi(x) = |f(x)|$$

where for $\alpha \in \text{WO}$, $|\alpha|$ is the length of $\leq_\alpha$. Then let $\preccurlyeq$ be defined by $x \preccurlyeq y \iff \psi(x) \leq \psi(y)$ and let φ be the norm of A such that $\preccurlyeq\, = \,\leq^\varphi$.

To see that φ is actually a Π^1_1-norm, we use the well-known fact that there exist relations $Q_{\Pi^1_1}, Q_{\Sigma^1_1}$ in Π^1_1, Σ^1_1 respectively, such that for $\beta \in \text{WO}$,

$$(\alpha \in \text{WO} \wedge |\alpha| \leq |\beta|) \iff Q_{\Pi^1_1}(\alpha, \beta) \iff Q_{\Sigma^1_1}(\alpha, \beta). \qquad \dashv$$

We now give two theorems which establish the prewellordering property for some pointclasses closed under $\exists^{\mathbb{R}}$.

Theorem 2.4. (Moschovakis, see [Mos70A]) Assume Γ is adequate, $A \in \Gamma$ and A admits a Γ-norm. Then $\exists^{\mathbb{R}} A$ admits an $\exists^{\mathbb{R}}\forall^{\mathbb{R}}\Gamma$-norm.

Corollary 2.5. If Γ is adequate, $\text{PWO}(\Gamma)$, and $\forall^{\mathbb{R}}\Gamma \subseteq \Gamma$ imply

$$\text{PWO}(\exists^{\mathbb{R}}\Gamma).$$

Corollary 2.6. $\text{PWO}(\Sigma^1_2)$.

Proof of Theorem 2.4. Let $A \subseteq \mathfrak{X} \times \mathbb{R}$, $A \in \Gamma, B = \exists^{\mathbb{R}} A = \{x : \exists\alpha(x, \alpha) \in A\}$. Let φ be a Γ-norm on A, define $\preccurlyeq$ on B by

$$x \preccurlyeq y \iff \min\{\varphi(x, \alpha) : \langle x, \alpha\rangle \in A\} \leq \min\{\varphi(y, \beta) : \langle y, \beta\rangle \in A\}$$

and let

$$\psi : B \twoheadrightarrow \lambda$$

be the unique norm such that $\leq^\psi\, = \,\preccurlyeq$.

To check that this is actually a $\exists^{\mathbb{R}}\forall^{\mathbb{R}}\Gamma$-norm, let $\leq^\varphi_\Gamma, \leq^\varphi_{\check{\Gamma}}, <^\varphi_\Gamma, <^\varphi_{\check{\Gamma}}$ be the relations associated with φ and notice that for $y \in B$,

$$\begin{aligned}(x \in B \wedge \psi(x) \leq \psi(y)) &\iff (\exists\alpha)(\langle x, \alpha\rangle \in A \wedge \forall\beta(\neg\langle y, \beta\rangle <^\varphi_{\check{\Gamma}} \langle x, \alpha\rangle)) \\ &\iff (\forall\beta)(\langle y, \beta\rangle \in A \Longrightarrow \exists\alpha(\langle x, \alpha\rangle \leq^\varphi_{\check{\Gamma}} \langle y, \beta\rangle)).\end{aligned}$$

$\dashv$ (Theorem 2.4)

Let $\leq$ be a wellordering of $\mathbb{R}$ of order-type $\aleph_1$; put

$$\text{In.Segment}_\leq(\gamma, \alpha) \iff \{\beta : \beta \leq \alpha\} = \{(\gamma)_n : n = 0, 1, 2, \dots\}.$$

We call $\leq$ **Γ-good** if both the relation $\leq$ and the relation $\text{In.Segment}_\leq$ are in $\Gamma \cap \check{\Gamma}$.

Theorem 2.7 (Essentially Addison [Add59]). Assume Γ is adequate, closed under both $\exists^\omega, \forall^\omega$ and $\check{\Gamma} \subseteq \exists^{\mathbb{R}}\Gamma$ and there is some wellordering $\leq$ of $\mathbb{R}$ which is Γ-good; then $\text{PWO}(\exists^{\mathbb{R}}\Gamma)$.

PROOF. If $A = \exists^{\mathbb{R}} B$ with $B \in \Gamma$, define on A

$$x \preccurlyeq y \Longleftrightarrow \min\{\alpha : \langle x, \alpha\rangle \in B\} \leq \min\{\beta : \langle y, \beta\rangle \in B\},$$

where the minima are taken in the good wellordering $\leq$ and let ψ be the associated norm. The computation is easy. ⊣

COROLLARY 2.8. **V=L** implies that for every $k \geq 2$, $\mathrm{PWO}(\Sigma^1_k)$.

PROOF. **V=L** implies there exists a wellordering of $\mathbb{R}$ which is Δ^1_2-good. ⊣

COROLLARY 2.9 (Silver, [Sil71]). If μ is a normal κ-additive measure on some cardinal, then $\mathbf{V}=\mathbf{L}[\mu]$ implies that for all $k \geq 2$, $\mathrm{PWO}(\Sigma^1_k)$.

PROOF. $\mathbf{V}=\mathbf{L}[\mu] \Longrightarrow$ there exists a wellordering of $\mathbb{R}$ which is Δ^1_3-good. ⊣

2.3. The First Periodicity Theorem.

THEOREM 2.10 (Martin, Moschovakis, [Mar68, AM68]). Assume that Γ is adequate and that $\mathrm{Det}(\underset{\sim}{\Gamma} \cap \underset{\sim}{\check{\Gamma}})$ holds. Then if $A \in \Gamma$ and A admits a Γ-norm, $\forall^{\mathbb{R}} A$ admits a $\forall^{\mathbb{R}}\exists^{\mathbb{R}}\Gamma$-norm.

COROLLARY 2.11. If Γ is adequate with $\mathrm{PWO}(\Gamma)$, $\mathrm{Det}(\underset{\sim}{\Gamma} \cap \underset{\sim}{\check{\Gamma}})$ and $\exists^{\mathbb{R}}\Gamma \subseteq \Gamma$, then $\mathrm{PWO}(\forall^{\mathbb{R}}\Gamma)$.

PROOF OF THEOREM 2.10. Assume the hypotheses hold and let ψ be a Γ-norm on $A \subseteq \mathfrak{X} \times \mathbb{R}$, $A \in \Gamma$. Let $B = \forall^{\mathbb{R}} A = \{x : \forall \alpha(\langle x, \alpha\rangle \in A)\}$. Instead of giving directly a norm φ on A we shall define the associated prewellordering $\leq^{\varphi}$, for simplicity $\preccurlyeq$.

Given $x, y \in \mathfrak{X}$ consider the game $\mathrm{G}(x, y)$

$$\begin{array}{cc} \mathrm{I} & \mathrm{II} \\ \alpha(0) & \beta(0) \\ \alpha(1) & \beta(1) \\ \alpha(2) & \beta(2) \\ \vdots & \vdots \\ \alpha & \beta \end{array}$$

Player I plays α, player II plays β and player II wins if

$$\langle y, \beta\rangle \notin A \vee (\langle y, \beta\rangle \in A \wedge \langle x, \alpha\rangle \in A \wedge \psi(x, \alpha) \leq \psi(y, \beta)).$$

Now for $x, y \in B$ put

$$x \preccurlyeq y \Longleftrightarrow \text{player II has a winning strategy in } \mathrm{G}(x, y).$$

The idea is that $x \preccurlyeq y$ if there is some procedure (strategy) τ which to each α assigns bit-by-bit some $[\alpha] * \tau$ so that $\psi(x, \alpha) \leq \psi(y, [\alpha] * \tau)$; in some sense, $\sup_\alpha\{\psi(x, \alpha)\} \leq \sup_\beta\{\psi(y, \beta)\}$ *effectively*.

CLAIM 2.12. $\forall x \in B(x \preccurlyeq x)$.

Proof of Claim 2.12. In the game $G(x, x)$, where $x \in B$, player II has always a trivial winning strategy, namely copying down the moves of player I.

⊣ (Claim 2.12)

Claim 2.13. $\forall x, y, z \in B(x \preccurlyeq y \wedge y \preccurlyeq z \Longrightarrow x \preccurlyeq z)$.

Proof of Claim 2.13. Let $x, y, z \in B \wedge x \preccurlyeq y \wedge y \preccurlyeq z$. Then player II has winning strategies in both $G(x, y)$ and $G(y, z)$. Fix one in each game and consider the diagram:

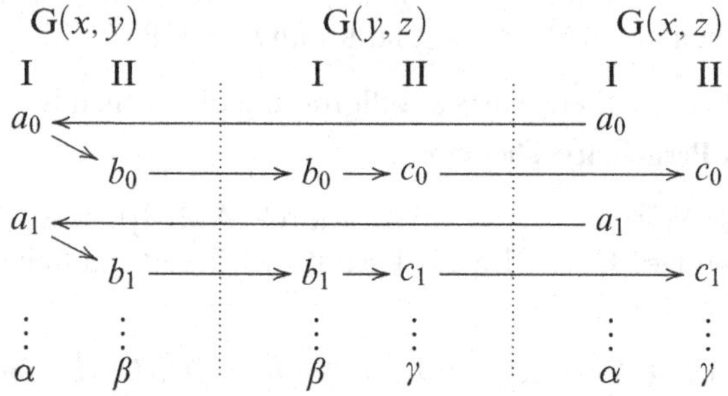

We describe a strategy for player II in $G(x, z)$ as follows: Player I plays in $G(x, z)$ a_0. Then player I copies in $G(x, y)$ a_0 and player II answers in $G(x, y)$ by his winning strategy to give b_0. Player I plays in $G(y, z)$ this b_0 and player II answers in $G(y, z)$ by his winning strategy to give c_0. Player II's answer in $G(x, z)$ is this c_0, and so on as in the diagram. After the game is over, reals α, β, γ result, and, since player II played with his winning strategy in $G(x, y)$ and $G(y, z)$, it follows that $\psi(x, \alpha) \leq \psi(y, \beta) \leq \psi(z, \gamma)$ (recall that $x, y, z \in B$). Thus $\psi(x, \alpha) \leq \psi(z, \gamma)$ and player II wins $G(x, z)$.

We described a winning strategy for player II in $G(x, z)$, so $x \preccurlyeq z$.

⊣ (Claim 2.13)

Claim 2.14. For any x, y, $G(x, y)$ is determined. If $x, y \in B$, then $x \prec y \Longleftrightarrow$ player I has a winning strategy in $G(y, x)$, where $x \prec y \Longleftrightarrow x \preccurlyeq y \wedge \neg(y \preccurlyeq x)$. Thus, $\forall x, y \in B(x \preccurlyeq y \text{ or } y \preccurlyeq x)$.

Proof of Claim 2.14. Let $x, y \in \mathfrak{X}$ be given. If $y \notin B$, pick β_0 such that $\langle y, \beta_0 \rangle \notin A$; player II then wins by playing β_0. If $y \in B$ then we have

$$\begin{aligned} \text{player II wins } G(x, y) &\Longleftrightarrow (\langle x, \alpha \rangle \in A \wedge \psi(x, \alpha) \leq \psi(y, \beta)) \\ &\Longleftrightarrow \langle x, \alpha \rangle \leq^{\psi}_{\Gamma} \langle y, \beta \rangle \\ &\Longleftrightarrow \langle x, \alpha \rangle \leq^{\psi}_{\check{\Gamma}} \langle y, \beta \rangle. \end{aligned}$$

Thus the game is in $\underset{\sim}{\Gamma} \cap \underset{\sim}{\check{\Gamma}}$ and is determined.

Now assume $x, y \in B$. If $x \prec y$, then $\neg(y \preccurlyeq x)$, so player II does not have a winning strategy in $G(y, x)$ and player I has a winning strategy in $G(y, x)$.

Conversely assume that player I has a winning strategy in $G(y, x)$. We shall show that $x \prec y$, *i.e.*, player II has a winning strategy in $G(x, y)$, but player II has no winning strategy in $G(y, x)$. The last statement is obvious so we proceed to invent a winning strategy for player II in $G(x, y)$.

Fix a strategy for player I in $G(y, x)$. Consider the diagram

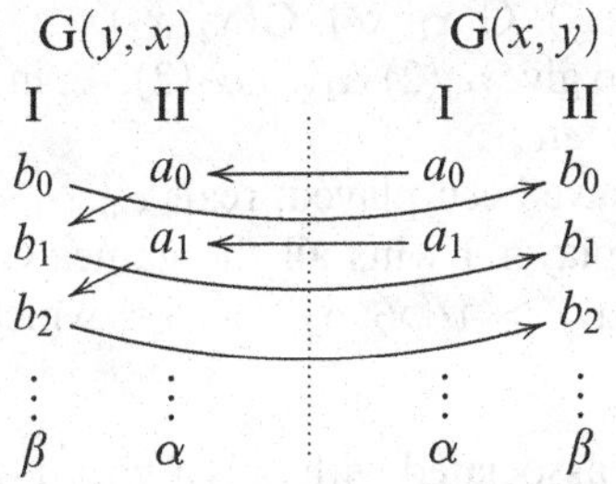

We describe a strategy for player II in $G(x, y)$ as follows: Player I plays in $G(x, y)$ a_0. Then player I plays b_0 by his strategy in $G(y, x)$ and player II plays this b_0 in $G(x, y)$ answering a_0. Then player I plays a_1 in $G(x, y)$. Player II copies a_0 in $G(y, x)$ and player I answers by his strategy in $G(y, x)$ to give b_1. Player II plays this b_1 in $G(x, y)$ answering to a_1. Then player I gives a_2 in $G(x, y)$, player II copies a_1 in $G(y, x)$ and player I answers in $G(y, x)$ by his strategy to give b_2. Player II then copies b_2 in $G(x, y)$, *etc.* After the end of the game reals α, β result as in the picture. Since player I played by his winning strategy in $G(y, x)$, we have $\psi(y, \beta) > \psi(x, \alpha)$; thus player II wins $G(x, y)$. Thus the above described strategy is a winning one for player II in $G(x, y)$ and we are done. ⊣ (Claim 2.14)

Claim 2.15. The relation $\preccurlyeq$ is wellfounded.

Proof of Claim 2.15. We have to show that there is no infinite descending chain $x_0 \succ x_1 \succ x_2 \succ \cdots$. Assume not, towards a contradiction. If $x_0 \succ x_1 \succ x_2 \succ \cdots$, then player I wins $G(x_i, x_{i+1})$ for each $i \geq 0$. Fix winning strategies for player I in each one of these games.

Consider the diagram:

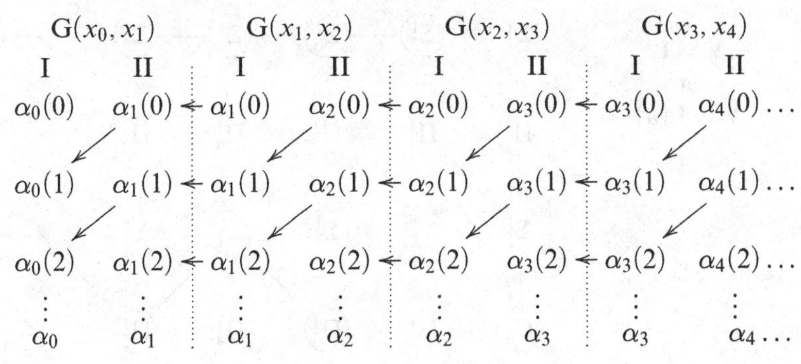

Let first player I play $\alpha_0(0)$, $\alpha_1(0)$, $\alpha_2(0), \ldots$ in $G(x_0, x_1)$, $G(x_1, x_2)$, $G(x_2, x_3), \ldots$, respectively by following his winning strategies. Then let player II play $\alpha_1(0)$, $\alpha_2(0)$, $\alpha_3(0), \ldots$ in $G(x_0, x_1)$, $G(x_1, x_2)$, $G(x_2, x_3), \ldots$, respectively. Player I answers by his winning strategies to give $\alpha_0(1), \alpha_1(1), \alpha_2(1), \ldots$ in $G(x_0, x_1)$, $G(x_1, x_2)$, $G(x_2, x_3), \ldots$, respectively. Then player II plays $\alpha_1(1)$, $\alpha_2(1)$, $\alpha_3(1), \ldots$ in $G(x_0, x_1)$, $G(x_1, x_2)$, $G(x_2, x_3), \ldots$, respectively. Player I answers by his strategies to give $\alpha_0(2)$,$\alpha_1(2)$,$\alpha_2(2), \ldots$ in $G(x_0, x_1)$, $G(x_1, x_2)$, $G(x_2, x_3), \ldots$, respectively, *etc.*

After all these moves have been played, reals $\alpha_0, \alpha_1, \alpha_2, \ldots$ are created as in the picture and since player I wins all the games $G(x_i, x_{i+1}), i \geq 0$, we have $\psi(x_0, \alpha_0) > \psi(x_1, \alpha_1) > \psi(x_2, \alpha_2) > \cdots$, which is a contradiction. ⊣ (Claim 2.15)

Claim 2.16. The norm associated with $\preccurlyeq$ is a $\forall^{\mathbb{R}}\exists^{\mathbb{R}}\Gamma$-norm on B.

Proof of Claim 2.16. Notice first that for $y \in B, x \in B \wedge x \preccurlyeq y \iff$ player II has a winning strategy in $G(x, y) \iff \exists\tau\forall\alpha(\langle x, \alpha\rangle \leq^{\psi}_{\Gamma} \langle y, [\alpha] * \tau\rangle)$. But since $G(x, y)$ is determined, player II has a winning strategy in $G(x, y) \iff$ player I has no winning strategy in $G(x, y) \iff \forall\sigma\exists\beta(\langle x, \sigma * [\beta]\rangle \leq^{\psi}_{\Gamma} \langle y, \beta\rangle)$. Thus for $y \in B$,

$$x \in B \wedge x \preccurlyeq y \iff \exists\tau\forall\alpha(\langle x, \alpha\rangle \leq^{\psi}_{\Gamma} \langle y, [\alpha] * \tau\rangle)$$
$$\iff \forall\sigma\exists\beta(\langle x, \sigma * [\beta]\rangle \leq^{\psi}_{\Gamma} \langle y, \beta\rangle),$$

and we are done. ⊣ (Claim 2.16)

⊣ (Theorem 2.10)

2.4. The zig-zag picture. It is not hard to verify that the Reduction property cannot hold both for Σ^1_n and Π^1_n for any n, hence the same is true for the Prewellordering property. If we make a diagram of the classes Σ^1_n, Π^1_n and circle those which have the Prewellordering property, we get the following two pictures, under the hypotheses $\mathbf{V} = \mathbf{L}$ (or $\mathbf{V} = \mathbf{L}[\mu]$) and PD:

$\mathbf{V} = \mathbf{L}$ or $\mathbf{V} = \mathbf{L}[\mu]$

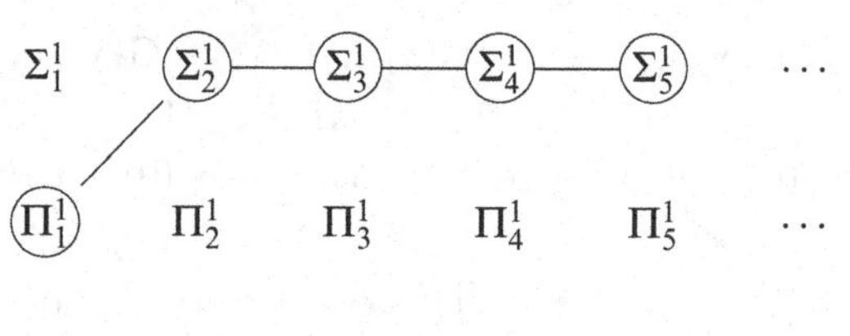

PD

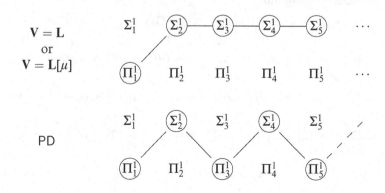

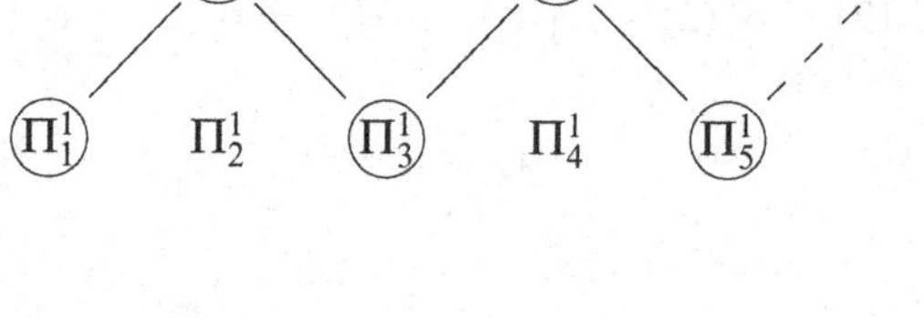

This second picture is the motivation for the name "Periodicity Theorem." We will see later that (assuming PD) we can construct models in which this picture has any finite predetermined number of "teeth."

§3. Scales. In this section we define scales and the property Scale(Γ), we prove that Scale(Γ) $\Longrightarrow$ Unif(Γ) for suitable Γ (in particular $\Pi^1_1, \Pi^1_3, \Pi^1_5, \dots$) and we establish Scale(Γ) for $\Gamma = \Pi^1_1, \Sigma^1_2, \Pi^1_3, \Sigma^1_4, \dots$ under the hypothesis PD. The elementary theory of scales is quite similar to that of norms, so the structure of this section is parallel to that of §2.

3.1. Definitions and basic properties. A **scale** on a pointset A is a sequence of norms $\langle \varphi_n : n \in \omega \rangle$ on A with the following **limit property**:

If $x_0, x_1, x_2, \dots \in A$, if $\lim_{i \to \infty} x_i = x$, if for each n and all large i, $\varphi_n(x_i) = \lambda_n$, then $x \in A$ and for each n, $\varphi_n(x) \le \lambda_n$.

It is easy to see that assuming AC every pointset admits a scale (take every φ_n to be equal to a fixed 1-1 mapping from the pointset onto an ordinal). In order to get something interesting we place definability conditions on a scale as follows:

Let Γ be a pointclass, $\langle \varphi_n : n \in \omega \rangle$ a scale on a pointset A. We call $\langle \varphi_n : n \in \omega \rangle$ a **Γ-scale** if there exist relations $S_\Gamma, S_{\check{\Gamma}}$ in Γ and $\check{\Gamma}$ respectively such that, for every y,

$$y \in A \Longrightarrow \forall x\{[x \in A \wedge \varphi_n(x) \le \varphi_n(y)] \Longleftrightarrow S_\Gamma(n, x, y) \Longleftrightarrow S_{\check{\Gamma}}(n, x, y)]. \quad (3)$$

Finally we put:

$$\text{Scale}(\Gamma) \Longleftrightarrow \text{Every pointset } A \text{ in } \Gamma \text{ admits a } \Gamma\text{-scale.}$$

The notion of a scale and the associated scale property were introduced in connection with the uniformization problem in [Mos71A]. If Γ, Γ^* are pointclasses, put:

$$\begin{aligned}\text{Unif}(\Gamma, \Gamma^*) \Longleftrightarrow\ & \text{For every } P \in \Gamma, P \subseteq \mathfrak{X} \times \mathcal{Y} \text{ we can find}\\ & P^* \in \Gamma^*, P^* \subseteq \mathfrak{X} \times \mathcal{Y} \text{ such that } P^* \subseteq P\\ & \text{and } \forall x(\exists y P(x, y) \Longleftrightarrow \exists! y P^*(x, y)).\end{aligned}$$

(In this case we say that P^* **uniformizes** P.) In this definition we allow $P \subseteq \mathcal{Y}$ in which case $P^* \subseteq P$ and $\exists y P(y) \Longleftrightarrow \exists! y P^*(y)$, i.e., P^* is a singleton contained in P, if $P \ne \varnothing$.

We abbreviate

$$\text{Unif}(\Gamma) \Longleftrightarrow \text{Unif}(\Gamma, \Gamma).$$

THEOREM 3.1. *Assume Γ is adequate and closed under $\forall^{\mathbb{R}}$. Then,*

$$\text{Scale}(\Gamma) \Longrightarrow \text{Unif}(\Gamma).$$

Proof. Let $P \in \Gamma, P \subseteq \mathfrak{X} \times \mathcal{Y}$, where for simplicity and without loss of generality we can assume $\mathcal{Y} = \mathbb{R}$. Let $\langle \varphi_n : n \in \omega \rangle$ be a Γ-scale for P with associated relations $S_\Gamma, S_{\check{\Gamma}}$. Fix $x \in \mathfrak{X}$ and define inductively the following (where we agree that $\min(\varnothing) = 0$):

$$\begin{cases} P_0^x = \{\alpha : P(x,\alpha)\} \\ \lambda_0^x = \min\{\varphi_0(x,\alpha) : \alpha \in P_0^x\} \\ k_0^x = \min\{\alpha(0) : \alpha \in P_0^x \wedge \varphi_0(x,\alpha) = \lambda_0^x\}, \end{cases}$$

$$\begin{cases} P_{n+1}^x = \{\alpha : \alpha \in P_n^x \wedge \alpha(n) = k_n^x \wedge \varphi_n(x,\alpha) = \lambda_n^x\} \\ \lambda_{n+1}^x = \min\{\varphi_{n+1}(x,\alpha) : \alpha \in P_{n+1}^x\} \\ k_{n+1}^x = \min\{\alpha(n+1) : \alpha \in P_{n+1}^x \wedge \varphi_{n+1}(x,\alpha) = \lambda_{n+1}^x\}. \end{cases}$$

Finally put

$$P_\infty^x = \bigcap_{n \in \omega} P_n^x.$$

We have now the following:

(1) $P_0^x \supseteq P_1^x \supseteq P_2^x \supseteq \cdots$.
(2) $\exists \alpha P(x,\alpha) \Longrightarrow \forall n (P_n^x \neq \varnothing)$.
(3) Assume $\exists \alpha P(x,\alpha)$. Let $\alpha^x = \langle k_0^x, k_1^x, k_2^x, \dots \rangle$. Then $P_\infty^x = \{\alpha^x\}$.

Proof of (3). If $\alpha \in P_\infty^x$, then for each $n, \alpha(n) = k_n^x$, *i.e.*, $\alpha = \alpha^x$. Conversely, if $\exists \alpha P(x,\alpha)$ pick reals $\alpha_i \in P_i^x, i = 0, 1, \dots$. Then for $i > n, \alpha_i(n) = k_n^x$, thus $\alpha_i \to \alpha^x$ and $\langle x, \alpha_i \rangle \to \langle x, \alpha^x \rangle$. Also for $i > n, \varphi_n(x,\alpha_i) = \lambda_n^x$, hence by the limit property of scales, $\langle x, \alpha^x \rangle \in P$ and $\varphi_n(x,\alpha^x) \leq \lambda_n^x$. Then certainly $\alpha^x \in P_0^x$. But also $\alpha^x(0) = k_0^x$ and $\varphi_0(x,\alpha^x) \leq \lambda_0^x$, *i.e.*, $\varphi_0(x,\alpha^x) = \lambda_0^x$. Thus $\alpha^x \in P_1^x$. A similar argument shows inductively that for all $n, \alpha^x \in P_n^x$. ⊣ (3)

Put now

$$P^*(x,\alpha) \Longleftrightarrow \exists \alpha P(x,\alpha) \wedge \alpha = \alpha^x.$$

Clearly $P^* \subseteq P$ and

$$\exists \alpha P(x,\alpha) \Longrightarrow \exists! \alpha P^*(x,\alpha).$$

To complete the proof it will be enough to show that $P^* \in \Gamma$. It is easier to show that the complement of P^* is in $\check{\Gamma}$. And this follows from the computation

$$
\begin{aligned}
&\neg P^*(x,\alpha)\\
&\iff\\
&\neg P(x,\alpha) \vee \Big\{P(x,\alpha) \wedge \exists n \exists \beta \Big\{P(x,\beta)\\
&\qquad \wedge \Big((\forall i < n)\big(\alpha(i) = \beta(i)\\
&\qquad\qquad \wedge \varphi_i(x,\alpha) = \varphi_i(x,\beta)\big)\\
&\qquad \wedge \big(\varphi_n(x,\beta) < \varphi_n(x,\alpha)\\
&\qquad\qquad \vee (\varphi_n(x,\beta) = \varphi_n(x,\alpha)\\
&\qquad\qquad\quad \wedge \beta(n) < \alpha(n))\big)\Big)\Big\}\Big\}\\
&\iff\\
&\neg P(x,\alpha) \vee \Big\{\exists n \exists \beta \{((\forall i < n)(S_{\breve{\Gamma}}(i,x,\beta,x,\alpha)\\
&\qquad \wedge S_{\breve{\Gamma}}(i,x,\alpha,x,\beta) \wedge \alpha(i) = \beta(i))\\
&\qquad \wedge (S_{\breve{\Gamma}}(n,x,\beta,x,\alpha) \wedge \neg S_{\Gamma}(n,x,\alpha,x,\beta)\\
&\qquad\qquad \vee (\beta(n) < \alpha(n) \wedge S_{\breve{\Gamma}}(n,x,\beta,x,\alpha)\\
&\qquad\qquad\quad \wedge S_{\breve{\Gamma}}(n,x,\alpha,x,\beta)))\}\Big\}.
\end{aligned}
$$

⊣

Again we should mention the trivial observation that if Γ is adequate, then Scale(Γ) implies Scale($\underset{\sim}{\Gamma}$).

3.2. Establishing the Scale property.

Theorem 3.2. Scale(Π^1_1).

Corollary 3.3 (The classical Novikoff-Kondô-Addison Theorem). Unif(Π^1_1).

Proof of Theorem 3.2. Let $A \subseteq \mathfrak{X}, A \in \Pi^1_1$. Then for some recursive function $f : \mathfrak{X} \to \mathbb{R}$

$$x \in A \iff f(x) \in \mathrm{WO}.$$

For $\alpha \in \mathrm{WO}$ put $\alpha \restriction n = \{m : m <_\alpha n\}$ (where $m <_\alpha n \iff m \leq_\alpha n \wedge m \neq n$) and $|\alpha \restriction n|$ = the length of $\alpha \restriction n$ ($\alpha \restriction n$ is an initial segment of $\leq_\alpha$). If $n \notin \mathrm{Field}(\leq_\alpha)$, *i.e.*, if $\alpha(\ulcorner n, n \urcorner) \neq 0$, then of course $|\alpha \restriction n| = 0$. Define now the following prewellorderings on A

$$
\begin{aligned}
x \leq_n y \iff & |f(x)| < |f(y)| \vee \\
& (|f(x)| = |f(y)| \wedge |f(x) \restriction n| \leq |f(y) \restriction n|).
\end{aligned}
$$

Let φ_n be the norm on A such that $\leq^{\varphi_n} = \leq_n$. We will show that $\langle \varphi_n : n \in \omega \rangle$ is a Π^1_1-scale on A.

CLAIM 3.4. $\langle \varphi_n : n \in \omega \rangle$ is a scale.

PROOF OF CLAIM 3.4. Assume $x_i \in A$ for every $i, x_i \to x$ and for some $\langle \lambda_n : n \in \omega \rangle, \varphi_n(x_i) = \lambda_n$ for all sufficiently large i. This implies that $|f(x_i)| = \lambda'$ for some λ' and all sufficiently large i and also that for some $\langle \lambda'_n : n \in \omega \rangle, |f(x_i) \restriction n| = \lambda'_n$ for all sufficiently large i. We show first that $x \in A$, *i.e.*, that $f(x) \in \mathrm{WO}$, by proving that the mapping

$$n \mapsto \lambda'_n$$

is order preserving on the field of $\leq_{f(x)}$. In fact, let $n <_{f(x)} m$, *i.e.*, $f(x)(\ulcorner n, m \urcorner) = 0 \wedge f(x)(\ulcorner m, n \urcorner) \neq 0$. Since f is continuous and $x_i \to x$, clearly for all sufficiently large i, $f(x_i)(\ulcorner n, m \urcorner) = 0$ and $f(x_i)(\ulcorner m, n \urcorner) \neq 0$, *i.e.*, $n <_{f(x_i)} m$. But then $|f(x_i) \restriction n| < |f(x_i) \restriction m|$. Taking i sufficiently large this shows that $\lambda'_n < \lambda'_m$ and we are done.

Finally we have to show that for each n

$$|f(x)| < \lambda' \vee (|f(x)| = \lambda' \wedge |f(x) \restriction n| \leq \lambda'_n).$$

Since $n \mapsto \lambda'_n$ is order preserving on the field of $\leq_{f(x)}$ it follows that $|f(x) \restriction n| \leq \lambda'_n$ for each n. But $\lambda'_n \leq \lambda'$ for each n (since for large enough i, $\lambda'_n = |f(x_i) \restriction n|$ while $\lambda' = |f(x_i)|$). Thus

$$|f(x)| = \sup\{|f(x) \restriction n| : n \in \omega\} \leq \lambda'. \qquad \dashv \text{(Claim 3.4)}$$

CLAIM 3.5. $\langle \varphi_n : n \in \omega \rangle$ is a Π^1_1-scale.

This is proved by a computation similar to that in the proof of Theorem 2.3. ⊣ (Theorem 3.2)

It should be pointed out here that the proofs of Theorems 3.2 and 3.1 taken together constitute the classical proof of the Novikoff-Kondô-Addison theorem.

We proceed now to prove the analogue of Theorem 2.4.

THEOREM 3.6 (Moschovakis, [Mos71A]). Assume Γ is adequate, $A \in \Gamma$ and A admits a Γ-scale; then $\exists^{\mathbb{R}} A$ admits an $\exists^{\mathbb{R}}\forall^{\mathbb{R}}\Gamma$-scale.

COROLLARY 3.7. If Γ adequate and $\forall^{\mathbb{R}}\Gamma \subseteq \Gamma$, then $\mathrm{Scale}(\Gamma)$ implies $\mathrm{Scale}(\exists^{\mathbb{R}}\Gamma)$.

COROLLARY 3.8. $\mathrm{Scale}(\Sigma^1_2)$.

PROOF OF THEOREM 3.6. If λ is any ordinal and $n \in \omega$, consider the **lexicographical wellordering** of ${}^n\lambda$,

$$\langle \xi_1, \dots, \xi_n \rangle \leq \langle \eta_1, \dots, \eta_n \rangle \iff \xi_1 < \eta_1 \vee (\xi_1 = \eta_1 \wedge \xi_2 < \eta_2) \vee \dots \vee (\xi_1 = \eta_1 \wedge \dots \wedge \xi_{n-1} = \eta_{n-1} \wedge \xi_n \leq \eta_n).$$

This wellorders ${}^n\lambda$ with ordinal λ^n and we let $\ulcorner\xi_1,\dots,\xi_n\urcorner$ = ordinal of $\langle\xi_1,\dots,\xi_n\rangle$ in the lexicographical wellordering. Each $\vartheta < \lambda^n$ can be written uniquely as $\vartheta = \ulcorner\xi_1,\dots,\xi_n\urcorner$ for $\xi_i < \lambda$.

Now let $A \subseteq \mathfrak{X} \times \mathbb{R}$, $A \in \Gamma$, and assume $\langle\varphi_n : n \in \omega\rangle$ is a Γ-scale on A. If $B = \exists^{\mathbb{R}} A = \{x : \exists\alpha(\langle x,\alpha\rangle \in A)\}$, define for $x \in B$

$$\psi'_n(x) = \min\{\ulcorner\varphi_0(x,\alpha),\alpha(0),\varphi_1(x,\alpha),\alpha(1),\dots,\varphi_n(x,\alpha),\alpha(n)\urcorner : \langle x,\alpha\rangle \in A\}.$$

Let $\leq_n$ be the prewellordering (on B)

$$x \leq_n y \iff \psi'_n(x) \leq \psi'_n(y)$$

and let ψ_n be the associated norm. We will prove that $\langle\psi_n : n \in \omega\rangle$ is a $\exists^{\mathbb{R}}\forall^{\mathbb{R}}\Gamma$-scale on A.

CLAIM 3.9. $\langle\psi_n : n \in \omega\rangle$ is a scale.

PROOF OF CLAIM 3.9. Assume $x_i \in B$, and $x_i \to x$ and for some $\langle\lambda_n : n \in \omega\rangle$ we have for each n, $\psi_n(x_i) = \lambda_n$, for all large enough i. In fact we may assume without loss of generality that for $i \geq n$, $\psi_n(x_i) = \lambda_n$. This implies that for some $\langle\lambda_n : n \in \omega\rangle$ and all $i \geq n$, $\psi'_n(x_i) = \lambda'_n$. Each λ'_n can be written uniquely as

$$\lambda'_n = \ulcorner\lambda_0^n, k_0^n, \dots, \lambda_n^n, k_n^n\urcorner.$$

We claim that for $i \geq n$, $\lambda_n^i = \lambda_n^n$ and $k_n^i = k_n^n$. Because if $i \geq n$,

$$\psi'_n(x_i) = \lambda'_n = \min\{\langle\varphi_0(x_i,\alpha),\alpha(0),\dots,\varphi_n(x_i,\alpha),\alpha(n)\rangle : \langle x_i,\alpha\rangle \in A\}$$

and

$$\psi'_i(x_i) = \lambda'_i = \min\{\ulcorner\varphi_0(x_i,\alpha),\alpha(0),\dots,\varphi_n(x_i,\alpha),\alpha(n),\dots,\varphi_i(x_i,\alpha),\alpha(i)\urcorner : \langle x_i,\alpha\rangle \in A\}.$$

Thus if α_i is such that $\langle x_i,\alpha_i\rangle \in A$ and $\psi'_i(x_i) = \lambda'_i = \ulcorner\varphi_0(x_i,\alpha_i), \alpha_i(0),\dots,\varphi_i(x_i,\alpha_i),\alpha_i(i)\urcorner$, then we must have $\ulcorner\varphi_0(x_i,\alpha_i),\alpha_i(0),\dots, \varphi_n(x_i,\alpha_i),\alpha_i(n)\urcorner = \lambda'_n$ (since to minimize a sequence lexicographically one has to minimize first all its initial segments). Thus $\lambda_n^i = \varphi_n(x_i,\alpha_i) = \lambda_n^n$ and $k_n^i = \alpha_i(n) = k_n^n$, and the claim is proved.

Now find α_i such that $\langle x_i,\alpha_i\rangle \in A$ and $\psi'_i(x_i) = \lambda'_i = \ulcorner\varphi_0(x_i,\alpha_i),\alpha_i(0), \dots,\varphi_i(x_i,\alpha_i),\alpha_i(i)\urcorner$. As shown above, for $i \geq n$, $\alpha_i(n) = k_n^n$ and therefore $\alpha_i \to \alpha = \langle k_0^0, k_1^1, k_2^2, \dots\rangle$. Thus $\langle x_i,\alpha_i\rangle \to \langle x,\alpha\rangle$. Moreover, for $i \geq n$, $\varphi_n(x_i,\alpha_i) = \lambda_n^n$, so that by the limit property of scales

$$\langle x,\alpha\rangle \in A \wedge \varphi_n(x,\alpha) \leq \lambda_n^n.$$

But then $x \in B$ and

$$\begin{aligned}\psi'_n(x) &= \min\{\ulcorner\varphi_0(x,\alpha),\alpha(0),\dots,\varphi_n(x,\alpha),\alpha(n)\urcorner : \langle x,\alpha\rangle \in A\}\\ &\leq \ulcorner\lambda_0^0,\alpha(0),\lambda_1^1,\alpha(1),\dots,\lambda_n^n,\alpha(n)\urcorner\\ &= \ulcorner\lambda_0^n,k_0^n,\lambda_1^n,k_1^n,\dots,\lambda_n^n,k_n^n\urcorner = \lambda'_n.\end{aligned}$$

Thus $x \in B$ and $\psi_n(x) \leq \lambda_n$ and we are done. ⊣ (Claim 3.9)

CLAIM 3.10. $\langle\psi_n : n \in \omega\rangle$ is a $\exists^{\mathbb{R}}\forall^{\mathbb{R}}\Gamma$-scale.

This is very similar to the computation done in the proof of Theorem 2.4 and we omit the details. ⊣ (Theorem 3.6)

The direct analogue of Theorem 2.7 is true, but we have no use for it. The useful analogue of Theorem 2.7 gives uniformization directly.

THEOREM 3.11 (Essentially Addison, [Add59]). Assume Γ is adequate, closed under both $\exists^\omega$, $\forall^\omega$, $\check{\Gamma} \subseteq \exists^{\mathbb{R}}\Gamma$, $\exists^{\mathbb{R}}(\Gamma\cap\check{\Gamma}) = \Gamma$ and there is a wellordering $\leq$ of $\mathbb{R}$ which is Γ-good. Then $\mathsf{Unif}(\exists^{\mathbb{R}}\Gamma, \exists^{\mathbb{R}}\Gamma)$.

PROOF. Since for Γ, Γ' adequate

$$\mathsf{Unif}(\Gamma,\Gamma') \Longrightarrow \mathsf{Unif}(\exists^{\mathbb{R}}\Gamma, \exists^{\mathbb{R}}\Gamma')$$

it will be enough to prove $\mathsf{Unif}(\Gamma\cap\check{\Gamma}, \Gamma\cap\check{\Gamma})$. But if $P \subseteq \mathfrak{X}\times\mathcal{Y}$ is in $\Gamma\cap\check{\Gamma}$, where we can assume without loss of generality $\mathcal{Y} = \mathbb{R}$, let

$$P^*(x,\alpha) \Longleftrightarrow P(x,\alpha) \wedge (\forall\beta < \alpha)\neg P(x,\beta).$$

Then P^* uniformizes P and clearly $P^* \in \Gamma\cap\check{\Gamma}$. ⊣

COROLLARY 3.12 (Addison, [Add59]). $\mathbf{V} = \mathbf{L} \Longrightarrow \mathsf{Unif}(\Sigma^1_k, \Sigma^1_k)$ $(k \geq 2)$.

COROLLARY 3.13 (Silver, [Sil71]). $\mathbf{V} = \mathbf{L}[\mu] \Longrightarrow \mathsf{Unif}(\Sigma^1_k, \Sigma^1_k)$ $(k \geq 2)$.

3.3. The Second Periodicity Theorem.

THEOREM 3.14 (Moschovakis, [Mos71A]). Assume Γ is adequate and $\mathsf{Det}(\underset{\sim}{\Gamma}\cap\underset{\sim}{\check{\Gamma}})$. Then if $A \in \Gamma$ and A admits a Γ-scale, $\forall^{\mathbb{R}}A$ admits a $\forall^{\mathbb{R}}\exists^{\mathbb{R}}\Gamma$-scale.

COROLLARY 3.15. Suppose that Γ is adequate, and that Scale(Γ), $\mathsf{Det}(\underset{\sim}{\Gamma}\cap\underset{\sim}{\check{\Gamma}})$, $\exists^{\mathbb{R}}\Gamma \subseteq \Gamma$ hold. Then Scale($\forall^{\mathbb{R}}\Gamma$) holds.

COROLLARY 3.16. PD implies $\mathsf{Unif}(\Pi^1_{2n+1})$, $n \geq 1$.

PROOF OF THEOREM 3.14. Enumerate in a 1-1 recursive way all the finite sequences of integers, say $u_0, u_1, u_2, \dots$ such that $u_0 = \langle\,\rangle$ (the empty sequence) and if u_i is a proper initial segment of u_j then $i < j$.

Put $B = \forall^{\mathbb{R}}A \subseteq \mathfrak{X}$ and define

$$x \in B_n \Longleftrightarrow (\forall\alpha \supseteq u_n)A(x,\alpha)$$

(where $\alpha \supseteq u_n \iff \alpha$ extends u_n). Notice that $B_0 = B$ and for every $n, B \subseteq B_n$. We shall define norms on each B_n by considering games as follows: For each $n \in \omega$ and $x, y \in \mathfrak{X}$ let $\mathrm{G}_n(x, y)$ be the following game

$$\begin{array}{cc} u_n & u_n \\ \mathrm{I} & \mathrm{II} \\ \alpha' & \beta' \end{array}$$

Player I plays α', player II plays β' and if we call $\alpha = u_n{}^\frown\alpha'$, $\beta = u_n{}^\frown\beta'$ then player II wins if and only if $A(y, \beta)$ doesn't hold, or

$$(A(y,\beta) \wedge A(x,\alpha) \wedge \ulcorner\varphi_0(x,\alpha),\dots,\varphi_n(x,\alpha)\urcorner \leq \ulcorner\varphi_0(y,\beta),\dots,\varphi_n(y,\beta)\urcorner)$$

holds, where $\langle \varphi_n : n \in \omega\rangle$ is a Γ-scale on A.

Finally put for $x, y \in B_n$

$$x \leq_n y \iff \text{player II has a winning strategy in } \mathrm{G}_n(x, y).$$

Proofs similar to those in 2.3 show that each $\leq_n$ is a prewellordering on B_n and that for some S, S' in $\forall^{\mathbb{R}}\exists^{\mathbb{R}}\Gamma, \exists^{\mathbb{R}}\forall^{\mathbb{R}}\check{\Gamma}$ respectively we have

$$y \in B_n \implies \forall x((x \in B_n \wedge x \leq_n y) \iff S(n,x,y) \iff S'(n,x,y)).$$

Call ψ_n the norm associated with $\leq_n$. Suppose we can prove that if $x_i \in B, x_i \to x$ and $\psi_n(x_i) = \lambda_n$ for all large enough i, then $x \in B$ and $\psi_n(x) \leq \lambda_n$. Then if we put for $x \in B$

$$\psi_n'(x) = \ulcorner\psi_0(x), \psi_n(x)\urcorner$$

and let ψ_n'' be the norm on B such that

$$\psi_n''(x) \leq \psi_n''(y) \iff \psi_n'(x) \leq \psi_n'(y)$$

it is easy to check that $\langle \psi_n'' : n \in \omega\rangle$ is a $\forall^{\mathbb{R}}\exists^{\mathbb{R}}\Gamma$-scale on B.

Thus, to complete the proof, we assume $x_i \in B, x_i \to x$ and $\psi_n(x_i) = \lambda_n$ for all $i \geq n$, and try to show that $x \in B$ and $\psi_n(x) \leq \lambda_n$.

Claim 3.17. $x \in B$.

Proof of Claim 3.17. We have to show that for every α, $\langle x, \alpha\rangle \in A$. Fix α. Let n_i be such that $u_{n_i} = \bar{\alpha}(i)$. Notice that $n_0 = 0 \wedge n_0 < n_1 < n_2 \cdots$. Consider the subsequence $\langle x_{n_0}, x_{n_1}, x_{n_2}, \dots\rangle$. Then $\psi_{n_i}(x_{n_i}) = \psi_{n_i}(x_{n_{i+1}})$. Thus player II has a winning strategy in all the games $\mathrm{G}_{n_i}(x_{n_{i+1}}, x_{n_i})$ (since $x_{n_i} \geq_{n_i} x_{n_{i+1}}$).

Fix strategies for player II in each one of the games $\mathrm{G}_{n_i}(x_{n_{i+1}}, x_{n_i})$ and consider the diagram below:

$G_{n_0}(x_{n_1}, x_{n_0})$		$G_{n_1}(x_{n_2}, x_{n_1})$		$G_{n_2}(x_{n_3}, x_{n_2})$		$G_{n_3}(x_{n_4}, x_{n_3})$		
						$\alpha(0)$	$\alpha(0)$	$\cdots$
				$\alpha(0)$	$\alpha(0)$	$\alpha(1)$	$\alpha(1)$	$\cdots$
		$\alpha(0)$	$\alpha(0)$	$\alpha(1)$	$\alpha(1)$	$\alpha(2)$	$\alpha(2)$	$\cdots$
I	II	I	II	I	II	I	II	
$\alpha(0) \to \alpha_0(0)$		$\alpha(1) \to \alpha_1(1)$		$\alpha(2) \to \alpha_2(2)$		$\alpha(3) \to \alpha_3(3)$		$\cdots$
$\alpha_1(1) \to \alpha_0(1)$		$\alpha_2(2) \to \alpha_1(2)$		$\alpha_3(3) \to \alpha_2(3)$		$\alpha_4(4) \leftharpoondown \alpha_3(4)$		$\cdots$
$\alpha_1(2) \to \alpha_0(2)$		$\alpha_2(3) \to \alpha_1(3)$		$\alpha_3(4) \to \alpha_2(4)$		$\alpha_4(5) \leftharpoondown \alpha_3(5)$		$\cdots$
$\vdots$	$\vdots$	$\vdots$	$\vdots$	$\vdots$	$\vdots$	$\vdots$	$\vdots$	
α_1	α_0	α_2	α_1	α_3	α_2	α_4	α_3	

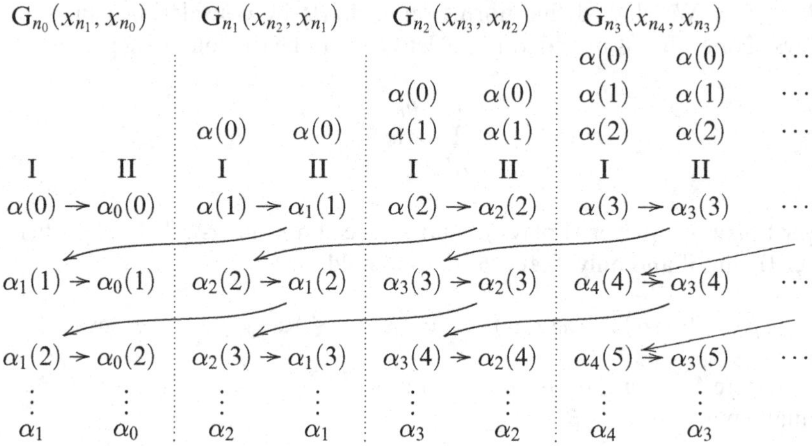

Player I plays $\alpha(0)$ in $G_{n_0}(x_{n_1}, x_{n_0})$ and player II answers with his strategy to give $\alpha_0(0)$. Then player I plays $\alpha(1)$ in $G_{n_1}(x_{n_2}, x_{n_1})$ and player II answers with his strategy to give $\alpha_1(1)$, *etc.* After these moves have been played, player I plays $\alpha_1(1)$ in $G_{n_0}(x_{n_1}, x_{n_0})$ and player II answers with his strategy to give $\alpha_0(1)$. Then player I plays $\alpha_2(2)$ in $G_{n_1}(x_{n_2}, x_{n_1})$ and player II answers by his strategy to give $\alpha_1(2)$, *etc.* Finally reals $\alpha_0, \alpha_1, \alpha_2, \ldots$ are formed (where, *e.g.*, $\alpha_4 = \langle \alpha(0), \alpha(1), \alpha(2), \alpha(3), \alpha_4(4), \ldots \rangle$). Clearly $\alpha_i \to \alpha$. Moreover since player II always wins we have

$$\varphi_0(x_{n_1}, \alpha_1) \leq \varphi_0(x_{n_0}, \alpha_0)$$
$$\varphi_0(x_{n_2}, \alpha_2) \leq \varphi_0(x_{n_1}, \alpha_1)$$
$$\varphi_0(x_{n_3}, \alpha_3) \leq \varphi_0(x_{n_2}, \alpha_2)$$

$$\vdots$$

Thus for all large enough $i, \varphi_0(x_{n_i}, \alpha_i)$ is constant. Looking at such i's and since again player II always wins we have

$$\varphi_1(x_{n_{i+1}}, \alpha_{i+1}) \leq \varphi_1(x_{n_i}, \alpha_i)$$
$$\varphi_1(x_{n_{i+2}}, \alpha_{i+2}) \leq \varphi_1(x_{n_{i+1}}, \alpha_{i+1})$$

$$\vdots$$

Thus again $\varphi_1(x_{n_i}, \alpha_i)$ becomes eventually constant. Then we look at φ_2 *etc.* Thus for the sequence $\langle \langle x_{n_i}, \alpha_i \rangle : i \in \omega \rangle$ we have $\langle x_{n_i}, \alpha_i \rangle \in A, \langle x_{n_i}, \alpha_i \rangle \to \langle x, \alpha \rangle$ and for each $n, \varphi_n(x_{n_i}, \alpha_i)$ becomes eventually constant. So $\langle x, \alpha \rangle \in A$.
$\dashv$ (Claim 3.17)

CLAIM 3.18. For each $n, \psi_n(x) \leq \lambda_n$.

PROOF OF CLAIM 3.18. Since for $i \geq n, \psi_n(x_i) = \lambda_n$, it will be enough to prove that for each $n, x \leq_n x_n$, i.e., that player II has a winning strategy in $\mathrm{G}_n(x, x_n)$.

Since $\psi_k(x_k) = \psi_k(x_m)$, for $k \leq m$, we have $x_m \leq_k x_k$ for $m \geq k$, thus player II has a winning strategy in each one of the games $\mathrm{G}_k(x_m, x_k)$ for $m \geq k$. Fix strategies for player II in each one of these games. Fix n. In order to invent a strategy for player II in $\mathrm{G}_n(x, x_n)$ consider the diagram

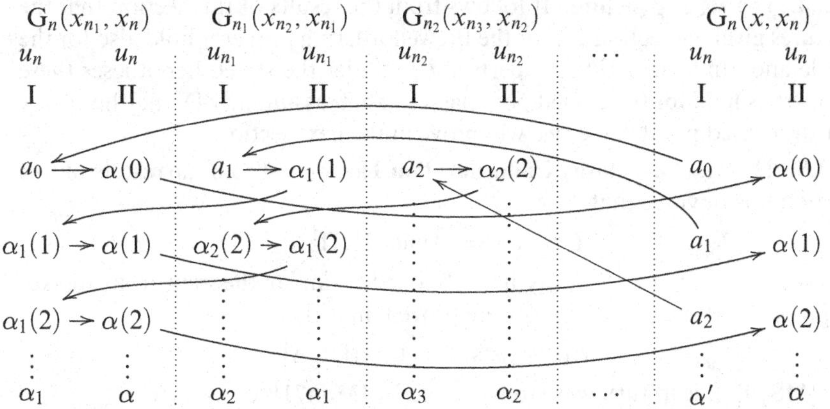

Let player I play a_0 in $\mathrm{G}_n(x, x_n)$. Let $u_{n_1} = u_n{}^\frown a_0$. Then $n_1 > n$. Consider the game $\mathrm{G}_n(x_{n_1}, x_n)$ and let player I play in $\mathrm{G}_n(x_{n_1}, x_n)$ a_0 and player II answer in this game by his winning strategy to give $\alpha(0)$. This $\alpha(0)$ is player II's answer to a_0 in $\mathrm{G}_n(x, x_n)$. Then player I plays a_1 in $\mathrm{G}_n(x, x_n)$. Let $u_{n_2} = u_n{}^\frown a_0{}^\frown a_1$. Then $n_2 > n_1$. Consider the game $\mathrm{G}_{n_1}(x_{n_2}, x_{n_1})$ and let player I play a_1 in $\mathrm{G}_{n_1}(x_{n_2}, x_{n_1})$ and player II answer in this game by his winning strategy to give $\alpha_1(1)$. Then player I plays in $\mathrm{G}_n(x_{n_1}, x_n)$ this $\alpha_1(1)$ and player II answers by his strategy in $\mathrm{G}_n(x_{n_1}, x_n)$ to give $\alpha(1)$ which is player II's next move in $\mathrm{G}_n(x, x_n)$ *etc.* After all these moves have been played $\alpha', \alpha, \alpha_1, \alpha_2, \ldots$ are created (*e.g.*, $\alpha_2 = u_{n_1}{}^\frown\langle a_1, \alpha_2(2), \alpha_2(3), \ldots\rangle = u_n{}^\frown\langle a_0, a_1, \alpha_2(2), \alpha_2(3), \ldots\rangle$). Clearly $\alpha_n \to \alpha' = u_n{}^\frown\langle a_0, a_1, a_2, a_3, \ldots\rangle$ and, since player II always wins, we have $\varphi_0(x_n, \alpha) \geq \varphi_0(x_{n_1}, \alpha_1) \geq \varphi_0(x_{n_2}, \alpha_2) \geq \cdots$. Thus after a while $\varphi_0(x_{n_i}, \alpha_i)$ becomes constant. Then we look at $\varphi_1(x_{n_i}, \alpha_i)$ for such i's; it is nonincreasing with i, thus becomes eventually constant, *etc.* Thus for each $n, \varphi_n(x_{n_i}, \alpha_i)$ becomes eventually constant. But also $\langle x_{n_i}, \alpha_i\rangle \to \langle x, \alpha'\rangle$, so $\langle x, \alpha'\rangle \in A$ and $\varphi_n(x, \alpha') \leq \lim_i \varphi_n(x_{n_i}, \alpha_i)$. Then, as we saw above, $\varphi_0(x_n, \alpha) \geq \lim_i \varphi_0(x_{n_i}, \alpha_i) \geq \varphi_0(x, \alpha')$. If $\varphi_0(x_n, \alpha) > \varphi_0(x, \alpha')$ clearly player II wins the game $\mathrm{G}_n(x, x_n)$ and we are done. If $\varphi_0(x_n, \alpha) = \varphi_0(x, \alpha')$, then for all $i, \varphi_0(x_{n_i}, \alpha_i) = \varphi_0(x_n, \alpha) = \varphi_0(x, \alpha')$. But then if $n \geq 1$, $\varphi_1(x_n, \alpha) \geq \varphi_1(x_{n_1}, \alpha_1) \geq \varphi_1(x_{n_2}, \alpha_2) \geq \cdots$, thus $\varphi_1(x_n, \alpha) \geq \varphi_1(x, \alpha')$. If again $\varphi_1(x_n, \alpha) > \varphi_1(x, \alpha')$ we are done, otherwise $\varphi_1(x_n, \alpha) = \varphi_1(x_{n_i}, \alpha_i) =$

$\varphi_1(x, \alpha')$ and then we look (if $n \geq 2$) at φ_2, *etc.* In any case this shows that

$$\ulcorner\varphi_0(x_n, \alpha), \varphi_1(x_n, \alpha), \dots, \varphi_n(x_n, \alpha)\urcorner \geq$$
$$\ulcorner\varphi_0(x, \alpha'), \dots, \varphi_n(x, \alpha')\urcorner, \textit{ i.e.}, \text{ player II wins.}$$

Thus we have described a winning strategy for player II in $G_n(x, x_n)$, so $x \leq_n x_n$. ⊣ (Claim 3.18)

⊣ (Theorem 3.14)

3.4. The zig-zag picture. It follows from the results of this section that the pictures given in Section 2.4 for the Prewellordering property hold also for the Scale and Uniformization properties, *i.e.*, under the stated hypotheses these properties hold for the circled pontclasses. That, assuming PD, they hold *only* for the circled pointclasses we will prove in the next section.

3.5. The Martin-Solovay Uniformization Theorem. From the results of this section it is obvious that

$$\mathsf{Det}(\utilde{\mathbf{\Delta}}^1_2) \Longrightarrow \mathsf{Unif}(\Pi^1_2, \Pi^1_3).$$

However Martin and Solovay had obtained a similar theorem from weaker hypotheses before these results were proved, namely

$$\forall\alpha(\alpha^{\#} \text{ exists}) \Longrightarrow \mathsf{Unif}(\Pi^1_2, \Delta^1_4),$$

see [MS69]; this in turn was strengthened by [Man71] to

$$\forall\alpha(\alpha^{\#} \text{ exists}) \Longrightarrow \mathsf{Unif}(\Pi^1_2, \Pi^1_3).$$

These proofs (and in fact the statements of the theorems) involve the **theory of indiscernibles** with which we are not concerned here. We will state one by-product of this work which will be useful later.

THEOREM 3.19. Assume that there exists a measurable cardinal or (the weaker hypothesis) that for each α, $\alpha^{\#}$ exists, let $u_1 = \aleph_1, u_2, u_3, \dots, u_\omega$ be the first $\omega + 1$ uniform indiscernibles. Then:

1. $u_n \leq \aleph_n$, $u_\omega \leq \aleph_\omega$ and $\text{cf}(u_{n+1}) = \text{cf}(u_2)$.
2. If AC holds, then $u_\omega < \aleph_3$.
3. Every Π^1_2 set A admits a Π^1_3 scale on u_ω, *i.e.*, a scale $\langle\varphi_n : n \in \omega\rangle$ with each $\varphi_n : A \to u_\omega$.

((1) and (2) are due to Solovay, (3) is implicit in [MS69, Man71])

§4. Bases. One of the most interesting corollaries of uniformization results is the computation of **bases** for pointclasses. If Γ is a pointclass and C a set of reals, put

$$\text{Basis}(\Gamma, C) \Longleftrightarrow \text{ for each } A \in \Gamma, A \subseteq \mathbb{R}, A \neq \varnothing \Longrightarrow A \cap C \neq \varnothing.$$

If Λ is a pointclass, we often abbreviate

$$\text{Basis}(\Gamma, \Lambda) \Longleftrightarrow \text{Basis}(\Gamma, \{\alpha : \text{ the set } \{\langle n, m\rangle : \alpha(n) = m\} \in \Lambda\}).$$

4.1. Computation of bases. It is immediately obvious that

$$\mathsf{Unif}(\Gamma, \Gamma') \text{ implies } \text{Basis}(\Gamma, \{\alpha : \{\alpha\} \in \Gamma'\}) \tag{4.1.1}$$

and it is easy to see that

if Γ is adequate and Basis(Γ, C) holds, then

$$\text{Basis}(\exists^{\mathbb{R}}\Gamma, \{\alpha : (\exists\beta)(\beta \in C \wedge \alpha \text{ is recursive in } \beta)\}). \tag{4.1.2}$$

From this and the results in §3 it is clear that

$$\mathsf{Det}(\underset{\sim}{\mathbf{\Delta}}^1_{2n}) \text{ implies } \text{Basis}(\Sigma^1_{2n+2}, \Delta^1_{2n+2}) \tag{4.1.3}$$

and

$$\mathsf{PD} \text{ implies } \text{Basis}(\Sigma^1_{2n}, \Delta^1_{2n}), \quad n \geq 2.$$

On the other hand we have

THEOREM 4.4. $\mathsf{Det}(\underset{\sim}{\mathbf{\Delta}}^1_{2n})$ implies $\neg$ Basis$(\Sigma^1_{2n+1}, \Delta^1_{2n+1})$.

PROOF. Since $\mathsf{Det}(\underset{\sim}{\mathbf{\Delta}}^1_{2n})$, we have PWO$(\Pi^1_{2n+1})$. But this has a consequence that $\{\alpha : \alpha \in \Delta^1_{2n+1}\} \in \Pi^1_{2n+1}$. (This is announced in [AM68].) From this the result follows immediately. (For another proof see [MS69].) ⊣

The periodicity phenomenon is again clear in (4.1.3) and Theorem 4.4.

THEOREM 4.5 (Martin, Solovay, Mansfield, [MS69, Man71]). If $\alpha^\#$ exists for all reals α, then there exists a fixed Π^1_3 singleton α_0 (*i.e.*, $\{\alpha_0\} \in \Pi^1_3$) such that

$$\text{Basis}(\Sigma^1_3, \{\beta : \beta \text{ is recursive in } \alpha_0\}).$$

THEOREM 4.6 (Moschovakis, [Mos71A]). If $\mathsf{Det}(\underset{\sim}{\mathbf{\Delta}}^1_{2n})$ holds, then there exists a fixed Π^1_{2n+1} singleton α_0 such that

$$\text{Basis}(\Sigma^1_{2n+1}, \{\beta : \beta \text{ is recursive in } \alpha_0\}).$$

PROOF. By (4.1.2) it will be enough to find a Π^1_{2n+1} singleton α_0 such that every Π^1_{2n} set contains a real recursive in α_0.

Let $B \subseteq \omega \times \mathbb{R}$ be a universal Π^1_{2n} set. Uniformize B by some $B^* \in \Pi^1_{2n+1}$. Then $B^* \subseteq B$ and $\exists\alpha B(n, \alpha) \iff \exists!\alpha B^*(n, \alpha)$. Define $B^{**}(n, \alpha) \iff B^*(n, \alpha) \vee (\forall\beta(\neg B(n, \beta)) \wedge \alpha = \lambda t 0\}$. Then $B^{**} \in \Pi^1_{2n+1}$ and $\forall n \exists! \alpha B^{**}(n, \alpha)$. Put

$$\alpha \in C \iff \forall n B^{**}(n, (\alpha)_n).$$

For this proof choose $(\alpha)_n$ so that α is completely determined by $\{(\alpha)_n : n \in \omega\}$. Thus C is a singleton and $C \in \Pi^1_{2n+1}$. If $C = \{\alpha_0\}$ we show that every Π^1_{2n} set A contains a real recursive in α_0. In fact if $A \in \Pi^1_{2n}$ we have $a \in A \iff \langle n_0, \alpha\rangle \in B$, for some n_0. Then $A \neq \varnothing \implies \exists\alpha(\langle n_0, \alpha\rangle \in B)$, so $\langle n_0, (\alpha)_{n_0}\rangle \in B$ *i.e.*, $(\alpha)_{n_0} \in A$. ⊣

A well known basis theorem says that for some fixed Σ^1_1 subset of ω, say A_0, we have

$$\text{Basis}(\Sigma^1_1, \{\alpha : \alpha \text{ is recursive in } A_0\}).$$

The following question is open: Does this generalize (under any reasonable hypothesis) to $\Sigma^1_{2n+1}, n \geq 1$?[1]

4.2. Independence results. It is clear that the weakest basis result one can expect for a ("lightface") pointclass Γ is

$$\text{Basis}(\Gamma, \{\alpha : \alpha \text{ is ordinal definable}\}).$$

But even such a weak result is not provable in ZFC for Γ beyond Σ^1_2 as the next theorem shows.

THEOREM 4.7 (Lévy, [Lév66]). *In ZFC alone we cannot prove*

$$\text{Basis}(\Pi^1_2, \{\alpha : \alpha \text{ is ordinal definable}\}).$$

PROOF. It is enough to show that if M is a countable model of ZF+V=L and α is a real Cohen generic over M, then in $N = M[\alpha]$ there is a Π^1_2 set containing no ordinal definable real. In fact, *in* N, consider the set $A = \{\beta : \beta \notin L\}$. Then $A = \Pi^1_2$ and $A \neq \varnothing$. But A cannot contain an ordinal definable real, since all such reals belong already to $M = \mathbf{L}^N$ (because the notion of forcing is homogeneous). ⊣

Of course we have a basis theorem for Σ^1_3 assuming, for example, that there exists a measurable cardinal (Theorem 4.6). Unfortunately we cannot go further even with this stronger hypothesis.

THEOREM 4.8. (Lévy's method for Theorem 4.7 using a key result of Silver [Sil71].) *In ZFC + "there exists a measurable cardinal", we cannot prove*

$$\text{Basis}(\Pi^1_3, \{\alpha : \alpha \text{ is ordinal definable}\}).$$

PROOF. Repeat the proof of 4.7, but now start with an M which is a countable model of ZF+V=L[μ], where μ is a normal measure on a cardinal κ. ⊣

§5. Partially playful universes.

We outline here a construction which (granting PD) yields for each $n \geq 3$ a model M^n of ZF+AC such that

$$M^n \models \text{Det}(\underset{\sim}{\Delta}^1_{n-1}),$$
$$M^n \models \mathbb{R} \text{ admits a } \Sigma^1_{n+1}\text{-good wellordering}.$$

In particular the zig-zag picture of §2.4 for the scale property in M^n has only finitely many teeth, *i.e.*, the scale property settles on the Σ side for $k \geq n+1$. The results are due to Moschovakis.

[1] Martin and Solovay have shown in 1972 that this generalization is false for $n \geq 1$, granting Det($\underset{\sim}{\Delta}^1_{2n}$). They also show that in Theorem 4.6, α_0 can be *any* Π^1_{2n+1} singleton which is not Δ^1_{2n+1}. This turns out to be the correct generalization of the Kleene Basis Theorem for Σ^1_1. See [MS].

Fix $n \geq 3$, let k be the largest even integer less than n such that $k = n - 1$ or $k = n - 2$ and assume $\mathsf{Det}(\underset{\sim}{\mathbf{\Delta}}^1_k)$. By the Second Periodicity Theorem we have

$$\mathsf{Unif}(\Pi^1_{n-1}, \Pi^1_n)$$

whether n is odd or even, so let $P_{n-1}(m, \alpha, \beta)$ be the standard Π^1_{n-1} universal relation and let $P^*_{n-1}(m, \alpha, \beta)$ be the Π^1_n relation that comes out of the proof of the Second Periodicity Theorem such that

$$P^*_{n-1}(m, \alpha, \beta) \Longrightarrow P_{n-1}(m, \alpha, \beta), \tag{$*$}$$

$$(\exists \beta) P_{n-1}(m, \alpha, \beta) \Longrightarrow (\exists! \beta) P^*_{n-1}(m, \alpha, \beta). \tag{$**$}$$

Finally define

$$F^*_n(m, \alpha) = \begin{cases} \text{the unique } \beta & \text{such that } P^*_{n-1}(m, \alpha, \beta), \\ & \text{if } (\exists \beta) P_{n-1}(m, \alpha, \beta), \\ \lambda t 0 & \text{if } \forall \beta \neg P_{n-1}(m, \alpha, \beta). \end{cases}$$

Clearly F^*_n is a function whose graph is Π^1_n.

Let M be a model of ZF, transitive and containing all ordinals (for brevity, **standard** model). We call M **Σ^1_n-correct** if for every Σ^1_n formula $\vartheta(\alpha_1, \ldots, \alpha_\ell)$,

$$\alpha_1, \ldots, \alpha_\ell \in M \Longrightarrow (\vartheta(\alpha_1, \ldots \alpha_\ell) \Longleftrightarrow M \models \vartheta(\alpha_1, \ldots, \alpha_\ell)).$$

LEMMA 5.1. Assume $\mathsf{Det}(\underset{\sim}{\mathbf{\Delta}}^1_k)$. A standard model M of ZF+DC is Σ^1_n-correct if and only if M is closed under F^*_n, *i.e.*,

$$\alpha \in M \Longrightarrow F^*_n(m, \alpha) \in M.$$

PROOF. Assume first that M is Σ^1_n-correct. Notice that if for some $\alpha \in M$ and some m_1, m_2,

$$\forall \beta (P_{n-1}(m_1, \alpha, \beta) \Longleftrightarrow \neg P_{n-1}(m_2, \alpha, \beta)),$$

then the same equivalence holds in M (it is expressible by a Π^1_n formula); thus $\langle m_1, m_2, \alpha \rangle$ codes a $\underset{\sim}{\mathbf{\Delta}}^1_{n-1}$ set in M if and only if it does in the world. This applies to $\underset{\sim}{\mathbf{\Delta}}^1_k$ sets, since $k \leq n - 1$, and it is now easy to verify that

$$M \models \mathsf{Det}(\underset{\sim}{\mathbf{\Delta}}^1_k).$$

Hence the Second Periodicity Theorem holds in M, so that $(*)$ and $(**)$ hold. Now if for some $m, \alpha \in M$, $(\exists \beta) P_{n-1}(m, \alpha, \beta)$, then $M \models (\exists \beta) P_{n-1}(m, \alpha, \beta)$, hence for some $\beta \in M$, $M \models P^*_{n-1}(m, \alpha, \beta)$, hence $P^*_{n-1}(m, \alpha, \beta)$ in the world and $\beta = F^*_n(m, \alpha) \in M$.

To prove the converse, assume that M is closed under F^*_n and then show by induction on $i \leq n$ that M is Σ^1_i-correct. *This part of the proof does not need the assumption that* $M \models$ DC. We omit the details. ⊣

(Actually neither direction of the equivalence needs the assumption $M \models$ DC, but the proof is a bit more complicated.)

Define by induction on the ordinal ξ,

$$\begin{aligned} M_0^n &= \varnothing, \\ M_{\xi+1}^n &= M_\xi^n \cup \{x \subseteq M_\xi^n : x \text{ is definable in } \langle M_\xi^n, \in\restriction M_\xi^n\rangle\} \\ &\qquad \cup \{F_n^*(m,\alpha) : \alpha \in M_\xi^n\}, \\ M_\eta^n &= \bigcup_{\xi<\eta} M_\xi^n \text{ if } \eta = \bigcup \eta > 0, \end{aligned}$$

and put

$$M^n = \bigcup_\xi M_\xi^n.$$

THEOREM 5.2. Let $n \geq 3$, assume $\mathsf{Det}(\underset{\sim}{\mathbf{\Delta}}_k^1)$, with k = largest even integer $< n$, let M_ξ^n, M^n be defined as above.

1. M^n is a standard model of ZF, it is closed under F_n^* and it is Σ_n^1-correct.
2. The relation "$x \in M_\xi^n$" is definable by a formula which is absolute for all Σ_n^1-correct models of ZF that are closed under F_n^*.
3. $M^n \models \forall x \exists \xi (x \in M_\xi^n)$
4. $M^n \models$ AC.
5. M^n is the smallest standard model of ZF+DC which is Σ_n^1-correct.
6. $M^n \models \mathsf{Det}(\underset{\sim}{\mathbf{\Delta}}_k^1)$ and if $\mathsf{Det}(\underset{\sim}{\mathbf{\Delta}}_{n-1}^1)$ holds in the world it also holds in M^n.
7. $M^n \models \mathsf{Scale}(\Pi_i^1)$ for i odd, $i \leq n$, $M^n \models \mathsf{Scale}(\Sigma_i^1)$ for i even, $i \leq n+1$.
8. $M^n \models$ Generalized Continuum Hypothesis.
9. $M^n \models \mathbb{R}$ admits a Σ_{n+1}^1-good wellordering.
10. $M^n \models \mathsf{Scale}(\Sigma_i^1)$ for $i \geq n+1$.

OUTLINE OF PROOF. 1-5 are easy by standard methods, 6 follows by the remarks on absoluteness made in the proof of Lemma 5.1 and 7 follows from this. The key to 8 and 9 is a version of the Gödel Condensation Lemma that is apropriate to Σ_n^1-correct models. First notice that there is a finite subset Φ_0 of the axioms in ZF+DC+$\mathsf{Det}(\underset{\sim}{\mathbf{\Delta}}_k^1)$ such that the function $\xi \mapsto M_\xi^n$ is absolute for transitive sets which are models of Φ_0 and closed under F_n^* and hence the transitive models of $\Phi_0 + \forall x \exists \xi (x \in M_\xi^n)$ which are closed under F_n^* are precisely of the form M_ξ^n. Now get the Condensation Lemma as usually, except that in taking elementary submodels close under F_n^*. We omit the details. ⊣

We draw the zig-zag picture for the scale property for the models M^3, M^4:

V=M^3
or
V=M^4

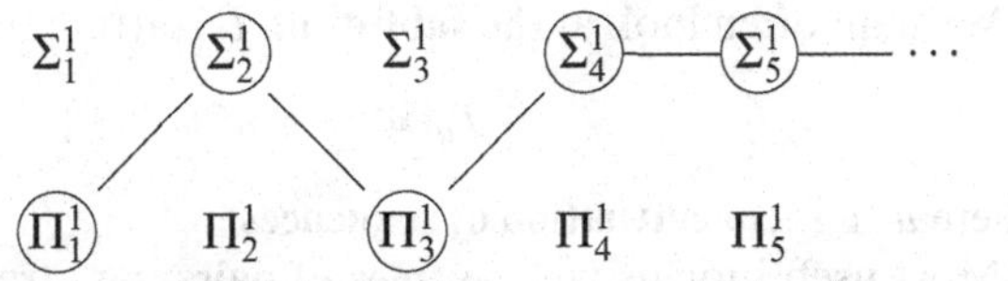

Another interesting model, M^ω, can be obtained by closing under *all* $F^*_n, n = 3, 4, 5, \ldots$. This satisfies PD and has the same zig-zag picture as **V** (assuming PD of course), but in $M^\omega, \mathbb{R}$ admits a very simple (hyperanalytic) good wellordering.

Kechris has shown by indiscernibility considerations that

$$M^3 \subsetneqq M^4 \subsetneqq M^5 \subsetneqq \cdots$$

and in fact for each $n \geq 3$, there is an $\alpha \in M^{n+1} \setminus M^n$.

§6. **Trees.** We show here that the existence of a scale on a set A yields a **representation** for A in terms of a **tree on ordinals** which is very similar to the classical representation for $\underset{\sim}{\Sigma}^1_1$ sets. This is the key to the applications of scales described in the remainder of this paper.

6.1. Notation for trees. A **tree** on some set C is a set T of finite sequences from C such that

$$\text{if } \langle c_0, c_1, \ldots, c_k \rangle \in T \wedge i \leq k, \text{ then } \langle c_0, \ldots, c_i \rangle \in T;$$

in particular every non-empty tree contains the **empty sequence** $\langle\,\rangle$.

A **branch through** (or **of**) a tree T on C is any function $f \in {}^\omega C$ such that

$$\text{for all } n, (\langle f(0), \ldots, f(n) \rangle \in T).$$

Put

$$[T] = \text{the set of all branches through } T$$

and call T **wellfounded** if $[T] = \varnothing$, *i.e.*, if T has no infinite branches. The idea here is a bit clearer if we consider the relation $\succ$ of **proper extension** on finite sequences.

$$\langle c_0, \ldots, c_k \rangle \succ \langle d_0, \ldots, d_\ell \rangle \iff k < \ell \wedge c_0 = d_0 \wedge \ldots \wedge c_k = d_k;$$

T is wellfounded if and only if $\prec \restriction T$ has no infinite descending chains, *i.e.*, if and only if $\prec \restriction T$ is wellfounded. We can now assign an **ordinal rank** to every sequence of a wellfounded tree in the canonical way we do this for any wellfounded relation,

$$|u|_T = \sup\{|v|_T + 1 : v \in T, u \succ v\}$$

(where $\sup(\varnothing) = 0$) and define the **rank of** T,

$$|T| = \sup\{|u|_T : u \in T\} = |\langle\,\rangle|_T.$$

By convention let also $|u|_T = -1$, if $u \notin T$.

We shall often look at **the subtree of** T starting from some sequence,

$$T_u = \{v : u^\frown v \in T\},$$

where $u^\frown v$ is **concatenation of sequences**.

Most useful for us will be **trees of pairs**, *i.e.*, trees on sets $C = A \times B$—usually $C = \omega \times \kappa$ for some ordinal κ. A typical member of a tree T on $A \times B$ is a sequence

$$\langle\langle a_0, b_0\rangle, \langle a_1, b_1\rangle, \dots, \langle a_n, b_n\rangle\rangle$$

and a branch through T is a function $f \in {}^\omega(A \times B)$. It will be convenient to represent each branch f by the pair $\langle g, h\rangle, g \in {}^\omega A, h \in {}^\omega B$ which determines it,

$$f(n) = \langle g(n), h(n)\rangle.$$

For each fixed $g \in {}^\omega A$ now, we can define a new tree $T(g)$ on B by

$$T(g) = \{\langle b_0, \dots, b_n\rangle : \langle\langle g(0), b_0\rangle, \dots, \langle g(n), b_n\rangle\rangle \in T\}.$$

In the typical case when T is a tree on $\omega \times \kappa$, for each $\alpha \in \mathbb{R}$ we will have a tree on κ

$$T(\alpha) = \{\langle \xi_0, \dots, \xi_n\rangle : \langle\langle \alpha(0), \xi_0\rangle, \dots, \langle \alpha(n), \xi_n\rangle\rangle \in T\};$$

notice that the function

$$\alpha \mapsto T(\alpha)$$

is **continuous** in a strong sense, *i.e.*,

$$\langle \xi_0, \dots, \xi_n\rangle \in T(\alpha) \wedge \bar{\alpha}(n+1) = \bar{\beta}(n+1) \Longrightarrow \langle \xi_0, \dots, \xi_n\rangle \in T(\beta).$$

6.2. κ-scales and their trees. Let $\langle \varphi_n : n \in \omega\rangle$ be a scale on A; we call $\langle \varphi_n : n \in \omega\rangle$ a **κ-scale**, if every φ_n is a function on A *into* κ, *i.e.*, if the length of each prewellordering $\leq^{\varphi_n}$ is $\leq \kappa$. With each κ-scale $\langle \varphi_n : n \in \omega\rangle$ on A we define the associated tree T on $\omega \times \kappa$ by

$$T = \{\langle\langle \alpha(0), \varphi_0(\alpha)\rangle, \langle \alpha(1), \varphi_1(\alpha)\rangle, \dots, \langle \alpha(n), \varphi_n(\alpha)\rangle\rangle : \alpha \in A\}.$$

THEOREM 6.1. Let A be a pointset, $A \subseteq \mathbb{R}$, $\langle \varphi_n : n \in \omega\rangle$ a κ-scale on A, T the associated tree. Then

$$\begin{aligned} \alpha \in A &\iff T(\alpha) \text{ is not wellfounded} \\ &\iff (\exists f)(\alpha, f) \in [T]. \end{aligned}$$

(this is an idea implicit in many of the classical proofs.)

PROOF. If $\alpha \in A$, then $\langle \varphi_0(\alpha), \varphi_1(\alpha), \varphi_2(\alpha), \dots \rangle$ is a branch through $T(\alpha)$. Conversely, suppose $\langle \xi_0, \xi_1, \xi_2 \dots \rangle$ is a branch through $T(\alpha)$, *i.e.*, for each n,

$$\langle \langle \alpha(0), \xi_0 \rangle, \dots, \langle \alpha(n), \xi_n \rangle \rangle \in T;$$

by the definition of T, there must exist reals $\alpha_0, \alpha_1, \dots$ in A, so that for each n,

$$\langle \langle \alpha_n(0), \varphi_0(\alpha_n) \rangle, \langle \alpha_n(1), \varphi_1(\alpha_n) \rangle, \dots, \langle \alpha_n(n), \varphi_n(\alpha_n) \rangle \rangle$$
$$= \langle \langle \alpha(0), \xi_0 \rangle, \langle \alpha(1), \xi_1 \rangle, \dots, \langle \alpha(n), \xi_n \rangle \rangle.$$

This implies immediately that $\lim_n \alpha_n = \alpha$ and for $m \leq n, \varphi_m(\alpha_n) = \xi_m$, so that by the basic property of scales $\alpha \in A$. ⊣

Kechris has shown that a converse to Theorem 6.1 is true, namely: if $\alpha \in A \iff T(\alpha)$ *is not wellfounded*, where T is a tree on $\omega \times \kappa$, then A admits a κ^ω-scale. This shows a connection between the notion of scale and some ideas of Mansfield in [Man70].

6.3. Computing lengths of scales.

THEOREM 6.2. If $A \subseteq \mathfrak{X} \times \mathbb{R}$ admits a κ-scale, $\kappa \geq \omega$, then $\exists^{\mathbb{R}} A$ admits a κ^ω-scale.

PROOF. See the proof of Theorem 3.6. ⊣

THEOREM 6.3. Every $\utilde{\mathbf{\Sigma}}^1_1$ set admits an ω^ω-scale.

PROOF. Every closed set admits an ω-scale. ⊣

Now define

$$\utilde{\boldsymbol{\delta}}^1_n = \sup\{\xi : \xi \text{ is the length of a } \utilde{\mathbf{\Delta}}^1_n \text{ prewellordering of } \mathbb{R}\}.$$

Classically it is known that $\utilde{\boldsymbol{\delta}}^1_1 = \aleph_1$.

Clearly every $\utilde{\mathbf{\Pi}}^1_{2n+1}$-norm on a set has length $\leq \utilde{\boldsymbol{\delta}}^1_{2n+1}$. Thus:

THEOREM 6.4. Assume $\mathsf{Det}(\utilde{\mathbf{\Delta}}^1_{2n})$. Then every $\utilde{\mathbf{\Pi}}^1_{2n+1}$ set admits a $\utilde{\boldsymbol{\delta}}^1_{2n+1}$-scale (by the Periodicity Theorem 3.2).

COROLLARY 6.5. (a) Every $\utilde{\mathbf{\Pi}}^1_1$ set admits a $\aleph_1$-scale.
(b) Every $\utilde{\mathbf{\Sigma}}^1_2$ set admits a $\aleph_1^\omega$-scale.

COROLLARY 6.6. Assume $\mathsf{Det}(\utilde{\mathbf{\Delta}}^1_{2n})$. Then every $\utilde{\mathbf{\Sigma}}^1_{2n+2}$ set admits a $(\utilde{\boldsymbol{\delta}}^1_{2n+1})^\omega$-scale.

From §3.5 we also have

THEOREM 6.7 (Martin, Solovay, [MS69]). If $\alpha^\#$ exists for all reals α, then every $\utilde{\mathbf{\Pi}}^1_2$ set admits a u_ω-scale.

COROLLARY 6.8 (Martin, [Mar70B]). If $\alpha^\#$ exists for all reals α, then every $\utilde{\mathbf{\Sigma}}^1_3$ set admits a $(u_\omega)^\omega$-scale. If we also assume AC, then every $\utilde{\mathbf{\Sigma}}^1_3$ set admits a κ-scale, with $\kappa < \aleph_3$.

The reader should have noticed that in this section a considerable change in our attitude towards scales has happened. We started worrying not only about definability of a scale but also about its length. Later sections will show why.

§7. Computing lengths of wellfounded relations.

7.1. The Kunen-Martin theorem. Recall that for a wellfounded relation $<$ we put for $x \in \text{Field}(<)$,

$$|x|_< = \sup\{|y|_< + 1 : y < x\}$$

and we define the **length of** $<$ by

$$|<| = \sup\{|x|_< : x \in \text{Field}(<)\}.$$

THEOREM 7.1. Let $< \subseteq \mathbb{R} \times \mathbb{R}$ be a wellfounded relation and assume $<$ (as a pointset) admits a κ-scale. Then $|<| < \kappa^+$. (Kunen, Martin, independently, unpublished; the proof below is Kunen's)

PROOF. To the wellfounded relation $<$ associate a tree T of reals as follows:

$$T = \{\langle \alpha_0, \alpha_1, \dots, \alpha_n \rangle : \alpha_0, \alpha_1, \dots, \alpha_n \in \text{Field}(<) \wedge \alpha_0 > \alpha_1 > \cdots > \alpha_n\}.$$

Notice that T is also wellfounded and in fact $| < | \leq |T|$. To prove the last statement one can show by $<$-induction that for any $\alpha \in \text{Field}(<)$ and any $\alpha_0, \dots, \alpha_n$ such that $\alpha_0 > \alpha_1 \cdots > \alpha_n > \alpha$, we have $|\alpha|_< = |\langle \alpha_0, \alpha_1, \dots, \alpha_n, \alpha \rangle|_T$.

We shall define a mapping

$$f : T \twoheadrightarrow S,$$

where S is a set of finite sequences from an ordinal $\lambda < \kappa^+$ such that

$$\langle \alpha_0, \alpha_1, \dots, \alpha_n \rangle \succ \langle \beta_0, \dots, \beta_m \rangle \Longrightarrow f(\langle \alpha_0, \dots, \alpha_n \rangle) \succ f(\langle \beta_0, \dots, \beta_m \rangle)$$

at least when $m \geq 1$. Of course $u \succ v$ means u is a proper initial segment of v. Then we will show that $\prec \restriction S$ is wellfounded, therefore $|T| \leq |\prec \restriction S| + 1$; but $|\prec \restriction S| < \kappa^+$ and the proof will be complete.

Let $\langle \varphi_n : n \in \omega \rangle$ be a κ-scale on $>$. To simplify the definition of f (though it is not essential) we put for $\alpha > \beta$

$$\psi_n(\alpha, \beta) = \ulcorner \alpha(0), \beta(0), \varphi_0(\alpha, \beta), \dots, \alpha(n), \beta(n), \varphi_n(\alpha, \beta) \urcorner.$$

Then notice the following limit property of $\langle \psi_n : n \in \omega \rangle$:

If $\alpha_i > \beta_i$ for all i, and for each n, $\psi_n(\alpha_i, \beta_i)$ is eventually constant, then $\lim_i(\alpha_i, \beta_i) = (\alpha, \beta)$ exists and $\alpha > \beta$.

Define now f by induction on the length of sequences:

$$f(\langle\,\rangle) = \langle\,\rangle$$
$$f(\langle\alpha_0\rangle) = \langle\,\rangle$$
$$f(\langle\alpha_0,\alpha_1\rangle) = \langle\psi_0(\alpha_0,\alpha_1)\rangle$$
$$f(\langle\alpha_0,\alpha_1,\alpha_2\rangle) = \langle\psi_0(\alpha_0,\alpha_1),\psi_1(\alpha_0,\alpha_1),\psi_1(\alpha_1,\alpha_2),\psi_0(\alpha_1,\alpha_2)\rangle,$$

and in general

$$f(\langle\alpha_0,\dots,\alpha_{n-1},\alpha_n\rangle) = f(\langle\alpha_0,\dots,\alpha_{n-1}\rangle) ^\frown\langle\psi_{n-1}(\alpha_0,\alpha_1),\psi_{n-1}(\alpha_1,\alpha_2),\dots,\psi_{n-1}(\alpha_{n-1},\alpha_n), \psi_{n-2}(\alpha_{n-1},\alpha_n),\dots,\psi_0(\alpha_{n-1},\alpha_n)\rangle.$$

The idea is to include in $f(\langle\alpha_0,\dots,\alpha_n\rangle)$ all $\psi_j(\alpha_i,\alpha_{i+1})$ for $i \le n-1, j \le n-1$. The diagram below explains the way we have done it:

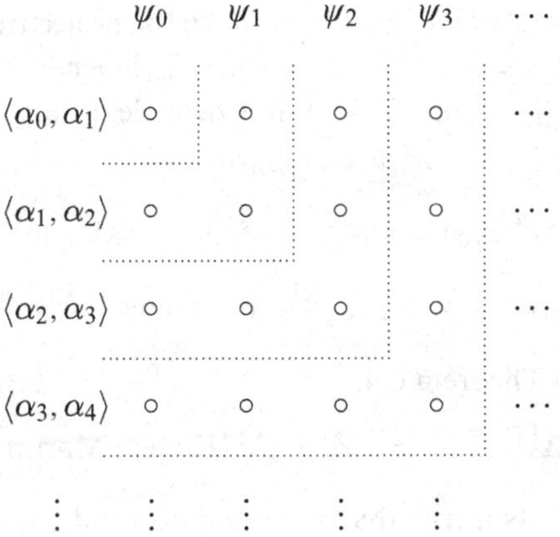

Clearly f is an $\prec$-preserving (on $T - \{\langle\,\rangle\}$) map from T onto a set of finite sequences S on $\kappa^\omega = \lambda$. Thus it will be enough to show that $\prec\restriction S$ is wellfounded. Assume not, towards a contradiction. Then we have

$$f(\langle\alpha_0^0\rangle) \succ f(\langle\alpha_0^1,\alpha_1^1\rangle) \succ f(\langle\alpha_0^2,\alpha_1^2,\alpha_2^2\rangle) \succ \cdots$$

for some $\alpha_0^0, \alpha_0^1, \alpha_1^1, \dots$, such that $\alpha_0^1 > \alpha_1^1, \alpha_0^2 > \alpha_1^2 > \alpha_2^2$ *etc.* Then in the diagram

$$\begin{array}{llllll}
\psi_0(\alpha_0^1,\alpha_1^1) & & & & & \\
\psi_0(\alpha_0^2,\alpha_1^2) & \psi_1(\alpha_0^2,\alpha_1^2) & \psi_1(\alpha_1^2,\alpha_2^2) & \psi_0(\alpha_1^2,\alpha_2^2) & & \\
\psi_0(\alpha_0^3,\alpha_1^3) & \psi_1(\alpha_0^3,\alpha_1^3) & \psi_1(\alpha_1^3,\alpha_2^3) & \psi_0(\alpha_1^3,\alpha_2^3) & \psi_2(\alpha_0^3,\alpha_1^3) & \dots \\
\vdots & \vdots & \vdots & \vdots & \vdots &
\end{array}$$

each column consists of identical ordinals, thus for each n and for each j, $\psi_n(\alpha^i_j, \alpha^i_{j+1})$ becomes constant for large enough i. Thus for each j, $\langle \alpha^i_j, \alpha^i_{j+1}\rangle \to \langle \alpha_j, \alpha_{j+1}\rangle$ and $\alpha_j > \alpha_{j+1}$, *i.e.*, $\alpha_0 > \alpha_1 > \alpha_2 > \cdots$, a contradiction. ⊣

Corollary 7.2. Every $\utilde{\Sigma}^1_1$ wellfounded relation has length $< \aleph_1$ (classical result).

Corollary 7.3. Every $\utilde{\Sigma}^1_2$ wellfounded relation has length $< \aleph_2$. Thus if there exists a $\utilde{\Sigma}^1_2$ wellordering of $\mathbb{R}$, the Continuum Hypothesis holds. (Martin; by an unpublished forcing argument before scales were introduced.)

Corollary 7.4. Assume $\forall\alpha(\alpha^\#$ exists). Then every $\utilde{\Sigma}^1_3$ wellfounded relation has length $< (u_\omega)^+$. If we also assume AC, then every $\utilde{\Sigma}^1_3$ wellfounded relation has length $< \aleph_3$. (That $\utilde{\delta}^1_3 \leq (u_\omega)^+$ was already shown by Martin, [Mar70B].)

7.2. Projective ordinals. We introduced in §6 the **projective ordinals** $\utilde{\delta}^1_n$ and we mentioned that $\utilde{\delta}^1_1 = \aleph_1$ (this follows also independently from our first corollary in 7.1). By the results in § 7.1, it is then clear that

$$\utilde{\delta}^1_2 \leq \aleph_2 \text{ (Martin)} \tag{7.2.1}$$

$$\forall\alpha(\alpha^\# \text{ exists}) + \mathsf{AC} \Longrightarrow \utilde{\delta}^1_3 \leq \aleph_3 \text{ (Martin)} \tag{7.2.2}$$

$$\mathsf{Det}(\utilde{\Delta}^1_{2n}) \Longrightarrow \utilde{\delta}^1_{2n+2} \leq (\utilde{\delta}^1_{2n+1})^+ \text{ (Kunen, Martin)} \tag{7.2.3}$$

(To prove 7.2.3 recall Theorem 6.4.)

$$\mathsf{Det}(\utilde{\Delta}^1_2) + \mathsf{AC} \Longrightarrow \utilde{\delta}^1_4 \leq \aleph_4 \text{ (Kunen, Martin)} \tag{7.2.4}$$

Open Problem 7.5. Is it true that assuming AC (and any other reasonable hypotheses), $\utilde{\delta}^1_n \leq \aleph_n, n \geq 5$?

We shall mention some other known results about the projective ordinals in the last section.

§8. Construction principles. A **construction principle** for a pointclass Γ asserts, roughly speaking, that every set in Γ can be expressed, in some canonical way, in terms of sets in a simpler pointclass Γ'. A classical example is the result that every analytic ($\utilde{\Sigma}^1_1$) set can be expressed both as a union and an intersection of $\aleph_1$ Borel sets.

8.1. Inductive analysis of projection of trees. Let T be a tree on $\omega \times \kappa$. We write

$$A = \mathrm{p}[T]$$

iff $\alpha \in A \iff \exists f(\langle \alpha, f\rangle \in [T]) \iff T(\alpha)$ is not wellfounded.

THEOREM 8.1. Let T be a tree on $\omega \times \kappa$ and $A = \mathrm{p}[T]$. Put for $1 \leq \xi < \kappa^+$ and u a finite sequence from κ,

$$A_u^\xi = \{\alpha : |T(\alpha)_u| < \xi\},$$

where for any tree J we abbreviate

$$|J| < \xi \iff J \text{ is wellfounded and } |J| < \xi.$$

Then, if $\mathrm{lh}(u) = n$ we have

$$A_u^0 = \{\alpha : \langle\langle\alpha(0), u_0\rangle, \dots, \langle\alpha(n), u_{n-1}\rangle\rangle \notin T\}$$

$$A_u^{\xi+1} = A_u^\xi \cup \bigcap_{\eta<\kappa} A_{u\frown\eta}^\xi$$

$$A_u^\lambda = \bigcup_{\xi<\lambda} A_u^\xi, \text{ if } \lambda = \bigcup \lambda > 0$$

and

$$\mathbb{R} \setminus A = \bigcup_{\xi<\kappa^+} A_{\langle\,\rangle}^\xi.$$

(For $\kappa = \omega$ this is apparently due to Sierpinski, see [Kur66, p. 32]; Martin [Mar70B] first applied these methods to $\kappa > \omega$.)

PROOF. Notice that $\alpha \notin A \iff T(\alpha)$ is wellfounded

$$\iff (\exists\xi < \kappa^+)(|T(\alpha)| < \xi) \iff (\exists\xi < \kappa^+)(|T(\alpha)_{\langle\,\rangle}| < \xi). \qquad \dashv$$

If λ is an ordinal put

$\mathcal{B}(\lambda) =$ the smallest Boolean algebra containing all closed sets and closed under unions of length $< \lambda$.

Let also

$$\mathcal{B}_n = \mathcal{B}(\underset{\sim}{\delta}_n^1).$$

THEOREM 8.2. Let T be a tree on $\omega \times \kappa$ and $A = \mathrm{p}[T]$. Then A is both the union and the intersection of κ^+ sets in $\mathcal{B}(\kappa^+)$. (Sierpinski for $\kappa = \omega$.)

PROOF. In the notation of 8.1, we clearly have A_u^1 clopen, for all u. Thus $A_u^\xi \in \mathcal{B}(\kappa^+)$, for any ξ and u. So A is the intersection of κ^+ sets in $\mathcal{B}(\kappa^+)$.

Now put

$$B_\xi = \{\alpha : |T(\alpha)| < \xi\} \cup \{\alpha : (\exists u)(|T(\alpha)_u| = \xi)\}.$$

Since

$$B_\xi = A_{\langle\,\rangle}^\xi \cup \bigcup_u (A_u^{\xi+1} \setminus A_u^\xi),$$

clearly each B_ξ is in $\mathcal{B}(\kappa^+)$. It is then easy to check that

$$\mathbb{R} \setminus A = \bigcap_{\xi<\kappa^+} B_\xi;$$

thus A is the union of κ^+ sets in $\mathcal{B}(\kappa^+)$. ⊣

COROLLARY 8.3. (a) Each $\utilde{\mathbf{\Sigma}}^1_1$ set is both the union and the intersection of $\aleph_1$ Borel sets. (Classical)

(b) Each $\utilde{\mathbf{\Sigma}}^1_2$ set is both the union and the intersection of $\aleph_2$ sets in $\mathcal{B}(\aleph_2)$.

(c) If $\alpha^\#$ exists for all reals α, each $\utilde{\mathbf{\Sigma}}^1_3$ set is both the union and the intersection of u^+_ω sets in $\mathcal{B}(u^+_\omega)$. Thus, if also AC holds, each $\utilde{\mathbf{\Sigma}}^1_3$ set is both the union and the intersection of $\aleph_3$ sets in $\mathcal{B}(\aleph_3)$. (Martin [Mar70B])

(d) $\mathsf{Det}(\utilde{\mathbf{\Delta}}^1_{2n})$ implies that every $\utilde{\mathbf{\Sigma}}^1_{2n+2}$ set is both the union and the intersection of $(\utilde{\boldsymbol{\delta}}^1_{2n+1})^+$ sets in $\mathcal{B}((\utilde{\boldsymbol{\delta}}^1_{n+1})^+)$. Thus $\mathsf{Det}(\utilde{\mathbf{\Delta}}^1_2)+\mathsf{AC} \Longrightarrow$ Every $\utilde{\mathbf{\Sigma}}^1_4$ set is both the union and the intersection of $\aleph_4$ sets in $\mathcal{B}(\aleph_4)$.

In §§ 8.2 and 8.3 will see how most of the results of this corollary can be improved.

8.2. An extension of Suslin's Theorem.

THEOREM 8.4. (Martin, [Mar70B] for $n = 1$; Moschovakis [Mos71A] in general.) Assume $\mathsf{Det}(\utilde{\mathbf{\Delta}}^1_{2n})$. Then

$$\utilde{\mathbf{\Delta}}^1_{2n+1} \subseteq \mathcal{B}_{2n+1}.$$

PROOF. (In this proof we use essentially [Mos70A, Lemmas 9 & 10].)

Let $A \subseteq \mathbb{R}, A \in \utilde{\mathbf{\Delta}}^1_{2n+1}$. Find a Π^1_{2n+1}-scale on $\mathbb{R} \setminus A$, say $\langle \varphi_m : m \in \omega\rangle$. Since $\mathbb{R} \setminus A \in \utilde{\mathbf{\Delta}}^1_{2n+1}$, each prewellordering $\leq^{\varphi_m}$ is actually a $\utilde{\mathbf{\Delta}}^1_{2n+1}$ prewellordering, thus it has length $\xi_m < \utilde{\boldsymbol{\delta}}^1_{2n+1}$. Since obviously $\mathrm{cf}(\utilde{\boldsymbol{\delta}}^1_{2n+1}) > \omega$, $\sup_m \xi_m < \utilde{\boldsymbol{\delta}}^1_{2n+1}$, say $\xi_m < \kappa < \utilde{\boldsymbol{\delta}}^1_{2n+1}$, for all m, and $\langle \varphi_m : m \in \omega\rangle$ is a κ-scale.

For each α put

$$\begin{aligned} J(\alpha) = \{\langle \xi_0, \dots, \xi_k\rangle : \xi_i < \kappa \wedge ((\exists \alpha_0 \dots \alpha_k \in \mathbb{R} \setminus A)(\forall i \leq k(\bar{\alpha}_i(i) = \bar{\alpha}(i)) \\ \wedge (\forall i \leq k \forall j)(i \leq j \leq k \Longrightarrow \varphi_i(\alpha_j) = \xi_i)))\}. \end{aligned}$$

The mapping $\alpha \mapsto J(\alpha)$, from $\mathbb{R}$ into trees on κ, is continuous, *i.e.*, for each $\langle \xi_0, \dots, \xi_k\rangle$, $\{\alpha : \langle \xi_0, \dots, \xi_k\rangle \in J(\alpha)\}$ is open, and

$$\alpha \notin A \Longleftrightarrow J(\alpha) \text{ is not wellfounded;}$$

thus $\alpha \in A \Longleftrightarrow J(\alpha)$ *is wellfounded.* Put $\lambda = \sup\{J(\alpha) + 1 : \alpha \in A\}$. Suppose we can prove $\lambda < \utilde{\boldsymbol{\delta}}^1_{2n+1}$. Then as in 8.1 we can define $A^\xi_u = \{\alpha : |J(\alpha)_u| < \xi\}$ for $1 \leq \xi < \lambda$ and u a finite sequence from κ and prove that $A^\xi_u \in \mathcal{B}_{2n+1}$ (since $\kappa, \lambda < \utilde{\boldsymbol{\delta}}^1_{2n+1}$). Since $A = \bigcup_{\xi<\lambda} A^\xi_{\langle\,\rangle}$ we have $A \in \mathcal{B}_{2n+1}$.

We prove now $\lambda < \underset{\sim}{\delta}^1_{2n+1}$. Define the tree

$$J^*(\alpha) = \{\langle \alpha_0, \dots, \alpha_k \rangle : \langle \alpha_0, \dots, \alpha_k \rangle \in \mathbb{R} \setminus A \wedge \forall i \leq k(\bar{\alpha}_i(i) = \bar{\alpha}(i)) \wedge \forall i \leq k \forall j \forall \ell (i \leq j \leq \ell \leq k \Longrightarrow (\varphi_i(\alpha_j) = \varphi_i(\alpha_\ell)))\}.$$

Then there is an obvious surjective map f from $J^*(\alpha)$ onto $J(\alpha)$, namely

$$f(\langle \alpha_0, \dots, \alpha_k \rangle) = \langle \varphi_0(\alpha_0), \dots, \varphi_k(\alpha_k) \rangle,$$

which is clearly $\prec$-preserving. Thus $J^*(\alpha)$ is wellfounded for any $\alpha \in A$ and it is easy to check that for $\alpha \in A$

$$|\langle \alpha_0, \dots, \alpha_k \rangle|_{J^*(\alpha)} = |f(\langle \alpha_0, \dots, \alpha_k \rangle)|_{J(\alpha)}.$$

Thus $|J(\alpha)| = |J^*(\alpha)|$, for every $\alpha \in A$. It will then be enough to show $\sup\{|J^*(\alpha)| + 1 : \alpha \in A\} < \underset{\sim}{\delta}^1_{2n+1}$. To prove this define

$$\begin{aligned} &\langle \alpha, \langle \alpha_0, \dots, \alpha_k \rangle \rangle > \langle \beta, \langle \beta_0, \dots, \alpha_m \rangle \rangle \\ &\quad \Longleftrightarrow \alpha = \beta \in A \wedge \langle \alpha_0, \dots, \alpha_k \rangle, \langle \beta_0, \dots, \beta_m \rangle \in J^*(\alpha) \\ &\quad \wedge \langle \alpha_0, \dots, \alpha_k \rangle \succ \langle \beta_0, \dots, \beta_m \rangle. \end{aligned}$$

Then $<$ is a $\underset{\sim}{\mathbf{\Delta}}^1_{2n+1}$ wellfounded relation, if we code the finite sequences by single reals. But a simple variation of Lemma 10 in [Mos70A] shows that every $\underset{\sim}{\mathbf{\Delta}}^1_{2n+1}$ wellfounded relation has length $< \underset{\sim}{\delta}^1_{2n+1}$, thus $| < | < \underset{\sim}{\delta}^1_{2n+1}$ and we are done, since $|J^*(\alpha)| \leq | < |$, for each $\alpha \in A$. ⊣

Corollary 8.5 (Moschovakis, [Mos71A]). Assume $\mathrm{Det}(\underset{\sim}{\mathbf{\Delta}}^1_{2n})$. Then every $\underset{\sim}{\mathbf{\Sigma}}^1_{2n+2}$ set is the union of $\underset{\sim}{\delta}^1_{2n+1}$ sets in $\mathcal{B}_{2n+1}$.

Proof. It follows from $\mathrm{PWO}(\underset{\sim}{\mathbf{\Pi}}^1_{2n+1})$ that every $\underset{\sim}{\mathbf{\Pi}}^1_{2n+1}$ set is the union of $\underset{\sim}{\delta}^1_{2n+1} \underset{\sim}{\mathbf{\Delta}}^1_{2n+1}$ sets (see remarks before 6.4). It also follows from $\mathrm{PWO}(\underset{\sim}{\mathbf{\Pi}}^1_{2n+1})$ that a 1-1 continuous image of a $\underset{\sim}{\mathbf{\Delta}}^1_{2n+1}$ set is also a $\underset{\sim}{\mathbf{\Delta}}^1_{2n+1}$ set. We can get then the result easily by applying the uniformization theorem and 8.4. ⊣

Let AD be the statement "Every set is determined". Martin [Mar70B] has shown that AD implies that $\underset{\sim}{\mathbf{\Delta}}^1_{2n+1} \supseteq \mathcal{B}_{2n+1}$; thus AD implies that $\underset{\sim}{\mathbf{\Delta}}^1_{2n+1} = \mathcal{B}_{2n+1}$. Moschovakis has shown in [Mos70A], that AD implies that every union of $\underset{\sim}{\delta}^1_{2n+1}$ sets in $\underset{\sim}{\mathbf{\Delta}}^1_{2n+1}$ is in $\underset{\sim}{\mathbf{\Sigma}}^1_{2n+2}$; thus AD implies

$$A \in \underset{\sim}{\mathbf{\Sigma}}^1_{2n+2} \Longleftrightarrow A \text{ is the union of } \underset{\sim}{\delta}^1_{2n+1} \text{ sets in } \mathcal{B}_{2n+1}.$$

8.3. Unions of Borel sets.

Theorem 8.6 (Martin, [Mar70B]). Assume AC. If a pointset A admits a κ-scale with $\kappa < \aleph_{n+1}$, $0 < n \in \omega$, then A is the union of $\aleph_n$ Borel sets.

Corollary 8.7. (a) Assume AC$+\forall\alpha(\alpha^{\#}$ exists). Then every $\underset{\sim}{\mathbf{\Sigma}}^1_3$ set is the union of $\aleph_2$ Borel sets.

(b) Assume AC$+\mathrm{Det}(\underset{\sim}{\mathbf{\Delta}}^1_2)$. Then every $\underset{\sim}{\mathbf{\Sigma}}^1_4$ set is the union of $\aleph_3$ Borel sets.

Proof of the Theorem. Assume $A \subseteq \mathbb{R}$ admits a κ-scale with $\kappa < \aleph_{n+1}$. Then for some tree T, $\alpha \in A \iff T(\alpha)$ is not wellfounded, where we may assume that T is actually a tree on $\omega \times \aleph_n$ (we replace if necessary the tree coming from the scale on A by an isomorphic tree, noticing that $|\kappa| \leq \aleph_n$). Since $n > 0$, $\mathrm{cf}(\aleph_n) > \omega$, thus if $T(\alpha)$ is not wellfounded there is a $\xi < \aleph_n$ such that $T^\xi(\alpha)$ is not wellfounded, where

$$T^\xi = \{\langle\langle k_0, \xi_0\rangle, \dots, \langle k_m, \xi_m\rangle\rangle \in T : \xi_0, \dots, \xi_m \leq \xi\}.$$

Thus

$$\alpha \in A \iff \exists \xi_n < \aleph_n[T^{\xi_n}(\alpha) \text{ is not wellfounded}].$$

If $\xi_n < \aleph_n$ we can replace T^{ξ_n} by an isomorphic tree on $\omega \times \aleph_{n-1}$, say $T_1^{\xi_n}$. Then

$$\alpha \in A \iff (\exists \xi_n < \aleph_n)(T_1^{\xi_n}(\alpha) \text{ is not wellfounded}).$$

If $n - 1 > 0$ we have again

$$\begin{aligned} &T_1^{\xi_n}(\alpha) \text{ is not wellfounded} \\ &\iff (\exists \xi_{n-1} < \aleph_{n-1})((T_1^{\xi_n})^{\xi_{n-1}} \text{ is not wellfounded}) \end{aligned}$$

and we proceed similarly. After at most n steps we get

$$\begin{aligned} \alpha \in A \iff &\exists \xi_n < \aleph_n \exists \xi_{n-1} < \aleph_{n-1} \dots \exists \xi_1 < \aleph_1 \\ &(T^{\xi_n, \dots, \xi_1}(\alpha) \text{ is not wellfounded}) \end{aligned}$$

where $T^{\xi_n,\dots,\xi_1}$ is a tree on $\omega \times \omega$. But then $\{\alpha : T^{\xi_n,\dots,\xi_1}(\alpha) \text{ is not wellfounded}\}$ is a $\utilde{\Sigma}^1_1$ set, thus it is the union of $\aleph_1$ Borel sets and the proof is complete. $\dashv$

Open Problem 8.8. Prove assuming AC (and any other reasonable hypotheses) that every $\utilde{\Sigma}^1_{n+1}$ set is the union of $\aleph_n$ Borel sets, $n \geq 4$. Notice that a solution to the problem at the end of §7 solves this problem too.

§9. Constructibility in the tree associated with a scale.

In §6 we associated with each κ-scale $\langle \varphi_n : n \in \omega \rangle$ on a set $A \subseteq \mathbb{R}$ a tree T on $\omega \times \kappa$. We introduce and study here the models $\mathbf{L}[T]$, where T comes from a complete Π^1_{2n+1} set, granting $\mathsf{Det}(\utilde{\Delta}^1_{2n})$—these are the basic tools for the results in the next two sections. The key theorem in the present section is that the tree that comes from a complete Π^1_1 set is in fact constructible.

9.1. The models $\mathbf{L}[T^{2n+1}]$.

Theorem 9.1 (Folk-type result). Let T be the tree associated with a κ-scale on some set A, let $Q \subseteq \mathbb{R} \times \mathbb{R}$ and assume that for some recursive $f : \mathbb{R} \times \mathbb{R} \to \mathbb{R}$,

$$Q(\alpha, \beta) \iff f(\alpha, \beta) \in A.$$

Then for some tree $S \in \mathbf{L}[T]$ on $\omega \times \kappa$

$$\exists\beta Q(\alpha,\beta) \iff S(\alpha) \text{ is not wellfounded.} \quad (1)$$
$$\iff (\exists\beta \in \mathbf{L}[T,\alpha])Q(\alpha,\beta) \quad (2)$$

PROOF. We have

$$\exists\beta Q(\alpha,\beta) \iff \exists\beta(f(\alpha,\beta) \in A)$$
$$\iff \exists\beta\exists\gamma(f(\alpha,\beta) = \gamma \wedge \gamma \in A)$$
$$\iff \exists\beta\exists\gamma(\forall n R(\bar{\alpha}(n), \bar{\beta}(n), \bar{\gamma}(n)) \wedge T(\gamma) \text{ is not wellfounded})$$

where R is recursive. Put

$$S' = \{\langle\langle a_0, \langle b_0, c_0, \xi_0\rangle\rangle, \dots, \langle a_k, \langle b_k, c_k, \xi_k\rangle\rangle\rangle : \langle\langle c_0, \xi_0\rangle, \dots, \langle c_k, \xi_k\rangle\rangle \in T$$
$$\wedge R(\ulcorner a_0, \dots, a_k\urcorner, \ulcorner b_0, \dots, b_k\urcorner, \ulcorner c_0, \dots, c_k\urcorner)\}.$$

Then S' is a tree on $\omega \times (\omega \times \omega \times \kappa)$, $S' \in \mathbf{L}[T]$ and

$$\exists\beta Q(\alpha,\beta) \iff S'(\alpha) \text{ is not wellfounded}$$
$$\iff \exists\beta \in \mathbf{L}[T,\alpha]Q(\alpha,\beta)$$

where the last equivalence follows from the usual absoluteness of wellfoundedness. To get instead of S' a tree on $\omega \times \kappa$ fix a 1-1 mapping from $\omega \times \omega \times \kappa$ onto κ *in* $\mathbf{L}[T]$ and replace S' by an isomorphic tree on $\omega \times \kappa$, call it S. ⊣

Fix now a *complete* Π^1_{2n+1} set $\mathcal{P}_{2n+1} \subseteq \mathbb{R}$. ($\mathcal{P}_{2n+1}$ is such that for every $A \subseteq \mathfrak{X}, A \in \Pi^1_{2n+1}$ we can find $f : \mathfrak{X} \to \mathbb{R}$ recursive such that $x \in A \iff f(x) \in \mathcal{P}_{2n+1}$.) Assuming $\mathrm{Det}(\utilde{\mathbf{\Delta}}^1_{2n})$, let $\langle\varphi_n : n \in \omega\rangle$ be a fixed for the discussion Π^1_{2n+1} scale on $\mathcal{P}_{2n+1}$ and let T^{2n+1} be the associated tree. T^{2n+1} is a tree on $\omega \times \utilde{\delta}^1_{2n+1}$.

THEOREM 9.2. Assume $\mathrm{Det}(\utilde{\mathbf{\Delta}}^1_{2n})$. Then for every Σ^1_{2n+2} set A we can find a tree $S \in \mathbf{L}[T^{2n+1}]$ such that

$$\alpha \in A \iff S(\alpha) \text{ is not wellfounded.}$$

THEOREM 9.3. Assume $\mathrm{Det}(\utilde{\mathbf{\Delta}}^1_{2n})$. Then Σ^1_{2n+2} formulas are absolute for $\mathbf{L}[T^{2n+1}]$.

PROOF. We show successively that for $2n+2 \geq k \geq 2$, Σ^1_k formulas are absolute for $\mathbf{L}[T^{2n+1}]$. For $k = 2$ this is Shoenfield's theorem, while for $k \geq 3$ we proceed using (3) and (4) of Theorem 9.1. ⊣

Unfortunately, except for the case $n = 0$, which we shall study in the rest of this section, there is practically nothing known about the *internal structure* of $\mathbf{L}[T^{2n+1}]$.[2]

[2]It has been recently shown by Harrington and Kechris that for all $n \geq 0$, $\mathbb{R} \cap \mathbf{L}[T^{2n+1}]$ is the largest countable Σ^1_{2n+2} set of reals (as conjectured by Moschovakis) and that additionally,

$$\mathbf{L}[T^{2n+1}] \models \text{“}\mathbb{R} \text{ has a } \Delta^1_{2n+2}\text{-good wellordering”}.$$

9.2. Absoluteness of closed games. Let $\mathcal{S}$ be a set of even finite sequences from a set A. We define the game $\mathrm{G}_{\mathcal{S}}$ as follows:

I	II	
a_0	b_0	I plays $a_0, a_1, \ldots$ and II plays $b_0, b_1 \ldots$, $a_i, b_i \in A$. Then I wins iff for some n,
a_1	b_1	$\langle a_0, b_0, \ldots, a_n, b_n \rangle \in \mathcal{S}$. Clearly the game is open in I.
$\vdots$	$\vdots$	

The following is a folk-type result.

THEOREM 9.4. Let $M \models \mathsf{ZF}+\mathsf{DC}$ and $M \supseteq \mathrm{Ord}$. Let $A, \mathcal{S} \in M$, and assume A is wellorderable in M. Then

player I has a winning strategy in $\mathrm{G}_{\mathcal{S}}$ iff $M \models$ player I has a winning strategy in $\mathrm{G}_{\mathcal{S}}$

and similarly for player II. Moreover the player who has a winning strategy has a winning strategy (for the game in the world) which lies in M.

PROOF. For each $\langle a_0, b_0, \ldots, a_n, b_n \rangle$ consider the subgame $\mathrm{G}_{\mathcal{S}}(a_0, b_0, \ldots, a_n, b_n)$ defined by;

I	II	
α	β	I plays α, II plays β and I wins iff for some m $\langle a_0, b_0, \ldots, a_n, b_n \rangle ^\frown \langle \alpha(0), \beta(0), \ldots, \alpha(m), \beta(m) \rangle \in \mathcal{S}$.

Then define

$$\mathcal{S}^0 = \mathcal{S}$$

$$\mathcal{S}^\xi = \{\langle a_0, b_0, \ldots, a_n, b_n \rangle : \exists a_{n+1} \in A \forall b_{n+1} \in A \\ \exists \eta < \xi(\langle a_0, b_0, \ldots, a_{n+1}, b_{n+1} \rangle \in \mathcal{S}^\eta).$$

Then for each ξ, $\langle a_0, b_0, \ldots, a_n, b_n \rangle \in \mathcal{S}^\xi \Longrightarrow$ player I has a winning strategy in $\mathrm{G}_{\mathcal{S}}(a_0, b_0, \ldots, a_n, b_n)$. Using this we show:

CLAIM 9.5. Player II has a winning strategy in G if and only if $\forall \xi (\langle\,\rangle \notin \mathcal{S}^\xi)$.

PROOF OF CLAIM 9.5. If player II has a winning strategy in $\mathrm{G}_{\mathcal{S}} = \mathrm{G}_{\mathcal{S}}(\langle\,\rangle)$, then player I has no winning strategy in $\mathrm{G}_{\mathcal{S}}(\langle\,\rangle)$, thus for all $\xi, \langle\,\rangle \notin \mathcal{S}^\xi$. Conversely assume that for each $\xi, \langle\,\rangle \notin \mathcal{S}^\xi$. We describe a winning strategy for player II in $\mathrm{G}_{\mathcal{S}}$ as follows: If player I plays a_0, player II plays the least b_0 (in a fixed wellordering of A) such that $\forall \xi(\langle a_0, b_0 \rangle \notin \mathcal{S}^\xi)$. Such a b_0 exists, because otherwise for all b, there exists a ξ such that $\langle a_0, b \rangle \in \mathcal{S}^\xi$. Let $g(b) =$ least such ξ and find $\xi_0 >$ all $g(b), b \in A$. Then $\forall b \exists \xi < \xi_0 (a_0, b) \in \mathcal{S}^\xi$, thus $\langle\,\rangle \in \mathcal{S}^0$, a contradiction. Similarly if player I plays a_1, player II picks the least b_1 such that $\forall \xi(\langle a_0, b_0, a_1, b_1 \rangle \notin \mathcal{S}^\xi)$, *etc.* ⊣ (Claim 9.5)

These results suggest that $\mathbf{L}[T^{2n+1}]$ is a correct higher level analogue of $\mathbf{L}$. Their proof uses determinacy of all hyperprojective sets. See [HK77].

Since the above equivalence was proved under the assumption "ZF+DC+A is wellorderable" and since $\xi \mapsto S^\xi$ is clearly an absolute map and $M \supseteq \mathrm{Ord}$, it is immediate the "player II has a winning strategy" is absolute for M, thus the same is true for "player I has a winning strategy." Moreover the argument above clearly provides a winning strategy for player II which lies in M and wins in the world, thus it will be enough in order to complete the proof to show that when player I has a winning strategy we can find one (who wins in the world also) in M. Notice that

$$\text{Player I has a winning strategy} \iff \exists\xi(\langle\,\rangle \in g^S)$$

and check that the following is a winning strategy for player I which lies in M. Put $\xi_0 =$ least ξ such that $\langle\,\rangle \in S^\xi$. If $\xi_0 = 0$, player I has already won. If $\xi_0 > 0$, let player I play the least a_0 such that for every b, $\exists\xi < \xi_0(\langle a_0, b\rangle \in S^\xi)$. If now player II plays b_0, let $\xi_1 =$ least $\xi < \xi_0$ such that $\langle a_0, b_0\rangle \in S^\xi$. If $\xi_1 = 0$, player I has already won, otherwise let player I play the least a_1 such that for all b, $\exists\xi < \xi_1(\langle a_0, b_0, a_1, b\rangle \in S^\xi)$ *etc.* (Notice that $\xi_0 > \xi_1 > \cdots$, so this cannot go on.) ⊣ (Theorem 9.4)

9.3. Proof that $T^1 \in$ L. Suppose $A \subseteq \aleph_1$. We let

$$\mathrm{Code}(A) = \{\alpha \in \mathrm{WO} : |\alpha| \in A\}.$$

Similarly, if T is a tree on $\omega \times \aleph_1$, we let

$$\mathrm{Code}(T) = \{\langle k_0, \alpha_0, \dots, k_n, \alpha_n\rangle : \langle\langle k_0, |\alpha_0|\rangle, \dots, \langle k_n, |\alpha_n|\rangle\rangle \in T\}.$$

We say that $\mathrm{Code}(T)$ is in Γ iff

$$\{\ulcorner k_0, \alpha_0, \dots, k_n, \alpha_n\urcorner : \langle k_0, \alpha_0, \dots, \alpha_n\rangle \in \mathrm{Code}(T)\} \in \Gamma$$

where $\ulcorner k_0, \alpha_0, \dots, k_n, \alpha_n\urcorner = \langle n, k_0, \dots, k_n, \alpha_0(0), \dots, \alpha_n(0), \alpha_0(1), \dots, \alpha_n(1), \dots\rangle \in \mathbb{R}$.

Lemma 9.6 (Kechris). Let $A \subseteq \mathbb{R}, A \in \Pi^1_1$ and assume $\langle\varphi_n : n \in \omega\rangle$ is a Π^1_1-scale on A. Then the tree T associated with $\langle\varphi_n : n \in \omega\rangle$ is Σ^1_2 in the codes.

Proof. We have

$$\begin{aligned}\langle k_0, \alpha_0, \dots, k_n, \alpha_n\rangle \in \mathrm{Code}(T) \iff & \alpha_0, \dots, \alpha_n \in \mathrm{WO} \\ & \wedge(\exists\alpha)(\alpha \in A \wedge \varphi_0(\alpha) = |\alpha_0| \wedge \alpha(0) = k_0 \\ & \wedge \dots \wedge \varphi_n(\alpha) = |\alpha_n| \wedge \alpha(n) = k_n).\end{aligned}$$

The result follows immediately if we can show that for each n,

$$\alpha \in A \wedge \beta \in \mathrm{WO} \wedge \varphi_n(\alpha) = |\beta|$$

is a Σ^1_2 relation in α, β, n. But each φ_n is a Π^1_1-norm, thus every initial segment of $\leq^{\varphi_n} = \leq_n$ will have countable length (since $\underset{\sim}{\delta}^1_1 = \aleph_1$). From this we have

$$\begin{aligned}\alpha \in A \wedge \beta \in \mathrm{WO} \wedge \varphi_n(\alpha) = |\beta| \iff & \alpha \in A \wedge \beta \in \mathrm{WO} \\ & \wedge \exists\delta((\forall m)(m \leq_\beta m \implies (\delta)_m \leq_n \alpha) \\ & \wedge (\forall\gamma)(\gamma \leq_n \alpha \implies (\exists m)(m \leq_\beta m \wedge (\delta)_m \leq_n \alpha \wedge \alpha \leq_n (\delta)_m) \\ & \wedge \forall m, \ell(m <_\beta \ell \iff (\delta)_m <_n (\delta)_\ell))\end{aligned}$$

(where of course $\delta <_n \varepsilon \iff \varphi_n(\delta) < \varphi_n(\varepsilon)$) and the proof is complete. ⊣

THEOREM 9.7. Suppose $A \subseteq \aleph_1$ and $\mathrm{Code}(A)$ is Σ^1_2. Then A is in $\mathbf{L}$. Similarly if T is a tree on $\omega \times \aleph_1$ and $\mathrm{Code}(A)$ is Σ^1_2, then $T \in \mathbf{L}$.

(For $A \subseteq \aleph_1$ with $\mathrm{Code}(A) \in \Pi^1_1$ the result is implicit in methods (using forcing) of Solovay. The game method used below as a substitute of forcing was used by Moschovakis to prove a version of the corollary below and traces to [Mos70B]. The present version of the theorem is due to Kechris.)

COROLLARY 9.8 (Moschovakis). $T^1 \in \mathbf{L}$.

PROOF OF THEOREM 9.7. We give the proof for $A \subseteq \aleph_1$, the case of a tree being similar.

Thus let $A \subseteq \aleph_1$ and $\mathrm{Code}(A) = P \in \Sigma^1_2$. Then

$$\xi \in A \iff (\exists\alpha)(\alpha \in P \wedge |\alpha| = \xi).$$

Let $\alpha \in P \iff \exists\beta\, Q(\alpha, \beta) \iff \exists\beta(f(\alpha, \beta) \in \mathrm{WO})$, where $Q \in \Pi^1_1$ and $f : \mathbb{R} \times \mathbb{R} \to \mathbb{R}$ is recursive and for all α, β, $f(\alpha, \beta) \in \mathrm{LOR}$. Then

$$\xi \in A \iff \exists\alpha\exists\beta(f(\alpha, \beta) \in \mathrm{WO} \wedge |\alpha| = \xi).$$

Consider the following game G_ξ:

$$\begin{array}{cccccc} \mathrm{I} & & & \mathrm{II} & & \\ \xi_0 & a_0 & b_0 & \eta_0 & \vartheta_0 & k_0 \\ \xi_1 & a_1 & b_1 & \eta_1 & \vartheta_1 & k_1 \\ \vdots & \vdots & \vdots & \vdots & \vdots & \vdots \\ & \alpha & \beta & & & \end{array}$$

Player I and player II play as in the diagram natural numbers and ordinals $< \aleph_1$ and player II wins iff *for every n*,

either for some $i \leq n$, $\xi_i \geq \xi$ *or* all the following are true:

(a) The mapping $i \mapsto \eta_i$ $(i \leq n)$ is order preserving on the part of $\leq_{f(\alpha,\beta)}$ already determined by $\langle\langle a_0, \dots, a_n\rangle, \langle b_0, \dots, b_n\rangle\rangle$ (notice that f is continuous).

(b) The mapping $i \mapsto \vartheta_i$ $(i \leq n)$ is order preserving on the part of $\leq_\alpha$ already determined by $\langle a_0, \dots, a_n\rangle$ and $\vartheta_i < \xi$, for each $i \leq n$.

If $k_i \leq n$, then $\vartheta_{k_i} = \xi_i$ and if $\ulcorner k_i, k_i \urcorner = j \leq n$, then $a_j = 0$.
Notice now the following:

Claim 9.9. For $\xi < \aleph_1$, $\xi \in A$ if and only if player II has a winning strategy in G_ξ.

Proof of Claim 9.9. Assume $\xi \in A$; let α, β be such that $f(\alpha, \beta) \in \mathrm{WO}$ and $|\alpha| = \xi$ and let $i \to \eta_i$ be an order preserving map on $\leq_{f(\alpha,\beta)}$ into $\aleph_1$ and $i \to \vartheta_i$ a mapping from ω into ξ such that its restriction to Field $(\leq_\alpha)$ is an order preserving bijection onto ξ, with inverse g. Consider the following strategy for player II in G_ξ and verify easily that it is winning: If player I plays ξ_0, player II plays $\alpha(0), \beta(0), \eta_0, \vartheta_0, g(\xi_0)$ (unless $\xi_0 \geq \xi$ in which case player II plays anything). If player I plays ξ_1, player II gives $\alpha(1), \beta(1), \eta_1, \vartheta_1, g(\xi_1)$ *etc.*

Conversely assume player II has a winning strategy. Let player I play $\xi_0, \xi_1, \dots$ enumerating without repetitions ξ, *i.e.*, $\xi = \{\xi_0, \xi_1, \xi_2, \dots\}$. Then player II plays by his winning strategy and produces $\alpha, \beta, \langle \eta_0, \eta_1, \dots \rangle$, $\langle \vartheta_0, \vartheta_1, \dots \rangle, \langle k_0, k_1, \dots \rangle$, such that $i \to \eta_i$ is order preserving on $\leq_{f(\alpha,\beta)}$, thus $f(\alpha, \beta) \in \mathrm{WO}$, $i \to \vartheta_i$ is order preserving on $\leq_\alpha$ into ξ, thus $\alpha \in \mathrm{WO}$, and finally $\xi_i \to k_i$ is an inverse to $i \to \vartheta_i$ on Field$(\leq_\alpha)$, thus $|\alpha| = \xi$ and the proof is complete. ⊣ (Claim 9.9)

It is clear now that $\mathrm{G}_\xi = \mathrm{G}_{\mathcal{S}_\xi}$ where $\mathcal{S}_\xi$ is a set of finite sequences and moreover the map $\xi \mapsto \mathcal{S}_\xi$ is absolute for $\mathbf{L}$. Thus *for* $\xi < \aleph_1$,

$$\begin{aligned} \xi \in A &\iff \text{player II has a winning strategy in } \mathrm{G}_{\mathcal{S}_\xi} \\ &\iff \mathbf{L} \models \text{player II has a winning strategy in } \mathrm{G}_{\mathcal{S}_\xi}. \end{aligned}$$

So A is definable in $\mathbf{L}$, therefore $A \in \mathbf{L}$. Notice that the definition of A involves as the *only* parameter $\aleph_1$, thus $A = \tau^{\mathbf{L}}(\aleph_1)$ for some term τ. ⊣

The following converse to Theorem 9.7 was proved by Kechris:

$$\forall \alpha (\alpha^\# \text{ exists}) \Longrightarrow \text{Every } A \in \mathbf{L}, A \subseteq \aleph_1 \text{ has } \mathrm{Code}(A) \in \utilde{\mathbf{\Sigma}}^1_2.$$

The proof (as also the proof of Theorem 9.7) relativizes to any real and thus gives the elegant characterization:

If $\tilde{\mathbf{L}} = \bigcup_{\alpha \in \mathbb{R}} \mathbf{L}[\alpha]$ and $\forall \alpha (\alpha^\# \text{ exists})$, then

$$A \subseteq \aleph_1 \Longrightarrow (A \in \tilde{\mathbf{L}} \iff \mathrm{Code}(A) \in \utilde{\mathbf{\Sigma}}^1_2).$$

Since T^{2n+1} is a tree definable by some formula of set theory, it makes sense to talk of the tree $(T^{2n+1})^M$ for any model M of ZF+DC+Det$(\utilde{\mathbf{\Delta}}^1_{2n})$. In particular, for the models M^n introduced in §5, we have (by Theorems 5.2 and 9.3) that

$$M = \mathbf{L}[(T)^M]$$

if $M = M^{2n+1}$ or $M = M^{2n+2}$ and $T = T^{2n+1}$. Thus the models of §5 are ordinary relative constructibility models, but we do not have an independent characterization of the trees $(T^{2n+1})^M$ for these M's.

Finally, we should mention that in contrast to $\mathbf{L} = \mathbf{L}[T^1]$, Kechris has noticed that

$$M^{2n} \subsetneq \mathbf{L}[T^{2n-1}], \quad n \geq 2;$$

this is because $M^{2n} \subseteq \mathbf{L}[\alpha]$ for some α.

§10. Lebesgue measurability and the property of Baire. We prove here Solovay's results that $\boldsymbol{\Sigma}^1_2$ sets are Lebesgue measurable and have the property of Baire, if $\forall\alpha(\aleph_1^{L[\alpha]} < \aleph_1)$. These will appear as corollaries of more general results about **approximations** of sets which admit definable scales, but we do not know any other applications of these general theorems. The forcing-free proofs given here are an adaptation by Moschovakis of the original Solovay proofs which used the forcing method.

10.1. Sets which are ∞-Boolean over a model of ZFC. We define a set of **codes** $\mathrm{C}(\kappa)$ and for each $a \in \mathrm{C}(\kappa)$ a set B_a in the Boolean algebra $\mathcal{B}(\kappa)$, by the induction:

1. $\langle 1, \ulcorner n_0, \dots, n_k \urcorner \rangle \in \mathrm{C}(\kappa)$ and

$$\mathrm{B}_{\langle 1, \ulcorner n_0, \dots, n_k \urcorner \rangle} = \{\alpha : \bar{\alpha}(k+1) = \ulcorner n_0, \dots, n_k \urcorner\},$$

2. If $a \in \mathrm{C}(\kappa)$, then $\langle 2, a \rangle \in \mathrm{C}(\kappa)$ and

$$\mathrm{B}_{\langle 2, a \rangle} = \mathbb{R} \setminus \mathrm{B}_a.$$

3. If $f : \xi \to \mathrm{C}(\kappa)$ is a function with domain some $\xi < \kappa$ and values in $\mathrm{C}(\kappa)$, then $\langle 3, f \rangle \in \mathrm{C}(\kappa)$ and

$$\mathrm{B}_{\langle 3, f \rangle} = \bigcup_{\eta < \xi} B_{f(\eta)}.$$

Clearly, using AC,

$$\mathcal{B}(\kappa) = \{\mathrm{B}_a : a \in \mathrm{C}(\kappa)\}.$$

Moreover the relation "$a \in \mathrm{C}(\kappa)$" is definable by a formula which is absolute for models of ZFC.

In particular the set $\mathrm{C}(\omega + 1)$ gives canonical codes to the algebra $\mathcal{B}(\omega + 1)$ of Borel sets.

Suppose $\mathcal{M}$ is a (transitive, containing all ordinals) model of ZFC. A set of reals A is **κ-Boolean over** $\mathcal{M}$, if

$$A = B_a \text{ for some } a \in \mathrm{C}(\kappa) \cap \mathcal{M}.$$

A is ∞-**Boolean over** $\mathcal{M}$ if it is κ-Boolean over $\mathcal{M}$ for some κ. A is $\omega + 1$-Boolean over $\mathcal{M}$ if it is "**Borel rational over** $\mathcal{M}$" in Solovay's terminology, [Sol70].

We would expect that sets which are ∞-Boolean over "thin", definable models have some regularity properties. But first let us show how we can get such sets.

THEOREM 10.1. Let T be a tree on $\omega \times \kappa$ and $A = \mathrm{p}[T] = \{\alpha : T(\alpha)$ is not wellfounded$\}$. Then A is $\kappa^+ + 1$-Boolean over $\mathbf{L}[T]$.

PROOF. Immediate from Theorem 8.1. ⊣

COROLLARY 10.2. If $A \in \Sigma^1_2$, then A is $\aleph_1 + 1$-Boolean over $\mathbf{L}$.

Remark. We know already that a Σ^1_2 set is the union of $\aleph_1$ Borel sets, a seemingly better representation than the above. Its only disadvantage is that we cannot have $A = \bigcup_{\xi<\aleph_1} \mathrm{B}_{f(\xi)}$, where $f \in \mathbf{L}$, *i.e.*, we cannot control this representation from inside $\mathbf{L}$, unless $\aleph_1^{\mathbf{L}} = \aleph_1$.

10.2. Ideals of Borel sets which are suitable over a model of ZFC. We isolate in a definition the key properties of the ideals of sets of measure 0 and sets of the first category that we need.

Let $\mathcal{M}$ be a fixed model of ZFC, $\mathcal{J}$ a collection of Borel sets. We call $\mathcal{J}$ **suitable over** $\mathcal{M}$ if the following conditions hold:

1. $\mathcal{J} \neq \varnothing$ and $\mathcal{J}$ is closed under subsets and countable unions (*i.e.*, $\mathcal{J}$ is a σ-**ideal** in the Boolean algebra of Borel sets).
2. There is a definable functor

$$G : \mathcal{M} \to \mathcal{M}$$

(*i.e.*, an operation on $\mathcal{M}$ definable by a formula, perhaps with parameters from $\mathcal{M}$) such that if

$$f : \xi \to C(\omega + 1), \quad f \in \mathcal{M}$$

is a function in $\mathcal{M}$ which maps some ordinal ξ in the Borel codes of $\mathcal{M}$, then

$$G(f) = \langle \eta_i : i \in \omega \rangle$$

gives a countable sequence of ordinals less than ξ such that

$$\text{for all } \eta < \xi, B_{f(\eta)} \setminus \bigcup_i B_{f(\eta_i)} \in \mathcal{J}.$$

This last condition says in effect that the Boolean algebra $\mathcal{B}(\omega + 1)/\mathcal{J}$ has the **countable chain condition** for wellordered sequences of Borel sets in $\mathcal{M}$. It is the key property that we need.

If $\mathcal{J}$ is the ideal of Borel sets of measure 0 (where we can take the measure on $\mathbb{R}$ as coming from the measure on the true reals via the standard topological identification of $\mathbb{R}$ with the irrationals *or* we can take only the subsets of ${}^{\omega}2$ with the product measure) and if $\mathcal{J}$ is the ideal of Borel sets of the first category, then $\mathcal{J}$ is suitable over every $\mathcal{M}$ *which admits a definable* (with parameters from $\mathcal{M}$) *wellordering*, in particular any $\mathbf{L}[X]$. This is not hard to prove from standard analytic and topological facts about these ideals, *e.g.*, see [Sol70, Kur66].

Let $\mathcal{J}$ be suitable over $\mathcal{M}$ and put

$$\mathrm{Alg}(\mathcal{M}, \mathcal{J}) = \{\alpha : \text{ for some } a \in C(\omega+1) \cap \mathcal{M}, \mathrm{B}_a \in \mathcal{J} \text{ and } \alpha \in B_a\}.$$

The reals in $\mathrm{Alg}(\mathcal{M}, \mathcal{J})$ are called **$\mathcal{J}$- algebraic over** $\mathcal{M}$. If we think of sets in $\mathcal{J}$ as **small** sets, they are reals which can be *approximated* in $\mathcal{M}$, in the sense that they belong to a small set with code in $\mathcal{M}$. Let

$$\mathrm{Trans}(\mathcal{M}, \mathcal{J}) = \{\alpha : \alpha \notin \mathrm{Alg}(\mathcal{M}, \mathcal{J})\}$$

be the set of reals **$\mathcal{J}$-transcendental** over $\mathcal{M}$. In the case of measure 0 or of the first category these are Solovay's **random** and **Cohen generic** (over $\mathcal{M}$) reals, respectively. This is the key notion needed for the statement and proof of the approximation theorem we want.

THEOREM 10.3. Let $\mathcal{M}$ be a model of ZFC. Let $\mathcal{J}$ be a σ-ideal of Borel sets which is suitable over $\mathcal{M}$. Then for every set A which is ∞-Boolean over $\mathcal{M}$, there is a Borel set A^*, rational over $\mathcal{M}$, such that

$$A \Delta A^* = (A \setminus A^*) \cup (A^* \setminus A) \subseteq \mathrm{Alg}(\mathcal{M}, \mathcal{J}).$$

If in addition $|\mathcal{M} \cap \mathbb{R}| < \aleph_1$, then

$$A \Delta A^* \in \mathcal{J}.$$

PROOF. Let A be κ-Boolean over $\mathcal{M}$. We assign *by induction in* $\mathcal{M}$ on the codes $\mathrm{C}(\kappa) \cap \mathcal{M}$, to each $a \in \mathrm{C}(\kappa) \cap \mathcal{M}$ a code $a^* \in \mathrm{C}(\omega+1) \cap \mathcal{M}$ such that $\mathrm{B}_a \Delta \mathrm{B}_{a^*} \subseteq \mathrm{Alg}(\mathcal{M}, \mathcal{J})$, *i.e.*, for any $\alpha \in \mathrm{Trans}(\mathcal{M}, \mathcal{J})$

$$\alpha \in B_a \Longleftrightarrow \alpha \in B_{a^*} \tag{$*$}$$

Case 1. $a = \langle 1, \ulcorner n_0, \dots, n_k \urcorner \rangle$. Put $a^* = a$.
Case 2. $a = \langle 2, b \rangle$. Put $a^* = \langle 2, b^* \rangle$.
Case 3. $a = \langle 3, f \rangle$. Put $f^*(\xi) = (f(\xi))^*$.

Then let $\langle \eta_i : i \in \omega \rangle = G(f^*)$ and $f_1(i) = f^*(\eta_i)$. Clearly $f_1 \colon \omega \to \mathrm{C}(\omega+1) \cap \mathcal{M}$ and we define

$$a^* = \langle 3, f_1 \rangle.$$

The verification that $(*)$ holds is trivial, except for the third case. Thus let $a = \langle 3, f \rangle \in \mathcal{M}$, where $f : \xi \to \mathrm{C}(\kappa) \cap \mathcal{M}$, and by induction hypothesis assume that for $\alpha \in \mathrm{Trans}(\mathcal{M}, \mathcal{J})$

$$\alpha \in B_{f(\eta)} \Longleftrightarrow \alpha \in B_{(f(\eta))^*} \Longleftrightarrow \alpha \in B_{f^*(\eta)}.$$

We have to show that for $\alpha \in \mathrm{Trans}(\mathcal{M}, \mathcal{J})$,

$$\alpha \in B_a \iff \alpha \in B_{\langle 3, f_1 \rangle}.$$

Let $\alpha \in \mathrm{Trans}(\mathcal{M}, \mathcal{J})$ and $\alpha \in \mathrm{B}_a$. Then $\alpha \in \mathrm{B}_{f(\eta)}$ for some $\eta > \xi$. Thus $\alpha \in \mathrm{B}_{f^*(\eta)}$. Then by 2 above $\alpha \notin \mathrm{B}_{f^*(\eta)} \setminus \bigcup_i \mathrm{B}_{f^*(\eta_i)}$, so $\alpha \in \bigcup_i \mathrm{B}_{f^*(\eta_i)} = \mathrm{B}_{\langle 3, f_1 \rangle}$.

Conversely assume $\alpha \in \mathrm{Trans}(\mathcal{M}, \mathcal{J})$ and $\alpha \in \mathrm{B}_{\langle 3, f_1 \rangle}$. Then $\alpha \in \bigcup_i \mathrm{B}_{f^*(\eta_i)}$, so for some i, $\alpha \in \mathrm{B}_{f^*(\eta_i)}$, thus $\alpha \in \mathrm{B}_{f(\eta_i)}$ and $\alpha \in \bigcup_{\eta < \xi} \mathrm{B}_{f(\xi)} = \mathrm{B}_a$. ⊣

Corollary 10.4. (a) Assume $\forall \alpha (\aleph_1^{\mathbf{L}[\alpha]} < \aleph_1)$. Then every $\utilde{\Sigma}^1_2$ set is Lebesgue Measurable and has the property of Baire. (Solovay, unpublished.)

(b) Assume $\mathrm{Det}(\utilde{\Delta}^1_{2n})$ and that for each α, $|\mathbb{R} \cap \mathbf{L}[T^{2n+1}, \alpha]| = \aleph_0$. Then every $\utilde{\Sigma}^1_{2n+2}$ set is Lebesgue Measurable and has the property of Baire. (Solovay, unpublished.)

It should be pointed out that it has been known for some time (see [MS64]) that PD implies that all projective sets are Lebesgue measurable and have the property of Baire. The proof of (b) above seems to use "less determinacy" (none if $n = 0$) and has a different flavor.

§11. Perfect subsets of pointsets. The main results here are that if PD holds and for each n, $|\mathbf{L}[T^{2n+1}] \cap \mathbb{R}| = \aleph_0$, then every uncountable projective set has a perfect subset and there exist largest countable Σ^1_{2n+2} sets. The first result does not need the hypothesis $|\mathbf{L}[T^{2n+1}]| \cap \mathbb{R}| = \aleph_0$, see [Dav64], [Myc64], but the proof given here (due to Solovay and Mansfield) uses "less determinacy", none if $n = 0$.

11.1. The theorem on perfect sets. Solovay in [Sol66] proved that, if $\forall \alpha$ $[\aleph_1^{\mathbf{L}[\alpha]} < \aleph_1]$, then every uncountable $\utilde{\Sigma}^1_2$ set contains a perfect subset. His method was one of the earliest applications of forcing to the proof of positive results. A few months later Mansfield obtained a similar theorem (see [Man69]) and in [Man70] he generalized the result to Theorem 11.1 below. His proof also used forcing. Finally Solovay obtained a new forcing-free proof of Mansfield's result. This is essentially the proof reproduced below with one alteration: Solovay's "inductive analysis" was replaced by the notion of "derivation on a tree"; as a result the proof becomes astonishingly similar to Cantor's proof of the Cantor-Bendixson theorem.

Theorem 11.1 (Mansfield, [Man70]). Assume T is a tree on $\omega \times \kappa$ and $A = \mathrm{p}[T]$. Then if A contains an element not in $\mathbf{L}[T]$, A contains a perfect set.

Proof. For any tree T on $\omega \times \kappa$ we define the **derivative** T' of T as follows: $\langle\langle k_0, \xi_0\rangle, \dots, \langle k_n, \xi_n\rangle\rangle \in T' \iff$ There are two, incompatible in the first coordinate, extensions of $\langle\langle k_0, \xi_0\rangle, \dots, \langle k_n, \xi_n\rangle\rangle$, both in T, *i.e.*, we can find $\langle\langle k_0', \xi_0'\rangle, \dots, \langle k_m', \xi_m'\rangle\rangle, \langle\langle k_0'', \xi_0''\rangle, \dots, \langle k_\ell'', \xi_\ell''\rangle\rangle \in T$ extending $\langle\langle k_0, \xi_0\rangle, \dots, \langle k_n, \xi_n\rangle\rangle$, such that $\langle k_0', \dots, k_m'\rangle$ is incompatible with $\langle k_0'', \dots, k_\ell''\rangle$. Notice that T' is a tree and $T' \subseteq T$.

Then we define á la Cantor the ξ**th-derivative** of T by

$$T^0 = T$$
$$T^{\xi+1} = (T^\xi)'$$
$$T^\lambda = \bigcap_{\xi<\lambda} T^\xi, \text{ if } \lambda = \bigcup \lambda > 0.$$

It is then clear that $\xi \to T^\xi$ is a function absolute for any model containing T, in particular for $\mathbf{L}[T]$. Moreover $T^0 \supseteq T^1 \supseteq T^2 \supseteq \cdots \supseteq T^\xi \supseteq T^{\xi+1} \supset \cdots$; thus let ξ_T be the least ξ such that $T^\xi = T^{\xi+1}$.

Case 1. $T^{\xi_T} = \varnothing$. Then consider $\alpha \in A$. Since $A = \mathrm{p}[T]$, we can find f such that $\langle \alpha, f\rangle \in [T]$. Since $\langle \alpha, f\rangle \notin [T^{\xi_T}] = \varnothing$, let $\xi < \xi_T$ be such that $\langle \alpha, f\rangle \in [T^\xi] \setminus [T^{\xi+1}]$. Let n be the least integer such that $\langle\langle \alpha(0), f(0)\rangle, \dots, \langle \alpha(n), f(n)\rangle\rangle \notin T^{\xi+1}$. Since $\langle\langle \alpha(0), f(0)\rangle, \dots, \langle \alpha(n), f(n)\rangle\rangle \in T^\xi$, it is clear that all branches of T^ξ extending $\langle\langle \alpha(0), f(0)\rangle, \dots, \langle \alpha(n), f(n)\rangle\rangle$ have the same real part, namely α, therefore $\mathrm{p}[T^\xi_{\langle\langle \alpha(0), f(0)\rangle, \dots, \langle \alpha(n), f(n)\rangle\rangle}] = \{\beta\}$, where $\alpha = \langle \alpha(0), \dots, \alpha(n)\rangle^\frown\beta$. But then clearly $\beta \in \mathbf{L}[T]$, since β is definable absolutely from elements of $\mathbf{L}[T]$; thus $\alpha \in \mathbf{L}[T]$. So in this case $A \subseteq \mathbf{L}[T]$.

Case 2. $T^{\xi_T} \neq \varnothing$. Then $T^{\xi_T} = (T^{\xi_T})' \neq \varnothing$, *i.e.*, every sequence in T^{ξ_T} has two extensions in T^{ξ_T} which are incompatible in the first coordinate. Then it is easy to show that $\mathrm{p}[T^{\xi_T}]$ ($\subseteq \mathrm{p}[T] = A$) contains a perfect set. ⊣

Corollary 11.2 (Solovay, [Sol66]). Every Σ^1_2 set with an element not in $\mathbf{L}$ contains a perfect set; thus, $\forall\alpha(\aleph_1^{\mathbf{L}[\alpha]} < \aleph_1) \implies$ Every uncountable $\boldsymbol{\Sigma}^1_2$ set contains a perfect subset.

Corollary 11.3. Assume $\mathsf{Det}(\boldsymbol{\Delta}^1_{2n})$. Then every Σ^1_{2n+2} set with an element not in $\mathbf{L}[T^{2n+1}]$ contains a perfect set. Thus if $|\mathbf{L}[T^{2n+1}, \alpha] \cap \mathbb{R}| = \aleph_0$, for all $\alpha \in \mathbb{R}$, every uncountable $\boldsymbol{\Sigma}^1_{2n+2}$ set contains a perfect subset.

11.2. Largest countable Σ^1_{2n} sets. The following is also a corollary of Theorem 11.1.

Theorem 11.4 (Solovay, [Sol66]). Assume $\aleph_1^{\mathbf{L}} < \aleph_1$. Then there exists a largest countable Σ^1_2 set of reals, namely $\{\alpha : \alpha \in \mathbf{L}\}$.

The next result extends Theorem 11.4 to higher levels.

Theorem 11.5 (Kechris, Moschovakis.). Assume $\mathsf{Det}(\boldsymbol{\Delta}^1_{2n})$. If the set $\mathbb{R} \cap \mathbf{L}[T^{2n+1}]$ is countable, then there exists a largest countable Σ^1_{2n+2} set.

PROOF. Notice first that $\mathsf{Unif}(\Pi^1_{2n+1})$ implies that for every countable Σ^1_{2n+2} set A we can find a countable Π^1_{2n+1} set B, so that every real in A is recursive in some real in B. Thus it will be enough to find a countable Σ^1_{2n+2} set C which contains all countable Π^1_{2n+1} sets. Then $C^* = \{\alpha : (\exists\beta)(\beta \in C \wedge \alpha \text{ is recursive in } \beta)\}$ is the largest countable Σ^1_{2n+2} set.

It will be convenient for this proof to choose a particular Π^1_{2n+1}-complete set $\mathcal{P}_{2n+1}$ and a Π^1_{2n+1}-scale on it as follows: Let $W_{2n+1} \subseteq \omega \times \mathbb{R}$ be universal for Π^1_{2n+1} subsets of $\mathbb{R}$ and put

$$\alpha \in \mathcal{P}_{2n+1} \iff (\alpha(0), \alpha') \in W_{2n+1},$$

where $\alpha' = \langle \alpha(1), \alpha(2), \dots \rangle$. Let also $\langle \varphi_n : n \in \omega \rangle$ be a Π^1_{2n+1}-scale on $\mathcal{P}_{2n+1}$. Let T^{2n+1} be the tree associated with this scale.

We define now C and then we show that it works:

$$\alpha \in C \iff \exists m \exists \xi (|\{m^\frown\beta \in \mathcal{P}_{2n+1} : \varphi_0(m^\frown\beta) \leq \xi\}| \leq \aleph_0 \wedge \varphi_0(m^\frown\alpha) \leq \xi).$$

CLAIM 11.6. $C \in \Sigma^1_{2n+2}$.

PROOF OF CLAIM 11.6. Notice that the statements "$\alpha \in C$" and "$\exists m \exists \beta (\beta \in \mathcal{P}_{2n+1} \wedge \varphi_0(m^\frown\alpha) \leq \varphi_0(\beta) \wedge \exists\gamma\forall\delta(\varphi_0(m^\frown\delta) \leq \varphi_0(\beta) \text{ implies } \exists k(\delta = (\gamma)_k)))$" are equivalent. ⊣ (Claim 11.6)

CLAIM 11.7. C contains every countable Π^1_{2n+1} set.

PROOF OF CLAIM 11.7. Let $B \in \Pi^1_{2n+1}$, $B \subseteq \mathbb{R}$, $|B| \leq \aleph_0$. Find m such that $\beta \in B \iff \langle m, \beta \rangle \in W_{2n+1} \iff m^\frown\beta \in \mathcal{P}_{2n+1}$. If $B \not\subseteq C$, let $\beta_0 \in B \setminus C$. Put $\xi = \varphi_0(m^\frown\beta)$. Then since $\beta_0 \notin C$, $|\{m^\frown\beta \in \mathcal{P}_{2n+1} : \varphi_0(m^\frown\beta) \leq \xi\}| > \aleph_0$; but $B \supseteq \{m^\frown\beta \in \mathcal{P}_{2n+1} : \varphi_0(m^\frown\beta) \leq \xi\}$, a contradiction. ⊣ (Claim 11.7)

CLAIM 11.8. $C \subseteq \mathbf{L}[T^{2n+1}]$; thus $|C| = \aleph_0$.

PROOF OF CLAIM 11.8. It is enough to show that if for some m, ξ, $|\{m^\frown\beta \in \mathcal{P}_{2n+1} : \varphi_0(m^\frown\beta) \leq \xi\}| \leq \aleph_0$, then $\{m^\frown\beta \in \mathcal{P}_{2n+1} : \varphi_0(m^\frown\beta) \leq \xi\} \subseteq \mathbf{L}[T^{2n+1}]$.

$$(T^{2n+1})_{m,\xi} = \{\langle\langle k_0, \xi_0\rangle, \dots, \langle k_\ell, \xi_\ell\rangle\rangle \in T^{2n+1} : k_0 = m \wedge \xi_0 \leq \xi\}.$$

Clearly $(T^{2n+1})_{m,\xi} \in \mathbf{L}[T^{2n+1}]$ and the limit property of scales shows that

$$\alpha \in \mathrm{p}[(T^{2n+1})_{m,\xi}] \iff \alpha \in \mathcal{P}_{2n+1} \wedge \varphi_0(\alpha) \leq \xi \wedge \alpha(0) = m.$$

Thus $\{m^\frown\beta \in \mathcal{P}_{2n+1} : \varphi_0(m^\frown\beta) \leq \xi\} = \mathrm{p}[(T^{2n+1})_{m,\xi}]$, so by Theorem 11.1

$$\begin{aligned} &|\{m^\frown\beta \in \mathcal{P}_{2n+1} : \varphi_0(m^\frown\beta) \leq \xi\}| \leq \aleph_0 \\ &\quad \implies \{m^\frown\beta \in \mathcal{P}_{2n+1} : \varphi_0(m^\frown\beta) \leq \xi\} \subseteq \mathbf{L}[(T^{2n+1})_{m,\xi}] \subseteq \mathbf{L}[T]. \end{aligned}$$

⊣ (Claim 11.8)

⊣

Open problem[4]. It is a well known result, that every countable Σ^1_1 set contains only Δ^1_1 reals. Thus there is no largest countable Σ^1_1 set (since $\{\alpha : \alpha \in \Delta^1_1\} \in \Pi^1_1 \setminus \Sigma^1_1$). Does any of these results generalize to Σ^1_{2n+1} ($n \geq 1$), under any reasonable hypotheses?

§12. A summary of results about projective ordinals. We give here a list of theorems about the projective ordinals. Proofs are omitted but many results follow from what we have already done.

1. $\underset{\sim}{\delta}^1_1 = \aleph_1$ (classical)
2. (a) $\underset{\sim}{\delta}^1_2 \leq \aleph_2$ (Martin, unpublished), If $\alpha^\#$ exists for all reals α, then $\underset{\sim}{\delta}^1_2 \leq u_2$ (Martin, unpublished),
   (b) If $\alpha^\#$ exists for all reals α, then $\underset{\sim}{\delta}^1_2 \geq u_2$ (Kechris, Martin, unpublished). Thus if $\alpha^\#$ exists for all reals α, then $\underset{\sim}{\delta}^1_2 = u_2$.
   (c) If $\alpha^\#$ exists for all reals α, then for $n \geq 3$, $\underset{\sim}{\delta}^1_n = u_{\underset{\sim}{\delta}^1_n}$ (Kechris), where $u_1, u_2, \ldots, u_\xi, \ldots$ is the increasing enumeration of the uniform indiscernibles.
3. If $\alpha^\#$ exists for all reals α and AC holds, then $\underset{\sim}{\delta}^1_3 \leq \aleph_3$ (Martin, [Mar70B]).
4. PD $\Longrightarrow \underset{\sim}{\delta}^1_{2n+2} \leq (\underset{\sim}{\delta}^1_{2n+1})^+$ (Kunen, Martin, unpublished).
5. Det($\underset{\sim}{\Delta}^1_2$)+AC $\Longrightarrow \underset{\sim}{\delta}^1_4 \leq \aleph_4$ (Kunen, Martin, unpublished).
6. (a) PD implies $\underset{\sim}{\delta}^1_{2n+1} < \underset{\sim}{\delta}^1_{2n+2}$ (Moschovakis, [Mos70A]).
   (b) PD implies $\underset{\sim}{\delta}^1_{2n} < \underset{\sim}{\delta}^1_{2n+1}$ (Kechris).
7. PD implies that every Π^1_{2n+1}-norm on a universal Π^1_{2n+1}-set has length $\underset{\sim}{\delta}^1_{2n+1}$ (Moschovakis, [Mos70A]).

If we now assume full determinacy (AD), the results have a different flavor. Assume AD; then:

1′. (a) $\underset{\sim}{\delta}^1_n$ is a cardinal and for n odd regular (Moschovakis, [Mos70A]).
   (b) $\underset{\sim}{\delta}^1_n$ is regular, for n even (Kunen, unpublished).
2′. $\underset{\sim}{\delta}^1_2 = \aleph_2$ ($= u_2$) (Martin).
3′. (a) $u_\omega = \aleph_\omega$; $\mathrm{cf}(\aleph_n) = \aleph_2$ ($n \geq 2$) (Martin, [Mar70B]).
   (b) $u_n = \aleph_n$ (Kunen, Solovay, unpublished).
4′. $\underset{\sim}{\delta}^1_3 = \aleph_{\omega+1}$ (Martin, [Mar70B]).
5′. (a) $\underset{\sim}{\delta}^1_{2n+2} = (\underset{\sim}{\delta}^1_{2n+1})^+$ (Kunen, Martin, unpublished).
   (b) $\underset{\sim}{\delta}^1_4 = \aleph_{\omega+2}$ (Kunen, Martin, unpublished).
6′. $\underset{\sim}{\delta}^1_{2n+1} = (\lambda_n)^+$, where λ_n is a cardinal and $\mathrm{cf}(\lambda_n) = \omega$; thus $\underset{\sim}{\delta}^1_{2n+1} \geq \aleph_{\omega n+1}$ (Kechris).

[4]Kechris [Kec75] has proved from PD that for each $n \geq 1$ there is no largest countable Σ^1_{2n+1} set. Martin [Mar73] then showed from PD that every countable Σ^1_{2n+1} set of reals contains only Δ^1_{2n+1} reals (Moschovakis has earlier shown this result for countable Δ^1_{2n+1} sets).

7′. (a) $\aleph_1 (= \underset{\sim}{\delta}^1_1), \aleph_2 (= \underset{\sim}{\delta}^1_2)$ are measurable (Solovay; for $\aleph_1$ see [Sol67]), for $\aleph_2$ unpublished).
(b) $\underset{\sim}{\delta}^1_{2n+1}$ is measurable (Martin, unpublished).
(c) $\underset{\sim}{\delta}^1_{2n}$ is measurable (Kunen, unpublished).

Postscript. While this paper was being typed, we received a preprint from Martin titled "Projective sets and cardinal numbers: some questions related to the continuum problem". This appears to contain most of the results of Martin that we have listed as "unpublished" or credited to [Mar70B].

REFERENCES

John W. Addison
[Add59] *Some consequences of the axiom of constructibility*, ***Fundamenta Mathematicae***, vol. 46 (1959), pp. 123–135.

John W. Addison and Yiannis N. Moschovakis
[AM68] *Some consequences of the axiom of definable determinateness*, ***Proceedings of the National Academy of Sciences of the United States of America***, no. 59, 1968, pp. 708–712.

Morton Davis
[Dav64] *Infinite games of perfect information*, ***Advances in game theory*** (Melvin Dresher, Lloyd S. Shapley, and Alan W. Tucker, editors), Annals of Mathematical Studies, vol. 52, 1964, pp. 85–101.

Leo A. Harrington and Alexander S. Kechris
[HK77] *Ordinal quantification and the models* $\mathbf{L}[T^{2n+1}]$, Mimeographed note, January 1977.

Alexander S. Kechris
[Kec75] *The theory of countable analytical sets*, ***Transactions of the American Mathematical Society***, vol. 202 (1975), pp. 259–297.

Alexander S. Kechris and Yiannis N. Moschovakis
[Cabal i] *Cabal seminar* 76–77, Lecture Notes in Mathematics, no. 689, Berlin, Springer, 1978.

Kazimierz Kuratowski
[Kur66] ***Topology***, vol. 1, Academic Press, New York and London, 1966.

Azriel Lévy
[Lév66] *Definability in axiomatic set theory*, ***Logic, methodology and philosophy of science. Proceedings of the 1964 international congress.*** (Amsterdam) (Yehoshua Bar-Hillel, editor), Studies in Logic and the Foundations of Mathematics, North-Holland, 1966, pp. 127–151.

Richard Mansfield
[Man69] ***The theory of*** Σ^1_2 ***sets***, Ph.D. thesis, Stanford University, 1969.
[Man70] *Perfect subsets of definable sets of real numbers*, ***Pacific Journal of Mathematics***, vol. 35 (1970), no. 2, pp. 451–457.
[Man71] *A Souslin operation on* Π^1_2, ***Israel Journal of Mathematics***, vol. 9 (1971), no. 3, pp. 367–379.

Donald A. Martin
[Mar68] *The axiom of determinateness and reduction principles in the analytical hierarchy*, ***Bulletin of the American Mathematical Society***, vol. 74 (1968), pp. 687–689.

[Mar70B] *Pleasant and unpleasant consequences of determinateness*, March 1970, unpublished manuscript.

[Mar73] *Countable Σ^1_{2n+1} sets*, 1973, circulated note.

DONALD A. MARTIN AND ROBERT M. SOLOVAY

[MS] *Basis theorems for Π^1_{2k} sets of reals*, unpublished.

[MS69] *A basis theorem for Σ^1_3 sets of reals*, ***Annals of Mathematics***, vol. 89 (1969), pp. 138–160.

YIANNIS N. MOSCHOVAKIS

[Mos70A] *Determinacy and prewellorderings of the continuum*, ***Mathematical logic and foundations of set theory. Proceedings of an international colloquium held under the auspices of the Israel Academy of Sciences and Humanities, Jerusalem, 11–14 November 1968*** (Y. Bar-Hillel, editor), Studies in Logic and the Foundations of Mathematics, North-Holland, Amsterdam-London, 1970, pp. 24–62.

[Mos70B] *The Suslin-Kleene theorem for countable structures*, ***Duke Mathematical Journal***, vol. 37 (1970), no. 2, pp. 341–352.

[Mos71A] *Uniformization in a playful universe*, ***Bulletin of the American Mathematical Society***, vol. 77 (1971), pp. 731–736.

[Mos80] ***Descriptive set theory***, Studies in Logic and the Foundations of Mathematics, no. 100, North-Holland, Amsterdam, 1980.

JAN MYCIELSKI

[Myc64] *On the axiom of determinateness*, ***Fundamenta Mathematicae***, vol. 53 (1964), pp. 205–224.

JAN MYCIELSKI AND STANISLAW SWIERCZKOWSKI

[MS64] *On the Lebesgue measurability and the axiom of determinateness*, ***Fundamenta Mathematicae***, vol. 54 (1964), pp. 67–71.

JACK H. SILVER

[Sil71] *Measurable cardinals and Δ^1_3 wellorderings*, ***Annals of Mathematics***, vol. 94 (1971), no. 2, pp. 141–446.

ROBERT M. SOLOVAY

[Sol66] *On the cardinality of Σ^1_2 set of reals*, ***Foundations of Mathematics: Symposium papers commemorating the 60th birthday of Kurt Gödel*** (Jack J. Bulloff, Thomas C. Holyoke, and S. W. Hahn, editors), Springer-Verlag, 1966, pp. 58–73.

[Sol67] *Measurable cardinals and the axiom of determinateness*, Lecture notes prepared in connection with the Summer Institute of Axiomatic Set Theory held at UCLA, Summer 1967.

[Sol70] *A model of set theory in which every set is Lebesgue measurable*, ***Annals of Mathematics***, vol. 92 (1970), pp. 1–56.

DEPARTMENT OF MATHEMATICS
CALIFORNIA INSTITUTE OF TECHNOLOGY
PASADENA, CA 91125, USA
E-mail: kechris@caltech.edu

DEPARTMENT OF MATHEMATICS
UNIVERSITY OF CALIFORNIA
LOS ANGELES, CA 90095, USA
E-mail: ynm@math.ucla.edu

PROPAGATION OF THE SCALE PROPERTY USING GAMES

ITAY NEEMAN

The aim of this short paper is to introduce the reader to the notion of a scale and to some of the basic techniques involved in the propagation of the scale property through the use of infinite games. None of the results presented is due to the author. For a full history see Moschovakis [Mos80]. We work throughout the paper with the space ${}^{\omega}\omega$. For $s \in {}^{<\omega}\omega$ we use N_s to denote the set $\{x \in {}^{\omega}\omega : x \text{ extends } s\}$. The sets N_s, $s \in {}^{<\omega}\omega$, form the basic open subsets of ${}^{\omega}\omega$. Following standard abuse we refer to the space ${}^{\omega}\omega$, equipped with the topology generated by these basic open sets, as $\mathbb{R}$.

Given a set $A \subset \mathbb{R}$ let $\mathrm{G}(A)$ denote the following game: Players I and II alternate playing $x(n)$ for $n \in \omega$ subject to the order displayed in Diagram 1, with $x(n) \in \omega$ for each n. If, after ω moves, the real $x = \langle x(n) : n < \omega \rangle$ belongs to A then player I wins. Otherwise player II wins. $\mathrm{G}(A)$ is **determined** if one of the two players has a winning strategy in the game.

I	$x(0)$		$x(2)$		......	
II		$x(1)$		$x(3)$		......

Diagram 1. The game $\mathrm{G}(A)$.

For $B \subset \mathbb{R} \times \mathbb{R}$ and $x \in \mathbb{R}$ let $B_x = \{y \in \mathbb{R} : \langle x, y \rangle \in B\}$. This is the x**-section** of B. Define $\Game B$ to be the set $\{x \in \mathbb{R} :$ player I has a winning strategy in $\mathrm{G}(B_x)\}$. We sometimes write $(\Game y)B(x, y)$, or $(\Game y)\langle x, y \rangle \in B$, for the statement $x \in \Game B$. This is deliberately meant to conjure up the notation used for statements involving the quantifiers $(\forall y)$ and $(\exists y)$. $(\Game y)$ really is a quantifier, giving precise meaning to the chain $(\exists y(0))(\forall y(1))(\exists(y(2))\cdots\cdots$ of quantifiers over ω.

Let $B \subset \mathbb{R} \times \mathbb{R}$ be open. Note that for each $\langle x, y \rangle \in B$ there exists some $n < \omega$ so that $N_{x \restriction n} \times N_{y \restriction n} \subset B$. Let $n(x, y)$ denote the least such n. We refer to $n(x, y)$ as the time of entry of $\langle x, y \rangle$ into B. For $\langle x, y \rangle \notin B$ we set $n(x, y) = \omega$.

Supported by the National Science Foundation under Grant No. DMS-0094174.

The Cabal Seminar. Volume I: Games, Scales and Suslin Cardinals
Edited by A. S. Kechris, B. Löwe, J. R. Steel
Lecture Notes in Logic, 31

F $y(0)$			S $y(1)$	F $y(2)$			$\cdots$
	S $y^*(0)$	F $y^*(1)$			S $y^*(2)$	$\cdots$	

Diagram 2. The game $\mathrm{H}(x^*, x)$.

Let $A = \Game B$. For $x, x^* \in \mathbb{R}$ define $\mathrm{H}(x^*, x)$ to be the following game: Players "first" and "second" (denoted F and S respectively) alternate moves subject to the format in Diagram 2. The moves are played sequentially from left to right, and are presented in two separate lines only for future convenience. The letters F and S indicate which player is responsible for each move. Each of the moves is a natural number. An infinite run leading to reals $y = \langle y(i) : i < \omega \rangle$ and $y^* = \langle y^*(i) : i < \omega \rangle$ is won by S if

$$\langle x, y \rangle \notin B, \quad \text{or} \quad \langle x, y \rangle \in B \;\&\; \langle x^*, y^* \rangle \in B \;\&\; n(x^*, y^*) \leq n(x, y).$$

Otherwise the run is won by F. With our convention that $n(x, y) = \omega$ for $\langle x, y \rangle \notin B$, a run $\langle y^*, y \rangle$ of $\mathrm{H}(x^*, x)$ is won by S just in case that $n(x^*, y^*) \leq n(x, y)$.

The game $\mathrm{H}(x^*, x)$ thus involves a simultaneous play of both $\mathrm{G}(B_x)$, taking place on the upper line in Diagram 2, and $\mathrm{G}(B_{x^*})$, taking place on the lower line. We think of the former as owned by F, and of the latter as owned by S. Each plays for I on the line she owns, and for II on the line owned by her opponent.

To win, S must make sure that the play on her line (namely the lower line, where she plays for I) does not lag behind the play on her opponent's line: if the play on her opponent's line enters B then she has to make sure that the play on her line enters B, at the same time or earlier.

Define a relation $\preceq$ on $\mathbb{R}$ by setting $x^* \preceq x$ iff S has a winning strategy in $\mathrm{H}(x^*, x)$. This winning strategy should be viewed as a "translation mechanism." It translates a strategy for I in $\mathrm{G}(B_x)$ into a strategy for I in $\mathrm{G}(B_{x^*})$, making sure that the translated strategy never lags behind the original strategy.

Claim 1. The relation $\preceq$ is reflexive.

Proof. S can win $\mathrm{H}(x, x)$ simply by copying the moves played by F. More precisely, the strategy defined by the conditions $y^*(n) = y(n)$ for even n and $y(n) = y^*(n)$ for odd n is winning for S in $\mathrm{H}(x, x)$. ⊣

Claim 2. The relation $\preceq$ is transitive.

Proof. Suppose that $x^{**} \preceq x^* \preceq x$. Let τ_1 be a winning strategy for S in $\mathrm{H}(x^*, x)$ and let τ_2 be a winning strategy for S in $\mathrm{H}(x^{**}, x^*)$. Let τ be the strategy in $\mathrm{H}(x^{**}, x)$ obtained by composing τ_1 and τ_2. A typical play according to τ is illustrated in Diagram 3. The play starts in the upper left corner with a move by F, and proceeds along the arrows obtaining additional moves through plays by F in $\mathrm{H}(x^{**}, x)$, uses of τ_1, and uses of τ_2, as indicated in the diagram. Note that a play $\langle y^{**}, y \rangle$ is according to τ iff there is a real y^* so that $\langle y^{**}, y^* \rangle$ is according to τ_2 and $\langle y^*, y \rangle$ is according to τ_1. It is easy using this characterization to check that τ is winning for S in $\mathrm{H}(x^{**}, x)$. ⊣

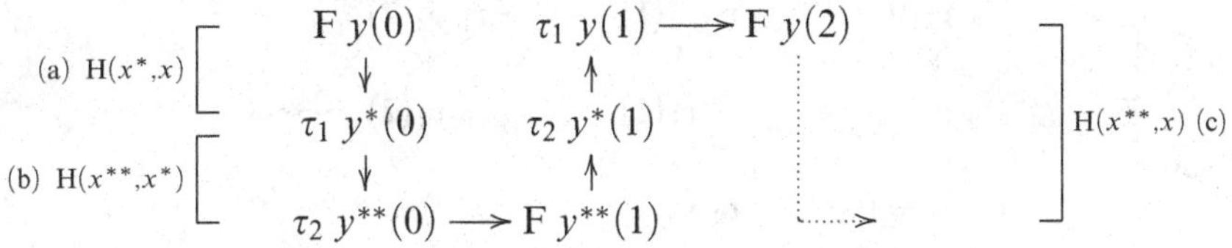

Diagram 3. Composing τ_1 and τ_2.

LEMMA 3. For $x, x^* \in \mathbb{R}$ set $x \prec x^*$ iff $x^* \not\preceq x$. Suppose that each of the games $\mathrm{H}(x^*, x)$, $x, x^* \in \mathbb{R}$, is determined. Then the relation $\prec$ is wellfounded.

PROOF. Suppose for contradiction that $\langle x_i : i < \omega \rangle$ is a sequence of reals so that $x_{i+1} \prec x_i$, meaning $x_i \not\preceq x_{i+1}$, for each i. Using the assumption of determinacy it follows that F has a winning strategy in $\mathrm{H}(x_i, x_{i+1})$. Let τ_i be such a strategy. Construct reals y_i, $i < \omega$, following Diagram 4. The construction proceeds column by column from left to right, setting $y_0(k) = 0$ for each even k and using the strategies τ_i, as indicated in the diagram, to produce all other objects.

For each $i < \omega$, the lines corresponding to y_i and y_{i+1} together form a play of $\mathrm{H}(x_i, x_{i+1})$, according to τ_i. (For $i = 1$ the progress of this play is indicated in the diagram through squiggly arrows, and the play itself is indicated in boldface.) Using the fact that τ_i is winning for F in $\mathrm{H}(x_i, x_{i+1})$ it follows that $\langle x_{i+1}, y_{i+1} \rangle \in B$ and that $n(x_{i+1}, y_{i+1}) < n(x_i, y_i)$ for each $i < \omega$. But this gives an infinite descending sequence of natural numbers, and hence a contradiction. ⊣

REMARK 4. It follows from Lemma 3 that $\preceq$ is total: if $x \not\preceq x^*$ and $x^* \not\preceq x$ then $\langle x, x^*, x, x^*, \dots \rangle$ is an infinite descending sequence in $\prec$, a contradiction.

We know now that the relation $\preceq$ is a prewellorder, meaning that the relation $x \sim y$ iff $x \preceq y$ & $y \preceq x$ is an equivalence relation and that $\preceq$ induces a wellordering of the equivalence classes of $\sim$. Let $\varphi \colon \mathbb{R} \to \mathrm{Ord}$ be the rank function associated to $\preceq$. Precisely, φ is defined through the condition $\varphi(x) = \sup\{\varphi(\bar{x}) + 1 : \bar{x} \prec x\}$ (with the supremum of the empty set taken to be 0). We shall see that the relationship between A and $x \mapsto \varphi(x)$ is analogous to the relationship between the open set B and the function $\langle x, y \rangle \mapsto n(x, y)$ defined earlier.

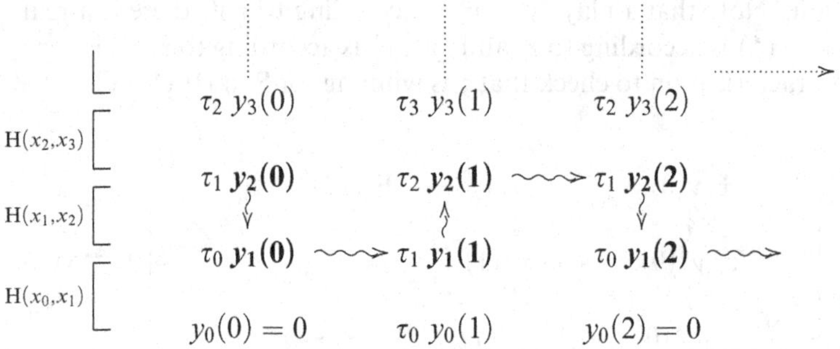

Diagram 4. The wellfoundedness of $\prec$.

Note that

$$\text{if } \langle x, y\rangle \in B \text{ and } n(x^*, y^*) \leq n(x, y) \text{ then } \langle x^*, y^*\rangle \in B. \tag{1}$$

We now establish a similar relationship between the set $A = \Game B$ and the function $x \mapsto \varphi(x)$:

LEMMA 5. Suppose that $x \in A$ and $\varphi(x^*) \leq \varphi(x)$. Then $x^* \in A$.

PROOF. Since $x \in A = \Game B$, player I has a winning strategy in $G(B_x)$. Let σ be such a strategy. Since $\varphi(x^*) \leq \varphi(x)$, S has a winning strategy in $H(x^*, x)$. Let τ be such a strategy. σ and τ combine naturally to give rise to a strategy σ^* for player I in $G(B_{x^*})$. σ^* is characterized by the condition that y^* is according to σ^* iff there is a real y so that y is according to σ and $\langle y^*, y\rangle$ is according to τ.

A typical run y^* of σ^* is presented on the lower line of Diagram 5, and the associated real y is presented on the upper line. Since σ is winning for I in $G(B_x)$ the real y must belong to B_x. In other words $\langle x, y\rangle \in B$. Since τ is winning for S in $H(x^*, x)$, $n(x^*, y^*) \leq n(x, y)$. By (1) above $\langle x^*, y^*\rangle \in B$. This shows that σ^* is winning for I in $G(B_{x^*})$, and hence $x^* \in A$. ⊣

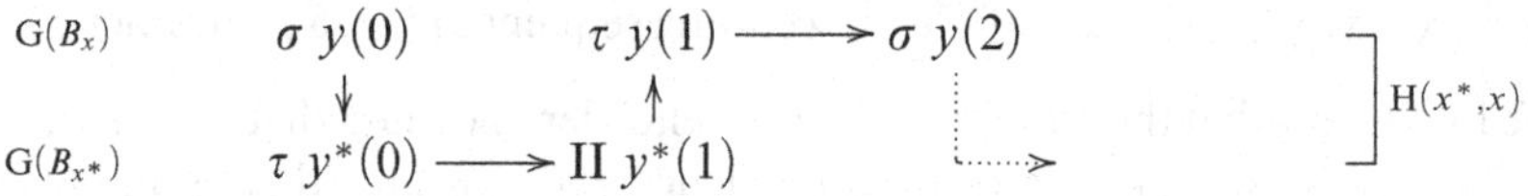

Diagram 5. Composing σ and τ.

The last lemma shows that A forms an initial segment of $\mathbb{R}$ in the prewellorder given by φ, just as B forms an initial segment of $\mathbb{R}\times\mathbb{R}$ in the prewellorder given by $\langle x, y\rangle \mapsto n(x, y)$. Note that the complement of B is a single equivalence class at the very top of the prewellorder given by $\langle x, y\rangle \mapsto n(x, y)$. Precisely,

$$\text{if } \langle x, y\rangle \notin B \text{ then } n(x^*, y^*) \le n(x, y) \text{ for all } \langle x^*, y^*\rangle \in \mathbb{R} \times \mathbb{R}. \tag{2}$$

The next claim establishes the same property for φ and A.

Claim 6. Suppose that each of the games $\mathrm{G}(B_x)$, $x \in \mathbb{R}$, is determined. Let $x, x^* \in \mathbb{R}$, and suppose that $x \notin A$. Then $x^* \preceq x$.

Proof. Since $x \notin A = \Game B$, and $\mathrm{G}(B_x)$ is determined, player II must have a winning strategy in $\mathrm{G}(B_x)$. Let σ be such a strategy. Let τ be the strategy for S in $\mathrm{H}(x^*, x)$ which follows σ on the upper line, and plays 0s on the lower line. Plays $\langle y^*, y\rangle$ according to τ are characterized by the condition that y is according to σ and $y^*(k) = 0$ for each even k. From the first clause and the fact that σ is winning for II in $\mathrm{G}(B_x)$ is follows that $\langle x, y\rangle \notin B$, so that $\langle y^*, y\rangle$ is won by S in $\mathrm{H}(x^*, x)$. ⊣

Recall that $B \subset \mathbb{R}^2$ is open, that is $\mathbf{\Sigma}^0_1$, and $n(x, y)$ is equal to the least n so that $N_{x\restriction n} \times N_{y\restriction n} \subset B$ if $\langle x, y\rangle \in B$, and to ω if $\langle x, y\rangle \in \mathbb{R}^2 - B$. Define $\sqsubseteq$ by setting $\langle x^*, y^*\rangle \sqsubseteq \langle x, y\rangle$ iff $n(x^*, y^*) \le n(x, y)$, and define $\sqsubset$ by setting $\langle x^*, y^*\rangle \sqsubset \langle x, y\rangle$ iff $n(x^*, y^*) < n(x, y)$. Both are relations on $\mathbb{R}^2$, equivalently subsets of $\mathbb{R}^2 \times \mathbb{R}^2$. It is easy to see that

$$\text{both } \sqsubset \text{ and } \sqsubseteq \cap (B \times \mathbb{R}^2) \text{ are } \mathbf{\Sigma}^0_1. \tag{3}$$

By $\sqsubseteq \cap (B \times \mathbb{R}^2)$ we mean $\{\langle \bar{x}, \bar{y}, x, y\rangle : n(\bar{x}, \bar{y}) \le n(x, y) \ \&\ \langle \bar{x}, \bar{y}\rangle \in B\}$. The restriction to $\langle \bar{x}, \bar{y}\rangle \in B$ is important. The full relation $\sqsubseteq$ is not $\mathbf{\Sigma}^0_1$. (We could, for symmetry, also restrict $\sqsubset$ to $B \times \mathbb{R}^2$ in (3). But this would not make any difference: $\langle \bar{x}, \bar{y}\rangle \sqsubset \langle x, y\rangle$ already implies that $\langle \bar{x}, \bar{y}\rangle \in B$.)

For a pointclass[1] Γ let $\Game\Gamma$ be the pointclass $\{\Game D : D \in \Gamma\}$. The set $A = \Game B$ is $\Game\mathbf{\Sigma}^0_1$. The next lemma establishes the parallel of property (3) for A and φ.

Lemma 7. Suppose that all length ω games with $\mathbf{\Sigma}^0_1$ payoff are determined. Then $\prec$ and $\preceq \cap (A \times \mathbb{R})$ are both $\Game\mathbf{\Sigma}^0_1$.

Proof. Using determinacy, $\bar{x} \prec x$ iff F has a winning strategy in $\mathrm{H}(x, \bar{x})$. A run $\langle y, \bar{y}\rangle$ of $\mathrm{H}(x, \bar{x})$ is won by F iff $n(\bar{x}, \bar{y}) < n(x, y)$, and this is a $\mathbf{\Sigma}^0_1$ condition by property (3). Hence the set $\{\langle \bar{x}, x\rangle : \text{F has a winning strategy in } \mathrm{H}(x, \bar{x})\}$ is $\Game\mathbf{\Sigma}^0_1$.

It remains to prove that $\preceq \cap (A \times \mathbb{R})$ is $\Game\mathbf{\Sigma}^0_1$.

Define $\mathrm{H}'(x^*, x)$ to be played according to the rules of $\mathrm{H}(x^*, x)$ but with the modified payoff condition that S wins just in case that $\langle x^*, y^*\rangle \in B$ and $n(x^*, y^*) \le n(x, y)$. Note that this modified condition is $\mathbf{\Sigma}^0_1$ by property (3). Hence:

[1]By a **pointclass** we always mean a class of sets closed under recursive substitutions, conjunctions, disjunctions, and bounded number quantifications.

(i) The set $\{\langle x^*, x\rangle : \text{S has a winning strategy in } \mathrm{H}'(x^*, x)\}$ is $\Game\utilde{\Sigma}^0_1$.

Note further that the modified condition is harder for S than the original condition. Hence:

(ii) If S has a winning strategy in $\mathrm{H}'(x^*, x)$ then she also has a winning strategy in $\mathrm{H}(x^*, x)$.

We intend to show that for $x^* \in A$ the games are in fact equivalent.

Claim 8. Let $x^* \in A$. Then S has a winning strategy in $\mathrm{H}'(x^*, x^*)$.

Proof. Suppose for contradiction that S does not have a winning strategy in $\mathrm{H}'(x^*, x^*)$. By determinacy then F has a winning strategy. Let τ be such a strategy.

Let σ be a winning strategy for I in $\mathrm{G}(B_{x^*})$. Player I has a winning strategy in this game since x^* is assumed to be in A.

Diagram 6 shows how to compose σ and infinitely many copies of τ to produce a sequence of reals y_i, $i < \omega$, with the property that y_0 is according to σ, and for each i, $\langle y_i, y_{i+1}\rangle$ is according to τ. (Each copy of $\mathrm{H}'(x^*, x^*)$ in the diagram is labelled by a roman letter, and the copies of τ have superscripts indicating which of the copies of $\mathrm{H}'(x^*, x^*)$ they belong to. The copy of $\mathrm{H}'(x^*, x^*)$ labelled (b) is highlighted in squiggly arrows in the diagram, and the moves by F in this game, made by the copy of τ labelled (b), are indicated in boldface.)

Since σ is winning for I in $\mathrm{G}(B_{x^*})$, $\langle x^*, y_0\rangle \in B$. Since τ is winning for F in $\mathrm{H}'(x^*, x^*)$, either $\langle x^*, y_i\rangle \notin B$ or $n(x^*, y_i) \not\leq n(x^*, y_{i+1})$, for each $i < \omega$. Notice that the latter disjunct implies that $\langle x^*, y_{i+1}\rangle \in B$, by property (2). It therefore follows by induction that $\langle x^*, y_i\rangle \in B$ and $n(x^*, y_{i+1}) < n(x^*, y_i)$ for each $i < \omega$. But this gives an infinite descending sequence of natural numbers, a contradiction. ⊣

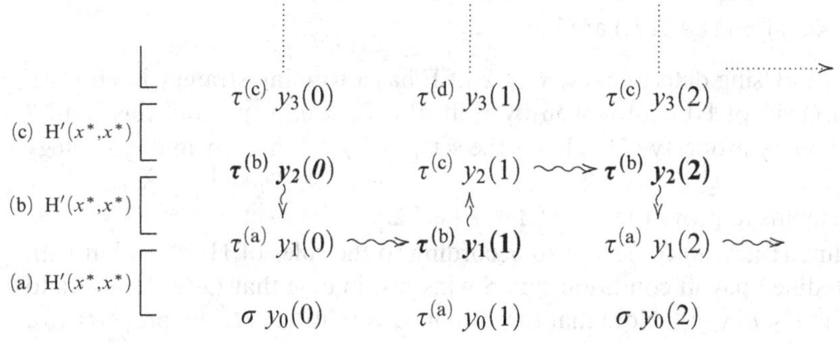

Diagram 6. F cannot have a winning strategy in $\mathrm{H}'(x^*, x^*)$ for $x^* \in A$.

CLAIM 9. Suppose that S has a winning strategy in $\mathrm{H}(x^*, x)$ and that $x^* \in A$. Then S has a winning strategy in $\mathrm{H}'(x^*, x)$.

PROOF. Let τ_1 be a winning strategy for S in $\mathrm{H}(x^*, x)$, and using the previous claim let τ_2 be a winning strategy for S in $\mathrm{H}'(x^*, x^*)$. Let τ be obtained by composing τ_1 and τ_2 in the manner of Diagram 3. (But here the games labelled (b) and (c) in the diagram are $\mathrm{H}'(x^*, x^*)$ and $\mathrm{H}'(x^*, x)$ respectively.) Then τ is winning for S in $\mathrm{H}'(x^*, x)$. ⊣

From Claim 9 and condition (ii) above it follows that for $x^* \in A$, S has a winning strategy in $\mathrm{H}(x^*, x)$ iff she has a winning strategy in $\mathrm{H}'(x^*, x)$. From this and condition (i) it follows that the set $\{\langle x^*, x\rangle : x^* \in A$ and S has a winning strategy in $\mathrm{H}(x^*, x)\}$ is $\Game\boldsymbol{\Sigma}^0_1$. ⊣ (Lemma 7)

LIST 10. The following list summarizes the properties of φ and A obtained so far (with Γ standing for $\Game\boldsymbol{\Sigma}^0_1$):

1. If $x \in A$ and $\varphi(\bar{x}) \leq \varphi(x)$ then $\bar{x} \in A$.
2. If $x \notin A$ and $\bar{x} \notin A$ then $\varphi(x) = \varphi(\bar{x})$.
3. Both the sets $\{\langle \bar{x}, x\rangle : \varphi(\bar{x}) < \varphi(x)\}$ and $\{\langle \bar{x}, x\rangle : \bar{x} \in A \;\&\; \varphi(\bar{x}) \leq \varphi(x)\}$ are in Γ.

Conditions (1) and (2) merely note that $\mathbb{R} - A$ forms a single equivalence class of the prewellorder induced by φ, located above all elements of A in this prewellorder. Notice that any function $\varphi\colon A \to \mathrm{Ord}$ can be extended to a function on $\mathbb{R}$ satisfying conditions (1) and (2) simply by setting $\varphi(x) = \sup\{\varphi(\bar{x}) + 1 : \bar{x} \in A\}$ for $x \in \mathbb{R} - A$. We use the same letter to refer both to the function defined on A and to an extension of the function to $\mathbb{R}$ subject to conditions (1) and (2).

Condition (3) in List 10 is the crucial one, connecting φ to the pointclass Γ. The next definition abstracts an equivalent condition, that refers only to the restriction of φ to A. The equivalence is proved in Claim 12.

DEFINITION 11. A function $\varphi\colon A \to \mathrm{Ord}$ is a Γ **norm** on A just in case that there are sets U and V in Γ and $\neg\Gamma$ respectively, so that, for every $x \in A$,

$$\{\bar{x} : \varphi(\bar{x}) \leq \varphi(x)\} = \{\bar{x} : \langle \bar{x}, x\rangle \in U\} = \{\bar{x} : \langle \bar{x}, x\rangle \in V\}.$$

CLAIM 12. Let $\varphi\colon A \to \mathrm{Ord}$ and extend φ to $\mathbb{R}$ in line with conditions (1) and (2) in List 10. Then φ is a Γ norm iff it satisfies condition (3) in the list.

PROOF. Assuming condition (3), let $U = \{\langle \bar{x}, x\rangle : \bar{x} \in A \;\&\; \varphi(\bar{x}) \leq \varphi(x)\}$, and let $V = \{\langle \bar{x}, x\rangle : \varphi(x) \not< \varphi(\bar{x})\}$. Assuming the condition in Definition 11 note that $\varphi(\bar{x}) < \varphi(x)$ iff $\bar{x} \in A \;\&\; \langle x, \bar{x}\rangle \notin V$, and that $\bar{x} \in A \;\&\; \varphi(\bar{x}) \leq \varphi(x)$ iff $(x \in A \;\&\; \langle \bar{x}, x\rangle \in U) \vee (\bar{x} \in A \;\&\; \langle x, \bar{x}\rangle \notin V)$. ⊣

The condition in Definition 11 states that for each $x \in A$, the initial segment $\{\bar{x} : \varphi(\bar{x}) \leq \varphi(x)\}$ belongs to both $\Gamma(x)$ and $\neg\Gamma(x)$, and that this holds uniformly in x. Working with norms in this paper it is more convenient to use the equivalent characterization in List 10, and we shall do this without further comment throughout the paper.

A pointclass Γ is said to have the **prewellordering property** if every set in Γ admits a Γ norm. The prewellordering property has various applications to questions in descriptive set theory, see for example the theorem on reduction in [KM78B, §2].

The sequence of results given above shows how to produce a $\Game\utilde{\Sigma}^0_1$ norm on a given $\Game\utilde{\Sigma}^0_1$ set, starting from $\utilde{\Sigma}^0_1$ norms on $\utilde{\Sigma}^0_1$ sets. But there is nothing specific to $\utilde{\Sigma}^0_1$ sets in any of the proofs. They generalize routinely to yield the following theorem:

THEOREM 13. Let Γ be a pointclass. Suppose that every length ω game with payoff in Γ is determined. Suppose that Γ has the prewellordering property. Then $\Game\Gamma$ has the prewellordering property.

PROOF. Let $B \subset \mathbb{R}^2$ belong to Γ and let $\vartheta\colon \mathbb{R} \to \mathrm{Ord}$ be a Γ norm on B. Follow the sequence of definitions and claims above, only replacing the uses of $n(x, y)$ by uses of $\vartheta(x, y)$. It is easy to check that the proofs adapt, showing that the resulting function φ is a $\Game\Gamma$ norm on $A = \Game B$. Let us only note that the definability expressed by condition (3) in List 10, for the norm ϑ, is such that all the games that come up in the adapted proofs have payoff sets in Γ, and are therefore determined. (Their determinacy is needed in the proofs.) ⊣

There is one crucial property of the norm $\langle x, y\rangle \mapsto n(x, y)$ that was not considered in the discussion so far. It is easy to check that this norm and the $\utilde{\Sigma}^0_1$ set B satisfy

let $\langle x_i, y_i\rangle$, $i < \omega$, be elements of B. Suppose that $\lim_{i\longrightarrow\infty} x_i$ and $\lim_{i\longrightarrow\infty} y_i$ exist and let x_∞ and y_∞ respectively denote the limits. Suppose that $n(x_i, y_i)$ is eventually constant as $i \longrightarrow \infty$. Then $\langle x_\infty, y_\infty\rangle \in B$ and $n(x_\infty, y_\infty) \leq$ eventual value of $n(x_i, y_i)$. (4)

As stated this additional property is not true at the level of $\Game\utilde{\Sigma}^0_1$ sets, but we can obtain a parallel property at that level by using countably many norms.

Given $x, x^* \in \mathbb{R}$, $p \in {}^{<\omega}\omega$ of even length, say $2k$, and $h, h^* \in \omega$, define $H_p(x^*, h^*, x, h)$ to be played as follows: Players F and S alternate moves subject to the format in Diagram 7. The moves are played sequentially from left to right, starting from the vertical line, and each of the moves is a natural number. At the end of an infinite run we set $y = p^\frown\langle h\rangle^\frown\langle y(i) : 2k < i < \omega\rangle$ and $y^* = p^\frown\langle h^*\rangle^\frown\langle y^*(i) : 2k < i < \omega\rangle$. The run is won by S if

$$\langle x, y\rangle \notin B, \quad \text{or} \quad \langle x, y\rangle \in B \;\&\; \langle x^*, y^*\rangle \in B \;\&\; n(x^*, y^*) \leq n(x, y),$$

and otherwise the run is won by F.

$H_p(x^*, h^*, x, h)$ may thus be viewed as a version of $\mathrm{H}(x^*, x)$ with $y{\restriction}2k+1$ set equal to $p^\frown\langle h\rangle$ and $y^*{\restriction}2k+1$ set equal to $p^\frown\langle h^*\rangle$. For reason of notational

$p(0)\ \cdots\ p(2k-1)$	h			S $y(2k+1)$	F $y(2k+2)$
$p(0)\ \cdots\ p(2k-1)$		h^*	F $y^*(2k+1)$		$\ldots$

Diagram 7. The game $H_p(x^*, h^*, x, h)$.

convenience we refer to $\langle y^*, y\rangle$, rather than the sequence of actual moves, as a **run** of $H_p(x^*, h^*, x, h)$.

Define $\preceq_p$ by setting $\langle x^*, h^*\rangle \preceq_p \langle x, h\rangle$ iff S has a winning strategy in the game $H_p(x^*, h^*, x, h)$. The previous proofs adapt to show that $\preceq_p$ is a prewellorder. Let $\varphi_p \colon \mathbb{R} \times \omega \to \mathrm{Ord}$ be the associated rank function, defined by $\varphi_p(x, h) = \sup\{\varphi_p(\bar{x}, \bar{h}) + 1 : \langle \bar{x}, \bar{h}\rangle \prec_p \langle x, h\rangle\}$. φ_p is then a $\Game\Sigma^0_1$ norm on the set $\{\langle x, h\rangle : p^\frown\langle h\rangle$ is a winning position for I in $\mathrm{G}(B_x)\}$.

For $p \in {}^{<\omega}\omega$ of even length and $x \in \mathbb{R}$ let $\psi_p(x) = \min\{\varphi_p(x, h) : h \in \omega\}$ and let $h_p(x)$ be the smallest number h realizing the minimum, that is $h_p(x) = \min\{h : \varphi_p(x, h) = \psi_p(x)\}$.

Exercise 14. Let $x \in A$. Let $p \in {}^{<\omega}\omega$ be a node of even length, say $2k$. Suppose that $p(2i) = h_{p\restriction 2i}(x)$ for each $i < k$. Show that p is a winning position for I in $\mathrm{G}(B_x)$.

Hint for the case $k = 1$. Let σ be a winning strategy for I in $\mathrm{G}(B_x)$. Let h be the first move played by σ. By assumption $p(0) = h_\varnothing(x)$, and it follows from the definition of $h_\varnothing(x)$ that $\varphi_\varnothing(x, p(0)) \le \varphi(x, h)$. So S has a winning strategy, τ say, in the game $H_\varnothing(x, p(0), x, h)$. Diagram 8 shows how to win $\mathrm{G}(B_x)$ from p, against an opponent who plays for II, using a composition of σ and τ. ⊣

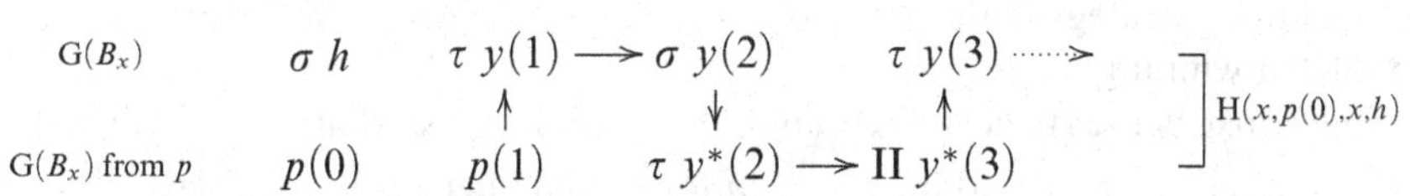

Diagram 8. Hint for Exercise 14.

Claim 15. The norm $\psi_\varnothing$ is equivalent to the earlier norm φ on A, in the sense that $\psi_\varnothing(x^*) \le \psi_\varnothing(x)$ iff $\varphi(x^*) \le \varphi(x)$.

Proof. $\psi_\varnothing(x^*) \le \psi_\varnothing(x)$ iff $(\forall n)\,(\exists n^*)$ so that S has a winning strategy in $H_\varnothing(x^*, n^*, x, n)$. Prepending moves corresponding to the quantifier string $(\forall n)(\exists n^*)$ to the game $H_\varnothing(x^*, n^*, x, n)$ we obtain precisely the game $\mathrm{H}(x^*, x)$. So $\psi_\varnothing(x^*) \le \psi_\varnothing(x)$ iff S has a winning strategy in $\mathrm{H}(x^*, x)$ iff $\varphi(x^*) \le \varphi(x)$. ⊣

EXERCISE 16. Let $A_p = \{x : p \text{ is a winning position for I in } \mathrm{G}(B_x)\}$. Show that ψ_p is a $\Game\underset{\sim}{\Sigma}^0_1$ norm on A_p. In fact, show that the sets $\{\langle p, \bar{x}, x\rangle : \psi_p(\bar{x}) < \psi_p(x)\}$ and $\{\langle p, \bar{x}, x\rangle : \bar{x} \in A_p \ \& \ \psi_p(\bar{x}) \leq \psi_p(x)\}$ are in $\Game\underset{\sim}{\Sigma}^0_1$.

HINT. That $\{\langle p, \bar{x}, x\rangle : \psi_p(\bar{x}) < \psi_p(x)\}$ is $\Game\underset{\sim}{\Sigma}^0_1$ follows directly from (determinacy and) the definitions. For $\{\langle p, \bar{x}, x\rangle : \bar{x} \in A_p \ \& \ \psi_p(\bar{x}) \leq \psi_p(x)\}$ use an argument similar to that in the proof of Lemma 7. ⊣

We now approach the parallel of property (4):

THEOREM 17. Let $\langle x_i : i < \omega\rangle$ be a sequence of reals in the set $A = \Game B$. Suppose that $\lim_{i \longrightarrow \infty} x_i$ exists and let x_∞ denote this limit. Suppose that for each p, both $\psi_p(x_i)$ and $h_p(x_i)$ are eventually constant as $i \longrightarrow \infty$. Let λ_p and h_p respectively be their eventual values. Then:

1. x_∞ belongs to A.
2. For each p, $\langle \psi_p(x_\infty), h_p(x_\infty)\rangle \leq_{\text{lex}} \langle \lambda_p, h_p\rangle$.

PROOF. We show that $\varphi_\varnothing(x_\infty, h_\varnothing) \leq \lambda_\varnothing$. A similar proof establishes that $\varphi_p(x_\infty, h_p) \leq \lambda_p$ for each p. Condition (2) of the theorem follows directly from this. Condition (1) follows from the instance $\psi_\varnothing(x_\infty) \leq \psi_\varnothing(x_i)$ (for all sufficiently large i) using Lemma 5 and Claim 15.

For each p let $k(p) < \omega$ be large enough that $\psi_p(x_i) = \lambda_p$ and $h_p(x_i) = h_p$ for all $i \geq k(p)$. Choose $k(p)$ inductively so that $k(p) > k(\bar{p})$ whenever $\bar{p}$ is a strict initial segment of p.

Let $p_0 = \varnothing$ and let $k_0 = k(\varnothing)$.

Suppose for contradiction that $\varphi_\varnothing(x_\infty, h_\varnothing) \not\leq \lambda_\varnothing$. Since $\lambda_\varnothing = \psi_\varnothing(x_{k_0}) = \min\{\varphi_\varnothing(x_{k_0}, h) : h \in \omega\}$ this means that there is some $h \in \omega$ so that $\varphi_\varnothing(x_\infty, h_\varnothing) \not\leq \varphi_\varnothing(x_{k_0}, h)$. Fix such an h.

Using determinacy the fact that $\varphi_\varnothing(x_\infty, h_\varnothing) \not\leq \varphi_\varnothing(x_{k_0}, h)$ implies that F has a winning strategy in the game $H_\varnothing(x_\infty, h_\varnothing, x_{k_0}, h)$ which we denote $H^{(\infty)}$. Fix such a winning strategy σ.

Now construct sequences $\langle k_n\rangle$, $\langle p_n\rangle$, $\langle \tau^{(n)}\rangle$, and $\langle y_n\rangle$ so that:

(a) $p_0 \subset p_1 \subset p_2 \cdots$ and $\mathrm{lh}(p_n) = 2n$ for each n. Let $y_\infty = \bigcup_{n<\omega} p_n$.
(b) $y_0(0) = h$, $y_1(0) = h_\varnothing$, and the pair $\langle y_\infty, y_0\rangle$ is a run of $H^{(\infty)}$ played according to σ.
(c) $p_{n+1} = p_n{}^\frown\langle h_{p_n}, y_{n+1}(2n+1)\rangle$.
(d) $k_{n+1} = k(p_{n+1})$.
(e) $\tau^{(n)}$ is a winning strategy for S in the game $H_{p_n}(x_{k_{n+1}}, h_{p_n}, x_{k_n}, y_n(2n))$ which we denote $H^{(n)}$.
(f) The pair $\langle y_{n+1}, y_n\rangle$ is a run of $H^{(n)}$ played according to $\tau^{(n)}$.

Diagram 9 illustrates the construction. The construction begins on the upper left corner, with the assignments $y_0(0) = h$ and $y_1(0) = h_\varnothing$. The construction continues following the arrows in the diagram, assigning to each entry a value either by setting it equal to h_{p_n} for some n or by using one of the strategies, as indicated. The symbol $''$ in an entry indicates copying the value of the entry above it.

Note that the values of $y_{n+1}(2n)$ (set equal to h_{p_n}) and $y_{n+1}(2n+1)$ (determined using σ) can be determined *before* k_{n+1} is known. (They do not depend on $\tau^{(n)}$.) Once these assignments are made we set $p_{n+1} = p_n{}^\frown\langle y_{n+1}(2n), y_{n+1}(2n+1)\rangle$, and set $k_{n+1} = k(p_{n+1})$. Since $y_{n+1}(2n) = h_{p_n}$, and since both k_n and k_{n+1} are greater than or equal to $k(p_n)$,

$$\begin{aligned}\varphi_{p_n}(x_{k_{n+1}}, y_{n+1}(2n)) &= \lambda_{p_n} \\ &= \varphi_{p_n}(x_{k_n}, h_{p_n}) \\ &\leq \varphi_{p_n}(x_{k_n}, y_n(2n)).\end{aligned}$$

Thus S has a winning strategy in $H^{(n)} = H_{p_n}(x_{k_{n+1}}, y_{n+1}(2n), x_{k_n}, y_n(2n))$, allowing us to pick $\tau^{(n)}$ subject to condition (e) above and continue with the construction, following the arrows.

The sequences y_∞ (whose entries are indicated in boldface in Diagram 9) and y_0 together form a run of $H_\varnothing(x_\infty, h_\varnothing, x_{k_0}, h)$, played according to σ. Since σ is winning for F in the game,

(i) $n(x_{k_0}, y_0) < n(x_\infty, y_\infty)$.

The sequences y_{n+1} and y_n together form a run of $H^{(n)}$ according to $\tau^{(n)}$. Since $\tau^{(n)}$ is winning for S in the game,

(ii) $n(x_{k_{n+1}}, y_{n+1}) \leq n(x_{k_n}, y_n)$.

It follows from this, and the wellfoundedness of ω, that $n(x_{k_n}, y_n)$ is eventually constant as $n \longrightarrow \infty$. Now $\lim_{n \longrightarrow \infty} x_{k_n} = x_\infty$ by the assumptions of the theorem, and $\lim_{n \longrightarrow \infty} y_n = y_\infty$ by construction. Using property (4) above it follows that

(iii) $n(x_\infty, y_\infty) \leq$ the eventual value of $n(x_{k_n}, y_n)$.

But from this and condition (ii) it follows that $n(x_\infty, y_\infty) \leq n(x_{k_0}, y_0)$, and this contradicts condition (i). ⊣

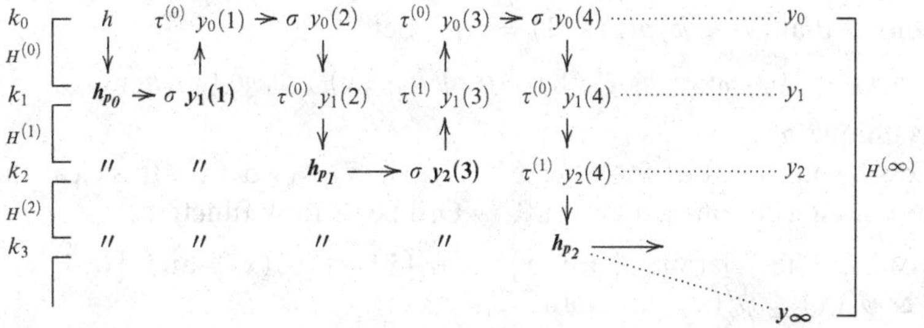

Diagram 9. The proof of Theorem 17.

Definition 18. Let $A \subset \mathbb{R}$. A countable sequence $\langle \vartheta_\ell \rangle$ of norms on A is called a **scale** on A if it satisfies the following limit condition:

(∗) Let x_i for $i < \omega$ belong to A. Suppose that $\lim_{i \longrightarrow \infty} x_i$ exists and let x_∞ denote the limit. Suppose that for each ℓ, $\vartheta_\ell(x_i)$ is eventually constant as $i \longrightarrow \infty$, and let λ_ℓ denote the eventual value. Then: $x_\infty \in A$; and for each ℓ, $\vartheta_\ell(x_\infty) \le \lambda_\ell$.

A scale $\langle \vartheta_\ell \rangle$ is called a Γ **scale** if both the sets $\{\langle \ell, \bar{x}, x\rangle : \vartheta_\ell(\bar{x}) < \vartheta_\ell(x)\}$ and $\{\langle \ell, \bar{x}, x\rangle : \bar{x} \in A \ \&\ \vartheta_\ell(\bar{x}) \le \vartheta_\ell(x)\}$ belong to Γ, or, equivalently, if there are sets U and V in Γ and $\neg\Gamma$ respectively so that

$$\{\bar{x} : \vartheta_\ell(\bar{x}) \le \vartheta_\ell(x)\} = \{\bar{x} : \langle \bar{x}, x, \ell\rangle \in U\} = \{\bar{x} : \langle \bar{x}, x, \ell\rangle \in V\}$$

for all $\ell < \omega$ and all $x \in A$.

Γ has the **scale property** if each set in Γ admits a Γ scale.

Remark 19. Many applications involve scales $\langle \vartheta_\ell \rangle$ so that the sets $\{\langle \bar{x}, x\rangle : \vartheta_\ell(\bar{x}) < \vartheta_\ell(x)\}$ and $\{\langle \bar{x}, x\rangle : \bar{x} \in A \ \&\ \vartheta_\ell(\bar{x}) \le \vartheta_\ell(x)\}$ belong to Γ for each individual ℓ, meaning that each ϑ_ℓ is a Γ norm, but the joins $\{\langle \ell, \bar{x}, x\rangle : \vartheta_\ell(\bar{x}) < \vartheta_\ell(x)\}$ and $\{\langle \ell, \bar{x}, x\rangle : \bar{x} \in A \ \&\ \vartheta_\ell(\bar{x}) \le \vartheta_\ell(x)\}$ do not belong to Γ. We call such scales **weakly** Γ, and say that Γ is **weakly scaled** if every set in Γ admits a weakly Γ scale.

Theorem 17 shows that $\langle \psi_p, h_p \rangle$ is a scale on $A = \Game B$. But it need not be a $\Game\utilde{\Sigma}^0_1$ scale. The problem is with the shift from A to A_p in the definability of the norms ψ_p in Exercise 16. We now solve this problem by restricting to p which are winning for I in $G(B_x)$, through a use of Exercise 14.

Call $p \in {}^{<\omega}\omega$ of even length $2k$ **correct** for $x \in A$ just in case that $(\forall i < k)$ $p(2i) = h_{p \restriction 2i}(x)$.

Remark 20. The assumption in Theorem 17, that $\psi_p(x_i)$ and $h_p(x_i)$ are eventually constant as $i \longrightarrow \infty$ for each p, can be weakened to apply only to p which are correct for x_i for almost all i (meaning all but finitely many i), as these are the only p which come up during the proof of the theorem.

Note that for every $r \in {}^{<\omega}\omega$ of length k there is a unique p which is correct for x and so that $(\forall i < k) p(2i + 1) = r(i)$. Let

$$\vec{\alpha}_r(x) = \langle \psi_{p\restriction 0}(x), h_{p\restriction 0}(x), \psi_{p\restriction 2}(x), h_{p\restriction 2}(x), \dots, \psi_p(x), h_p(x)\rangle$$

for this unique p.

Set $\bar{x} \trianglelefteq_r x$ iff $x \notin A$ or $\bar{x} \in A \ \&\ x \in A \ \&\ \vec{\alpha}_r(\bar{x}) \le_{\text{lex}} \vec{\alpha}_r(x)$. It is clear that $\trianglelefteq_r$ is a prewellorder on $\mathbb{R}$. Let $\psi'_r : \mathbb{R} \to \text{Ord}$ be its rank function.

Claim 21. The relations $\{\langle r, \bar{x}, x\rangle : \psi'_r(\bar{x}) < \psi'_r(x)\}$ and $\{\langle r, \bar{x}, x\rangle : \bar{x} \in A \ \&\ \psi'_r(\bar{x}) \le \psi'_r(x)\}$ are both $\Game\utilde{\Sigma}^0_1$.

Proof. This is a simple calculation using Exercise 16, a parallel of the same exercise for the norm φ_p, and the fact that if p is correct for x then $x \in A_p$, given by Exercise 14. ⊣

CLAIM 22. Let x_i be a sequence of reals in A and suppose that for each r, $\psi'_r(x_i)$ is eventually constant as $i \longrightarrow \infty$. Suppose that p is correct for x_i for almost all i. Then $\psi_p(x_i)$ and $h_p(x_i)$ are eventually constant as $i \longrightarrow \infty$.

PROOF. Let k be such that $\mathrm{lh}(p) = 2k$. Let $r = \langle p(1), \dots, p(2k-1) \rangle$. Let n be large enough that p is correct for x_i for all $i > n$. Then $\vec{\alpha}_r(x_i) = \langle \psi_{p\restriction 0}(x_i), h_{p\restriction 0}(x_i), \psi_{p\restriction 2}(x_i), h_{p\restriction 2}(x_i), \dots, \psi_p(x_i), h_p(x_i) \rangle$ for all $i > n$, and the fact that $\psi_p(x_i)$ and $h_p(x_i)$ are eventually constant follows from the fact that $\vec{\alpha}_r(x_i)$ is eventually constant. ⊣

COROLLARY 23. The sequence $\langle \psi'_r : r \in {}^{<\omega}\omega \rangle$ is a $\Game\Sigma^0_1$ scale on $A = \Game B$.

PROOF. The definability required by Definition 18 is given by Claim 21. The limit condition is given by Theorem 17 using Remark 20 and Claim 22. ⊣

Our work so far produced a $\Game\underset{\sim}{\Sigma}^0_1$ scale on a given $\Game\underset{\sim}{\Sigma}^0_1$ set, starting from a norm satisfying property (4) above. But in fact the argument can be adapted to start with a scale, rather than a single norm, and to use the limit condition in Definition 18, rather than property (4). The result is the following theorem:

THEOREM 24. Let Γ be a pointclass. Let $B \subset \mathbb{R}^2$ belong to Γ. Let $\langle \vartheta_\ell : \ell < \omega \rangle$ be a Γ scale (respectively weakly Γ scale) on B. Suppose that Γ determinacy holds. Define H_p, $\preceq_p$, φ_p, h_p, ψ_p, and ψ'_r as above, but replacing the condition "$n(x^*, y^*) \leq n(x, y)$" with the condition

$$\langle \vartheta_0(x^*), \dots, \vartheta_{\mathrm{lh}(p)/2}(x^*) \rangle \leq_{\mathrm{lex}} \langle \vartheta_0(x), \dots, \vartheta_{\mathrm{lh}(p)/2}(x) \rangle$$

throughout.

Then $\langle \psi'_r : r \in {}^{<\omega}\omega \rangle$ is a $\Game\Gamma$ scale (respectively weakly $\Game\Gamma$ scale) on the set $\Game B$.

PROOF. The proofs given above generalize routinely to these settings. Let us only make the following comments:

First, note that all the games that come up during the proofs have payoff sets in Γ, since the norms ϑ_ℓ are all Γ norms. The games are therefore determined. This is important since the proofs require their determinacy.

Second, condition (i) in the proof of Theorem 17 is revised in the general settings to state that

(i)′ $\vartheta_0(x_{k_0}, y_0) < \vartheta_0(x_\infty, y_\infty)$.

Third, condition (ii) in the proof of Theorem 17 is revised in the general settings to state that

(ii)′ $\langle \vartheta_0(x_{k_{n+1}}, y_{n+1}), \dots, \vartheta_n(x_{k_{n+1}}, y_{n+1}) \rangle \leq_{\mathrm{lex}} \langle \vartheta_0(x_{k_n}, y_n), \dots, \vartheta_n(x_{k_n}, y_n) \rangle$.

It follows from the revised condition, and from the wellfoundedness of the ordinals, that $\vartheta_\ell(x_{k_n}, y_n)$ is eventually constant as $n \longrightarrow \infty$, for each $\ell < \omega$. Using the limit condition in Definition 18 then

(iii)′ $\vartheta_\ell(x_\infty, y_\infty) \leq$ eventual value of $\vartheta_\ell(x_{k_n}, y_n)$ as $n \longrightarrow \infty$, for each ℓ.

From this and condition (ii)′ it certainly follows that $\vartheta_0(x_\infty, y_\infty) \leq \vartheta_0(x_{k_0}, y_0)$, contradicting condition (i)′. ⊣

Theorems 24 and 13 are *propagation theorems*. They show that some desirable properties, the prewellordering property in the case of Theorem 13 and scale properties in the case of Theorem 24, propagate from a pointclass Γ to the pointclass $\Game\Gamma$. Infinite games are central to the proofs of both theorems, and both theorems require determinacy.

We noted above that the prewellordering property can be used to settle the classical problem of reduction. The scale property too has applications to classical problems, specifically to the problem of uniformization. Roughly speaking the problem involves definably selecting elements from non-empty sets of reals. The following claim is an indication of how scales connect with such selections.

Claim 25. Let $E \subset \mathbb{R}$ be non-empty and let $\langle \vartheta_\ell : \ell < \omega \rangle$ be a scale on E. For $x \in \mathbb{R}$ set $\vec{\alpha}(x) = \langle \vartheta_0(x), x(0), \vartheta_1(x), x(1), \ldots \rangle$. Let $\langle \lambda_0, h_0, \lambda_1, h_1, \ldots \rangle$ be the lexicographic infimum of the set $\{\vec{\alpha}(x) : x \in E\}$. (The infimum is characterized precisely by the condition that for each n, $\langle \lambda_0, h_0, \ldots, \lambda_{n-1}, h_{n-1} \rangle$ is the lexicographically smallest element of $\{\vec{\alpha}(x) \restriction n : x \in E\}$.)

Then the real $y = \langle h_n : n < \omega \rangle$ belongs to E.

Proof. For each n pick $x_n \in E$ so that $\vec{\alpha}(x_n) \restriction n = \langle \lambda_0, h_0, \ldots, \lambda_{n-1}, h_{n-1} \rangle$. Note that if $n \longrightarrow \infty$, then $x_n \longrightarrow y$, and that $\vartheta_\ell(x_n)$ is equal to λ_ℓ for all $n > \ell$. By the limit condition in Definition 18 it follows that $y \in E$. ⊣

Thus there is a canonical way to select an element of E given a scale on E. This canonical selection process can be turned into a solution for the problem of uniformization, cf. [KM78B, §3.1].

In a similar fashion, scales can be used to select canonical winning strategies, producing $\Game\Gamma$ winning strategies in Γ games won by the player aiming to enter the Γ set:

Exercise 26. Let $E \subset \mathbb{R}$ belong to a pointclass Γ and let $\langle \vartheta_\ell : \ell < \omega \rangle$ be a Γ scale on E. Suppose that every length ω game with payoff in Γ is determined. Suppose that player I wins the game $G_\omega(E)$. Prove that player I has a winning strategy σ which belongs to the pointclass $\Game\Gamma$, meaning that the relation (p is winning for I) & $\sigma(p) = h$ is $\Game\Gamma$.

Hint. Given $p \in {}^{<\omega}\omega$ of even length, say $2k$, and $h, h^* \in \omega$, define $H_p(h^*, h)$ to be played according to Diagram 10, with the moves played sequentially from left to right, starting from the vertical line. At the end of an infinite run set $y = p^\frown\langle h\rangle^\frown\langle y(i) : 2k < i < \omega\rangle$ and $y^* = p^\frown\langle h^*\rangle^\frown\langle y^*(i) : 2k < i < \omega\rangle$. The run is won by S if

$$y \notin E, \quad \text{or}$$
$$y \in E \;\&\; y^* \in E \;\&\; \langle \vartheta_0(y^*), \ldots, \vartheta_{\mathrm{lh}(p)/2}(y^*)\rangle \leq_{\mathrm{lex}} \langle \vartheta_0(y), \ldots, \vartheta_{\mathrm{lh}(p)/2}(y)\rangle,$$

$$\begin{array}{llll|lll} p(0) \cdots p(2k-1) & h & & & & \mathrm{S}\ y(2k+1) & \mathrm{F}\ y(2k+2) \\ p(0) \cdots p(2k-1) & & h^* & & \mathrm{F}\ y^*(2k+1) & & \cdots \end{array}$$

Diagram 10. The game $H_p(h^*, h)$.

and otherwise the run is won by F.

Let Q be the set of positions $p \in {}^{<\omega}\omega$ of even length from which player I can continue to win $G_\omega(E)$. For $p \in Q$ and $h, h^* \in \omega$ set $h^* \preceq_p h$ iff S has a winning strategy in the game $H_p(h^*, h)$. Notice the similarity between the definitions here and the ones preceding Exercise 14 (with the modification indicated in Theorem 24). Adapting the proofs connected to the exercise show that:

(i) $\preceq_p$ is a prewellorder on ω.

Let $\sigma(p)$ be the smallest number in the set $\{h : h \text{ is } \preceq_p \text{ minimal}\}$. Show that:

(ii) The relation $p \in Q \mathbin{\&} \sigma(p) = h$ is $\Game\Gamma$.

(iii) If p is consistent with σ then p is a winning position for I in $G_\omega(E)$.

(iv) σ is a winning strategy for I in $G_\omega(E)$.

Items (i) and (ii) involve adaptations of proofs preceding Exercise 14. Item (iii), which precisely parallels Exercise 14, is a warm-up for item (iv). ⊣

Claim 25 and Exercise 26 are examples of basic applications of norms and scales in descriptive set theory. The papers in Part II of this volume contain many more applications, demonstrating the fundamental importance of scales in the study of consequences of determinacy.

The papers in Part I for the most part concentrate on establishing the scale property for various pointclasses. In many cases this is done through propagation, building on and expanding the introductory methods presented here.

REFERENCES

ALEXANDER S. KECHRIS AND YIANNIS N. MOSCHOVAKIS

[CABAL i] *Cabal seminar* 76–77, Lecture Notes in Mathematics, no. 689, Berlin, Springer, 1978.

[KM78B] *Notes on the theory of scales*, this volume, originally published in *Cabal Seminar* 76–77 [CABAL i], pp. 1–53.

YIANNIS N. MOSCHOVAKIS

[Mos80] ***Descriptive set theory***, Studies in Logic and the Foundations of Mathematics, no. 100, North-Holland, Amsterdam, 1980.

DEPARTMENT OF MATHEMATICS
UNIVERSITY OF CALIFORNIA LOS ANGELES
LOS ANGELES, CA 90095, USA
E-mail: ineeman@math.ucla.edu

SCALES ON Σ^1_1-SETS

JOHN R. STEEL

Let Λ be the class of sets which are $(\omega \cdot n)$-Π^1_1 for some $n < \omega$. We shall show that every set in Λ admits a very good scale all of whose norms are in Λ. Even for Σ^1_1 sets, this is the best possible definability bound on very good scales. Busch shows in [Bus76] that every Σ^1_1 set admits a not very good scale all of whose norms are $(\omega + 3)$-Π^1_1.

Some terminology: if L is an ordered set, then "$\leq_{\text{lex}}$" denotes the lexicographic ordering on ${}^{<\omega}L \cup {}^{\omega}L$. A tree on L is a subset of ${}^{<\omega}L$ closed under initial segment. $[T]$ is the set of infinite branches of the tree T. If T is a tree on $L \times M$ and $x \in {}^{\omega}L$, then $T(x) = \{\tau \in {}^{<\omega}M : \langle x \restriction \text{lh}(\tau), \tau \rangle \in T\}$, and $\text{p}[T] = \{x \in {}^{\omega}L : [T(x)] \neq \varnothing\}$. If M is wellordered and $x \in \text{p}[T]$, then f^T_x is the leftmost branch of $T(x)$. That is, $f^T_x \in [T(x)]$ and $\forall g \in [T(x)](f^T_x \leq_{\text{lex}} g)$. Finally, if $\langle A_\alpha : \alpha < \beta \rangle$ is a sequence of sets, then

$$\text{Diff}_{\alpha<\beta} A_\alpha = \{x : \exists \alpha < \beta(\alpha \text{ is odd} \wedge x \in A_\alpha \wedge \forall \gamma < \alpha(x \notin A_\gamma))\}.$$

A set $A \subseteq {}^{\omega}\omega$ is β-$\boldsymbol{\Pi}^1_1$ iff $A = \text{Diff}_{\alpha<\beta} A_\alpha$ for some sequence $\langle A_\alpha : \alpha < \beta \rangle$ of $\boldsymbol{\Sigma}^1_1$ sets. If the sequence of Σ^1_1 indices for the A_α's can be taken to be recursive, then A is β-Π^1_1.

Lemma 1.1. If A and B are $(\omega \cdot n)$-Π^1_1, where $n < \omega$, then $A \cap B$ is $(\omega \cdot (n^2))$-Π^1_1.

Proof. We claim first that there is a wellorder $\preccurlyeq$ of $\omega^2 \times \omega^2$ of order type ω^2 which extends the product partial order. (That is, if $\alpha \leq \beta$ and $\gamma \leq \delta$, then $\langle \alpha, \gamma \rangle \preccurlyeq \langle \beta, \delta \rangle$.) For simply let

$$\langle \omega \cdot i_0 + j_0, \omega \cdot k_0 + \ell_0 \rangle \preccurlyeq \langle \omega \cdot i_1 + j_1, \omega \cdot k_1 + \ell_1 \rangle \Longleftrightarrow$$
$$(2^{i_0} \cdot 3^{k_0} < 2^{i_1} \cdot 3^{k_1} \vee (2^{i_0} \cdot 3^{k_0} = 2^{i_1} \cdot 3^{k_1} \wedge 2^{j_0} \cdot 3^{\ell_0} \leq 2^{j_1} \cdot 3^{\ell_1})).$$

Notice that $\preccurlyeq \restriction (\omega \cdot n \times \omega \cdot n)$ has order type $\omega \cdot (n^2)$.

Now let $A = \text{Diff}_{\alpha<\omega\cdot n} A_\alpha$ and $B = \text{Diff}_{\alpha<\omega\cdot n} B_\alpha$, where the A_α's and B_α's are Σ^1_1. Fix an $\alpha < \omega \cdot (n^2)$, and let $\langle \beta, \gamma \rangle$ be the αth element of $\omega \cdot n \times \omega \cdot n$ under $\preccurlyeq$. If β and γ are both odd, set $C_{2\alpha} = \varnothing$ and $C_{2\alpha+1} = A_\beta \cap B_\gamma$; otherwise set $C_{2\alpha} = A_\beta \cap B_\gamma$ and $C_{2\alpha+1} = \varnothing$. We claim that $A \cap B = \text{Diff}_{\alpha<\omega\cdot(n^2)} C_\alpha$. For example, let $x \in A \cap B$. Let β be least so that $x \in A_\beta$ and γ be least so that

The Cabal Seminar. Volume I: Games, Scales and Suslin Cardinals
Edited by A. S. Kechris, B. Löwe, J. R. Steel
Lecture Notes in Logic, 31

$x \in B_\gamma$. Then β and γ are odd, so $x \in C_{2\alpha+1}$, where $\langle \alpha, \beta\rangle$ is the αth element of $\omega \cdot n \times \omega \cdot n$ under $\preccurlyeq$. If $x \in C_\delta$ for some $\delta < 2\alpha + 1$, then $x \in A_{\beta'} \cap B_{\gamma'}$ for some $\langle \beta', \gamma'\rangle \preccurlyeq \langle \beta, \gamma\rangle$. But $\beta \leq \beta'$ and $\gamma \leq \gamma'$ by the definitions of β and γ, so $\langle \beta, \gamma\rangle \preccurlyeq \langle \beta', \gamma'\rangle$, a contradiction. Thus $x \notin C_\delta$ for $\delta < 2\alpha + 1$, so that $x \in \text{Diff}_{\alpha<\omega\cdot(n^2)} C_\alpha$. It is equally easy to check that $\text{Diff}_{\alpha<\omega\cdot(n^2)} C_\alpha \subseteq A \cap B$. thus $A \cap B$ is $(\omega \cdot (n^2))$-Π^1_1. ⊣

With a little care, one can show that the intersection of two $(\omega \cdot n)$-Π^1_1 sets is $(\omega \cdot (2n - 1))$-Π^1_1.

Since the complement of an α-Π^1_1 set is $(\alpha+2)$-Π^1_1, the class $\Lambda = \bigcup_{n<\omega}((\omega \cdot n)$-$\Pi^1_1)$ is closed under complement, intersection and union. This means that "interweaving" Λ-norms produces a Λ-norm. In particular, interweaving the norms of the Busch scale on a Σ^1_1 set produces a very good scale on that set with norms in Λ. We give now a direct construction of such a scale.

THEOREM 1.2. Every Σ^1_1 set A admits a very good scale $\langle \varphi_n \rangle$ such that for all n,

$$\varphi_n \colon A \to \omega \cdot (n + 1)$$

and

$$\varphi_n \text{ is an } (\omega \cdot (n + 1))\text{-}\Pi^1_1 \text{ norm.}$$

PROOF. Let $A = \mathrm{p}[T]$, where T is a recursive tree on $\omega \times \omega$. We may assume that if $\langle \sigma, \tau \rangle \in T$, then $\forall i < \mathrm{lh}(\sigma) \exists k (\tau(i) = 2^{\sigma(i)} \cdot 3^k)$, so that any branch of $T(x)$ codes x. Let $\langle \sigma_i : i \leq n \rangle$ be a recursive enumeration of ${}^{<\omega}\omega$ such that for all n, $\{\sigma_i : i \leq n\}$ is a tree. For $n < \omega$, let

$$E_n = \{\tau \in {}^{<\omega}\omega : \tau \notin \{\sigma_i : i \leq n\} \wedge \tau{\restriction}(\mathrm{lh}(\tau) - 1) \in \{\sigma_i : i \leq n\}\}.$$

For $x \in A$ and $n < \omega$, let

$$\tau^x_n = \text{unique } \tau \in E_n (\tau \subseteq f^T_x)$$

and

$$\varphi_n(x) = \text{the ordinal rank of } \tau^x_n \text{ in } \leq_{\text{lex}}{\restriction}E_n.$$

Notice that $\leq_{\text{lex}}{\restriction}E_n$ has order type $\omega \cdot (n + 1)$, so that $\varphi_n \colon A \to \omega \cdot (n + 1)$. We now check that $\langle \varphi_n \rangle$ is a very good scale on A. Clearly, if $x \in A$ and $n \leq m$, then $\tau^x_n \subseteq \tau^x_m$. So if $x, y \in A$ and $\tau^x_n <_{\text{lex}} \tau^y_n$, then $\tau^x_m <_{\text{lex}} \tau^y_m$ for $m \geq n$, as required by the refinement property of very goodness. Now let $\langle x_i : i < \omega \rangle$ be a sequence of elements of A such that for all $n < \omega$, the sequence $\langle \varphi_n(x_i) : i < \omega \rangle$ is eventually constant, say with constant value (the ordinal rank in $\leq_{\text{lex}}{\restriction}E_n$ of) τ_n. Then for any n, $\tau_n = \tau^{x_i}_n$ and $\tau_{n+1} = \tau^{x_i}_{n+1}$ for sufficiently large i, so $\tau_n \subseteq \tau_{n+1}$. But $\tau_n \notin \{\sigma_i : i \leq n\}$, so $\mathrm{lh}(\tau_n) \to \infty$ as $n \to \infty$. Let $f = \bigcup_n \tau_n$. Now for any n, $\langle x_i{\restriction}n, f{\restriction}n \rangle \in T$ for sufficiently large i, so by our assumption on T, $\lim_{i\to\infty} x_i$ exists. Let $x = \lim_{i\to\infty} x_i$. Then

$\langle x, f\rangle \in [T]$, so $x \in A$; moreover $f_x^T \leq_{\text{lex}} f$, so $\tau_n^x \leq_{\text{lex}} \tau_n$ for all n. Thus $\langle \varphi_n\rangle$ is a very good scale on A.

It is easy to compute that φ_n is an $(\omega \cdot (n+1))$-Π_1^1 norm. ⊣

The universal Σ_1^1 set admits no very good scale whose norms are all $(\omega \cdot n)$-$\underset{\sim}{\Pi}_1^1$ for some fixed n, at least if $x^\#$ exists for all reals x. For if it did, the Third Periodicity Theorem [Mos80, p. 335] would give an $x \in {}^\omega\omega$ so that every $\Sigma_1^1(x)$ game has a winning strategy which is $\Game((\omega \cdot n)$-$\Pi_1^1(x))$. But every $\Game((\omega \cdot n)$-$\Pi_1^1(x))$ real is recursive in the type of the first $n+1$ indiscernibles for $\mathbf{L}[x]$, and hence a member of $\mathbf{L}[x]$, by Martin's analysis of $(\omega \cdot n)$-Π_1^1 games. This is impossible since there are $\Sigma_1^1(x)$ games without winning strategies in $\mathbf{L}[x]$.

THEOREM 1.3. *Let Λ be the class of sets which are $(\omega \cdot n)$-Π_1^1 for some $n < \omega$. Then every set in Λ admits a very good scale whose norms are in Λ.*

PROOF. Let $A = \text{Diff}_{\alpha<\omega\cdot n} A_\alpha$ where the A_α sets are Σ_1^1. For $\alpha < \omega \cdot n$, let $\langle \varphi_i^\alpha\rangle$ be a scale on A_α as given in Theorem 1.2 (or in [Bus76]), and let $\langle \vartheta_i^\alpha\rangle$ be a very good Π_1^1 scale on $\{x : \forall \beta < \alpha(x \notin A_\beta)\}$. Let $\langle F_i : i < \omega\rangle$ be an increasing sequence of finite sets whose union is $\{\alpha < \omega \cdot n : \alpha \text{ is odd}\}$. For $x \in A$, let

$$\alpha_x = \min\{\beta : x \in A_\beta\},$$

and let

$$\psi_i(x) = \begin{cases} \langle \alpha_x\rangle & \text{if } \alpha_x \notin F_i, \\ \langle \alpha_x, \varphi_0^{\alpha_x}(x), \vartheta_0^{\alpha_x}(x), \cdots, \varphi_i^{\alpha_x}(x), \vartheta_i^{\alpha_x}(x)\rangle & \text{if } \alpha_x \in F_i. \end{cases}$$

Here, of course, $\psi_i(x)$ is identified with an ordinal using $\leq_{\text{lex}} \restriction \omega_1^{\leq 2i+1}$.

It is routine to check that $\langle \psi_i\rangle$ is a very good scale on A. We shall check that ψ_i is a Λ norm; for notational simplicity, we consider only $i = 0$. First, notice that there is a relation $R \in \Lambda$ so that for $x, y \in A$, $R(x, y)$ iff $\alpha_x < \alpha_y$. For simply let

$$B_{2\cdot\alpha} = \{\langle x, y\rangle : y \in A_\alpha\}$$

and

$$B_{2\cdot\alpha+1} = \{\langle x, y\rangle : x \in A_\alpha\}$$

for $\alpha < \omega \cdot n$, and then set $R = \text{Diff}_{\alpha<\omega\cdot n} B_\alpha$. Let $E(x, y)$ iff $\neg(R(x, y) \vee R(y, x))$, so that $E \in \Lambda$ and for $x, y \in A$, $E(x, y) \iff \alpha_x = \alpha_y$. Now let

$$x \leq_0 y \iff x \in A \wedge (y \notin A \vee \psi_0(x) \leq \psi_0(y)),$$
$$x <_0 y \iff x \in A \wedge (y \notin A \vee \psi_0(x) < \psi_0(y)).$$

Then

$$x \leq_0 y \iff (x \in A \wedge (y \notin A \vee (y \in A \wedge (R(x,y) \vee (E(x,y) \wedge \bigwedge_{\beta \in F_0} (x \in A_\beta \setminus \bigcup_{\gamma<\beta} A_\gamma \implies \langle \psi_0^\beta(x), \vartheta_0^\beta(x) \rangle \leq_{\text{lex}} \langle \psi_0^\beta(y), \vartheta_0^\beta(y) \rangle))))))).$$

Our Lemma 1.1 implies that $\leq_0 \in \Lambda$. Similarly $<_0 \in \Lambda$, so that ψ_0 is a Λ norm. ⊣

According to the Third Periodicity Theorem, if every Γ set has a very good scale with norms in Γ, then every $\Game\Gamma$ set has a (very good) scale with norms in Γ. So we have

COROLLARY 1.4. Let $\Lambda = \bigcup_n (\omega \cdot n)$-$\Pi^1_1$. Then for all $k < \omega$, every $\Game^k\Lambda$ set admits a very good scale with norms in $\Game^k\Lambda$.

In particular, every $\Pi^1_2(= \Game\Sigma^1_1)$ set admits a scale with $\Game\Lambda$ norms. The Martin-Solovay scale [Kec78A, Theorem 6.3] also has this property. In fact, if one propagates the scale of Theorem 1.2 to Π^1_2 via the second periodicity construction, one obtains the Martin-Solovay scale. (Martin and the author proved this independently.) One can see this by looking at the auxiliary closed games associated to the $(\omega \cdot n)$-Π^1_1 games involved in propagating the scale of Theorem 1.2.

Theorem 1.3 generalizes easily to the class $\Lambda_\alpha = \bigcup_n (\omega^2 \cdot \alpha + (\omega \cdot n)$-$\underset{\sim}{\Pi}^1_1)$, where $\alpha < \omega_1$.

REFERENCES

DOUGLAS R. BUSCH

[Bus76] *λ-Scales, κ-Souslin sets and a new definition of analytic sets*, ***The Journal of Symbolic Logic***, vol. 41 (1976), p. 373.

ALEXANDER S. KECHRIS

[Kec78A] AD *and projective ordinals*, In Kechris and Moschovakis [CABAL i], pp. 91–132.

ALEXANDER S. KECHRIS AND YIANNIS N. MOSCHOVAKIS

[CABAL i] *Cabal seminar* 76–77, Lecture Notes in Mathematics, no. 689, Berlin, Springer, 1978.

YIANNIS N. MOSCHOVAKIS

[Mos80] ***Descriptive set theory***, Studies in Logic and the Foundations of Mathematics, no. 100, North-Holland, Amsterdam, 1980.

DEPARTMENT OF MATHEMATICS
UNIVERSITY OF CALIFORNIA
BERKELEY, CA 94720, USA
E-mail: steel@math.berkeley.edu

INDUCTIVE SCALES ON INDUCTIVE SETS

YIANNIS N. MOSCHOVAKIS

Let $\mathfrak{X} = X_1 \times \cdots \times X_n$ be any product of copies of ω and $\mathbb{R} = {}^\omega\omega$. A **pointset** $P \subseteq \mathfrak{X}$ is **inductive** if there is a projective set $Q \subseteq \mathfrak{X} \times \mathbb{R}$ such that

$$P(x) \iff ((\forall\alpha_1)(\exists\alpha_2)(\forall\alpha_3)(\exists\alpha_4)\dots)(\exists n)Q(x, \ulcorner\alpha_1, \dots, \alpha_n\urcorner), \quad (1)$$

where $\ulcorner\alpha_1, \dots, \alpha_n\urcorner$ is the usual (recursive) coding of tuples and the (open) **game quantifier** is interpreted in the obvious way. If Q is analytical in some $\beta_0 \in \mathbb{R}$, we call P **inductive in** β_0 and if we can choose β_0 recursive, we call P **absolutely inductive**. Sets which are both inductive and coinductive are **hyperprojective**; "**hyperprojective in** β_0" and "**absolutely hyperprojective**" sets are defined in the obvious way.

The theory of inductive sets (on arbitrary structures) is developed in some detail in [Mos74A]. Our purpose here is to outline a proof of the following:

Main Theorem. If every hyperprojective game is determined, then every absolutely inductive pointset admits an absolutely inductive scale; it follows that inductive sets admit inductive scales and hyperprojective sets admit hyperprojective scales.

Part of the interest in this result lies in the fact that the collection of inductive sets is the largest collection of pointsets for which we can presently establish the scale property, from any hypotheses.

§1. Proof of the Main Theorem. We assume that the tupling function $\ulcorner\alpha_1, \dots, \alpha_n\urcorner$ is defined for the empty tuple ($n = 0$),

$$\ulcorner\ \urcorner = t \mapsto 1$$

and that **concatenation** is given by a recursive function $^\frown$ on the codes,

$$\ulcorner\alpha_1, \dots, \alpha_n\urcorner {}^\frown \ulcorner\beta_1, \dots, \beta_m\urcorner = \ulcorner\alpha_1, \dots, \alpha_n, \beta_1, \dots, \beta_m\urcorner.$$

We will prove the main theorem in a sequence of simple lemmas.

Generalizing slightly the definition (1) above, for any given $Q \subseteq \mathfrak{X} \times \mathbb{R}$, put

$$R(x, \alpha) \iff ((\forall\alpha_1)(\exists\alpha_2)(\forall\alpha_3)(\exists\alpha_4)\dots)(\exists n)Q(x, \alpha {}^\frown \ulcorner\alpha_1, \dots, \alpha_n\urcorner) \quad (2)$$

The Cabal Seminar. Volume I: Games, Scales and Suslin Cardinals
Edited by A. S. Kechris, B. Löwe, J. R. Steel
Lecture Notes in Logic, 31

and for each ordinal ξ define by induction

$$R_\xi(x,\alpha) \iff Q(x,\alpha) \vee (\forall\beta)(\exists\gamma)(\exists\eta < \xi)R_\eta(x,\alpha^\frown\ulcorner\beta,\gamma\urcorner). \tag{3}$$

LEMMA 1.1. $R(x,\alpha) \iff (\exists\xi)R_\xi(x,\alpha)$.

PROOF. First check by a simple induction on ξ that

$$R_\xi(x,\alpha) \Longrightarrow R(x,\alpha).$$

For the converse, assume $(\forall\xi)\neg R_\xi(x,\alpha)$ and show by applying (3) repeatedly that in that case

$$((\exists\alpha_1)(\forall\alpha_2)(\exists\alpha_3)(\forall\alpha_4)\ldots)(\forall_n)\neg Q(x,\alpha^\frown\ulcorner\alpha_1,\ldots,\alpha_n\urcorner)$$

which is equivalent to $\neg R(x,\alpha)$. ⊣

It follows from results of [Mos74A] that if Q is analytical (or projective), then

$$R(x,\alpha) \iff (\exists\xi < \kappa)R_\xi(x,\alpha),$$

where κ is **the closure ordinal** for positive elementary inductive definitions on $\mathbb{R}$, or alternatively,

$$\kappa = \sup\{\text{rank}(\preceq) : \preccurlyeq \text{ is a hyperprojective prewellordering of } \mathbb{R}\}.$$

Suppose now that R and R_ξ are defined by (2) and (3) and suppose we are given a scale $\vec{\varphi}^0 = \langle\varphi_n^0 : n \in \omega\rangle$ on $Q = R_0$. We will define by induction on ξ a scale $\vec{\varphi}^\xi$ on each R_ξ.

Assuming that $\vec{\varphi}^\eta$ has been defined for each $\eta < \xi$, define first on

$$R^{<\xi} = \bigcup_\eta R_\eta$$

the sequence of norms $\vec{\psi}^\xi = \langle\psi_n^\xi : n \in \omega\rangle$ by

$$\psi_0^\xi(x,\alpha) = \min\{\eta : R_\eta(x,\alpha)\}, \tag{4}$$

$$\psi_{n+1}^\xi(x,\alpha) = \ulcorner\zeta,\varphi_n^\zeta(x,\alpha)\urcorner, \text{ where } \zeta = \psi_0^\xi(x,\alpha). \tag{5}$$

LEMMA 1.2. If each $\vec{\varphi}^\eta$ is a scale on $R_\eta (\eta < \xi)$, then (4) and (5) define a scale $\vec{\psi}^\xi$ on $R^{<\xi}$.

PROOF. If $\langle x_i,\alpha_i\rangle \to \langle x,\alpha\rangle$ and all norms $\psi_n^\xi(x_i,\alpha_i)$ are ultimately fixed, then in particular

$$\zeta = \psi_0^\xi(x_i,\alpha_i) = \min\{\eta : R_\eta(x_i,\alpha_i)\}$$

is ultimately fixed; hence $R_\zeta(x,\alpha)$ holds and for all n we have

$$\varphi_n^\zeta(x,\alpha) \le \varphi_n^\zeta(x_i,\alpha_i) \text{ (for all large } i). \tag{6}$$

Now if $\psi_0^\xi(x,\alpha) = \zeta$, then (6) implies directly that for all n, $\psi_{n+1}^\xi(x,\alpha) = \ulcorner\zeta, \varphi_n^\zeta(x,\alpha)\urcorner \leq \ulcorner\zeta, \varphi_n^\zeta(x_i,\alpha_i)\urcorner = \psi_{n+1}^\xi(x_i,\alpha_i)$ for all large i. If on the other hand $\psi_0^\xi(x,\alpha) = \zeta' < \zeta$, then the lexicographic ordering used in (5) implies that

$$\psi_{n+1}^\xi(x,\alpha) = \ulcorner\zeta', \varphi_n^{\zeta'}(x,\alpha)\urcorner < \ulcorner\zeta, \varphi_n^\zeta(x_i,\alpha_i)\urcorner = \psi_{n+1}^\xi(x_i,\alpha_i),$$

so that in either case, $\vec{\psi}^\xi$ has the critical lower semicontinuity property of scales. ⊣

To extend the construction and define a scale on R_ξ from given scales on all the $R_\eta, \eta < \xi$, we need the determinacy hypothesis.

Lemma 1.3. *Assume that every hyperprojective game is determined and that both $R^{<\xi}$ and the scale $\vec{\psi}^\xi$ on $R^{<\xi}$ constructed above is hyperprojective; then we can construct a hyperprojective scale on R_ξ.*

Proof. This is immediate from [KM78B, 3.2.2] and [KM78B, 3.3.1], taking Γ = all hyperprojective sets and using the rather obvious fact that Γ is adequate and closed under $\exists^{\mathbb{R}}, \forall^{\mathbb{R}}$. ⊣

To complete the construction, we now put a scale $\vec{\psi}^\kappa$ on

$$R = R^{<\kappa}$$

by taking $\xi = \kappa$ in Lemma 1.2.

To prove the main theorem, it will be enough to show that if P is defined from an analytical pointset Q by (1) and R is defined by (2), then this construction actually gives an inductive scale $\vec{\psi}^\kappa$ on R, since we can then obviously get an inductive scale on P by setting

$$\varphi_n(x) = \psi_n^\kappa(x, \ulcorner\ \urcorner).$$

The proof naturally rests on some elementary properties of inductive and hyperprojective sets, all proved in [Mos74A].

To begin with, the collection of absolutely inductive sets is adequate in the sense of [KM78B] and it is closed under $\exists^\omega, \forall^\omega, \exists^{\mathbb{R}}, \forall^{\mathbb{R}}$—this is almost obvious. The hyperprojective sets are additionally closed under $\neg$ and continuous substitutions.

It is a bit less obvious that *the inductive sets can be parameterized in* $\mathbb{R}$ *by absolutely inductive universal sets*; *i.e.*, for each $\mathfrak{X}$, there is an absolutely inductive $G^{\mathfrak{X}} = G \subseteq \mathbb{R} \times \mathfrak{X}$ such that for $P \subseteq \mathfrak{X}$,

$$P \text{ is inductive} \iff \text{for some } \alpha \in \mathbb{R}, P = G_\alpha = \{x : G(\alpha, x)\}.$$

This can be proved by checking that we can always take Q to be $\undertilde{\Sigma}_2^1$—see [Mos74A, 5D.2].

The most important fact about inductive sets that we need is the **Stage Comparison Theorem**, [Mos74A, 2A.2]:

THEOREM 1.4 (Stage Comparison Theorem). If Q is analytical in (2), then the obvious norm on R given by

$$\varphi(x, \alpha) = \min\{\xi : R_\xi(x, \alpha)\}$$

is an absolutely inductive norm.

In particular, the pointclasses of absolutely inductive and coinductive sets have **the prewellordering property**; see [KM78B, 2A].

For each $\mathfrak{X}$, fix an absolutely inductive $G^{\mathfrak{X}} \subseteq \mathbb{R} \times \mathfrak{X}$ which parametrizes the inductive subsets of $\mathfrak{X}$ as above and put

$$\ulcorner\alpha, \beta\urcorner \in C \iff (\forall x \in \mathfrak{X})(G_\alpha(x) \iff \neg G_\beta(x)).$$

Each $\gamma = \ulcorner\alpha, \beta\urcorner \in C$ **codes** a hyperprojective subset of $\mathfrak{X}$,

$$H_\gamma = G_\alpha = \mathfrak{X} \setminus G_\beta$$

and every hyperprojective subset of $\mathfrak{X}$ receives at least one code in C. It is not very hard to check that *if the system $\{G^{\mathfrak{X}}\}$ of universal sets is chosen carefully, then the hyperprojective sets are effectively closed in this coding under every operation which preserves both absolute inductiveness and absolute co-inductiveness*; for example, there is a recursive function

$$f : \mathbb{R} \times \mathbb{R} \to \mathbb{R}$$

such that if $\gamma, \delta \in C$, then

$$\varepsilon = f(\gamma, \delta) \in C$$

and

$$H_\varepsilon = H_\gamma \cap H_\delta.$$

The key to this argument is to choose universal sets so that for suitable recursive functions $S^{\mathfrak{X},\mathfrak{Y}} = S$,

$$G(\alpha, x, y) \iff G(S(\alpha, x), y)$$

and such that in addition,

$$P \text{ is absolutely inductive} \implies P = G_\alpha \text{ with a recursive } \alpha.$$

We get effective closure under $\wedge$ (as an example) using such universal sets by setting

$$P_0(\gamma, \delta, x) \iff G((\gamma)_0, x) \wedge G((\delta)_0, x),$$
$$P_1(\gamma, \delta, x) \iff G((\gamma)_1, x) \vee G(\delta)_1, x),$$

choosing recursive $\varepsilon_0, \varepsilon_1$ so that

$$P_0(\gamma, \delta, x) \iff G(\varepsilon_0, \gamma, \delta, x) \iff G(S(\varepsilon_0, \gamma, \delta), x)$$
$$P_1(\gamma, \delta, x) \iff G(\varepsilon_1, \gamma, \delta, x) \iff G(S(\varepsilon_1, \gamma, \delta), x)$$

and taking

$$f(\gamma,\delta) = \ulcorner S(\varepsilon_0,\gamma,\delta), S(\varepsilon_1,\gamma,\delta)\urcorner.$$

It is simple to compute that if $\gamma, \delta \in C$ and $\varepsilon = f(\gamma,\delta)$, then $\varepsilon \in C$ and $H_\varepsilon = H_\gamma \cap H_\delta$.

One can always find "good" universal sets with these properties—see [Mos74A, 9C.7].

After these preliminaries we can now describe the main construction for the proof.

If $\vec{\varphi} = \langle \varphi_n : n \in \omega \rangle$ is a scale on a set $A \subseteq \mathfrak{X}$, then a (hyperprojective) **code for** $\vec{\varphi}$ is any $\alpha \in C$ which codes a set $H_\alpha \subseteq \omega \times \mathfrak{X} \times \mathfrak{X}$ such that

$$y \in A \Longrightarrow (\forall x)((x \in A \wedge \varphi_n(x) \leq \varphi_n(y)) \Longleftrightarrow (n, x, y) \in H_\alpha).$$

Lemma 1.5. Assume that every hyperprojective game is determined and that

$$R(x,\alpha) \Longleftrightarrow ((\forall \alpha_1)(\exists \alpha_2)(\forall \alpha_3)(\exists \alpha_4)\dots)(\exists_n)Q(x, \alpha^\frown \ulcorner \alpha_1 \dots \alpha_n \urcorner)$$

with Q analytical, let

$$R_\xi(x,\alpha) \Longleftrightarrow (\forall \beta)(\exists \gamma)(Q(x,\alpha) \vee (\exists \eta < \xi)R_\eta(x, \alpha^\frown \ulcorner \beta, \gamma \urcorner))$$

as above and for $\langle x, \alpha \rangle \in R$, put

$$|x,\alpha| = \min\{\xi : R_\xi(x,\alpha)\}.$$

For each ξ, there is a scale $\vec{\varphi}^\xi = \langle \varphi_n^\xi : n \in \omega \rangle$ on R_ξ such that for some recursive function

$$\pi \colon \mathfrak{X} \times \mathbb{R} \to \mathbb{R},$$

if $R(x,\alpha)$ and $|x,\alpha| = \xi$, then $\pi(x,\alpha)$ is a code of $\vec{\varphi}^\xi$.

In particular, each $\vec{\varphi}^\xi$ is a hyperprojective scale on R_ξ.

Proof. The scales $\vec{\varphi}^\xi$ will be exactly those defined in Lemmas 1.2 and 1.3—we will have to make sure that the hypothesis of Lemma 1.3 is satisfied.

Since the definition of $\vec{\varphi}^\xi$ is by transfinite induction and we want π to be recursive, we must apply Kleene's well-known method of definition of **effective transfinite induction**, using the **recursion theorem**; *i.e.*, we will set

$$\pi(x,\alpha) = \{\varepsilon_0\}(x,\alpha)$$

where ε_0 is a fixed recursive real satisfying

$$\{\varepsilon_0\}(x,\alpha) = f(\varepsilon_0, x, \alpha)$$

with a recursive (possibly partial) $f \colon \mathbb{R} \times \mathfrak{X} \times \mathbb{R} \to \mathbb{R}$. (The necessary coding machinery for recursive partial functions on the reals can be found in [Mos70A] or [Mos80].)

Now the key to the construction is the fact that the transfer theorems for scales [KM78B, 3.2.2] and [KM78B, 3.3.1] have effective proofs. Combined with the remarks about effective closure of the hyperprojective sets above, this implies easily that there are recursive functions $f_1(\alpha,\gamma), f_2(\alpha,\gamma)$ such that *if* α *happens to code a scale on* $H_\gamma (\gamma \in C)$, *then* $f_1(\alpha,\gamma)$ *codes a scale on* $\exists^{\mathbb{R}} H_\gamma$ *and* $f_2(\alpha,\gamma)$ *codes a scale on* $\forall^{\mathbb{R}} H_\gamma$. Considering the positive analytical definition of R_ξ from $R^{<\xi}$, this in turn implies that *there is a recursive function* $f_3(\alpha,\gamma)$ *such that if* $R^{<\xi} = H_\alpha$ *and* γ *happens to code a scale on* $R^{<\xi}$, *then* $f_3(\alpha,\gamma)$ *codes a scale on* R_ξ. (Notice that the hypothesis here implies that both $R^{<\xi}$ and the scale on it are hyperprojective, so the appropriate games in Lemma 1.3 are determined.)

At the same time, the Stage Comparison Theorem implies easily that for a suitable recursive $f_4(x,\alpha)$, *if* $R(x,\alpha)$ *holds and* $|x,\alpha| = \xi$, *then* $f_4(x,\alpha)$ *is a hyperprojective code of* $R^{<\xi}$. Finally, looking at (4) and (5), it is not hard to believe that there is a further recursive function $f_5(\varepsilon,x,\alpha)$ with the following property: *if there are scales* $\vec{\varphi}^\eta$ *on* R_η *for each* $\eta < \xi$ *and if for each* $\langle x',\alpha'\rangle \in R$ *such that* $|x',\alpha'| = \eta < \xi$, $\{\varepsilon\}(x',\alpha')$ *is defined and codes* $\vec{\varphi}^\eta$, *then* $f_5(\varepsilon,x,\alpha)$ *codes the scale* $\vec{\psi}^\xi$ *defined from the* $\vec{\varphi}^\eta$ *by* (4), (5).

Now choose a recursive $\varepsilon^* \in R$ such that

$$\begin{aligned} f_5(\varepsilon,x,\alpha) &= \{\varepsilon^*\}(\varepsilon,x,\alpha) \\ &= \{S(\varepsilon^*,\varepsilon)\}(x,\alpha) \end{aligned}$$

using the appropriate recursive S for our coding of recursive partial functions and set

$$f(\varepsilon,x,\alpha) = f_3(f_4(x,\alpha), S(\varepsilon^*,\varepsilon)).$$

If we pick ε_0 by the recursion theorem and set

$$\pi(x,\alpha) = \{\varepsilon_0\}(x,\alpha) = f(\varepsilon_0,x,\alpha),$$

it is then trivial to verify by induction on ξ that π has the required property. ⊣

This lemma completes the proof of the main theorem, since in the scale $\vec{\psi}^\kappa$ we put on $R = R^{<\kappa}$ by (4) and (5) the first norm ψ_0^κ is absolutely inductive by the Stage Comparison Theorem and the later norms can be computed easily using the recursive function π.

§2. Corollaries and remarks. Of course the main corollary concerns uniformization.

COROLLARY 2.1. If every hyperprojective game is determined, then the collection of absolutely inductive pointsets has the uniformization property.

PROOF. See [KM78B, 3.1.1]. ⊣

To look a bit closer at one of the more interesting consequences of this uniformization result, let us go back and reconsider the open games on $\mathbb{R}$ that we used to define inductive sets. In a definition of the form

$$P(x) \iff ((\forall\alpha_1)(\exists\alpha_2)(\forall\alpha_3)(\exists\alpha_4)\ldots)(\exists n)Q(x,\ulcorner\alpha_1,\ldots,\alpha_n\urcorner)$$

we normally interpret the expression on the right as asserting the existence of a **winning strategy** for player II in the obvious game. This understanding of the definitions will force us to use the axiom of choice in the proof of Lemma 1.2 which is best avoided in the context of determinacy; instead we can reinterpret the infinite alternating string using **multiple-valued** strategies (**quasistrategies** in [Mos74A]) and if we do that, then the proof of the main theorem is easily given in ZF+DC.

On the other hand, once we have the theorem, we should point out that if $P(x)$ holds, then player II can win the game on the right using a (single-valued) strategy which is hyperprojective in x.

Corollary 2.2. Assume that every hyperprojective game is determined and suppose that

$$((\forall\alpha_1)(\exists\alpha_2)(\forall\alpha_3)(\exists\alpha_4)\dots)(\exists n)Q(\ulcorner\alpha_1,\dots,\alpha_n\urcorner),$$

where Q is hyperprojective in some β_0. Then there exists a function

$$\sigma : \mathbb{R} \to \mathbb{R}$$

which is hyperprojective in β_0 such that

$$((\forall\alpha_1)(\forall\alpha_3)(\forall\alpha_5)\dots)(\exists n)Q(\ulcorner\alpha_1,\alpha_2,\alpha_3,\alpha_4,\dots,\alpha_n\urcorner), \tag{7}$$

where for even n,

$$\alpha_n = \sigma(\ulcorner\alpha_1,\dots,\alpha_{n-1}\urcorner).$$

Proof. Put

$$P(\alpha,\beta) \iff ((\forall\alpha_{n+1})(\exists\alpha_{n+2})\dots)(\exists k)Q(\alpha^\frown\ulcorner\beta,\alpha_{n+1},\alpha_{n+2},\dots,\alpha_k\urcorner),$$

so that P is inductive in β_0. Let P^* uniformize P and set for even n

$$\sigma(\ulcorner\alpha_1,\dots,\alpha_{n-1}\urcorner) = \beta \iff P^*(\ulcorner\alpha_1,\dots,\alpha_{n-1}\urcorner,\beta) \qquad \dashv$$

A further consequence of this remark concerns the absoluteness properties of the partially playful universe associated with the inductive sets.

Corollary 2.3. Assume that every hyperprojective game is determined and let T be the tree on $\omega \times \kappa$ associated with some absolutely inductive scale on some universal inductive set, let $\mathbf{L}[T]$ be the associated model as in [KM78B]. If Q is analytical and for $x \in \mathbf{L}[T]$

$$((\forall\alpha_1)(\exists\alpha_2)(\forall\alpha_3)(\exists\alpha_4)\dots)(\exists n)Q(x,\ulcorner\alpha_1,\dots,\alpha_n\urcorner)$$

holds (in the world), then the relativization of this open game assertion to $\mathbf{L}[T]$ also holds.

Proof. $\mathbf{L}[T]$ is easily closed under functions hyperprojective in x, with $x \in \mathbf{L}[T]$. $\dashv$

The corollary is quite strong, as there are a lot of things that one can say with open game assertions—particularly if one assumes full determinacy and uses the coding Lemma 3 of [Mos70A].

The proof of the main theorem makes it clear that the result goes through for various pointclasses (included in the inductive sets) which enjoy some of the structural and closure properties of the inductive sets. In particular, the careful reader will notice that *closure of the inductive sets under* $\exists^{\mathbb{R}}$ *is not used in the proof*—although we did use (effective) closure of the hyperprojective sets under $\exists^{\mathbb{R}}$. Using this observation, anyone who is somewhat familiar with recursion in higher types (as presented in [KM77], for example) will verify easily

COROLLARY 2.4 (to the proof). If every game which is Kleene recursive in ${}^3E$ and some $\beta_0 \in \mathbb{R}$ is determined, then every pointset which is semirecursive in ${}^3E$ admits a scale which is semirecursive in ${}^3E$.

In particular, the collection of pointsets semirecursive in ${}^3E$ has the uniformization property (under this determinacy hypothesis).

In the other direction, it is hard to see how one could extend the class of sets now known to possess definable scales without using some entirely new principle of constructing scales. It appears that the question *whether coinductive sets admit definable scales* is one of the critical open problems in this theory.

REFERENCES

ALEXANDER S. KECHRIS AND YIANNIS N. MOSCHOVAKIS

[KM77] *Recursion in higher types*, ***Handbook of mathematical logic*** (K. J. Barwise, editor), North-Holland, 1977, pp. 681–737.

[CABAL i] *Cabal seminar* 76–77, Lecture Notes in Mathematics, no. 689, Berlin, Springer, 1978.

[KM78B] *Notes on the theory of scales*, this volume, originally published in *Cabal Seminar* 76–77 [CABAL i], pp. 1–53.

YIANNIS N. MOSCHOVAKIS

[Mos70A] *Determinacy and prewellorderings of the continuum*, ***Mathematical logic and foundations of set theory. Proceedings of an international colloquium held under the auspices of the Israel Academy of Sciences and Humanities, Jerusalem, 11–14 November 1968*** (Y. Bar-Hillel, editor), Studies in Logic and the Foundations of Mathematics, North-Holland, Amsterdam-London, 1970, pp. 24–62.

[Mos74A] ***Elementary induction on abstract structures***, North-Holland, 1974.

[Mos80] ***Descriptive set theory***, Studies in Logic and the Foundations of Mathematics, no. 100, North-Holland, Amsterdam, 1980.

DEPARTMENT OF MATHEMATICS
UNIVERSITY OF CALIFORNIA
LOS ANGELES, CA 90095, USA
E-mail: ynm@math.ucla.edu

SCALES ON COINDUCTIVE SETS

YIANNIS N. MOSCHOVAKIS

The main purpose of this note is to prove (under appropriate determinacy hypotheses) that *coinductive pointsets admit definable scales*. This solves the second Victoria Delfino problem [CABAL i, p. 279-282] and was announced in [MMS82], together with many stronger related results of Martin and Steel.

§1. Lemmas on preservation of scales. When this question was first posed in [Mos78], there was a feeling (certainly on my part) that its answer might require methods for constructing scales that would be radically different from the simple and well-understood ideas of the periodicity theorems. As it happens, this is not true: one can define a scale on a given coinductive set quite easily, by a judicious mix of the constructions in the second periodicity theorems [Mos80, 6C.1 & 6C.3]. In fact, the simplest way to explain the proof of this theorem is to reexamine the proofs of these two results and extract from them a bit more than was stated explicitly in [Mos80].

Suppose P is a pointset, i.e., $P \subseteq \mathfrak{X} = X_1 \times \cdots \times X_n$ with each $X_i = \omega$ or $X_i = {}^{\omega}\omega$. A **putative scale** on P is a sequence

$$\vec{\varphi} = \langle \varphi_i : i \in \omega \rangle$$

of norms on P, each norm

$$\varphi_i : P \to \mathrm{Ord}$$

mapping P into the ordinals.

A putative scale $\vec{\varphi}$ on P defines in a natural way a **notion of convergence** for sequences of points in P: we say that x_n **converges to** x **modulo** $\vec{\varphi}$ and write

$$x_n \to x \pmod{\vec{\varphi}}$$

if $\lim_{n\to\infty} x_n = x$ in the usual topology on $\mathfrak{X}$ and if in addition, for each fixed i, the sequence of ordinals $\varphi_i(x_0), \varphi_i(x_1), \varphi_i(x_2), \dots$ is ultimately constant.

We say that $\vec{\varphi}$ is a **semiscale** on P if for every sequence $\langle x_n : n \in \omega \rangle$ in P,

$$x_n \to x \pmod{\vec{\varphi}} \Longrightarrow x \in P.$$

During the preparation of this paper the author was partially supported by NSF Grant MCS78-02989.

The Cabal Seminar. Volume I: Games, Scales and Suslin Cardinals
Edited by A. S. Kechris, B. Löwe, J. R. Steel
Lecture Notes in Logic, 31

A norm $\varphi\colon P \to \mathrm{Ord}$ is **lower semicontinuous relative to the putative scale** $\vec{\varphi}$, if for each sequence $\langle x_n : n \in \omega\rangle$ in P,

$$x_n \to x \pmod{\vec{\varphi}} \Longrightarrow (x \in P \land \varphi(x) \le \lim_{n\to\infty} \varphi(x_n));$$

we call $\vec{\varphi}$ a **scale** if each φ_i is lower semicontinuous relative to $\vec{\varphi}$.

To give a topological interpretation of these definitions (for those familiar with **uniform spaces**), let $\tau(\vec{\varphi})$ be the topology on P generated by the neighbourhoods

$$N(x_0, k) = \{x : x\restriction k+1 = x_0\restriction k+1 \land \forall i \le k(\varphi_i(x) = \varphi_i(x_0))\}$$

relative to a putative scale $\vec{\varphi}$. It is easy to check that $\tau(\vec{\varphi})$ is a uniform topology in the sense of Kelley [Kel55] and that

$$\begin{aligned}\vec{\varphi} \text{ is a semiscale} &\iff P \text{ is complete},\\ \vec{\varphi} \text{ is a scale} &\iff \text{for each } i \text{ and } \lambda,\\ &\qquad\text{the "closed ball" } \{x : \varphi_i(x) \le \lambda\} \text{ is complete.}\end{aligned}$$

Of course, the theory of uniform spaces is tailored to the examples where the topology is determined by a sequence of pseudo-metrics with real values and is not useful in the present context of ordinal-valued norms.

If $\vec{\varphi}$ and $\vec{\varphi}'$ are both putative scales on P, we say that $\vec{\varphi}$ **is finer than** $\vec{\varphi}'$ if for every $x_0, x_1, \dots$ in P,

$$x_n \to x \pmod{\vec{\varphi}} \Longrightarrow x_n \to x \pmod{\vec{\varphi}'}.$$

For example, one can **refine** a putative scale $\vec{\varphi}$ by adding some norms or **interweaving**, *i.e.*, setting

$$\psi_n(x) = \ulcorner \varphi_0(x), \varphi_1(x), \dots, \varphi_n(x) \urcorner$$

where

$$\langle \xi_0, \dots, \xi_n\rangle \mapsto \ulcorner \xi_0, \dots, \xi_n \urcorner$$

is some order-preserving map (depending on n) of the $(n+1)$-tuples of ordinals (ordered lexicographically) into the ordinals.

Suppose that

$$Q(x) \iff \exists\alpha(P(x, \alpha))$$

and that we are given norms $\varphi_0, \dots, \varphi_n$ on P. We define the norm

$$\psi_n = \text{"inf"}\{\varphi_0, \dots, \varphi_n\}$$

on Q by

$$\psi_n(x) = \inf\{\ulcorner \varphi_0(x, \alpha), \alpha(0), \varphi_1(x, \alpha), \alpha(1), \dots, \varphi_n(x, \alpha), \alpha(n) \urcorner : P(x, \alpha)\}$$

where $\ulcorner \dots \urcorner$ again stands for an order-preserving map of $2n$-tuples into the ordinals. If $\vec{\varphi} = \langle \varphi_n : n \in \omega \rangle$ is a putative scale on P, then this definition associates with $\vec{\varphi}$ the infimum putative scale

$$\vec{\psi} = \text{“inf”}\, \vec{\varphi} = \langle \psi_n : n \in \omega \rangle$$

on Q—but *it is important that the definition of each* ψ_n *in* “inf” $\vec{\varphi}$ *depends only on* $\varphi_0, \dots, \varphi_n$ *and not on the whole sequence* $\vec{\varphi}$.

Now the proof of [Mos80, 6C.1] (stated there for “very good scales”) actually establishes the following result.

LEMMA 1.1 (The Infimum Lemma). Suppose $\vec{\varphi}$ is a putative scale on $P \subseteq \mathfrak{X} \times {}^{\omega}\omega$,

$$Q(x) \iff \exists \alpha (P(x, \alpha))$$

and $\vec{\psi} = \text{“inf”}\, \vec{\varphi}$.

1. If $\vec{\psi}'$ is any putative scale finer than $\vec{\psi}$ and if $\langle x_n : n \in \omega \rangle$ is any sequence of points in Q such that

   $$x_n \to x \pmod{\vec{\psi}'},$$

   then we can find a subsequence

   $$x_k^* = x_{n_k} \text{ with } (n_0 < n_1 < \dots)$$

   and irrationals $\alpha, \alpha_0, \alpha_1, \dots$, such that $P(x_k, \alpha_k)$ holds for each k and

   $$\langle x_k^*, \alpha_k \rangle \to \langle x, \alpha \rangle \pmod{\vec{\varphi}}.$$

2. If $\vec{\varphi}$ is a semiscale on P and for $i = 0, \dots, m$ each φ_i is lower semicontinuous relative to $\vec{\varphi}$, then ψ_m is lower semicontinuous relative to $\vec{\psi}$. In particular, if $\vec{\varphi}$ is a scale on P, then $\vec{\psi} = \text{“inf”}\, \vec{\varphi}$ is a scale on Q.

PROOF. For 1., notice that it is enough to consider the case $\vec{\psi} = \text{“inf”}\, \vec{\varphi}$ and choose n_k to be the least n such that for $i = 0, \dots, k$, and all $m \geq n$,

$$\psi_i(x_m) = \psi_i(x_n).$$

By definition,

$$\psi_i(x_k^*) = \ulcorner \varphi_0(x_k^*, \alpha_i^k), \alpha_i^k(0), \varphi_1(x_k^*, \alpha_i^k), \alpha_i^k(1), \dots, \varphi_i(x_k^*, \alpha_i^k), \alpha_i^k(i) \urcorner$$

for some α_i^k and for $k \geq i, \ell \geq i$,

$$\alpha_i^k(0) = \alpha_i(0), \dots, \alpha_i^k(i) = \alpha_i^\ell(i)$$

so we can take

$$\alpha_k = \alpha_k^k$$

and the result follows immediately.

2. is not hard to prove directly from the definition of the infimum norms and we will skip it. ⊣

We will also need the analogous modification of the second periodicity theorem [Mos80, 6C.3] which applies to arbitrary putative scales that need not be semiscales or scales.

Suppose then that

$$Q(x) \iff \forall \alpha P(x, \alpha)$$

and $\vec{\varphi}$ is a given putative scale on Q and let $u(0), u(1), \ldots$ be a canonical enumeration of all finite sequences of integers such that if $u(i)$ is a proper initial segment of $u(j)$, then $i < j$. As in [Mos80, 6C.3] (*but interweaving the norms in* $\vec{\varphi}$), we define for each $x, y \in \mathfrak{X}$ and each n a game $\mathrm{G}_n(x, y)$ where player I plays α', player II plays β', $\alpha = u(n)^\frown\alpha'$, $\beta = u(n)^\frown\beta'$, and

$$\text{Player II wins} \iff \ulcorner \varphi_0(x, \alpha), \ldots, \varphi_n(x, \alpha) \urcorner \leq \ulcorner \varphi_0(y, \beta), \ldots, \varphi_n(y, \beta) \urcorner.$$

The relation

$$x \leq_n y \iff x, y \in Q \wedge \text{player II wins } \mathrm{G}_n(x, y)$$

is a prewellordering (granting the determinacy of these games), so we can define on Q for each n the norm

$$\psi_n = \text{“sup”}\{\varphi_0, \ldots \varphi_n\}$$

by setting $\psi_n(x)$ to be the rank of x in the prewellordering $\leq_n$. The putative scale $\vec{\psi} = \langle \psi_n : n \in \omega \rangle$ on Q is the **(fake) supremum** putative scale associated with $\vec{\varphi}$ on P, in symbols

$$\vec{\psi} = \text{“sup”}\, \vec{\varphi}.$$

Notice again that the definition of each ψ_n in “sup” $\vec{\varphi}$ depends only on $\varphi_0, \ldots, \varphi_n$.

LEMMA 1.2 (The Fake Supremum Lemma). Suppose $\vec{\varphi}$ is a putative scale on $P \subseteq \mathfrak{X} \times {}^\omega\omega$,

$$Q(x) \iff \forall \alpha P(x, \alpha),$$

assume that each φ_i is a Γ-norm where Γ is an adequate pointclass satisfying $\mathsf{Det}(\underset{\sim}{\mathbf{\Delta}})$ and let $\vec{\psi} = \text{“sup”}\, \vec{\varphi}$.

1. If $\vec{\psi}'$ is any putative scale on Q which is finer than $\vec{\psi}$ and if $\langle x_n : n \in \omega \rangle$ is a sequence of points in Q such that

   $$x_n \to x \pmod{\vec{\psi}'},$$

   then for each α, there exists a subsequence

   $$x^*_k = x_{n_k} \text{ with } (n_0 < n_1 < \ldots)$$

   and irrationals $\alpha_0, \alpha_1, \ldots,$ such that

   $$\langle x^*_k, \alpha_k \rangle \to \langle x, \alpha \rangle \pmod{\vec{\varphi}}.$$

2. If $\vec{\varphi}$ is a semiscale on P and for $i = 0, \dots, m$ each φ_i is lower semicontinuous relative to $\vec{\varphi}$, then ψ_m is lower semicontinuous relative to $\vec{\psi}$.
In particular, if $\vec{\varphi}$ is a scale on P, then $\vec{\psi} =$ "sup" $\vec{\varphi}$ is a scale on Q.

Proof. The proof of this is exactly the proof of [Mos80, 6C.3] which was stated with additional hypotheses there. ⊣

§2. The main result. A pointset $Q \subseteq \mathfrak{X}$ is inductive if there is a projective $P \subseteq \mathfrak{X} \times {}^{\omega}\omega$ such that

$$Q(x) \iff ((\forall\alpha_0)(\exists\alpha_1)(\forall\alpha_2)\dots)(\exists t)P(x, \ulcorner\alpha_0, \dots, \alpha_{t-1}\urcorner). \tag{1}$$

This can be taken as the definition of inductive sets or it can be proved easily from the more natural definition in terms of **positive elementary induction on** ${}^{\omega}\omega$, see [Mos80, 7C]. (A quick summary of some of the basic properties of inductive sets can be found in [Mos78].)

For the proof in this paper, we will only need the following very simple normal form for inductive sets.

Lemma 2.1. A set $Q \subseteq \mathfrak{X}$ is inductive if and only if there is a number-theoretic relation R such that

$$Q(x) \iff ((\forall\alpha_0)(\exists\alpha_1)(\forall\alpha_2)\dots)(\exists t)R(\overline{x}(t), \ulcorner\overline{\alpha}_0(t), \dots, \overline{\alpha}_{t-1}\urcorner(t)); \tag{2}$$

consequently, Q is coinductive if for some R

$$Q(x) \iff ((\exists\alpha_0)(\forall\alpha_1)(\exists\alpha_2)\dots)(\forall t)R(\overline{x}(t), \ulcorner\overline{\alpha}_0(t), \dots, \overline{\alpha}_{t-1}\urcorner(t)).$$

Proof. Suppose, for example, that (1) above holds with some $P(x, \alpha)$ in $\underset{\sim}{\Sigma}^1_2$, so that

$$P(x, \alpha) \iff (\exists\beta)(\forall\gamma)(\exists s)R(\overline{x}(s), \overline{\alpha}(s), \overline{\beta}(s), \overline{\gamma}(s))$$

with a number-theoretic R. An easy game-argument shows that

$$\begin{aligned} Q(x) \iff & ((\forall\alpha_0)(\exists\alpha_1)(\exists u_1)(\forall\alpha_2)(\exists\alpha_3)(\exists u_3)\dots)(\exists t) \\ & (u_t = 1 \wedge (\forall i < t,\ i \text{ odd})(u_i = 0) \wedge \\ & \wedge \text{ if } s = \alpha_{t+2}(0), \\ & \qquad \text{then } R(\overline{x}(s), \ulcorner\overline{\alpha}(s), \dots, \overline{\alpha}_{t-1}(s)\urcorner, \overline{\alpha}_t(s), \overline{\alpha}_{t+1}(s))). \end{aligned} \tag{3}$$

(From a strategy that establishes $Q(x)$ by (1), get one that establishes $Q(x)$ by (3) and vice versa; notice that both (1) and (3) are interpreted via open games.) Now (3) gives a normal form for $Q(x)$ as in (2) (with a different R) by trivial recursive manipulations of the infinite quantifier string and the matrix. ⊣

Let $\underset{\sim}{\Sigma}_0^*$ be the pointclass of all Boolean combinations of inductive and coinductive pointsets, and for $n \geq 1$, by induction,

$$Q \in \underset{\sim}{\Sigma}_n^* \iff \text{for some } P \text{ in } \underset{\sim}{\Sigma}_{n-1}^*, \text{ and all } x,\ Q(x) \iff (\exists\alpha)\neg P(x,\alpha).$$

THEOREM 2.2. If every game with payoff in $\bigcup_{n\in\omega}\underset{\sim}{\Sigma}_n^*$ is determined, then every coinductive pointset Q admits a scale $\vec{\psi} = \langle \psi_i : i \in \omega\rangle$, such that each ψ_i is a $\underset{\sim}{\Sigma}_i^*$-norm.

PROOF. Suppose by Lemma 2.1 that

$$Q(x) \iff ((\exists\alpha_0)(\forall\alpha_1)\dots)(\forall t)R(\overline{x}(t), \ulcorner\overline{\alpha}_0(t),\dots,\overline{\alpha}_{t-1}\urcorner(t))$$

and for each *even* n define

$$Q_n(x,\alpha_0,\dots,\alpha_{n-1}) \iff ((\exists\alpha_n)(\forall\alpha_{n+1})\dots)(\forall t)R(\overline{x}(t), \ulcorner\overline{\alpha}_0(t),\dots,\overline{\alpha}_{t-1}(t)\urcorner).$$

For *odd* n, let

$$Q_n(x,\alpha_0,\dots,\alpha_{n-1}) \iff (\forall\alpha_n)Q_{n+1}(x,\alpha_0,\dots,\alpha_{n-1},\alpha_n).$$

We will define a putative scale $\vec{\psi}^n = \{\psi_i^n : i \in \omega\}$ on each Q_n and then show that $\vec{\psi}^0$ on $Q = Q_0$ is actually a scale.

The definition of ψ_i^n is by induction on i, simultaneously for all n. We will use "vector notation" for tuples, *i.e.*,

$$\vec{\alpha} = \langle\alpha_0,\dots,\alpha_{n-1}\rangle, \vec{\beta} = \langle\beta_0,\dots,\beta_{n-1}\rangle.$$

CASE 1. n is even and $i = 0$.
If $Q_n(x,\vec{\alpha})$ holds, put

$$\psi_0^n(x,\vec{\alpha}) = 0.$$

This norm just records (by being defined) that $Q_n(x,\vec{\alpha})$ holds. It is a $\underset{\sim}{\Sigma}_0^*$-norm since all Q_n are clearly coinductive and (in the notation of [Mos80, 4B]),

$$\langle x,\vec{\alpha}\rangle \leq^*_{\psi_0^n} \langle y,\vec{\beta}\rangle \iff Q_n(x,\vec{\alpha}),$$
$$\langle x,\vec{\alpha}\rangle <^*_{\psi_0^n} \langle y,\vec{\beta}\rangle \iff Q_n(x,\vec{\alpha}) \wedge \neg Q_n(y,\vec{\beta}).$$

It is not hard to check that in general we cannot expect ψ_0^n to be either inductive or coinductive.

CASE 2. n is odd.
Now

$$Q_n(x,\vec{\alpha}) \iff (\forall\alpha_{n+1})Q_{n+1}(x,\vec{\alpha},\alpha_{n+1})$$

and assuming that ψ_j^{n+1} is defined for $j \leq i$, we put

$$\psi_i^n = \text{"sup"}\{\psi_0^{n+1},\dots,\psi_i^{n+1}\}.$$

CASE 3. n is even and $i > 0$.
Now

$$Q_n(x,\vec{\alpha}) \iff (\exists\alpha_{n+1})Q_{n+1}(x,\vec{\alpha},\alpha_{n+1})$$

and assuming that ψ_j^{n+1} is defined for all $j < i$, we put

$$\psi_i^n = \text{“inf”}\{\psi_0^{n+1}, \dots, \psi_{i-1}^{n+1}\}.$$

This completes the definition of a putative scale $\vec{\psi}^n$ on each Q_n and the definability assertion in the theorem is immediate from the proofs of [Mos80, 6C.1 & 6C.3], by induction on i.

Notice that by the construction, immediately, for each even n, $\vec{\psi}^n$ is finer than “inf” $\vec{\psi}^{n+1}$ and for each odd n, $\vec{\psi}^n =$ “sup” $\vec{\psi}^{n+1}$, so the basic lemmas apply.

To check first that $\vec{\psi}^0$ is a semiscale on $Q = Q_0$, suppose

$$x_m \to x \pmod{\vec{\psi}^0};$$

to prove that $Q(x)$ holds, we must describe a strategy for $\exists$ to win the closed game which interprets

$$((\exists\alpha_0)(\forall\alpha_1)(\exists\alpha_2)\dots)(\forall t)R(\overline{x}(t), \ulcorner\overline{\alpha}_0(t), \dots, \overline{\alpha}_{t-1}(t)\urcorner).$$

By the Infimum Lemma, choose a subsequence $\langle x_m^0 : m \in \omega\rangle$ of $\langle x_m : m \in \omega\rangle$ and irrationals $\alpha_0, \alpha_{0,1}^0, \alpha_{0,2}^0, \dots$ such that

$$\langle x_m^0, \alpha_{0,m}^0\rangle \to \langle x, \alpha_0\rangle \pmod{\vec{\psi}^1}$$

and play this α_0. If your opponent responds with some α_1, choose by the Fake Supremum Lemma a subsequence $\langle\langle x_m^1, \alpha_{0,m}^1\rangle : m \in \omega\rangle$ of $\langle\langle x_m^0, \alpha_{0,m}^0\rangle : m \in \omega\rangle$ and irrationals $\langle\alpha_{1,m}^1 : m \in \omega\rangle$ such that

$$\langle x_m^1, \alpha_{0,m}^1, \alpha_{1,m}^1\rangle \to \langle x, \alpha_0, \alpha_1\rangle \pmod{\vec{\psi}^2}$$

and then apply the Infimum Lemma again to $\langle\langle x_m^1, \alpha_{0,m}^1, \alpha_{1,m}^1\rangle : m \in \omega\rangle$ to get a subsequence $\langle\langle x_m^2, \alpha_{0,m}^2, \alpha_{1,m}^2\rangle : m \in \omega\rangle$ and irrationals α_2 and $\langle\alpha_{2,m}^2 : m \in \omega\rangle$ such that

$$\langle x_m^2, \alpha_{0,m}^2, \alpha_{1,m}^2, \alpha_{2,m}^2\rangle \to \langle x, \alpha_0, \alpha_1, \alpha_2\rangle \pmod{\vec{\psi}^3}$$

and play this α_2.

It is clear that $\exists$ can continue to play in this manner indefinitely, so we have defined a strategy for him. To see that he wins, notice that by the construction, for each n we know that for each m,

$$Q_n(x_m^{n-1}, \alpha_{0,m}^{n-1}, \alpha_{1,m}^{n-1}, \dots, \alpha_{n-1,m}^{n-1})$$

so that by the definition of Q_n, taking $t = n$ in the matrix, we have for each m

$$R(\overline{x}_m^{n-1}(n), \ulcorner\overline{\alpha}_{0,m}^{n-1}(n), \dots, \overline{\alpha}_{n-1,m}^{n-1}(n)\urcorner);$$

letting $m \to \infty$, since

$$\langle x_m^{n-1}, \alpha_{0,m}^{n-1}, \dots, \alpha_{n-1,m}^{n-1}\rangle \to \langle x, \alpha_0, \dots, \alpha_{n-1}\rangle,$$

we have for each n

$$R(\overline{x}(n), \ulcorner\overline{\alpha}_0(n), \dots, \overline{\alpha}_{n-1}(n)\urcorner)$$

which is precisely the condition for $\exists$ to win. Thus $\vec{\psi}^0$ is a semiscale on Q_0.

Now the same argument shows easily that each $\vec{\psi}^n$ is a semiscale in Q_n, with just some added notation. To show that these are actually scales, notice first that for n even, ψ_0^n is surely semicontinuous since it is constant; the Infimum Lemma and the Fake Supremum Lemma then imply immediately that all ψ_i^n are lower semicontinuous relative to $\vec{\psi}^n$, by induction on i. ⊣

REFERENCES

Alexander S. Kechris and Yiannis N. Moschovakis

[Cabal i] *Cabal seminar* 76–77, Lecture Notes in Mathematics, no. 689, Berlin, Springer, 1978.

John Kelley

[Kel55] ***General topology***, The University series in higher mathematics, Van Nostrand, Princeton, NJ, 1955.

Donald A. Martin, Yiannis N. Moschovakis, and John R. Steel

[MMS82] *The extent of definable scales*, ***Bulletin of the American Mathematical Society***, vol. 6 (1982), pp. 435–440.

Yiannis N. Moschovakis

[Mos78] *Inductive scales on inductive sets*, this volume, originally published in Kechris and Moschovakis [Cabal i], pp. 185–192.

[Mos80] ***Descriptive set theory***, Studies in Logic and the Foundations of Mathematics, no. 100, North-Holland, Amsterdam, 1980.

DEPARTMENT OF MATHEMATICS
UNIVERSITY OF CALIFORNIA
LOS ANGELES, CA 90095, USA
E-mail: ynm@math.ucla.edu

THE EXTENT OF SCALES IN $\mathbf{L}(\mathbb{R})$

DONALD A. MARTIN AND JOHN R. STEEL

We shall show that every set of reals which is Σ_1-definable over $\mathbf{L}(\mathbb{R})$ from parameters in $\mathbb{R} \cup \{\mathbb{R}\}$ admits a scale in $\mathbf{L}(\mathbb{R})$, and in fact that the class of sets so definable has the scale property. Kechris and Solovay observed some time ago that no other sets admit scales in $\mathbf{L}(\mathbb{R})$, so that our result determines precisely the extent of scales in $\mathbf{L}(\mathbb{R})$.

Some preliminaries: We work in ZF + DC, and state our additional determinacy hypotheses as we need them. We let $\mathbb{R} = {}^{\omega}\omega$, the Baire space, and call its elements reals. Variables $z, y, x, \dots$ range over $\mathbb{R}$, while $\alpha, \beta, \gamma, \dots$ range over the class Ord of ordinals. Let * be a recursive homeomorphism from ${}^{\omega}\mathbb{R}$ to $\mathbb{R}$ such that $\langle x_n : n < \omega\rangle^* \restriction i$ depends only on $\langle x_n : n < i\rangle$. We use $\langle x_n : n < i\rangle^*$ to denote the common value of $\langle x_n : n < \omega\rangle^* \restriction i$ for all extensions $\langle x_n : n < \omega\rangle$ of $\langle x_n : n < i\rangle$.

A pointclass is a class of relations on $\mathbb{R}$ closed under recursive substitutions; a boldface pointclass is a pointclass closed under continuous substitutions. Let $\Sigma_n(M, X)$ be the class of relations on $\mathbb{R}$ which are Σ_n-definable over M from parameters in X. We are mainly interested in the pointclass $\Sigma_1(\mathbf{L}(\mathbb{R}), \{\mathbb{R}\})$, and its boldface counterpart $\Sigma_1(\mathbf{L}(\mathbb{R}), \mathbb{R} \cup \{\mathbb{R}\})$. It is easy to show that $\Sigma_1(\mathbf{L}(\mathbb{R}), \{\mathbb{R}\}) = (\Sigma^2_1)^{\mathbf{L}(\mathbb{R})}$ and $\Sigma_1(\mathbf{L}(\mathbb{R}), \mathbb{R} \cup \{\mathbb{R}\}) = (\undertilde{\Sigma}^2_1)^{\mathbf{L}(\mathbb{R})}$ (*cf.* [Ste83B, Lemma 1.12]). We shall have no use for this fact, however.

For any X, ${}^{<\omega}X$ is the set of finite sequences from X. A tree on X is a subset of ${}^{<\omega}X$ closed under initial segment. If T is a tree on X, $[T] = \{f \in {}^{\omega}X : \forall n (f \restriction n \in T)\}$; T is well founded just in case $[T] = \varnothing$. We sometimes regard a sequence of tuples as a tuple of sequences (of the same length), so that a tree on $X_1 \times \cdots \times X_k$ becomes a subset of ${}^{<\omega}X_1 \times \cdots \times {}^{<\omega}X_k$. If T is a tree on $X \times Y$ and $f \in {}^{\omega}X$, then $T(f) = \{u \in {}^{<\omega}Y : (f \restriction \mathrm{lh}(u), u) \in T\}$. Notice that $T(f)$ is a tree on Y depending continuously on f.

A quasi-strategy for player I in a game on X is a nonempty tree S on X such that if $u \in S$ and $\mathrm{lh}(u)$ is even, then $\exists a \in X \forall b \in X (u^\frown\langle a\rangle^\frown\langle b\rangle \in S)$. We say S is a winning quasi-strategy for player I in the game with payoff A just in

The preparation of the this paper was partially supported by NSF Grant Number MCS78-02989

The Cabal Seminar. Volume I: Games, Scales and Suslin Cardinals
Edited by A. S. Kechris, B. Löwe, J. R. Steel
Lecture Notes in Logic, 31

case $[S] \subseteq A$. If G is the game on $\mathbb{R}$ with payoff A, and Γ is a pointclass, we call G a Γ game iff $\{\langle x_n : n < \omega\rangle^* : \langle x_n : n < \omega\rangle \in A\}$ is in Γ. $\mathrm{Det}_{\mathbb{R}}(\Gamma)$ is the assertion that all Γ games on $\mathbb{R}$ are determined. Similarly, if $B \subseteq \mathbb{R} \times \mathbb{R}$, let

$$\begin{aligned} \Game^{\mathbb{R}} B &= \{x : \text{player I has a winning quasi-strategy in the game with payoff} \\ &\qquad \{\langle y_n : n < \omega\rangle : B(x, \langle y_n : n < \omega\rangle^*\}\}, \\ &= \{x : \exists y_0 \forall y_1 \exists y_2 \cdots B(x, \langle y_n : n < \omega\rangle^*)\}, \end{aligned}$$

and if Γ is a pointclass, let $\Game^{\mathbb{R}}\Gamma = \{\Game^{\mathbb{R}} B : B \in \Gamma\}$.

For whatever else we use from Descriptive Set Theory, and in particular for the notions of a scale and of the scale property, we refer the reader to [Mos80].

Out main theorem builds directly on a scale construction due to [Mos83]. We shall describe briefly the slight generalization of this construction we need. Suppose that $\alpha \in \mathrm{Ord}$, and that for each $x \in \mathbb{R}$ we have a game G_x in which player I's moves come from $\mathbb{R} \times \alpha$ while player II's moves come from $\mathbb{R}$. Thus a typical run of G_x has the form

player I	x_0, β_0		x_2, β_1		$\cdots$
player II		x_1		x_3	

where $x_i \in \mathbb{R}$ and $\beta_i < \alpha$ for all i. Suppose that G_x is closed and continuously associated to x in the strong sense that for some tree T on $\omega \times \omega \times \alpha$

$$\text{player I wins } \mathrm{G}_x \text{ iff } (x, \langle x_n : n < \omega\rangle^*, \langle \beta_n : n < \omega\rangle) \in [T].$$

Let $P_k(x, u)$ iff u is a position of length k from which player I has a winning quasi-strategy in G_x, and

$$P(x) \text{ iff } P_0(x, \varnothing).$$

If sufficiently many games are determined, then Moschovakis' construction yields a scale on P. [As in [Mos81], we extend the concept of a scale to relations with arguments from Ord as well as $\mathbb{R}$ by giving Ord the discrete topology. Following [Mos83] then, we define putative scales $\vec{\varphi}^k$ on P_k for all k simultaneously. If $P_k(x, u)$, then $\varphi_0^k(x, u) = 0$. (Otherwise $\varphi_0^k(x, u)$ is undefined.) Now either $(P_k(x, u) \iff \forall y P_{k+1}(x, u^\frown\langle y\rangle))$ or $(P_k(x, u) \iff \exists y P_{k+1}(x, u^\frown\langle y\rangle))$ or $(P_k(x, u) \iff \exists \beta P_{k+1}(x, u^\frown\langle \beta\rangle))$, depending on whose turn it is to play what at move k. In the first two cases, we define φ_{i+1}^k from $\langle \varphi_0^{k+1}, \ldots, \varphi_i^{k+1}\rangle$ as in [Mos83]. In the third case, if $P_k(x, u)$ then

$$\varphi_{i+1}^k(x, u) = \langle \beta, \varphi_0^{k+1}(x, u^\frown\langle\beta\rangle), \ldots, \varphi_i^{k+1}(x, u^\frown\langle\beta\rangle)\rangle,$$

where β is least so that $P_{k+1}(x, u^\frown\langle\beta\rangle)$. (More precisely, $\varphi_{i+1}^k(x, u)$ is the ordinal of this tuple in the lexicographic order.) As in [Mos83], one can show that each $\vec{\varphi}^k$ is in fact a scale on P_k, and thus $\vec{\varphi}^0$ is a scale on P, as desired.] The prewellordering $\leq_i$ induced by the ith norm of this scale on P is definable from the P_k's for $k \leq i$ by means of recursive substitution, the Boolean operations, and quantification over $\mathbb{R} \times \alpha$. In particular, if the map $k \mapsto P_k$ is $\Sigma_1(\mathbf{L}(\mathbb{R}), \{\alpha, \mathbb{R}\})$, then so is the map $i \mapsto \leq_i$. The determinacy required to construct this scale on P is closely related to the definability of the scale constructed. In particular, if the map $k \mapsto P_k$ is in $\mathbf{L}(\mathbb{R})$, then the determinacy of all games on ω in $\mathbf{L}(\mathbb{R})$ suffices.

In the circumstances just described we call the map $x \mapsto G_x$ a **closed game representation** of P. One can show directly that every $\Sigma_1(\mathbf{L}(\mathbb{R}), R \cup \{\mathbb{R}\})$ set admits a closed game representation, and hence a scale, in $\mathbf{L}(\mathbb{R})$. This is done in [Ste83B]. We shall take a slightly more circuitous route here, our reward being some additional information concerning $\mathbf{\Pi}^1_1$ games on $\mathbb{R}$.

Lemma 1. Every $\Game^{\mathbb{R}}\mathbf{\Pi}^1_1$ set admits a closed game representation in $\mathbf{L}(\mathbb{R})$.

Proof. Let P be $\Game^{\mathbb{R}}\mathbf{\Pi}^1_1$, and let T be a tree on $\omega \times \omega \times \omega$ so that

$$P(x) \text{ iff player I has a winning quasi-strategy in } \mathrm{G}_x,$$

where G_x is the game on $\mathbb{R}$ whose payoff for player I is $\{\langle x_n : n < \omega\rangle : T(x, \langle x_n : n < \omega\rangle^*) \text{ is wellfounded}\}$. We shall associate to each G_x auxilliary closed games $\mathrm{G}^*_{x,\alpha}$ for $\alpha \in \mathrm{Ord}$ as in Martin's proof of $\mathbf{\Pi}^1_1$ determinacy. A typical run of $\mathrm{G}^*_{x,\alpha}$ has the form

player I	x_0, β_0		x_2, β_1		$\dots$
player II		x_1		x_3	

where $X_i \in \mathbb{R}$ and $\beta_i < \alpha$ for all i. Let $\langle \sigma_i : i < \omega\rangle$ be an enumeration of ${}^{<\omega}\omega$ such that if $\sigma_i \subseteq \sigma_j$ then $i \leq j$. We say player I wins the run of $\mathrm{G}^*_{x,\alpha}$ just displayed just in case whenever $\sigma_i, \sigma_j \in T(x, \langle x_n : n < \omega\rangle^*)$ and $\sigma_i \subsetneq \sigma_j$, we have $\beta_i > \beta_j$.

Claim. Player I has a winning quasi-strategy in G_x iff $\exists\alpha$(player I has a winning quasi-strategy in $\mathrm{G}^*_{x,\alpha}$).

Proof. If player I has a winning quasi-strategy in $\mathrm{G}^*_{x,\alpha}$, then player I has a winning quasi-strategy in G_x, since G_x requires less of him. So suppose S is a winning quasi-strategy for player I in G_x. Let

$$\begin{aligned} U = \{\langle u, v\rangle \in {}^{<\omega}\mathbb{R} \times {}^{<\omega}\omega : \mathrm{lh}(u) = \mathrm{lh}(v) \text{ and } u \in S \\ \text{and } \langle x\restriction \mathrm{lh}(u), u^*, v\rangle \in T\}. \end{aligned} \tag{1}$$

Then U is a tree on $\mathbb{R} \times \omega$, and since S is a winning quasi-strategy for player I, U is wellfounded. Let $\|\langle u, v\rangle\|_U$ be the rank of $\langle u, v\rangle$ in U if $\langle u, v\rangle \in U$, and 0 otherwise. Let $\alpha = \|\langle\varnothing,\varnothing\rangle\|_U + 1$. Let

$$S_0^* = \{\langle x_0, \beta_0, x_1, \dots, x_{2k}, \beta_k, x_{2k+1}\rangle : \langle x_0, x_1, \dots, x_{2k+1}\rangle \in S \text{ and } \forall i \leq k(\beta_i = \|\langle\langle x_0, \dots, x_\ell\rangle, \sigma_i\rangle\|_U, \text{ where } \ell = \text{lh}(\sigma_i) - 1\},$$

and let S^* be the closure of S_0^* under initial segment. It is easy to verify that S^* is a winning quasi-strategy for player I in $\mathrm{G}^*_{x,\alpha}$. This proves the claim. $\dashv_{\text{Claim}}$

If player I has a winning quasi-strategy in G_x, let α_x be the least α such that player I has a winning quasi-strategy in $\mathrm{G}^*_{x,\alpha}$. Let $\lambda = \sup\{\alpha_x$: player I has a winning quasi-strategy in $\mathrm{G}_x\}$. Since any winning quasi-strategy for player I in $\mathrm{G}^*_{x,\alpha}$ is a winning quasi-strategy for player I in all $\mathrm{G}^*_{x,\beta}$ with $\beta \geq \alpha$, the map $x \mapsto \mathrm{G}^*_{x,\lambda}$ is a closed game representation of P. The map $x \mapsto \mathrm{G}^*_{x,\lambda}$ is clearly definable over $\mathbf{L}(\mathbb{R})$ (from T and λ). $\dashv$

The proof of Lemma 1 easily gives the following corollary.

Corollary 2. If G is a $\underset{\sim}{\Pi}^1_1$ game on $\mathbb{R}$ and player I has a winning quasi-strategy for G, then player I has a winning quasi-strategy for G in $\mathbf{L}(\mathbb{R})$. Thus $\Game^{\mathbb{R}}\underset{\sim}{\Pi}^1_1 = (\Game^{\mathbb{R}}\underset{\sim}{\Pi}^1_1)^{\mathbf{L}(\mathbb{R})}$.

Corollary 2 is the best absoluteness result for $\mathbf{L}(\mathbb{R})$ of its kind, at least if $\mathsf{Det}_{\mathbb{R}}(\Pi^1_1)$ holds. For in that case, Corollary 6 below implies that there is a Σ^1_1 game on $\mathbb{R}$ for which player I has a winning quasi-strategy in $\mathbf{V}$, but no winning quasi-strategy in $\mathbf{L}(\mathbb{R})$.

In proofs of determinacy from large cardinal hypotheses one associates to a complicated game G a closed game G^* in which one or both players make additional ordinal moves. One then uses measures on the possible additional moves to "integrate" strategies for G^*, and thereby show that whoever wins G^* wins G (and vice-versa, since G^* is determined). We could have proved Lemma 1 this way, but would have required the existence of $\mathbb{R}^\#$ in order to do so. The proof of Lemma 1 we did give shows that whoever wins G wins G^* (and vice-versa, *if* G is determined). In the large cardinals/determinacy problem this simpler method for proving the equivalence of G with G^* may give some guidance in discovering G^*, although integration is unavoidable in actually proving the determinacy of G.

Lemma 3. $\Game^{\mathbb{R}}\Pi^1_1 = \Sigma_1(\mathbf{L}(\mathbb{R}), \{\mathbb{R}\})$.

Proof. Since quantification over $\mathbb{R}$ is bounded, $(\Game^{\mathbb{R}}\Pi^1_1)^{\mathbf{L}(\mathbb{R})} \subseteq \Sigma_1(\mathbf{L}(\mathbb{R}), \{\mathbb{R}\})$. By Corollary 2 then, $\Game^{\mathbb{R}}\Pi^1_1 \subseteq \Sigma_1(\mathbf{L}(\mathbb{R}), \{\mathbb{R}\})$. For the other inclusion, let $\varphi(v_0, v_1)$ be a Σ_1 formula. We must associate recursively to each real x a $\Pi^1_1(x)$ game G_x on $\mathbb{R}$ so that $\mathbf{L}(\mathbb{R}) \models \varphi[x, \mathbb{R}]$ iff player I has a winning quasi-strategy in G_x. Our plan is to force player I in G_x to describe a countable, wellfounded model of $\mathsf{ZF}^- + \mathbf{V}{=}\mathbf{L}(\mathbb{R}) + \varphi[x, \mathbb{R}]$, which contains all the reals played in the course of G_x, and minimally so. If $\mathbf{L}(\mathbb{R}) \models \varphi[x, \mathbb{R}]$, then player

I will be able to win G_x by describing an elementary submodel of $\mathbf{L}_\alpha\mathbb{R}$, where α is least so that $\mathbf{L}_\alpha\mathbb{R} \models \mathsf{ZF}^- + \varphi[x, \mathbb{R}]$. On the other hand, if player I has a winning quasi-strategy in G_x, we shall be able to piece together the models he describes in different runs of G_x according to his strategy, and thereby produce a model of the form $\mathbf{L}_\alpha\mathbb{R}$ satisfying $\varphi[x, \mathbb{R}]$.

By "$\mathbf{V}{=}\mathbf{L}(\mathbb{R})$" we mean some (easily constructed) sentence ϑ in the language of set theory such that if M is a transitive model of ZF^-, then $M \models \vartheta$ iff $M = \mathbf{L}_\alpha(\mathbb{R}^M)$ for some α.

Player I describes his model in the language $\mathcal{L}$ which has, in addition to $\in$ and $=$, constant symbols $\boldsymbol{x}_i$ for $i < \omega$. He uses $\boldsymbol{x}_i$ to denote the ith real played in the course of G_x. Let us fix recursive maps

$$m, n \colon \{\vartheta : \vartheta \text{ is an } \mathcal{L}\text{-formula}\} \to \{2n : 1 \le n < \omega\}.$$

which are one-one, have disjoint recursive ranges, and are such that whenever $\boldsymbol{x}_i$ occurs in ϑ, $i < \min(m(\vartheta), n(\vartheta))$. These maps give stages sufficiently late in G_x for player I to decide certain statements about his model.

Player I's description must extend the following $\mathcal{L}$-theory T. The axioms of T include

$$\mathsf{ZF}^- + \mathsf{DC} + \mathbf{V}{=}\mathbf{L}(\mathbb{R}) \tag{2}$$

$$\boldsymbol{x}_i \in \mathbb{R} \tag*{$(3)_i$}$$

$$\varphi(\boldsymbol{x}_0, \mathbb{R}) \wedge \forall\beta(\mathbf{L}_\beta(\mathbb{R}) \not\models \mathsf{ZF}^- + \varphi[\boldsymbol{x}_0, \mathbb{R}]) \tag{4}$$

Finally, T has axioms which guarantee that for any model $\mathfrak{U}$ of T, the definable closure of $\{\boldsymbol{x}_i^{\mathfrak{U}} : i < \omega\}$ is an elementary submodel of $\mathfrak{U}$. Let $\vartheta_0(v_0, v_1, v_2)$ be a Σ_1 formula such that whenever M is transitive and $M \models \mathsf{ZF}^- + \mathbf{V}{=}\mathbf{L}(\mathbb{R})$, ϑ_0^M defines the graph of a map from $\mathrm{Ord}^M \times \mathbb{R}^M$ onto M. It is easy to construct such a formula; *cf.* [Ste83B, Lemma 1.4]. Now, for any $\mathcal{L}$-formula $\varphi(v)$ of one free variable, T has axioms

$$\exists v\varphi(v) \Longrightarrow \exists v\exists\alpha(\varphi(v) \wedge \vartheta_0(\alpha, \boldsymbol{x}_{m(\varphi)}, v)), \tag*{$(5)_\varphi$}$$

$$\exists v(\varphi(v) \wedge v \in \mathbb{R}) \Longrightarrow \varphi(\boldsymbol{x}_{n(\varphi)}). \tag*{$(6)_\varphi$}$$

This completes the list of axioms of T.

A typical run of G_x has the form

player I	i_0, x		i_1, x_2	
				$\dots$
player II		x_1		x_3

where for all k, $i_k \in \{0, 1\}$ and $x_k \in \mathbb{R}$. If $u = \langle\langle i_k, x_{2k}, x_{2k+1}\rangle : k < n\rangle$ is a position in G_x, then

$$T^*(u) = \{\vartheta : \vartheta \text{ is a sentence of } \mathcal{L} \text{ and } n(\vartheta) < n \text{ and } i_{n(\vartheta)} = 0\},$$

and if p is a full run of G_x,

$$T^*(p) = \bigcup_{n<\omega} T^*(p{\restriction}n).$$

Now let $p = \langle\langle i_k, x_{2k}, x_{2k+1}\rangle : k < \omega\rangle$ be a run of G_x; we say that p is a winning run for player I iff

(a) $x_0 = x$
(b) $T^*(p)$ is a complete, consistent extension T such that for all i, m, n "$\boldsymbol{x}_i(\boldsymbol{n}) = \boldsymbol{m}$" $\in T^*(p)$ iff $x_i(n) = m$, and
(c) There is no sequence $\langle\vartheta_i : i < \omega\rangle$ of $\mathcal{L}$-formulae of one free variable such that for all i, "$\imath v\vartheta_{i+1}(v) \in \imath v\vartheta_i(v)$" $\in T^*(p)$.

In condition (c) we have used the "unique v" operator as an abbreviation: $\psi(\imath v\vartheta(v))$ abbreviates "$\exists v(\psi(v) \wedge \forall u(\vartheta(u) \iff u = v))$".

It is clear that G_x is a $\Pi^1_1(x)$ game on $\mathbb{R}$, uniformly in x; its complexity comes from the wellfoundedness condition (c). So we need only prove the following claim.

Claim. $\mathbf{L}(\mathbb{R}) \models \varphi[x, \mathbb{R}]$ iff player I has a winning quasi-strategy in G_x.

Proof. ($\Rightarrow$) Let α be least so that $\mathbf{L}_\alpha(\mathbb{R}) \models \mathsf{ZF}^- + \varphi[x, \mathbb{R}]$. We call a position $u = \langle\langle i_k, x_{2k}, x_{2k+1}\rangle : k < n\rangle$ **honest** iff

(i) $n > 0 \Longrightarrow x_0 = x$,
(ii) if we let $I_u(\boldsymbol{x}_i) = x_i$ for $i < 2n$, then all axioms of $T \cup T^*(u)$ thereby interpreted in $\langle\mathbf{L}_\alpha(\mathbb{R}), I_u\rangle$ are true in the structure.

It is easy to check that if u is an honest position of length n, then $\exists i, x\forall y$ $(u^\frown\langle i, x, y\rangle)$ is honest. [If $n = n(\vartheta)$ for some sentence ϑ of $\mathcal{L}$, put $i = 0$ iff $\langle\mathbf{L}_\alpha(\mathbb{R}), I_u\rangle \models \vartheta$. Otherwise let i be random. If $n = m(\varphi)$ for φ an $\mathcal{L}$-formula of one free variable, choose x so that $\langle\mathbf{L}_\alpha(\mathbb{R}), I_{u^\frown\langle i,x,y\rangle}\rangle \models (5)_\varphi$. If $n = n(\varphi)$, choose x for the sake of axiom $(6)_\varphi$, and otherwise let x be random.] Now if p is a run of G_x such that $p{\restriction}n$ is honest for all n, and $I_p = \bigcup_n I_{p\restriction n}$, then $\langle\mathbf{L}_\alpha(\mathbb{R}), I_p\rangle \models T \cup T^*(p)$. It follows at once that p is a winning run for player I. Thus $\{u : u \text{ is honest}\}$ determines a winning quasi-strategy for player I in G_x.

($\Leftarrow$) Let S be a winning quasi-strategy for player I in G_x. If p is a run of G_x, let $\mathbb{R}_p$ be the set of reals played during p, that is, let $\mathbb{R}_p = \operatorname{ran}(I_p)$. If $p \in [S]$ then $T^*(p)$ is consistent, and by $(4)_\varphi$ has, up to isomorphism, a unique model $\mathcal{M}_p$ such that every element of $\mathcal{M}_p$ is $\mathcal{L}$-definable over $\mathcal{M}_p$. By player I's payoff condition (c), $\mathcal{M}_p$ is wellfounded, so we may assume that $\mathcal{M}_p$ is transitive. By $(5)_\varphi$ and player I's payoff condition (b), $\mathbb{R}^{\mathcal{M}_p} = \mathbb{R}_p$. Thus $\mathcal{M}_p = \langle\mathbf{L}_{\alpha_p}(\mathbb{R}), I_p\rangle$ for some α_p such that no reals beyond those in $\mathbb{R}_p$ appear in $\mathbf{L}_{\alpha_p}(\mathbb{R}_p)$. Notice that by $(3)_\varphi$, if $q \in [S]$ and $\mathbb{R}_q = \mathbb{R}_p$, then $\alpha_q = \alpha_p$.

Claim (Subclaim). If $p, q \in [S]$ and $I_p(\boldsymbol{x}_i) = I_q(\boldsymbol{x}_{n_i})$ for $i \leq k$, then for any formula $\vartheta(v_0, \ldots, v_k)$ of the language of set theory, $\vartheta(\boldsymbol{x}_0, \ldots, \boldsymbol{x}_k) \in T^*(p)$ iff $\vartheta(\boldsymbol{x}_{n_0}, \ldots, \boldsymbol{x}_{n_k}) \in T^*(q)$.

Proof. Suppose not, and let $u = p \restriction \ell$ and $v = q \restriction \ell$ where ℓ is so large that (say) $\vartheta(\boldsymbol{x}_0, \ldots, \boldsymbol{x}_k) \in T^*(p \restriction \ell)$ and $\neg\vartheta(\boldsymbol{x}_{n_0}, \ldots, \boldsymbol{x}_n) \in T^*(q \restriction \ell)$. Since player II is free in G_x to play whatever reals he pleases, we can find $s, t \in [S]$ so that $u \subseteq s, v \subseteq t$, and $\mathbb{R}_s = \mathbb{R}_t$. But then $\alpha_s = \alpha_t$. Since $\vartheta(v_0, \ldots, v_k)$ involves no constant symbols $\boldsymbol{x}_i$, and since $I_s(\boldsymbol{x}_i) = I_t(\boldsymbol{x}_{n_i})$ for $i \leq k$, we have $\mathcal{M}_s \models \vartheta(\boldsymbol{x}_0, \ldots, \boldsymbol{x}_k)$ iff $\mathcal{M}_t \models \vartheta(\boldsymbol{x}_{n_0}, \ldots, \boldsymbol{x}_{n_k})$. But $\mathcal{M}_s \models T^*(u)$ and $\mathcal{M}_t \models T^*(v)$, a contradiction.

By the subclaim, if $p, q \in [S]$ and $\mathbb{R}_p \subseteq \mathbb{R}_q$, then there is a unique elementary embedding $j_{p,q} : \mathbf{L}_{\alpha_p}(\mathbb{R}_p) \to \mathbf{L}_{\alpha_q}(\mathbb{R}_q)$. The uniqueness of the embeddings implies that if $p, q, s \in [S]$ and $\mathbb{R}_p \subseteq \mathbb{R}_q \subseteq \mathbb{R}_s$, then $j_{p,s} = j_{q,s} \circ j_{p,q}$. Thus we can form the direct limit M of the $\mathbf{L}_{\alpha_p}(\mathbb{R}_p)$'s under these embeddings. Now whenever $\{p_i : i < \omega\} \subseteq [S]$ we can find a $q \in [S]$ so that $\mathbb{R}_{p_i} \subseteq \mathbb{R}_q$ for all i, and therefore M is wellfounded. So we may assume M is transitive. Since $\mathbb{R} \subseteq \bigcup\{\mathbb{R}_p : p \in [S]\}$, $\mathbb{R} \subseteq M$, and so $M = \mathbf{L}_\alpha(\mathbb{R})$ for some α. But $M \models \varphi[x, \mathbb{R}]$.

This completes the proof of the claim, and thereby Lemma 3. ⊣

Our proof of Lemma 3 sharpens some earlier arguments of Solovay [Sol78B] which implies, assuming $\mathsf{AD}_{\mathbb{R}}$, that every set of reals in $\mathbf{L}(\mathbb{R})$ is $\Game^{\mathbb{R}}\undertilde{\mathbf{\Delta}}^1_3$.

Theorem 4. Assume all games in $\mathbf{L}(\mathbb{R})$ are determined. Then the pointclasses $\Sigma_1(\mathbf{L}(\mathbb{R}), \{\mathbb{R}\})$ and $\Sigma_1(\mathbf{L}(\mathbb{R}), \mathbb{R} \cup \{\mathbb{R}\})$ have the scale property.

Proof. It is enough to show $\Sigma_1(\mathbf{L}(\mathbb{R}), \{\mathbb{R}\})$ has the scale property. Let P be $\Sigma_1(\mathbf{L}(\mathbb{R}), \{\mathbb{R}\})$. Lemmas 1 and 3 yield a map $\langle x, \alpha \rangle \mapsto G^*_{x,\alpha}$ which is $\Delta_1(\mathbf{L}(\mathbb{R}), \{\mathbb{R}\})$ such that

$$P(x) \text{ iff } \exists\alpha(\text{player I has a winning quasi-strategy in } G^*_{x,\alpha}).$$

Let $P^\alpha(x)$ iff player I has a winning quasi-strategy in $G^*_{x,\alpha}$. Our slight generalisation of [Mos83], and the uniformity in its proof, yield a $\Delta_1(\mathbf{L}(\mathbb{R}), \{\mathbb{R}\})$ map $\alpha \mapsto \vec{\psi}^\alpha$ such that $\vec{\psi}^\alpha$ is a scale on P^α for all α. Now for $x \in P$ let

$$\varphi_0(x) = \min\{\alpha : P^\alpha(x)\},$$

and

$$\varphi_{i+1}(x) = \langle \varphi_0(x), \psi_i^{\varphi_0(x)}(x) \rangle,$$

where of course we use the lexicographic order to identify $\varphi_{i+1}(x)$ with an ordinal. It is easy to check that $\vec{\varphi}$ is a $\Sigma_1(\mathbf{L}(\mathbb{R}), \{\mathbb{R}\})$ scale on P. ⊣

Combining Theorem 4 with the result of Kechris and Solovay we mentioned earlier, we have

COROLLARY 5. Assume all games in $\mathbf{L}(\mathbb{R})$ are determined. Then P admits a scale in $\mathbf{L}(\mathbb{R})$ iff P is $\Sigma_1(\mathbf{L}(\mathbb{R}), \{\mathbb{R}\})$.

PROOF. One direction is Theorem 4; we prove the other for the sake of completeness. For $x, y \in \mathbb{R}$, let

$$C(x, y) \text{ iff } \mathbf{L}(\mathbb{R}) \models y \text{ is ordinal definable from } x.$$

By the reflection theorem, C is $\Sigma_1(\mathbf{L}(\mathbb{R}), \{\mathbb{R}\})$. Now $\neg C$ has no uniformization, hence no scale, in $\mathbf{L}(\mathbb{R})$. [If D uniformizes $\neg C$ and $D \in \mathbf{L}(\mathbb{R})$, then we can find $x_0 \in \mathbb{R}$ so that $\mathbf{L}(\mathbb{R}) \models D$ is ordinal definable from x_0. Then $D(x_0, y_0) \Longrightarrow C(x_0, y_0)$, a contradiction since $\exists y \neg C(x_0, y)$.] Now the class Γ of sets admitting scales in $\mathbf{L}(\mathbb{R})$ is a boldface pointclass. Since $\neg C \notin \Gamma$ and C is $\Sigma_1(\mathbf{L}(\mathbb{R}), \mathbb{R} \cup \{\mathbb{R}\})$, Wadge's lemma implies that $\Gamma \subseteq \Sigma_1(\mathbf{L}(\mathbb{R}), \mathbb{R} \cup \{\mathbb{R}\})$. ⊣

As promised, our proof of Theorem 4 gives some information about Π^1_1 games on $\mathbb{R}$.

COROLLARY 6. If $\mathsf{Det}_{\mathbb{R}}(\Pi^1_1)$ holds, then $\Game^{\mathbb{R}}\Sigma^1_1 \neq (\Game^{\mathbb{R}}\Sigma^1_1)^{\mathbf{L}(\mathbb{R})}$. Thus $\mathsf{Det}_{\mathbb{R}}(\Pi^1_1)$ is false in $\mathbf{L}(\mathbb{R})$.

PROOF. If $\mathsf{Det}_{\mathbb{R}}(\Pi^1_1)$ holds, then $\Game^{\mathbb{R}}\Sigma^1_1$ is the dual of $\Game^{\mathbb{R}}\Pi^1_1$, so that $\Game^{\mathbb{R}}\Sigma^1_1 = \Pi_1(\mathbf{L}(\mathbb{R}), \{\mathbb{R}\})$. But $(\Game^{\mathbb{R}}\Sigma^1_1)^{\mathbf{L}(\mathbb{R})} \subseteq \Sigma_1(\mathbf{L}(\mathbb{R}), \{\mathbb{R}\})$ by a direct computation. ⊣

COROLLARY 7. Assume AD. Then the following are equivalent:

(a) $\mathsf{Det}_{\mathbb{R}}(\Pi^1_1)$
(b) $\mathbb{R}^\#$ exists

PROOF. If $\mathbb{R}^\#$ exists, then $\mathsf{Det}_{\mathbb{R}}(\Pi^1_1)$ follows by the argument of [Mar70A]. (AD is not necessary here.) If $\mathsf{Det}_{\mathbb{R}}(\Pi^1_1)$ holds, then $\exists A \subseteq \mathbb{R}(A \notin \mathbf{L}(\mathbb{R}))$ by Corollary 6. But in the presence of AD this last statement implies that $\mathbb{R}^\#$ exists; *cf.* [SVW82]. ⊣

It seems likely that $\mathsf{Det}_{\mathbb{R}}(\Pi^1_1)$ and the existence of $\mathbb{R}^\#$ are provably equivalent in $\mathsf{ZF} + \mathsf{DC}$. We have not tried to show this.

We have seen that the complete $\Pi_1(\mathbf{L}(\mathbb{R}), \{\mathbb{R}\})$ set admits no scale in $\mathbf{L}(\mathbb{R})$. On the other hand, this set admits a scale just beyond $\mathbf{L}(\mathbb{R})$: Solovay has shown that if $\mathbb{R}^\#$ exists and $\mathbf{L}(\mathbb{R}) \models \mathsf{AD}$, then every set of reals in $\mathbf{L}(\mathbb{R})$ admits a scale each of whose norms is in $\mathbf{L}(\mathbb{R})$. (By [SVW82, Theorem 1.3.3], one needs $R^\#$ in order to construct definable scales beyond those in $\mathbf{L}(\mathbb{R})$.) This is a special case of a more general result of [Mar83B]: if Γ is a reasonably closed pointclass with the scale property, $\mathsf{Det}_{\mathbb{R}}(\Gamma)$ holds, and $\Game^{\mathbb{R}}\Gamma$ games on ω are determined, then $\Game^{\mathbb{R}}\Gamma$ has the scale property. (Solovay's result follows by taking Γ to be the class of sets which are $\omega \cdot n$-$\underset{\sim}{\Pi}^1_1$ for some $n < \omega$.) We shall conclude by proving part of Martin's theorem in a way which avoids the determinacy of games on $\mathbb{R}$. Our proof generalizes the proof of Lemma 1.

THEOREM 8. Assume AD. Then whenever A admits a scale, $\Game^{\mathbb{R}}A$ admits a scale.

PROOF. Let $\vec{\varphi}$ be a scale on A whose norms map into κ, where $\kappa < \vartheta$. Let R be the tree of $\vec{\varphi}$, that is

$$R = \{\langle s, t, u\rangle \in {}^{<\omega}\omega \times {}^{<\omega}\omega \times {}^{<\omega}\kappa :$$
$$\mathrm{lh}(s) = \mathrm{lh}(t) = \mathrm{lh}(u) \wedge \exists x, y(A(x, y) \wedge \forall i < \mathrm{lh}(u)(u(i) = \varphi_i(x, y)))\}.$$

Thus $A = \{(x, y) : [R(x, y)] \neq \varnothing\}$.

We now obtain a "homogeneous" tree representation of $\neg A$ in the standard way (*cf.* [Kec81A]). For any $\langle s, t\rangle \in {}^{<\omega}\omega \times {}^{<\omega}\omega$ such that $\mathrm{lh}(s) = \mathrm{lh}(t)$, let $W(s, t)$ be the Brouwer-Kleene order of ${}^{<\omega}\kappa$ restricted to

$$\mathrm{fld}(W(s, t)) = \{u \in {}^{<\omega}\kappa : \mathrm{lh}(u) \leq \mathrm{lh}(s) \wedge (s{\restriction}\mathrm{lh}(u), t{\restriction}\mathrm{lh}(u), u) \in R\}.$$

$W(s, t)$ is a wellorder because the sequences in its field have length $\leq \mathrm{lh}(s)$. By AD we have a $\lambda > \kappa$ such that $\lambda \to (\lambda)^\lambda$. For any $C \subseteq \lambda$, let $[C]^{W(s,t)}$ be the set of order-preserving maps from $\mathrm{fld}(W(s, t))$ into C. Now let T be the tree on $\omega \times \omega \times \bigcup_{\langle s,t\rangle}[\lambda]^{W(s,t)}$ given by

$$T = \{\langle s, t, \langle f_1, \dots, f_\ell\rangle\rangle : \mathrm{lh}(s) = \mathrm{lh}(t) = \ell \wedge \forall i \leq \ell \left(f_i \in [\lambda]^{W(s{\restriction}i, t{\restriction}i)}\right)$$
$$\wedge \forall i, j \leq \ell(i \leq j \Longrightarrow f_i = f_j{\restriction}W(s{\restriction}i, t{\restriction}i))\}.$$

A path through T is equivalent to a pair $\langle x, y\rangle$ of reals together with a map of $R(x, y)$ into λ preserving the Brouwer-Kleene order on $R(x, y)$. Thus

$$A = \{\langle x, y\rangle : [T(x, y)] = \varnothing\}.$$

For any $\langle s, t\rangle \in {}^{<\omega}\omega \times {}^{<\omega}\omega$ such that $\mathrm{lh}(s) = \mathrm{lh}(t)$, let

$$T(s, t) = \{u : \langle s, t, u\rangle \in T\}.$$

Each $T(s, t)$ carries a canonical measure $\mu_{s,t}$ given by the strong partition property of λ. [Fix $\alpha \leq \lambda$. For $C \subseteq \alpha$, let

$$C\uparrow = \{h \in {}^{\alpha}[\lambda] : \exists g \in [C]^{\omega\alpha} \forall \beta < \alpha(h(\beta) = \sup\{g(\omega\beta + n) : n \in \omega\})\}.$$

Then since $\lambda \to (\lambda)^\lambda$, the sets of the form $C\uparrow$ for C closed and unbounded in λ are the base for an ultrafilter μ_α on ${}^{\alpha}[\lambda]$. If α is the order type of $W(s, t)$, then μ_α is isomorphic in an obvious way to an ultrafilter $\mu_{s,t}$ on $[\lambda]^{W(s,t)}$. But an inspection of the definition of T shows that we may identify $T(s, t)$ with ${}^{W(s,t)}[\lambda]$. (See [Kec81A] for further details on this construction.)]. The measures $\mu_{s,t}$ are compatible, in the sense that if $\mu_{s,t}(X) = 1$, and $i < \mathrm{lh}(s)$, then $\mu_{s{\restriction}i,t{\restriction}i}(\{u{\restriction}i : u \in X\}) = 1$. They also have a "homogeneity" property: if $[T(x, y)] \neq \varnothing$ and $\mu_{x{\restriction}i,y{\restriction}i}(Y_i) = 1$ for all $i < \omega$, then $\exists f \in [T(x, y)]\forall i(f{\restriction}i \in Y_i)$. We shall now produce a closed game representation of $\Game^{\mathbb{R}} A$. For $x \in \mathbb{R}$ and $\alpha \in \mathrm{Ord}$, we define a closed game $\mathrm{G}^*_{x,\alpha}$. A typical run of $\mathrm{G}^*_{x,\alpha}$ has the form

$$\begin{array}{lcccc} \text{player I} & x_0, \beta_0 & & x_2, \beta_1 & \\ & & & & \dots \\ \text{player II} & & x_1 & & x_3 \end{array}$$

where $x_i \in \mathbb{R}$ and $\beta_i < \alpha$ for all i. Player I wins this run of $\mathrm{G}^*_{x,\alpha}$ just in case, if we set $y = \langle x_n : n < \omega \rangle^*$, and for all $i < \omega$ pick maps $F_i : T(X \restriction i, y \restriction i) \to \mathrm{Ord}$ such that $\beta_i = [F_i]_{\mu_{x\restriction i, y\restriction i}}$, then whenever $i < k$,

$$\mu_{x\restriction k, y\restriction k}(\{u : F_k(u) < F_i(u \restriction i)\} = 1.$$

The compatibility of the measures implies that player I's payoff condition does not depend on which map F_i is chosen to represent β_i in the ultrapower by $\mu_{x\restriction i, y\restriction i}$.

Claim. $x \in \Game^{\mathbb{R}} A$ iff $\exists \alpha$(player I has a winning quasi-strategy in $\mathrm{G}^*_{x,\alpha}$).

Proof. Let S^* be a winning quasi-strategy for player I in $\mathrm{G}^*_{x,\alpha}$. Let S result from S^* by omitting the ordinals from player I's moves. Let $\langle x_n : n < \omega \rangle \in [S]$. Then the homogeneity property of the $\mu_{s,t}$'s guarantees that $[T(x, \langle x_n : n < \omega \rangle^*)] = \varnothing$. Thus $(x, \langle x_n : n < \omega \rangle^*) \in A$, and S witnesses that $x \in \Game^{\mathbb{R}} A$.

Conversely, let S be a winning quasi-strategy for player I in the game whose payoff is $\{\langle x_n : n < \omega \rangle : A(x, \langle x_n : n < \omega \rangle^*)\}$. Let U be the tree on $\mathbb{R} \times \bigcup_{\langle s,t \rangle} {}^{W(s,t)}[\lambda]$ given by

$$U = \{\langle u, v \rangle : \mathrm{lh}(u) = \mathrm{lh}(v) \wedge u \in S \wedge \langle x \restriction \mathrm{lh}(u), u^*, v \rangle \in T\}.$$

Then U is wellfounded since S is a winning quasi-strategy for player I. For $u \in S$, let $F_u : T(x \restriction \mathrm{lh}(u), u^*) \to \mathrm{Ord}$ be defined by

$$F_u(v) = \text{ rank of } \langle u, v \rangle \text{ in } U$$

We can now define a winning quasi-strategy S^* for player I as follows: let

$$\begin{aligned} S_0^* = \{\langle x_0, \beta_0, x_1, \dots, x_{2k}, \beta_k, x_{2k+1} \rangle &: \langle x_0, \dots, x_{2k+1} \rangle \in S \\ &\wedge \forall i \leq k(\beta_i = [F_i]_\mu \text{ where } F_i = F_{\langle x_n : n < i \rangle} \text{ and } \mu = \mu_{x\restriction i, \langle x_n : n < i \rangle^*})\}. \end{aligned}$$

and let S^* be the closure of S_0^* under initial segment. It is easy to check that S^* is a winning quasi-strategy for player I in $\mathrm{G}^*_{x,\alpha}$ for all sufficiently large α. This completes the proof of the claim.

Since a winning quasi-strategy for player I in $\mathrm{G}^*_{x,\alpha}$ is also a winning quasi-strategy for player I in $\mathrm{G}^*_{x,\beta}$ for all $\beta \geq \alpha$, we can find $\lambda \in \mathrm{Ord}$ such that $\forall x(x \in \Game^{\mathbb{R}} A$ iff player I has a winning quasi-strategy in $\mathrm{G}^*_{x,\lambda})$. This $x \to \mathrm{G}^*_{x,\lambda}$ is a closed game representation of $\Game^{\mathbb{R}} A$, and $\Game^{\mathbb{R}} A$ admits a scale. ⊣

REFERENCES

ALEXANDER S. KECHRIS
[Kec81A] *Homogeneous trees and projective scales*, In Kechris et al. [CABAL ii], pp. 33–74.

ALEXANDER S. KECHRIS, DONALD A. MARTIN, AND YIANNIS N. MOSCHOVAKIS
[CABAL ii] *Cabal seminar* 77–79, Lecture Notes in Mathematics, no. 839, Berlin, Springer, 1981.
[CABAL iii] *Cabal seminar* 79–81, Lecture Notes in Mathematics, no. 1019, Berlin, Springer, 1983.

ALEXANDER S. KECHRIS AND YIANNIS N. MOSCHOVAKIS
[CABAL i] *Cabal seminar* 76–77, Lecture Notes in Mathematics, no. 689, Berlin, Springer, 1978.

DONALD A. MARTIN
[Mar70A] *Measurable cardinals and analytic games*, ***Fundamenta Mathematicae***, (1970), no. LXVI, pp. 287–291.
[Mar83B] *The real game quantifier propagates scales*, this volume, originally published in Kechris et al. [CABAL iii], pp. 157–171.

YIANNIS N. MOSCHOVAKIS
[Mos80] ***Descriptive set theory***, Studies in Logic and the Foundations of Mathematics, no. 100, North-Holland, Amsterdam, 1980.
[Mos81] *Ordinal games and playful models*, In Kechris et al. [CABAL ii], pp. 169–201.
[Mos83] *Scales on coinductive sets*, this volume, originally published in Kechris et al. [CABAL iii], pp. 77–85.

ROBERT M. SOLOVAY
[Sol78B] *The independence of* DC *from* AD, In Kechris and Moschovakis [CABAL i], pp. 171–184.

JOHN R. STEEL
[Ste83B] *Scales on Σ^1_1-sets*, this volume, originally published in Kechris et al. [CABAL iii], pp. 72–76.

JOHN R. STEEL AND ROBERT VAN WESEP
[SVW82] *Two consequences of determinacy consistent with choice*, ***Transactions of the American Mathematical Society***, (1982), no. 272, pp. 67–85.

DEPARTMENT OF MATHEMATICS
UNIVERSITY OF CALIFORNIA
LOS ANGELES, CA 90095, USA
E-mail: dam@math.ucla.edu

DEPARTMENT OF MATHEMATICS
UNIVERSITY OF CALIFORNIA
BERKELEY, CA 94720, USA
E-mail: steel@math.berkeley.edu

THE LARGEST COUNTABLE THIS, THAT, AND THE OTHER

DONALD A. MARTIN

§1. Introduction. The set ${}^{\omega}\omega \cap \mathbf{L}$ has, if it is countable, an implicit descriptive set-theoretic characterization as the largest countable Σ^1_2 set [Mos80, p. 538]. In this paper, we give an explicit descriptive set-theoretic characterization of this and (under determinacy hypotheses) other related sets.

A subset A of ${}^{k}\omega \times ({}^{\omega}\omega)^m$ is α-$\underset{\sim}{\mathbf{\Pi}}^1_1$, where α is an ordinal, if there are $\underset{\sim}{\mathbf{\Pi}}^1_1$ sets A_β, $\beta < \alpha$ such that

$$x \in A \iff \min\{\beta : \beta = \alpha \vee x \notin A_\beta\} \text{ is odd.}$$

If α is small, say recursive, the lightface notion, α-Π^1_1, has an obvious definition.

A is $\Game\Gamma$ if there is a $B \in \Gamma$ such that

$$x \in A \iff \text{player I has a winning strategy for } \mathrm{G}_x(B),$$

where $\mathrm{G}_x(B)$ is the game in which the players cooperatively produce a $y \in {}^{\omega}\omega$ and player I wins just in case $\langle x, y\rangle \in B$.

Let C_{2n} be the largest countable Σ^1_{2n} subset of ${}^{\omega}\omega$, for $n \geq 1$. See [Mos80, p. 346]. In §2 and §3, we prove the following theorem (where $\Game^{i+1}\Gamma = \Game(\Game^{i}\Gamma)$):

THEOREM 1.1 (Projective Determinacy, PD).

$$\mathrm{C}_{2k} = \bigcup_{n\in\omega} \Game^{2k-1}(\omega \cdot n)\text{-}\Pi^1_1.$$

COROLLARY 1.2 ($0^{\#}$ exists).

$$\mathbf{L} = \bigcup_{n\in\omega} \Game(\omega \cdot n)\text{-}\Pi^1_1.$$

We also use our proof to characterize $0^{\#}$ as the recursive join of a sequence of complete $\Game\omega \cdot n$-Π^1_1 sets.

Let Σ^*_0 be the class of unions of inductive (see [Mos80, p. 410]) and coinductive sets. (We take "inductive" in the lightface sense.) Let Π^*_n be the class of complements of Σ^*_n sets and let Σ^*_{n+1} be the class of projections of Π^*_n sets. In §4, we prove

Research partially supported by NSF grant.

The Cabal Seminar. Volume I: Games, Scales and Suslin Cardinals
Edited by A. S. Kechris, B. Löwe, J. R. Steel
Lecture Notes in Logic, 31

THEOREM 1.3 (Det(Σ_n^*) for all n). The largest countable inductive set is $\bigcup_{n\in\omega}\Sigma_n^*$.

[Mos83] shows that every coinductive set admits a scale whose complexity is the join of the Σ_n^*. It follows from the theorem that there is a coinductive set which has no Σ_n^* scale for any n.

§2. Every $\Game^{2k-1}(\omega\cdot n)$-$\Pi_1^1$ set belongs to C_{2k}. We first fix a coding of $(\omega\cdot n)$-Π_1^1 sets. Let each $z\in{}^\omega\omega$ code, in some effective manner, a sequence $\langle A_\beta^z : \beta<\omega^2\rangle$ of $\underset{\sim}{\Pi}_1^1$ sets. For each z let $A^{z,n}$ be the $\omega\cdot n$-$\underset{\sim}{\Pi}_1^1$ set determined by the sequence $\langle A_\beta^z : \beta<\omega\cdot n\rangle$. For each z, let G_z^n be the game with player I's winning set $A^{z,n}$. Do this coding so that $\{z : G_z^n$ is a win for player I$\}$ is a complete $\Game(\omega\cdot n)$-Π_1^1 set, uniformly in n.

LEMMA 2.1. For each z and each sequence $\gamma_1,\dots,\gamma_n$ of ordinals, there is an open game $G^*_{z,\gamma_1,\dots,\gamma_n}$, played on ordinals strictly less than $\sup\{\gamma_1,\dots,\gamma_n\}$ such that

1. The winning conditions for $G^*_{z,\gamma_1,\dots,\gamma_n}$ are definable in $\mathbf{L}[z]$ from z and $\gamma_1,\dots,\gamma_n$ by a definition independent of z and $\gamma_1,\dots,\gamma_n$.
2. If $\gamma_1<\cdots<\gamma_n$ are indiscernibles for $\mathbf{L}[z]$, then whoever has a winning strategy for $G^*_{z,\gamma_1,\dots,\gamma_n}$ has a winning strategy for G_z^n.
3. There is a Π_1^1 relation $R(n,z,y)$ such that, if $y_1,\dots,y_n$ are codes for countable ordinals $\gamma_1,\dots,\gamma_n$ respectively, then

$$\text{player I wins } G^*_{z,\gamma_1,\dots,\gamma_n} \iff R(n,z,\langle y_1,\dots,y_n\rangle).$$

PROOF. For simplicity, we suppress z and often n from the notation. Let $\langle A_\beta : \beta<\omega\cdot n\rangle$ be the $\underset{\sim}{\Pi}_1^1$ sets given by z. We can associate with each $x\in{}^\omega\omega$ and each $\beta<\omega\cdot n$ a linear ordering R_x^β of ω in an effective manner, such that $R_x^\beta\restriction k$ depends only on $x\restriction k$, such that 0 is maximum in R_x^β, and such that $x\in A_\beta$ if and only if R_x^β is a well-ordering.

Let $h\colon\omega\to\omega\cdot n\times\omega$ be an effective bijection such that

(a) if $h(k)=\langle\beta,t\rangle$, then β is even if and only if k is even;
(b) if $h(k_1)=\langle\beta,t_1\rangle$, $h(k_2)=\langle\beta,t_2\rangle$, and $t_1<t_2$, then $k_1\le k_2$;
(c) if $h(k_1)=\langle\omega\cdot i+j_1,0\rangle$, $h(k_2)=\langle\omega\cdot i+j_2,0\rangle$, and $j_1<j_2$, then $k_1<k_2$.

We play $G^*_{\gamma_1,\dots,\gamma_n}$ as follows:

Player I	Player II
m_0,ϱ_0	
	m_1,ϱ_1
m_2,ϱ_2	
	m_3,ϱ_3
$\vdots$	

The **rules** of the game are:

(i) $m_i \in \omega$.
(ii) If $h(k) = \langle \omega \cdot i + j, t \rangle$, then $\varrho_k < \gamma_{i+1}$.
(iii) Let $F_\beta(t) = \varrho_k$, where $h(k) = \langle \beta, t \rangle$. F_β must embed the ordering R_x^β into the ordinals, where $x = \langle m_i : i \in \omega \rangle$.

The first player to violate the rules loses. Otherwise, player II wins.

The players are thus trying to verify that the R_x^β are well-orderings. Player I is responsible for even β, player II for odd β. Condition (b) ensures that the maximum value of F_β is played before any other values, and that $R_x^\beta \lceil t + 1$ is known before $F_\beta(t)$ is played. Condition (c) ensures that $F_\beta(0)$ is played before $F_\delta(0)$ whenever $\delta > \beta$ and β and δ are in the same ω-block.

Condition (1) in the statement of the lemma is clear. Condition (3) is also easily verified, since given codes for $\gamma_1, \ldots, \gamma_n$ (*i.e.*, well-orderings of those order types), we can construe $\mathrm{G}^*_{\gamma_1,\ldots,\gamma_n}$ as an ordinary open game. Since $\Game\Sigma_1^0 = \Pi_1^1$, this yields (3).

By (1), it suffices to verify (2) in the special case $\gamma_i = \omega_i$, where we may assume $z^\#$ exists.

Assume for definiteness that player I has a winning strategy σ^* for $\mathrm{G}^* = \mathrm{G}^*_{\omega_1,\ldots,\omega_n}$.

Let us say that player II **plays well** if player II obeys all rules, $F_{\omega\cdot i+2j+1} > \gamma_i$ for all $i \geq 1$ and $j \in \omega$, $F_{\omega\cdot i+2j+1}(t) > F_{\omega\cdot i+j'}(0)$ for all $j' \leq 2j$, and $F_{\omega\cdot i+2j+1}(t)$ is an indiscernible for $\mathbf{L}[z]$ for all $i \in \omega$ and $j \in \omega$.

We define a strategy σ for player I for G_z^n by assuming player II plays well and player I plays σ^*.

Note that, in positions in G^* such that player II has played well, player I's moves m_i and $F_\beta(t)$ given by σ^* are independent of player II's moves $F_{\beta'}(t)$ for $\beta' > \beta$.

Let x be any play of G ($= \mathrm{G}_z^n$) consistent with σ. Let β_0 be the least β such that $\beta = \omega \cdot n$ or $x \notin A_\beta$. By induction, we define F_β for $\beta < \beta_0$. Assume $F_{\beta'}$ is defined for $\beta' < \beta$.

Suppose β is even. Let F_β be given by letting x be played, letting player II play $F_{\beta'}$ for odd $\beta' < \beta$, letting player II play well otherwise, and letting player I play according to σ^*.

Suppose $\beta = \omega \cdot i + j$ with j odd. Let η be the supremum of $\{\gamma_{i'} : i' < i\} \cup \{F_{\omega\cdot i+j'}(0) : j' < j\}$. Let $F_\beta \colon \omega \to \omega_{i+1}$ embed R_x^β into the indiscernibles for $\mathbf{L}[z]$ between η and γ_{i+1}.

Suppose $\beta_0 = \omega \cdot n$. Then the F_β extend x to a play of G^* according to σ^* with all rules obeyed. Since this play is a win for player II, we have a contradiction.

Suppose β_0 is even. Let F_{β_0} be given by letting x be played, letting player II play F_β for odd $\beta < \beta_0$, letting player II play well otherwise, and letting player

I play σ^*. Eventually F_{β_0} must violate rule (iii). Extending to a position where player II has played well, we get a contradiction.

Thus β_0 is odd, and so x is a win for player I. Hence σ is a winning strategy for player I for G. ⊣

LEMMA 2.2 ($0^\#$ exists). If S is a $\Game\omega \cdot n$-Π^1_1 set of integers, then $S \in \mathbf{L} = \mathrm{C}_2$.

PROOF. There is recursive $f : \omega \to {}^\omega\omega$ such that

$$m \in S \iff \text{player I wins } \mathrm{G}^n_{f(m)}.$$

Let

$$B = \{S' : \exists \gamma_1, \dots, \gamma_n < \omega_1 \forall m (\text{player I wins } \mathrm{G}^*_{f(m),\gamma_1,\dots,\gamma_n} \iff m \in S')\}.$$

By part (2) of Lemma 2.1, $S \in B$.

It is clear that $S' \in B$ implies that $S' \in \mathbf{L}$, but we prove directly that B is a countable Σ^1_2 set. By part (3) of Lemma 2.1, $S' \in B$ if and only if there are codes $y_1, \dots, y_n$ for countable ordinals such that for all m, we have

$$R(n, f(m), y_1, \dots, y_n) \iff m \in S'.$$

Thus $B \in \Sigma^1_2$. Since B has a Σ^1_2 well-ordering, B is countable. ⊣

LEMMA 2.3 (PD). For $i \in \omega$, every $\Game^{2i+1}(\omega \cdot n)$-$\Pi^1_1$ set $S \subseteq \omega$ belongs to C_{2i+2}.

PROOF. There is a recursive function f such that $m \in S$ if and only if player I wins the following game $\mathrm{G}(m)$ of length $\omega \cdot 2i$:

$x_1, \dots, x_{2i}$ is a win for player I just in case player I has a winning strategy for $\mathrm{G}^n_{f(m,x_1,\dots,x_{2i})}$.

Given degrees of unsolvability $\boldsymbol{d}_1, \dots, \boldsymbol{d}_{2i}$, consider the game $\mathrm{G}(m, \boldsymbol{d}_1, \dots, \boldsymbol{d}_{2i})$:

Player I	Player II
σ_1	
	x_1
σ_2	
	x_2
$\vdots$	$\vdots$
σ_{2i}	
	x_{2i}

σ_j must be a strategy for player I for an ordinary game on ω and must satisfy $\deg(\sigma_j) \leq \boldsymbol{d}_j$. x_j must be a play consistent with σ_j such that $\deg(x_j) \leq \boldsymbol{d}_j$. Player I wins if and only if $x_1, \dots, x_{2i}$ is a win for player I in $\mathrm{G}^n_{f(m,x_1,\dots,x_{2i})}$.

SUBLEMMA 2.3.1. *For almost every $\boldsymbol{d}_1, \dots, \boldsymbol{d}_{2i}$ (with respect to the iterated product of the usual measure on the degrees), for all m*

$$\textit{player I wins } \mathrm{G}(m) \iff \textit{player I wins } \mathrm{G}(m, \boldsymbol{d}_1, \dots, \boldsymbol{d}_{2i}).$$

PROOF OF SUBLEMMA. Suppose that player I has a winning strategy σ for $G(m)$. For $1 \le j \le 2i$ and degrees $\boldsymbol{d}_1, \ldots, \boldsymbol{d}_{j-1}$, let $\boldsymbol{d}_j \in A(\boldsymbol{d}_1, \ldots, \boldsymbol{d}_{j-1})$ just in case, for all $\langle x_1, \ldots, x_{j-1}\rangle$ consistent with σ such that $\deg(x_{j'}) \le \boldsymbol{d}_{j'}$ for all $j' < j$, the degree of σ restricted to the next ω moves from the position $\langle x_1, \ldots, x_{j-1}\rangle$ is $\le \boldsymbol{d}_j$. Let $\boldsymbol{d}_1, \ldots, \boldsymbol{d}_{2i} \in A$ just in case $\boldsymbol{d}_j \in A(\boldsymbol{d}_1, \ldots, \boldsymbol{d}_{j-1})$ for $1 \le j \le 2i$. Clearly σ gives a winning strategy for $G(m, \boldsymbol{d}_1, \ldots, \boldsymbol{d}_{2i})$ whenever $\langle \boldsymbol{d}_1, \ldots, \boldsymbol{d}_{2i}\rangle \in A$. Since A has iterated product measure 1, we have proved the first half of the sublemma.

A similar argument shows that if II has a winning strategy for $G(m)$ then II has a winning strategy for $G(m, \boldsymbol{d}_1, \ldots, \boldsymbol{d}_{2i})$ for almost every $\langle \boldsymbol{d}_1, \ldots, \boldsymbol{d}_{2i}\rangle$. ⊣

For each $\boldsymbol{d}_1, \ldots, \boldsymbol{d}_{2i}$, and each $\gamma_1, \ldots, \gamma_n < \omega_1$, let $G(m, \boldsymbol{d}_1, \ldots, \boldsymbol{d}_{2i}, \gamma_1, \ldots, \gamma_n)$ be the game played like $G(m, \boldsymbol{d}_1, \ldots, \boldsymbol{d}_{2i})$ which player I wins just in case player I wins $G^*_{f(m, x_1, \ldots, x_{2i}), \gamma_1, \ldots, \gamma_n}$.

Let $B = \{S' : \text{for almost all } \boldsymbol{d}_1, \ldots, \boldsymbol{d}_{2i} \text{ there exist } \gamma_1, \ldots, \gamma_n < \omega_1 \text{ such that for all } m \text{ player I wins } G(m, \boldsymbol{d}_1, \ldots, \boldsymbol{d}_{2i}, \gamma_1, \ldots, \gamma_n) \iff m \in S'\}$.

We first note that B is (projectively) well-orderable. For $S' \in B$, let $\phi_{S'}(\boldsymbol{d}_1, \ldots, \boldsymbol{d}_{2i})$ be the lexicographically least $\langle \gamma_1, \ldots, \gamma_n\rangle$ such that for all m, player I wins $G(m, \boldsymbol{d}_1, \ldots, \boldsymbol{d}_{2i}, \gamma_1, \ldots, \gamma_n)$ if and only if $m \in S'$, if it is defined. It is defined for almost all $\boldsymbol{d}_1, \ldots, \boldsymbol{d}_{2i}$. If $\phi_{S'_1}(\boldsymbol{d}_1, \ldots, \boldsymbol{d}_{2i}) = \phi_{S'_2}(\boldsymbol{d}_1, \ldots, \boldsymbol{d}_{2i})$ almost everywhere, then $S'_1 = S'_2$.

Next we show that B is Σ^1_{2i+1}:

$S' \in B$ if and only if

$$\exists \boldsymbol{d}_1 \forall \boldsymbol{d}'_1 \ge \boldsymbol{d}_1 \forall \boldsymbol{d}_2 \exists \boldsymbol{d}'_2 \ge \boldsymbol{d}_2 \ldots \forall \boldsymbol{d}_{2i} \exists \boldsymbol{d}'_{2i} \ge \boldsymbol{d}_{2i} \exists y_1, \ldots, y_m$$
$$y_1, \ldots, y_m \text{ are codes for countable ordinals and for all } m$$
$$\Big(\exists \sigma_1 \text{ with } \deg(\sigma_1) \le \boldsymbol{d}'_1 \forall x_1 \text{ with } \deg(x_1) \le \boldsymbol{d}'_1$$
$$\ldots \exists \sigma_{2i} \text{ with } \deg(\sigma_{2i}) \le \boldsymbol{d}'_{2i} \forall x_{2i} \text{ with } \deg(x_{2i}) \le \boldsymbol{d}'_{2i}$$
$$\big(R(n, f(m, x_1, \ldots, x_{2i}), y_1, \ldots, y_n) \iff m \in S'\big)\Big).$$

Finally we show that $S \in B$. Let $\boldsymbol{d}_1, \ldots, \boldsymbol{d}_{2i}$ be such that for all m, player I wins $G(m)$ if and only if player I wins $G(m, \boldsymbol{d}_1, \ldots, \boldsymbol{d}_{2i})$. By Sublemma 2.3.1, almost all $\langle \boldsymbol{d}_1, \ldots, \boldsymbol{d}_{2i}\rangle$ have this property. Now let $\gamma_1 < \cdots < \gamma_n$ be countable indiscernibles for $\mathbf{L}[w]$ with $\deg(w) \ge \boldsymbol{d}_j$ for $j = 1, \ldots, 2i$. By Lemma 2.1, if $x_1, \ldots, x_{2i}$ is the result of a play of $G(m, \boldsymbol{d}_1, \ldots, \boldsymbol{d}_{2i})$, then

$$x_1, \ldots, x_{2i} \text{ is a win for player I in } G(m, \boldsymbol{d}_1, \ldots, \boldsymbol{d}_{2i})$$
$$\iff x_1, \ldots, x_{2i} \text{ is a win for player I in } G(m, \boldsymbol{d}_1, \ldots, \boldsymbol{d}_{2i}, \gamma_1, \ldots, \gamma_n).$$

Now for all m, we have $m \in S$ if and only if player I wins $G(m)$ if and only if player I wins $G(m, \boldsymbol{d}_1, \ldots, \boldsymbol{d}_{2i}, \gamma_1, \ldots, \gamma_n)$. ⊣

§3. Characterization of C_{2i} and of $0^{\#}$.

LEMMA 3.1 (Moschovakis; Sufficient determinacy). Suppose every set in Γ admits a very good scale $\langle \phi_i : i \in \omega \rangle$ such that each ϕ_i is a Γ_i norm.

1. Every set in $\Game\Gamma$ admits a very good scale $\langle \psi_i : i \in \omega \rangle$ such that each ψ_i is a $\Game\Gamma_i$ norm.
2. Every Γ game won by player I has a winning strategy σ such that σ restricted to positions of length i belongs to $\Game\Gamma_i$.

An inspection of the proof of [Mos80, Theorem 6E.1] and [Mos80, Theorem 6E.15] will reveal that they prove the lemma.

LEMMA 3.2 ([Ste83B]). Every Σ^1_1 set admits a very good scale $\langle \phi_i : i \in \omega \rangle$ such that each ϕ_i is $(\omega \cdot (i+1))$-Π^1_1.

LEMMA 3.3 (PD). Every Π^1_{2i+1} (resp. Σ^1_{2i+2}) game won by player II has a winning strategy τ such that τ restricted to positions of length n is $\Game^{2i+1}(\omega \cdot (n+1))$-$\Pi^1_1$ (resp. $\Game^{2i+2}(\omega \cdot (n+1))$-$\Pi^1_1$).

PROOF. Lemma 3.1 and Lemma 3.2 allow one to prove this lemma by induction. ⊣

LEMMA 3.4. For each $i \geq 1$, every member of C_{2i} is $\Game^{2i-1}(\omega \cdot n)$-$\Pi^1_1$ for some n.

PROOF. Let A be any countable Σ^1_{2i} subset of ${}^{\omega}2$. Let $A = \{x : \exists y(\langle x, y\rangle \in B)\}$ for $B \in \Pi^1_{2i-1}$.

Consider the following game:

Player I	Player II
m_0, s_0	
	ε_0
m_1, s_1	
	ε_1
$\vdots$	

Each m_j must be a natural number. Each s_j must be an element of ${}^{<\omega}2$ with $s_{j+1} \supseteq s_j{}^\frown\varepsilon_j$. ε_j must be 0 or 1. If all rules are obeyed, player I wins just in case $\langle x, y\rangle \in B$ where $x = \bigcup_{j\in\omega} s_j$ and $y = \langle m_i : i \in \omega\rangle$.

If player I has a winning strategy, then A has a perfect subset, contradicting its countability.

By Lemma 3.3, let τ be a winning strategy for player II such that τ restricted to positions of length n is $\Game^{2n-1}(\omega \cdot (n+1))$-$\Pi^1_1$.

Suppose $x, y \in B$. Then there is a position p in our game with last move ε_{j-1} (we permit p to be the initial position, *i.e.*, $j = 0$) such that

1. $m_{j'} = y(j')$ for all $j' < j$;

2. $s_{j-1}{}^\frown\varepsilon_{j-1} \subseteq x$ (if $j > 0$);
3. for any s_j such that $s_{j-1}{}^\frown\varepsilon_{j-1} \subseteq s_j \subseteq x$, if player I plays $\langle y(j), s_j\rangle$ at p, then τ calls for player II to play $\varepsilon_j = 1 - x(\mathrm{lh}(s_j))$.

If no such position existed, then there would be a play of the game consistent with τ with result $\langle x, y\rangle$.

For such a position p, x is clearly recursive in τ restricted to positions of length $2j+1$. Thus every $x \in A$ belongs to $\Game^{2i-1}(\omega \cdot n)\text{-}\Pi^1_1$ for some n. ⊣

THEOREM 3.5 (PD). $C_{2i} = \bigcup_{n\in\omega} \Game^{2i-1}(\omega \cdot n)\text{-}\Pi^1_1$.

COROLLARY 3.6 ($0^\#$ exists). ${}^\omega\omega \cap \mathbf{L} = \bigcup_{n\in\omega} \Game(\omega\cdot n)\text{-}\Pi^1_1$.

THEOREM 3.7 ($0^\#$ exists). Let B_n be a complete $\Game(\omega \cdot n)\text{-}\Pi^1_1$ subset of ω, uniformly in n. $0^\#$ is recursively isomorphic with $\{\langle n, m\rangle : m \in B_n\} = B$.

PROOF. There is a one-one recursive function f such that $\langle n, m\rangle \in B$ if and only if player I has a winning strategy for $G^*_{f(n,m),\gamma_1,\dots,\gamma_n}$, for any indiscernibles $\gamma_1 < \cdots < \gamma_n$ for $\mathbf{L}$. This follows from Lemma 2.1, part (2). By Lemma 2.1, part (1), B is one-one reducible to $0^\#$.

For the other direction, let $\phi(x_1,\dots,x_n)$ be any formula in the language of set theory. We play a game G_ϕ as follows: player I chooses (dividing up ω) $x_\beta \in {}^\omega\omega$ for each $\beta < \omega\cdot(n+1)$. Player II chooses $y_\beta \in {}^\omega\omega$ for each $\beta < \omega\cdot(n+1)$. If some x_β or y_β is not a code for a countable ordinal, then player I wins if, for the least such β, x_β is a code for a countable ordinal, and player II wins otherwise. If x_β (resp. y_β) codes an ordinal, let $|x_\beta|$ (resp. $|y_\beta|$) be the ordinal coded. If all $|x_\beta|$ and $|y_\beta|$ are defined, let

$$\gamma_i = \sup\{\max\{|x_{\omega\cdot i+j}|, |y_{\omega\cdot i+j}|\} : j \in \omega\}.$$

Player I wins just in case $\mathbf{L}_{\gamma_n} \models \phi[\gamma_1,\dots,\gamma_{n-1}]$. G_ϕ is $(\omega\cdot(n+1))\text{-}\Pi^1_1$, so it is enough to prove that player I wins G_ϕ if and only if $\ulcorner\phi\urcorner \in 0^\#$.

Assume for definiteness that player I has a winning strategy σ for G_ϕ. By boundedness, there is a closed, unbounded $C \subseteq \omega_1$, such that, for all $\beta < \omega\cdot(n+1)$, all $\alpha \in C$, and all plays consistent with σ, if for all $\beta' < \beta$, the ordinal $|y_\beta|$ is defined and $\sup\{|y_{\beta'}| : \beta' < \beta\} < \alpha$, then $|x_\beta| < \alpha$.

Let $\overline{\gamma_1} < \cdots < \overline{\gamma_n}$ be indiscernibles for $\mathbf{L}$ which belong to C. Let player II play y_β such that each $\langle |y_{\omega\cdot i+j}| : j \in \omega\rangle$ is an increasing sequence with limit $\overline{\gamma_i}$. If player I plays σ, then $\gamma_i = \overline{\gamma_i}$ and so $\mathbf{L}_{\overline{\gamma_n}} \models \phi[\overline{\gamma_1},\dots,\overline{\gamma_{n-1}}]$ and so $\mathbf{L} \models \phi[\overline{\gamma_1},\dots,\overline{\gamma_{n-1}}]$, *i.e.*, $\ulcorner\varphi\urcorner \in 0^\#$. ⊣

§4. The largest countable inductive set.

LEMMA 4.1 (Inductive Determinacy). If $A \subseteq \omega$ is Σ^*_n, then A belongs to the largest countable inductive set.

Proof. Let A be Π^*_n if n is even and Σ^*_n if n is odd. Then

$$m \in A \iff (Q_1 x_1)(Q_2 x_2) \dots (\exists x_n)(\langle m, x_1, \dots, x_n \rangle \in B \cap C)$$

where the quantifiers are alternating and B is inductive and C is coinductive. By [Mos80, p. 419], let φ be an inductive norm on B and let ψ be an inductive norm on $\neg C$.

For degrees $\boldsymbol{d}_1, \dots, \boldsymbol{d}_n$, let

$$\begin{aligned} &E_{\boldsymbol{d}_1,\dots,\boldsymbol{d}_n} = \\ &\quad \{S \subseteq \omega : \\ &\quad (Q_1 x_1 \text{ of degree } \leq \boldsymbol{d}_1)(Q_2 x_2 \text{ of degree } \leq \boldsymbol{d}_2) \dots (\exists x_n \text{ of degree } \leq \boldsymbol{d}_n) \\ &\quad \exists \gamma \forall m (m \in S \iff (\varphi(m, x_1, \dots, x_n) < \gamma \wedge \psi(m, x_1, \dots, x_n) \not< \gamma))\}. \end{aligned}$$

Since the inductive sets are closed under number quantification, $E_{\boldsymbol{d}_1,\dots,\boldsymbol{d}_n}$ is inductive uniformly in reals $x_1, \dots, x_n$ of degrees $\boldsymbol{d}_1, \dots, \boldsymbol{d}_n$ respectively.

Let

$$E = \{S : \text{For almost all } (\boldsymbol{d}_1, \dots, \boldsymbol{d}_n), S \in E_{\boldsymbol{d}_1,\dots,\boldsymbol{d}_n}\}.$$

E is inductive, since $S \in E$ just in case

$$\exists \boldsymbol{d}_1 \forall \boldsymbol{d}'_1 \geq \boldsymbol{d}_1 \forall \boldsymbol{d}_2 \exists \boldsymbol{d}'_2 \geq \boldsymbol{d}_2 \dots Q^*_n \boldsymbol{d}'_n \geq \boldsymbol{d}_n S \in E_{\boldsymbol{d}'_1,\dots,\boldsymbol{d}'_n},$$

and the inductive sets are closed under quantification over ${}^\omega\omega$.

E is (inductively) well-orderable, since the function $\varphi_S(\boldsymbol{d}_1, \dots, \boldsymbol{d}_n) =$ the least γ which witnesses $S \in E_{\boldsymbol{d}_1,\dots,\boldsymbol{d}_n}$ embeds E in the ultrapower of the ordinals by the iterated product measure (where we only need the measure defined on inductive sets).

Let us finally show that $A \in E$. For almost all $(\boldsymbol{d}_1, \dots, \boldsymbol{d}_n)$,

$$\begin{aligned} &(Q_1 x_1 \text{ of degree } \leq \boldsymbol{d}_1) \dots (\exists_n x_n \text{ of degree } \leq \boldsymbol{d}_n) \\ &\qquad\qquad (\forall m)((m, x_1, \dots, x_n) \in B \cap C \iff m \in A). \end{aligned}$$

Let $\gamma(m, \boldsymbol{d}_1, \dots, \boldsymbol{d}_n) = \sup\{\varphi(m, x_1, \dots, x_n) : (m, x_1, \dots, x_n) \in A \wedge \forall i \leq n \deg(x_i) \leq \boldsymbol{d}_i\}$. Let $\gamma(\boldsymbol{d}_1, \dots, \boldsymbol{d}_n) = \sup\{\gamma(m, \boldsymbol{d}_1, \dots, \boldsymbol{d}_n) : m \in \omega\}$. $\gamma(\boldsymbol{d}_1, \dots, \boldsymbol{d}_n)$ witnesses that $A \in E_{\boldsymbol{d}_1,\dots,\boldsymbol{d}_n}$. ⊣

Lemma 4.2 (Determinacy for all Σ^*_n games). Every member of any countable inductive set is Σ^*_n for some n.

Proof. By [Mos83] every coinductive set admits a very good scale $\langle \varphi_i : i \in \omega \rangle$ such that φ_i is Σ^*_1. By Lemma 3.3, every inductive game won by player II has a winning strategy τ such that τ restricted to positions of length i is Σ^*_j for some j.

Using a game like that in the proof of Lemma 3.4, except that there are no m_i, we see that every member of a countable inductive set is Σ^*_n for some n. ⊣

Theorem 4.3. The largest countable inductive set is $\bigcup_{n \in \omega} \Sigma^*_n$.

COROLLARY 4.4 (Determinacy of all Σ_n^* games). There is a coinductive set which does not admit a $\underset{\sim}{\Sigma}_n^*$ scale for any n.

PROOF. For each x, let C_x be the largest countable set inductive in x. Let $C = \{\langle x, y\rangle : y \in C_x\}$. $\neg C$ is coinductive. Suppose $\neg C$ admits a $\underset{\sim}{\Sigma}_n^*$ scale. This scale is Σ_n^* in x for some x. Thus $\neg C_x$ has a member Δ_n^* in x. This contradicts the relativization of the theorem. ⊣

The results of this section hold for wider classes than the inductive sets. Suppose Γ has the scale property and is closed under trivial operations and integer and real quantification. With the obvious definitions, Lemma 4.1 holds for Γ (by the same proof). Hence Corollary 4.4 holds for Γ. Lemma 4.2 does not hold in general. We must replace Σ_n^* by, roughly speaking, the nth norm on the largest countable $\widetilde{\Gamma}$ set.

REFERENCES

ALEXANDER S. KECHRIS, DONALD A. MARTIN, AND YIANNIS N. MOSCHOVAKIS

[CABAL iii] *Cabal seminar* 79–81, Lecture Notes in Mathematics, no. 1019, Berlin, Springer, 1983.

YIANNIS N. MOSCHOVAKIS

[Mos80] ***Descriptive set theory***, Studies in Logic and the Foundations of Mathematics, no. 100, North-Holland, Amsterdam, 1980.

[Mos83] *Scales on coinductive sets*, this volume, originally published in Kechris et al. [CABAL iii], pp. 77–85.

JOHN R. STEEL

[Ste83B] *Scales on* Σ_1^1*-sets*, this volume, originally published in Kechris et al. [CABAL iii], pp. 72–76.

DEPARTMENT OF MATHEMATICS
UNIVERSITY OF CALIFORNIA
LOS ANGELES, CA 90095, USA
E-mail: dam@math.ucla.edu

SCALES IN $\mathbf{L}(\mathbb{R})$

JOHN R. STEEL

§0. Introduction. We now know the extent of scales in $\mathbf{L}(\mathbb{R})$: $(\boldsymbol{\Sigma}_1^2)^{\mathbf{L}(\mathbb{R})}$ sets admit $(\boldsymbol{\Sigma}_1^2)^{\mathbf{L}(\mathbb{R})}$ scales, while properly $(\boldsymbol{\Pi}_1^2)^{\mathbf{L}(\mathbb{R})}$ sets admit no scales whatsoever in $\mathbf{L}(\mathbb{R})$. It follows that $(\boldsymbol{\Delta}_1^2)^{\mathbf{L}(\mathbb{R})}$ sets admit $(\boldsymbol{\Delta}_1^2)^{\mathbf{L}(\mathbb{R})}$ scales, but this is by no means a local result, in that the simplest possible scale on a given $(\boldsymbol{\Delta}_1^2)^{\mathbf{L}(\mathbb{R})}$ set may be substantially more complicated than the set itself. Here we shall consider the problem of finding scales of minimal complexity on sets in $\mathbf{L}(\mathbb{R})$, and obtain a fairly complete solution. Given a set A in $\mathbf{L}(\mathbb{R})$, we shall identify by means of reflection properties of the Levy hierarchy for $\mathbf{L}(\mathbb{R})$ the first level $\boldsymbol{\Sigma}_n(\mathbf{L}_\alpha(\mathbb{R}))$ of this hierarchy at which a scale on A is definable. This level occurs very near the least α such that $A \in \mathbf{L}_\alpha(\mathbb{R})$ and for some Σ_1 formula $\varphi(v)$ and real x, $\mathbf{L}_{\alpha+1}(\mathbb{R}) \models \varphi[x]$ while $\mathbf{L}_\alpha(\mathbb{R}) \models \neg\varphi[x]$. That is, in $\mathbf{L}(\mathbb{R})$ the construction of new scales is closely tied to the verification of new Σ_1 statements about reals.

Scales are important in Descriptive Set Theory because they provide the only known general method which will take arbitrary definitions in a given logical form of sets of reals, and produce definitions of members of those sets. This is something a descriptive set theorist will often want to do. It is a pleasing consequence of our work and the earlier work upon which it builds that there is no better general method in $\mathbf{L}(\mathbb{R})$. We shall see that there are no simpler uniformizations of arbitrary $\boldsymbol{\Sigma}_n(\mathbf{L}_\alpha(\mathbb{R}))$ relations on reals than those given by scales.

Our work knits together earlier work of Kechris and Solovay, Martin [Mar83], Martin-Steel [MS83], and Moschovakis [Mos83]. We shall credit this work in the appropriate places. What is new is our systematic use of the Levy hierarchy for $\mathbf{L}(\mathbb{R})$, especially its reflection properties and fine structure.

The paper is organized as follows. In §1 we exposit rather carefully the basic fine structure theory of $\mathbf{L}(\mathbb{R})$. Although there is nothing really new here, we have included this section as a service to the scrupulous reader. In §2 we present the heart of our analysis of the complexity of scales in $\mathbf{L}(\mathbb{R})$. §3 is devoted to the one case in this analysis not covered by §2; we have isolated this case because of it is technically more involved than the others, and the

The Cabal Seminar. Volume I: Games, Scales and Suslin Cardinals
Edited by A. S. Kechris, B. Löwe, J. R. Steel
Lecture Notes in Logic, 31

casual reader might want to skip it. Finally, in §4 we refine the results of §§2 and 3 slightly, and use these results to prove some theorems concerning Suslin cardinals and the pointclass $\boldsymbol{S}(\kappa)$ of κ-Suslin sets.

Some preliminaries and notation: Except in §4, we work in ZF + DC, and state our additional determinacy hypotheses as we need them. (This is done chiefly as a service to the readers and authors of [KW83], who must keep close watch on the determinacy we assume in Theorems 2.1 and 3.7.) We let $\mathbb{R} = {}^{\omega}\omega$, the Baire space, and call its elements reals. Variables $z, y, x, w, \ldots$ range over $\mathbb{R}$, while $\alpha, \beta, \gamma, \delta, \ldots$ range over the class Ord of ordinals. If $0 \le k \le \omega$ and $1 \le \ell \le \omega$, then ${}^{k}\omega \times {}^{\ell}({}^{\omega}\omega)$ is recursively homeomorphic to $\mathbb{R}$, and we sometimes tacitly identify the two. A pointclass is a class of subsets of $\mathbb{R}$ closed under recursive substitutions; a boldface pointclass is a pointclass closed under continuous substitutions. If Γ is a pointclass, then $\check{\Gamma} = \{\mathbb{R} \setminus A : A \in \Gamma\}$ is the dual of Γ, and $\exists^{\mathbb{R}}\Gamma = \{\exists^{\mathbb{R}} A : A \in \Gamma\}$ where $\exists^{\mathbb{R}} A = \{x : \exists y(\langle x, y\rangle \in A)\}$ and $\forall^{\mathbb{R}}\Gamma = (\exists^{\mathbb{R}}\check{\Gamma})\check{}$.

By "Det(Γ)" we mean the assertion that all games whose payoff set is in Γ are determined. For whatever else we use from Descriptive Set Theory, and in particular for the notions of a scale and of the scale property, we refer the reader to [Mos80].

Our general set theoretic notation is standard. The least ordinal satisfying P is denoted by $\mu\alpha\, P(\alpha)$. For any set X, $X^{<\omega}$ is the set of finite sequences of elements of X, $[S]^{<\omega}$ is the set of finite subsets of X, and $\wp(X)$ is the set of all subsets of X. $\mathbf{V}_\alpha$ is the set of sets of rank less than α. If $X \cup \{\mathbf{V}_{\omega+1}\} \subseteq M$, where M is any set, then $\Sigma_n(M, X)$ is the class of relations on M definable over $(M, \in)$ by a Σ_n formula from parameters in $X \cup \{\mathbf{V}_{\omega+1}\}$. Thus we are always allowed $\mathbf{V}_{\omega+1}$ itself (not necessarily its elements) as a parameter in definitions. $\Sigma_\omega(M, X) = \bigcup_{n<\omega} \Sigma_n(M, X)$. We write "$\Sigma_n(M)$" for "$\Sigma_n(M, \varnothing)$", and "$\boldsymbol{\Sigma}_n(M)$" for "$\boldsymbol{\Sigma}_n(M, M)$". Similar conventions apply to the Π_n and Δ_n notations. If $X \cup \{\mathbf{V}_{\omega+1}\} \subseteq M \subseteq N$, then "$M \prec_n^X N$" means that for all $\vec{a} \in (X \cup \{\mathbf{V}_{\omega+1}\})^{<\omega}$ and all Σ_n formulae φ, $M \models \varphi[\vec{a}]$ iff $N \models \varphi[\vec{a}]$. We write "$M \prec_n N$" for "$M \prec_n^M N$".

In the sequel we shall sometimes refer to "the pointclass $\Sigma_n(M, X)$", or assert "$\Sigma_n(M, X)$ has the scale property". In such contexts we are actually referring to $\Sigma_n(M, X) \cap \wp(\mathbb{R})$; the context should make this clear.

§1. The fine structure of L(ℝ). This section contains a straightforward generalization to $\mathbf{L}(\mathbb{R})$ of Jensen's fine structure theory of $\mathbf{L}$. Although the reader could no doubt supply the generalization himself, we have assumed he would rather not, and so have given a fairly complete exposition. ZF + DC will suffice for the results of this section; no determinacy is required.

Because it is convenient, we shall use Jensen $\mathbf{J}$-hierarchy for $\mathbf{L}(\mathbb{R})$. We presume the basic facts about rudimentary functions set forth in [Jen72]. Let rud(M) be the closure of $M \cup \{M\}$ under the rudimentary functions. By [Jen72, Corollary 1.7], $\text{rud}(M) \cap \wp(M) = \underset{\sim}{\Sigma}_\omega(M)$ for transitive sets M. Now let

$$\mathbf{J}_1(\mathbb{R}) = \mathbf{V}_{\omega+1} = \{a : \text{rank}(a) \leq \omega\}$$
$$\mathbf{J}_{\alpha+1}(\mathbb{R}) = \text{rud}(\mathbf{J}_\alpha(\mathbb{R})), \text{ for } \alpha > 0,$$

and

$$\mathbf{J}_\lambda(\mathbb{R}) = \bigcup_{\alpha<\lambda} \mathbf{J}_\alpha(\mathbb{R}), \text{ for } \lambda \text{ limit.}$$

Of course, $\mathbf{L}(\mathbb{R}) = \bigcup_{\alpha\in\text{Ord}} \mathbf{J}_\alpha(\mathbb{R})$. For all $\alpha \geq 1$, $\mathbf{J}_\alpha(\mathbb{R})$ is transitive and $\mathbf{J}_{\alpha+1}(\mathbb{R}) \cap \wp(\mathbf{J}_\alpha(\mathbb{R})) = \underset{\sim}{\Sigma}_\omega(\mathbf{J}_\alpha(\mathbb{R}))$. Thus $\mathbf{J}_\alpha(\mathbb{R}) \in \mathbf{J}_{\alpha+1}(\mathbb{R})$ for $\alpha \geq 1$, and $\mathbf{J}_\alpha(\mathbb{R}) \subseteq \mathbf{J}_\beta(\mathbb{R})$ for $1 \leq \alpha \leq \beta$. If $\alpha > 1$, $\text{rank}(\mathbf{J}_\alpha(\mathbb{R})) = \text{Ord} \cap \mathbf{J}_\alpha(\mathbb{R}) = \omega\alpha$.

It is useful to refine this hierarchy. Recall from [Jen72] the rudimentary functions $F_1, \ldots, F_8$ from which all rudimentary functions can be generated by composition. Let $F_9(a, b) = \langle a, b\rangle$, $F_{10}(a, b) = a[b] = \{c : \langle b, c\rangle \in a\}$, and $F_{11}(a, b, c) = \langle a, b, c\rangle$. Define

$$S(M) = M \cup \{M\} \cup \bigcup_{i=1}^{11} F_i[M \cup \{M\}].$$

Then if M is transitive, so is $S(M)$ (which is why we included F_9, F_{10}, and F_{11}); moreover,

$$\text{rud}(M) = \bigcup_{n<\omega} S^n(M).$$

Now let

$$S_\omega(\mathbb{R}) = \mathbf{J}_1(\mathbb{R}),$$
$$S_{\alpha+1}(\mathbb{R}) = S(S_\alpha(\mathbb{R})) \text{ for } \alpha \geq \omega,$$

and

$$S_\lambda(\mathbb{R}) = \bigcup_{\alpha<\lambda} S_\alpha(\mathbb{R}) \text{ for } \lambda \text{ limit.}$$

Clearly the $S_\alpha(\mathbb{R})$'s are transitive and cumulative, $S_\alpha(\mathbb{R}) \in S_{\alpha+1}(\mathbb{R})$, and $S_{\omega\alpha}(\mathbb{R}) = \mathbf{J}_\alpha(\mathbb{R})$ for all α.

Lemma 1.1. The sequences $\langle S_\gamma(\mathbb{R}) : \gamma < \omega\alpha\rangle$ and $\langle \mathbf{J}_\beta(\mathbb{R}) : \beta < \alpha\rangle$ are uniformly $\Sigma_1(\mathbf{J}_\alpha(\mathbb{R}))$ for $\alpha > 1$.

Proof. The reason is that the two sequences are defined by local Σ_0 recursions. Notice that

$$M = S_\gamma(\mathbb{R}) = \exists f\, (\Phi(f)\ \&\ f(\gamma) = M),$$

where

$$\Phi(f) \iff \big[\, f \text{ is a function } \& \operatorname{dom} f \in \mathrm{Ord} \;\&\; f(\omega) = \mathbf{V}_{\omega+1} \;\&$$
$$\forall \alpha \in \operatorname{dom} f \; \big(\alpha + 1 \in \operatorname{dom} f \Rightarrow f(\alpha+1) = S(f(\alpha))\big) \;\&$$
$$\forall \lambda \in \operatorname{dom} f \; \big(\lambda \text{ limit } \Rightarrow f(\lambda) = \bigcup_{\alpha<\lambda} f(\alpha)\big)\big].$$

The formula Φ is Σ_0 since rudimentary functions have Σ_0 graphs and we are always allowed $\mathbf{V}_{\alpha+1}$ as a parameter. So it is enough to show that for $\gamma < \omega\alpha$

$$M = S_\gamma(\mathbb{R}) \iff \exists f \in \mathbf{J}_\alpha(\mathbb{R}) \; \big(\Phi(f) \;\&\; f(\gamma) = M\big).$$

That is, we must show that $\langle S_\delta(\mathbb{R}) : \delta \leq \gamma \rangle \in \mathbf{J}_\alpha(\mathbb{R})$ for all γ and α such that $\gamma < \omega\alpha$. This can be proved by induction on γ. For limit γ, say $\gamma = \omega\beta$, we use the fact that $\langle S_\delta(\mathbb{R}) : \delta < \gamma \rangle$ is $\Sigma_1(\mathbf{J}_\beta(\mathbb{R}))$ by induction hypothesis.

A similar argument shows that $\langle \omega\beta : \beta < \alpha \rangle$ is uniformly $\Sigma_1(\mathbf{J}_\alpha(\mathbb{R}))$, and therefore $\langle \mathbf{J}_\beta(\mathbb{R}) : \beta < \alpha \rangle$ is uniformly $\Sigma_1(\mathbf{J}_\alpha(\mathbb{R}))$. ⊣

In a similar vein,

Lemma 1.2. There is a Π_2 sentence ϑ such that for all transitive sets M with $\mathbf{V}_{\omega+1} \in M$,

$$M \models \vartheta \text{ iff } \exists \alpha > 1 \; \big(M = \mathbf{J}_\alpha(\mathbb{R})\big).$$

Proof. Let $\Phi(f, u)$ be the Σ_0 formula resulting from the Σ_0 formula $\Phi(f)$ of Lemma 1.1 by replacing "$\mathbf{V}_{\omega+1}$" with the variable u. Let

$$\psi(u) \iff \forall a \exists f \exists \gamma \geq \omega \; \big(\Phi(f, u) \;\&\; a \in f(\gamma)\big) \;\&$$
$$\forall \gamma \in \mathrm{Ord} \, \exists \beta \in \mathrm{Ord}(\gamma < \beta) \;\&$$
$$\forall \gamma \in \mathrm{Ord} \, \exists f \, \big(\Phi(f, u) \;\&\; \gamma \in \operatorname{dom} f\big).$$

Then ψ is Π_2, and it is easy to check that for transitive M with $\mathbf{V}_{\omega+1} \in M$,

$$M \models \psi\,[\mathbf{V}_{\omega+1}] \text{ iff } \exists \alpha > 1 \; \big(M = \mathbf{J}_\alpha(\mathbb{R})\big).$$

The desired sentence ϑ is therefore

$$\forall u \; \big(\forall x (x \in u \iff \operatorname{rank}(x) \leq \omega) \rightarrow \psi(u)\big).$$

⊣

We shall call the Π_2 sentence ϑ provided by Lemma 1.2 "V=L(ℝ)".

Corollary 1.3. If $M \prec_1 \mathbf{J}_\alpha(\mathbb{R})$ and $\mathbf{V}_{\omega+1} \in M$, then $M \cong \mathbf{J}_\beta(\mathbb{R})$ for some $\beta \leq \alpha$.

Of course, if there is any point to this paper there is no definable wellorder of $\mathbf{J}_\alpha(\mathbb{R})$. The next lemma is as much as we can get in the direction of such a wellorder.

LEMMA 1.4. There are uniformly $\Sigma_1(\mathbf{J}_\alpha(\mathbb{R}))$ maps f_α such that

$$f_\alpha\colon [\omega\alpha]^{<\omega} \times \mathbb{R} \twoheadrightarrow \mathbf{J}_\alpha(\mathbb{R}).$$

PROOF. Fix a rudimentary function h such that for all finite $F \subseteq \mathrm{Ord}$,

$$h[\{F\} \times \mathbb{R}] = \{\langle G_1, G_2, G_3\rangle : G_i \subseteq F \text{ for } 1 \leq i \leq 3\}.$$

It is easy to construct such an h. We now define by induction on $\gamma \geq \omega$ functions

$$g_\gamma\colon [\gamma]^{<\omega} \times \mathbb{R} \twoheadrightarrow S_\gamma(\mathbb{R})$$

such that $g_\gamma \subseteq g_\delta$ if $\gamma \leq \delta$. We leave g_ω to the reader's discretion. If λ is a limit, let

$$g_\lambda = \bigcup_{\beta<\lambda} g_\beta.$$

Finally, suppose g_γ is given. Let F and x be given, and set

$$g_{\gamma+1}(F, x) = g_\gamma(F, x)$$

if $\gamma \notin F$. If $\gamma \in F$, let $h(F - \{\gamma\}, (x)_0) = \langle G_1, G_2, G_3\rangle$, and for $1 \leq i \leq 3$, let

$$a_i = \begin{cases} g_\gamma(G_i, (x)_i) & \text{if } (x)_4(i) = 0 \\ S_\gamma(\mathbb{R}) & \text{if } (x)_4(i) \neq 0. \end{cases}$$

Let also $j \in \{1, \dots, 11\}$ be such that $j = (x)_4(4) \pmod{11}$. Then we set

$$g_{\gamma+1}(F, x) = \begin{cases} F_j(a_1, a_2) & \text{if } j \neq 11 \\ F_j(a_1, a_2, a_3) & \text{if } j = 11. \end{cases}$$

By an easy induction we have $g_\gamma\colon [\gamma]^{<\omega} \times \mathbb{R} \twoheadrightarrow S_\gamma(\mathbb{R})$ for all γ. But now notice that there is a rudimentary function G such that $g_{\gamma+1} = G(g_\gamma, S_\gamma(\mathbb{R}))$ for all γ. By the proof of Lemma 1.1, $\langle g_\gamma : \gamma < \omega\alpha\rangle$ is uniformly $\Sigma_1(\mathbf{J}_\alpha(\mathbb{R}))$. Thus the desired f_α is given by

$$f_\alpha = \bigcup_{\gamma<\omega\alpha} g_\gamma.$$

[Jen72, Lemma 2.10] implies that for all α there is a $\underset{\sim}{\Sigma}_1(\mathbf{J}_\alpha(\mathbb{R}))$ map of α onto $[\omega\alpha]^{<\omega}$, and therefore a $\underset{\sim}{\Sigma}_1(\mathbf{J}_\alpha(\mathbb{R}))$ map of $\omega\alpha \times \mathbb{R}$ onto $\mathbf{J}_\alpha(\mathbb{R})$. However, the parameters involved in the Σ_1 definitions of these map, and the consequent lack of uniformity in their definitions, make the maps described in Lemma 1.4 more useful.

One can identify finite sets of ordinals with descending sequences of ordinals. The tree of all such descending sequences is wellfounded, and so its Brouwer-Kleene order is a wellorder. This gives us a wellorder of $[\mathrm{Ord}]^{<\omega}$. More explicitly, for $F, G \in [\mathrm{Ord}]^{<\omega}$ let

$$F \leq_{\mathrm{BK}} G \text{ iff } \big(\exists\alpha \in G\ (G = F \setminus \alpha) \vee \max(G \mathbin{\triangle} F) \in G\big).$$

Then $\leq_{\text{BK}}$ is a Σ_0 wellorder of $[\text{Ord}]^{<\omega}$. ⊣

We begin now to make our way toward the Σ_n selection theorem.

One central feature of fine structure theory is that it allows us to form Skolem hulls in the most effective way possible. This involves Σ_n selection (or uniformization). Now in $\mathbf{L}(\mathbb{R})$ we cannot hope to select from arbitrary set in a definable way; for example, the relation

$$R(\alpha, x) \iff \alpha < \omega_1 \;\&\; x \in \mathbb{R} \;\&\; x \text{ codes } \alpha$$

cannot be uniformized at all in $\mathbf{L}(\mathbb{R})$. However, we shall throw all of $\mathbb{R}$ into the hulls we build anyway, so that in view of Lemma 1.4 it suffices to select effectively from subsets of $[\text{Ord}]^{<\omega}$. Here we have a chance.

Definition 1.5. $\mathbf{J}_\alpha(\mathbb{R})$ **satisfies** Σ_n **selection** iff whenever $R \subseteq \mathbf{J}_\alpha(\mathbb{R}) \times [\omega\alpha]^{<\omega}$ is $\utilde{\Sigma}_n(\mathbf{J}_\alpha)$, there is a $\utilde{\Sigma}_n(\mathbf{J}_\alpha(\mathbb{R}))$ set $S \subseteq R$ such that

$$\forall a(\exists F\; R(a, F) \to \exists! F\; S(a, F)).$$

Definition 1.6. Let $h\colon \mathbf{J}_\alpha(\mathbb{R}) \times \mathbb{R} \to \mathbf{J}_\alpha(\mathbb{R})$ be a partial $\utilde{\Sigma}_n(\mathbf{J}_\alpha(\mathbb{R}))$ map. Then h is a Σ_n **Skolem function** for $\mathbf{J}_\alpha(\mathbb{R})$ iff whenever S is $\Sigma_n(\mathbf{J}_\alpha(\mathbb{R}), \{a\})$ and $S \neq 0$, then $\exists x \in \mathbb{R}\,(h(a, x) \in S)$.

Recall [Jen72, Corollary 1.13], according to which the satisfaction relation restricted to Σ_n formula is uniformly $\Sigma_n(M)$ for transitive, rudimentarily closed M.

Lemma 1.7. If $\mathbf{J}_\alpha(\mathbb{R})$ satisfies Σ_n selection, then there is a Σ_n Skolem function for $\mathbf{J}_\alpha(\mathbb{R})$.

Proof. Let f_α be the map given by Lemma 1.4, and let $\langle \varphi_i : i < \omega \rangle$ be an enumeration of the Σ_n formulae of two free variables. Let

$$R(a, x, F) \text{ iff } \mathbf{J}_\alpha(\mathbb{R}) \models \varphi_{x(0)}\,[a, f_\alpha(F, \lambda i.x(i+1))].$$

Since R is $\Sigma_n(\mathbf{J}_\alpha(\mathbb{R}))$, we have a partial $\utilde{\Sigma}_n(\mathbf{J}_\alpha(\mathbb{R}))$ map g uniformizing R. Let

$$h(a, x) = f_\alpha\big(g(a, x), \lambda i.x(i+1)\big).$$

It is easy to check that h is the desired Skolem function. ⊣

Definition 1.8. Let $X \subseteq M$ and $1 \leq n \leq \omega$. Then

$$\text{Hull}_n^M(X) = \{a \in M : \{a\} \text{ is } \Sigma_n(M, X)\}.$$

We write "$\text{Hull}_n^\alpha(X)$" for "$\text{Hull}_n^{\mathbf{J}_\alpha(\mathbb{R})}(X)$". The connection between these hulls and Σ_n Skolem functions is explained in the next lemma.

Lemma 1.9. Let $H = \text{Hull}_n^\alpha(X)$ where $n < \omega$ and $\mathbb{R} \subseteq X$. Suppose that for some $p \in H$ there is a Σ_n Skolem function for $\mathbf{J}_\alpha(\mathbb{R})$ which is in fact $\Sigma_n(\mathbf{J}_\alpha(\mathbb{R}), \{p\})$. Then

(a) $X \prec_n \mathbf{J}_\alpha(\mathbb{R})$,

(b) $H = h[X^{<\omega}]$ for some partial $\Sigma_n(H)$ map h, and
(c) if $X = Y \cup \mathbb{R}$ where Y is finite, then $H = h[\mathbb{R}]$ for some partial $\Sigma_n(H)$ map h.

PROOF. (a) Notice that if $\{c\}$ is $\Sigma_n(\mathbf{J}_\alpha(\mathbb{R}), H)$, then $c \in H$. Now let $a_1, \dots, a_n \in H$ and suppose $\mathbf{J}_\alpha(\mathbb{R}) \models \varphi[a_1, \dots, a_n, b]$ where φ is Σ_n. We want to see that $\mathbf{J}_\alpha(\mathbb{R}) \models \varphi[a_1, \dots, a_n, c]$ for some $c \in H$. Let $\tilde{h}$ be a Σ_n Skolem function for $\mathbf{J}_\alpha(\mathbb{R})$ which is $\Sigma_n(\mathbf{J}_\alpha(\mathbb{R}), \{p\})$ with $p \in H$. Then for some $x \in \mathbb{R}$, if we set $c = \tilde{h}(\langle a_1, \dots, a_n\rangle, x)$ then $\mathbf{J}_\alpha(\mathbb{R}) \models \varphi[a_1, \dots, a_n, c]$. Since $\{c\}$ is $\Sigma_n(\mathbf{J}_\alpha(\mathbb{R}), \{a_1, \dots, a_n, x, p\})$, $c \in H$.

(b) Let $\tilde{h}$ be as in (a), and set

$$h(\langle a_1, \dots, a_n\rangle) = \tilde{h}(\langle a_1, \dots, a_{n-1}, p\rangle, a_n).$$

One can easily check that $h[X^{<\omega}] = H$.

(c) This follows easily from (b). ⊣

Incidentally, we don't need Σ_n selection to show that for $\mathbb{R} \subseteq X$,

$$\mathrm{Hull}^\alpha_{n+1}(X) \prec_n \mathbf{J}_\alpha(\mathbb{R})$$

for all $n < \omega$, and therefore

$$\mathrm{Hull}^\alpha_\omega(X) \prec_\omega \mathbf{J}_\alpha(\mathbb{R}).$$

The crude Skolem functions obtained by selecting the BK-least F in the proof of Lemma 1.7 will suffice for this.

LEMMA 1.10. For all $\alpha > 1$, $\mathbf{J}_\alpha(\mathbb{R})$ satisfies Σ_1 selection.

PROOF. Let $R \subseteq \mathbf{J}_\alpha(\mathbb{R}) \times [\omega\alpha]^{<\omega}$, and

$$R(a, F) \text{ iff } \exists b\big(\mathbf{J}_\alpha(\mathbb{R}) \models \varphi[a, F, b, p]\big),$$

where φ is Σ_0 and $p \in \mathbf{J}_\alpha(\mathbb{R})$. Define

$$\begin{aligned} S(a, F) \text{ iff } \exists \gamma < \omega\alpha\ \big[&\exists b \in S_\gamma(\mathbb{R})\, \varphi(a, F, b, p)\ \&\\ &\forall \delta < \gamma \forall b \in S_\delta(\mathbb{R}) \neg\varphi(a, F, b, p)\ \&\\ &\forall G \in S_\gamma(\mathbb{R}) \forall b \in S_\gamma(\mathbb{R})\big(G \leq_{\mathrm{BK}} F \rightarrow \neg\varphi(a, G, b, p)\big)\big]. \end{aligned}$$

By Lemma 1.1, S is $\Sigma_1(\mathbf{J}_\alpha(\mathbb{R}), \{p\})$, and clearly S uniformizes R. ⊣

Lemma 1.10 fails for $\alpha = 1$ since there is a Σ^1_1 subset of $\mathbb{R} \times \{0, 1\}$ with no Σ^1_1 uniformization.

Before going on to Σ_n selection, we make some simple use of our ability to construct Σ_1 Skolem hulls.

LEMMA 1.11. Suppose that $\alpha > 1$, and for no $\beta < \alpha$ do we have $\mathbf{J}_\beta(\mathbb{R}) \prec^{\mathbb{R}}_1 \mathbf{J}_\alpha(\mathbb{R})$. Then

(a) There is a $\Sigma_1(\mathbf{J}_\alpha(\mathbb{R}))$ partial map $h\colon \mathbb{R} \twoheadrightarrow \mathbf{J}_\alpha(\mathbb{R})$,
(b) If $\beta < \alpha$, then there is a total map $h\colon \mathbb{R} \twoheadrightarrow \mathbf{J}_\beta(\mathbb{R})$ such that $h \in \mathbf{J}_\alpha(\mathbb{R})$.

PROOF. (a) Let $H = \mathrm{Hull}_1^{\alpha}(\mathbb{R})$. From the proofs of Lemmas 1.7 and 1.10, we see that $\mathbf{J}_\alpha(\mathbb{R})$ has a Σ_1 Skolem function which is (*lightface*) $\Sigma_1(\mathbf{J}_\alpha(\mathbb{R}))$. By Lemma 1.9 then, $H \prec_1 \mathbf{J}_\alpha(\mathbb{R})$ and H is the image of $\mathbb{R}$ under a $\Sigma_1(H)$ map $\tilde{h}$. Let $\pi \colon H \cong \mathbf{J}_\beta(\mathbb{R})$ be the collapse map. Since $\pi \restriction (\mathbb{R} \cup \{\mathbb{R}\})$ is the identity, $\mathbf{J}_\beta(\mathbb{R}) \prec_1^{\mathbb{R}} \mathbf{J}_\alpha(\mathbb{R})$, and therefore $\beta = \alpha$. But then $\pi[\tilde{h}]$ is the desired h.

(b) Suppose $\beta < \alpha$. By increasing β if necessary we may assume that for some Σ_1 formula φ and real x, $\mathbf{J}_{\beta+1}(\mathbb{R}) \models \varphi[x]$ but $\mathbf{J}_\beta(\mathbb{R}) \not\models \varphi[x]$. Let

$$S_{\omega\beta+n} \models \varphi[x].$$

Now rudimentary functions are simple [Jen72, Lemma 1.2], and so we can find a Σ_0 formula ψ such that for any transitive M and $x \in M$,

$$\psi(M, x) \iff S^n(M) \models \varphi[x].$$

Clearly, there is a first order formula ϑ such that for any transitive M and $x \in M$

$$(M \models \vartheta[x]) \iff \psi(M, x).$$

Let ϑ be Σ_n. Let $H := \mathrm{Hull}_{n+1}(\mathbb{R})$. Then $H \prec_n \mathbf{J}_\beta(\mathbb{R})$. Let $\pi \colon H = \mathbf{J}_\gamma(\mathbb{R})$ be the collapse map. Since $\pi(x) = x$ and $\pi(\mathbb{R}) = \mathbb{R}$, $\mathbf{J}_\gamma(\mathbb{R}) \models \vartheta[x]$, and so $\gamma = \beta$. It is easy to get a $\utilde{\Sigma}_\omega(H)$ map of $\mathbb{R}$ onto H, which then collapses to a $\utilde{\Sigma}_\omega(\mathbf{J}_\beta(\mathbb{R}))$ map of $\mathbb{R}$ onto $\mathbf{J}_\beta(\mathbb{R})$. This map belongs to $\mathbf{J}_\alpha(\mathbb{R})$. ⊣

Let Θ be the least ordinal not the surjective image of $\mathbb{R}$, and $\utilde{\delta}_1^2$ the least ordinal not the image of $\mathbb{R}$ under a surjection f such that $\{\langle x, y\rangle : f(x) \leq f(y)\}$ is $\utilde{\Delta}_1^2$. A Skolem hull argument like those just given shows that $\Theta^{\mathbf{L}(\mathbb{R})} = \mu\alpha\,[\wp(\mathbb{R}) \cap \mathbf{L}(\mathbb{R}) \subseteq \mathbf{J}_\alpha(\mathbb{R})]$. For $(\utilde{\delta}_1^2)^{\mathbf{L}(\mathbb{R})}$ we have

LEMMA 1.12. Let σ be least such that $\mathbf{J}_\sigma(\mathbb{R}) \prec_1^{\mathbb{R}} \mathbf{L}(\mathbb{R})$. Then

(a) $(\utilde{\Sigma}_1^2)^{\mathbf{L}(\mathbb{R})} = \utilde{\Sigma}_1(\mathbf{J}_\sigma(\mathbb{R})) \cap \wp(\mathbb{R})$

(b) $(\utilde{\Delta}_1^2)^{\mathbf{L}(\mathbb{R})} = \mathbf{J}_\sigma(\mathbb{R}) \cap \wp(\mathbb{R})$

(c) $(\utilde{\delta}_1^2)^{\mathbf{L}(\mathbb{R})} = \sigma$.

PROOF. (a) Clearly, any $(\utilde{\Sigma}_1^2)^{\mathbf{L}(\mathbb{R})}$ set of reals is $\Sigma_1(\mathbf{L}(\mathbb{R}), \mathbb{R})$, and therefore $\Sigma_1(\mathbf{J}_\sigma(\mathbb{R}), \mathbb{R})$. For the converse, let A be a $\utilde{\Sigma}_1(\mathbf{J}_\sigma(\mathbb{R}))$ set of reals. By Lemma 1.11, A is $\Sigma_1(\mathbf{J}_\sigma(\mathbb{R}), \mathbb{R})$. Let

$$\begin{aligned} x \in A &\iff \mathbf{J}_\sigma(\mathbb{R}) \models \varphi[x, y] \\ &\iff \mathbf{L}(\mathbb{R}) \models \varphi[x, y] \end{aligned}$$

where φ is Σ_1. Since $\mathbf{J}_\sigma(\mathbb{R})$ is the surjective image of $\mathbb{R}$ under a map in $\mathbf{L}(\mathbb{R})$,

$$\begin{aligned} x \in A \iff \exists E \in \mathbf{L}(\mathbb{R})\ (&\mathfrak{M} = (\mathbb{R}, E) \text{ is a wellfounded extensional model} \\ &\text{of } V = \mathbf{L}(\mathbb{R})\ \&\ \mathbb{R}^{\mathfrak{M}} \text{ collapses to } \mathbb{R}\ \&\ \mathfrak{M} \models \varphi[x^{\mathfrak{M}}, y^{\mathfrak{M}}]), \end{aligned}$$

where we use $x^{\mathfrak{M}}$ and $y^{\mathfrak{M}}$ for the inverse images of x and y under the collapse of $\mathfrak{M}$. Thus A is $(\utilde{\Sigma}_1^2)^{\mathbf{L}(\mathbb{R})}$.

(b) By Lemma 1.11, $\mathbf{J}_\sigma(\mathbb{R}) \prec_1 \mathbf{L}(\mathbb{R})$, and so $\mathbf{J}_\sigma(\mathbb{R})$ is admissible. If A is $(\underset{\sim}{\Delta}_1^2)^{\mathbf{L}(\mathbb{R})}$, then A is $\underset{\sim}{\Delta}_1(\mathbf{J}_\sigma(\mathbb{R}))$ by (a), and so $A \in \mathbf{J}_\sigma(\mathbb{R})$ by admissibility.

(c) $(\underset{\sim}{\delta}_1^2)^{\mathbf{L}(\mathbb{R})} \leq \sigma$ by (b) and the fact that $\mathbf{J}_\sigma(\mathbb{R})$ is admissible. But if $\beta < \sigma$, then Lemma 1.11 (b) guarantees a map $h \colon \mathbb{R} \twoheadrightarrow \beta$ in $\mathbf{J}_\sigma(\mathbb{R})$. Since $\{\langle x, y\rangle : h(x) \leq h(y)\} \in \mathbf{J}_\sigma(\mathbb{R})$, it is $(\underset{\sim}{\Delta}_1^2)^{\mathbf{L}(\mathbb{R})}$ and therefore $\beta < (\underset{\sim}{\delta}_1^2)^{\mathbf{L}(\mathbb{R})}$. Thus $\sigma \leq (\underset{\sim}{\delta}_1^2)^{\mathbf{L}(\mathbb{R})}$. ⊣

Lemma 1.2 indicates an analogy between the pointclass $(\underset{\sim}{\Sigma}_1^2)^{\mathbf{L}(\mathbb{R})}$ and the class $\Sigma_2^1 \cap \wp(\omega) = (\Sigma_2^1)^{\mathbf{L}} \cap \wp(\omega)$. This analogy is strengthened by the result of [MS83] that

$$(\underset{\sim}{\Sigma}_1^2)^{\mathbf{L}(\mathbb{R})} = \Game^{\mathbb{R}}\underset{\sim}{\Pi}_1^1,$$

while of course $\Sigma_2^1 \cap \wp(\omega) = \Game\Pi_1^1$. One further evidence of the analogy is that if $A \in (\underset{\sim}{\Delta}_1^2)^{\mathbf{L}(\mathbb{R})}$, then $A \leq_{\mathrm{W}} B$ for some $B \subseteq \mathbb{R}$ such that $\{B\}$ is $\underset{\sim}{\Pi}_1^1$. ($A \leq_{\mathrm{W}} B$ iff $\exists f$ (f continuous & $A = f^{-1}(B)$)). To see this, let $A \in \mathbf{J}_\alpha(\mathbb{R})$ where for some formula $\varphi(v)$ and real x, $\mathbf{J}_{\alpha+1}(\mathbb{R}) \models \varphi[x]$ but $\mathbf{J}_\alpha(\mathbb{R}) \models \neg\varphi[x]$. Take $B = \{\langle \vartheta, y\rangle : y \in \mathbb{R} \;\&\; \mathbf{J}_{\alpha+1}(\mathbb{R}) \models \vartheta[y]\}$. It is easy to show that $\{B\}$ is $\Pi_1^1(x)$, and $A \leq_{\mathrm{W}} B$.

We proceed to the Σ_n selection theorem. Our proof is a transcription of Jensen's proof for $\mathbf{L}$, as simplified by Sy Friedman (cf. [Sim78]).

Definition 1.13. For any $\alpha \in \mathrm{Ord}$ and $n \in \omega$

$$\rho_n^\alpha = \text{the least } \beta \text{ such that there is a } \underset{\sim}{\Sigma}_n(\mathbf{J}_\alpha(\mathbb{R})) \text{ partial map}$$
$$f \colon \mathbf{J}_\beta(\mathbb{R}) \twoheadrightarrow \mathbf{J}_\alpha(\mathbb{R})$$

if $n \geq 1$, and $\rho_n^\alpha = \alpha$ if $n = 0$.

By Lemma 1.4, there is in fact a $\underset{\sim}{\Sigma}_n(\mathbf{J}_\alpha(\mathbb{R}))$ partial map from $\omega\rho_n^\alpha \times \mathbb{R}$ onto $\mathbf{J}_\alpha(\mathbb{R})$. We shall write "$\rho_n^\alpha = \mathbb{R}$" to mean $\rho_n^\alpha = 1$.

Lemma 1.14. Suppose $n \geq 1$ and $\mathbf{J}_\alpha(\mathbb{R})$ satisfies Σ_n selection. Then

$$\rho_n^\alpha = \text{the least } \beta \text{ such that some } \underset{\sim}{\Sigma}_n(\mathbf{J}_\alpha(\mathbb{R})) \text{ subset of } \mathbf{J}_\beta(\mathbb{R})$$
$$\text{is not a member of } \mathbf{J}_\alpha(\mathbb{R})$$

Proof. Let β be as on the right hand side. If $h \colon \mathbf{J}_{\rho_n^\alpha}(\mathbb{R}) \twoheadrightarrow \mathbf{J}_\alpha(\mathbb{R})$ is partial Σ_n, then $\{a \in \operatorname{dom} h : a \notin h(a)\}$ witnesses that $\beta \leq \rho_n^\alpha$. Now let $p \in \mathbf{J}_\alpha(\mathbb{R})$ be such that $\mathbf{J}_\alpha(\mathbb{R})$ has a Σ_n Skolem function which is $\Sigma_n(\mathbf{J}_\alpha(\mathbb{R}), \{p\})$ and there is a $\Sigma_n(\mathbf{J}_\alpha(\mathbb{R}), \{p\})$ subset of $\mathbf{J}_\beta(\mathbb{R})$ not in $\mathbf{J}_\alpha(\mathbb{R})$. Let

$$H = \mathrm{Hull}_n^\alpha(\mathbf{J}_\beta(\mathbb{R}) \cup \{p\}).$$

Then $H \prec_n \mathbf{J}_\alpha(\mathbb{R})$ and there is a $\underset{\sim}{\Sigma}_n(H)$ partial map h of $\mathbf{J}_\beta(\mathbb{R})$ onto H, by Lemma 1.9. Let $\pi \colon H \cong \mathbf{J}_\delta(\mathbb{R})$ be the collapse. Then $\delta = \alpha$, as our new subset of $\mathbf{J}_\beta(\mathbb{R})$ at α is constructed at δ. Thus $\pi[h]$ witnesses that $\rho_n^\alpha \leq \beta$. ⊣

The next lemma gives the basic quantifier complexity reduction behind the proof of the Σ_n selection theorem.

LEMMA 1.15. Let $n \geq 1$ and assume $\mathbf{J}_\alpha(\mathbb{R})$ satisfies Σ_m selection for all $m \leq n$. Let Q be a $\underset{\sim}{\Sigma}_n(\mathbf{J}_\alpha(\mathbb{R}))$ relation and f a $\underset{\sim}{\Sigma}_n(\mathbf{J}_\alpha(\mathbb{R}))$ partial function. Define S by

$$S(a, \beta) \text{ iff } \beta < p_n^\alpha \ \& \ \forall b \in \mathbf{J}_\beta(\mathbb{R}) \ (b \in \operatorname{dom} f \to Q(a, \beta, f(b))).$$

Then S is $\underset{\sim}{\Sigma}_{n+1}(\mathbf{J}_\alpha(\mathbb{R}))$.

PROOF. We proceed by induction on n. Suppose $Q(a, b, c) \iff \exists d P(a, b, c, d)$, where P is $\underset{\sim}{\Pi}_{n-1}(\mathbf{J}_\alpha(\mathbb{R}))$. Let $\rho_n = p_n^\alpha$ and $\rho_{n-1} = \rho_{n-1}^\alpha$. Let $h\colon \mathbf{J}_{\rho_{n-1}}(\mathbb{R}) \twoheadrightarrow \mathbf{J}_\alpha(\mathbb{R})$ be a partial $\underset{\sim}{\Sigma}_{n-1}(\mathbf{J}_\alpha(\mathbb{R}))$ map.

Case 1. For some $\delta < \rho_n$ and total $\underset{\sim}{\Sigma}_n(\mathbf{J}_\alpha(\mathbb{R}))$ map $g\colon \mathbf{J}_\delta(\mathbb{R}) \to \rho_{n-1}$, the range of g is cofinal in ρ_{n-1}.

PROOF. Suppose $S(a, \beta)$. Then if $b \in \mathbf{J}_\beta(\mathbb{R})$ and $b \in \operatorname{dom}(f)$, we can find $c \in \mathbf{J}_{\rho_{n-1}}(\mathbb{R})$ so that $P(a, \beta, f(b), h(c))$. Moreover, $c \in \mathbf{J}_{g(d)}(\mathbb{R})$ for some $d \in \mathbf{J}_\delta(\mathbb{R})$. Let

$$\begin{aligned} H(b, d) \iff\ & b \in \mathbf{J}_\beta(\mathbb{R}) \cap \operatorname{dom} f \ \& \ d \in \mathbf{J}_\delta(\mathbb{R}) \\ & \& \ \forall e \forall \eta \ ((e = f(b) \ \& \ \eta = g(d)) \\ & \quad \to \exists c \in \mathbf{J}_\eta(\mathbb{R}) \ P(a, \beta, e, h(c))). \end{aligned}$$

We have just seen that $S(a, \beta) \to \forall b \, \exists d \, H(b, d)$. Our induction hypothesis implies (or if $n = 1$, it is trivial that) the consequent of the final implication defining H is $\underset{\sim}{\Pi}_n$. Thus H is $\underset{\sim}{\Sigma}_n$ & $\underset{\sim}{\Pi}_n$, and being bounded in $\mathbf{J}_{\rho_n}(\mathbb{R})$, $H \in \mathbf{J}_\alpha(\mathbb{R})$. But then $S(a, \beta)$ is equivalent to the existence of such an H in $\mathbf{J}_\alpha(\mathbb{R})$; that is,

$$\begin{aligned} S(a, \beta) \iff\ & \beta < \rho_n \ \& \ \exists H \in \mathbf{J}_\alpha(\mathbb{R}) \\ & \Big(\forall b \in \mathbf{J}_\beta(\mathbb{R}) \ (b \in \operatorname{dom} f \to \exists d \, H(b, d)) \\ & \quad \& \ \forall b \in \mathbf{J}_\beta(\mathbb{R}) \, \forall d \, \forall \eta \, \forall e \\ & \quad\quad (H(b, d) \ \& \ \eta = g(d) \ \& \ e = f(b) \to \\ & \quad\quad\quad \exists c \in \mathbf{J}_\eta(\mathbb{R}) \ P(a, \beta, e, h(c)))\Big). \end{aligned}$$

Using the induction hypothesis again, we have that S is $\underset{\sim}{\Sigma}_{n+1}(\mathbf{J}_\alpha(\mathbb{R}))$.

Case 2. Otherwise.

PROOF. Suppose $S(a, \beta)$. Let $D = \operatorname{dom} f \cap \mathbf{J}_\beta(\mathbb{R})$; then $D \in \mathbf{J}_\alpha(\mathbb{R})$ since $\beta < \rho_n$. Define

$$R(b, \eta) \iff b \in D \ \& \ \eta < \rho_{n-1} \ \& \ \exists c \in \mathbf{J}_\eta(\mathbb{R}) \ P(a, \beta, f(b), h(c)),$$

so that R is $\underset{\sim}{\Sigma}_n(\mathbf{J}_\alpha(\mathbb{R}))$. By Σ_n selection for $\mathbf{J}_\alpha(\mathbb{R})$, we have a total $\underset{\sim}{\Sigma}_n(\mathbf{J}_\alpha(\mathbb{R}))$ map $g\colon D \to \rho_{n-1}$ which uniformizes R. By case hypothesis, the range of g is bounded in ρ_{n-1}. But then

$$S(a,\beta) \iff \exists\eta < \rho_{n-1}\,\forall b \in \mathbf{J}_\beta(\mathbb{R})\,\forall d$$
$$(d = f(b) \to \exists c \in \mathbf{J}_\eta(\mathbb{R})\ P(a,\beta,d,h(c))).$$

By induction hypothesis (or obviously in the case $n = 1$) S is $\underset{\sim}{\Sigma}_{n+1}(\mathbf{J}_\alpha(\mathbb{R}))$.

⊣

THEOREM 1.16. Suppose $\alpha \in \mathrm{Ord}$, $n \geq 1$ and $\rho^\alpha_{n-1} \neq 1$. Then $\mathbf{J}_\alpha(\mathbb{R})$ satisfies Σ_n selection.

PROOF. Suppose $R(a,F) \iff \exists b\, Q(a,F,b)$, where Q is $\underset{\sim}{\Pi}_{n-1}(\mathbf{J}_\alpha(\mathbb{R}))$ and $R \subseteq \mathbf{J}_\alpha(\mathbb{R}) \times [\omega\alpha]^{<\omega}$. Let $f\colon \mathbf{J}_{\rho^\alpha_{n-1}}(\mathbb{R}) \twoheadrightarrow \mathbf{J}_\alpha(\mathbb{R})$ be a partial $\underset{\sim}{\Sigma}_{n-1}(\mathbf{J}_\alpha(\mathbb{R}))$ map. Define

$$S(a,F) \iff \exists\beta < \rho^\alpha_{n-1}$$
$$\big[\exists b \in \mathbf{J}_{\beta+1}(\mathbb{R})\ \big(b \in \operatorname{dom} f \,\&\, f(b)_0 = F \,\&\, Q(a, f(b)_0, f(b)_1)\big)$$
$$\&\ \forall b \in \mathbf{J}_\beta(\mathbb{R})\ \big(b \in \operatorname{dom} f \to \neg Q(a, f(b)_0, f(b)_1)\big)$$
$$\forall b \in \mathbf{J}_{\beta+1}(\mathbb{R})\ \big(b \in \operatorname{dom} f \to$$
$$(F \leq_{\mathrm{BK}} f(b)_0 \vee \neg Q(a, f(b)_0, f(b)_1))\big)\big].$$

Here, of course, $f(b) = \langle f(b)_0, f(b)_1\rangle$. Lemma 1.15 easily implies that S is $\underset{\sim}{\Sigma}_n(\mathbf{J}_\alpha(\mathbb{R}))$, and clearly S uniformizes R. (By Lemma 1.10, we may assume $n \geq 2$, and then Lemma 1.4 easily implies that ρ^α_{n-1} is a limit ordinal, so that "$\forall b \in \mathbf{J}_{\beta+1}(\mathbb{R})$" is a bounded quantification over $\mathbf{J}_{\rho^\alpha_{n-1}}(\mathbb{R})$.) ⊣

We shall see in §4 that the hypothesis $\rho^\alpha_{n-1} \neq 1$ of the Σ_n selection theorem is essential. Of course, if $\mathbf{L}(\mathbb{R}) = \mathbf{L}$ we can weaken it to "$\alpha \neq 1$ or $n \neq 1$"; that is, we need only rule out $\Sigma_1(\mathbf{J}_1(\mathbb{R}))$. On the other hand, if $\mathsf{Det}(\mathbf{L}(\mathbb{R}))$ then $\mathbf{J}_1(\mathbb{R})$ satisfies Σ_n selection iff n is even, and a similar periodicity of order two takes over on arbitrary $\mathbf{J}_\alpha(\mathbb{R})$ above the least n such that $\rho^\alpha_n = 1$.

§2. Scales on $\Sigma_1(\mathbf{J}_\alpha(\mathbb{R}))$ sets. Our positive results on the existence of scales are refinements of [MS83, Theorem 1], which builds directly on a scale construction due to Moschovakis [Mos83]. We shall describe briefly the slight generalization of this construction we need. Suppose that $\alpha \in \mathrm{Ord}$, and that for each $x \in \mathbb{R}$ we have a game G_x in which player I's moves come from $\mathbb{R} \times \alpha$ while player II's moves come from $\mathbb{R}$. Thus a typical run of G_x has the form

I	x_0, β_0		x_2, β_1	...
II		x_1		x_3

where $x_i \in \mathbb{R}$ and $\beta_i < \alpha$ for all i. Suppose that G_x is closed and continuously associated to x in the strong sense that for some $Q \subseteq (\omega^{<\omega})^{<\omega} \times \alpha^{<\omega}$,

$$\text{player I wins } \mathrm{G}_x \text{ iff } \forall n\, Q(\langle x\restriction n, x_0\restriction n, \dots, x_n\restriction n\rangle, \langle \beta_0, \dots, \beta_n\rangle).$$

Thus by Gale-Stewart, one of the players in G_x has a winning quasi-strategy. (We are not assuming the Axiom of Choice, so we don't get full strategies.) Let

$$P_k(x, u) \text{ iff } u \text{ is a position of length } k \text{ from which player I has a winning quasi-strategy}$$

and

$$P(x) \text{ iff } P_0(x, \varnothing).$$

If sufficiently many games are determined , then Moschovakis' construction yields a scale on P. The prewellordering $\leq_i$ induced by the ith norm in this scale is definable from the P_k's for $k \leq i$ by means of recursive substitution, the Boolean operations, and quantification over $\mathbb{R} \times \alpha$. In particular, if $\alpha < \omega\gamma$ and $\langle P_k : k \leq i\rangle \in \mathbf{J}_\gamma(\mathbb{R})$, then $\leq_i \in \mathbf{J}_\gamma(\mathbb{R})$. The determinacy required to construct the Moschovakis scale is closely related to the definability of the scale constructed. In particular, if $\alpha < \omega\gamma$ and $P_k \in \mathbf{J}_\gamma(\mathbb{R})$ for all $k \leq \omega$, then $\mathsf{Det}(\mathbf{J}_\gamma(\mathbb{R}))$ suffices.

In the circumstances just described we call the map $x \mapsto \mathrm{G}_x$ a **closed game representation** of P. Which sets P admit game representations, and therefore scales? According to [MS83], every $(\underset{\sim}{\Sigma}^2_1)^{\mathbf{L}(\mathbb{R})}$ set admits a closed game representation in $\mathbf{L}(\mathbb{R})$. Our first theorem refines that result.

THEOREM 2.1. *If $\alpha > 1$ and $\mathsf{Det}(\mathbf{J}_\alpha(\mathbb{R}))$, then the pointclass $\Sigma_1(\mathbf{J}_\alpha(\mathbb{R}))$ has the scale property.*

PROOF. Let us first assume that α is a limit ordinal, and deal with the general case later. Let $\varphi_0(v)$ be a Σ_1 formula, and let

$$P(x) \text{ iff } \mathbf{J}_\alpha(\mathbb{R}) \models \varphi_0[x],$$

for $x \in \mathbb{R}$, For $\beta < \alpha$, let

$$P^\beta \text{ iff } \mathbf{J}_\beta(\mathbb{R}) \models \varphi_0[x].$$

Thus $P = \bigcup_{\beta<\alpha} P^\beta$. For each $\beta < \alpha$ we will construct a closed game representation $x \mapsto \mathrm{G}^\beta_x$ of P^β. Let

$$P^\beta_k(x, u) \iff u \text{ is a position of length } k \text{ from which player I has a winning quasi-strategy in } \mathrm{G}^\beta_x.$$

We shall arrange that for each $P^\beta_k \in \mathbf{J}_\alpha(\mathbb{R})$, and that the map $(\beta, k) \mapsto P^\beta_k$ is $\Sigma_1(\mathbf{J}_\alpha(\mathbb{R}))$. This will suffice for Theorem 2.1. For let $\{\varphi^\beta_k\}$ be the Moschovakis scale on P^β, and let $\leq^\beta_k$ be the prewellorder of $\mathbb{R}$ induced by

φ_k^β. Then $\leq_k^\beta \in \mathbf{J}_\alpha(\mathbb{R})$ by the remarks above. A simple inspection of [Mos83] shows that those remarks are true uniformly in β, so that the map $(\beta, k) \mapsto \leq_k^\beta$ is $\Sigma_1(\mathbf{J}_\alpha(\mathbb{R}))$. We can then define a scale $\{\psi_k\}$ on P:

$$\psi_0(x) = \mu\beta\, P^\beta(x),$$

and

$$\psi_{k+1}(x) = \langle \psi_0(x), \varphi_k^{\psi_0(x)}(x) \rangle.$$

(In the definition of ψ_{k+1}, we use the lexicographic order to assign ordinals to pairs of ordinals.) It is easy to check that $\{\psi_k\}$ is a $\Sigma_1(\mathbf{J}_\alpha(\mathbb{R}))$ scale on P.

So let β and x be given; we want to define G_x^β. Our plan is to force player I in G_x^β to describe a countable model of $V = \mathbf{L}(\mathbb{R}) + \varphi_0(x) + \forall\gamma\, (\mathbf{J}_\gamma(\mathbb{R}) \not\models \varphi_0\,[x])$ which contains all the reals played by player II, while using ordinals less than $\omega\beta$ to prove that his model is wellfounded. If $\mathbf{J}_\beta(\mathbb{R}) \models \varphi_0\,[x]$, then player I will be able to win G_x^β by describing an elementary submodel of $\mathbf{J}_\gamma(\mathbb{R})$, where γ is least so that $\mathbf{J}_\gamma(\mathbb{R}) \models \varphi_0\,[x]$. On the other hand, if player I has a winning quasi-strategy in G_x^β, we will be able to piece together the models he describes in different runs of G_x^β according to his strategy, and thereby produce a model of the form $\mathbf{J}_\gamma(\mathbb{R})$ so that $\gamma \leq \beta$ and $\mathbf{J}_\gamma(\mathbb{R}) \models \varphi_0\,[x]$.

Player I describes his model in the language $\mathcal{L}$ which has, in addition to $\in$ and $=$, constant symbols $\underline{x}_i$ for $i < \omega$. He uses $\underline{x}_i$ to denote the ith real played in the course of G_x^β. Let us fix recursive maps

$$m, n \colon \{\vartheta : \vartheta \text{ is an } \mathcal{L}\text{-formula}\} \to \{2n : 1 \leq n < \omega\}$$

which is one-one, have disjoint recursive ranges, and are such that whenever $\underline{x}_i$ occurs in ϑ, then $i < \min(m(\vartheta), n(\vartheta))$. These maps give stages sufficiently late in G_x^β for player I to decide certain statements about his model.

Player I's description must extend the following $\mathcal{L}$-theory T. The axioms of T include

(1) Extensionality
(2) $V = \mathbf{L}(\mathbb{R})$
$(3)_\varphi$ $\exists v \varphi(v) \to \exists v\, (\varphi(v)\ \&\ \forall u \in v\, \neg\varphi(u))$
$(4)_i$ $\underline{x}_i \in \mathbb{R}$.

Of course, these axioms are true in any $\mathbf{J}_\gamma(\mathbb{R})$ as long as the $\underline{x}_i$'s are interpreted as reals. The next axiom restricts the models of this form to at most one possibility

(5) $\varphi_0(\underline{x}_0)\ \&\ \forall\delta\, (\mathbf{J}_\delta(\mathbb{R}) \not\models \varphi\,[\underline{x}_0])$.

Finally, T has axioms which guarantee that any model $\mathfrak{A}$ can be regarded as a definable closure of $\{\underline{x}_i^{\mathfrak{A}} : i < \omega\}$. Recall from Lemma 1.4 the uniformly definable maps $f_\gamma \colon [\omega\gamma]^{<\omega} \times \mathbb{R} \twoheadrightarrow \mathbf{J}_\gamma(\mathbb{R})$; let $\vartheta_0(v_0, v_1, v_2)$ be a Σ_1 formula which for all γ defines the graph of f_γ over $\mathbf{J}_\gamma(\mathbb{R})$. Now, for any $\mathcal{L}$-formula $\varphi(v)$ of one free variable, T has axioms

$$(6)_\varphi \quad \exists v \varphi(v) \to \exists v \exists F\, (\varphi(v) \,\&\, \vartheta_0(F, \underline{x}_{m(\varphi)}, v))$$
$$(7)_\varphi \quad \exists v\, (\varphi(v) \,\&\, v \in \mathbb{R}) \to \varphi(\underline{x}_{n(\varphi)}).$$

This completes the list of axioms of T.

A typical run of G_x^β has the form

$$\begin{array}{lccccc} \mathrm{I} & i_0, x_0, \eta_0 & & i_1, x_2, \eta_1 & & \\ & & & & & \cdots \\ \mathrm{II} & & x_1 & & x_3 & \end{array}$$

where for all k, $i_k \in \{0, 1\}$, $x_k \in \mathbb{R}$, and $\eta_k < \omega\beta$. If u is a position of length n, say $u = \langle\langle i_k, x_{2k}, \eta_k, x_{2k+1}\rangle : k < n\rangle$, then we let

$$T^*(u) = \{\vartheta : \vartheta \text{ is a sentence of } \mathcal{L} \,\&\, i_{n(\vartheta)} = 0\},$$

and if p is a full run of G_x^β,

$$T^*(p) = \bigcup_{n<\omega} T^*(p \restriction n).$$

Now let $p = \langle\langle i_k, x_{2k}, \eta_k, x_{2k+1}\rangle : k < \omega\rangle$ be a run of G_x; we say that p is a winning run for player I iff

(a) $x_0 = x$
(b) $T^*(p)$ is complete, consistent extension of T such that for all i, m, n, "$\underline{x}_i(\underline{n}) = \underline{m}$" $\in T^*(p)$ iff $x_i(n) = m$, and
(c) If φ and ψ are $\mathcal{L}$-formulae of one free variable, and "$\imath v\, \varphi(v) \in \mathrm{Ord}$ & $\imath v\, \psi(v) \in \mathrm{Ord}$" $\in T^*(p)$, then "$\imath v\, \varphi(v) \le \imath v\, \psi(v)$" $\in T^*(p)$ iff $\eta_{n(\varphi)} \le \eta_{n(\psi)}$.

In condition (c) we have used the "unique v" operator as an abbreviation: if ϑ and φ are $\mathcal{L}$-formulae, then $\vartheta(\imath v\, \varphi(v))$ abbreviates

$$\text{“}\exists v(\vartheta(v) \,\&\, \forall u\, (\varphi(u) \iff u = v))\text{”}.$$

It is clear that G_x is closed and continuously associated to x. In order to show that $x \mapsto \mathrm{G}_x$ is the desired closed game representation of P^β, we want to characterize the winning positions for player I in G_x as those in which player I has been "honest" about the minimal model of $\varphi_0(x)$. More precisely, let us call a position $u = \langle\langle i_k, x_{2k}, \eta_k, x_{2k+1}\rangle : k < n\rangle$ of length n *(β, x)-honest* iff $\mathbf{J}_\beta(\mathbb{R}) \models \varphi_0[x]$, and if $\gamma \le \beta$ is least such that $\mathbf{J}_\gamma(\mathbb{R}) \models \varphi_0[x]$, then

(i) $n > 0 \to x_0 = x$,
(ii) if we let $I_u(\underline{x}_i) = x_i$ for $i < 2n$, then all axioms of $T^*(u) \cup T$ thereby interpreted in $(\mathbf{J}_\gamma(\mathbb{R}), I_u)$ are true in this structure, and
(iii) If $\vartheta_0, \dots, \vartheta_m$ enumerate those $\mathcal{L}$-formulae ϑ of one free variable such that $n(\vartheta) < n$ and

$$(\mathbf{J}_\gamma(\mathbb{R}), I_u) \models \imath v\, \vartheta(v) \in \mathrm{Ord}$$

and if $\delta_i < \omega\gamma$ is such that

$$(\mathbf{J}_\gamma(\mathbb{R}), I_u) \models \iota v\, \vartheta_i(v) = \delta_i,$$

then the map

$$\delta_i \mapsto \eta_{n(\vartheta_i)}$$

is well-defined and extendible to an order preserving map of $\omega\gamma$ into $\omega\beta$.

According to the following claim, honesty is the only rational policy for player I in G^β_x.

Claim. Let u be a position in G^β_x. Then player I has a winning quasi-strategy in G^β_x starting from u iff u is (β, x)-honest.

PROOF. $(\rightarrow)$ Let u be a (β, x)-honest position of length n; then it is easy to check that $\exists i, x, \eta\, \forall y\, (u^\frown\langle i, x, \eta, y\rangle)$ is (β, x)-honest. [If $n = n(\vartheta)$ for some sentence ϑ of $\mathcal{L}$, put $i = 0$ iff $(\mathbf{J}_\gamma(\mathbb{R}), I_u) \models \vartheta$, otherwise i is random. If $n = m(\varphi)$ for φ an $\mathcal{L}$ formula of one free variable, choose x so that $(\mathbf{J}_\gamma(\mathbb{R}), I_{u^\frown\langle i,x,\eta,y\rangle}) \models$ axiom $(6)_\varphi$ of T. If $n = n(\varphi)$, choose x for the sake of $(7)_\varphi$, and otherwise let x be random. Finally, if $n = \varphi(n)$ for φ of one free variable, and $\delta < \omega\gamma$ is such that $(\mathbf{J}_\gamma(\mathbb{R}), I_u) \models \iota v\, \varphi(v) = \delta$, then set $\eta = f(\delta)$ where f witnesses part (iii) of honesty for u. Otherwise, choose η randomly.] Moreover, no (β, x)-honest position is an immediate loss for player I in G^β_x. Thus if u is (β, x)-honest , player I can win G^β_x from u by keeping to (β, x)-honest positions.

$(\Leftarrow)$ The proof we give here is due to Hugh Woodin; our original proof followed more closely the corresponding proof in [MS83].

Let ϑ be the sentence we want to prove, that is, let $\vartheta =$ "For all β, x, and u, if player I has a winning quasi-strategy in G^β_x starting from u, then u is (β, x)-honest." We show that $M \models \vartheta$ whenever M is a countable, transitive model of (a sufficiently large fragment of) $\mathsf{ZF} + \mathsf{DC}$. But then $\mathsf{ZF} + \mathsf{DC} \vdash \vartheta$, and we're done.

So, let M be such a model, and let

$$M \models S \text{ is a winning quasi-strategy for player I in } \mathrm{G}^\beta_x \text{ starting from } u.$$

In M, define

$$P = \{v : v \text{ is a position in } \mathrm{G}^\beta_x \text{ extending } u \text{ and allowed by } S\},$$

and let $\mathbb{P} = (P, \supseteq)$. Let G be $\mathbb{P}$-generic over M, and let $p = \bigcup G$. We can think of p as a "generic run" of G^β_x according to S. In particular, p satisfies the requirements for being a winning run for player I, since these requirements are closed.

Thus $T^*(p)$ is a complete, consistent extension of T. Let $\mathfrak{B} \models T^*(p)$, and let $\mathfrak{A}$ be the substructure of $\mathfrak{B}$ whose universe is

$$|\mathfrak{A}| = \{b \in |\mathfrak{B}| \,:\, \exists i \in \omega \,\exists \psi \,(\psi \text{ is a formula of } \mathcal{L} \text{ with no constant symbols but } \underline{x}_i \text{ and } b = \imath v\,(\mathfrak{B} \models \psi\,[v]))\}.$$

Then by axioms (3) and (6) of T, $\mathfrak{A} \prec \mathfrak{B}$.

Let $p = \langle\langle i_k, x_{2k}, \eta_k, x_{2k+1}\rangle \,:\, k < \omega\rangle$. If $b \in \mathrm{Ord}^{\mathfrak{A}}$, and $b = \imath v\,(\mathfrak{B} \models \psi\,[v])$, then let

$$f(b) = \eta_{n(\psi)}.$$

By requirement (c) on player I in G^β_x, f is a well-defined order-preserving map of $\mathrm{Ord}^{\mathfrak{A}}$ into $\omega\beta$. Thus we may assume that $\mathfrak{A} = (\mathbf{J}_\gamma(\mathbb{R}^{\mathfrak{A}}), I)$ for some $\gamma \leq \beta$, where by requirement (b) on player I in G^β_x, $I(\underline{x}_i) = x_i$. By axioms (7) of T, $R^{\mathfrak{A}} = \{x_i \,:\, i < \omega\}$; on the other hand, since p is a generic run and player II is free to play whatever reals he pleases, $\{x_i \,:\, i < \omega\} = \mathbb{R}^M$.

Thus $\mathfrak{A} = (\mathbf{J}_\gamma(\mathbb{R}^M), I)$, where $I = \bigcup_k I_{p\restriction k}$. Moreover, by axiom (5) of T and requirement (a) on player I in G^β_x, γ is least such that $\mathbf{J}_\gamma(\mathbb{R}) \models \varphi_0\,[x]$. It is now easy to verify conditions (i)–(iii) of (β, x)-honesty for u in $M[G]$; the map f defined above witnesses (iii). Since honesty is absolute, u is (β, x)-honest in M.

This proves the claim. ⊣

Notice that the empty position is (β, x)-honest iff $\mathbf{J}_\beta(\mathbb{R}) \models \varphi_0\,[x]$. The claim therefore implies that player I has a winning quasi-strategy in G^β_x iff $\mathbf{J}_\beta(\mathbb{R}) \models \varphi_0\,[x]$. If we let

$$\begin{aligned} P^\beta_n(x, u) &\iff \text{player I has a winning quasi-strategy in } \mathrm{G}^\beta_x \\ &\ \text{from the length } n \text{ position } u \\ &\iff u \text{ has length } n \text{ and is } (\beta, x)\text{-honest,} \end{aligned}$$

then it is easy to see that $P^\beta_n \in \mathbf{J}_{\beta+1}(\mathbb{R})$ for all β, n, and in fact that the map $(\beta, n) \mapsto P^\beta_n$ is $\Sigma_1(\mathbf{J}_\alpha(\mathbb{R}))$. (To see that $P^\beta_n \in \mathbf{J}_{\beta+1}(\mathbb{R})$, notice that the sentences which come up in evaluating the honesty of positions of length n have quantifier complexity bounded by some recursive function on n. Also, notice that if there is any order preserving map as required by part (iii) of honesty, then there is a $\underset{\sim}{\Sigma}_1(\mathbf{J}_\beta(\mathbb{R}))$ such map.) Thus the claim gives us Theorem 2.1 in the case that is a limit.

Minor modifications make this proof work for arbitrary α. Again, let

$$P(x) \iff \mathbf{J}_\alpha(\mathbb{R}) \models \varphi_0\,[x],$$

where φ_0 is Σ_1. Using the simplicity of rudimentary function S as in the proof of Lemma 1.11(b), we can find first order formulae $\psi_n(v)$ for $n < \omega$ so that for all $\beta \in \text{Ord}$ and $x \in \mathbb{R}$,

$$\mathbf{J}_\beta(\mathbb{R}) \models \psi_n[x] \text{ iff } S_{\omega\beta+\eta}(\mathbb{R}) \models \varphi_0[x].$$

For each $\beta < \alpha$ and $n < \omega$ we can define a game $G_x^{\beta,n}$ so that player I wins, $G_x^{\beta,n}$ iff $\exists \gamma \leq \beta\, (\mathbf{J}_\gamma(\mathbb{R}) \models \psi_n[x])$ iff $S_{\omega\beta+n}(\mathbb{R}) \models \varphi_0[x]$. The rules of $G_x^{\beta,n}$ are exactly those of G_x^β except that axiom (5) of T is replaced by

(5)' $\psi_n(\underset{\sim}{x}_0)\ \&\ \forall\delta\, (\mathbf{J}_\delta(\mathbb{R}) \models -\psi_n(\underset{\sim}{x}_0))$.

We can define (β, n, x)-honesty and prove the analogue of our claim just as before. This analogous claim yields Theorem 2.1 in the general case at once. ⊣

Of course, the existence of a $\Sigma_1(\mathbf{J}_\alpha(\mathbb{R}))$ set of reals universal for $\Sigma_1(\mathbf{J}_\alpha(\mathbb{R}), \mathbb{R})$ sets of reals immediately gives the boldface version of Theorem 2.1.

Corollary 2.2. If $\alpha > 1$ and $\text{Det}(\mathbf{J}_\alpha(\mathbb{R}))$, then the boldface pointclass $\Sigma_1(\mathbf{J}_\alpha(\mathbb{R}), \mathbb{R})$ has the scale property.

The hypothesis that $\alpha > 1$ is necessary in Theorem 2.1 and Corollary 2.2, since $\Sigma_1(\mathbf{J}_1(\mathbb{R})) \cap \wp(\mathbb{R}) = \Sigma_1^1$, and Σ_1^1 doesn't have the scale property.

Theorem 2.1 immediately implies the special cases from which it was abstracted.

Corollary 2.3. If $\text{Det}(\mathbf{L}(\mathbb{R}))$, then

(a) The pointclass of inductive sets has the scale property ([Mos78]),
(b) The pointclass Σ_ω^* has the scale property; in particular, every coinductive set admits a Σ_ω^* scale ([Mos83]),
(c) The pointclass $(\Sigma_1^2)^{\mathbf{L}(\mathbb{R})}$ has the scale property ([MS83]).

Proof. Let

$$\kappa^{\mathbb{R}} = \mu\alpha\, [\mathbf{J}_\alpha(\mathbb{R}) \text{ is admissible}].$$

Then inductive $= \Sigma_1(\mathbf{J}_{\kappa^{\mathbb{R}}}(\mathbb{R}))$ and $\Sigma_\omega^* = \Sigma_1(\mathbf{J}_{\kappa^{\mathbb{R}}+1}(\mathbb{R}))$, so we have (a) and (b). Let

$$\sigma = \mu\alpha\, [\mathbf{J}_\alpha(\mathbb{R}) \prec_1 \mathbf{L}(\mathbb{R})].$$

Then by Lemma 1.12, $(\Sigma_1^2)^{\mathbf{L}(\mathbb{R})} = \Sigma_1(\mathbf{J}_\sigma(\mathbb{R}))$, so we have (c). ⊣

We shall use Theorem 2.1 together with some further work to give a complete description of those levels of the Levy hierarchy for $\mathbf{L}(\mathbb{R})$ (that is, those pointclasses of the form $\underset{\sim}{\Sigma}_n(\mathbf{J}_\gamma(\mathbb{R})))$ which have the scale property. For this, it is convenient to have the following definition.

Definition 2.4. Let $\alpha, \beta \in \text{Ord}$ and $\alpha \leq \beta$. The interval $[\alpha, \beta]$ is a Σ_1**-gap** iff

(i) $\mathbf{J}_\alpha(\mathbb{R}) \prec_1^{\mathbb{R}} \mathbf{J}_\beta(\mathbb{R})$
(ii) $\forall \alpha' < \alpha\, (\mathbf{J}_{\alpha'}(\mathbb{R}) \not\prec_1^{\mathbb{R}} \mathbf{J}_\alpha(\mathbb{R}))$

(iii) $\forall \beta' > \beta\,(\mathbf{J}_\beta(\mathbb{R}) \not\prec_1^{\mathbb{R}} \mathbf{J}_{\beta'}(\mathbb{R}))$.

That is, a Σ_1-gap is a maximal interval of ordinals in which no new Σ_1 facts about elements of $\mathbb{R} \cup \{\mathbf{V}_{\omega+1}\}$ are verified. If $[\alpha, \beta]$ is a Σ_1-gap, we say α begins the gap and β ends it. Notice that we allow $\alpha = \beta$. We shall also allow $[(\underset{\sim}{\delta}_1^2)^{\mathbf{L}(\mathbb{R})}, \Theta^{\mathbf{L}(\mathbb{R})}]$ as a Σ_1-gap.

LEMMA 2.5. The Σ_1-gaps partition $\Theta^{\mathbf{L}(\mathbb{R})}$.

PROOF. Given $\gamma < \Theta^{\mathbf{L}(\mathbb{R})}$, let $\alpha \leq \gamma$ be least so that $\mathbf{J}_\alpha(\mathbb{R}) \prec_1^{\mathbb{R}} \mathbf{J}_\gamma(\mathbb{R})$, and let $\beta = \sup\{\delta < \Theta^{\mathbf{L}(\mathbb{R})} : \mathbf{J}_\gamma(\mathbb{R}) \prec_1^{\mathbb{R}} \mathbf{J}_\delta(\mathbb{R})\}$. Then $[\alpha, \beta]$ is a Σ_1-gap, and $\gamma \in [\alpha, \beta]$. Any two distinct Σ_1-gaps are disjoint because we have restricted the parameters involved to $\mathbb{R} \cup \{\mathbf{V}_{\omega+1}\}$. ⊣

Let us work our way now through an arbitrary Σ_1-gap, looking for levels of the Levy hierarchy having the scale property. If α begins a Σ_1-gap, then by Lemma 1.11 there is a partial $\Sigma_1(\mathbf{J}_\alpha(\mathbb{R}))$ map from $\mathbb{R}$ onto $\mathbf{J}_\alpha(\mathbb{R})$, and therefore $\underset{\sim}{\Sigma}_1(\mathbf{J}_\alpha(\mathbb{R})) = \Sigma_1(\mathbf{J}_\alpha(\mathbb{R}), \mathbb{R})$.

COROLLARY 2.6. If α begins a Σ_1-gap, $\alpha > 1$, and $\mathrm{Det}(\mathbf{J}_\alpha(\mathbb{R}))$, then $\underset{\sim}{\Sigma}_1(\mathbf{J}_\alpha(\mathbb{R}))$ has the scale property.

The next classes to consider are the $\underset{\sim}{\Sigma}_n(\mathbf{J}_\alpha(\mathbb{R}))$, $n > 1$, for α which begin a Σ_1-gap. The scale property here depends upon the admissibility of $\mathbf{J}_\alpha(\mathbb{R})$.

LEMMA 2.7. Suppose α begins a Σ_1-gap and $\mathbf{J}_\alpha(\mathbb{R})$ is not admissible. Then for all $n \geq 1$

$$\underset{\sim}{\Sigma}_{n+1}(\mathbf{J}_\alpha(\mathbb{R})) = \exists^{\mathbb{R}}(\underset{\sim}{\Pi}_n(\mathbf{J}_\alpha(\mathbb{R})))$$

and

$$\underset{\sim}{\Pi}_{n+1}(\mathbf{J}_\alpha(\mathbb{R})) = \forall^{\mathbb{R}}(\underset{\sim}{\Sigma}_n(\mathbf{J}_\alpha(\mathbb{R}))).$$

PROOF. The two conclusions are of course equivalent. Let S be $\underset{\sim}{\Sigma}_{n+1}(\mathbf{J}_\alpha(\mathbb{R}))$; say

$$S(u) \iff \exists v\, P(u, v)$$

where P is $\underset{\sim}{\Sigma}_n(\mathbf{J}_\alpha(\mathbb{R}))$. Let

$$f\colon \mathbb{R} \twoheadrightarrow \mathbf{J}_\alpha(\mathbb{R})$$

be a partial $\Sigma_1(\mathbf{J}_\alpha(\mathbb{R}))$ map; There is such a map since α begins a Σ_1-gap. Then clearly

$$S(u) \iff \exists x \in \mathbb{R}\,\big(x \in \operatorname{dom} f \;\&\; \forall v\,(v = f(x) \rightarrow P(u, v))\big).$$

If $n \geq 2$, this implies $S \in \exists^{\mathbb{R}}(\underset{\sim}{\Pi}_n(\mathbf{J}_\alpha(\mathbb{R})))$, as desired. For $n = 1$, we need to know that "$x \in \operatorname{dom} f$" $\in \exists^{\mathbb{R}}(\underset{\sim}{\Pi}_1(\mathbf{J}_\alpha(\mathbb{R})))$. This is a direct consequence of inadmissibility. For by inadmissibility we have a total $\underset{\sim}{\Sigma}_1(\mathbf{J}_\alpha(\mathbb{R}))$ map

$$h\colon D \rightarrow \omega\alpha$$

such that $D \in \mathbf{J}_\alpha(\mathbb{R})$ and h has range cofinal in $\omega\alpha$. Since α begins a Σ_1-gap, and easy Skolem hull argument gives a total

$$g : \mathbb{R} \twoheadrightarrow D$$

such that $g \in \mathbf{J}_\alpha(\mathbb{R})$. Let $k = h \circ g$. Let

$$Q(u) \iff \mathbf{J}_\alpha(\mathbb{R}) \models \varphi[u, p],$$

where φ is Σ_1, be any $\underset{\sim}{\Sigma}_1(\mathbf{J}_\alpha(\mathbb{R}))$ set. Then

$$\begin{aligned} Q(u) &\iff \exists x \in \mathbb{R}\,(S_{k(x)}(\mathbb{R}) \models \varphi[u, p]) \\ &\quad\ \exists x \in \mathbb{R}\,\forall S\,\forall \gamma\,[(\gamma = k(x)\ \&\ S = S_\gamma(\mathbb{R})) \rightarrow S \models \varphi[u, p]] \end{aligned}$$

so that $Q \in \exists^{\mathbb{R}}(\Pi_1(\mathbf{J}_\alpha(\mathbb{R})))$. ⊣

The second periodicity theorem, Corollary 2.6 and Lemma 2.7 yield at once

Corollary 2.8. Suppose α begins a Σ_1-gap, $\alpha > 1$, $\mathbf{J}_\alpha(\mathbb{R})$ is not admissible, and $\mathsf{Det}(\mathbf{J}_{\alpha+1}(\mathbb{R}))$. Then for all $n < \omega$, the classes $\underset{\sim}{\Sigma}_{2n+1}(\mathbf{J}_\alpha(\mathbb{R}))$ and $\underset{\sim}{\Pi}_{2n+2}(\mathbf{J}_\alpha(\mathbb{R}))$ have the scale property.

Martin [Mar83] shows that at admissible α beginning a gap the scale property fails above $\Sigma_1(\mathbf{J}_\alpha(\mathbb{R}))$ in a strong way.

Theorem 2.9 (Martin). Suppose α begins a Σ_1-gap, $\mathbf{J}_\alpha(\mathbb{R})$ is admissible, and $\mathsf{Det}(\mathbf{J}_{\alpha+1}(\mathbb{R}))$. Then there is a $\Pi_1(\mathbf{J}_\alpha(\mathbb{R}))$ subset of $\mathbb{R} \times \mathbb{R}$ with no uniformization in $\mathbf{J}_{\alpha+1}(\mathbb{R})$.

Corollary 2.10 (Martin). If α begins a Σ_1-gap, $\mathbf{J}_\alpha(\mathbb{R})$ is admissible, and $\mathsf{Det}(\mathbf{J}_{\alpha+1}(\mathbb{R}))$, then none of the classes $\underset{\sim}{\Sigma}_n(\mathbf{J}_\alpha(\mathbb{R}))$ or $\underset{\sim}{\Pi}_n(\mathbf{J}_\alpha(\mathbb{R}))$, for $n > 1$, have the scale property.

We are ready to venture inside our Σ_1 gaps. There we shall find no new scales.

Theorem 2.11. Let $[\alpha, \beta]$ be a Σ_1-gap, and assume $\mathsf{Det}(\mathbf{J}_{\alpha+1}(\mathbb{R}))$. Then there is a $\Pi_1(\mathbf{J}_\alpha(\mathbb{R}))$ subset of $\mathbb{R} \times \mathbb{R}$ with no $\underset{\sim}{\Sigma}_1(\mathbf{J}_\beta(\mathbb{R}))$ uniformization.

Proof. For $x, y \in \mathbb{R}$, let

$$\mathrm{C}_\alpha(x, y) \iff \exists \gamma < \omega\alpha\, \exists F \subseteq \gamma\, (F \text{ is finite } \&\ \{y\} \text{ is } \Sigma_1(S_\gamma(\mathbb{R}), \{x, F\})).$$

By Lemma 1.1, C_α is $\Sigma_1(\mathbf{J}_\alpha(\mathbb{R}))$. Suppose for a contradiction that $\mathbb{R}$ is a $\underset{\sim}{\Sigma}_1(\mathbf{J}_\beta(\mathbb{R}))$ relation uniformizing $\neg\mathrm{C}_\alpha$. By Lemma 1.4, we can fix an $x \in \mathbb{R}$ and Σ_0 formula φ so that for some finite $F \subseteq \omega\beta$

$$R(u, v) \iff \exists w\, (\mathbf{J}_\beta(\mathbb{R}) \models \varphi[u, v, w, x, F]).$$

Now $\{y : \mathrm{C}_\alpha(x, y)\}$ has a wellorder definable over $\mathbf{J}_\alpha(\mathbb{R})$, and so by $\mathsf{Det}(\mathbf{J}_{\alpha+1}(\mathbb{R}))$ is countable. Thus we have a unique $y \in \mathbb{R}$ so that $R(x, y)$. Let

$\vartheta(u, v)$ be the Σ_1 formula

$$\exists \gamma\, \exists F \subseteq \gamma\, \exists S\, \big(F \text{ is finite } \&\ S = S_\gamma(\mathbb{R})$$
$$\&\ \forall z \in \mathbb{R}\, \big(z = v \iff \exists w \in S\, \varphi(u, z, w, u, F)\big)\big).$$

Then $\mathbf{J}_\beta(\mathbb{R}) \models \vartheta[x, y]$, so $\mathbf{J}_\alpha(\mathbb{R}) \models \vartheta[x, y]$. But inspecting ϑ, we see that this means $\mathrm{C}_\alpha(x, y)$, a contradiction. ⊣

Corollary 2.12. If $[\alpha, \beta]$ is a Σ_1-gap and $\alpha < \gamma < \beta$, then none of the classes $\underset{\sim}{\Sigma}_n(\mathbf{J}_\gamma(\mathbb{R}))$ or $\underset{\sim}{\Pi}_n(\mathbf{J}_\gamma(\mathbb{R}))$, for $n < \omega$, have the scale property.

Theorem 2.11 extends in a simple way an example due to Solovay. For $x, y \in \mathbb{R}$, let $\mathbf{OD}(x, y)$ iff y is ordinal definable from x. Solovay showed that if

$$\forall x \exists y\, \neg \mathbf{OD}(x, y),$$

then $\neg\mathbf{OD}$ has no uniformization ordinal definable from a real. [If R is such a uniformization, ordinal definable from x, and $R(x, y)$, then $\mathbf{OD}(x, y)$, a contradiction.] Let $\delta = (\underset{\sim}{\delta}^2_1)^{\mathbf{L}(\mathbb{R})}$ and $\Theta = \Theta^{\mathbf{L}(\mathbb{R})}$. By the reflection theorem and the stability of δ, $\mathbf{OD}^{\mathbf{L}(\mathbb{R})}$ is just the set C_δ defined in Theorem 2.11. In $\mathbf{L}(\mathbb{R})$, every set is ordinal definable from a real, so Solovay's example interpreted in $\mathbf{L}(\mathbb{R})$ gives Theorem 2.11 for the gap $[\delta, \Theta]$ (as noticed by Kechris and Solovay). Theorem 2.11 comes from simply localizing this example. It is interesting that the non-uniformizable $\Pi_1(\mathbf{J}_\alpha(\mathbb{R}))$ relation referred to in Martin's Theorem 2.9 is just $\neg\mathrm{C}_\alpha$; all counterexamples to uniformization in $\mathbf{L}(\mathbb{R})$ come from localizing a single example. The next easy proposition sheds more light on the C_α's.

Proposition 2.13. Suppose $\mathbf{J}_\alpha(\mathbb{R})$ is admissible and $\mathrm{Det}(\mathbf{J}_\alpha(\mathbb{R}))$. Let C_α be the set defined in the proof of Theorem 2.11. Then for all x, $\{y : \mathrm{C}_\alpha(x, y)\}$ is the largest countable $\Sigma_1(\mathbf{J}_\alpha(\mathbb{R}), \{x\})$ set of reals.

Proof (Sketch). Let P be countable and $\Sigma_1(\mathbf{J}_\alpha(\mathbb{R}), \{x\})$. Since $\Sigma_1(\mathbf{J}_\alpha(\mathbb{R}), \{x\})$ has the scale property, P is Suslin via a tree on $\omega \times \alpha$ which is $\Delta_1(\mathbf{J}_\alpha(\mathbb{R}), \{x\})$ (here we use admissibility). Since P is countable, P is Suslin via $T \restriction \delta$ for some $\delta < \alpha$. Taking the appropriate derivatives of $T \restriction \delta$, we eventually isolate any $y \in P$ at some stage $\beta < \alpha$. But then $\{y\}$ is $\Sigma_1(\mathbf{J}_\beta(\mathbb{R}), \{x, F\})$ for the appropriate F. ⊣

Let us return to the question of scales in Σ_1-gaps. We have only one question left, but its answer is sufficiently long-winded to warrant a new section.

§3. Scales at the end of a Σ_1-gap.

The results of §2 leave open the question whether any of the classes $\underset{\sim}{\Sigma}_n(\mathbf{J}_\beta(\mathbb{R}))$ have scale property when β ends a Σ_1 gap. Notice that there are new pointclasses here, that is, $\rho^\beta_n = \mathbb{R}$ for some $n < \omega$. This follows at once from part (b) of Lemma 1.11. It turns out that the scale property for these classes hinges on a reflection property of β

somewhat subtler than the "Σ_1 reflection on reals" involved in Theorem 2.1 and Corollary 2.6.

For $a \in \mathbf{J}_\alpha(\mathbb{R})$, let $\Sigma^n_{a,\alpha}$ be the Σ_n-type realized by a in $\mathbf{J}_\alpha(\mathbb{R})$, that is

$$\Sigma^n_{a,\alpha} = \{\vartheta(v) : \vartheta \text{ is either } \Sigma_n \text{ or } \Pi_n \;\&\; \mathbf{J}_\alpha(\mathbb{R}) \models \vartheta[a]\}.$$

Definition 3.1. An ordinal β is **strongly Π_n-reflecting** iff every Σ_n type realized in $\mathbf{J}_\beta(\mathbb{R})$ is realized in $\mathbf{J}_\alpha(\mathbb{R})$ for some $\alpha < \beta$; that is, iff

$$\forall b \in \mathbf{J}_\beta(\mathbb{R})\, \exists \alpha < \beta\, \exists a \in \mathbf{J}_\alpha(\mathbb{R})\; (\Sigma^n_{b,\beta} = \Sigma^n_{a,\alpha}).$$

Definition 3.2. Let $[\alpha, \beta]$ be a Σ_n-gap. We call $[\alpha, \beta]$ **strong** iff β is strongly Π_n-reflecting, where n is least such that $\rho^\beta_n = \mathbb{R}$. Otherwise, $[\alpha, \beta]$ is **weak**.

Martin's proof of Theorem 2.9 easily gives the following generalization.

Theorem 3.3. *Let $[\alpha, \beta]$ be a strong Σ_1-gap, and assume $\mathrm{Det}(\mathbf{J}_{\alpha+1}(\mathbb{R}))$. Then there is a $\Pi_1(\mathbf{J}_\alpha(\mathbb{R}))$ relation (namely, the $\neg C_\alpha$ of Theorem 2.11) which has no uniformization in $\mathbf{J}_{\beta+1}(\mathbb{R})$.*

Corollary 3.4. *If $[\alpha, \beta]$ is a strong Σ_1-gap and $\mathrm{Det}(\mathbf{J}_{\alpha+1}(\mathbb{R}))$, then none of the classes $\boldsymbol{\Sigma}_n(\mathbf{J}_\beta(\mathbb{R}))$ or $\boldsymbol{\Pi}_n(\mathbf{J}_\beta(\mathbb{R}))$, for $n < \omega$, have the scale property.*

Thus at strong gaps $[\alpha, \beta]$, the scale property first re-appears on $\boldsymbol{\Sigma}_1(\mathbf{J}_{\beta+1}(\mathbb{R}))$. At weak gaps it re-appears on $\boldsymbol{\Sigma}_n(\mathbf{J}_\beta(\mathbb{R}))$ where n is least so that $\rho^\beta_n = \mathbb{R}$; that is, as soon as possible given Theorem 2.11. Before we show this, let us consider examples of the two different kinds if gap.

Example 3.5. Let $[\alpha, \beta]$ be the first Σ_1 gap such that $\beta \geq \alpha + \omega_1$. A Σ_1 Skolem hull argument shows easily that $\beta = \alpha + \omega_1$ and $\rho^\beta_1 = \mathbb{R}$. Since $\mathrm{cf}(\beta) > \omega$, any Σ_1-type realized in $\mathbf{J}_\beta(\mathbb{R})$ is realized in some $\mathbf{J}_\delta(\mathbb{R})$ for $\delta < \beta$. Thus β is strongly Π_1 reflecting.

Example 3.6. Let β be least such that $\rho^\beta_1 \neq \mathbb{R}$, and let $\alpha < \beta$ be least so that $\mathbf{J}_\alpha(\mathbb{R}) \prec^{\mathbb{R}}_1 \mathbf{J}_\beta(\mathbb{R})$. The minimality of β implies that $[\alpha, \beta]$ is a Σ_1 gap, and that $\rho^\beta_2 = \mathbb{R}$. Now for any n, let ϑ_n be a Σ_2 sentence so that for all γ,

$$\mathbf{J}_\gamma(\mathbb{R}) \models \vartheta_n \text{ iff } \exists \alpha_0 \ldots \exists \alpha_n\, (\alpha_0 < \cdots < \alpha_n < \gamma$$
$$\text{and } \mathbf{J}_{\alpha_0}(\mathbb{R}) \prec_1 \cdots \prec_1 \mathbf{J}_{\alpha_n}(\mathbb{R}) \prec_1 \mathbf{J}_\gamma(\mathbb{R})).$$

Every Σ_2-type realized in $\mathbf{J}_\beta(\mathbb{R})$ contains all of the ϑ_n's, but no Σ_2-type realized in any $\mathbf{J}_\delta(\mathbb{R})$ for $\delta < \beta$ includes all of the ϑ_n's. Thus β is not strongly Π_2 reflecting.

There are simpler weak gaps than that given by Example 3.6. For example, the first gap of the form $[\alpha, \alpha + 1]$ is weak. However, Example 3.6 seems to better illustrate the general situation. Indeed, the reader might gain some feeling for the proof of our next theorem by trying to show that for the β described in Example 3.6, $\boldsymbol{\Sigma}_2(\mathbf{J}_\beta(\mathbb{R}))$ has the scale property.

THEOREM 3.7. Let $[\alpha, \beta]$ be a weak Σ_1 gap, and suppose $\text{Det}(\mathbf{J}_\alpha(\mathbb{R}))$. Then if n is least so that $\rho_n^\beta = \mathbb{R}$, the class $\underset{\sim}{\Sigma}_n(\mathbf{J}_\beta(\mathbb{R}))$ has the scale property.

PROOF. Let α, β, and n satisfy the hypothesis. The case $n = 1$ is somewhat special, so we assume first that $n > 1$.

We shall follow the proof of Theorem 2.1 in outline, but some new complications arise in this situation. In order to eventually obtain a $\underset{\sim}{\Sigma}_n(\mathbf{J}_\beta(\mathbb{R}))$ scale, we must regard $\mathbf{J}_\beta(\mathbb{R})$ as the union of a canonical sequence of Σ_{n-1} substructures of $\mathbf{J}_\beta(\mathbb{R})$, each of which collapse to some $\mathbf{J}_\delta(\mathbb{R})$ with $\delta < \beta$. Our non-reflecting Σ_n type gives such a sequence.

Let $h\colon \mathbf{J}_\beta(\mathbb{R}) \times \mathbb{R} \to \mathbf{J}_\beta(\mathbb{R})$ be a Σ_{n-1} Skolem function. We want to standardize the parameter from which such a function is Σ_{n-1} definable. Fix ϑ_{sk} a Σ_{n-1} formula, w_1 a real, and F' a finite subset of $\omega\beta$ such that

$$h(u, x) = v \text{ iff } \mathbf{J}_\beta(\mathbb{R}) \models \vartheta_{\text{sk}}[u, x, v, w_1, F'].$$

Now fix $F \leq_{\text{BK}} F'$ to be the BK-least finite subset of $\omega\beta$ so that

$$\{\langle u, x, v\rangle : \mathbf{J}_\beta(\mathbb{R}) \models \vartheta_{\text{sk}}[u, x, v, w_1, F]\}$$

is a Σ_{n-1} Skolem function for $\mathbf{J}_\beta(\mathbb{R})$.

We next standardize the parameter satisfying a non-reflected Σ_n type. Let $b \in \mathbf{J}_\beta(\mathbb{R})$ be such that $\Sigma^n_{b,\beta}$ is not realized in any $\mathbf{J}_\gamma(\mathbb{R})$ for $\gamma < \beta$. Let $b = f(G', w_2)$, where f is $\Sigma_1(\mathbf{J}_\beta(\mathbb{R}))$, G' is a finite subset of $\omega\beta$, and $w_2 \in \mathbb{R}$. Set

$$\Sigma = \Sigma^n_{\langle G', w_2\rangle, \beta}.$$

Clearly Σ is not realized in $\mathbf{J}_\gamma(\mathbb{R})$ for any $\gamma < \beta$. Now fix $G \leq_{\text{BK}} G'$ to be BK-least so that

$$\Sigma = \Sigma^n_{\langle G, w_2\rangle, \beta}.$$

We can now define our canonical sequence $\langle H_i : i < \omega\rangle$ of Σ_{n-1} hulls. Simultaneously we shall define sequence $\langle K_i : i < \omega\rangle$ of finite subsets of $\omega\beta$ and $\langle \vartheta_i : i < \omega\rangle$ of Σ_n formulae in Σ. Let $K_0 = \varnothing$. Given $K_0, \dots, K_i$, let

$$\begin{aligned} H_i &= \text{Hull}^\beta_{n-1}(\{F, G, K_0 \dots K_i\} \cup \mathbb{R}) \\ &= \{a : \{a\} \text{ is } \Sigma_{n-1}^{\mathbf{J}_\beta(\mathbb{R})}(\{F, G, K_0 \dots K_i\} \cup \mathbb{R})\}. \end{aligned}$$

Because we have thrown F into H_i, $H_i \prec_{n-1} \mathbf{J}_\beta(\mathbb{R})$ and H_i is the image of $\mathbb{R}$ under a partial $\underset{\sim}{\Sigma}_{n-1}(H_i)$ function. Since $\rho^\beta_{n-1} \neq \mathbb{R}$, H_i must collapse to some $\mathbf{J}_\gamma(\mathbb{R})$ for $\gamma < \beta$. But then $\pi(\langle G, w_2\rangle)$ does not realize Σ in $\mathbf{J}_\gamma(\mathbb{R})$, where π is the collapse map, and therefore $\langle G, w_2\rangle$ does not realize Σ in H_i. Since $H_i \prec_{n-1} \mathbf{J}_\beta(\mathbb{R})$, there must be a Σ_n formula $\vartheta(v)$ in Σ so that $H_i \not\models \vartheta[\langle G, w_2\rangle]$. Let

$$\vartheta_i = \text{the least } \Sigma_n \text{ formula } \vartheta(v) \in \Sigma \text{ such that } H_i \not\models \vartheta[\langle G, w_2\rangle],$$

where by "least" we mean least in some fixed ordering $<_\Sigma$ of Σ of order type ω. Now every element of $\mathbf{J}_\beta(\mathbb{R})$ is Σ_1 definable from a real and a finite subset of $\omega\beta$, so that for some finite $K \subseteq \omega\beta$, $\text{Hull}^\beta_{n-1}(\{F, K\} \cup \mathbb{R}) \models \vartheta_i [\langle G, w_2\rangle]$. Let K_{i+1} be the BK-least finite subset K of $\omega\beta$ such that

$$\text{Hull}^\beta_{n-1}(\{F, G, K_0, \dots, K_i, K\} \cup \mathbb{R}) \models \vartheta_i [\langle G, w_2\rangle].$$

This completes the definitions of the H_i's, K_i's, and ϑ_i's.

We record some simple properties of these sequences in a claim.

CLAIM 3.8.

(a) $\forall i\, (H_i \prec_{n-1} H_{i+1} \prec_{n-1} \mathbf{J}_\beta(\mathbb{R}))$.
(b) $\forall i\, (H_i \not\models \vartheta_i [\langle G, w_2\rangle]\ \&\ H_{i+1} \models \vartheta_i [\langle G, w_2\rangle])$.
(c) $\forall i\, \exists \gamma < \beta\, (H_i \cong \mathbf{J}_\gamma(\mathbb{R}))$.
(d) $\bigcup_{i<\omega} H_i = \mathbf{J}_\beta(\mathbb{R})$.

PROOF. (a) and (b) are obvious, and (c) was proved in the course of defining ϑ_i.

For (d), notice that

$$\bigcup_{i<\omega} H_i \models \vartheta [\langle G, w_2\rangle], \quad \forall \vartheta \in \Sigma,$$

because the ϑ_i's were chosen least in an ordering of type ω. Thus if π collapses $\bigcup_{i<\omega} H_i$ to a trasnsitive set, then

$$\pi\colon \bigcup_{i<\omega} H_i \cong \mathbf{J}_\beta(\mathbb{R}),$$

as otherwise $\pi(\langle G, w_2\rangle)$ would realize Σ in some $\mathbf{J}_\gamma(\mathbb{R})$ for $\gamma < \beta$. We shall show $\pi(F) = F$, $\pi(G) = G$, and $\pi(K_i) = K_i$ for all i. Of course $\pi(x) = x$ for $x \in \mathbb{R}$. Since $\bigcup_{i<\omega} H_i = \text{Hull}^{\bigcup_i H_i}_{n-1}(\{F, G, K_0, K_1, \dots\} \cup \mathbb{R})$, it follows that $\mathbf{J}_\beta(\mathbb{R}) = \text{Hull}^\beta_{n-1}(\pi[\{F, G, \dots\} \cup \mathbb{R}])$, so $\mathbf{J}_\beta(\mathbb{R}) = \text{Hull}^\beta_{n-1}(\{F, G, \dots\} \cup \mathbb{R})$, that is, $\mathbf{J}_\beta(\mathbb{R}) = \bigcup_{i<\omega} H_i$.

To see that $\pi(F) = F$, notice that ϑ_{sk} defines a Σ_{n-1} Skolem function from $\langle F, w_1\rangle$ over $\bigcup_i H_i$, and thus from $\langle \pi(F), w_1\rangle$ over $\mathbf{J}_\beta(\mathbb{R})$. But $\pi(F) \leq_{\text{BK}} F$, while F is BK-least so that ϑ_{sk} defines a Σ_{n-1} Skolem function from $\langle F, w_1\rangle$ over $\mathbf{J}_\beta(\mathbb{R})$. Thus $\pi(F) = F$.

Similarly, $\langle \pi(G), w_2\rangle$ realizes Σ in $\mathbf{J}_\beta(\mathbb{R})$ and $\pi(G) \leq_{\text{BK}} G$, so the BK-minimality of G implies $\pi(G) = G$.

Finally, $\pi(K_i) = K_i$ by induction on i. Certainly $\pi(K_0) = K_0$. If $\pi(K_j) = K_j$ for $j \leq i$, then

$$\pi[H_{i+1}] = \text{Hull}^\beta_{n-1}(\{F, G, K_0 \dots K_i, \pi(K_{i+1})\} \cup \mathbb{R}).$$

Since $\pi[H_{i+1}] \models \vartheta_i [\langle G, w_2\rangle]$, $\pi(K_{i+1}) \leq_{\text{BK}} K_{i+1}$, and K_{i+1} is BK-least so that $\text{Hull}^\beta_{n-1}(\{F, G, K_0 \dots K_{i+1}\} \cup \mathbb{R}) \models \vartheta_i [\langle G, w_2\rangle]$, we have $\pi(K_{i+1}) = K_{i+1}$. $\dashv$ (Claim 3.8)

Let P be a $\underset{\sim}{\Sigma}_n(\mathbf{J}_\beta(\mathbb{R}))$ set of reals. We want a closed game representation of P simple enough to yield a $\underset{\sim}{\Sigma}_n(\mathbf{J}_\beta(\mathbb{R}))$ scale on P. By part (d) of Claim 1, we can fix $e \in \omega$ so that P is $\Sigma_n(\mathbf{J}_\beta(\mathbb{R}), \{F, G, K_0 \dots K_e\} \cup \mathbb{R})$. Let

$$P(x) \text{ iff } \mathbf{J}_\beta(\mathbb{R}) \models \varphi_0\,[x, a]$$

where φ_0 is Σ_n, and a is a parameter of the form $\langle F, G, K_0, \dots, K_e, w_3\rangle$ for $w_3 \in \mathbb{R}$. For $i \geq e$, let

$$P^i(x) \text{ iff } H_i \models \varphi_0\,[x, a].$$

We shall construct closed game representations $x \mapsto \mathrm{G}^i_x$ of the P^i's in such a way that if

$$P^i_k(x, u) \text{ iff } u \text{ is a winning position for player I in } \mathrm{G}^i_x \text{ of length } k,$$

then P^i_k is $\underset{\sim}{\Sigma}_\omega(H_{\max(i,k)})$. Suppose we have done this. Inspecting [Mos83] again, we see that this gives a scale $\{\varphi^i_k\}_{k\in\omega}$ on P^i so that the prewellordering $\leq^i_k$ of $\mathbb{R}$ induced by φ^i_k is $\underset{\sim}{\Sigma}_\omega(H_{\max(i,k)})$. But $H_{\max(i,k)}$ collapses to some $\mathbf{J}_\gamma(\mathbb{R})$ with $\gamma < \beta$, and $\leq^i_k$ is fixed by the collapse map, so $\leq^i_k \in \mathbf{J}_\beta(\mathbb{R})$. Similarly, each $P^i \in \mathbf{J}_\beta(\mathbb{R})$. But now $\rho^\beta_n = \mathbb{R}$, so any countable subset of $\mathbf{J}_\beta(\mathbb{R})$ is $\underset{\sim}{\Sigma}_n(\mathbf{J}_\beta(\mathbb{R}))$, so the map $(i, k) \mapsto (P^i, \leq^i_k)$ is $\underset{\sim}{\Sigma}_n(\mathbf{J}_\beta(\mathbb{R}))$. We then have the obvious scale $\{\psi_i\}_{i<\omega}$ on P:

$$\psi_0(x) = \mu i\,[P^i(x)],$$

and

$$\psi_{k+1}(x) = \langle \psi_0(x), \varphi^{\psi_0(x)}_k\rangle,$$

where again ψ_{k+1} is defined using the lexicographical order. Since $(i, k) \mapsto (P^i, \leq^i_k)$ is $\underset{\sim}{\Sigma}_n(\mathbf{J}_\beta(\mathbb{R}))$, the scale $\{\psi_i\}_{i<\omega}$ is $\underset{\sim}{\Sigma}_n(\mathbf{J}_\beta(\mathbb{R}))$.

It suffices, then, to construct the representations $x \mapsto \mathrm{G}^i_x$ in such a way that P^i_k is first-order definable over $H_{\max(i,k)}$ for all i, k. Our plan is to force player I in G^i_x to describe the truth in $\mathbf{J}_\beta(\mathbb{R})$ about F, G, and the K_j's. Since

$$\mathbf{J}_\beta(\mathbb{R}) = \mathrm{Hull}^\beta_{n-1}(\{F, G\} \cup \{K_j \,:\, j < \omega\} \cup \mathbb{R})$$

it is enough that player I describe only Σ_{n-1} truths; moreover, this restriction is important if P^i_k is to be first-order definable over $H_{\max(i,k)}$. For this reason we also restrict player I to playing at move k only ordinals in H_k and formulae involving no K_j for $j > k$. Our main problem is how, within these restrictions, to prevent player I from describing the Σ_{n-1} truths about some parameters different from F, G, or the K_j's.

Player I's description is given in the language $\mathcal{L}$, which has $\in$, $=$, constant symbols $\underline{x}_i$ for $i \in \omega$, and constant symbols $\underline{F}$, $\underline{G}$, and $\underline{K}_i$ for $i \in \omega$. If φ is an $\mathcal{L}$-formula containing no constants $\underline{K}_i$ for $i > m$, we say φ has support m.

Let $\mathcal{B}_{n-1}$ be the class of boolean combinations of Σ_{n-1} formulae of $\mathcal{L}$. I will actually describe the truth of $\mathcal{B}_{n-1}$ formulae. We shall use the "unique v" operator applied to Σ_{n-1} formulae to abbreviate formulae in $\mathcal{B}_{n-1}$. Let σ, τ, and φ be Σ_{n-1} formulae, and let ψ be Π_1. The reader can easily check that $\varphi(\iota v\, \sigma(v))$, $\psi(\iota v\, \sigma(v))$, and $\psi(\iota v\, \sigma(v, \iota u\, \tau(u)))$ can be considered abbreviations of $\mathcal{B}_{n-1}$ formulae. There are the sorts of abbreviations we shall use.

For expository reasons, we allow player I to play finitely many sentences and finitely many reals in a single move of G^i_x. A typical run of G^i_x then has the form

$$\begin{array}{lccc} \mathrm{I} & T_0, s_0, \eta_0, m_0 & & T_1, s_1, \eta_1, m_1 & & \dots \\ \mathrm{II} & & y_0 & & y_1 & \end{array}$$

where T_k is a finite set of sentences in $\mathcal{B}_{n-1}$, all of which have support k, $s_k \in \mathbb{R}^{<\omega}$, $\eta_k \in \mathrm{Ord} \cap H_k$, $m_k \in \omega$, and $m_k > k$, and $y_k \in \mathbb{R}$. The roles of these objects have been explained, with the exception of the m_k's. We shall explain their role shortly.

Given the run of G^i_x displayed above, set

$$\langle x_i : i < \omega \rangle = s_0 {}^\frown \langle y_0 \rangle {}^\frown s_1 {}^\frown \langle y_1 \rangle {}^\frown \cdots,$$

and

$$T^* = \bigcup_{k<\omega} T_k.$$

We say this run is a win for player I just in case it meets the following requirements. Let $n \colon \mathcal{B}_{n-1} \hookrightarrow \omega$ be such that any $\vartheta \in \mathcal{B}_{n-1}$ has support $n(\vartheta)$ and involves no $\underline{x}_j$ for $j \geq n(\vartheta)$.

1. $s_0(0) = x$, $s_0(1) = w_1$, $s_0(2) = w_2$, and $s_0(3) = w_3$.
2. (a) T^* is consistent,
   (b) if $\vartheta \in \mathcal{B}_{n-1}$ and ϑ is a sentence, then either $\vartheta \in T_{n(\vartheta)}$ or $\neg\vartheta \in T_{n(\vartheta)}$,
   (c) if $\vartheta \in T_k$, then ϑ has support k and involves no constant $\underline{x}_j$ for $j \geq \mathrm{dom}(s_0 {}^\frown \langle y_0 \rangle {}^\frown \cdots \langle y_{k-1} \rangle {}^\frown s_k)$,
   (d) The Axiom of Extensionality is in T_0.
   (e) if $\tau(v)$ is Σ_{n-1} and $\exists u\, (u = \iota v\, \tau(v)) \in T_k$, then "$\exists \alpha\, (\iota v\, \tau(v) \in \mathbf{J}_\alpha(\mathbb{R}))$" $\in T_{k+1}$,
   (f) $\underline{x}_k(\underline{n}) = \underline{m} \in T^*$ iff $x_k(n) = m$.

Requirement 2(e) could be replaced by "$\mathbf{V}{=}\mathbf{L}(\mathbb{R}) \in T^*$" in the case $n > 2$.

Next we require player I to verify that $\varphi_0(x, a)$ holds in H_i. Let $\varphi_0 = \exists v \psi_0$ where ψ_0 is Π_1.

3. For some Σ_{n-1} formula $\tau(v)$ with support i,

$$\psi_0(\underline{x}_0, \langle \underline{F}, \underline{G}, \underline{K}_0 \dots \underline{K}_e, \underline{x}_3 \rangle, \iota v\, \tau(v)) \in T_i.$$

Player I must verify that any real he describes is one of the x_k's:

4. If $\tau(v)$ is Σ_{n-1} and "$\iota v\,\tau(v) \in \mathbb{R}$" $\in T_k$, then for some j,
$$\text{“}\iota v\,\tau(v) = \underline{x}_j\text{”} \in T_{k+1}.$$

Player I must verify that the model he is describing is well-founded:

5. If $\sigma(v)$ and $\tau(v)$ are Σ_{n-1}, and "$\iota v\,\sigma(v) \in \text{Ord}\ \&\ \iota v\,\tau(v) \in \text{Ord}$" $\in T^*$, then
$$\text{“}\iota v\,\sigma(v) \leq \iota v\,\tau(v)\text{”} \in T^* \text{ iff } \eta_{n(\sigma)} \leq \eta_{n(\tau)}.$$

Finally, player I must verify that $\underline{F}$ denotes F, $\underline{G}$ denotes G, and $\underline{K}_j$ denotes K_j for $j < \omega$. In order to do so, player I must verify certain Σ_n and Π_n sentences which come up as he plays. Now in order to verify $\forall v \vartheta(v)$, where ϑ is Σ_{n-1}, player I must simply refrain from putting $\neg\vartheta(\iota v\,\sigma(v))$ into T^* for any Σ_{n-1} formula $\sigma(v)$. This is a closed requirement on player I's play. Dually, in order to verify $\exists v \vartheta(v)$, where ϑ is Π_1, player I must put $\vartheta(\iota v\,\sigma(v))$ into T^* for some Σ_{n-1} formula $\sigma(v)$. Unfortunately, this is an open requirement on player I's play. In simple cases, such as requirement 3, we can bound in advance the move at which player I is expected to verify his Σ_n sentence, so that the requirement becomes closed. In view of our restriction on sentences in T_k to those with support k, this amounts to bounding in advance the hull which is to satisfy player I's Σ_n sentence. In some cases we can't do this, and so we require player I himself to provide the bound. This is the role of m_k; it is player I's prediction at move k of the later move at which he will discharge an accrued obligation to verify certain Σ_n sentences. (In 7(c)(ii) below, m_k plays a similar but slightly different role.)

One may object that this violates the spirit of our restrictions on the first k moves: since the m_j for $j < k$ can refer to arbitrarily large hulls, P^i_k won't be first-order definable over $H_{\max(i,k)}$. The Coding Lemma (of all things!) will overcome this objection.

Requirements 6(a) and (b) force player I to assert that ϑ_{sk} defines a Σ_{n-1} Skolem function from $\underline{F}$ and $\underline{x}_1$, while 6(c) forces him to assert that nothing $\leq_{\text{BK}} \underline{F}$ has this property of $\underline{F}$.

6. (a) The sentence
$$\forall v, w, u, x\ \big[(\vartheta_{\text{sk}}(u, x, v, \underline{F}, \underline{x}_1)\ \&\ \vartheta_{\text{sk}}(u, x, w, \underline{F}, \underline{x}_1)) \rightarrow v = w\big]$$
is in T_0,

(b) if $\vartheta(v_0, v_1)$ and $\tau(u)$ are Σ_{n-1}, and $\exists v\, \vartheta(v, \iota u\,\tau(u)) \in T_k$, then for some j
$$\exists v\ (\vartheta(v, \iota u\,\tau(u))\ \&\ \vartheta_{\text{sk}}(\iota u\,\tau(u), \underline{x}_j, v, \underline{F}, \underline{x}_1))$$
is in T_{k+1},

(c) if "$\iota v\,\sigma(v) <_{\text{BK}} \underline{F}$" $\in T_k$, then either the sentence
$$\exists v, w, u, x\ (\vartheta_{\text{sk}}(u, x, v, \iota v\,\sigma(v), \underline{x}_1)\ \&\ \vartheta_{\text{sk}}(u, x, w, \iota v\,\sigma(v), \underline{x}_1), \&\ v \neq w)$$

is in T_{m_k}, or for some Σ_{n-1} formulae $\vartheta(v_0, v_1)$ and $\tau(u)$, the sentence

$$\exists v\, \vartheta(v, \iota u\, \vartheta(u)) \;\&\; \neg\exists x\, \exists v\, (\vartheta_{\text{sk}}(\iota u\, \tau(u), x, v, \iota v\sigma(v), \underline{x}_1) \;\&\; \vartheta(v, \iota u\, \tau(u)))$$

is in T_{m_k}.

A similar requirement will fix the meaning of $\underline{G}$. Recall the Σ_n formula $\vartheta_j(v)$ in Σ used to define K_{j+1}, and the order $<_\Sigma$ of Σ.

7. (a) If $\forall u\, \vartheta(u, v) \in \Sigma$, where ϑ is Σ_{n-1}, and if $\tau(u)$ is any Σ_{n-1} formula, then

$$\neg\vartheta(\iota u\, \tau(u), \langle \underline{G}, \underline{x}_2 \rangle) \notin T_k$$

(b) if $\exists u\, \vartheta(u, v) \in \Sigma$, where ϑ is Π_1, and $\exists u\, \vartheta(u, v) <_\Sigma \vartheta_k$, then for some Σ_{n-1} formula $\tau(u)$

$$\vartheta(\iota u\, \tau(u), \underline{G}, \underline{x}_2) \in T_k,$$

(c) if "$\iota v\, \sigma(v) <_{\text{BK}} \underline{G}$" $\in T_k$, where $\sigma(v)$ is Σ_{n-1}, then either

(i) there is a Σ_{n-1} formula $\vartheta(u, v)$ such that $\forall u\, \vartheta(u, v) \in \Sigma$, but for some Σ_{n-1} formula $\tau(u)$,

$$\neg\vartheta(\iota u\, \tau(u), \langle \iota v\, \sigma(v), \underline{x}_2 \rangle) \in T_{m_k}$$

or

(ii) there is a Π_1 formula $\vartheta(u, v)$ such that $\exists u\, \vartheta(u, v)$ is one of the first m_k elements of Σ under $<_\Sigma$, and for all Σ_{n-1} formulae $\tau(u)$ and all $j \in \omega$

$$\vartheta(\iota u\, \tau(u), \langle \iota v\, \sigma(v). \underline{x}_2 \rangle) \notin T_j.$$

Finally, we have a requirement fixing the meaning of $\underline{K}_j$. Recall the sequence $\langle \vartheta_j : j < \omega \rangle$ of Σ_n formulae in Σ. Let $\vartheta_j = \exists u\, \vartheta'_j$, where ϑ'_j is Π_{n-1}.

8. (a) "$\underline{K}_0 = \varnothing$" $\in T_0$

(b) For any $j > 0$, there is a Σ_{n-1} $\tau(u)$ with support j such that

$$\vartheta'_{j-1}(\iota u\, \tau(u), \langle \underline{G}, \underline{x}_2 \rangle) \in T_j,$$

(c) For any $j > 0$, and any k, if

$$\text{"}\iota v\, \sigma(v) <_{\text{BK}} \underline{K}_j\text{"} \in T_k,$$

then for no Σ_{n-1} $\tau(u, v)$ with support $j - 1$ and no l do we have

$$\vartheta'_{j-1}(\iota u\, \tau(u, \iota v\, \sigma(v)), \langle \underline{G}, \underline{x}_2 \rangle) \in T_l.$$

This completes the description of the payoff for player I in G^i_x.

We next want to characterize the winning positions for player I in G^i_x as the honest ones. Let

$$u = \langle \langle T_k, s_k, \eta_k, m_k, y_k \rangle : k < s \rangle$$

be a position of length s. Let

$$\langle x_j : j < l \rangle = s_0{}^\frown\langle y_0\rangle^\frown \cdots {}^\frown s_k{}^\frown\langle y_k\rangle,$$

and let

$$I_u(\underline{x}_j) = x_j \quad \text{for } j < l,$$
$$I_u(\underline{F}) = F, \qquad I_u(\underline{G}) = G,$$

and

$$I_u(\underline{K}_j) = K_j \quad \text{for all } j < \omega.$$

Let us call u **reasonable** if it is not an immediate loss for player I because of $\langle T_k : k < s\rangle$, in the sense that all assertions in 1–4 and 6–8 about the T_k's hold so far for the T_k with $k < s$. [Take for example 7(c). If u is unreasonable on its account, there must be a $k < s$ such that $m_k < s$, and a formula "$\iota v\, \sigma(v) <_{\mathrm{BK}} \underline{G}$" in T_k such that 7(c)(i) fails for T_{m_k} while there are enough $j < s$ to witness the failure of 7(c)(ii).]

Finally, we say that u is (i, x)**-honest** if the following conditions are met:

1. $H_i \models \varphi_0\,[x, a]$.
2. u is reasonable.
3. $s > 0 \rightarrow s_0(0) = x \;\&\; s_0(1) = w_1 \;\&\; s_0(2) = w_2 \;\&\; s_0(3) = w_3$.
4. $(\mathbf{J}_\beta(\mathbb{R}), I_u) \models \bigcup_{k<s} T_k$.
5. if $\langle \sigma_0, \ldots, \sigma_m\rangle$ enumerates those Σ_{n-1} formulae σ such that $n(\sigma) < s$ and

$$(\mathbf{J}_\beta(\mathbb{R}), I_u) \models \iota v\, \sigma(v) \in \mathrm{Ord},$$

and if $\delta_1 < \omega\beta$ is such that

$$(\mathbf{J}_\beta(\mathbb{R}), I_u) \models \iota v\, \sigma_i(v) = \delta_i$$

then the map

$$\delta_i \mapsto \eta_{n(\sigma_i)}$$

is well-defined and extendible to an order-preserving map of $\mathrm{Ord} \cap H_s$ into $\mathrm{Ord} \cap H_s$.

The final two requirements for (i, x)-honesty guarantee that player I has made commitments m_k, $k < s$, which can be kept.

6. If $k < s$ and "$\iota v\, \sigma(v) <_{\mathrm{BK}} \underline{F}$" $\in T_k$, then either

$$\exists v, w, u, x\, (\vartheta_{\mathrm{sk}}(u, x, v, \iota v\, \sigma(v), \underline{x}_1) \;\&\; \vartheta_{\mathrm{sk}}(u, x, w, \iota v\, \sigma(v), \underline{x}_1) \;\&\; v \neq w)$$

is true in (H_{m_k}, I_u), or for some Σ_{n-1} formula $\vartheta(v_0, v_1)$

$$\exists u\, [\exists v\, \vartheta(v, u) \;\&\; \neg\exists x \exists v\, (\vartheta(v, u) \;\&\; \vartheta_{\mathrm{sk}}(u, x, v, \iota v\, \sigma(v), \underline{x}_1))]$$

is true in (H_{m_k}, I_u).

7. If $k < s$ and "$\imath v\, \sigma(v) <_{\mathrm{BK}} \underline{G}$" $\in T_k$, then either there is a Π_n formula $\vartheta(v) \in \Sigma$ such that
$$(H_{m_k}, I_u) \models \neg\vartheta(\langle \imath v\, \sigma(v), \underline{x}_2\rangle),$$
or there is a Σ_n formula $\vartheta(v) \in \Sigma$ which is one of the first m_k elements of Σ such that
$$(\mathbf{J}_\beta(\mathbb{R}), I_u) \models \neg\vartheta(\langle \imath v\, \sigma(v), \underline{x}_2\rangle).$$

Because of conditions 6 and 7 in the definition of (i, x)-honesty, the following claim requires a proof.

CLAIM 3.9. $\{(x,u) : u$ is an (i,x)-honest position of length $s\}$ is $\utilde{\Sigma}_\omega(H_{\max(i,s)})$.

PROOF. Conditions 1, 2, and 3 are trivially $\utilde{\Sigma}_\omega(H_{\max(i,s)})$. Condition 4 of honesty is first order over $H_{\max(i,s)}$ because of our restrictions on the sentences in T_k for $k < s$. Condition 5 is first order because if there is any map as required by 5, there is a $\utilde{\Sigma}_1(H_s)$ such map, and H_s is $\utilde{\Sigma}_{n+5}$ over $H_{\max(i,s)}$.

We turn now to condition 7; since the proof that 6 is $\utilde{\Sigma}_\omega(H_{\max(i,s)})$ is similar but simpler, we shall omit it. Let
$$A = \{\langle K, \vartheta, m\rangle : K <_{\mathrm{BK}} G \;\&\; K \in H_s \;\&\; \vartheta \in \Sigma \;\&\; (H_m, \in) \models \vartheta[\langle K, w_2\rangle]\}.$$
It suffices to show that A is $\utilde{\Sigma}_\omega(H_s)$; from this it easily follows that condition 7 is $\utilde{\Sigma}_\omega(H_{\max(i,s)})$.

Since there is a $\utilde{\Sigma}_{n-1}(H_s)$ partial map of $\mathbb{R}$ onto H_s, there is a $\utilde{\Sigma}_n(H_s)$ total map
$$f : \mathbb{R} \twoheadrightarrow \{K \in H_s : K <_{\mathrm{BK}} G\} \times \Sigma.$$
For $y, z \in \mathbb{R}$, let
$$y \leq^* z \text{ iff } f(y)_0 <_{\mathrm{BK}} f(z)_0 \vee \big(f(y)_0 = f(z)_0 \;\&\; f(y)_1 \leq_\Sigma f(z)_1\big).$$
Thus $\leq^*$ is a $\utilde{\Sigma}_n(H_s)$ prewellorder of $\mathbb{R}$. Since $H_s \cong \mathbf{J}_\gamma(\mathbb{R})$ for some $\gamma\omega$, $\leq^* \in \mathbf{J}_\beta(\mathbb{R})$. Now for $m < \omega$, let
$$A_m = \{\langle K, \vartheta\rangle : \langle K, \vartheta, m\rangle \in A\}.$$
Now A_m is $\utilde{\Sigma}_\omega(H_{\max(m,s)})$, so $f^{-1}(A_m)$ is $\utilde{\Sigma}_\omega(H_{\max(m,s)})$, and so $f^{-1}(A_m) \in \mathbf{J}_\beta(\mathbb{R})$ for all m. By the Coding Lemma [Mos80, 7D.5], $f^{-1}(A_m)$ is $\utilde{\Sigma}^1_1(\leq^*)$ for each m. [Since $\leq^*, f^{-1}(A_m) \in \mathbf{J}_\beta(\mathbb{R})$, $\mathsf{Det}(\mathbf{J}_\beta(\mathbb{R}))$ suffices for this application of the Coding Lemma. Now $\mathbf{J}_\alpha(\mathbb{R}) \prec_1^{\mathbb{R}} \mathbf{J}_\beta(\mathbb{R})$, so if there is a non-determined game in $\mathbf{J}_\beta(\mathbb{R})$, there is a non-determined game in $\mathbf{J}_\alpha(\mathbb{R})$. However, we have assumed $\mathsf{Det}(\mathbf{J}_\alpha(\mathbb{R}))$.] Since $\utilde{\Sigma}^1_1(\leq^*)$ is closed under countable unions, we have $B \in \utilde{\Sigma}^1_1(\leq^*)$, where
$$B = \bigcup_m (f^{-1}(A_m) \times \{m\}).$$

But then B is $\utilde{\Sigma}_\omega(H_s)$. Since

$$(K, \vartheta, m) \in A \text{ iff } \exists y\,(\langle y, m\rangle \in B \;\&\; f(y)_0 = K \;\&\; f(y)_1 = \vartheta)$$

we have that A is $\utilde{\Sigma}_\omega(H_s)$, as desired. ⊣ (Claim 3.9)

Our original proof of Claim 3.9 required more that $\mathrm{Det}(\mathbf{J}_\alpha(\mathbb{R}))$, since we had applied the Coding Lemma directly to A. Kechris had the idea of applying the Coding Lemma to the A_m's individually. This reduces the determinacy needed in the hypotheses of Theorem 3.7 to $\mathrm{Det}(\mathbf{J}_\alpha(\mathbb{R}))$, a reduction which is crucial for the use of Theorem 3.7 in Kechris-Woodin [KW83].

The next claim finishes our proof of Theorem 3.7 in the case that $n > 1$.

Claim 3.10. For all i, x, and u, u is (i, x)-honest iff u is a winning position for player I in G^i_x.

Proof (Sketch). (⇒) It is enough to show that whenever v is (i, x)-honest, then $\exists T, s, \eta, m\, \forall y\, (v^\frown\langle T, s, \eta, m, y\rangle$ is (i, x)-honest). For if so, then player I can win G^i_x by keeping to (i, x)-honest positions. So let v be an (i, x)-honest position of length k. Because v satisfies 2 and 4 of honesty, player I has told as much of the truth about $\mathbf{J}_\beta(\mathbb{R})$ in his first k moves as we required him to tell. Because v satisfies 6 and 7 of honesty, player I has made predictions m_i for $i < k$ which he can fulfill. Thus player I can choose T and s to insure continued satisfaction of 2 and 4 by the new position; that is, he can tell as much more of the truth as he must. (If $k = i$, we need that v satisfies 1 as well.) Continued satisfaction of 1 and 3 are trivial to insure, while continued satisfaction of 6 and 7 can be insured by choosing m_k large enough. Finally, player I must choose η so as to insure continued satisfaction of 5. Now let v determine the map

$$g(\delta_i) = \eta_{n(\sigma_i)}, \quad \text{for } 0 \le i \le m,$$

as in 5 of honesty for v. We may assume that $i \le j \to \delta_i \le \delta_j$. It is enough to see that g can be extended to an order-preserving map of $\mathrm{Ord} \cap H_{k+1}$ into $\mathrm{Ord} \cap H_{k+1}$. Now by 5 for v and the fact that $H_k \prec_1 \mathbf{J}_\beta(\mathbb{R})$,

$$H_k \models \exists f : [\delta_i, \delta_{i+1}] \xrightarrow{\text{op}} [g(\delta_1), g(\delta_{i+1})]$$

for all $i < m$. Since $H_k \prec_1 H_{k+1}$,

$$H_{k+1} \models \exists f : [\delta_i, \delta_{i+1}] \xrightarrow{\text{op}} [g(\delta_1), g(\delta_{i+1})]$$

for all $i < m$. Similarly,

$$H_{k+1} \models \exists f : [0, \delta_0] \xrightarrow{\text{op}} [0, g(\delta_0)].$$

Thus in order to obtain the desired extensions of g, it is enough to show that there is an order-preserving $f : (\mathrm{Ord} \cap H_{k+1}) - \delta_m \to (\mathrm{Ord} \cap H_{k+1}) - g(\delta_m)$.

But such an f fails to exist just in case for some $n < \omega$

$$H_{k+1} \models (g(\delta_m) - \delta_m) \cdot n \text{ doesn't exist},$$

as the following diagram makes clear (see page 160).

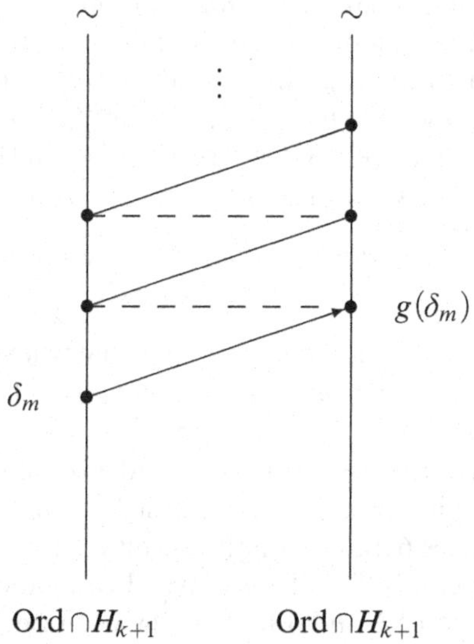

This is a Π_1 property of H_{k+1}, and so would pass downward to H_k, which is impossible since g can be extended over H_k. Thus there is such an f.

($\Leftarrow$) Let S be a winning strategy for player I starting from u. As in Theorem 2.1, let $\langle\langle T_k, s_k, \eta_k, m_k, y_k\rangle : k < \omega\rangle$ be a "generic play" according to S, and let $\langle x_k : k < \omega\rangle = s_0{}^\frown\langle y_0\rangle{}^\frown s_1{}^\frown\langle y_1\rangle \cdots$. Since $\bigcup_k T_k$ is consistent, it has a model $\mathfrak{B}$. By payoff requirements 6(a) and (b), $\mathfrak{A} \prec \mathfrak{B}$ where

$$\mathfrak{A} = \mathrm{Hull}^{\mathfrak{B}}(\{\underline{x}_j^{\mathfrak{B}} : j < \omega\} \cup \{\underline{F}^{\mathfrak{B}}, \underline{G}^{\mathfrak{B}}\} \cup \{\underline{K}_j^{\mathfrak{B}} : j < \omega\}).$$

If $\mathfrak{A} \models \iota v\, \sigma(v) \in \mathrm{Ord}$, then set

$$f(\iota v\, \sigma(v)^{\mathfrak{A}}) = \eta_{n(\sigma)}.$$

By payoff requirement 5, f is well-defined and order-preserving. Thus

$$\mathfrak{A} \cong (\mathbf{J}_\gamma(\mathbb{R}^{\mathfrak{A}}), I)$$

for some $\gamma \leq \beta$ and interpretation of constants I. By payoff requirements 7(a) and (b), $\mathfrak{A}$ realizes Σ, so $\gamma = \beta$. By 2(f), 4, and genericity, $\mathbb{R}^{\mathfrak{A}} = \mathbb{R}^M$, where M is our ground model. Clearly

$$I(\underline{x}_j) = x_j \quad \text{for } j < \omega,$$

while payoff requirements 6, 7, and 8 insure that

$$I(\underline{F}) = F, \quad I(\underline{G}) = G,$$

and

$$I(\underline{K}_j) = K_j \quad \text{for } j < \omega.$$

Therefore

$$\mathfrak{A} = (\mathbf{J}_\beta(\mathbb{R}), I)$$

where I is the natural interpretation. It is easy to verify conditions 1 through 4 and conditions 6 and 7 of (i, x)-honesty for u at this point. For condition 5, notice that the order-preserving map f defined above actually maps any H_s into H_s (since I is the natural interpretation). Thus f witnesses the satisfaction of condition 5. ⊣ (Claim 3.10)

Of course, Claim 3.10 applied to the empty position implies that $x \mapsto \mathrm{G}^i_x$ is a closed game representation of P^i. Claims 3.9 and 3.10 imply that the sets P^i_s are simply definable enough to yield $\underset{\sim}{\Sigma}_n(\mathbf{J}_\beta(\mathbb{R}))$ scale on P.

We are left with the proof of the theorem in the case $n = 1$. Since the basic plan in the case is the same as the plan in the case $n > 1$, we shall just sketch a few of the details.

Assume first that β is a limit ordinal and $n = 1$. We shall need no analogue of F. Let G be a finite subset of β and w_1 a real such that Σ, the Σ_1-type of $\langle G, w_1\rangle$ over $\mathbf{J}_\beta(\mathbb{R})$, is not realized in $\mathbf{J}_\gamma(\mathbb{R})$ for any $\gamma < \beta$. Further, let G be BK-least so that $\langle G, w_1\rangle$ realizes Σ. Define $\beta_i < \beta$ and $\vartheta_i(v) \in \Sigma$ by: $\beta_0 = 0$, and

$$\vartheta_i = \text{least } \Sigma_1 \text{ formula } \vartheta(v) \in \Sigma \text{ such that } \mathbf{J}_{\beta_i}(\mathbb{R}) \not\models \vartheta[\langle G, w_1\rangle]$$

and

$$\beta_{i+1} = \text{least } \beta > \beta_i \text{ such that } \mathbf{J}_\beta(\mathbb{R}) \models \vartheta[\langle G, w_1\rangle].$$

The β_i's are analogues of the K_i's. Let also

$$H_i = \mathrm{Hull}^{\beta_{i+1}}_\omega(\{G, \beta_0 \ldots \beta_i\} \cup \mathbb{R}).$$

Of course, $H_i \cong \mathbf{J}_\gamma(\mathbb{R})$ for some $\gamma < \beta$, moreover, H_i is the image of $\mathbb{R}$ under a $\underset{\sim}{\Sigma}_1(H_{i+1})$ map. As before, $\bigcup_{i<\omega} H_i = \mathbf{J}_\beta(\mathbb{R})$.

Let

$$P(x) \iff \mathbf{J}_\beta(\mathbb{R}) \models \varphi[x],$$

where φ is Σ_1 and we have dropped the parameter for simplicity. Let

$$P^i(x) \iff H_i \models \varphi[x].$$

Again, it is enough to construct representations $x \mapsto \mathrm{G}^i_x$ of the P^i in such a way that the corresponding P^i_k are first-order over $H_{\max(i,k)}$. For this, it is convenient to have in player I's language $\mathcal{L}$, besides the $\underline{x}_i$, constants $\underline{\beta}_i$

and $\underline{\mathbf{J}_{\beta_i}(\mathbb{R})}$ for $i < \omega$. In G^i_x, player I must play a Σ_0-complete. consistent set of Σ_0 sentences of $\mathcal{L}$, mentioning at move k no sentences involving $\underline{\beta}_j$ or $\underline{\mathbf{J}_{\beta_i}(\mathbb{R})}$ for $j > k$. For each j, player I must play at move j the Σ_0 sentence "$\underline{\mathbf{J}_{\beta_j}(\mathbb{R})} \models V = \mathbf{L}(\mathbb{R})$". At move $i+1$, player I must assert that some object definable over $\mathbf{J}_{\beta_{i+1}}(\mathbb{R})$ from G, $\beta_0, \ldots, \beta_i$, and some real (which he has played) witnesses that $\varphi(\underline{x}_0)$. Player I must play ordinals $\eta_k \in H_k$ to prove well-foundness. Finally, player I must prove that he is interpreting his constant symbols correctly; for $\underline{G}$ this involves commitments m_k made at move k as in the case $n > 1$. The Coding Lemma argument of Claim 3.9 goes through because each H_k is the image of $\mathbb{R}$ under a map in $\mathbf{J}_\beta(\mathbb{R})$. Claim 3.10 is proved as in the case $n > 1$.

Finally, we have the case $\beta = \gamma + 1$ and $n = 1$. Let $\langle G, w\rangle$ realize over $\mathbf{J}_\beta(\mathbb{R})$ a non-reflected Σ_1-type. Now $G = h(p, \mathbf{J}_\gamma(\mathbb{R}))$ for some rud function h and $p \in \mathbf{J}_\gamma(\mathbb{R})$. Using the simplicity of rud functions (as in Theorem 2.1), we can recursively associate to any Σ_1 formula ϑ and any $n \in \omega$ a formula $\psi_{\vartheta,n}$ so that for all α and all $a \in \mathbf{J}_\alpha(\mathbb{R})$,

$$\mathbf{J}_\alpha(\mathbb{R}) \models \psi_{\vartheta,n}[a] \text{ iff } S_{\omega\alpha+n}(\mathbb{R}) \models \vartheta[h(a, \mathbf{J}_\alpha(\mathbb{R}))].$$

It follows that if Σ is the full elementary type realized by $\langle p, w\rangle$ over $\mathbf{J}_\gamma(\mathbb{R})$, then Σ is not realized in $\mathbf{J}_\delta(\mathbb{R})$ for any $\delta < \gamma$. Again, we may replace p by some finite $G \subseteq \omega\gamma$ (changing w) and take G to be BK-least so that $\langle G, w\rangle$ realizes Σ over $\mathbf{J}_\gamma(\mathbb{R})$. Then $\mathbf{J}_\gamma(\mathbb{R}) = \mathrm{Hull}(\{G\} \cup \mathbb{R})$.

Let

$$P(x) \iff \mathbf{J}_\beta(\mathbb{R}) \models \varphi[x]$$

where φ is Σ_1 and we have ignored the parameter. Let

$$P^i(x) \iff S_{\omega\gamma+1}(\mathbb{R}) \models \varphi[x].$$

As above, we can find first-order formulae ψ_i such that

$$P^i(x) \iff \mathbf{J}_\gamma(\mathbb{R}) \models \psi_i[x].$$

We construct the desired representation $x \mapsto \mathrm{G}^i_x$ of P^i as follows. Let $\mathcal{L}$ have $\in$, $=$, $\underline{x}_i$ for $i < \omega$, and $\underline{G}$. In G^i_x player I must produce a (fully) complete, consistent $\mathcal{L}$ theory containing $\psi_i(\underline{x}_0)$, while using ordinals less than $\omega\gamma$ to prove well-foundness. At move k he can put only Σ_k sentences into his hierarchy. His theory must be "Skolemized", and whenever he describes a real he must play it bodily on the board, as before. Finally, player I must prove that he is interpreting $\underline{G}$ by G; this again involves commitments of the form m_k. The Coding Lemma argument of Claim 3.9 goes through because for each k we have a map $f : \mathbb{R} \to \mathbf{J}_\gamma(\mathbb{R})$ such that $f \in \mathbf{J}_\beta(\mathbb{R})$ and $G \in f[\mathbb{R}]$ and $f[\mathbb{R}] \prec_k \mathbf{J}_\gamma(\mathbb{R})$. Claim 3.10 is proved as in the case $n > 1$.

This completes the proof of Theorem 3.7. ⊣

In each case in the proof of Theorem 3.7, we expressed our arbitrary $\boldsymbol{\Sigma}_n(\mathbf{J}_\beta(\mathbb{R}))$ set of reals as a countable union of sets belonging to $\mathbf{J}_\beta(\mathbb{R})$. Thus we have

COROLLARY 3.11. If $[\alpha, \beta]$ is a weak Σ_1-gap, and n is least such that $\rho_n^\beta = \mathbb{R}$, then $P \in \boldsymbol{\Sigma}_n(\mathbf{J}_\beta(\mathbb{R})) \cap \wp(\mathbb{R})$ iff $P = \bigcup_i P_i$, where $P_i \in \mathbf{J}_\beta(\mathbb{R})$ for all i.

COROLLARY 3.12. If $[\alpha, \beta]$ is a weak Σ_1-gap, and n is least such that $\rho_n^\beta = \mathbb{R}$, then each of the classes $\boldsymbol{\Sigma}_{n+2k}(\mathbf{J}_\beta(\mathbb{R}))$ and $\boldsymbol{\Pi}_{n+2k+1}(\mathbf{J}_\beta(\mathbb{R}))$, for $k < \omega$, has the scale property.

PROOF. By the second periodicity theorem, it is enough to show that

$$\boldsymbol{\Sigma}_{n+k+1}(\mathbf{J}_\beta(\mathbb{R})) \cap \wp(\mathbb{R}) = \exists^{\mathbb{R}}(\boldsymbol{\Pi}_{n+k}(\mathbf{J}_\beta(\mathbb{R}))) \cap \wp(\mathbb{R}).$$

The proof of this is like the proof of Lemma 2.7, the key fact being that

$$\Sigma_n(\mathbf{J}_\beta(\mathbb{R})) \cap \wp(\mathbb{R}) \subseteq \exists^{\mathbb{R}}(\Pi_n(\mathbf{J}_\beta(\mathbb{R}))).$$

This fact follows trivially from Corollary 3.11. ⊣

Our description of the levels of the Levy hierarchy for $\mathbf{L}(\mathbb{R})$ having the scale property is now complete. It is interesting that the negative results on scales in $\mathbf{L}(\mathbb{R})$ (Corollaries 2.10, 2.12, and 3.4) all come from negative results on uniformization (Theorems 2.9, 2.11, and 3.3). In $\mathbf{L}(\mathbb{R})$, the best way to uniformize arbitrary relations in a given Levy class is by means of scales on those relations, as we claimed in the introduction.

It follows easily from Wadge's Lemma that if $A \in \boldsymbol{\Pi}_1(\mathbf{J}_\alpha(\mathbb{R})) - \boldsymbol{\Sigma}_1(\mathbf{J}_\alpha(\mathbb{R}))$ and A admits a $\boldsymbol{\Sigma}_n(\mathbf{J}_\beta(\mathbb{R}))$ scale, then so does every $\boldsymbol{\Pi}_1(\mathbf{J}_\alpha(\mathbb{R}))$ set. This observation and an inspection of our results above yield the promised characterization of the smallest Levy class $\boldsymbol{\Sigma}_n(\mathbf{J}_\beta(\mathbb{R}))$ at which a scale on a given set A is definable. Let $\langle \beta, n \rangle$ be lexicographically least so that A admits a $\boldsymbol{\Sigma}_n(\mathbf{J}_\beta(\mathbb{R}))$ scale; then $A \in \boldsymbol{\Sigma}_n(\mathbf{J}_\beta(\mathbb{R}))$ and either $\boldsymbol{\Sigma}_n(\mathbf{J}_\beta(\mathbb{R}))$ or $\boldsymbol{\Sigma}_{n+1}(\mathbf{J}_\beta(\mathbb{R}))$ has the scale property. Our results on the scale property therefore characterize $\langle \beta, n \rangle$ by means of reflection properties.

§4. Suslin cardinals.

We shall use results of §2 and §3 to extend some work of [Kec81] and [KSS81] (see also [Ste81]) in the global theory of boldface pointclasses. To simplify the exposition, we assume ZF + AD + DC throughout this section. The results of [Kec81] and [KSS81] are formulated in terms of a hierarchy slightly finer and considerably more general than the Levy hierarchy for $\mathbf{L}(\mathbb{R})$. We now define this hierarchy. Let Sep(Γ), PWO(Γ), and Scale(Γ) mean, respectively, that Γ has the separation, prewellordering, and scale properties. If $\boldsymbol{\Gamma}$ and $\boldsymbol{\Lambda}$ are nonselfdual boldface pointclasses, let

$$\{\boldsymbol{\Gamma}, \breve{\boldsymbol{\Gamma}}\} <_{\mathrm{w}} \{\boldsymbol{\Lambda}, \breve{\boldsymbol{\Lambda}}\} \text{ iff } \boldsymbol{\Gamma} \subseteq \boldsymbol{\Lambda} \cap \breve{\boldsymbol{\Lambda}}.$$

Wadge and Martin have shown that $<_W$ is a wellorder. The set of pairs $\{\boldsymbol{\Gamma}, \check{\boldsymbol{\Gamma}}\}$ such that $\exists^{\mathbb{R}}\boldsymbol{\Gamma} \subseteq \boldsymbol{\Gamma}$ or $\exists^{\mathbb{R}}\check{\boldsymbol{\Gamma}} \subseteq \check{\boldsymbol{\Gamma}}$ has order type Θ under $\leq_W$; for $0 \leq \alpha < \Theta$, let P_α be the αth such pair. Now let

$$\boldsymbol{\Sigma}^1_\alpha = \text{unique } \boldsymbol{\Gamma} \in P_\alpha \left[\exists^{\mathbb{R}}\boldsymbol{\Gamma} \subset \boldsymbol{\Gamma} \,\&\, (\forall^{\mathbb{R}}\boldsymbol{\Gamma} \not\subseteq \boldsymbol{\Gamma} \vee \text{Sep}(\boldsymbol{\Gamma}))\right].$$

Since exactly one of $\text{Sep}(\boldsymbol{\Gamma})$ and $\text{Sep}(\check{\boldsymbol{\Gamma}})$ holds for every nonselfdual boldface $\boldsymbol{\Gamma}$, $\boldsymbol{\Sigma}^1_\alpha$ is well defined. Let

$$\boldsymbol{\Pi}^1_\alpha = (\boldsymbol{\Sigma}^1_\alpha)\check{},$$
$$\boldsymbol{\Delta}^1_\alpha = \boldsymbol{\Sigma}^1_\alpha \cap \boldsymbol{\Pi}^1_\alpha,$$

and

$$\boldsymbol{\delta}^1_\alpha = \sup\{\lambda : \exists f\ (f : \mathbb{R} \twoheadrightarrow \lambda \,\&\, \{\langle x, y\rangle : f(x) \leq f(y)\} \in \boldsymbol{\Delta}^1_\alpha)\}.$$

The sequence $\langle \boldsymbol{\Sigma}^1_\alpha : \alpha < \vartheta\rangle$ is just the natural extension of the projective hierarchy cofinally through all boldface pointclasses. In particular, $\boldsymbol{\Sigma}^1_0 = \boldsymbol{\Sigma}^0_1$, and for $1 \leq n < \omega$, $\boldsymbol{\Sigma}^1_n$ is just the class usually given that name. For $\alpha < \Theta^{\mathbf{L}(\mathbb{R})}$, $\boldsymbol{\Sigma}^1_\alpha$ is essentially the αth level of the restriction to sets of reals of the Levy hierarchy for $\mathbf{L}(\mathbb{R})$, as we now show.

First, more terminology. If $\boldsymbol{\Gamma}$ is a boldface pointclass, then

$$\bigcup_\alpha \boldsymbol{\Gamma} = \Big\{\bigcup_{\beta<\alpha} A_\beta : \forall\beta < \alpha\,(A_\beta \in \boldsymbol{\Gamma})\Big\}$$

and

$$\bigcap_\alpha \boldsymbol{\Gamma} = \Big(\bigcup_\alpha \boldsymbol{\Gamma}\Big)\check{}.$$

We can classify limit ordinals λ according to the closure properties of $\boldsymbol{\Sigma}^1_\lambda$:

$$\lambda \text{ is type I} \iff \bigcap_2 \boldsymbol{\Sigma}^1_\lambda \subseteq \boldsymbol{\Sigma}^1_\lambda \,\&\, \bigcap_\omega \boldsymbol{\Sigma}^1_\lambda \not\subseteq \boldsymbol{\Sigma}^1_\lambda,$$
$$\lambda \text{ is type II} \iff \bigcap_2 \boldsymbol{\Sigma}^1_\lambda \not\subseteq \boldsymbol{\Sigma}^1_\lambda,$$
$$\lambda \text{ is type III} \iff \bigcap_\omega \boldsymbol{\Sigma}^1_\lambda \subseteq \boldsymbol{\Sigma}^1_\lambda \,\&\, \forall^{\mathbb{R}}\boldsymbol{\Sigma}^1_\lambda \not\subseteq \boldsymbol{\Sigma}^1_\lambda,$$
$$\lambda \text{ is type IV} \iff \forall^{\mathbb{R}}\boldsymbol{\Sigma}^1_\lambda \subseteq \boldsymbol{\Sigma}^1_\lambda.$$

The type of λ is just the type, in the sense of [KSS81], of the projective-like hierarchy immediately above $\boldsymbol{\Sigma}^1_\lambda$. Some facts from [KSS81] and [Ste81]: λ is of type I iff $\text{cf}(\lambda) = \omega$ iff $\boldsymbol{\Sigma}^1_\lambda = \bigcup_\omega(\bigcup_{\alpha<\omega} \boldsymbol{\Sigma}^1_\alpha)$. If λ is of type II,III, or IV, then $\boldsymbol{\Delta}^1_\lambda = \bigcup_{\alpha<\lambda} \boldsymbol{\Sigma}^1_\alpha$. If λ is of type I, then $\text{PWO}(\boldsymbol{\Sigma}^1_\lambda)$; otherwise (provided $\lambda < \Theta^{\mathbf{L}(\mathbb{R})}$) $\text{PWO}(\boldsymbol{\Pi}^1_\lambda)$. (At this point we should warn the reader that our definition of the extended projective hierarchy differ slightly from that of [Kec81]. The difference is that the classes $\underset{\sim}{\Sigma}^1_\lambda$ for λ of type I or II are omitted from the hierarchy of [Kec81].)

Let $\langle \delta_\alpha : \alpha < \Theta^{\mathbf{L}(\mathbb{R})} \rangle$ enumerate in increasing order those ordinals δ such that $\rho^\delta_n = \mathbb{R}$ for some n, with $\delta_0 = 1$. Let n_α be the least n such that $\rho^{\delta_\alpha}_n = \mathbb{R}$.

LEMMA 4.1. For all $\alpha < \Theta^{\mathbf{L}(\mathbb{R})}$,

(a) $\mathrm{PWO}(\underset{\sim}{\Sigma}_{n_\alpha}(\mathbf{J}_{\delta_\alpha}(\mathbb{R})))$,

(b) $\underset{\sim}{\Sigma}_{n_\alpha}(\mathbf{J}_{\delta_\alpha}(\mathbb{R})) \cap \wp(\mathbb{R}) \subseteq \bigcup_{\delta_\alpha}(\mathbf{J}_{\delta_\alpha}(\mathbb{R}) \cap \wp(\mathbb{R}))$.

PROOF. Let $\delta = \delta_\alpha$, $n = n_\alpha$, and $\rho = \rho^\delta_{n-1}$. Fix a partial $\Sigma_{n-1}(\mathbf{J}_\delta(\mathbb{R}))$ map $f : \mathbf{J}_\rho(\mathbb{R}) \twoheadrightarrow \mathbf{J}_\delta(\mathbb{R})$. Suppose

$$x \in A \iff \mathbf{J}_\delta(\mathbb{R}) \models \exists v \varphi\,[x],$$

where φ is Π_1, and for $x \in A$ let

$$\psi(x) = \mu\gamma \le \rho\,\big[\exists u \in \mathbf{J}_\gamma(\mathbb{R})(u \in \operatorname{dom} f \;\&\; \mathbf{J}_\delta(\mathbb{R}) \models \varphi\,[x, f(u)])\big].$$

By Lemma 1.15, ψ is a $\underset{\sim}{\Sigma}_n(\mathbf{J}_\delta(\mathbb{R}))$ norm, so we have (a). Fix $\gamma < \rho$. Then $\operatorname{dom} f \cap \mathbf{J}_\gamma(\mathbb{R}) \in \mathbf{J}_\delta(\mathbb{R})$, from which it easily follows that $\{x : \psi(x) \le \gamma\} \in \mathbf{J}_\delta(\mathbb{R})$. This proves (b). ⊣

THEOREM 4.2. For all $\alpha < \Theta^{\mathbf{L}(\mathbb{R})}$,

(a) if $\omega\alpha = 0$ or $\omega\alpha$ is of type I, then for all $k < \omega$

$$\underset{\sim}{\Sigma}^1_{\omega\alpha+k} = \underset{\sim}{\Sigma}_{n_\alpha+k}(\mathbf{J}_{\delta_\alpha}(\mathbb{R}))\ ;$$

(b) if $\omega\alpha$ is of type II or III, then for all $k < \omega$

$$\underset{\sim}{\Sigma}^1_{\omega\alpha+k+1} = \underset{\sim}{\Sigma}_{n_\alpha+k}(\mathbf{J}_{\delta_\alpha}(\mathbb{R}))\ ;$$

(c) if $\omega\alpha$ is of type IV, then

$$\underset{\sim}{\Pi}^1_{\omega\alpha} = \underset{\sim}{\Sigma}_{n_\alpha}(\mathbf{J}_{\delta_\alpha}(\mathbb{R})),$$

and for $k < \omega$ such that $k \neq 0$,

$$\underset{\sim}{\Sigma}^1_{\omega\alpha+k+1} = \underset{\sim}{\Sigma}_{n_\alpha+k}(\mathbf{J}_{\delta_\alpha}(\mathbb{R})).$$

PROOF. By induction on α. The case $\alpha = 0$ is clear. Set $\Lambda = \bigcup_{\beta<\omega\alpha} \underset{\sim}{\Sigma}^1_\beta = \mathbf{J}_{\delta_\alpha}(\mathbb{R}) \cap \wp(\mathbb{R})$.

(a) Since $\underset{\sim}{\Sigma}^1_{\omega\alpha} = \bigcup_\omega \Lambda$, $\underset{\sim}{\Sigma}^1_{\omega\alpha} \subseteq \underset{\sim}{\Sigma}_{n_\alpha}(\mathbf{J}_{\delta_\alpha}(\mathbb{R}))$. Let $A \in \underset{\sim}{\Sigma}_{n_\alpha}(\mathbf{J}_{\delta_\alpha}(\mathbb{R})) \cap \wp(\mathbb{R})$. By Lemma 4.1(b) and the fact that $\operatorname{cf}(\omega\alpha) = \omega$

$$A = \bigcup_{n<\omega}\Big(\bigcup_{\beta<\delta_\alpha} A^n_\beta\Big),$$

where for some fixed $\gamma_n < \omega\alpha$, $A^n_\beta \in \underset{\sim}{\Sigma}^1_{\gamma_n}$ for all $\beta < \delta_\alpha$. By Kechris [Kec81], we may assume that $\underset{\sim}{\Sigma}^1_{\gamma_n}$ is closed under wellordered unions, so that $A \in \bigcup_\omega \Lambda = \underset{\sim}{\Sigma}^1_{\omega\alpha}$, as desired. Thus (a) is true for $k = 0$, and the case $k > 0$ follows easily.

(b) $\underset{\sim}{\Sigma}^1_{\omega\alpha} \neq \Sigma_{n_\alpha}(\mathbf{J}_{\delta_\alpha}(\mathbb{R}) \cap \wp(\mathbb{R}))$, by Lemma 4.1(a) and the fact that PWO($\underset{\sim}{\Pi}^1_{\omega\alpha}$). Thus $\underset{\sim}{\Sigma}^1_{\omega\alpha+1} \subseteq \Sigma_{n_\alpha}(\mathbf{J}_{\delta_\alpha}(\mathbb{R}) \cap \wp(\mathbb{R}))$. But $\underset{\sim}{\Sigma}^1_{\omega\alpha+1}$ is closed under wellordered unions by Kechris' theorem, so Lemma 4.1(b) gives $\underset{\sim}{\Sigma}_{n_\alpha}(\mathbf{J}_{\delta_\alpha}(\mathbb{R})) \cap \wp(\mathbb{R}) \subseteq \underset{\sim}{\Sigma}^1_{\omega\alpha+1}$. Thus (b) is true for $k = 0$, and for $k > 0$ again follows easily.

(c) The Coding lemma implies $\bigcup_{\omega\alpha} \Pi^1_a \subseteq \Pi^1_a$, which with the proof of Lemma 4.1(b) implies the first statement. The second statement is immediate once we notice that $\underset{\sim}{\Sigma}^1_{\omega\alpha+1} = \exists^{\mathbb{R}} \bigcup_2(\underset{\sim}{\Sigma}^1_{\omega\alpha} \cup \underset{\sim}{\Pi}^1_{\omega\alpha})$. ⊣

If T is a tree on $\omega \times \kappa$, then $[T]$ is the set of infinite branches of T, and $\mathrm{p}([T])$ the projection of this set, that is,

$$\mathrm{p}([T]) = \{x \in \mathbb{R} : \exists f \forall n\, ((x \upharpoonright n, f \upharpoonright n) \in T)\}.$$

We let

$$\boldsymbol{S}(\kappa) = \{\mathrm{p}([T]) : T \text{ is a tree on } \omega \times \kappa\},$$

and call the members of $\boldsymbol{S}(\kappa)$ κ-Suslin sets. If $\kappa > \omega$, then the κ-Suslin sets are precisely those admitting scales all of whose norms map into κ. We say κ is Suslin iff $\boldsymbol{S}(\kappa) \setminus \bigcup_{\alpha<\kappa} \boldsymbol{S}(\alpha) \neq \varnothing$. Suslin ordinals are cardinals. We shall locate the pointclasses $\boldsymbol{S}(\kappa)$ among the $\underset{\sim}{\Sigma}^1_\nu$'s, and the Suslin cardinals among the $\underset{\sim}{\delta}^1_\nu$'s.

Let ν_α be the αth ordinal ν such that either Scale($\underset{\sim}{\Sigma}^1_\nu$) or Scale($\underset{\sim}{\Pi}^1_\nu$). By the second periodicity theorem, $\nu_{\alpha+1} = \nu_\alpha + 1$ as long as ν_α is not a limit ordinal of type IV. If ν_α is of type IV, then $\nu_{\alpha+1}$ is a limit ordinal of type I by the results of §§2 and 3.

THEOREM 4.3. Let $\lambda < (\underset{\sim}{\delta}^2_1)^{\mathrm{L}(\mathbb{R})}$ be a limit ordinal, and let $\nu = \sup\{\nu_\alpha : \alpha < \lambda\}$. Then

(a) If ν is type I, then for all $n < \omega$

$$\mathrm{Scale}(\underset{\sim}{\Sigma}^1_{\nu+2n}), \quad \mathrm{Scale}(\underset{\sim}{\Pi}^1_{\nu+2n+1}),$$
$$\boldsymbol{S}(\kappa_{\lambda+n}) = \underset{\sim}{\Sigma}^1_{\nu+n+1},$$
$$\kappa_{\lambda+2n+1} = \underset{\sim}{\delta}^1_{\nu+2n+1} = (\kappa_{\lambda+2n})^+,$$
$$\mathrm{cf}(\kappa_{\lambda+2n}) = \omega\ ;$$

(b) if ν is type II or III, then for all $n < \omega$

$$\mathrm{Scale}(\underset{\sim}{\Pi}^1_{\nu+2n}), \quad \mathrm{Scale}(\underset{\sim}{\Sigma}^1_{\nu+2n+1}),$$
$$\boldsymbol{S}(\kappa_{\lambda+n}) = \underset{\sim}{\Sigma}^1_{\nu+n+1},$$
$$\kappa_{\lambda+2n+2} = \underset{\sim}{\delta}^1_{\nu+2n+2} = (\kappa_{\lambda+2n+1})^+,$$
$$\mathrm{cf}(\kappa_{\lambda+2n+1}) = \omega\ ;$$

(c) if ν is type IV, then

$$\text{Scale}(\underset{\sim}{\mathbf{\Pi}}^1_\nu), \quad \boldsymbol{S}(\kappa_\lambda) = \underset{\sim}{\mathbf{\Pi}}^1_\nu, \quad \kappa_\lambda = \underset{\sim}{\boldsymbol{\delta}}^1_\nu,$$

and if $\mu = \nu_{\lambda+1}$, then for all $n < \omega$

$$\text{Scale}(\underset{\sim}{\mathbf{\Sigma}}^1_{\mu+2n}), \quad \text{Scale}(\underset{\sim}{\mathbf{\Pi}}^1_{\mu+2n+1}),$$
$$\boldsymbol{S}(\kappa_{\lambda+n+1}) = \underset{\sim}{\mathbf{\Sigma}}^1_{\mu+n+1},$$
$$\kappa_{\lambda+2n+2} = \underset{\sim}{\boldsymbol{\delta}}^1_{\mu+2n+1} = (\kappa_{\lambda+2n+1})^+,$$
$$\text{cf}(\kappa_{\lambda+2n+1}) = \omega.$$

Proof (Sketch). By induction on λ.

(a) By Theorem 4.2 and the results of §§2 and 3, $\underset{\sim}{\mathbf{\Sigma}}^1_\nu = \Sigma_1(\mathbf{J}_\alpha(\mathbb{R}))$ for some α beginning a Σ_1 gap. Thus Scale($\underset{\sim}{\mathbf{\Sigma}}^1_\nu$), and by induction $\kappa_\lambda = \sup\{\delta^1_{\nu_\beta} : \beta < \lambda\}$. Thus $\kappa_\lambda < \underset{\sim}{\boldsymbol{\delta}}^1_{\nu+1}$, and $\boldsymbol{S}(\kappa_\lambda) \subseteq \underset{\sim}{\mathbf{\Sigma}}^1_{\nu+1}$. But $\underset{\sim}{\mathbf{\Sigma}}^1_{\nu+1} \subseteq \boldsymbol{S}(\kappa_\lambda)$ since $\underset{\sim}{\mathbf{\Sigma}}^1_\nu \subseteq \boldsymbol{S}(\kappa_\lambda)$ and $\boldsymbol{S}(\kappa)$ is closed under $\exists^{\mathbb{R}}$ and countable intersection. Thus $\boldsymbol{S}(\kappa_\lambda) = \underset{\sim}{\mathbf{\Sigma}}^1_{\nu+1}$. The remaining assertions follow by arguments like those for the projective hierarchy ($\underset{\sim}{\mathbf{\Sigma}}^1_\nu$ being analogous to $\underset{\sim}{\mathbf{\Sigma}}^1_0$).

(b) By Theorem 4.2 and the results of §§2 and 3, $\underset{\sim}{\mathbf{\Sigma}}^1_{\nu+1} = \Sigma_1(\mathbf{J}_\alpha(\mathbb{R}))$ for some α beginning a Σ_1 gap. Thus Scale($\underset{\sim}{\mathbf{\Sigma}}^1_{\nu+1}$), and $\kappa_\lambda = \sup\{\delta^1_{\nu_\beta} : \beta < \lambda\} = \underset{\sim}{\boldsymbol{\delta}}^1_\nu$. We have $\boldsymbol{S}(\kappa_\lambda) = \underset{\sim}{\mathbf{\Sigma}}^1_{\nu+1}$ for the same reasons as in (a). The remaining assertions follow by the arguments for the projective hierarchy, except for Scale($\underset{\sim}{\mathbf{\Pi}}^1_\nu$). This can be proved using the ideas of the proof of PWO($\underset{\sim}{\mathbf{\Pi}}^1_\nu$); cf. [Kec81, Theorem 3.1(iii)] and [Ste81, Theorem 3.1], .

(c) By Theorem 4.2 and §§2 and 3, $\underset{\sim}{\mathbf{\Pi}}^1_\nu = \Sigma_1(\mathbf{J}_\alpha(\mathbb{R}))$, where α begins a Σ_1 gap. This proves the first set of assertions. The second set follows from our analysis of the gaps in §§2 and 3, and the arguments for the projective hierarchy ($\underset{\sim}{\mathbf{\Sigma}}^1_\mu$ being analogous to $\underset{\sim}{\mathbf{\Sigma}}^1_0$). ⊣

Corollary 4.4. (a) The sequences $\langle \kappa_\alpha : \alpha \leq (\underset{\sim}{\boldsymbol{\delta}}^2_1)^{\mathbf{L}(\mathbb{R})} \rangle$ and $\langle \nu_\alpha : \alpha \leq (\underset{\sim}{\boldsymbol{\delta}}^2_1)^{\mathbf{L}(\mathbb{R})} \rangle$ are continuous at limits.

(b) For any $\kappa \leq (\underset{\sim}{\boldsymbol{\delta}}^2_1)^{\mathbf{L}(\mathbb{R})}$, either $\boldsymbol{S}(\kappa)$ or its dual has the scale property.

As a final application of §2 and §3, we prove

Corollary 4.5. In $\mathbf{L}(\mathbb{R})$, the reliable cardinals are precisely the Suslin cardinals.

This result should be stamped "Made in Los Angeles". The reader who cares can untangle some of the credits for it from what follows.

Recall that an ordinal λ is reliable iff there is a scale $\{\varphi_i\}$ on a set $A \subseteq \mathbb{R}$ so that

$$\varphi_i \colon A \to \lambda$$

for all i, and

$$\lambda = \{\varphi_0(x) : x \in A\}.$$

The interest of the notion of reliability stems from results of [Bec80] and [Mos81], which to date have been proved only for reliable ordinals.

One direction of Corollary 4.5 is easy.

LEMMA 4.6. Every Suslin cardinal is reliable.

PROOF. Clearly ω is reliable. Let κ be an uncountable Suslin cardinal, let $A \in \boldsymbol{S}(\kappa) - \bigcup_{\alpha<\kappa} \boldsymbol{S}(\alpha)$, and let $\{\varphi_i\}$ be a scale on A all of whose norms map onto (perhaps improper) initial segments of κ. Let

$$B = \{x \in \mathbb{R} : \lambda i.x(i+1) \in A\},$$

and for $x \in B$

$$\psi_0(x) = \varphi_{x(0)}(\lambda i.x(i+1)),$$
$$\psi_{n+1}(x) = \varphi_n(\lambda i.x(i+1)).$$

Then $\{\psi_i\}$ is a scale on B all of whose norms map into κ. Since $A \notin \bigcup_{\alpha<\kappa} \boldsymbol{S}(\alpha)$, $\kappa = \{\psi_0(x) : x \in B\}$. ⊣

There are reliable ordinals which are not cardinals (cf. [Bec80]). The first step toward showing that in $\mathbf{L}(\mathbb{R})$ all reliable cardinals are Suslin, and the realization that it is a first step, are due to Kechris.

LEMMA 4.7 (Kechris). Let λ be reliable, and let $\kappa = \sup\{\gamma \le \lambda : \gamma \text{ is Suslin}\}$. Then there is a strictly increasing sequence of sets in $\boldsymbol{S}(\kappa)$ of length λ.

PROOF. Let $\{\varphi_i\}$ be a scale on A witnessing the reliability of λ. For $\alpha < \lambda$, let

$$A_\alpha = \{x \in A : \varphi_0(x) \le \alpha\}.$$

Then $\alpha < \beta \rightarrow A_\alpha \subsetneq A_\beta$; moreover, each A_α is in $\boldsymbol{S}(\kappa)$ via a subtree of the tree of $\{\varphi_i\}$. But $\boldsymbol{S}(\lambda) = \boldsymbol{S}(\kappa)$ by the definition of κ. ⊣

The next and most substantial step toward Corollary 4.5 is due to Martin (exploiting an idea of Jackson).

THEOREM 4.8. [JM83] For $1 \le n < \omega$, there is no strictly increasing sequence of $\boldsymbol{\Sigma}^1_{2n}$ sets of length $(\boldsymbol{\delta}^1_{2n-1})^+$.

By Lemma 4.7 and Theorem 4.8 and standard facts about the Suslin cardinals below $\boldsymbol{\delta}^1_\omega$, every reliable cardinal below $\boldsymbol{\delta}^1_\omega$ is Suslin. But Martin's argument gives more than Theorem 4.8, and in fact, with some care we can prove

THEOREM 4.9. Suppose $\alpha < \Theta^{\mathbf{L}(\mathbb{R})}$ and $\mathrm{Scale}(\boldsymbol{\Pi}^1_\alpha)$. Then there is no strictly increasing sequence of $\exists^{\mathbb{R}}\boldsymbol{\Pi}^1_\alpha$ sets of length $(\boldsymbol{\delta}^1_\alpha)^+$.

Proof. We shall assume that the reader is familiar with the proof of Theorem 4.8 given in [JM83]. Our only problem in extending that proof is to show that the coding of ordinals below $(\underset{\sim}{\delta}^1_{2n-1})^+$ which it employs generalizes suitably.

Since Scale($\underset{\sim}{\Pi}^1_\alpha$), every $\exists^{\mathbb{R}}\underset{\sim}{\Pi}^1_\alpha$ set has a scale whose norms map into $\underset{\sim}{\delta}^1_\alpha$. Let $U \subseteq \mathbb{R}^3$ be universal $\exists^{\mathbb{R}}\underset{\sim}{\Pi}^1_\alpha$, and $\{\varphi_i\}$ a scale on U mapping into $\underset{\sim}{\delta}^1_\alpha$. Define a tree T on $\omega \times \omega \times \underset{\sim}{\delta}^1_\alpha$ by

$$T = \Big\{(s,t,u) \;:\; \mathrm{lh}(s) = \mathrm{lh}(t) = \mathrm{lh}(u) \;\&$$
$$\exists x \supseteq s\; \exists y \supseteq t\; \Big[\bigwedge_{1<\mathrm{lh}(s)} U(x,(y)_i,(y)_{i+1}) \;\&$$
$$\bigwedge_{1<\mathrm{lh}(s)} \varphi_{(i)_0}(x,(y)_{(i)_1},(y)_{(i)_1+1}) = U(i)\Big]\Big\}.$$

Here we let $\langle\cdot,\cdot\rangle$ be a bijection of $\omega\times\omega$ onto ω, and for any i, $i = \langle (i)_0,(i)_1\rangle$, and for any y, i, n, $(y)_i(n) = y(\langle i,n\rangle)$. For $x \in \mathbb{R}$, let

$$T_x = \{(t,u) \;:\; (\restriction \mathrm{lh}(t), t, u) \in T\}.$$

T_x is the "Kunen tree" associated to x. If U_x is a wellfounded relation, then T_x is a wellfounded tree, and $|U_x| \le |T_x|$ (where $|R|$ denotes the rank of the relation R). In order to carry out the argument of [JM83] in the present situation, we need only to verify

(a) Suppose that $f : \mathbb{R} \to \mathbb{R}$ is continuous and that the following holds for all $x \in \mathbb{R}$ and $\beta < \underset{\sim}{\delta}^1_\alpha$: if for all $\gamma < \beta$, $T_{(x)_0}\restriction\gamma$ is wellfounded, then $T_{f(x)}\restriction\beta$ is wellfounded. Then, if there is a $\delta < (\underset{\sim}{\delta}^1_\alpha)^+$ such that for all $x \in \mathbb{R}$, $T_{(x)_0}$ is wellfounded, then $|T_{f(x)}| < \delta$;

(b) Suppose that $f : \mathbb{R} \to \mathbb{R}$ is continuous and that the following holds for all $x \in \mathbb{R}$ and $\beta < \underset{\sim}{\delta}^1_\alpha$: if for all $\gamma < \beta$, $T_{(x)_0}\restriction\gamma$ is wellfounded, then $T_{f(x)}\restriction\beta$ is wellfounded. Then there are unboundedly many $\delta < (\underset{\sim}{\delta}^1_\alpha)^+$ such that for all $x \in \mathbb{R}$, we have

$$|T_x| < \delta \to |T_{f(x)}| < \delta \;;$$

(c) For all sufficiently large $\delta < (\underset{\sim}{\delta}^1_\alpha)^+$, $\{x \;:\; |T_x| < \delta\}$ is not $\underset{\sim}{\Pi}^1_{\alpha+1}$.

(If R is a tree on $\omega^k \times \alpha$, then $R\restriction\gamma = \{(\sigma_1 \dots \sigma_k) \in R \;:\; \mathrm{ran}(\tau) \subseteq \gamma\}$.)

Now (c) can be proved exactly as in [JM83]. Since (a) and (b) have similar proofs, we shall just prove (b).

Lemma 4.10. Suppose $\alpha < \Theta^{\mathbf{L}(\mathbb{R})}$, α is a successor, and Scale($\underset{\sim}{\Pi}^1_\alpha$). Let R be a tree on $\omega \times \beta$, where $\beta < \underset{\sim}{\delta}^1_\alpha$, and let $f : \mathbb{R} \to \mathbb{R}$ be continuous and such that $\forall x(R_x$ is wellfounded $\to R_{f(x)}$ is wellfounded). Then there is a club $C \subseteq \underset{\sim}{\delta}^1_\alpha$ so that for $\delta \in C$, $\forall x(|R_x| < \delta \to |R_f(x)| < \delta)$.

Proof. By a result of Martin, $\bigcup_\delta \underset{\sim}{\mathbf{\Delta}}^1_\alpha \subseteq \underset{\sim}{\mathbf{\Delta}}^1_\alpha$ for all $\delta < \underset{\sim}{\boldsymbol{\delta}}^1_\alpha$ (cf. [Kec78, Theorem 3.7]). It follows (as in [Kec78, Theorem 3.8]) that for all $u \in \beta^{<\omega}$, and $\delta < \underset{\sim}{\boldsymbol{\delta}}^1_\alpha$,

$$\{x : |R^u_x| < \delta\} \in \underset{\sim}{\mathbf{\Delta}}^1_\alpha,$$

where R^u_x is the subtree of R_x below u; the proof is by induction on δ.

It is enough to show that $\forall\delta < \underset{\sim}{\boldsymbol{\delta}}^1_\alpha\, \exists\eta < \underset{\sim}{\boldsymbol{\delta}}^1_\alpha\, \forall x \in \mathbb{R}\,(|R_x| < \delta \to |R_{f(x)}| < \eta)$. So fix $\delta < \underset{\sim}{\boldsymbol{\delta}}^1_\alpha$, and let $A = \{x : |R_x| < \delta\}$. Now our hypotheses on α and Theorem 4.3 together imply that $\underset{\sim}{\mathbf{\Sigma}}^1_\alpha = \boldsymbol{S}(\gamma)$ for some $\gamma < \underset{\sim}{\boldsymbol{\delta}}^1_\alpha$. Thus we can fix a tree Q on $\omega \times \gamma$ so that $A = \mathrm{p}([Q])$. Define a tree P on $\omega \times \gamma \times \omega \times \beta$ by

$$\begin{aligned} P = \{(s,t,u,v) :\ & \mathrm{lh}(s) = \mathrm{lh}(t) = \mathrm{lh}(u) = \mathrm{lh}(v)\ \& \\ & (s,t) \in Q\ \&\ (u,v) \in R\ \&\ \exists x \supseteq s\, \exists y \supseteq u\, (f(x) = y)\}. \end{aligned}$$

Then P is wellfounded, and if $x \in A$ then $R_{f(x)}$ can be embedded in P. Thus $\eta = |P|$ is as desired. ⊣

We can now complete the proof of (b) in the case α is a successor as in [JM83]. Since α is a successor and Scale($\underset{\sim}{\mathbf{\Pi}}^1_\alpha$), we have $\bigcup_\omega \underset{\sim}{\mathbf{\Pi}}^1_\alpha \subseteq \underset{\sim}{\mathbf{\Pi}}^1_\alpha$. But then the ω-club subsets of $\underset{\sim}{\boldsymbol{\delta}}^1_\alpha$ generate a normal ultrafilter μ on $\underset{\sim}{\boldsymbol{\delta}}^1_\alpha$. Set $\kappa = \underset{\sim}{\boldsymbol{\delta}}^1_\alpha$. Since $\underset{\sim}{\mathbf{\Sigma}}^1_\alpha = \boldsymbol{S}(\gamma)$ for some $\gamma < \kappa$, we can apply Kunen's argument ([Kec78, Theorem 14.3]) to show that ${}^\kappa\kappa/_\mu$ has order type κ^+. For $\beta < \kappa$ let

$$C_\beta = \{\delta < \kappa : \forall x(|T_x \restriction \beta| < \delta \to |T_{f(x)} \restriction \beta| < \delta)\},$$

so that C_β is club in κ by Lemma 4.10. Let $D = \{\delta < \kappa : \forall\beta < \delta\, (\delta \in C_\beta)\}$, and let

$$C = \{[h]_\mu : h\colon \kappa \to D\ \&\ [h]_\mu \geq \kappa\}.$$

Clearly C is unbounded in κ^+. Suppose $|T_x| < \delta$, where $\delta \in C$. Let $\delta = [h]$, where $h\colon \kappa \to D$. Then

$$|T_x \restriction \beta| < h(\beta) \quad (\mu\text{-a.e.}),$$

so

$$|T_{f(x)} \restriction \beta| < h(\beta) \quad (\mu\text{-a.e.}),$$

so

$$|T_x| < [h]_\mu = \delta,$$

as desired. ⊣

Finally, let α be a limit. Since Scale($\underset{\sim}{\mathbf{\Pi}}^1_\alpha$), Theorems 4.3 and 4.2 tell us that $\exists^{\mathbb{R}}\underset{\sim}{\mathbf{\Pi}}^1_\alpha = \underset{\sim}{\mathbf{\Sigma}}_1(\mathbf{J}_\rho(\mathbb{R}))$ for some ρ such that $\mathrm{cf}(\rho) > \omega$. Let $\lambda = \mathrm{cf}(\rho)$, and let

$\mathcal{U}$ be the supercompactness measure on $P_{\omega_1}(\lambda)$ given by of [HK81, Theorem 6.2.1]. Let μ be the measure on λ defined by

$$\mu(A) = 1 \text{ iff } \{X : \sup X \in A\} \in \mathcal{U}.$$

Then μ is weakly normal (that is, if $h(\beta) < \beta$ a.e., then $\exists \gamma < \lambda\,(h(\beta) < \gamma$ a.e.)) and $\mu(A) = 1$ for every ω-club A. Let

$$g : \lambda \xrightarrow{\text{cofinal}} \rho$$

be strictly increasing, continuous, have range cofinal in ρ, and be such that for all $\beta < \lambda$, $\mathbf{J}_{g(\beta)}(\mathbb{R}) \not\prec_1^{\mathbb{R}} \mathbf{J}_{g(\beta)+1}(\mathbb{R})$. For $\beta < \lambda$, let also

$$h(\beta) = \sup\{|\leq| : \leq \text{ is a prewellorder of } \mathbb{R} \text{ in } \mathbf{J}_{g(\beta)}(\mathbb{R})\}.$$

Then h is strictly increasing, continuous, and has range cofinal in $\underset{\sim}{\boldsymbol{\delta}}^1_\alpha$.

The following claim is the crucial new ingredient we need in the case α is a limit ordinal. The proof of part (2), its non-trivial part, is due to Kechris.

Claim. (1) $[h]_\mu = \underset{\sim}{\boldsymbol{\delta}}^1_\alpha$,
(2) $[\lambda\beta.h(\beta)^+] = (\underset{\sim}{\boldsymbol{\delta}}^1_\alpha)^+$.

Proof. (1) $\underset{\sim}{\boldsymbol{\Delta}}^1_\alpha = \bigcup_{\beta<\alpha} \underset{\sim}{\boldsymbol{\Delta}}^1_\beta = \mathbf{J}_\rho(\mathbb{R}) \cap \wp(\mathbb{R})$, and therefore every $\underset{\sim}{\boldsymbol{\Delta}}^1_\alpha$ set is γ-Suslin for some $\gamma < \underset{\sim}{\boldsymbol{\delta}}^1_\alpha$. If $\underset{\sim}{\boldsymbol{\delta}}^1_\alpha$ is regular (i.e. $\lambda = \underset{\sim}{\boldsymbol{\delta}}^1_\alpha$), then the proof of [Ste81, Theorem 3.2] implies that $\bigcup_\omega \underset{\sim}{\boldsymbol{\Pi}}^1_\alpha \subseteq \underset{\sim}{\boldsymbol{\Pi}}^1_\alpha$, so that the ω-club subsets of $\underset{\sim}{\boldsymbol{\delta}}^1_\alpha$ generate a normal ultrafilter, which must be μ. But then $[h]_\mu = \underset{\sim}{\boldsymbol{\delta}}^1_\alpha$. So suppose $\lambda < \underset{\sim}{\boldsymbol{\delta}}^1_\alpha$. The proof of [HK81, Theorem 6.2.1] shows in this case that $\mu \in \mathbf{J}_\rho(\mathbb{R})$. Thus if $\gamma < \underset{\sim}{\boldsymbol{\delta}}^1_\alpha$, then ${}^\lambda\gamma/\mu$ has order type less than $\underset{\sim}{\boldsymbol{\delta}}^1_\alpha$. By the weak normality of μ, $[h]_\mu$ is just the supremum of these order types for $\gamma < \underset{\sim}{\boldsymbol{\delta}}^1_\alpha$. Thus $[h]_\mu = \underset{\sim}{\boldsymbol{\delta}}^1_\alpha$.

(2) That $[\lambda\beta.h(\beta)^+]_{\mu^+} \geq (\underset{\sim}{\boldsymbol{\delta}}^1_\alpha)^+$ is easy to see. For the other inequality, let $\ell : \lambda \to \underset{\sim}{\boldsymbol{\delta}}^1_\alpha$ and $\ell(\beta) < h(\beta)^+$ for all β. We shall construct a wellfounded tree W on $\underset{\sim}{\boldsymbol{\delta}}^1_\alpha$ so that for μ-a.e. $\beta < \lambda$, $\ell(\beta) \leq |W \restriction h(\beta)|$. This implies that

$$[\ell]_\mu \leq [\lambda\beta.|W\restriction h(\beta)|]_\mu = |[\lambda\beta.W\restriction h(\beta)]_\mu|.$$

But by (a) of our claim, $[\lambda\beta.W\restriction h(\beta)]_\mu$ is a wellfounded tree on $\underset{\sim}{\boldsymbol{\delta}}^1_\alpha$. Thus $[\ell]_\mu < (\underset{\sim}{\boldsymbol{\delta}}^1_\alpha)^+$, as desired.

In order to construct W, we construct first a tree R on $\omega \times \omega \times \omega \times \underset{\sim}{\boldsymbol{\delta}}^1_\alpha$ so that for all limit ordinals $\delta < \lambda$,

$$h(\beta)^+ = \sup\{|R_{xy}\restriction h(\beta)| : R_{xy} \text{ is wellfounded}\}. \qquad (*)$$

For this, let

$$V = \{\langle x, y\rangle : \mathbf{J}_\beta(\mathbb{R}) \models \sigma_{x(0)}[\lambda i.x(i+1), y]\},$$

where σ_i is the ith Σ_1 formula of two free variables, and let ϑ be a Σ_1 formula so that (identifying $\mathbb{R}^2$ with $\mathbb{R}$)

$$V(x) \text{ iff } \mathbf{J}_\rho(\mathbb{R}) \models \vartheta[x].$$

Let also

$$L(x, y) \text{ iff } \exists\delta < \rho\ (\mathbf{J}_\delta(\mathbb{R}) \models \vartheta[x] \ \&\ -\vartheta[y]).$$

Since L is $\Sigma_1(\mathbf{J}_\rho(\mathbb{R}))$, Theorem 2.1 gives us a scale $\{\varphi_i\}$ on L. Notice that the scale $\{\varphi_i\}$ we get from Theorem 2.1 has the property that if $L(x, y)$, then $\varphi_0(x, y) = \mu\delta\,[\mathbf{J}_\delta(\mathbb{R}) \models \vartheta[x]]$, and for all $i > 0$. $\varphi(x, y)$ is the order type of a prewellorder of $\mathbb{R}$ which is $\underset{\sim}{\Sigma}_\omega(\mathbf{J}_{\varphi_0(x,y)}(\mathbb{R}))$. Let

$$M(x, y) \text{ iff } \forall n < \omega\ L((x)_n, y),$$

and let $\{\psi_i\}$ be the scale on M given by

$$\psi_{\langle m,n\rangle}(x, y) = \varphi_m((x)_n, y).$$

Finally, let

$$R = \Big\{(s, t, u, v) : \text{lh}(s) = \text{lh}(t) = \text{lh}(u) = \text{lh}(v)\ \&$$
$$\exists x \supseteq s\, \exists y \supseteq t\, \exists z \supseteq u\ \Big(M(x, \langle y, z\rangle)\ \&\ \bigwedge_{i<\text{lh}(s)} v(i) = \psi_i(x, \langle y, z\rangle)\Big)\Big\}.$$

In order to show $(*)$, fix a limit ordinal $\beta < \lambda$, and fix an $x \in \mathbb{R}$ so that $(x)_n \in V$ for all n, and in fact

$$g(\beta) = \mu\gamma\, \big[\forall n < \omega\ (\mathbf{J}_\gamma(\mathbb{R}) \models \vartheta[(x)_n])\big].$$

We can find such an x by the universality of V and the fact that g grows as fast as it does. Then for all y, z,

$$\begin{aligned} M(x, \langle y, z\rangle) &\text{ iff } \langle y, z\rangle \in \text{p}([R_x]),\\ &\text{ iff } \langle y, z\rangle \in \text{p}([R_x \restriction h(\beta)]), \end{aligned}$$

where the second equivalence follows from our observations on thee "local" nature of $\{\varphi_i\}$. Thus for any y, R_{xy} is wellfounded iff $R_{xy}\restriction h(\beta)$ is wellfounded. Now $\{\langle y, z\rangle : M(x, \langle y, z\rangle)\}$ is clearly universal for the class $\Pi_1(\mathbf{J}_{g(\beta)}(\mathbb{R}), \mathbb{R})$, and so if we let

$$y \in N \text{ iff } \exists z\, M(x, \langle y, z\rangle),$$

then N is in the class $\exists^{\mathbb{R}}\Pi_1(\mathbf{J}_{g(\beta)}(\mathbb{R}), \mathbb{R})$, but not in its dual. From Theorem 4.2 we see easily that $\exists^{\mathbb{R}}\Pi_1(\mathbf{J}_{g(\beta)}(\mathbb{R}), \mathbb{R})$ is a class of the form $\underset{\sim}{\Sigma}^1_{\gamma+1}$, where Scale($\underset{\sim}{\Pi}^1_{\gamma+1}$) and, by Theorem 4.3, $\underset{\sim}{\delta}^1_{\gamma+1} = h(\beta)^+$. Now for $\eta < \underset{\sim}{\delta}^1_{\gamma+1}$, $\{y : |R_{xy}\restriction h(\beta)| < \eta\}$ is $\underset{\sim}{\Delta}^1_{\gamma+1}$, by the argument in the proof of Lemma 4.10. Since $N = \{y : R_{xy} \text{ is wellfounded}\} = \{y : R_{xy}\restriction h(\beta) \text{ is wellfounded}\}$, and N is not $\underset{\sim}{\Delta}^1_{\gamma+1}$, we have $(*)$.

We can now complete the proof of (2) of the claim in the case $\underset{\sim}{\delta}^1_\alpha$ is singular, that is $\lambda < \underset{\sim}{\delta}^1_\alpha$. Fix a norm $\varphi\colon \mathbb{R} \twoheadrightarrow \lambda$ so that the induced prewellordering $\leq_\varphi$ is in $\mathbf{J}_\rho(\mathbb{R})$. By the Coding lemma there is a relation P such that $\forall w \exists x, y\, P(w, x, y)$, and for $\beta < \lambda$ a limit

$$(\varphi(w) = \beta \;\&\; P(w, x, y)) \rightarrow (R_{xy} \text{ is wellfounded } \;\&\; \ell(\beta) < \langle R_{xy} \restriction h(\beta) \rangle).$$

and $P \in \mathbf{J}_\rho(\mathbb{R})$. Thus P is γ-Suslin for some $\gamma < \underset{\sim}{\delta}^1_\alpha$; let $P = \mathrm{p}([S])$, where S is a tree on $\omega^3 \times \gamma$ and $\gamma < \underset{\sim}{\delta}^1_\alpha$. Finally, let

$$W = \{(q, r, s, t, u) : \mathrm{lh}(q) = \mathrm{lh}(r) = \cdots = \mathrm{lh}(u) \\ \& \; (q, r, s, t) \in S \;\&\; (r, s, u) \in R\}.$$

Then W is wellfounded. If $\beta < \lambda$ is a limit ordinal, $\gamma < h(\beta)$, $\varphi(w) = \beta$, and $P(w, x, y)$ then the tree $R_{xy} \restriction h(\beta)$ can be embedded into $W_{wxy} \restriction h(\beta)$. Since $|R_{xy} \restriction h(\beta)| > \ell(\beta)$, $|W \restriction h(\beta)| > \ell(\beta)$, as desired.

We shall just outline the proof of Claim (b) in the case that $\underset{\sim}{\delta}^1_\alpha$ is regular. Fix a complete $\underset{\sim}{\Pi}^1_\alpha$ set A, and a $\underset{\sim}{\Pi}^1_\alpha$-scale $\{\tau_i\}$ on A. Since $\bigcup_\omega \underset{\sim}{\Pi}^1_\alpha \subseteq \underset{\sim}{\Pi}^1_\alpha$, (by [Ste81]), any $\underset{\sim}{\Sigma}^1_\alpha$ subset of A is bounded in τ_0—here we use the regularity of $\underset{\sim}{\delta}^1_\alpha$. Thus μ is just the ω-club measure on $\lambda = \underset{\sim}{\delta}^1_\alpha$, and $g(\beta) = h(\beta) = \beta$ for μ-a.e. β. Consider the following Solovay game.

I	w
II	z, x, y

Player II wins if $\exists n\, ((w)_n \notin A \;\&\; \forall m < n\, ((z)_m \in A))$ or $\forall n\, ((w)_n, (z)_n \in A)$ and, if $\beta = \sup\{\tau((w)_n), \tau_0((z)_n) : n < w\}$, then R_{xy} is wellfounded and $l(\beta) < |R_{xy} \restriction \beta|$. By boundedness and the property $(*)$ of R, player II has a winning strategy σ. We can now use σ and the scale $\{\tau_i\}$ to construct W just as we used the tree S to construct W in the case $\underset{\sim}{\delta}^1_\alpha$ is singular. Where we used before that E was a tree on an ordinal less that $\underset{\sim}{\delta}^1_\alpha$, we now use that for μ-a.e. β, whenever $\tau_0((w)_n) < \beta$ for all n, then $\tau_i((w)_n) < \beta$ for all i, n.

This proves the claim. ⊣

Finally, we prove boundedness property (b) of the Kunen tree coding in the case α is a limit. Suppose $f : \mathbb{R} \rightarrow \mathbb{R}$ is continuous and $\forall \eta < \underset{\sim}{\delta}^1_\alpha\, \forall x \in \mathbb{R}\, (T_x \restriction \eta$ is wellfounded $\rightarrow T_{f(x)}$ is wellfounded$)$. Now for μ-a.e. $\beta < \lambda$, $h(\beta)^+ = \underset{\sim}{\delta}^1_{\gamma+1}$ for some γ of type I such that $\mathrm{Scale}(\underset{\sim}{\Pi}^1_{\gamma+1})$. By Lemma 4.10, the set

$$C_\beta = \{\delta < h(\beta)^+ : \forall x \in \mathbb{R}\, (|T_x \restriction h(\beta)| < \delta \rightarrow T_{f(x)} \restriction h(\beta) < \delta)\}$$

is club in $h(\beta)^+$ for such β. Let

$$C' = [\lambda\beta.C_\beta]_\mu - \underset{\sim}{\delta}^1_\alpha,$$

so that by our claim C' is club in $(\underset{\sim}{\delta}^1_\alpha)^+$. Now there is easy to show that there is a function $F\colon (\underset{\sim}{\delta}^1_\alpha)^\alpha \rightarrow (\underset{\sim}{\delta}^1_\alpha)^+$ so that whenever $\beta < (\underset{\sim}{\delta}^1_\alpha)^+$ and W is a tree on $\underset{\sim}{\delta}^1_\alpha$ so that $|W| = \beta$, then $|[\lambda\gamma.W\restriction h(\gamma)]_\mu| = F(\beta)$. (If μ is normal, then F is the identity.) Let

$$C = C' \cap \{\beta : F[\beta] \subseteq \beta\}.$$

Suppose $[p]_\mu \in C$ and $|T_x| < [p]$. Then $F(|T_x|) < [p]_\mu$, so $|T_x\restriction h(\beta)| < p(\beta)$ for μ-a.e. β, so $|T_{f(x)}\restriction f(\beta)| < p(\beta)$ for μ-a.e. β, so $F(|T_{f(x)}|) < [p]$. But it is easy to check that $\beta \leq F(\beta)$ for all β, so in fact $|T_{f(x)}| < [p]$. Thus C is as demanded by boundedness requirement (b), and we have (modulo [JM83], of course) proved Theorem 4.9.

From Theorem 4.9 and Lemma 4.7 we have at once the following theorem.

THEOREM 4.11. Let $\lambda < \Theta^{\mathbf{L}(\mathbb{R})}$ be a reliable ordinal. There is a Suslin cardinal κ so that $\kappa \leq \lambda < \kappa^+$.

PROOF. Let $\kappa = \sup\{\gamma \leq \lambda : \gamma \text{ is Suslin}\}$. By Corollary 4.4, κ is Suslin. If $S(\kappa) = \underset{\sim}{\Pi}^1_\alpha$ where α is type IV, then Scale($\underset{\sim}{\Pi}^1_\alpha$) and $\kappa = \underset{\sim}{\delta}^1_\alpha$ by Theorem 4.3. But $\exists^{\mathbb{R}}\underset{\sim}{\Pi}^1_\alpha = \underset{\sim}{\Pi}^1_\alpha$, so Lemma 4.7 and Theorem 4.9 combined tell us that $\lambda < \kappa^+$. Otherwise, inspection of Theorem 4.3 shows that $S(\kappa) = \underset{\sim}{\Sigma}^1_{\alpha+1} = \exists^{\mathbb{R}}\underset{\sim}{\Pi}^1_\alpha$ for some α. If Scale($\underset{\sim}{\Pi}^1_\alpha$), then $\kappa = \underset{\sim}{\delta}^1_\alpha$ by Theorem 4.3, and again Lemma 4.7 and Theorem 4.9 imply that $\lambda < \kappa^+$. If Scale($\underset{\sim}{\Pi}^1_\alpha$) fails, then Scale($\underset{\sim}{\Pi}^1_{\alpha+1}$) must hold, and $\underset{\sim}{\delta}^1_{\alpha+1} = \kappa^+$ is Suslin. By the definition of κ, $\lambda < \kappa^+$. ⊣

Theorem 4.11 clearly implies that all reliable cardinals below $\Theta^{\mathbf{L}(\mathbb{R})}$ are Suslin. This completes the proof of Corollary 4.5.

REFERENCES

HOWARD S. BECKER

[Bec80] *Thin collections of sets of projective ordinals and analogs of* L, ***Annals of Mathematical Logic***, vol. 19 (1980), pp. 205–241.

LEO A. HARRINGTON AND ALEXANDER S. KECHRIS

[HK81] *On the determinacy of games on ordinals*, ***Annals of Mathematical Logic***, vol. 20 (1981), pp. 109–154.

STEPHEN JACKSON AND DONALD A. MARTIN

[JM83] *Pointclasses and wellordered unions*, In Kechris et al. [CABAL iii], pp. 55–66.

RONALD B. JENSEN

[Jen72] *The fine structure of the constructible hierarchy*, ***Annals of Mathematical Logic***, vol. 4 (1972), pp. 229–308.

ALEXANDER S. KECHRIS

[Kec78] AD *and projective ordinals*, In Kechris and Moschovakis [CABAL i], pp. 91–132.

[Kec81] *Suslin cardinals, κ-Suslin sets, and the scale property in the hyperprojective hierarchy*, this volume, originally published in Kechris et al. [CABAL ii], pp. 127–146.

ALEXANDER S. KECHRIS, DONALD A. MARTIN, AND YIANNIS N. MOSCHOVAKIS

[CABAL ii] *Cabal seminar* 77–79, Lecture Notes in Mathematics, no. 839, Berlin, Springer, 1981.

[CABAL iii] *Cabal seminar* 79–81, Lecture Notes in Mathematics, no. 1019, Berlin, Springer, 1983.

ALEXANDER S. KECHRIS AND YIANNIS N. MOSCHOVAKIS

[CABAL i] *Cabal seminar* 76–77, Lecture Notes in Mathematics, no. 689, Berlin, Springer, 1978.

ALEXANDER S. KECHRIS, ROBERT M. SOLOVAY, AND JOHN R. STEEL

[KSS81] *The axiom of determinacy and the prewellordering property*, In Kechris et al. [CABAL ii], pp. 101–125.

ALEXANDER S. KECHRIS AND W. HUGH WOODIN

[KW83] *Equivalence of determinacy and partition properties*, ***Proceedings of the National Academy of Sciences of the United States of America***, vol. 80 (1983), no. 6 i., pp. 1783–1786.

DONALD A. MARTIN

[Mar83] *The largest countable this, that, and the other*, this volume, originally published in Kechris et al. [CABAL iii], pp. 97–106.

DONALD A. MARTIN AND JOHN R. STEEL

[MS83] *The extent of scales in* $\mathbf{L}(\mathbb{R})$, this volume, originally published in Kechris et al. [CABAL iii], pp. 86–96.

YIANNIS N. MOSCHOVAKIS

[Mos78] *Inductive scales on inductive sets*, this volume, originally published in Kechris and Moschovakis [CABAL i], pp. 185–192.

[Mos80] ***Descriptive set theory***, Studies in Logic and the Foundations of Mathematics, no. 100, North-Holland, Amsterdam, 1980.

[Mos81] *Ordinal games and playful models*, In Kechris et al. [CABAL ii], pp. 169–201.

[Mos83] *Scales on coinductive sets*, this volume, originally published in Kechris et al. [CABAL iii], pp. 77–85.

STEPHEN G. SIMPSON

[Sim78] ***A short course in admissible recursion theory, Generalized recursion theory II***, Studies in Logic, vol. 94, North Holland, Amsterdam, 1978.

JOHN R. STEEL

[Ste81] *Closure properties of pointclasses*, In Kechris et al. [CABAL ii], pp. 147–163.

DEPARTMENT OF MATHEMATICS
UNIVERSITY OF CALIFORNIA
BERKELEY, CA 94720, USA
E-mail: steel@math.berkeley.edu

SCALES IN $\mathbf{K}(\mathbb{R})$

JOHN R. STEEL

§1. Introduction. In this paper, we shall extend the fine-structural analysis of scales in $\mathbf{L}(\mathbb{R})$ [Ste83A] and $\mathbf{L}(\mu, \mathbb{R})$ [Cun90] to models of the form $\mathbf{L}(\vec{E}, \mathbb{R})$, constructed over the reals from a coherent sequence $\vec{E}$ of extenders. We shall show that in the natural hierarchy in an iterable model of the form $\mathbf{L}(\vec{E}, \mathbb{R})$ satisfying AD, the appearance of scales on sets of reals not previously admitting a scale is tied to the verification of new Σ_1 statements about $\vec{E}$ and individual reals in exactly the same way as it is in the special case $\vec{E} = \varnothing$ of [Ste83A]. For example, we shall show:

THEOREM 1.1. Let $\mathcal{M}$ be a passive, countably iterable premouse over $\mathbb{R}$, and suppose $\mathcal{M} \models \mathsf{AD}$; then the pointclass consisting of all $\Sigma_1^{\mathcal{M}}$ sets of reals has the scale property.

A premouse is said to be **countably iterable** if all its countable elementary submodels are $(\omega_1 + 1)$-iterable. It is easy to show, using a simple Löwenheim-Skolem argument, that if $\mathcal{M}$ and $\mathcal{N}$ are ω-sound, countably iterable premice over $\mathbb{R}$ which project to $\mathbb{R}$, then either $\mathcal{M}$ is an initial segment of $\mathcal{N}$, or vice versa. We shall write $\mathbf{K}(\mathbb{R})$ for the "union" of all such premice over $\mathbb{R}$, regarded as itself a premouse over $\mathbb{R}$. This is a small abuse of notation, since our $\mathbf{K}(\mathbb{R})$ is determined by its sets of reals, but since we are concerned with the scale property, sets of reals are all that matter here. In fact, as in [Ste83A] and the work of [Mos83] and [MS83] on which it rests, our existence results for scales require determinacy hypotheses, and so we are really only concerned here with the longest initial segment of $\mathbf{K}(\mathbb{R})$ satisfying AD .

Section 2 is devoted to preliminaries. In section 3 we show that for any $\mathbb{R}$-mouse $\mathcal{M}$ satisfying "Θ exists", $\mathbf{HOD}^{\mathcal{M}}$ is a T-mouse, for some $T \subseteq \Theta^{\mathcal{M}}$.[1] We use this representation of $\mathbf{HOD}^{\mathcal{M}}$ in the proof of Theorem 1.1, which is given in section 4. There we also extend the proof of Theorem 1.1 so as to obtain a complete description of those pointclasses which have the scale property and are definable over initial segments of $\mathbf{K}(\mathbb{R})$ satisfying AD.

[1] What we actually show is slightly weaker than this in some very technical respects.

The Cabal Seminar. Volume I: Games, Scales and Suslin Cardinals
Edited by A. S. Kechris, B. Löwe, J. R. Steel
Lecture Notes in Logic, 31

§2. Preliminaries.

2.1. Potential $\mathbb{R}$-premice. We shall be interested in premice built over $\mathbb{R}$, which we take to be $\mathbf{V}_{\omega+1}$ in this context, but nevertheless refer to as the set of all reals on occasion. In most respects, the basic theory of premice built over $\mathbb{R}$ is a completely routine generalization of the theory of ordinary premice (built over $\emptyset$); however, because $\mathbb{R}$-premice do not in general satisfy the axiom of choice, one must be careful at a few points. Here are some details.

Let M be a transitive, rud-closed set, and $X \in M$. Let E be an extender over M. We say that E is **(M, X)-complete** iff whenever a is a finite subset of $\mathrm{lh}(E)$ and $f: X \to E_a$ and $f \in M$, then $\bigcap \mathrm{ran}(f) \in E_a$. In the contrapositive: whenever $g: [\mathrm{crit}(E)]^{<\omega} \to \wp(X)$ is in M, then the following implication holds: if for E_a-almost every u there is an $i \in X$ such that $i \in g(u)$, then there is an $i \in X$ such that for E_a-almost every u, we have $i \in g(u)$. It is clear that if E is (M, X)-complete and (M, Y)-complete, then E is $(M, X \times Y)$-complete. Thus if E is (M, X)-complete and $\alpha < \mathrm{crit}(E)$, then E is $(M, X \times \alpha)$-complete.

For any transitive set M, let $o(M)$ be the least ordinal not in M.

DEFINITION 2.1. Let M be transitive and $X \in M$; we say M is **wellordered mod** X iff $\forall Y \in M \exists \alpha \in o(M) \exists g \in M(g: X \times \alpha \twoheadrightarrow Y)$.

Our $\mathbb{R}$-premice will be wellordered mod $\mathbb{R}$, moreover, if we take an ultrapower of such a premouse M by an extender E, then E will be $(M, \mathbb{R})$-complete. In this context we have

PROPOSITION 2.2. Let M be transitive, rud-closed, and wellordered mod X, where X is transitive. Let E be an extender over M; then the following are equivalent:

1. E is (M, X)-complete,
2. $\mathrm{ult}(M, E)$ satisfies the Łoś theorem for Σ_0 formulae, and the canonical embedding from M to $\mathrm{ult}(M, E)$ is the identity on $X \cup \{X\}$.

PROOF. We shall just sketch (1) $\Rightarrow$ (2), which is the direction we use anyway.

The usual proof of Łoś's theorem works except at the point where one would invoke the axiom of choice in M. At this point we have assumed

$$\text{for } E_a\text{-almost every } u \ (M \models \exists v \in g(u)\varphi[v, f_1(u), \ldots, f_k(u)]),$$

where φ is Σ_0, and $g, f_1, \ldots, f_k$ are in M, and we wish to find f in M such that

$$\text{for } E_a\text{-almost every } u \ (M \models \varphi[f(u), f_1(u), \ldots, f_k(u)]).$$

Now since M is wellordered mod X, we can fix $h \in M$ so that $h: X \times \alpha \twoheadrightarrow \bigcup \mathrm{ran}(g)$. For $u \in \mathrm{dom}(g)$ and $\beta < \alpha$, set

$$f^*(u, \beta) = \{i \in X : M \models \varphi[h(i, \beta), f_1(u), \ldots, f_k(u)]\}.$$

For $u \in \mathrm{dom}(g)$, let

$$t(u) = f^*(u, \beta_u), \text{ where } \beta_u \text{ is least s.t. } f^*(u, \beta_u) \neq \emptyset,$$

and let $t(u) = \varnothing$ if $f^*(u, \beta) = \varnothing$ for all β. Because M is rud-closed, the functions f^* and t are in M. But now for E_a-almost every u there is an $i \in t(u)$, and so by (M, X)-completeness we can fix i_0 such that for E_a-almost every u, we have $i_0 \in t(u)$. The desired function f is then given by

$$f(u) = h(i_0, \beta_u).$$

To see that the canonical embedding j is the identity on $X \cup \{X\}$, suppose $g \in M$ maps $[\text{crit}(E)]^{|a|}$ to X. We have that for E_a-almost every u there is an $i \in X$ such that $i = g(u)$, hence we can fix $i \in X$ such that $g(u) = i$ for E_a-almost every u. It follows that $[g] = j(i)$. ⊣

For X transitive and appropriate $\vec{E}$, we define $\mathbf{J}_\alpha^{\vec{E}}(X)$ by:

$$\mathbf{J}_0^{\vec{E}}(X) = X,$$

$$\mathbf{J}_{\alpha+1}^{\vec{E}}(X) = \text{ rud-closure of } \mathbf{J}_\alpha^{\vec{E}}(X) \cup \{\mathbf{J}_\alpha^{\vec{E}}(X), E_\alpha\},$$

and taking unions at limits. Here the appropriate $\vec{E}$ are those such that each E_α is either the emptyset or an extender over $\mathbf{J}_\alpha^{\vec{E}}(X)$ which is $(\mathbf{J}_\alpha^{\vec{E}}(X), X)$-complete. We write

$$\mathcal{J}_\alpha^{\vec{E}}(X) = (\mathbf{J}_\alpha^{\vec{E}}(X), \in, \vec{E} \restriction \alpha, E_\alpha, X)$$

for the structure for the language of set theory expanded by predicate symbols $\dot{E}$ for $\vec{E}$, $\dot{F}$ for E_α, and a constant symbol $\dot{\mathbb{R}}$ for X (chosen because $X = \mathbb{R} \cap \mathbf{J}_\alpha^{\vec{E}}(X)$ is the case of greatest interest).[2] This language of relativised premice we call $\mathcal{L}^*$.

Definition 2.3. An appropriate $\vec{E}$ is X**-acceptable at** α iff $\forall \beta < \alpha \forall \kappa$ $(\wp(\mathbf{J}_\kappa^{\vec{E}}(X) \cap (\mathbf{J}_{\beta+1}^{\vec{E}}(X) \setminus \mathbf{J}_\beta^{\vec{E}}(X)) \neq \varnothing) \Rightarrow (\mathbf{J}_{\beta+1}^{\vec{E}}(X) \models \exists f\colon \mathbf{J}_\kappa^{\vec{E}}(X) \twoheadrightarrow \mathbf{J}_\beta^{\vec{E}}(X))$.

The following proposition is a uniform, local version of the fact that that every set in $L[\vec{E}, X]$ is ordinal-definable from parameters in $X \cup \{X\}$.

Proposition 2.4.

1. There is a fixed Σ_1 formula φ_0 of our expanded language such that whenever X is transitive, $\vec{E}$ is appropriate for X, and $\alpha < \text{lh}(\vec{E})$, then φ defines over $\mathcal{J}_\alpha^{\vec{E}}(X)$ a map $h\colon (X^{<\omega} \times [\alpha]^{<\omega}) \twoheadrightarrow \mathbf{J}_\alpha^{\vec{E}}(X)$. We write $h_\alpha^{\vec{E},X}$ for the map h so defined.
2. We can (and do) take the maps $h_\alpha^{\vec{E},\mathbb{R}}$ to have domain $\mathbb{R} \times [\alpha]^{<\omega}$ (replacing φ_0 with φ_1, another Σ_1 formula).
3. If $\vec{E}$ is appropriate for $\mathbb{R}$, then for any α, there is a map from $\mathbb{R} \times \alpha$ onto $\mathbf{J}_\alpha^{\vec{E}}(\mathbb{R})$ which is Σ_1 definable from parameters over $\mathcal{J}_\alpha^{\vec{E}}(\mathbb{R})$.

[2] If $\alpha = 0$, take $\dot{\mathbb{R}}$ to name 0.

The proof is a routine extension of Jensen's [Jen72]. See [Ste83A] for the case $\vec{E} = \varnothing$. We shall need the *uniformly* Σ_1 maps of assertion (2) later on.

Using Proposition 2.4, we can reformulate $\mathbb{R}$-acceptability with $\mathbb{R} \times \kappa$ replacing $\mathbf{J}^{\vec{E}}_{\kappa}(\mathbb{R})$, etc. Let us call λ an **$\mathbb{R}$-cardinal** iff there is no map $f : \kappa \times \mathbb{R} \twoheadrightarrow \lambda$ with $\kappa < \lambda$. It is then easy to see that if $\vec{E}$ is $\mathbb{R}$-acceptable at α, and $\mathbf{J}^{\vec{E}}_{\alpha}(\mathbb{R}) \models \lambda$ is an $\mathbb{R}$-cardinal, then for all $\kappa < \lambda$, $\wp(\mathbf{J}^{\vec{E}}_{\kappa}(\mathbb{R})) \cap \mathbf{J}^{\vec{E}}_{\alpha}(\mathbb{R}) \subseteq \mathbf{J}^{\vec{E}}_{\lambda}(\mathbb{R})$.

Definition 2.5. Θ is the least ordinal which is not the surjective image of $\mathbb{R}$.

Let $\mathcal{M}$ be an $\mathbb{R}^{\mathcal{M}}$-premouse satisfying "Θ exists", and let $\theta = \Theta^{\mathcal{M}}$. It is easy to see, using Proposition 2.4, that θ is regular in $\mathcal{M}$. By acceptability, $\mathcal{M}|\theta$ satisfies "every set is the surjective image of $\mathbb{R}$". We can conclude then that the structure $\mathcal{M}|\theta$ is admissible.

It is easy to show, without using the axiom of choice, that for $\lambda > 1$, λ is an $\mathbb{R}$-cardinal iff λ is a cardinal and $\lambda \geq \Theta$. Thus if $\vec{E}$ is $\mathbb{R}$-acceptable at α, then $\mathcal{J}^{\vec{E}}_{\alpha}(\mathbb{R})$ satisfies: "whenever $\kappa \geq \Theta$ and κ^+ exists, then $\wp(\mathbf{J}^{\vec{E}}_{\kappa}(\mathbb{R})) \subseteq \mathbf{J}^{\vec{E}}_{\kappa^+}(\mathbb{R})$".

Definition 2.6. Let X be transitive; then a **fine extender sequence over** X is a sequence $\vec{E}$ such that for each $\alpha \in \mathrm{dom}(\vec{E})$, $\vec{E}$ is X-acceptable at α, and either $E_\alpha = \varnothing$ or E_α is a (κ, α) pre-extender over $\mathbf{J}^{\vec{E}}_{\alpha}(X)$ for some κ such that $\mathbf{J}^{\vec{E}}_{\alpha}(X) \models \wp(\mathbf{J}^{\vec{E}}_{\kappa}(X))$ exists, and E_α is $(\mathbf{J}^{\vec{E}}_{\alpha}(X), X)$-complete, and: E_α satisfies clauses (1), (2), and (3) of [Ste07B, Definition 2.4].

Of course, [Ste07B, Definition 2.4] was formulated there for the case $X = \varnothing$, but it is now easy to see what its clauses should mean in the general case.

Definition 2.7. A **potential premouse over** X (or X**-ppm**) is a structure of the form $\mathcal{J}^{\vec{E}}_{\alpha}(X)$, where $\vec{E}$ is a fine extender sequence over X. If $\mathcal{M} = \mathcal{J}^{\vec{E}}_{\alpha}(X)$ is an X-ppm, we write $\mathcal{J}^{\mathcal{M}}_{\beta}$, or simply $\mathcal{M}|\beta$, for the structure $\mathcal{J}^{\vec{E}}_{\beta}(X)$. We say $\mathcal{N}$ is an **initial segment** of $\mathcal{M}$, and write $\mathcal{N} \trianglelefteq \mathcal{M}$, iff $\mathcal{N} = \mathcal{M}|\beta$ for some β.

Active potential premice are in general not amenable structures, but we can code an active X-ppm $\mathcal{M}$ by an amenable structure $\mathfrak{C}_0(\mathcal{M})$. That involves replacing the last extender $F = \dot{F}^{\mathcal{M}}$ with $F{\restriction}\nu(F)$ in the case $\nu(F)$ (the sup of the generators of F) is a limit ordinal, and with a certain predicate F^* coding the fragments of F in the case $\nu(F)$ is a successor ordinal. The details are given in [Ste07B, 2.11]. The "Σ_0-code" $\mathfrak{C}_0(\mathcal{M})$ is also a structure for the language $\mathcal{L}^*$, and it is really this interpretation of $\mathcal{L}^*$ which is of importance in what follows. Whenever we speak of definability over a ppm $\mathcal{M}$, we shall in reality mean definabilty over $\mathfrak{C}_0(\mathcal{M})$.

2.2. Cores, Projecta, Soundness. These notions carry over routinely to X-ppm. One need only remember that $X \cup \{X\}$ is contained in all cores of

X-ppm. For example, we define the first projectum, standard parameter, and core of an X-ppm $\mathcal{M}$ by:[3]

$$\begin{aligned}\rho_1(\mathcal{M}) = &\text{ least } \alpha \text{ such that for some boldface } \utilde{\Sigma}_1^{\mathfrak{C}_0(\mathcal{M})} \text{ set } A,\\ &A \subseteq \mathcal{M}|\alpha \text{ and } A \notin \mathcal{M},\\ p_1(\mathcal{M}) = &\text{ least } p \in {}^{<\omega}[o(\mathcal{M})] \text{ such that there is an}\\ &A \subseteq \mathcal{M}|\rho_1(\mathcal{M}) \text{ s.t. } A \text{ is } \Sigma_1^{\mathfrak{C}_0(\mathcal{M})} \text{ in } p \text{ but } A \notin \mathcal{M},\\ \mathfrak{C}_1(\mathcal{M}) = &\ \Sigma_1 \text{ Skolem hull of } \mathcal{M} \text{ generated by } \mathcal{M}|\rho_1(\mathcal{M}) \cup p_1(\mathcal{M}).\end{aligned}$$

Here we order finite sets of ordinals by listing their elements in decreasing order, and comparing the resulting finite sequences lexicographically. It is possible that $\mathcal{M}|\rho_1(\mathcal{M}) = \mathcal{M}$; if this is not the case, then $\rho_1(\mathcal{M})$ is an X-cardinal of $\mathcal{M}$. In the case $\rho_1(\mathcal{M}) = 1$, we shall generally write $\rho_1(\mathcal{M}) = X$ instead. The core $\mathfrak{C}_1(\mathcal{M})$ is taken to be transitive, and a structure for the language of X-ppm, and so taken, it is in fact an X-ppm.

The definitions of solidity and universality for the standard parameter go over to X-ppm in the obvious way. We say $p_1(\mathcal{M})$ is **universal** if $\wp(\mathcal{M}|\rho_1(\mathcal{M}))\cap \mathcal{M} \subseteq \mathfrak{C}_1(\mathcal{M})$. We say $p = p_1(\mathcal{M})$ is **solid** if for each $\alpha \in p$, letting b be the set of Σ_1 sentences in our expanded language augmented further by names for all elements of $(p \setminus (\alpha+1)) \cup \alpha$ which are true in $\mathcal{M}$, we have $b \in \mathcal{M}$. We call such b the **solidity witnesses** for $p_1(\mathcal{M})$. If $p_1(\mathcal{M})$ is solid and universal, we go on to define $\rho_2(\mathcal{M})$, $p_2(\mathcal{M})$, and $\mathfrak{C}_2(\mathcal{M})$; the reader should consult [MS94] for all further details here regarding the $\rho_n(\mathcal{M})$, $p_n(\mathcal{M})$, and $\mathfrak{C}_n(\mathcal{M})$ for $n > 1$. In this paper, when we need to go into fine-structural details, we shall stick to the representative case $n = 1$.

DEFINITION 2.8. An X-ppm $\mathcal{M}$ is ***n*-solid** iff $\mathfrak{C}_n(\mathcal{M})$ exists, and $p_n(\mathfrak{C}_n(\mathcal{M}))$ is solid and universal. $\mathcal{M}$ is ***n*-sound** iff $\mathcal{M}$ is n-solid, and $\mathcal{M} = \mathfrak{C}_n(\mathcal{M})$. $\mathcal{M}$ is an ***X*-premouse** iff every proper initial segment of $\mathcal{M}$ is ω-sound (i.e., n-sound for all $n < \omega$).

2.3. Ultrapowers, Iteration Trees. It is easy to adapt the material of [MS94, §§4& 5], or [Ste07B, §§2&3], to X-premice. The definitions make sense, and the theorems continue to hold true, if one replaces "premouse" with "X-premouse" everywhere. The following propositions summarize some of the basic facts:

PROPOSITION 2.9. Let $\mathcal{M}$ be an n-sound premouse over X, and let E be an extender over $\mathcal{M}$ which is $(\mathcal{M}, X)$-complete; then

[3]Here and elsewhere, we write $\mathcal{M}$ for the universe of the structure $\mathcal{M}$, if no confusion can come from doing so.

1. For any generalized $r\Sigma_n$ formula φ, functions f_i definable over $\mathcal{M}$ from parameters using Σ_n Skolem terms $\tau_i \in \mathrm{Sk}_n$, and $a \in [\mathrm{lh}(E)]^{<\omega}$ such that $\mathrm{dom}(f_i) = [\mathrm{crit}(E)]^{|a|}$ for all i, the following are equivalent:
   (a) $\mathrm{ult}_n(\mathcal{M}, E) \models \varphi[[a, f_0], \ldots, [a, f_k]]$, and
   (b) for E_a-almost every u, $\mathcal{M} \models \varphi[f_0(u), \ldots, f_k(u)]$.
2. The canonical embedding $i_E^{\mathcal{M}}$ from $\mathcal{M}$ to $\mathrm{ult}_n(\mathcal{M}, E)$ is an n-embedding.
3. Suppose also E is close to $\mathcal{M}$, $\rho_{n+1}(\mathcal{M}) \leq \mathrm{crit}(E)$, and $p_{n+1}(\mathcal{M})$ is solid and universal; then

$$\rho_{n+1}(\mathcal{M}) = \rho_{n+1}(\mathrm{ult}_n(\mathcal{M}, E)),$$

   and

$$i_E^{\mathcal{M}}(p_{n+1}(\mathcal{M})) = p_{n+1}(\mathrm{ult}_n(\mathcal{M}, E)),$$

   and $p_{n+1}(\mathrm{ult}_n(\mathcal{M}, E))$ is solid and universal.

Here the ultrapower $\mathrm{ult}_0(\mathcal{M}, E)$ is formed using the functions $f_i \in \mathcal{M}$. Thus we have already proved the $n = 0$ case of Proposition 2.9 (1). There are no new ideas in the rest of the proof.

PROPOSITION 2.10. Suppose $\mathcal{T}$ is an n-maximal iteration tree on the n-sound X-premouse $\mathcal{M}$, and that $\alpha T \beta$ and $D^{\mathcal{T}} \cap (\alpha, \beta]_T = \varnothing$; then the canonical embedding $i^{\mathcal{T}}_{\alpha,\beta} \circ i^*_\alpha$ from $\mathcal{M}^*_\alpha$ to $\mathcal{M}^{\mathcal{T}}_\beta$ is a k-embedding, where $k = \deg^{\mathcal{T}}(\beta)$. Moreover, if $\mathrm{crit}(i^*_\alpha) \geq \rho_{k+1}(\mathcal{M}^*_\alpha)$, then $i^{\mathcal{T}}_{\alpha,\beta} \circ i^*(p_{k+1}(\mathcal{M}^*_\alpha)) = p_{k+1}(\mathcal{M}^{\mathcal{T}}_\beta)$.

The notions of [MS94] and [Ste07B] associated with iterability relativise to X-premice in an obvious way.

DEFINITION 2.11. Let $\mathcal{M}$ be a k-sound X-premouse; then we say $\mathcal{M}$ is **countably k-iterable** iff whenever $\overline{\mathcal{M}}$ is a countable $\overline{X}$-premouse, and there is a $\pi\colon \overline{\mathcal{M}} \to \mathcal{M}$ which is fully $\mathcal{L}^*$-elementary, then $\overline{\mathcal{M}}$ is $(k, \omega_1 + 1)$-iterable. We say $\mathcal{M}$ is **countably iterable** just in case it is countably ω-iterable.

The comparison process yields

THEOREM 2.12. Let $\mathcal{M}$ and $\mathcal{N}$ be X-premice. and suppose that $\mathcal{M}$ is $m+1$-sound and countably m-iterable, where $\rho_{m+1}(\mathcal{M}) = X$, and $\mathcal{N}$ is $n+1$-sound and countably n-iterable, where $\rho_{n+1}(\mathcal{N}) = X$; then either $\mathcal{M} \trianglelefteq \mathcal{N}$ or $\mathcal{N} \trianglelefteq \mathcal{M}$.

We can therefore define

DEFINITION 2.13. For any transitive set X, $\mathbf{K}(X)$ is the unique X premouse $\mathcal{M}$ whose proper initial segments are precisely all those countably iterable X-premice $\mathcal{N}$ such that $\rho_\omega(\mathcal{N}) = X$.

§3. Some local HOD's. Our analysis of scales proceeds by getting optimal closed game representations. In the relevant closed game, player I attempts to verify that a given $\mathbb{R}$-mouse $\mathcal{M}$ satisfies $\varphi(x)$, where φ is Σ_1 and $x \in \mathbb{R}$. He does so by describing an $\dot{\mathbb{R}}^{\mathcal{N}}$-premouse $\mathcal{N}$ satisfying $\varphi(x)$; one can think of him as claiming that his $\mathcal{N}$ is an elementary submodel of $\mathcal{M}$. Player II helps keep player I honest about this by playing reals which player I must then put into $\mathcal{N}$. In order to ensure that player I is indeed being honest about what is true in $\mathcal{M}$, we must ask him in addition to verify that his $\mathcal{N}$ is iterable. Of course, the obvious way to verify this is to play an elementary $\pi\colon \mathcal{N} \to \mathcal{M}$, but this leads to a payoff condition for player I which is not closed in the appropriate topology.[4] Our main new idea here is just that player I can verify iterability by elementarily embedding $\mathbf{HOD}^{\mathcal{N}}$ into $\mathbf{HOD}^{\mathcal{M}}$. The key here is that $\mathbf{HOD}^{\mathcal{M}}$ is definably wellordered,[5] so that the embedding is essentially an ω-sequence of ordinals, and the elementarity condition is closed in the appropriate topology.

We must consider here iterations of $\mathcal{N}$ involving Σ_n-ultrapowers of the form ult_n. We shall reduce such ultrapowers to Σ_n-ultrapowers of $\mathbf{HOD}^{\mathcal{N}}$, and to do so we need a fine-structure theory for $\mathbf{HOD}^{\mathcal{N}}$. Fortunately, we can restrict ourselves to $\mathcal{N}$ satisfying "Θ exists", and be content with a fine-structural analysis of $\mathbf{HOD}^{\mathcal{N}}$ *above* $\Theta^{\mathcal{N}}$: that is, a representation of $\mathbf{HOD}^{\mathcal{N}}$ as an X-mouse, for some $X \subseteq \Theta^{\mathcal{N}}$. Now in the case $\mathcal{N} = \mathbf{L}(\mathbb{R})$, the following theorem of Woodin does the job:

THEOREM 3.1 (Woodin). There is a partial order $\mathbb{P}$ on $\Theta^{\mathbf{L}(\mathbb{R})}$ such that

$$\mathbf{HOD}^{\mathbf{L}(\mathbb{R})} = \mathbf{L}(\mathbb{P}),$$

and moreover $\mathbf{L}(\mathbb{R})$ is an inner model of a $\mathbb{P}$-generic extension of $\mathbf{HOD}^{\mathbf{L}(\mathbb{R})}$.

Here $\mathbb{P}$ is a modification of the Vopěnka partial order designed to add a generic enumeration of $\mathbb{R}$.

We shall extend Woodin's argument so as to show that if $\mathcal{M}$ is an $\mathbb{R}$-premouse satisfying "Θ exists", then there is a $\mathbb{P} \subseteq \Theta^{\mathcal{M}}$ such that $\mathbf{HOD}^{\mathcal{M}}$ is the universe of a $\mathbb{P}$-premouse $\mathcal{H}$. The main new thing here is to show that the projecta and standard parameters of levels of $\mathcal{H}$ match those of the corresponding levels of $\mathcal{M}$, and indeed establish level-by-level intertranslatability of the theories of initial segments of $\mathcal{H}$ and $\mathcal{M}$ respectively. This we get from the fact that

---

[4]Player I may play reals as well as ordinals in our game, and the elements of $\text{ran}(\pi)$ can be coded by pairs $\langle x, \alpha\rangle$ where $x \in \mathbb{R}$ and $\alpha \in \text{Ord}$. However, the elementarity requirement on π would not be closed in the appropriate topology on $(\text{Ord} \times \mathbb{R})^\omega$, which is the product of ω copies of the discrete topology on Ord and the *Baire* (not discrete) topology on $\mathbb{R}$.

[5]The definition which guarantees a set is in $\mathbf{HOD}^{\mathcal{M}}$ must use the language $\mathcal{L}^*$ of premice. $\dot{E}$ and $\dot{F}$ are allowed, but names for individual reals are not! Further, this definition must be interpreted over some *proper* initial segment of $\mathcal{M}$; there may be sets of ordinals in $\mathcal{M}$ which are definable over $\mathcal{M}$ itself, yet not in $\mathbf{HOD}^{\mathcal{M}}$.

$\mathcal{M}$ is a symmetric[6] inner model of a $\mathbb{P}$-extension of $\mathcal{H}$, using the level-by-level definability of forcing.

In turning to the details, it will be convenient to replace $\mathbb{P}$ with a superficially more powerful set. Let us fix for the remainder of this section an $\mathbb{R}^{\mathcal{M}}$-premouse $\mathcal{M}$ such that

- $\mathcal{M} \models$ "Θ exists".

We set $\theta = \Theta^{\mathcal{M}}$, and we also fix an $n_0 < \omega$ such that

- $\mathcal{M}$ is n_0-sound and $\rho_{n_0}(\mathcal{M}) \geq \theta$.

Finally, letting $o(\mathcal{M}) = \omega\gamma_0$, we assume that

- for all $\langle \xi, k\rangle <_{\text{lex}} \langle \gamma_0, n_0\rangle$, $\mathcal{M}|\xi$ is countably k-iterable.[7]

We get a certain amount of condensation from these assumptions.

LEMMA 3.2. For any $\xi < \gamma_0$,

$$\mathcal{H}_1^{\mathcal{M}|\xi}(\mathbb{R}^{\mathcal{M}}) \cong \mathcal{M}|\tau$$

for some $\tau < \theta$.

In fact, if $\langle \xi, k\rangle \leq_{\text{lex}} \langle \gamma_0, n_0\rangle$ and $k \geq 1$, then for any finite $F \subseteq \omega\xi$, $\mathcal{H}_k^{\mathcal{M}|\xi}(\mathbb{R}^{\mathcal{M}} \cup F) \in \mathcal{M}|\theta$, since the theory of the hull is in $\mathcal{M}|\theta$, and the latter is an admissible structure. However, $\mathcal{H}_k^{\mathcal{M}|\xi}(\mathbb{R}^{\mathcal{M}} \cup F)$ may not be sound, and hence may not be of the form $\mathcal{M}|\tau$.

COROLLARY 3.3. $\mathcal{M}|\theta$ is a Σ_1-elementary submodel of $(\mathbf{J}_{\gamma_0}^{\dot{E}^{\mathcal{M}}}, \in, \dot{E}^{\mathcal{M}} \restriction \gamma_0, \varnothing)$.

Set

$$T^{\mathcal{M}} = \{\langle \varphi, \vec{\alpha}\rangle : \vec{\alpha} \in {}^{<\omega}\theta \text{ and } \mathcal{M}|\theta \models \varphi[\vec{\alpha}]\}.$$

Since θ is a cardinal in $\mathcal{M}$, we can use Gödel's pairing function to identify $T^{\mathcal{M}}$ with a subset of θ. Letting $\omega\eta = o(\mathcal{M})$, we have that

$$\mathbf{HOD}^{\mathcal{M}|\theta} \cap \mathbf{V}_\theta = \mathbf{J}_\eta(T^{\mathcal{M}}) \cap \mathbf{V}_\theta.$$

Since $\mathcal{M}|\theta$ is a Σ_1-elementary submodel of $\mathcal{M}$ with its last extender removed, we have

$$\mathbf{HOD}^{\mathcal{M}} \cap \mathbf{V}_\theta = \mathbf{J}_\eta(T^{\mathcal{M}}) \cap \mathbf{V}_\theta.$$

We can construct a $T^{\mathcal{M}}$-premouse whose universe is the whole of $\mathbf{HOD}^{\mathcal{M}}$ by simply constructing from $T^{\mathcal{M}}$ together with the extenders from the $\mathcal{M}$-sequence having critical points above θ. More precisely, letting

$$\mathcal{M} = \mathcal{J}_{\gamma_0}^{\vec{E}}(\mathbb{R}^{\mathcal{M}}),$$

[6] We shall explain the meaning of this shortly.

[7] The analysis of $\mathbf{HOD}^{\mathcal{M}}$ we are developing will be used to show that $\mathcal{M}$ is countably n_0-iterable, given that $\mathbf{HOD}^{\mathcal{M}}$ is.

we define an appropriate sequence $\vec{F}$ over $\theta \cup \{T^{\mathcal{M}}\}$ by setting

$$F_\alpha = E_{\theta+\alpha} \cap \mathbf{J}_\alpha^{\vec{F}\restriction\alpha}(\theta \cup \{T^{\mathcal{M}}\})$$

for all α such that $\theta+\alpha \leq \gamma_0$. It is not hard to see that the sequence $\vec{F}$ is indeed appropriate for $\theta \cup \{T^{\mathcal{M}}\}$; the main point is that $\text{crit}(E_{\theta+\alpha}) > \Theta^{\mathcal{M}|(\theta+\alpha)} = \theta$, which implies that F_α is sufficiently complete. We set

$$\mathcal{H}_\alpha = \mathcal{J}_\alpha^{\vec{F}}(\theta \cup \{T^{\mathcal{M}}\}),$$

for all α such that $\theta + \alpha \leq \gamma_0$.

It will be convenient to ignore the $\mathcal{H}_\alpha$ for small α. Therefore, we add to our assumptions on $\mathcal{M}$ that there is some $\lambda \leq \gamma_0$ such that $\mathcal{M}|\lambda \models \mathsf{ZF}$, and let

$$\lambda_0 = \text{ least } \lambda \text{ such that } \mathcal{M}|\lambda \models \mathsf{ZF}.$$

Notice that our new assumption on $\mathcal{M}$ holds if some $E_{\theta+\alpha} \neq \varnothing$, and in this case $\text{crit}(E_{\theta+\alpha}) > \lambda_0$. Since in our closed game representation we only need the $\mathcal{H}_\alpha$ to "verify" extenders from the $\mathcal{M}$-sequence with index above θ, we can afford to ignore $\mathcal{H}_\alpha$ except when λ_0 exists and $\lambda_0 \leq \theta + \alpha$. Note that $\theta + \alpha = \alpha$ for the α we do not ignore, so that we are already rewarded for our ignorance. We set

$$\mathcal{H} = \mathcal{H}_{\gamma_0}.$$

We shall show that for $\alpha \geq \lambda_0$, $\mathcal{H}_\alpha$ is a $T^{\mathcal{M}}$-premouse and $\mathcal{M}|\alpha$ is an inner model of a generic extension of $\mathcal{H}_\alpha$.[8] The relevant partial order is the same Vopěnka-like partial order used by Woodin.

Let us work in $\mathcal{M}$ for a while. Fix a bijection $\pi \colon \theta \to \mathcal{O}$, where $\mathcal{O}$ is the collection of all subsets of $\mathbb{R}^n = \{s : s \colon n \to \mathbb{R}\}$ which are definable from ordinal parameters over $\mathcal{M}|\theta$. We choose π so that it is definable over $\mathcal{M}|\theta$. We write A^* for $\pi(A)$ henceforth. Let

$$A \in \text{Vop}_n \Leftrightarrow \exists n < \omega(A^* \subseteq \mathbb{R}^n \wedge A^* \neq \varnothing),$$

and

$$A \in \text{Vop}_\omega \Leftrightarrow \exists n < \omega(A \in \text{Vop}_n).$$

For A in Vop_ω, we write $s(A)$ for the unique $n < \omega$ such that $A \in \text{Vop}_n$. For $A, B \in \text{Vop}_\omega$, we put

$$A \leq^v B \Leftrightarrow s(B) \leq s(A) \wedge \forall s \in A^*(s\restriction s(B) \in B^*).$$

We also use Vop_ω to denote the partial order $(\text{Vop}_\omega, \leq^v)$. Clearly, Vop_ω is coded into $T^{\mathcal{M}}$ in a simple way, and hence $\text{Vop}_\omega \in \mathcal{H}|2$.

The standard Vopěnka argument shows that for any $n < \omega$, Vop_n is a complete Boolean algebra in $\mathcal{H}$, and each $s \in \mathbb{R}^n$ determines an $\mathcal{H}$-generic filter $G_s = \{A \in \text{Vop}_n : s \in A\}$ on Vop_n.[9] It is easy to see that the inclusion

[8] The first assertion is true for smaller α as well.

[9] Note that any subset of $\mathbb{R}$ which is $\mathbf{OD}^{\mathcal{M}}$ is actually $\mathbf{OD}^{\mathcal{M}|\theta}$, since $\mathcal{M}|\theta$ is a Σ_1 elementary submodel of $\mathcal{M}$.

map is a complete embedding of Vop_n into Vop_ω. Motivated by this, we define for $h\colon \omega \to \mathbb{R}$ and $A \in \mathrm{Vop}_\omega$:

$$A \in G_h \Leftrightarrow h{\restriction}s(A) \in A^*.$$

LEMMA 3.4. If h is $\mathcal{M}$-generic over $\mathrm{Col}(\omega, \mathbb{R})$, then G_h is $\mathcal{H}$-generic over Vop_ω.

PROOF. Let $\mathcal{D}$ be dense in Vop_ω, and $\mathcal{D} \in \mathcal{M}$. Let $s \in \mathbb{R}^n$ be a condition in $\mathrm{Col}(\omega, \mathbb{R})$. It will be enough to find a t extending s in $\mathrm{Col}(\omega, \mathbb{R})$ such that $G_t \cap \mathcal{D} \neq \varnothing$. Let

$$X = \{u \in \mathbb{R}^n : \exists B \in \mathcal{D} \exists t (u \subseteq t \wedge t \in B^*)\}.$$

We want to see $s \in X$, so it will be enough to see $X = \mathbb{R}^n$. Suppose not; then since X is clearly **OD** in $\mathcal{M}$, there is an $A \in \mathrm{Vop}_n$ such that

$$A^* = \mathbb{R}^n \setminus X.$$

Since $\mathcal{D}$ is dense, we can find $B \in \mathcal{D}$ such that $B \leq A$. But now pick any $t \in B$, and it is clear that $t{\restriction}n \in X$, a contradiction. ⊣

We can recover h from G_h in a simple way. For $b \in \mathbf{V}_\omega$ and $n < \omega$, let $A_{b,n} \in \mathrm{Vop}_{n+1}$ be such that $A^*_{b,n} = \{s \in \mathbb{R}^{n+1} : b \in s(n)\}$. We assume that the map $\langle b, n\rangle \mapsto A_{b,n}$ is definable over $\mathcal{M}_\theta$, and hence in $\mathcal{H}$, as any natural such map will be.[10] Then clearly,

$$b \in h(n) \Leftrightarrow A_{b,n} \in G_h.$$

We define Vop_ω-terms for the $h(n)$ and $\mathrm{ran}(h)$ by

$$\sigma_n = \{\langle A, \check{b}\rangle : A \leq^v A_{b,n}\},$$

and

$$\dot{R} = \{\langle A, \sigma_n\rangle : A \in \mathrm{Vop}_\omega \wedge n < \omega\}.$$

These terms are of course in $\mathcal{H}$. It is easy to see

LEMMA 3.5.
1. For any $h\colon \omega \to \mathbb{R}$, $\sigma_n^{G_h} = h(n)$ for all n, and $\dot{R}^{G_h} = \mathrm{ran}(h)$.
2. If h is $\mathcal{M}$-generic for $\mathrm{Col}(\omega, \mathbb{R}^{\mathcal{M}})$, then $\dot{R}^{G_h} = \mathbb{R}^{\mathcal{M}}$.
3. For any condition $A \in \mathrm{Vop}_\omega$, there is an $\mathcal{H}$-generic filter G on Vop_ω such that $A \in G$ and $\dot{R}^G = \mathbb{R}^{\mathcal{M}}$.

By Lemma 3.5, truth in $\mathcal{H}(\mathbb{R}^{\mathcal{M}})$ can be reduced to truth in $\mathcal{H}$ via the forcing relation for Vop_ω. In order to see that $\mathcal{H}(\mathbb{R}^{\mathcal{M}})$ determines $\mathcal{M}$ we need to know that the extenders on $\vec{F}$ generate the corresponding extenders on $\vec{E}$. For this, we need that the forcing relation for Vop_ω is locally definable. We also need

[10]The $*$ map is one-one on the separative quotient of Vop_ω, so the question as to what to choose for $A_{b,n}$ disappears if we replace Vop_ω with its separative quotient.

this local definability to show that the reduction of $\mathcal{M}$-truth to $\mathcal{H}$-truth is local, and thereby that $\mathcal{H}$ is a $T^{\mathcal{M}}$-premouse.

We shall use the usual Shoenfield terms for our forcing language. Besides these terms, the language of forcing over an amenable structure

$$(\mathbf{J}_\xi^A(X), \in, A, X, B)$$

has $\in$, $=$, a constant symbol $\dot{R}$ for X, and predicate symbols $\dot{E}$ and $\dot{F}$ for A and B. A filter G over a poset $\mathbb{P} \in \mathbf{J}_\xi^A(X)$ is generic over this structure just in case it meets all $\mathbb{P}$-dense sets $D \in \mathbf{J}_\xi^A(X)$. We let $\mathbf{J}_\xi^A(X)[G] = \{\tau^G : \tau \in \mathbf{J}_\xi^A(X)\}$ be the set of G-interpretations of terms, and say

$$p \Vdash \varphi \Leftrightarrow \forall G(G \text{ is generic over } \mathbf{J}_\xi^A(X) \Rightarrow (\mathbf{J}_\xi^A(X)[G], \in, A, X, B) \models \varphi).$$

We use $\boldsymbol{S}_\alpha^A(X)$ for the αth level of Jensen's $\boldsymbol{S}$-hierarchy on $\mathbf{J}_\xi^A(X)$. Let $\Sigma_{0,n}$ be the collection of Σ_0 sentences of the forcing language containing at most n bounded quantifiers.

LEMMA 3.6. Let $(\mathbf{J}_\xi^A(X), \in, B)$ be amenable, and let $\mathbb{P}$ be a poset, with $\mathbb{P} \in \mathbf{J}_\nu^A(X)$, where $\nu < \xi$ and $\mathbf{J}_\nu^A(X) \models \mathsf{ZFC}$. For $\nu \leq \alpha < \omega\xi$, let

$$F_{\alpha,n} = \{\langle p, \varphi \rangle : p \in \mathbb{P} \wedge \varphi \in (\Sigma_{0,n} \cap \boldsymbol{S}_\alpha^A(X)) \wedge p \Vdash \varphi\}.$$

Then

1. $\forall \alpha < \omega\xi \forall n < \omega(F_{\alpha,n} \in \mathbf{J}_\xi^A(X))$; moreover the function $\langle \alpha, n \rangle \mapsto F_{\alpha,n}$ is $\Sigma_1^{(\mathbf{J}_\xi^A(X), \in, A, X, B)}$ in the parameter $\mathbb{P}$ (uniformly in ξ).
2. If $(\mathbf{J}_\xi^A(X)[G], \in, A, X, B) \models \varphi$, where φ is Σ_0 and G is generic over $\mathbf{J}_\xi^A(X)$, then $\exists p \in G(p \Vdash \varphi)$.

A proof of Lemma 3.6 can be organized as follows. Let $F_{\alpha,n}^* = \{\langle p, \varphi \rangle \in F_{\alpha,n} : \dot{F}$ does not occur in $\varphi\}$. One first proves the lemma with $F_{\alpha,n}^*$ replacing $F_{\alpha,n}$. This amounts to observing that the standard inductive definition of forcing for Σ_0 sentences is a "local Σ_0-recursion" of the same sort that defines the function $\alpha \mapsto \boldsymbol{S}_\alpha^A(X)$ itself in a Σ_1 way over $(\mathbf{J}_\xi^A(X), \in, A, X)$. Of course, one needs that everything true is forced to verify that the inductive definition works. The reason for restricting ourselves to $\alpha \geq \nu$ is that we need a starting point for the induction, and when $\alpha = \nu$, Lemma 3.6 literally is a standard basic forcing lemma.[11] Finally, one can show that $F_{\alpha,0}$ is uniformly rudimentary in $\langle F_{\alpha,0}^*, B \cap \boldsymbol{S}_\alpha^A(X) \rangle$ (since $p \Vdash \dot{F}(\tau) \Leftrightarrow \forall q \leq p \exists r \leq q \exists x \in B(r \Vdash \tau = \check{x})$). Also, $F_{\alpha,n+1}$ is uniformly rudimentary in $F_{\alpha,n}$. This completes our pseudo-proof of Lemma 3.6.

As a consequence of Lemma 3.6, we get the level-by-level adequacy of the Shoenfield terms:

[11] For small α, we have the problem that there may be sentences in $\boldsymbol{S}_\alpha^A(X)$ involving terms of rank greater than α.

LEMMA 3.7. Let G be $\mathcal{H}$-generic over Vop_ω; then for all ξ such that $\lambda_0 \leq \xi \leq \gamma_0$, $\mathbf{J}_\xi^{\vec{F}}(T^{\mathcal{M}})[G] = \mathbf{J}_\xi^{\vec{F}}(\langle T^{\mathcal{M}}, G\rangle)$.

If G is a Vopěnka-generic over $\mathcal{H}$ such that $\dot{R}^G = \mathbb{R}^{\mathcal{M}}$, then $\mathcal{H}[G]$ can recover $\mathcal{M}$:

LEMMA 3.8. If G is Vop_ω-generic over $\mathcal{H}$, $\dot{R}^{\mathcal{M}} = \mathbb{R}^{\mathcal{M}}$, and $\lambda_0 \leq \xi \leq \gamma_0$, then $\mathcal{M}|\xi$ is Δ_1-definable over $\mathcal{H}_\xi[G]$ from the parameter G; moreover, this definition is uniform in such G and ξ.

PROOF (SKETCH.) For $\xi = \lambda_0$ this is clear. In general, what we need to see is that if $F_\xi \neq \varnothing$, then F_ξ determines the corresponding extender E_ξ on the $\mathcal{M}$-sequence in a $\Delta_1^{\mathcal{H}_\xi[G]}$ way. We may assume by induction that $\mathbf{J}_\xi^{\vec{E}}(\mathbb{R}^{\mathcal{M}}) \subseteq \mathcal{H}_\xi[G]$. Since Vop_ω has cardinality strictly less than $\mathrm{crit}(F_\xi)$ in $\mathcal{H}_\xi$, F_ξ lifts to an extender F^* over $\mathbf{J}_\xi^{\vec{E}}(\mathbb{R}^{\mathcal{M}})$ defined by: for $a \in [\mathrm{lh}(F_\xi)]^{<\omega}$ and $Z \subseteq [\mathrm{crit}(F_\xi)]^{|a|}$ such that $Z \in \mathbf{J}_\xi^{\vec{E}}(\mathbb{R}^{\mathcal{M}})$,

$$Z \in F_a^* \Leftrightarrow \exists Y(Y \in (F_\xi)_a \wedge Y \subseteq Z).$$

Clearly, any extender over $\mathbf{J}_\xi^{\vec{E}}(\mathbb{R}^{\mathcal{M}})$ whose restriction to sets in $\mathcal{H}_\xi$ is F_ξ must then be equal to F^*. Thus $E_\xi = F^*$, and hence E_ξ is Δ_1 over $\mathcal{H}_\xi[G]$ in the parameter G. The uniformity in ξ and G is obvious upon inspection of the definition we have given. (The uniformity is needed to pass through limit stages.) ⊣

THEOREM 3.9. $\mathcal{H}$ is a $T^{\mathcal{M}}$-premouse; moreover for all $k \leq n_0$, $\mathcal{H}$ is k-sound, $\rho_k(\mathcal{H}) = \rho_k(\mathcal{M})$, and $p_k(\mathcal{H}) = p_k(\mathcal{M}) \setminus \{\theta\}$.

PROOF OF THEOREM 3.9. We show by induction on ξ such that $\lambda_0 \leq \xi \leq \gamma_0$, that $\mathcal{H}_\xi$ is a $T^{\mathcal{M}}$-premouse, and if $0 \leq k \leq \omega$, and $k \leq n_0$ if $\xi = \gamma_0$, then $\mathcal{H}_\xi$ is k-sound, $\rho_k(\mathcal{H}_\xi) = \rho_k(\mathcal{M}|\xi)$, and $p_k(\mathcal{H}_\xi) = p_k(\mathcal{M}|\xi)$.

This is clear for $\xi = \lambda_0$. Now let $\xi > \lambda_0$. We first show that $\mathcal{H}_\xi$ is a premouse. Since all proper initial segments of $\mathcal{H}_\xi$ are ω-sound $T^{\mathcal{M}}$-premice by our induction hypotheses, it suffices to show that $\mathcal{H}_\xi$ is a $T^{\mathcal{M}}$-ppm. If $F_\xi = \varnothing$, this is trivial, so assume $F_\xi = E_\xi \cap \mathcal{H}_\xi$ where E_ξ is an extender over $\mathcal{M}|\xi$. We must verify that $\vec{F}$ has the properties of a fine extender sequence at ξ, that is, that it satisfies clauses (1)-(3) of [Ste07B, Definition 2.4]. Let us write $F = F_\xi$ and $E = E_\xi$. Notice that $\xi = \mathrm{lh}(F) = \mathrm{lh}(E) = o(\mathcal{M}|\xi) = o(\mathcal{H}_\xi)$. Set $\kappa = \mathrm{crit}(F) = \mathrm{crit}(E)$.

CLAIM 3.10. If $a \in [\xi]^{<\omega}$ and $f \in \mathcal{M}|\xi$ and $f : [\kappa]^{|a|} \to \mathcal{H}_\xi$, then there is a $Z \in F_a$ such that $f \restriction Z \in \mathcal{H}_\xi$.

PROOF. We need to take a little care with the standard argument because $F_a \notin \mathcal{M}$ is possible. Note that by Proposition 2.4, f is definable over some $\mathcal{M}|\gamma$, where $\gamma < \xi$, from ordinals and a real x_0. We can therefore fix a term $\dot{f}$

in $\mathcal{H}_\xi$ such that whenever h is $\mathcal{M}$-generic over $\mathrm{Col}(\omega, \mathbb{R})$, and $G = G_h$ is the associated Vopěnka generic, then $\dot{f}^G = f$. We claim there is an $A \in \mathrm{Vop}_\omega$ such that

$$\text{for } F_a\text{-almost every } u, \exists\eta(A \Vdash \dot{f}(u) = \eta),$$

and

$$\exists s \in A^*(s(0) = x_0).$$

If not, then for each $A \in \mathrm{Vop}_\omega$ such that $\exists s \in A^*(s(0) = x_0)$, the set Z_A of all $u \in [\kappa]^{|a|}$ such that A forces no value for $\dot{f}(u)$ is in F_a. The local definability of Vopěnka forcing implies that the function $A \mapsto Z_A$ is in $\mathcal{M}$ (in fact, in $\mathcal{H}_\xi$). But F is $(\mathcal{M}, \mathbb{R})$-complete, and hence we have a u such that $u \in Z_A$ whenever Z_A is defined. Now let h be $\mathcal{M}$-generic over $\mathrm{Col}(\omega, \mathbb{R})$ with $h(0) = x_0$. Then $\exists A \in G_h \exists\eta(A \Vdash \dot{f}(u) = \eta)$, so Z_A is defined and $u \notin Z_A$, a contradiction.

Now let A be as in our claim, and let Z be the set of all u such that $\exists\eta A \Vdash \dot{f}(u) = \eta$. It is clear that $f \upharpoonright Z$ can be computed inside $\mathcal{H}_\xi$ from $A, \dot{f}$, and the forcing relation. (Note that there is an $\mathcal{M}$-generic h such that $h(0) = x_0$ and $A \in G_h$.) ⊣ (Claim 3.10)

Claim 3.11. The extenders F and E have the same generators.

Proof. If $\eta < \mathrm{lh}(F)$ is not a generator of F, then there is an $f \in \mathcal{H}_\xi$ and finite $a \subseteq \eta$ such that $f(u) = v$ for $(F)_{a\cup\{\eta\}}$-almost every $u \cup \{v\}$. Since $f \in \mathcal{M}|\xi$ and $F \subseteq E$, this means that η is not a generator of E. Conversely, if η is not a generator of E, as witnessed by $f \in \mathcal{M}_\xi$ and $a \subseteq \eta$ finite, then we can apply Claim 3.10 to see that η is not a generator of F. ⊣ (Claim 3.11)

Claim 3.12. For all $\theta \leq \eta < \xi$, η is a cardinal of $\mathcal{H}_\xi$ iff η is a cardinal of $\mathcal{M}|\xi$.

Proof. If η is a cardinal of $\mathcal{M}|\xi$, then η is a cardinal of the smaller model $\mathcal{H}_\xi$. If $\eta > \theta$ is a cardinal of $\mathcal{H}_\xi$ and G is Vop_ω-generic over $\mathcal{H}_\xi$, then η is a cardinal of $\mathcal{H}_\xi[G]$. Choosing G so that $\dot{R}^G = \mathbb{R}^{\mathcal{M}}$, we see that η is a cardinal of $\mathcal{M}|\xi$. For $\eta = \theta$, we have that η is a cardinal of both models. ⊣ (Claim 3.12)

We can now verify the first clause in the definition of fine extender sequences, that $\xi = \nu(F)^+$ in $\mathrm{ult}(\mathcal{H}_\xi, F)$. We have $\nu(F) = \nu(E)$ by Claim 3.11, and $\xi = \nu(E)^+$ in $\mathrm{ult}(\mathcal{M}|\xi, E)$ because $\vec{E}$ is a fine extender sequence. Letting $i_E\colon \mathcal{M}|\xi \to \mathrm{ult}(\mathcal{M}|\xi, E)$ be the canonical embedding, we have $\mathrm{ult}(\mathcal{H}_\xi, F) = i_E(\mathcal{H}_\xi)$ by our first claim.[12] By Claim 3.12 and the elementarity of i_E, $\mathrm{ult}(\mathcal{M}|\xi, E)$ has the same cardinals as $i_E(\mathcal{H}_\xi)$, so $\xi = \nu(E)^+$ in $i_E(\mathcal{H}_\xi)$. Thus $\xi = \nu(F)^+$ in $\mathrm{ult}(\mathcal{H}_\xi, F)$, as desired.

[12]Where $i_E(\mathcal{H}_\xi)$ is the "union" of the $i_E(\mathcal{H}_\eta)$ for $\eta < \xi$.

To verify clause 2, coherence, notice that $i_E(\vec{F})\restriction\xi = F\restriction\xi$ by coherence for the $\vec{E}$ sequence and the fact that the F_α are uniformly locally definable from the E_α. Since by Claim 3.10, $i_F(\vec{F}) = i_E(\vec{F})$, we are done.

The initial segment condition for E easily implies the initial segment condition for F; we leave the details to the reader. We have therefore shown that $\mathcal{H}_\xi$ is a $T^{\mathcal{M}}$-premouse.

We now show by induction on k such that $k \leq n_0$ if $\xi = \gamma_0$ that $\rho_k(\mathcal{H}_\xi) = \rho_k(\mathcal{M}|\xi)$, $p_k(\mathcal{H}_\xi) = p_k(\mathcal{M}|\xi) \setminus \{\theta\}$, and $\mathcal{H}_\xi$ is k-sound. This is trivial if $k = 0$. Let us first consider the case $k = 1$.

The key is that the Σ_1 theories (in the language $\mathcal{L}^*$) of $\mathcal{H}_\xi$ and $\mathcal{M}|\xi$ are intertranslatable. First, let us translate the Σ_1 theory of $\mathcal{H}_\xi$ into that of $\mathcal{M}|\xi$. Here we shall expand the latter theory by allowing a name for θ; notice that $T^{\mathcal{M}}$ is Σ_1-definable over $\mathcal{M}|\xi$ from the parameter θ. We can then see that the universe of $\mathcal{H}_\xi$, together with the interpretations of $\dot{E}$ and $\dot{F}$ in $\mathcal{H}_\xi$, are Δ_1-definable over $\mathcal{M}|\xi$ from θ. Clearly $\dot{\mu}^{\mathcal{H}_\xi} = \dot{\mu}^{\mathcal{M}|\xi} = \kappa$, and $\dot{\nu}^{\mathcal{H}_\xi} = \dot{\nu}^{\mathcal{M}|\xi}$ by Claim 3.11 above. We leave it to the reader to show that $\dot{\gamma}^{\mathcal{H}_\xi}$ is Σ_1-definable over $\mathcal{M}|\xi$ from θ; this is a bit of a mess because of the "one ultrapower away" case in the initial segment condition, but otherwise routine.[13] These calculations constitute a proof of:

Claim 3.13. There is a recursive map $\varphi(v_1,\dots,v_n) \mapsto \varphi^*(v_0, v_1,\dots,v_n)$ associating to each Σ_1 formula of $\mathcal{L}^*$ a Σ_1 formula of $\mathcal{L}^*$ with one additional free variable, such that for all $\varphi(v_1,\dots,v_n)$ and $a_1,\dots,a_n$,

$$\mathcal{H}_\xi \models \varphi[a_1,\dots,a_n] \Leftrightarrow \mathcal{M}|\xi \models \varphi^*[\theta, a_1,\dots,a_n].$$

We translate in the other direction using the strong forcing relation for Σ_1 formulae and Lemma 3.6. Let $\mathcal{L}^{**}$ be the sublanguage of $\mathcal{L}^*$ with symbols $\in, =, \dot{E}, \dot{F}$. If

$$\varphi(v_1,\dots,v_n) = \exists u_1 \dots \exists u_k \psi(u_1,..,u_k, v_1,\dots,v_n)$$

where ψ is a Σ_0 formula of $\mathcal{L}^{**}$, then for $p \in \mathrm{Vop}_\omega$ and $\tau_1,\dots,\tau_n$ Shoenfield terms, we put

$$p \Vdash_s \varphi(\tau_1,\dots,\tau_n) \Leftrightarrow \exists r_1 \dots \exists r_k (p \Vdash \psi(r_1,\dots,r_k,\tau_1,\dots,\tau_n)).$$

Now if G is generic over $\mathcal{H}_\xi$ for Vop_ω and $\dot{R}^G = \mathbb{R}^{\mathcal{M}}$, then the universe of $\mathcal{M}|\xi$ is Δ_1 definable over $\mathcal{H}_\xi[G]$ from $\dot{R}^G$; moreover, the interpretations in $\mathcal{M}|\xi$ of the symbols of $\mathcal{L}^*$ are Δ_1 definable over $\mathcal{H}_\xi[G]$ from their interpretations in $\mathcal{H}_\xi$.[14] Since strong forcing equals truth, we get

---

[13]The extenders E and F fall under the same case in the initial segment condition, and $\dot{\gamma}^{\mathcal{H}_\xi} = \dot{\gamma}^{\mathcal{M}|\xi}$ unless E and F are type II, and their last initial segments are an ultrapower away from the corresponding sequence.

[14]Our previous comments regarding $\dot{\gamma}$ apply here too.

CLAIM 3.14. There is a recursive map $\varphi(v_1,\ldots,v_n) \mapsto \varphi^\dagger(w,x,y,z,v_1,\ldots,v_n)$ associating to each Σ_1 formula of $\mathcal{L}^*$ a Σ_1 formula of $\mathcal{L}^{**}$ with four additional free variables, such that whenever G is generic over $\mathcal{H}_\xi$ for Vop_ω and $\dot{R}^G = \mathbb{R}^{\mathcal{M}}$, and $\tau_1,\ldots,\tau_n$ are Shoenfield terms, then

$$\mathcal{M}|\xi \models \varphi[\tau_1^G,\ldots,\tau_n^G]$$

if and only if

$$\exists p \exists w,x,y(p\in G \wedge \langle x,y,z\rangle = \langle \dot{\gamma}^{\mathcal{H}_\xi}, \dot{\mu}^{\mathcal{H}_\xi}, \dot{\nu}^{\mathcal{H}_\xi}\rangle \wedge p \Vdash_s \varphi^\dagger(\dot{R},\check{x},\check{y},\check{z},\tau_1,\ldots,\tau_n)).$$

These translations give

CLAIM 3.15. $\rho_1(\mathcal{H}_\xi) = \rho_1(\mathcal{M}|\xi)$.

PROOF. We first show $\rho_1(\mathcal{H}_\xi) \geq \rho_1(\mathcal{M}|\xi)$. This follows at once from

SUBCLAIM 3.15A. Let $S \subseteq \mathcal{H}_\xi$ be $\Sigma_1^{\mathcal{M}|\xi}$-definable from parameters in $\mathcal{H}_\xi$, and suppose $S \in \mathcal{M}|\xi$; then $S \in \mathcal{H}_\xi$.

PROOF OF SUBCLAIM 3.15A. We may as well assume S is a set of ordinals; say $S \subseteq \rho < \omega\xi$. By Proposition 2.4 we can fix a real x_0 and an ordinal δ such that $S = h_\xi^{\vec{E},\mathbb{R}^{\mathcal{M}}}(\delta, x_0)$. Now for y a real, let

$$y \in \mathrm{Ord} \leftrightarrow \exists \eta \in S \exists Z(Z = h_\xi^{\vec{E},\mathbb{R}^{\mathcal{M}}}(\delta,y) \wedge \eta \notin Z).$$

Since S is $\Sigma_1^{\mathcal{M}|\xi}$ in parameters from $\mathcal{H}_\xi$, O is $\Sigma_1^{\mathcal{M}|\xi}$ in ordinal parameters, so by Lemma 3.2, O is ordinal definable over $\mathcal{M}|\theta$. Hence there is a condition $r \in \mathrm{Vop}_1$ such that $r^* = \mathbb{R}^{\mathcal{M}} \setminus O$. Notice that $x_0 \in r^*$.

Let φ_h be a Σ_1 formula defining $h_\xi^{\vec{E},\mathbb{R}^{\mathcal{M}}}$ over $\mathcal{M}|\xi$. Let $\alpha < \omega\xi$ be large enough that

$$(S_\alpha^{\vec{E}}(\mathbb{R}^{\mathcal{M}}), \in, \vec{E} \restriction \alpha^*) \models \varphi_h[\delta, x_0, S],$$

where α^* is largest such that $\omega\alpha^* \leq \alpha$. By the proof of Lemma 3.8, we can fix a term $\tau \in \mathcal{H}_\xi$ such that whenever G is Vop_ω-generic over $\mathcal{H}_\xi$ and $\dot{R}^G = \mathbb{R}^{\mathcal{M}}$, then

$$\tau^G = (S_\alpha^{\vec{E}}(\mathbb{R}^{\mathcal{M}}), \in, \vec{E} \restriction \alpha^*).$$

SUBCLAIM 3.15B. For any $\eta < \rho$

$$\eta \notin S \Leftrightarrow \exists p \leq^v r(p \Vdash [\tau \models (\varphi_h(\check{\delta}, \sigma_0, Z) \wedge \check{\eta} \notin Z)]).$$

PROOF OF SUBCLAIM 3.15B. Assume $\eta \notin S$. Let $f : \omega \twoheadrightarrow \mathbb{R}^{\mathcal{M}}$ be $\mathcal{M}$-generic over $\mathrm{Col}(\omega, \mathbb{R}^{\mathcal{M}})$, with $f(0) = x_0$. It follows that $r \in G_f$, $\tau^{G_f} = (S_\alpha^{\vec{E}}(\mathbb{R}^{\mathcal{M}}), \in, \vec{E} \restriction \alpha^*)$, and $\sigma_0^{G_f} = x_0$. But then $\tau^{G_f} \models \exists Z(\varphi_h[\delta, \sigma_0^{G_f}, Z] \wedge \eta \notin Z)$, since S is in fact the unique such Z. Hence we have some $p \in G_f$ forcing this fact, and we may as well take $p \leq^v r$, so that p witnesses the right hand side of our equivalence.

Conversely, let p be as on the right hand side of our equivalence. By Lemma 3.5, we can find and $f : \omega \twoheadrightarrow \mathbb{R}^{\mathcal{M}}$ which is $\mathcal{M}$-generic over $\mathrm{Col}(\omega, \mathbb{R}^{\mathcal{M}})$ such that $p \in G_f$. Then $\tau^{G_f} = (S_\alpha^{\vec{E}}(\mathbb{R}^{\mathcal{M}}))$, and what's forced by p is true in the generic extension, so $(S_\alpha^{\vec{E}}(\mathbb{R}^{\mathcal{M}})) \models \exists Z(\varphi_h[\delta, \sigma_0^{G_f}, Z] \wedge \eta \notin Z)$. If $\eta \in S$, this implies $\sigma_0^{G_f} \in O$ by the definition of O; however, $\sigma_0^{G_f} = f(0) \in r^* = \neg O$ because $r \in G_f$. Thus $\eta \notin S$, as desired. ⊣ (Subclaim 3.15B)

From Subclaim 3.15B and the definability of the forcing relation for Σ_0 sentences given by Lemma 3.6, we get that that $S \in \mathcal{H}_\xi$. ⊣ (Subclaim 3.15A)

Now let $S \subseteq \rho_1(\mathcal{H}_\xi)$ be boldface $\boldsymbol{\Sigma}_1^{\mathcal{H}_\xi}$ but not in $\mathcal{H}_\xi$. By Claim 3.13 and Subclaim 3.15A, $S \notin \mathcal{M}|\xi$, and this implies $\rho_1(\mathcal{M}|\xi) \leq \rho_1(\mathcal{H}_\xi)$.

We now show $\rho_1(\mathcal{H}_\xi) \leq \rho_1(\mathcal{M}|\xi)$. Let $S \subseteq (\mathbb{R}^{\mathcal{M}} \times \rho)$ be boldface $\boldsymbol{\Sigma}_1^{\mathcal{M}|\xi}$ but not a member of $\mathcal{M}|\xi$, where $\rho = \rho_1(\mathcal{M}|\xi)$. Let

$$\langle x, \eta \rangle \in S \Leftrightarrow \mathcal{M}|\xi \models \varphi[x, \eta, y, \beta],$$

where $y \in \mathbb{R}^{\mathcal{M}}$ and $\beta < \omega\xi$ are fixed parameters. Now consider the strong forcing relation:

$$\langle p, \eta \rangle \in F \Leftrightarrow (p \in \mathrm{Vop}_\omega \wedge p \Vdash_s \varphi(\sigma_0, \check{\eta}, \sigma_1, \check{\beta})).$$

F is $\Sigma_1^{\mathcal{H}_\xi}$ by 3.6, and a subset of $\theta \times \rho$. Since $\theta \leq \rho$, we will be done if we show $F \notin \mathcal{H}_\xi$. In fact, $F \notin \mathcal{M}|\xi$, for

$$\langle x, \eta \rangle \in S \Leftrightarrow \exists p[\langle p, \eta \rangle \in F \wedge (\exists s \in p^*(s(0) = x \wedge s(1) = y)],$$

so that if $F \in \mathcal{M}|\xi$, then $S \in \mathcal{M}|\xi$. Both directions of the equivalence displayed are proved by considering Vopěnka generics of the form G_f, where f is $\mathrm{Col}(\omega, \mathbb{R}^{\mathcal{M}})$ generic over $\mathcal{M}|\xi$ and $f(0) = x$ and $f(1) = y$. We leave the rest to the reader. ⊣ (Claim 3.15)

Claim 3.16. $\rho_1(\mathcal{H}_\xi) = \rho_1(\mathcal{M}|\xi) \setminus \{\theta\}$.

Proof. The proof of Claim 3.15 actually shows that for any finite $F \subseteq \omega\xi$ and any α such that $\theta < \alpha$,

$$\mathrm{Th}_1^{\mathcal{H}_\xi}(\alpha \cup F) \in \mathcal{H}_\xi \Leftrightarrow \mathrm{Th}_1^{\mathcal{M}|\xi}(\mathbb{R}^{\mathcal{M}} \cup \alpha \cup F) \in \mathcal{M}|\xi,$$

where $\mathrm{Th}_1^{\mathcal{P}}(X)$ denotes the Σ_1-theory in $\mathcal{P}$ of parameters in X. Letting

$$p_1(\mathcal{H}_\xi) = \langle \alpha_0, \ldots, \alpha_n \rangle,$$

we have by the solidity of $p_1(\mathcal{H}_\xi)$ that for $i \leq n$,

$$\alpha_i = \text{least } \beta \text{ such that } \mathrm{Th}_1^{\mathcal{H}_\xi}(\beta \cup \{\alpha_0, \ldots, \alpha_{i-1}\}) \notin \mathcal{H}_\xi.$$

Using the equivalence displayed above and the solidity of $p_1(\mathcal{M}|\xi)$, we then get by induction on i that α_i is the ith member of $p_1(\mathcal{M}|\xi)$. Thus $p_1(\mathcal{H}_\xi) \subseteq p_1(\mathcal{M}|\xi) \setminus \{\theta\}$. (Note that $\theta \notin p_1(\mathcal{H}_\xi)$, since θ is easily definable over $\mathcal{H}_\xi$.) A similar argument shows $p_1(\mathcal{M}_\xi) \setminus \{\theta\} \subseteq p_1(\mathcal{H}_\xi)$. ⊣ (Claim 3.16)

CLAIM 3.17. $\mathcal{H}_\xi$ is 1-sound.

PROOF. Let $\eta < \omega\xi$; we must show η is $\Sigma_1^{\mathcal{H}_\xi}$-definable, as a point, from parameters in $\rho \cup p$, where $\rho = \rho_1(\mathcal{H}_\xi)$ and $p = p_1(\mathcal{H}_\xi)$. Since $\mathcal{M}|\xi$ is 1-sound, we can find a finite $F \subseteq \rho \cup p$ and a real x and a Σ_1 formula $\varphi(t, u, v, w)$ such that

$$\eta = \text{ unique } \beta \text{ such that } \mathcal{M}|\xi \models \varphi[x, F, \theta, \beta].$$

We may assume that φ has been "uniformised", so that over any premouse it defines the graph of a partial function of its first three variables. Now, letting f be $\text{Col}(\omega, \mathbb{R}^{\mathcal{M}})$-generic over $\mathcal{M}|\xi$ with $f(0) = x$, we can find a $A \in G_f$ such that

$$A \Vdash_s \varphi(\sigma_0, \check{F}, \check{\theta}, \check{\eta}).$$

Since φ has been uniformised, this gives us a $\Sigma_1^{\mathcal{H}_\xi}$ definition of η from $A, \sigma_0, F,$ and θ. But $A < \theta$, and $\sigma_0 \subseteq \theta$ is coded into $T^{\mathcal{M}}$ in a simple way. Thus η is $\Sigma_1^{\mathcal{H}_\xi}$ definable from F, as desired. ⊣ (Claim 3.17)

This finishes the $k = 1$ case in our induction on $\langle \xi, k \rangle$. The case $k > 1$ can be handled quite similarly, using the master code structures. For example, let $\mathcal{P} = (\mathbf{J}_\rho^{\vec{F}}(T^{\mathcal{M}}), \in, \vec{F}\restriction\rho, A)$ and $\mathcal{Q} = (\mathbf{J}_\rho^{\vec{E}}(\mathbb{R}^{\mathcal{M}}), \in, \vec{E}\restriction\rho, B)$ be the first master code structures of $\mathcal{H}_\xi$ and $\mathcal{M}|\xi$ respectively. The arguments above show that $\alpha \mapsto A \cap \alpha$ is $\Delta_1^{\mathcal{Q}}$ and total on $\mathcal{Q}$, and this can be used to show as above that $\rho_1(\mathcal{Q}) \leq \rho_1(\mathcal{P})$, that is, $\rho_2(\mathcal{M}|\xi) \leq \rho_2(\mathcal{H}_\xi)$. In the other direction, one can show that $\mathcal{Q}$ is $\Delta_1^{\mathcal{P}[G]}$, uniformly in all Vopěnka-generic G such that $\dot{R}^G = \mathbb{R}^{\mathcal{M}}$ (as in Lemma 3.8), and using the definability of forcing over $\mathcal{P}$ given by Lemma 3.6, this implies $\rho_1(\mathcal{P}) \leq \rho_1(\mathcal{Q})$, that is $\rho_2(\mathcal{H}_\xi) \leq \rho_2(\mathcal{M}|\xi)$. We leave the remaining details to the reader. ⊣ (Theorem 3.9)

If $\mathcal{M}$ is a model of ZFC minus the powerset axiom, and $\mathcal{H}$ is the $T^{\mathcal{M}}$-premouse we have defined above, then it is easy to see that the universe of $\mathcal{H}$ is just $\mathbf{HOD}^{\mathcal{M}}$. Indeed, $\mathcal{H} \subseteq \mathbf{HOD}^{\mathcal{M}}$ is clear, and $\mathbf{HOD}^{\mathcal{M}} \subseteq \mathcal{H}$ follows at once from Subclaim 3.15A. In general, for arbitrary $\mathcal{M}$ satisfying the assumptions behind Theorem 3.9,

$$x \in \mathcal{H} \Leftrightarrow \exists \alpha (\omega\alpha < o(\mathcal{M}) \wedge \forall y \in \text{tcl}(x) \cup \{x\}(y \in \mathbf{OD}^{\mathcal{M}|\alpha}).$$

It is therefore tempting to write $\mathcal{H} = \mathbf{HOD}^{\mathcal{M}}$ in general, as this would be a reasonable general meaning for $\mathbf{HOD}^{\mathcal{M}}$. However, we shall stick to

$$\mathcal{H} = \mathcal{H}(\mathcal{M}) = \mathcal{H}^{\mathcal{M}}.$$

Finally, we come to the main reason we have isolated $\mathcal{H}$.

THEOREM 3.18. Let $\mathcal{M}$ be n_0-sound $\mathbb{R}^{\mathcal{M}}$-premouse, and satisfy "Θ exists". Suppose $\rho_{n_0}(\mathcal{M}) \geq \Theta^{\mathcal{M}}$, and $\mathcal{M}|\lambda \models \text{ZF}$, for some $\lambda \geq \Theta^{\mathcal{M}}$. Finally, suppose

$\mathcal{M}|\xi$ is countably k-iterable, for all $\langle \xi, k\rangle <_{\text{lex}} \langle \gamma_0, n_0\rangle$ such that $\Theta^{\mathcal{M}} \leq \xi$, where γ_0 is such that $\omega\gamma_0 = o(\mathcal{M})$. Then, letting $\mathcal{H} = \mathcal{H}(\mathcal{M})$, we have for all τ

$$\mathcal{H} \text{ is } (n_0, \tau)\text{-iterable } \Rightarrow \mathcal{M} \text{ is } (n_0, \tau)\text{-iterable above } \Theta^{\mathcal{M}}.$$

PROOF (SKETCH). Let Σ be an (n_0, τ)-iteration strategy for $\mathcal{H}$. We define an (n_0, τ)-iteration strategy Γ for $\mathcal{M}$ which operates on trees all of whose extenders have critical points above $\Theta^{\mathcal{M}}$. Given such a tree $\mathcal{T}$ played according to Γ, the construction insures that there is a tree $\mathcal{T}^*$ on $\mathcal{H}$ which is according to Σ with the same tree order, drop, and degree structure as $\mathcal{T}$, and such that

- $\mathcal{T}^*$ is according to Σ,
- $\mathcal{M}_\alpha^{\mathcal{T}^*} = \mathcal{H}(\mathcal{M}_\alpha^{\mathcal{T}})$, for all $\alpha < \text{lh}(\mathcal{T})$, and
- $E_\alpha^{\mathcal{T}^*} = E_\alpha^{\mathcal{T}} \cap \mathcal{M}_\alpha^{\mathcal{T}^*}$.

Because $\mathcal{M}_\alpha^{\mathcal{T}}$ is an inner model of a generic extension of $\mathcal{M}_\alpha^{\mathcal{T}^*}$ via a poset (i.e., $\text{Vop}_\omega^{\mathcal{H}}$) of size $\Theta^{\mathcal{M}}$, these conditions imply that

- $i_{\alpha,\beta}^{\mathcal{T}^*} = i_{\alpha,\beta}^{\mathcal{T}} \restriction \mathcal{M}_\alpha^{\mathcal{T}^*}$.

Given that $\mathcal{T}$ has limit length $< \tau$, and we have a $\mathcal{T}^*$ as above, we simply define $\Gamma(\mathcal{T}) = \Sigma(\mathcal{T}^*)$. Setting $b = \Gamma(\mathcal{T})$, it is routine to verify that the b-extensions of $\mathcal{T}$ and $\mathcal{T}^*$ satisfy the conditions above. The main point is that for $\alpha \in b$ sufficiently large,

$$\mathcal{M}_b^{\mathcal{T}^*} = i_{\alpha,b}^{\mathcal{T}^*}(\mathcal{M}_\alpha^{\mathcal{T}^*}) = i_{\alpha,b}^{\mathcal{T}}(\mathcal{H}(\mathcal{M}_\alpha^{\mathcal{T}})) = \mathcal{H}(\mathcal{M}_b^{\mathcal{T}}),$$

where we have applied the embeddings to classes of their domain models in the usual way.

It is also clear that the existence of $\mathcal{T}^*$ as above propagates through successor steps in the construction of $\mathcal{T}$. This completes our sketch. ⊣

§4. The scale property in $\mathbf{K}(\mathbb{R})$. Using the local **HOD**'s of the last section to verify iterability, in the same way that the ordinals were used to verify wellfoundedness in [Ste83A], we shall construct closed game representations of minimal complexity for sets in $\mathbf{K}(\mathbb{R})$. As explained in [Ste83A], an argument due to Moschovakis [Mos83] converts these closed game representations to scales of minimal complexity. Modulo the use of the local **HOD**'s to verify iterability, everything goes pretty much as it did in [Ste83A]. We shall therefore keep our notation as close as possible to that of [Ste83A], and omit many of the details treated more carefully there.

4.1. Scales on $\Sigma_1^{\mathcal{M}}$ sets, for $\mathcal{M}$ passive. In this subsection we prove Theorem 1.1. In fact, we prove the slightly stronger

THEOREM 4.1. Let $\mathcal{M}$ be a passive, countably iterable premouse over $\mathbb{R}^{\mathcal{M}}$, and suppose $\mathcal{M} \models \mathsf{AD}$; then

$$\mathcal{M} \models \text{“the pointclass } \Sigma_1^{\mathcal{M}} \text{ has the scale property.”}$$

It should be clear what it means for $\mathcal{M}$ to believe that a pointclass of $\mathcal{M}$-definable sets of reals has the scale property: the norms of the putative scale, which are $\mathcal{M}$-definable, must have the limit and lower semi-continuity properties of a scale with respect to all sequences of reals $\langle x_i : i < \omega \rangle \in \mathcal{M}$. If $\mathbb{R}^{\mathcal{M}} = \mathbb{R}$ and $\mathcal{M}$ believes that the pointclass $\Sigma_1^{\mathcal{M}}$ has the scale property, then indeed $\Sigma_1^{\mathcal{M}}$ does have the scale property, as every ω-sequence of reals is in $\mathcal{M}$. Thus Theorem 4.1 implies Theorem 1.1.

PROOF. Let us fix a passive, countably 0-iterable $\mathbb{R}^{\mathcal{M}}$-premouse $\mathcal{M}$ such that $\mathcal{M} \models \mathsf{AD}$. We want to show that $\mathcal{M}$ satisfies a certain sentence, so by taking a Skolem hull we may assume that $\mathcal{M}$ is countable.

For $x \in \mathbb{R}^{\mathcal{M}}$, let

$$P(x) \Leftrightarrow \mathcal{M} \models \varphi_0[x],$$

where φ_0 is a Σ_1 formula of $\mathcal{L}^*$. Since $\mathcal{M}$ is passive, we may (and do) assume that φ_0 does not contain $\dot{F}$, $\dot{\mu}$, $\dot{\nu}$, or $\dot{\gamma}$; that is, it contains only $\in$ and $\dot{E}$. We want to show that $\mathcal{M}$ believes that there is $\Sigma_1^{\mathcal{M}}$ scale on P.

Let us first assume that $o(\mathcal{M}) = \omega\alpha$, where α is a limit ordinal, and deal with the general case later. If $\mathcal{M}$ satisfies "Θ exists", then set $\alpha^* = \Theta^{\mathcal{M}}$, and otherwise set $\alpha^* = \alpha$. For $\beta < \alpha^*$ and $x \in \mathbb{R}^{\mathcal{M}}$, let

$$P^{\beta}(x) \Leftrightarrow \mathcal{M}|\beta \models \varphi_0[x],$$

so that $P = \bigcup_{\beta<\alpha*} P^{\beta}$. Here we use Lemma 3.2 to see that the union out to α^* suffices.[15]

For each $\beta < \alpha^*$, we will construct a closed game representation $x \mapsto G_x^{\beta}$ of P^{β}. Letting

$$P_k^{\beta}(x, u) \Leftrightarrow u \text{ is a position of length } k \text{ from which player I has a winning quasi-strategy in } G_x^{\beta},$$

we shall arrange that each $P_k^{\beta} \in \mathcal{M}$, and that the map $\langle \beta, k \rangle \mapsto P_k^{\beta}$ is $\Sigma_1^{\mathcal{M}}$. As explained in [Ste83A], Moschovakis' argument then gives that $\mathcal{M}$ believes there is a scale on P of the desired complexity.[16]

So let β and x be given; we want to define G_x^{β}. Our plan is to force player I to describe a model of $\mathbf{V} = \mathbf{K}(\mathbb{R}) + \varphi_0(x) + \forall\gamma(\mathcal{J}_{\gamma}^{\vec{E}}(\mathbb{R}) \not\models \varphi[x])$ which includes all the reals played by player II. Player I will verify that his model is wellfounded by embedding its ordinals into $\omega\beta$, and verify his model is iterable by embedding its local **HOD**'s into the local **HOD**'s of $\mathcal{M}|\beta$ corresponding to them under his embedding of the ordinals.

---

[15]We have restricted ourselves to $\beta < \alpha^*$ for a minor technical reason connected to the definability of "honesty".

[16]Moschovakis uses the "second periodicity" method to construct scales on the P_k^{β}. It is here that one needs $\mathcal{M} \models \mathsf{AD}$.

Player I describes his model in the language $\mathcal{L}$, which is $\mathcal{L}^*$ together with new constant symbols $\dot{x}_i$ for $i < \omega$. He uses $\dot{x}_i$ to denote the ith real played in the course of G_x^β. Let us fix recursive maps

$$m, n\colon \{\sigma : \sigma \text{ is an } \mathcal{L}\text{-formula}\} \to \{2n : 1 \leq n < \omega\}$$

which are one-one, have disjoint recursive ranges, and are such that whenever $\dot{x}_i$ occurs in σ, then $i < \min(m(\sigma), n(\sigma))$. These maps give stages sufficiently late in G_x^β for player I to decide certain statements about his model.

Let us call an ordinal ξ of an $\mathbb{R}^{\mathcal{P}}$-premouse $\mathcal{P}$ **relevant** iff $\mathcal{P}|\xi$ satisfies "Θ exists, and there is a $\lambda > \Theta$ such that $\mathbf{J}_\lambda^{\dot{E}}(\mathbb{R}) \models \mathsf{ZF}$." That is, the relevant ξ are just those for which, under the additional assumption that $\mathcal{P}$ is countably iterable, we have defined and proved the existence of $\mathcal{H}_\xi^{\mathcal{P}}$. Also, "$v$ is relevant" is the $\mathcal{L}$-formula which expresses that v is relevant viv-a-vis the universe as $\mathcal{P}$. Similarly for the $\mathcal{L}$-formula "$\mathcal{H}_v$ exists".

Player I's description must extend the following $\mathcal{L}$-theory T. The axioms of T include

(1) Extensionality plus $\mathbf{V} = \mathbf{K}(\mathbb{R})$

(2) $\forall v(v \text{ is relevant } \Rightarrow \mathcal{H}_v \text{ exists})$

$(3)_\varphi$ $\exists v\varphi(v) \Rightarrow \exists v(\varphi(v) \wedge \forall u \in v \neg\varphi(u))$

$(4)_i$ $\dot{x}_i \in \mathbb{R}$.

(5) $\varphi_0(\dot{x}_0) \wedge \forall\delta(\mathcal{J}_\delta^{\dot{E}}(\mathbb{R}) \not\models \varphi_0[\dot{x}_0])$

Finally, T has axioms which guarantee that in any model, the definable closure of the interpretations of the $\dot{x}_i$ constitute an elementary submodel. Recall from Proposition 2.4 the uniformly definable maps $h_\gamma\colon [\omega\gamma]^{<\omega} \twoheadrightarrow \mathcal{M}|\gamma$; let $\sigma_0(v_0, v_1, v_2)$ be a Σ_1 formula which for all γ defines the graph of h_γ over $\mathcal{M}|\gamma$. Now, for any $\mathcal{L}$-formula $\varphi(v)$ of one free variable, T has axioms

(6) $\forall v_0 \forall v_1 \forall y \forall z(\sigma_0(v_0, v_1, y) \wedge \sigma_0(v_0, v_1, z) \Rightarrow y = z)$

$(7)_\varphi$ $\exists v\varphi(v) \Rightarrow \exists v \exists F \in [\mathrm{Ord}]^{<\omega}(\varphi(v) \wedge \sigma_0(F, \dot{x}_{m(\varphi)}, v))$

$(8)_\varphi$ $\exists v(\varphi(v) \wedge v \in \mathbb{R}) \Rightarrow \varphi(\dot{x}_{n(\varphi)})$.

This completes the axioms of T.

A typical run of G_x^β has the form

I	i_0, x_0, η_0		i_1, x_2, η_1	$\dots$
II		x_1		x_3

where for all k, $i_k \in \{0, 1\}$, $x_k \in \mathbb{R}$, and $\eta_k < \omega\beta$. If $u = \langle (i_k, x_{2k}, \eta_k, x_{2k+1}) : k < n \rangle$ is a position of length n, then we set

$$T^*(u) = \{\sigma : \sigma \text{ is a sentence of } \mathcal{L} \wedge i_{n(\sigma)} = 0\},$$

and if p is a full run of G_x^β,

$$T^*(p) = \bigcup_{n<\omega} T^*(p \restriction n).$$

Now let $p = \langle (i_k, x_{2k}, \eta_k, x_{2k+1}) : k < \omega \rangle$ be a run of G_x^β; we say that p is winning for player I iff

(a) $x_0 = x$,

(b) $T^*(p)$ is a complete, consistent extension of T such that for all i, m, n, "$\dot{x}_i(n) = m$" $\in T^*(p)$ iff $x_i(n) = m$,

(c) if φ and ψ are $\mathcal{L}$-formulae of one free variable, and "$\iota v \varphi(v) \in \text{Ord} \wedge \iota v \psi(v) \in \text{Ord}$" $\in T^*(p)$, then "$\iota v \varphi(v) \leq \iota v \psi(v)$" $\in T^*(p)$ iff $\eta_{n(\varphi)} \leq \eta_{n(\psi)}$, and

(d) if ψ is an $\mathcal{L}$-formula of one free variable, and "$\iota v \psi(v) \in \text{Ord} \wedge \iota v \psi(v)$ is relevant" $\in T^*(p)$, then $\eta_{n(\psi)}$ is relevant vis-a-vis $\mathcal{M}$; moreover if $\sigma_1, \ldots, \sigma_n$ are $\mathcal{L}$ formulae of one free variable such that for all k, "$\iota v \sigma_k(v) < (\iota v \psi(v))$" $\in T^*(p)$, then for any $\mathcal{L}^*$ formula $\theta(v_1, \ldots, v_n)$,

$$\text{“}\mathcal{H}_{\iota v \psi(v)} \models \theta[\iota v \sigma_1(v), \ldots, \iota v \sigma_n(v)]\text{”} \in T^*(p)$$

if and only if

$$\mathcal{H}^{\mathcal{M}}_{\eta_{n(\psi)}} \models \theta[\eta_{n(\sigma_1)}, \ldots, \eta_{n(\sigma_n)}].$$

Clearly, G_x^β is a game on $\mathbb{R} \times \omega\beta$ whose payoff is continuously associated to x. It remains to show that the winning positions for player I in G_x^β are those in which he has been honest. More precisely, let us call a position $u = \langle (i_k, x_{2k}, \eta_k, x_{2k+1}) : k < n \rangle$ (β, x)-**honest** iff $\mathcal{M}|\beta \models \varphi_0[x]$, and letting $\gamma \leq \beta$ be least such that $\mathcal{M}|\gamma \models \varphi_0[x]$, we have

(i) $n > 0 \Rightarrow x_0 = x$,

(ii) if we let $I_u(\dot{x}_i) = x_i$ for $i < 2n$, then all axioms of $T^*(u) \cup T$ thereby interpreted in $(\mathcal{M}|\gamma, I_u)$ are true in this structure, and

(iii) if $\sigma_0, \ldots, \sigma_m$ enumerates those $\mathcal{L}$-formulae σ of one free variable such that $n(\sigma) < n$ and

$$(\mathcal{M}|\gamma, I_u) \models \iota v \sigma(v) \in \text{Ord},$$

and if $\delta_i < \omega\gamma$ is such that

$$(\mathcal{M}|\gamma, I_u) \models \iota v \sigma_i(v) = \delta_i,$$

then the map

$$\delta_i \mapsto \eta_{n(\sigma_i)}$$

is well-defined and extendible to an order preserving map $\pi \colon \omega\gamma \to \omega\beta$ with the additional property that whenever δ_i is relevant vis-a-vis $\mathcal{M}$, so that $\eta_{n(\sigma_i)}$ is as well, then $\pi \restriction \omega\delta_i$ is extendible to an elementary embedding from $\mathcal{H}^{\mathcal{M}}_{\delta_i}$ to $\mathcal{H}^{\mathcal{M}}_{\eta_{n(\sigma_i)}}$.[17]

It is not immediately clear that the set of (β, x)-honest positions even belongs to $\mathcal{M}$, because of condition (iii).

[17]Notice that this extension is determined by π, since every point in $\mathcal{H}_{\delta_i}$ is definable from an ordinal.

CLAIM 4.2. Letting $Q_k^\beta(x,u)$ iff u is a (β,x)-honest position of length k, we have that $Q_k^\beta \in \mathcal{M}$ for all β, and the map $(\beta,k) \mapsto Q_k^\beta$ is $\Sigma_1^{\mathcal{M}}$.

PROOF (SKETCH). It is enough to see that the truth of clause (iii) can be determined within $\mathcal{M}$. But note that (iii) is equivalent to the existence of a winning strategy for the closed player in a certain "embedding game" on $\omega\beta$. It is enough then to see that if the closed player wins the embedding game in $\mathbf{V}$, then he wins it in $\mathcal{M}$. (The converse is obvious.) So suppose the closed player wins the embedding game in $\mathbf{V}$. Let $A \in \mathcal{M}$ be a set of ordinals which codes up this game; since $\beta < \Theta^{\mathcal{M}}$, and $\mathcal{M} \models \mathsf{AD}$, we can find a model N of ZFC such that $A \in N$, and $(N, \in)$ is coded by a set of reals $B_N \in \mathcal{M}$. (E.g., let $N = L_\alpha[A]$, where α is the supremum of the order types of the $\Delta_n^{\mathcal{M}|\gamma}$ prewellorders of $\mathbb{R}^{\mathcal{M}}$, for an appropriate n and γ.) Since $N \models \mathsf{ZFC}$, the closed player wins the embedding game via a strategy $\Sigma \in N$. For $\gamma < \omega\beta$, let $f(\gamma) = \{z \in \mathbb{R}^{\mathcal{M}} : z \text{ codes } \gamma \text{ via } B_N\}$. We can arrange that $f \in \mathcal{M}$, and use f to show that $\Sigma \in \mathcal{M}$. ⊣ (Claim 4.2)

We now show that the winning positions are the honest ones.

CLAIM 4.3. For any position u in G_x^β, player I has a winning quasi-strategy starting from u iff u is (β,x)-honest; that is, $P_k^\beta(x,u) \Leftrightarrow Q_k^\beta(x,u)$ for all k.

PROOF. It is easy to see that player I can win from honest positions u by continuing to tell the truth, while continuing to play his η's according to some map π satisfying (iii) in the definition of honesty for u.

Conversely, let Σ be a winning quasi-strategy for player I in G_x^β from u. Since $\mathbb{R}^{\mathcal{M}}$ is countable, we can easily construct a complete run

$$p = \langle (i_k, x_{2k}, \eta_k, x_{2k+1}) : k < \omega \rangle$$

of G_x^β according to Σ such that

$$\mathbb{R}^{\mathcal{M}} = \{x_i : i < \omega\}.$$

Since $T^*(p)$ is consistent, it has a model $\mathcal{A}$, and by the axioms in groups (3), (6), (7), and (8), the $\mathcal{L}$-definable points of $\mathcal{A}$ constitute an elementary submodel $\mathcal{N} \prec \mathcal{A}$. Let

$$\pi(\iota v \psi(v)^{\mathcal{N}}) = \eta_{n(\psi)}$$

whenever "$\iota v \psi(v) \in \mathrm{Ord}$" $\in T^*(p)$. Then π witnesses that $\mathcal{N}$ is wellfounded, and so we may assume $\mathcal{N}$ is transitive. By axiom (1) of T, $\mathcal{N}$ is a premouse. Note that $\mathcal{N}$ is a premouse over $\{\dot{x}_i^{\mathcal{A}} : i < \omega\}$, and this set is $\mathbb{R}^{\mathcal{M}}$ since $\dot{x}_i^{\mathcal{A}} = x_i$ by (ii). The main thing we need to show is that $\mathcal{N}$ is an initial segment of $\mathcal{M}|\beta$. Since π guarantees $o(\mathcal{N}) \leq \omega\beta$, it suffices to show $\mathcal{N}$ is an initial segment of $\mathcal{M}$.

We show by induction on γ that if $\rho_\omega(\mathcal{N}|\gamma) = 1$, then $\mathcal{N}|\gamma = \mathcal{M}|\gamma$. This is clear if $\gamma = 1$.

Suppose first that $\mathcal{N}|\gamma \not\models \Theta$ exists . Note that then γ is not active in $\mathcal{N}$. If γ is a limit ordinal, then there are arbitrarily large $\xi < \gamma$ such that $\rho_\omega(\mathcal{N}|\xi) = 1$, and by induction $\mathcal{N}|\xi = \mathcal{M}|\xi$ for such ξ. This implies γ is not active in $\mathcal{M}$, and $\mathcal{N}|\gamma = \mathcal{M}|\gamma$, as desired. If $\gamma = \xi + 1$, then $\rho_\omega(\mathcal{N}|\xi) = 1$, so $\mathcal{N}|\xi = \mathcal{M}|\xi$. Since successor ordinals are not active, we get $\mathcal{N}|\gamma = \mathcal{M}|\gamma$, as desired. (Note that if $\omega\gamma = o(\mathcal{N})$, then it falls under this last case by axiom (5) of T, so we may assume $\omega\gamma < o(\mathcal{N})$ henceforth.)

Next, suppose $\mathcal{N}|\gamma \models \Theta$ exists, and let $\theta = \Theta^{\mathcal{N}|\gamma}$. Note $\mathcal{N}|\theta = \mathcal{M}|\theta$ by the argument of the last paragraph.

If there is no $\xi \in (\theta, \gamma)$ such that $\mathcal{N}|\xi \models \mathsf{ZF}$, then $\mathcal{N}|\gamma$ is just the constructible closure of $\mathcal{N}|\theta$ through γ steps. Also, no $\xi \in (\theta, \gamma]$ can be active in $\mathcal{M}$, and so $\mathcal{N}|\gamma = \mathcal{M}|\gamma$. So we may assume that there is a $\xi \in (\theta, \gamma)$ such that $\mathcal{N}|\xi \models \mathsf{ZF}$; that is, γ is relevant in $\mathcal{N}$. Let n_0 be the largest $n < \omega$ such that $\rho_n(\mathcal{N}|\gamma) \geq \theta$. By rule (iii), π determines an elementary embedding from $\mathcal{H}(\mathcal{N}|\xi)$ to $\mathcal{H}(\mathcal{M}|\pi(\xi))$, for each relevant ξ of $\mathcal{N}$. Using Theorem 3.18 and a simple induction, we can then see that $\mathcal{N}|\gamma$ is countably n_0-iterable. Now $\beta < \alpha*$, and hence we can find a $\xi \geq \beta$ such that $\rho_\omega(\mathcal{M}|\xi) = 1$. Applying the comparison theorem 2.12 to $\mathcal{N}|\gamma$ and $\mathcal{M}|\xi$, we get that $\mathcal{N}|\gamma = \mathcal{M}|\gamma$, as desired.

Thus $\mathcal{N}$ is an initial segment of $\mathcal{M}|\beta$. Clearly, $\mathcal{N} = \mathcal{M}|\gamma$, where γ is least such that $\mathcal{M}|\gamma \models \varphi_0[x]$. The theory $T^*(u)$ is true in $\mathcal{N} = \mathcal{M}|\gamma$. The remainder of (β, x) honesty for u is witnessed by π. ⊣ (Claim 4.3)

Claims 4.2 and 4.3 yield the desired scale, as we have explained. This completes the proof of Theorem 4.1 in the case that $o(\mathcal{M}) = \omega\alpha$ for α a limit. The case that α is a successor ordinal can be handled similarly, using Jensen's $\boldsymbol{S}$-hierarchy. See [Ste83A]. ⊣

4.2. Σ_1 gaps.

Definition 4.4. Let $\mathcal{M}$ and $\mathcal{N}$ be X-premice; then we write $\mathcal{M} \prec_1 \mathcal{N}$ iff $\mathcal{M}$ is an initial segment of $\mathcal{N}$, and whenever $\varphi(v_1, \ldots, v_n)$ is a Σ_1 formula of the language $\mathcal{L}^*$ in which $\dot{F}, \dot{\mu}, \dot{v}$, and $\dot{\gamma}$ do not occur, then for any $a_1, \ldots, a_n \in X \cup \{X\}$,

$$\mathcal{N} \models \varphi[a_1, \ldots, a_n] \Rightarrow \mathcal{M} \models \varphi[a_1, \ldots, a_n].$$

Notice here that such Σ_1 formulae go up from $\mathcal{M}$ to $\mathcal{N}$ simply because $\mathcal{M}$ is an initial segment of $\mathcal{N}$. This uses our restriction that the symbols of $\mathcal{L}^*$ which have to do with the last extender of a premouse do not occur in φ.

Definition 4.5. Let $\mathcal{M}$ be an X-premouse, and suppose $\omega\alpha \leq \omega\beta \leq o(\mathcal{M})$; then we call the interval $[\alpha, \beta]$ a **Σ_1-gap of** $\mathcal{M}$ iff

1. $\mathcal{M}|\alpha \prec_1 \mathcal{M}|\beta$,
2. $\forall\gamma < \alpha(\mathcal{M}|\gamma \not\prec_1 \mathcal{M}|\alpha)$, and
3. $\forall\gamma > \beta(\mathcal{M}|\beta \not\prec_1 \mathcal{M}|\gamma)$.

That is, a Σ_1-gap is a maximal interval of ordinals in which no new Σ_1 facts about members of $X \cup \{X\}$ and the extender sequence $\vec{E}$ are verified. If $[\alpha, \beta]$ is a Σ_1-gap, we say α begins the gap and β ends it. Notice that we allow $\alpha = \beta$. It is easy to see

Lemma 4.6. Let $o(\mathcal{M}) = \omega\alpha$; then the Σ_1-gaps of $\mathcal{M}$ partition $\alpha + 1$.

We shall use the Σ_1-gaps of $\mathbf{K}(\mathbb{R})$ to characterize the levels of the Levy hierarchy in the initial segment of $\mathbf{K}(\mathbb{R})$ satisfying AD which have the Scale Property. Until we get to the end-of-gap case, the proofs are quite easy, and completely parallel to those of [Ste83A], so we shall omit them.

If $\mathcal{M}$ is an $\mathbb{R}^{\mathcal{M}}$-premouse, then by **the pointclass** $\Sigma_n^{\mathcal{M}}$ we mean the collection of all $A \subseteq \mathbb{R}^{\mathcal{M}}$ such that A is Σ_n definable over $\mathfrak{C}_{n-1}(\mathcal{M})$ from arbitrary parameters in $\mathfrak{C}_{n-1}(\mathcal{M})$, using the language $\mathcal{L}^*$.

Theorem 4.7. Let $\mathcal{M}$ be a countably 0-iterable $\mathbb{R}^{\mathcal{M}}$-premouse, and suppose α begins a Σ_1 gap of $\mathcal{M}$, and that $\mathcal{M}|\alpha \models$ AD; then $\mathcal{M}$ believes that the pointclass $\Sigma_1^{\mathcal{M}|\alpha}$ has the Scale Property.

Theorem 4.8. Let $\mathcal{M}$ be a countably 0-iterable $\mathbb{R}^{\mathcal{M}}$-premouse, and suppose α begins a Σ_1-gap of $\mathcal{M}$, and that $\mathcal{M}|(\alpha+1) \models$ AD, and that $\mathcal{M}|\alpha$ is not an admissible structure. Then for all $n < \omega$,

(a) $\Sigma_{n+2}^{\mathcal{M}|\alpha} = \exists^{\mathbb{R}}(\Pi_{n+1}^{\mathcal{M}|\alpha})$
$\Pi_{n+2}^{\mathcal{M}|\alpha} = \forall^{\mathbb{R}}(\Sigma_{n+1}^{\mathcal{M}|\alpha})$,
and

(b) $\mathcal{M}$ believes that the pointclasses $\Sigma_{2n+1}^{\mathcal{M}|\alpha}$ and $\Pi_{2n+2}^{\mathcal{M}|\alpha}$ have the Scale Property.

As in $\mathbf{L}(\mathbb{R})$, our negative results on the Scale Property are localizations of the fact that the relation "x is not ordinal definable from y" has no ordinal definable uniformization.

Definition 4.9. If $\mathcal{M}$ is an $\mathbb{R}^{\mathcal{M}}$-premouse and $o(\mathcal{M}) = \omega\alpha$, then for $x, y \in \mathbb{R}^{\mathcal{M}}$, we put

$$C^{\mathcal{M}}(x, y) \Leftrightarrow \exists\gamma < \alpha(\{y\} \text{ is } \mathcal{M}|\gamma\text{-definable from parameters in } \{x\} \cup \omega\gamma).$$

We also set $\neg C^{\mathcal{M}} = (\mathbb{R}^{\mathcal{M}} \times \mathbb{R}^{\mathcal{M}}) \setminus C^{\mathcal{M}}$.

It is clear that $C^{\mathcal{M}}$ is $\Sigma_1^{\mathcal{M}}$, and indeed, it is so via a formula which does not refer to the last extender $\dot{F}^{\mathcal{M}}$.

Theorem 4.10 (Martin, [Mar83A]). Let $\mathcal{M}$ be a countably 0-iterable $\mathbb{R}^{\mathcal{M}}$-premouse, and suppose that α begins a Σ_1-gap of $\mathcal{M}$, that $\mathcal{M}|(\alpha+1) \models$ AD, and that the structure $\mathcal{M}|\alpha$ is admissible. Then

(a) there is a $\Pi_1^{\mathcal{M}|\alpha}$ relation on $\mathbb{R}^{\mathcal{M}}$, namely $\neg C^{\mathcal{M}|\alpha}$, which has no uniformization in $\mathcal{M}|(\alpha+1)$, and hence

(b) $\mathcal{M}$ believes that none of the pointclasses $\Sigma_n^{\mathcal{M}|\alpha}$ or $\Pi_n^{\mathcal{M}|\alpha}$, for $n > 1$, have the Scale Property.

In the interior of a Σ_1-gap, we find no new scales.

THEOREM 4.11 (Kechris, Solovay). Let $\mathcal{M}$ be a countably 0-iterable $\mathbb{R}^{\mathcal{M}}$-premouse, and suppose $[\alpha, \beta]$ is a Σ_1-gap of $\mathcal{M}$, and that $\mathcal{M}|\alpha \models \mathsf{AD}$. Then

(a) the relation $\neg C^{\mathcal{M}|\alpha}$ has no uniformizing function f such that $f \in \mathcal{M}|\beta$, and hence

(b) if $\alpha < \gamma < \beta$, then $\mathcal{M}$ believes that none of the pointclasses $\Sigma_n^{\mathcal{M}|\gamma}$ or $\Pi_n^{\mathcal{M}|\gamma}$, for $n < \omega$, have the Scale Property.

4.3. Scales at the end of a gap. We are left with the question as to which, if any, of the pointclasses $\Sigma_n^{\mathcal{M}|\beta}$ and $\Pi_n^{\mathcal{M}|\beta}$ have the Scale Property in the case that β ends a Σ_1 gap $[\alpha, \beta]$ of $\mathcal{M}$, and $\alpha < \beta$. As in [Ste83A], the answer turns on the following reflection property of β.

DEFINITION 4.12. For $\mathcal{M}$ a relativised premouse and $1 \leq n < \omega$ and $a \in \mathcal{M}$, we let $\Sigma_a^{n,\mathcal{M}}$ be the Σ_n-type realized by a in $\mathcal{M}$; that is

$$\Sigma_a^{n,\mathcal{M}} = \{\theta(v) : \theta \text{ is either } \Sigma_n \text{ or } \Pi_n \text{ and } \mathfrak{C}_{n-1}(\mathcal{M}) \models \theta[a]\}.$$

We are allowing formulae of the full language $\mathcal{L}^*$ of relativised premice, so that the last extender $\dot{F}^{\mathcal{M}}$ is (partially) described in $\Sigma_a^{n,\mathcal{M}}$.

DEFINITION 4.13. An ordinal β is **strongly Π_n-reflecting in $\mathcal{M}$** iff every Σ_n-type realized in $\mathfrak{C}_{n-1}(\mathcal{M}|\beta)$ is realized in $\mathfrak{C}_{n-1}(\mathcal{M}|\xi)$ for some $\xi < \beta$; that is

$$\forall a \in \mathfrak{C}_{n-1}(\mathcal{M}|\beta)\exists\xi < \beta\exists b \in \mathfrak{C}_{n-1}(\mathcal{M}|\xi)(\Sigma_a^{n,\mathcal{M}|\beta} = \Sigma_b^{n,\mathcal{M}|\xi}).$$

DEFINITION 4.14. Let $[\alpha, \beta]$ be a Σ_1 gap of $\mathcal{M}$, with $\alpha < \beta$; then we call $[\alpha, \beta]$ **strong** iff β is strongly Π_n-reflecting in $\mathcal{M}$, where n is least such that $\rho_n(\mathcal{M}|\beta) = \mathbb{R}^{\mathcal{M}}$. Otherwise, $[\alpha, \beta]$ is **weak**.

Martin's reflection argument of [Mar83A] yields

THEOREM 4.15 (Martin). Let $\mathcal{M}$ be a countably 0-iterable $\mathbb{R}^{\mathcal{M}}$-mouse which satisfies AD , and let $[\alpha, \beta]$ be a strong Σ_1-gap of $\mathcal{M}$ such that $\omega\beta < o(\mathcal{M})$; then

(a) there is a $\Pi_1^{\mathcal{M}|\alpha}$ relation on $\mathbb{R}^{\mathcal{M}}$ which has no uniformization which is definable over $\mathcal{M}|\beta$, and hence

(b) $\mathcal{M}$ believes that none of the pointclasses $\Sigma_n^{\mathcal{M}|\beta}$ or $\Pi_n^{\mathcal{M}|\beta}$ have the Scale Property.

Thus at the end of strong gaps $[\alpha, \beta]$, the Scale Property first re-appears with the pointclass $\Sigma_1^{\mathcal{M}|(\beta+1)}$. The weak gap case is settled, under stronger determinacy hypotheses than should be necessary, by

THEOREM 4.16. Let $\mathcal{M}$ be a countably ω-iterable $\mathbb{R}^{\mathcal{M}}$-mouse which satisfies AD, and $[\alpha, \beta]$ a weak gap of $\mathcal{M}$ and $\omega\beta < o(\mathcal{M})$; then letting n be least such that $\rho_n(\mathcal{M}|\beta) = \mathbb{R}^{\mathcal{M}}$, we have that $\mathcal{M}$ believes that $\Sigma_n^{\mathcal{M}|\beta}$ has the Scale Property.

Remark. The hypothesis that $\omega\beta < o(\mathcal{M})$ should not be necessary. The proof below needs it because at a certain point we apply the Coding Lemma to a bounded subset of $\Theta^{\mathcal{M}|\beta}$ such that A is merely definable over $\mathcal{M}|\beta$, and we need enough determinacy to do this. (See below.) Unfortunately, adding this determinacy as a hypothesis in Theorem 4.16 makes the theorem significantly less useful in core model induction arguments than it would be otherwise. We can eliminate the additional determinacy hypothesis in one case:

THEOREM 4.17. Let $\mathcal{M}$ be a countably ω-iterable $\mathbb{R}^{\mathcal{M}}$-mouse which satisfies AD, and $[\alpha, \beta]$ a weak gap of $\mathcal{M}$. Suppose that either $\Theta^{\mathcal{M}}$ does not exist, or there are no extenders on the $\mathcal{M}$-sequence with index above $\Theta^{\mathcal{M}}$; then letting n be least such that $\rho_n(\mathcal{M}|\beta) = \mathbb{R}^{\mathcal{M}}$, we have that $\mathcal{M}$ believes that $\Sigma_n^{\mathcal{M}|\beta}$ has the Scale Property.

One can combine Theorem 4.17 with the work of [SteB], and thereby obtain a construction of scales at the end of a weak gap in $\mathbf{K}(\mathbb{R})$ which is more useful in a core model induction.

PROOF OF THEOREM 4.16. (Sketch) One gets a proof by integrating our use of the local $\mathbf{HOD}^{\mathcal{M}}$'s into the proof of the corresponding result (theorem 3.7) of [Ste83A]. This is fairly routine, yet involves many details. We shall therefore just sketch one case in which some care with the details is needed. We want also to point out the place where the additional determinacy hypothesis is used.[18]

The case we consider is $n = 1$ and $\mathcal{M}|\beta$ is active of type II. Let us make these assumptions.

Let F^* be the amenable-to-$\mathcal{M}$ predicate coding $\dot{F}^{\mathcal{M}}$ which is described in [Ste07B]. For $\gamma < \beta$, let

$$\mathcal{M}||\gamma = (\mathbf{J}_\gamma^{\dot{E}^{\mathcal{M}}}, \in, \dot{E}^{\mathcal{M}|\gamma}, F^* \cap \mathbf{J}_\gamma^{\dot{E}^{\mathcal{M}}}).$$

The $\mathcal{M}||\gamma$ are just the initial segments of the Σ_0-code $\mathfrak{C}_0(\mathcal{M})$. They are structures for the language $\mathcal{L}^*$ of $\mathfrak{C}_0(\mathcal{M})$, and for φ a Σ_1 formula of $\mathcal{L}^*$ and $x \in \mathcal{M}$, we have

$$\mathfrak{C}_0(\mathcal{M}) \models \varphi[x] \Leftrightarrow \exists\gamma < \beta(\mathcal{M}||\gamma \models \varphi[x]).$$

Further, $\mathcal{M}||\gamma \in \mathcal{M}$ for all $\gamma < \beta$.

Let $\Sigma = \Sigma_a^{1,\mathcal{M}}$ be our nonreflecting Σ_1-type. We may assume $a = \langle G, w_1\rangle$, where G is a finite subset of β and $w_1 \in \mathbb{R}$, and that G is Brouwer-Kleene

[18] This hypothesis is needed in the other cases not covered by Theorem 4.17 as well.

minimal, in the sense that whenever $H \in [\beta]^{<\omega}$ and $H <_b G$ then $\langle H, w_1 \rangle$ does not realize Σ in $\mathfrak{C}_0(\mathcal{M})$. (Here $H <_b G$ iff $\max(H \triangle G) \in G$.) Let $\kappa = \mathrm{crit}(\dot{F}^{\mathcal{M}})$.

We define a canonical sequence of initial segments $\mathcal{M}||\beta_i$ of $\mathfrak{C}_0(\mathcal{M})$. Let $\beta_0 = \nu(\dot{F}^{\mathcal{M}})$. Given $\beta_i < \beta$, let

$$Y_i = \{a : a \text{ is definable over } \mathcal{M}||\beta_i \text{ from parameters in } \mathbb{R}^{\mathcal{M}} \cup \{G, \beta_0, \dots, \beta_{i-1}\}\},$$

and

$$\xi_i = \sup(Y_i \cap (\kappa^+)^{\mathcal{M}}).$$

Letting

$$\psi_i = \text{ least } \psi \in \Sigma \text{ such that } \mathcal{M}||\beta_i \not\models \psi[\langle G, w_1 \rangle],$$

we set β_{i+1} the least γ such that

$$\mathcal{M}||\gamma \models \psi_i[\langle G, w_1 \rangle] \wedge \dot{F}^{\mathcal{M}||\gamma} \text{ measures all sets } A \in \mathcal{M}|\xi_i.$$

Note that $Y_i \in \mathcal{M}$, so $\xi_i < (\kappa^+)^{\mathcal{M}}$, and ψ_i exists and $\beta_{i+1} < \beta$, for all i.

CLAIM 4.18. $\bigcup_{i<\omega} Y_i = \mathcal{M}$.

PROOF. Let $\pi \colon N \to \bigcup_i Y_i$ be the transitive collapse map. Let $\vec{E} = \bigcup_i \pi^{-1}(\dot{E}^{\mathcal{M}||\beta_i})$ and $W = \bigcup_i \pi^{-1}(\dot{F}^{\mathcal{M}||\beta_i})$. It is not hard to see that $(N, \in, \vec{E}, W) = \mathfrak{C}_0(\mathcal{N})$ for some premouse $\mathcal{N}$; note here that our construction ensures that W measures all subsets of its critical point which lie in N. Further, $\mathcal{N}$ is $(0, \omega_1 + 1)$-iterable because we can lift trees on it to trees on $\mathcal{M}$. (Here we need that π is a weak 0-embedding. That is true because π is Σ_1-elementary on the collapses of the β_i, and that in turn is true because each β_i is Σ_1-definable over $\mathfrak{C}_0(\mathcal{M})$ from $\langle G, w_1 \rangle$, so that Σ_1 facts about β_i in $\mathfrak{C}_0(\mathcal{M})$ get recorded in the type Σ.) Thus $\mathcal{N}$ can be compared with $\mathcal{M}$, and since $\rho_1(\mathcal{N}) = \rho_1(\mathcal{M}) = \mathbb{R}$, and $\mathcal{N}$ realizes Σ, we get $\mathcal{N} = \mathcal{M}$. The $<_b$-minimality of G then implies $\pi^{-1}(G) = G$, and the minimality of the β_i implies $\pi^{-1}(\beta_i) = \beta_i$. From this the claim follows easily. ⊣ (Claim 4.18)

It follows that $\sup(\{\beta_i : i < \omega\}) = \beta$ and $\sup(\{\xi_i : i < \omega\}) = (\kappa^+)^{\mathcal{M}}$.

Let $P \subseteq \mathbb{R}$ be $\Sigma_1^{\mathfrak{C}_0(\mathcal{M})}$. Fix a Σ_1 formula φ_0 of $\mathcal{L}^*$ and a parameter b such that

$$P(x) \Leftrightarrow \mathfrak{C}_0(\mathcal{M}) \models \varphi_0[x, b].$$

By the claim, we may assume

$$b = \langle G, \beta_0, \dots, \beta_e, w_2 \rangle,$$

for some e and some real w_2. For $e < i < \omega$, let

$$P^i(x) \Leftrightarrow \mathcal{M}||\beta_i \models \varphi_0[x, b].$$

We shall construct closed game representations $i \mapsto G_x^i$ of the P^i in such a way that if

$$P_k^i(x,u) \Leftrightarrow u \text{ is a winning position for player I in } G_x^i \text{ of length } k\ ,$$

then P_k^i is first order definable over $\mathcal{M}||\beta_{\sup(i,k)}$. Such a closed game representation yields scales on each P^i each of whose norms belongs to $\mathcal{M}$, and hence a $\Sigma_1^{\mathcal{M}}$ scale on P.

In G_x^i, player I describes $\mathfrak{C}_0(\mathcal{M})$ as the union of the $\mathcal{M}||\beta_k$. The language $\mathcal{L}$ in which he does this has $\in, =$, and constant symbols $\dot{G}$, and $\dot{\mathcal{M}}_k$, $\dot{\beta}_k$, and $\dot{x}_k$ for all $k < \omega$. If φ is an $\mathcal{L}$-formula involving no constants $\dot{\mathcal{M}}_k$ or $\dot{\beta}_k$ for $k \geq m$, then we say φ has support m. Player I will produce a Σ_0-complete theory in $\mathcal{L}$, restricting himself at move m to Σ_0 sentences with support m. Let B_0 be the collection of Σ_0 formulae of $\mathcal{L}$, and let $n\colon B_0 \hookrightarrow \omega$ be such that any $\theta \in B_0$ has support $n(\theta)$ and involves no $\dot{x}_k$ for $k \geq n(\theta)$.

A typical run of G_x^i has the form

$$\begin{array}{lllll} \text{I} & T_0, s_0, \eta_0, m_0 & & T_1, s_2, \eta_1, m_1 & \\ & & & & \cdots \\ \text{II} & & s_1 & & s_3 \end{array}$$

where for all k, T_k is a finite set of sentences in B_0, all of which have support k, $s_k \in \mathbb{R}^{<\omega}$, $\eta_k < \omega\beta$, and $m_k \in \omega$. Given such a run of G_x^i, let

$$\langle x_k \ :\ k < \omega \rangle = \text{ concatenation of } \langle s_k \ :\ k < \omega \rangle,$$

and

$$T^* = \bigcup_k T_k.$$

Let S_0 be the set of sentences in B_0 which involve no constants of the form $\dot{x}_i$ for $i \notin \{1,2\}$, and are true in the interpretation under which $\dot{x}_1$ denotes w_1, $\dot{x}_2$ denotes w_2, and $\dot{\beta}_k$ and $\dot{\mathcal{M}}_k$ denote β_k and $\mathcal{M}||\beta_k$ for all $k < \omega$. S_0 will enter as a real parameter in the payoff condition for G_x^i, and hence in the definition of our scale on P. We could avoid this by replacing S_0 with an appropriate finitely axiomatized subtheory, but since real parameters will enter elsewhere, there is no point in doing so. Notice that it is part of S_0 that each $\dot{\mathcal{M}}_k$ is an $\mathcal{L}^*$-structure. For θ any formula of $\mathcal{L}^*$, let $\theta^{\dot{\mathcal{M}}_k}$ be the natural B_0-formula expressing that $\dot{\mathcal{M}}_k \models \theta$.

We say that the run of G_x^i displayed above is a win for player I iff the following conditions hold:

(1) $x_0 = x$, $x_1 = w_1$, and $x_2 = w_2$.
(2) T^* is a consistent extension of S_0 such that for all k, m, n, "$\dot{x}_k(n) = m$" $\in T^*$ iff $x_k(n) = m$.
(3) If $\theta \in B_0$ is a sentence, then either $\theta \in T_{n(\theta)}$ or $(\neg\theta) \in T_{n(\theta)}$.

(4) If $\exists v(v \in \mathbb{R} \wedge \sigma) \in T_k$, then for some j, $\sigma(\dot{x}_j) \in T_{k+1}$.
(5) $(\varphi_0(\dot{x}_0, \langle \dot{G}, \dot{\beta}_0, \dots, \dot{\beta}_e, \dot{x}_2 \rangle))^{\dot{\mathcal{M}}_i} \in T_{i+1}$.
(6) If $\theta(v_1, \dots, v_{n+2})$ is an $\mathcal{L}^*$-formula, and $\sigma_1, \dots, \sigma_n$ are B_0 formulae of one free variable with support k, and "$\imath v\sigma_m(v) \in \mathrm{Ord}$"$\in T^*$ for all $m \leq n$, then

$$\theta^{\dot{\mathcal{M}}_k}(\imath v\sigma_1(v), \dots, \imath v\sigma_n(v), \dot{x}_1, \dot{x}_2) \in T^*$$
$$\Leftrightarrow \mathcal{M}||\beta_k \models \theta[\eta_{n(\sigma_1)}, \dots, \eta_{n(\sigma_n)}, w_1, w_2].$$

(8) If $(\imath v\sigma(v) <_b \dot{G}) \in T_k$, then either
   (a) there is a Π_1 formula $\varphi \in \Sigma$ such that

$$(\neg\varphi(\langle \imath v\sigma(v), \dot{w}_1 \rangle))^{\dot{\mathcal{M}}_{m_k}} \in T_{m_k+1},$$

   or
   (b) there is a Σ_1 formula φ which is one of the first m_k elements of Σ such that for all j and ℓ,

$$(\varphi(\langle \imath v\sigma(v), \dot{w}_1 \rangle))^{\dot{\mathcal{M}}_\ell} \notin T_j.$$

This completes the description of the payoff set for player I in G^i_x. We now show that player I wins G^i_x iff $\mathcal{M}||\beta_i \models \varphi_0[x, b]$, and that the P^i_k are appropriately definable. Both claims follow from the fact that $P^i_k(x, u)$ iff u is honest. Honesty is defined as follows: let I_u be the interpretation of $\mathcal{L}$ under which $\dot{x}_j$ denotes x_j whenever x_j is the jth real determined by u, and $\dot{G}$, $\dot{\beta}_k$ and $\dot{\mathcal{M}}_k$ denote G, β_k, and $\mathcal{M}||\beta_k$ for all k. For u a position in G^i_x, we say u is x-**honest** iff

(i) $T^*(u)$ is true in $(|\mathcal{M}|, \in, I_u)$,
(ii) $\mathcal{M}||\beta_i \models \varphi_0[x, b]$,
(iii) $x_0 = x$, $x_1 = w_1$,and $x_2 = w_2$, if u determines x_0, x_1, and x_2.
(iv) the commitments represented by the m_k can be kept,
(v) if $\sigma_0, \dots, \sigma_n$ enumerates those B_0-formulae σ of one free variable such that $n(\sigma) \in \mathrm{dom}(u)$ and

$$(|\mathcal{M}|, \in, I_u) \models \imath v\sigma(v) \in \mathrm{Ord},$$

and if $\delta_m < o(\mathcal{M})$ is such that

$$(|\mathcal{M}|, \in, I_u) \models \imath v\sigma_m(v) = \delta_m,$$

for all $m \leq n$, then the map

$$\delta_m \mapsto \eta_{n(\sigma_m)}$$

is well-defined and extendible to an order preserving map $\pi \colon o(\mathcal{M}) \to o(\mathcal{M})$ such that for all k, all formulae θ of $\mathcal{L}^*$, and all tuples $\overline{\gamma}$ of ordinals from $\mathcal{M}||\beta_k$, $\pi \restriction \mathcal{M}||\beta_k \subseteq \mathcal{M}||\beta_k$ and

$$\mathcal{M}||\beta_k \models \theta[\overline{\gamma}, w_1, w_2] \Leftrightarrow \mathcal{M}||\beta_k \models \theta[\pi(\overline{\gamma}), w_1, w_2].$$

CLAIM 4.19. For any position u of G_x^i, I wins G_x^i from u iff u is x-honest.

PROOF. If u is x-honest, then I can win G_x^i from u by continuing to tell the truth, while using the map π given by condition (v) to play further η's.

Now suppose I wins G_x^i from u. Let p be a run of G_x^i by such a strategy, with $u \subseteq p$, such that the associated sequence of reals $\langle x_k : k < \omega \rangle$ enumerates $\mathbb{R}^{\mathcal{M}}$. Let $T^* = T^*(p)$ be the B_0-theory played by I. Let $\mathcal{A}$ be the unique model of T^* which is pointwise definable from parameters in $\mathbb{R}^{\mathcal{M}}$. (There is such a model by rule (4).) By rule (6) of G_x^i, $\mathcal{A}$ is wellfounded, and so we assume it is transitive. Let

$$\dot{\mathcal{M}}_k^{\mathcal{A}} = (\mathcal{N}_k, F_k^*),$$

$$\beta_k^* = \dot{\beta}_k^{\mathcal{A}},$$

and

$$G^* = \dot{G}^{\mathcal{A}}.$$

Since $S_0 \subseteq T^*$, the $\mathcal{N}_k$ are premice, and $\mathcal{N}_k \trianglelefteq \mathcal{N}_{k+1}$ for all k. Let $\mathcal{N}$ be the union of the $\mathcal{N}_k$, and F^* the union of the F_k^*. We can define $\pi \colon o(\mathcal{N}) \to o(\mathcal{N})$ by

$$\pi(\iota v\sigma(v)^{\mathcal{A}}) = \eta_{n(\sigma)},$$

for all B_0-formulae σ such that $\iota v\sigma(v) \in \mathrm{Ord}$ is in T^*. Clearly, π is well-defined, and for any tuple $\overline{\gamma}$ of ordinals from $\mathcal{N}_k$ and any formula θ of $\mathcal{L}^*$,

$$(\mathcal{N}_k, F_k^*) \models \theta[\overline{\gamma}, w_1, w_2] \Leftrightarrow \mathcal{M}||\beta_k \models \theta[\pi(\overline{\gamma}), w_1, w_2].$$

As in the proof of Theorem 4.1, this implies that $\mathcal{N}$ is countably iterable, and that π extends to an embedding, which we also call π, such that

$$\pi \colon \mathcal{H}^{\mathcal{N}} \to \mathcal{H}^{\mathcal{M}}$$

is Σ_1-elementary (in $\mathcal{L}^*$).

Because S_0 is true in $\mathcal{A}$, F_{k+1}^* measures all subsets of its critical point in $\mathcal{H}_1^{(\mathcal{N}_k, F_k^*)}(\mathbb{R}^{\mathcal{M}} \cup \{\beta_0^*, \dots, \beta_{k-1}^*, G^*\})$. But the union of these hulls has the same universe as $\mathcal{N}$, and thus F^* is an extender over $\mathcal{N}$. Similarly, we get

$$F_k^* \cap \mathcal{H}^{\mathcal{N}_k} \in \mathcal{H}^{\mathcal{N}_{k+1}}$$

for all k, and because π is sufficiently elementary,

$$\pi(F_k^* \cap \mathcal{H}^{\mathcal{N}_k}) = F \cap \mathcal{H}^{\mathcal{M}|\beta_k}.$$

Letting $E^* = F^* \cap \mathcal{H}^{\mathcal{N}}$ and $E = F \cap \mathcal{H}^{\mathcal{M}}$, we then have that

$$\pi \colon (\mathcal{H}^{\mathcal{N}}, E^*) \xrightarrow{\Sigma_1} \mathcal{H}^{\mathcal{M}}$$

is Σ_1 elementary. It follows that $(\mathcal{H}^{\mathcal{N}}, E^*)$ is a countably iterable premouse, and hence that $(\mathcal{N}, F^*)$ is a countably iterable premouse.

It is part of S_0 that our non-reflecting type Σ is realized for the first time, and thus we have $(\mathcal{N}, F^*) = \mathcal{M}$. Because player I has kept the commitments he made according to rule (7) of G^i_x, we have $G^* = G$, and $\beta^*_k = \beta_k$ and $(\mathcal{N}_k, F^*_k) = \mathcal{M}||\beta_k$ for all k. It is now easy to verify that u was x-honest; the map π witnesses that condition (v) of x-honesty is met. ⊣ (Claim 4.19)

CLAIM 4.20. Let $k \leq i$; then $\{u : u \text{ is an } x\text{-honest position of length } k\}$ is a member of $\mathcal{M}|\beta$.

PROOF. It is clear that the set of u satisfying conditions (i)-(iii) of x-honesty is definable over $\mathcal{M}||\beta_k$, and hence in $\mathcal{M}|\beta$. The Coding Lemma argument of [Ste83A] shows that the set of u satisfying condition (iv) is also in $\mathcal{M}|\beta$. Here, as in [Ste83A], we can apply the Coding Lemma to sets belonging to $\mathcal{M}|\beta$, so we don't actually need determinacy beyond the sets in $\mathcal{M}|\beta$.

For (v), let

$$s = (\iota v \sigma(v)^{\mathcal{M}||\beta_k} \mapsto \eta_{n(\sigma)})$$

be a finite map coded into a position u of length k satisfying (i)-(iv). Note that $\mathrm{dom}(s) \subseteq Y_k$, so that if s can be extended to a π as demanded in (v), then as Y_k is Σ_1-definable over $\mathcal{M}||\beta_{k+1}$ from $\beta_0, \dots, \beta_{k-1}, G$, $\mathrm{ran}(s) \subseteq Y_k$ as well. (Note here that by the proof of Claim 4.2, π must fix G and the β_m for $m < \omega$.) So if we let

$$Z_k = \{t \colon Y_k \to Y_k : |t| < \omega \wedge \exists \pi \supseteq t (\pi \text{ is as in (v)}) \},$$

then it suffices to show that Z_k is definable over $\mathcal{M}||\beta_{\sup(i,k)}$.

We proceed as in the proof of Claim 4.2. $\mathcal{M}$. Note that $t \in Z_k$ iff the closed player has a winning strategy in a certain "embedding game" on $\theta^{\mathcal{M}|\beta}$. We claim that if the closed player wins the embedding game in $\mathbf{V}$, then he wins it in $\mathcal{M}$. (The converse is obvious.) So suppose the closed player wins the embedding game in $\mathbf{V}$. Let $A \in \mathcal{M}$ be a set of ordinals which codes up the payoff of game; since $\theta^{\mathcal{M}|\beta} < \theta^{\mathcal{M}}$, and $\mathcal{M} \models \mathsf{AD}$, we can find a model N of ZFC such that $A \in N$, and $(N, \in)$ is coded by a set of reals $B_N \in \mathcal{M}$. (E.g., let $N = L_\alpha[A]$, where α is the supremum of the order types of the $\Delta^{\mathcal{M}|\beta}_n$ prewellorders of $\mathbb{R}^{\mathcal{M}}$, for an appropriate n.) Since $N \models$ ZFC, the closed player wins the embedding game via a strategy $\Sigma \in N$. For $\gamma < \theta^{\mathcal{M}|\beta}$, let $f(\gamma) = \{z \in \mathbb{R}^{\mathcal{M}} : z \text{ codes } \gamma \text{ via } B_N\}$. We can arrange that $f \in \mathcal{M}$, and use f to show that $\Sigma \in \mathcal{M}$.

The argument of the last paragraph actually shows that there is a fixed $n < \omega$ such that for all t, $t \in Z_k$ iff player II has a $\Delta^{\mathcal{M}|\beta}_n$ winning strategy in the embedding game associated to t. It follows that

$$Z_k \in \mathcal{M}.$$

But $\mathcal{M} \models \mathsf{AD}$, and Z_k can be identified with a bounded subset of $\theta^{\mathcal{M}|\beta}$, since $Y_k \in \mathcal{M}|\beta$ and is the surjective image of $\mathbb{R}$ by a map in $\mathcal{M}|\beta$. It follows from the Coding Lemma that $Z_k \in \mathcal{M}|\beta$, and in fact Z_k is definable over $\mathcal{M}||\beta_{\sup(i,k)}$.[19] ⊣ (Claim 4.20)

⊣ (Theorem 4.16)

As the reader can see, we use the determinacy of sets *definable over* $\mathcal{M}|\beta$ in the proof of Claim 4.20 above. The determinacy of sets belonging to $\mathcal{M}|\beta$ is not enough for the proof because the payoff set A for the embedding game may not be a member of $\mathcal{M}|\beta$. One can get by with the determinacy of sets in $\mathcal{M}|\beta$ in the proof of Theorem 4.17 because in that case, the only "global" role of the ordinals played by player I in G_x^i is to verify that the model he is playing is wellfounded. This aspect of honesty can be explicitly defined; player I needs only to have spaced his ordinals adequately. See [Ste83A] for the details. It is still true that player I will have to verify that the $\mathcal{H}_\nu^{\mathcal{A}}$ for $\nu < o(\mathcal{A})$ are iterable, by embedding them into a corresponding $\mathcal{H}_\mu^{\mathcal{M}|\beta}$, but these embeddings no longer need to fit together into a single embedding, and thus this aspect of honesty does not lead out of $\mathcal{M}|\beta$. We leave the further details of the proof of Theorem 4.17 to the reader.

REFERENCES

DANIEL CUNNINGHAM

[Cun90] ***The real core model***, Ph.D. thesis, UCLA, 1990.

RONALD B. JENSEN

[Jen72] *The fine structure of the constructible hierarchy*, ***Annals of Mathematical Logic***, vol. 4 (1972), pp. 229–308.

AKIHIRO KANAMORI AND MATTHEW FOREMAN

[KF07] ***Handbook set theory into the 21st century***, Springer, 2007.

ALEXANDER S. KECHRIS, DONALD A. MARTIN, AND YIANNIS N. MOSCHOVAKIS

[CABAL iii] *Cabal seminar* 79–81, Lecture Notes in Mathematics, no. 1019, Berlin, Springer, 1983.

DONALD A. MARTIN

[Mar83A] *The largest countable this, that, and the other*, this volume, originally published in Kechris et al. [CABAL iii], pp. 97–106.

DONALD A. MARTIN AND JOHN R. STEEL

[MS83] *The extent of scales in* $\mathbf{L}(\mathbb{R})$, this volume, originally published in Kechris et al. [CABAL iii], pp. 86–96.

WILLIAM J. MITCHELL AND JOHN R. STEEL

[MS94] ***Fine structure and iteration trees***, Lecture Notes in Logic, vol. 3, Springer-Verlag, Berlin, 1994.

YIANNIS N. MOSCHOVAKIS

[Mos83] *Scales on coinductive sets*, this volume, originally published in Kechris et al. [CABAL iii], pp. 77–85.

[19] All we really need to get the desired scale is that $Z_k \in \mathcal{M}|\beta$.

JOHN R. STEEL
[SteB] *Scales in* $\boldsymbol{K}(\mathbb{R})$ *at the end of a weak gap*, Preprint, available at `http://www.math.berkeley.edu/~steel`.
[Ste83A] *Scales in* $\mathbf{L}(\mathbb{R})$, this volume, originally published in Kechris et al. [CABAL iii], pp. 107–156.
[Ste07B] *An outline of inner model theory*, In Kanamori and Foreman [KF07].

DEPARTMENT OF MATHEMATICS
UNIVERSITY OF CALIFORNIA
BERKELEY, CA 94720, USA
E-mail: steel@math.berkeley.edu

THE REAL GAME QUANTIFIER PROPAGATES SCALES

DONALD A. MARTIN

§1. Introduction. Moschovakis [Mos80, 6E] shows, assuming sufficient determinacy, that the game quantifier (for games on the integers) propagates scales. In [Mos83] he shows how to put scales on sets defined by applying the real game quantifier to a closed matrix. In this paper we show, from appropriate determinacy hypotheses, that the real game quantifier, and the real or integer game quantifier of any fixed countable length, propagate scales.

Throughout the paper, we work in ZF+DC.

If α is an ordinal, a **game type of length** α is a function $g\colon \alpha \to \{0,1\}$. If g is a game type of length α and $Y \subseteq {}^{\alpha}\omega$, the game G_g^Y is played as follows: White and Black produce, in α moves, an element y of ${}^{\alpha}\omega$. White chooses $y(\beta)$ if $g(\beta) = 0$ and Black chooses $y(\beta)$ if $g(\beta) = 1$. White wins just in case $y \in Y$.

If $Y \subseteq {}^{(\gamma+\alpha)}\omega$ and g is a game type of length α,

$$g(Y) = \{x \in {}^{\gamma}\omega : \text{White has a winning strategy for } \mathrm{G}_g^{Y_x}\}$$

where $Y_x = \{z : x^\frown z \in Y\}$. (Note that $g(Y)$ depends in general on γ, though our notation does not indicate this dependence.)

The notion of a scale on a subset of ${}^{\alpha}\omega$ is defined just like that of a scale on a subset of ${}^{\omega}\omega$, using the product topology on ${}^{\alpha}\omega$.

The purpose of this paper is to show that, if $Y \subseteq {}^{(\gamma+\alpha)}\omega$, α is countable, Y admits a scale, and g is a game type of length α, then $g(Y)$ admits a scale. We shall need to assume determinacy for certain games of length $\leq 4\alpha$ ($= \alpha$ if α is a limit ordinal), the games $\mathrm{G}_g^{Y_x}$ and certain other games whose winning conditions are related to the norms on Y. Our definability results for the scale on $g(Y)$ will be natural generalizations of those of Moschovakis, the principal difference being that we need to use a well-ordering of a subset of ω of order type α.

Our plan is to break down Moschovakis' scale-propagation technique into three basic pieces and then to assemble the pieces in a routine fashion. The three pieces are:

The Cabal Seminar. Volume I: Games, Scales and Suslin Cardinals
Edited by A. S. Kechris, B. Löwe, J. R. Steel
Lecture Notes in Logic, 31

1. **Infimum norms**. Here we show how to go from a sequence of norms on Y to a sequence of norms on $g(Y)$ when $\alpha = 1$ and $g(0) = 0$ (i.e., for one-move games where White moves). If the original sequence is a scale, the resulting sequence will be a scale, but our main result is similar to Moschovakis' Infimum Lemma for existential real quantification.
2. **Supremum norms**. This is like (1) except that $g(0) = 1$ (Black moves). Our method is similar to that of Moschovakis' Supremum Lemma.
3. **Game norms**. If g is any game type and φ is any norm on Y, a natural generalization of a result of Moschovakis [Mos80] gives a norm $g(\varphi)$ on $G(Y)$, granted appropriate determinacy.

In §2 we discuss infimum norms, in §3 we discuss supremum norms, in §4 we discuss game norms, and in §5 we assemble a scale on $g(Y)$ from a scale on Y, infimum norms, supremum norms, game norms, and a well-ordering of a subset of ω of order type α. In §6 we define canonical winning strategies. In §7 we prove results about propagation of the scale property.

§2. Infimum norms. Let $Y \subseteq {}^{(\gamma+1)}\omega$. Suppose $\vec{\varphi} = \langle \varphi_i : i \in \omega \rangle$ is a **putative scale** on Y, i.e. $\varphi_i \colon Y \to \text{Ord}$ for each i and $\varphi_{i+1}(y_1) \leq \varphi_{i+1}(y_2)$ implies $\varphi_i(y_1) \leq \varphi_i(y_2)$ for all $y_1, y_2 \in Y$ and every $i \in \omega$. (The last clause is merely for convenience, and differs from Moschovakis' definition [Mos80].)

Let $k \in \omega$. We define a putative scale $\vec{\psi} = \langle \psi_i : i \in \omega \rangle = \text{Inf}_k\, \vec{\varphi}$ on $g(Y)$, where $g \colon 1 \to \{0, 1\}$ and $g(0) = 0$. Our definition will have the property that each ψ_i is defined solely from $\varphi_0, \ldots, \varphi_i$.

For $i \leq k$, let $\psi_i(x) = \inf\{\varphi_i(x^\frown n) : n \in \omega \wedge x^\frown n \in Y\}$. For $i > k$ let

$$\psi_i(x) = \inf\{\ulcorner \varphi_k(x^\frown n), n, \varphi_i(x^\frown n) \urcorner : n \in \omega \wedge x^\frown n \in Y\}$$

where $\ulcorner \cdot, \cdot, \cdot \urcorner$ embeds the lexicographic order on an appropriate $\kappa \times \omega \times \kappa$ into the ordinals.

For each $i \in \omega$ and $x_1, x_2 \in {}^{\gamma}\omega$ we now define the **game for verifying** $x_1 \in g(Y)$ **and either** $x_2 \notin g(Y)$ **or** $\psi_i(x_1) \leq \psi_i(x_2)$.

$i \leq k$. Player II plays n_1 and player I plays n_2. Player II wins just in case $x_1{}^\frown n_1 \in Y$ and either $x_2{}^\frown n_2 \notin Y$ or $\varphi_i(x_1{}^\frown n_1) \leq \varphi_i(x_2{}^\frown n_2)$.

$i > k$. Player II plays n_1, player I plays n_2, player I plays $j_1 < 3$, and player II plays $j_2 \leq j_1$. If $j_1 = j_2 = 2$, player II wins if $x_1{}^\frown n_1 \in Y$ and either $x_2{}^\frown n_2 \notin Y$ or $\varphi_i(x_1{}^\frown n_1) \leq \varphi_i(x_2{}^\frown n_2)$. If $j_1 = j_2 = 1$, player II wins if $n_1 \leq n_2$. If $j_1 = j_2 = 0$, player II wins if $x_1{}^\frown n_1 \in Y$ and either $x_2{}^\frown n_2 \notin Y$ or $\varphi_k(x_1{}^\frown n_1) \leq \varphi_k(x_2{}^\frown n_2)$. If $1 = j_2 < j_1$, player II wins if $n_1 < n_2$. If $0 = j_2 < j_1$, player II wins if $x_1{}^\frown n_1 \in Y$ and either $x_2{}^\frown n_2 \notin Y$ or $\varphi_k(x_1{}^\frown n_1) < \varphi_k(x_2{}^\frown n_2)$.

It is easily checked that player II has a winning strategy for this game just in case $x_1 \in g(Y)$ and either $x_2 \notin g(Y)$ or $\psi_i(x_1) \leq \psi_i(x_2)$.

There is a similar game for verifying $x_1 \in g(Y)$ and either $x_2 \notin g(Y)$ or $\psi_i(x_1) < \psi_i(x_2)$. We omit the definition.

A **standard play** of the game for verifying $x_1 \in g(Y)$ and either $x_2 \notin g(Y)$ or $\psi_i(x_1) \leq \psi_i(x_2)$ is a play of the game in which either $i \leq k$ or $i > k$ and $j_1 = j_2 \neq 1$. The **terminal condition** of a standard play is $\langle x_1{}^\frown n_1, x_2{}^\frown n_2, k\rangle$ if $i > k$ and $j_1 = 0$ and $\langle x_1{}^\frown n_1, x_2{}^\frown n_2, i\rangle$ if $i \leq k$ or $j_1 = 2$. Notice that player II wins a standard play just in case $x_1{}^\frown n_1 \in Y$ and either $x_2{}^\frown n_2 \notin Y$ or $\varphi_j(x_1{}^\frown n_1) \leq \varphi_j(x_2{}^\frown n_2)$ where j is the third component of the terminal condition.

LEMMA 2.1. Let $\vec{\varphi}$ be a putative scale on Y. Let $\vec{\psi} = \mathrm{Inf}_k\, \vec{\varphi}$. Let $f : \omega \to \omega$ be such that

a. $f(i) \leq f(i+1)$ for all i;
b. $f(i) \leq i$ for all i;
c. $\mathrm{ran}(f)$ is unbounded.

Let $x_i \in g(Y)$ for each $i \in \omega$, where $g: 1 \to \{0, 1\}$ and $g(0) = 0$. Assume that $\psi_{f(i)}(x_i) \geq \psi_{f(i)}(x_{i+1})$ for all i. Let σ_i be a winning strategy for player II for the game for verifying $x_{i+1} \in g(Y)$ and either $x_i \notin g(Y)$ or $\psi_{f(i)}(x_{i+1}) \leq \psi_{f(i)}(x_i)$, for each $i \in \omega$.

There are numbers $n, n_1, n_2, \ldots$ such that $\lim_i n_i = n$ and, for every n_0 with $x_0{}^\frown n_0 \in Y$, there is an $f^* : \omega \to \omega$ satisfying (a), (b) and (c), and

d. $f^*(i) \leq f(i)$ for all i,
e. $f^*(i) \geq \min\{f(i), k\}$,

and such that $\langle x_{i+1}{}^\frown n_{i+1}, x_i{}^\frown n_i, f^*(i)\rangle$ is the terminal condition of a standard play consistent with σ_i for each i.

PROOF. Let n_{i+1} be player II's first move as given by σ_i, for $i = 0, 1, 2, \ldots$. Let n_0 be such that $x_0{}^\frown n_0 \in Y$.

Assume inductively that $x_i{}^\frown n_i \in Y$. Since $\langle n_{i+1}, n_i\rangle$ is consistent with σ_i for $f(i) \leq k$ and $\langle n_{i+1}, n_i, 0, 0\rangle$ is consistent with σ_i for $i > k$ we have $x_{i+1}{}^\frown n_{i+1} \in Y$. For $f(i) \leq k$ we have also $\varphi_{f(i)}(x_{i+1}{}^\frown n_{i+1}) \leq \varphi_{f(i)}(x_i{}^\frown n_i)$. For $f(i) > k$, we have $\varphi_k(x_{i+1}{}^\frown n_{i+1}) \leq \varphi_k(x_i{}^\frown n_i)$.

Since the $\varphi_k(x_i{}^\frown n_i)$ are non-increasing for $f(i) > k$, there must be an i_0 with $f(i_0) > k$ such that $i \geq i_0$ implies $\varphi_k(x_i{}^\frown n_i) = \varphi_k(x_{i_0}{}^\frown n_{i_0})$.

Let $i \geq i_0$. The play $\langle n_{i+1}, n_i, 1, 0\rangle$ is inconsistent with σ_i, since it is a win for player I. Thus $\langle n_{i+1}, n_i, 1, 1\rangle$ is consistent with σ_i. This means $n_{i+1} \leq n_i$. Let then i_1 be large enough that $i_1 \geq i_0$ and $n_i = n_{i_1}$ for all $i \geq i_1$. Let $n = n_{i_1}$. Let $f^*(i) = f(i)$ if $f(i) \leq k$ or $i \geq i_1$. Let $f^*(i) = k$ otherwise. We have already seen that $\langle n_{i+1}, n_i, f^*(i)\rangle$ is the terminal condition of a standard play consistent with σ_i when $i < i_1$. Suppose $i \geq i_1$. Consider the play $\langle n_{i+1}, n_i, 2, j_2\rangle$ consistent with σ_1. $j_2 = 0$ or $j_2 = 1$ would lose for player II, so $j_2 = 2$. Hence $\langle n_{i+1}, n_i, f(i)\rangle$ is the terminal condition of a standard play consistent with σ. ⊣

Lemma 2.1 is similar to Moschovakis' [Mos83] Infimum Lemma. Our situation is simpler than his in that our moves are integers instead of reals, but our situation is complicated by the fact that we are going to use our lemma in studying long games, so we need the σ_i and the functions f and f^* for bookkeeping and avoiding the axiom of choice.

Let us say that a putative scale $\vec{\varphi} = \langle \varphi_i : i \in \omega \rangle$ on Y is **i-lsc** if whenever $\langle y_j : j \in \omega \rangle$ converges to y, each $y_j \in Y$, and the norms $\varphi_i(y_j)$ are all eventually constant as j increases, then $y \in Y$ and $\varphi_i(y) \leq \lim_j \varphi_i(y_j)$. Note that $\vec{\varphi}$ is a scale just in case $\vec{\varphi}$ is i-lsc for every $i \in \omega$.

Lemma 2.2. Let $\vec{\varphi}$ be a putative scale on Y, let $\vec{\psi} = \mathrm{Inf}_k\, \vec{\varphi}$, and let $i_0 \in \omega$. If $\vec{\varphi}$ is i_0-lsc so is $\vec{\psi}$.

Proof. Let $x_j \in g(Y)$ for $j \in \omega$ and suppose $\langle x_j : j \in \omega \rangle$ converges to x and the norms $\psi_i(x_j)$ are eventually constant. By thinning the sequence if necessary, we may assume that $\psi_i(x_i) = \lim_j \psi_i(x_j)$ for each i. Let n_i be such that $x_i{}^\frown n_i \in Y$ and $\varphi_i(x_i{}^\frown n_i)$, for $i \leq k$, or $\ulcorner \varphi_k(x_i{}^\frown n_i), n_i, \varphi_i(x_i{}^\frown n_i) \urcorner$, for $i > k$, is as small as possible. If $i \leq k$, we have that $\varphi_i(x_i{}^\frown n_i) = \varphi_i(x_j{}^\frown n_j)$ for all $j \geq i$, since the minimality of $\varphi_j(x_j{}^\frown n_j)$ implies that of $\varphi_i(x_j{}^\frown n_j)$. If $i > k$, we have then that $\varphi_k(x_i{}^\frown n_i) = \varphi_k(x_k{}^\frown n_k)$. Hence, for all $j > k$, $n_j = n_{k+1} = n$. Hence $\langle x_i{}^\frown n_i : i \in \omega \rangle$ converges to $(x{}^\frown n)$. If $j > i > k$, we must have $\varphi_i(x_i{}^\frown n_i) = \varphi_i(x_j{}^\frown n_j)$. Thus all norms $\varphi_i(x_j{}^\frown n_j)$ are eventually constant. It follows that $x{}^\frown n \in Y$ and $\varphi_{i_0}(x{}^\frown n) \leq \lim_j \varphi_{i_0}(x_j{}^\frown n_j)$. Thus $x \in g(Y)$. If $i_0 \leq k$, then $\psi_{i_0}(x) \leq \varphi_{i_0}(x{}^\frown n) \leq \lim_j \varphi_{i_0}(x_j{}^\frown n_j) = \lim_j \psi_{i_0}(x_j)$. If $i > k$, it suffices to note that $\inf\{\varphi_k(x{}^\frown m) : m \in \omega\} \leq \varphi_k(x{}^\frown n) \leq \lim_j \varphi_k(x_j{}^\frown n_j)$ and so that

$$\begin{aligned}\psi_{i_0}(x) &= \inf\{\ulcorner \varphi_k(x{}^\frown m), m, \varphi_i(x{}^\frown m) \urcorner : m \in \omega\} \\ &\leq \ulcorner \varphi_k(x{}^\frown n), n, \varphi_{i_0}(x{}^\frown n) \urcorner \leq \lim_j \ulcorner \varphi_k(x_j{}^\frown n_j), n_j, \varphi_{i_0}(x_j{}^\frown n_j) \urcorner. \qquad \dashv\end{aligned}$$

§3. Supremum norms. Let $Y \subseteq {}^{(\gamma+1)}\omega$. Suppose $\vec{\varphi} = \langle \varphi_i : i \in \omega \rangle$ is a putative scale on Y. Let $k \in \omega$. We define a putative scale $\vec{\psi} = \langle \psi_k : k \in \omega \rangle = \mathrm{Sup}_k\, \vec{\varphi}$ on $g(Y)$, where $g\colon 1 \to \{0, 1\}$ and $g(0) = 1$.

If $i \leq k$, let

$$\psi_i(x) = \sup\{\varphi_i(x{}^\frown n) : n \in \omega\}.$$

If $i > k$, let

$$\psi_i(x) = \ulcorner \psi_k(x), \varphi_{p_0}(x{}^\frown m_0), \varphi_{p_1}(x{}^\frown m_1), \ldots, \varphi_{p_{i-k-1}}(x{}^\frown m_{i-k-1}) \urcorner,$$

where $n \mapsto (p_n, m_n)$ is a bijection between ω and $\omega \times \omega$ with $p_n \leq n$ for all n, with $\ulcorner \;\; , \ldots , \;\; \urcorner$ an appropriate embedding of the lexicographic ordering into the ordinals.

$\vec{\psi}$ is clearly a putative scale.

For each i and x_1, x_2, we define the **game for verifying that** $x_1 \in g(Y)$ **and either** $x_2 \notin g(Y)$ **or** $\psi_i(x_1) \leq \psi_i(x_2)$.

For $i \leq k$, player I plays n_1 and player II plays n_2. Player II wins if $x_1{}^\frown n_1 \in Y$ and either $x_2{}^\frown n_2 \notin Y$ or $\varphi_i(x_1{}^\frown n_1) \leq \varphi_i(x_2{}^\frown n_2)$.

If $i > k$, player I plays $j_1 \leq i - k$ and player II plays $j_2 \leq j_1$. If $j_1 = j_2 = 0$, player I plays n_1, player II plays n_2, and player II wins $\Leftrightarrow$ $x_1{}^\frown n_1 \in Y$ and either $x_2{}^\frown n_2 \notin Y$ or $\varphi_k(x_1{}^\frown n_1) \leq \varphi_k(x_2{}^\frown n_2)$. If $j_1 = j_2 = n+1$, player II wins $\Leftrightarrow$ $x_1{}^\frown m_n \in Y$ and either $x_2{}^\frown m_n \notin Y$ or $\varphi_{p_n}(x_1{}^\frown m_n) \leq \varphi_{p_n}(x_2{}^\frown m_n)$. The cases $j_1 > j_2 = 0$ and $j_1 > j_2 = n+1$ are similar, except that "$<$" replaces "$\leq$", and player II plays n_2 before player I plays n_1 when $j_2 = 0$.

There is a similar **game for verifying that** $x_1 \in g(Y)$ **and either** $x_2 \notin g(Y)$ **or** $\psi_i(x_1) < \psi_i(x_2)$.

A **standard play** of our game for verifying $x_1 \in g(Y)$ and either $x_2 \notin g(Y)$ or $\psi_i(x_1) \leq \psi_i(x_2)$ is a play in which $i \leq k$ or $j_1 = j_2$. The **terminal condition** of a standard play is $\langle x_1{}^\frown n_1, x_2{}^\frown n_2, i\rangle$ if $i \leq k$, $\langle x_1{}^\frown n_1, x_2{}^\frown n_2, k\rangle$ if $i > k$ and $j_1 = 0$, and $\langle x_1{}^\frown m_n, x_2{}^\frown m_n, p_n\rangle$ if $j_1 = n+1$.

Lemma 3.1. Let $\vec{\varphi}$ be a putative scale on Y. Let $\vec{\psi} = \mathrm{Sup}_k\,\vec{\varphi}$. Let f satisfy (a), (b), and (c) as in Lemma 2.1. Let $x_i \in g(Y)$ for each $i \in \omega$ (where $g\colon 1 \to \{0,1\}$ and $g(0) = 1$). Assume that $\psi_{f(i)}(x_i) \geq \psi_{f(i)}(x_{i+1})$ for all i. Let σ_i, $i = 0, 1, \ldots$, be winning strategies for the game for verifying $x_{i+1} \in g(Y)$ and either $x_i \notin g(Y)$ or $\psi_{f(i)}(x_{i+1}) \leq \psi_{f(i)}(x_i)$. For each $n \in \omega$ there are numbers $n_0, n_1, \ldots$ such that $\lim_i n_i = n$ and there is a function $f^*\colon \omega \to \omega$ such that f^* satisfies (a), (b), (c), (d), and (e) of Lemma 2.1 and such that $\langle x_{i+1}{}^\frown n_{i+1}, x_i{}^\frown n_i, f^*(i)\rangle$ is the terminal condition of a standard play according to σ_i for each i.

Proof. Since the norms $\psi_i(x_j)$ are eventually non-decreasing as j increases, they are eventually constant.

Let $j(i)$ be given by $p_{j(i)} = i$ and $m_{j(i)} = n$. For each $t > k$, let $u(t)$ be such that $u(t) > u(t-1)$ if $t > k+1$, $f(u(t)) > k + j(t)$, and $\psi_{k+j(t)}(x_j) = \psi_{k+j(t)}(x_{u(t)})$ for all $j \geq u(t)$.

Set $n_i = n$ for all $i \geq u(k+1)$. For $u(t+1) > i \geq u(t)$, let player I play $j_1 = j(t)+1$ against σ_i. This is legal, since $f(i) \geq f(u(t)) > k + j(t) = k + j_1 - 1$ so $f(i) - k \geq j_1$. Since $\psi_{k+j(t)}(x_{i+1}) = \psi_{k+j(t)}(x_i)$, $\psi_k(x_i) = \psi_k(x_{i+1})$ and $\varphi_{p_j}(x_i{}^\frown m_j) = \varphi_{p_j}(x_{i+1}{}^\frown m_j)$ for all $j < j(t)$. This σ_i, a winning strategy, cannot call for player II to play $j_2 < j(t)+1$. Hence $j_2 = j(t)+1$ and the terminal condition of this standard play is

$$\langle x_{i+1}{}^\frown m_{j(t)}, x_i{}^\frown m_{j(t)}, p_{j(t)}\rangle = \langle x_{i+1}{}^\frown n, x_i{}^\frown n, t\rangle.$$

Assume that n_{i+1} is defined, where $k < f(i)$ and $i < u(k+1)$. Let player I play $j_1 = 0$ against σ_i. player II must respond with $j_2 = 0$. Let player I play n_{i+1} as his n_1. Let n_i be player II's response via σ_i. The terminal condition of this standard play is then

$$\langle x_{i+1}{}^\frown n_{i+1}, x_i{}^\frown n_i, k\rangle.$$

Suppose n_{i+1} is defined and $f(i) \leq k$. Let player I play n_{i+1} and player II play n_i as given by σ_i. The terminal condition of this standard play is

$$\langle x_{i+1}{}^\frown n_{i+1}, x_i{}^\frown n_i, f(i)\rangle.$$

Now define $f^*(i) = f(i)$ if $f(i) \leq k$, $f^*(i) = k$ if $k < f(i)$ and $i < u(k+1)$, and $f^*(i) = t$ if $u(t) \leq i < u(t+1)$. Since $f(u(t)) > k + j(t) > m_{j(t)} = t$, $f^*(i)$ is always $\leq f(i)$. ⊣

Lemma 3.1 has the same relations to Moschovakis' [Mos83] Supremum Lemma as Lemma 2.1 has to the Infimum Lemma.

LEMMA 3.2. *Let $\vec{\varphi}$ be a putative scale on Y and let $\vec{\psi} = \mathrm{Sup}_k\, \vec{\varphi}$. For $i_0 \in \omega$, if $\vec{\varphi}$ is i_0-lsc, so is $\vec{\psi}$.*

PROOF. Let $\langle x_j : j \in \omega\rangle$ converge to x with each $x_j \in g(Y)$ and let the norms $\psi_i(x_j)$ be eventually constant. Let $n \in \omega$. For each i, the norms $\varphi_i(x_j{}^\frown n)$ are eventually constant as j increases. Thus $x^\frown n \in Y$ and $\varphi_{i_0}(x^\frown n) \leq \lim_j \varphi_{i_0}(x_j{}^\frown n)$. Since $\vec{\varphi}$ is a putative scale, $\varphi_i(x^\frown n) \leq \lim_j \varphi_i(x_j{}^\frown n)$ for each $i \leq i_0$. Since this holds for all n, it follows that $x \in g(Y)$ and, from the definition of ψ_{i_0}, that $\psi_{i_0}(x) \leq \lim_j \psi_{i_0}(x_j)$, provided that $\psi_i(x) \leq \lim_j \psi_i(x_j)$ for all i such that $i \leq k$ and $i \leq i_0$. Fix such an i and let $\psi_i(x_{j_0}) = \psi_i(x_j)$ for all $j \geq j_0$. We must show that $\sup\{\varphi_i(x^\frown n) : n \in \omega\} \leq \sup\{\varphi_i(x_{j_0}{}^\frown n) : n \in \omega\}$. We already know, since $i \leq i_0$, that $\varphi_i(x^\frown n) \leq \lim_j \varphi_i(x_j{}^\frown n)$. Let $j^* \geq j_0$ be such that $\varphi_i(x_{j^*}{}^\frown n) \geq \varphi_i(x^\frown n)$. For some m, $\varphi_i(x_{j_0}{}^\frown m) \geq \varphi_i(x_{j^*}{}^\frown n)$, and the proof is complete. ⊣

§4. Game norms. Let $Y \subseteq {}^{(\gamma+\alpha)}\omega$ and let $g : \alpha \to \{0, 1\}$. Let φ be a norm on Y. We define a norm $\psi = g(\varphi)$ on $g(Y)$. Let $x_1, x_2 \in {}^{\gamma}\omega$. We consider a game of length 2α, played as follows. If White moves at β, i.e. if $g(\beta) = 0$, then player II moves at 2β and chooses $z_1(\beta)$, and player I moves at $2\beta + 1$ and chooses $z_2(\beta)$. If Black moves at β, then player I chooses $z_1(\beta)$ at 2β and player II chooses $z_2(\beta)$ at $2\beta + 1$. Player II wins the game if $x_1{}^\frown z_1 \in Y$ and either $x_2{}^\frown z_2 \notin Y$ or $\varphi(x_1{}^\frown z_1) \leq \varphi(x_2{}^\frown z_2)$.

(Our game and the lemmas that follow are obvious generalizations of [Mos80, 6E].)

We define $x_1 \preccurlyeq x_2$ to hold if and only if $x_1, x_2 \in g(Y)$ and player II has a winning strategy for the associated game. If we can show that $\preccurlyeq$ is a prewellordering, then our norm ψ is essentially defined. *Note*: We shall let ψ have range an initial segment of the ordinals.

LEMMA 4.1. *Assume all the games used in defining $\preccurlyeq$ are determined. There is no infinite sequence $x_0, x_1, x_2, \ldots$ of members of $g(Y)$ such that $x_i \not\preccurlyeq x_{i+1}$ for all i.*

PROOF. Suppose such a sequence exists. Let σ be a strategy for White witnessing that $x_0 \in g(Y)$. Let σ_i, for each i, be a winning strategy for player I for witnessing that $x_i \not\preccurlyeq x_{i+1}$. We shall define $z_i \in {}^\alpha\omega$ such that z_0 is a play according to σ and $\langle z_i, z_{i+1}\rangle$ is a play according to σ_i, for each i. Suppose each $z_i \restriction \beta$ is defined for some $\beta < \alpha$, with $z_0 \restriction \beta$ consistent with σ and each $\langle z_i \restriction \beta, z_{i+1} \restriction \beta\rangle$ consistent with σ_i. If White moves at β, let $z_0(\beta)$ be given by σ and, inductively, let $z_{i+1}(\beta)$ be given by σ_i. If Black moves at β, let $z_i(\beta)$ be given by σ_i.

Since z_0 is according to σ, $x_0{}^\frown z_0 \in Y$. Since $\langle z_i, z_{i+1}\rangle$ is according to σ_i, we have by induction that $z_{i+1} \in Y$ and $\varphi(x_{i+1}{}^\frown z_{i+1}) < \varphi(x_i{}^\frown z_i)$. This is a contradiction. ⊣

LEMMA 4.2. Assume that all games used in defining $\preccurlyeq$ are determined. $\preccurlyeq$ is a prewellordering of $g(Y)$.

PROOF. $x \preccurlyeq x$ since otherwise $x_i = x$ contradicts Lemma 4.1. Similarly, a failure of ($x \preccurlyeq x'$ or $x' \preccurlyeq x$) contradicts Lemma 4.1. If $x_1 \preccurlyeq x_2$ and $x_2 \preccurlyeq x_3$, then composing the two strategies in the obvious way gives $x_1 \preccurlyeq x_3$. Lemma 4.1 implies directly that the relation $(x_1 \preccurlyeq x_2 \wedge x_2 \not\preccurlyeq x_1)$ is well-founded. ⊣

For any x_1 and x_2 we call the associated game **the game for verifying** $x_1 \in g(Y)$ **and either** $x_2 \notin g(Y)$ **or** $\psi(x_1) \leq \psi(x_2)$. Note that player II has a winning strategy for the game if and only if $x_1 \in g(Y)$ and either $x_2 \notin g(Y)$ or $\psi(x_1) \leq \psi(x_2)$.

We now define **the game for verifying** $x_1 \in g(Y)$ **and either** $x_2 \notin g(Y)$ **or** $\psi(x_1) < \psi(x_1)$. This game is played just as the one for $x_2 \in g(Y)$ and either $x_1 \notin g(Y)$ or $\psi(x_2) \leq \psi(x_1)$ except that player I wins if and only if $x_1{}^\frown z_1 \in Y$ and either $x_2{}^\frown z_2 \notin Y$ or $\varphi(x_1{}^\frown z_1) < \varphi(x_2{}^\frown z_2)$. *Note*: Unlike the games of §2 and §3, it is a winning strategy for player I, not for player II, which "verifies" the relation in question.

LEMMA 4.3. Assume all relevant games are determined. Player I has a winning strategy for the game for verifying $x_1 \in g(Y)$ and either $x_2 \notin g(Y)$ or $\psi(x_1) < \psi(x_2)$ if and only if $x_1 \in g(Y)$ and either $x_2 \notin g(Y)$ or $\psi(x_1) < \psi(x_2)$.

PROOF. If player I has a strategy for this game clearly $x_1 \in g(Y)$. If $x_2 \in g(Y)$ and $\psi(x_2) \leq \psi(x_1)$, we can play player II's strategy witnessing this against player I's strategy and get a contradiction.

Suppose $x_1 \in g(Y)$ and either $x_2 \notin g(Y)$ or $\psi(x_1) < \psi(x_2)$ and suppose also that player II has a winning strategy for the game. We can compose this strategy with player II's strategy witnessing $x_1 \preccurlyeq x_1$, getting a strategy witnessing that $x_2 \preccurlyeq x_1$. ⊣

§5. **The main construction.** Let $Y \subseteq {}^{(\gamma+\alpha)}\omega$ and let $g: \alpha \to \{0,1\}$ with α countable. Let $\vec{\varphi} = \langle \varphi_i : i \in \omega \rangle$ be a putative scale on Y. Let $\beta \mapsto k_\beta$ be a one-one function from α into ω.

For $\beta \le \alpha$, let $g_\beta(\delta) = g(\beta + \delta)$ for $\beta + \delta < \alpha$. Let $X_\beta = g_\beta(Y)$. $X_\beta \subseteq {}^{(\gamma+\beta)}\omega$. For $\beta_1 < \beta_2 \le \alpha$, let $g_{\beta_1,\beta_2} = g_{\beta_1} \restriction \delta$ where δ is the least ordinal such that $\beta_1 + \delta = \beta_2$. Note that $X_{\beta_1} = g_{\beta_1,\beta_2}(X_{\beta_2})$.

We define a relation $\lhd$ on $\omega \times (\alpha + 1)$ as follows:

If $\beta < \alpha$ and $k_\beta < n$, then $\langle m, \beta + 1 \rangle \lhd \langle n, \beta \rangle$ for all $m \le n$. If $\beta < \alpha$ and $k_\beta \ge n$, let δ be minimal such that $\delta = \alpha$ or $k_\delta < n$. $\langle n, \delta \rangle \lhd \langle n, \beta \rangle$. Otherwise $\lhd$ never obtains.

Lemma 5.1. $\lhd$ is a well-founded relation.

Proof. Suppose $\langle n_0, \beta_0 \rangle \rhd \langle n_1, \beta_1 \rangle \rhd \cdots$. Since $n_0 \ge n_1 \ge n_2 \ge \cdots$, $n_i = n$ for all $i \ge$ some i_0. Since $\beta_0 < \beta_1 < \cdots$, let $\beta < \sup\{\beta_i : i \in \omega\}$ be such that $k_\delta \ge n$ for all δ such that $\beta \le \delta < \sup\{\beta_i : i \in \omega\}$. Let i be such that $i \ge i_0$ and $\beta_i \ge \beta$. Since $n_i = n$ and $k_\delta \ge n$ for all δ with $\beta_i \le \delta \le \beta_{i+1}$, $\langle n_{i+1}, \beta_{i+1} \rangle \not\lhd \langle n_i, \beta_i \rangle$. ⊣

The construction which follows will depend on certain games' being determined. *Assume for the rest of this section that all relevant games are determined.*

By induction on $\lhd$ we define norms ψ_i^β on X_β for $i \in \omega$ and $\beta \le \alpha$.

Let $\psi_i^\alpha = \varphi_i$ for all $i \in \omega$.

If $k_\beta < i$ and $g(\beta) = 0$, let $\psi_i^\beta = (\mathrm{Inf}_{k_\beta} \vec{\psi}^{\beta+1})_i$. This is well defined since it depends only on $\psi_0^{\beta+1}, \ldots, \psi_i^{\beta+1}$.

If $k_\beta < i$ and $g(\beta) = 1$, let $\psi_i^\beta = (\mathrm{Sup}_{k_\beta} \vec{\psi}^{\beta+1})_i$.

If $k_\beta > i$, let $\psi_i^\beta = g_{\beta,\delta}(\psi_i^\delta)$ where $\delta > \beta$ is minimal such that $\delta = \alpha$ or $k_\delta < i$.

Lemma 5.2. Let $\beta < \alpha$ and $x_1, x_2 \in X_\beta$. For all $i \in \omega$, if $g(\beta) = 0$ then $\psi_i^\beta(x_1) \le \psi_i^\beta(x_2) \Leftrightarrow (\mathrm{Inf}_{k_\beta} \vec{\psi}^{\beta+1})_i(x_1) \le (\mathrm{Inf}_{k_\beta} \vec{\psi}^{\beta+1})_i(x_2)$, and if $g(\beta) = 1$ then $\psi_i^\beta(x_1) \le \psi_i^\beta(x_2) \Leftrightarrow (\mathrm{Sup}_{k_\beta} \vec{\psi}^{\beta+1})_i(x_1) < (\mathrm{Sup}_{k_\beta} \vec{\psi}^{\beta+1})_i(x_2)$.

Proof. For $i > k_\beta$ it is immediate from the definition that ψ_i^β is identical with the corresponding infimum or supremum norm. Assume then that $i \le k_\beta$. For some $\delta > \beta$, $\psi_i^\beta = g_{\beta,\delta}(\psi_i^\delta)$. It is easily seen that $g_{\beta,\delta}(\psi_i^\delta) = g_{\beta,\beta+1}(g_{\beta+1,\delta}(\psi_i^\delta)) = g_{\beta,\beta+1}(\psi_i^{\beta+1})$. Since $i \le k_\beta$, the verification game for $g_{\beta,\beta+1}(\psi_i^{\beta+1})$ is identical with that for $(\mathrm{Inf}_{k_\beta}(\vec{\psi}^{\beta+1}))_i$ or $(\mathrm{Sup}_{k_\beta}(\vec{\psi}^{\beta+1}))_i$, depending on whether $g(\beta) = 0$ or $g(\beta) = 1$. ⊣

Note: ψ_i^β may not be identical with the corresponding infimum or supremum norm when $i \le k_\beta$, simply because the infimum and supremum norms need not have range an initial segment of the ordinals. Had we wished, we could have modified the definitions to make this so.

LEMMA 5.3. *If $k_\delta \geq i$ for all δ such that $\beta_1 \leq \delta < \beta_2$ then $\psi_i^{\beta_1} = g_{\beta_1,\beta_2}\psi_i^{\beta_2}$.*

By induction on $\lhd$ we define **games for verifying** $x_1 \in X_\beta$ **and either** $x_2 \notin X_\beta$ **or** $\psi_i^\beta(x_1) \leq \psi_i^\beta(x_2)$, and also **games for verifying that** $x_1 \in X_\beta$ **and** $x_2 \notin X_\beta$ **or** $\psi_i^\beta(x_1) < \psi_i^\beta(x_2)$.

If $k_\beta \geq i$, let $\delta > \beta$ be minimal such that $\delta = \alpha$ or $k_\delta < i$. First play the game given by the fact that $\psi_i^\beta = g_{\beta,\delta}\psi_i^\delta$ (reversing the roles of player I and player II in the $<$ case). If $\langle z_1, z_2\rangle$ is a play of this game, then player II wins just in case $x_1{}^\frown z_1 \in X_\delta$ and either $(x_2{}^\frown z_2) \notin X_\delta$ or $\psi^\delta(x_1{}^\frown z_1) <$(resp. $\leq$) $\psi^\delta(x_2{}^\frown z_2)$. Now play the game for verifying this fact.

If $k_\beta < i$, we first play the finite game given by the fact that ψ_i^β is an infimum or supremum norm defined from $\vec{\psi}_i^{\beta+1}$. The winning conditions for this game involve an inequality on $\psi_{i'}^{\beta+1}$, for some $i' \leq i$ or else a numerical inequality. In the latter case, the game terminates. In the former case, continue by playing the appropriate game.

A **standard play** of the game for verifying $x_1 \in X_\beta$ and either $x_2 \notin X_\beta$ or $\psi_i^\beta(x_1) \leq \psi_i^\beta(x_2)$ is a play in which all the plays of constituent games for infimum or supremum norms are standard. The **terminal condition** of a standard play is defined in the obvious ways. We also speak of a **standard partial play** of such a game and of the **terminal condition** of such a partial play.

LEMMA 5.4. *If $\vec{\varphi}$ is 0-lsc so is $\vec{\psi}^0$.*

PROOF. Let $\langle x_j : j \in \omega\rangle$ converge to x and suppose that each $x_j \in g(Y)$ and that the $\psi_i^0(x_j)$ are eventually constant as j increases. Thinning the sequence if necessary, we may suppose that $\psi_i(x_i) = \psi_i(x_j)$ for each $j \geq i$. For a contradiction, let σ be a winning strategy for player I for the game for verifying $x_0 \in g(Y)$ and either $x \notin g(Y)$ or $(g(\varphi_0))(x_0) < (g(\varphi_0))(x)$. Note that $g(\varphi_0) = \psi_0^0$. Let σ_i be a winning strategy for player II for the game for verifying $x_{i+1} \in g(Y)$ and either $x_i \notin g(Y)$ or $\psi_i^0(x_{i+1}) \leq \psi_i^0(x_i)$.

We shall define $z, z_0, z_1, \ldots$, each $\in {}^\alpha\omega$, functions $f^\beta : \omega \to \omega$ for $\beta \leq \alpha$, and standard plays $w_0, w_1, \ldots$, with each w_i consistent with σ_i, such that the following conditions hold:

i. $\langle z_i : i \in \omega\rangle$ converges to z.
ii. $\langle z, z_0\rangle$ is a play consistent with σ.
iii. For each $\alpha \leq \beta$ and each $i \in \omega$, the terminal condition of the corresponding part of the play w_i is

$$\langle x_{i+1}{}^\frown z_{i+1}\restriction\beta, x_i{}^\frown z_i\restriction\beta, f^\beta(i)\rangle.$$

iv. $f^0(i) = i$ for all i.
v. $f^\beta(i) \leq f^\beta(i+1)$ for all $\beta \leq \alpha$ and all $i \in \omega$.
vi. Rangef^β is unbounded for all $\beta \leq \alpha$.
vii. $f^{\beta+1}(i) \geq \min\{f^\beta(i), k_\beta\}$.

viii. For each i, $f^{\beta}(i)$ is a non-increasing function of β.
ix. If $\lambda \leq \alpha$ is a limit ordinal, then $f^{\lambda}(i) = \inf\{f^{\beta}(i) : \beta < \lambda\}$ for all $i \in \omega$.

Suppose $z^{\beta} = z\upharpoonright\beta$, $z_0^{\beta} = z_0\upharpoonright\beta, \ldots, f^{\beta}$, and the appropriate parts, w_i^{β}, of the w_i are defined. Suppose $\langle z_i^{\beta} : i \in \omega\rangle$ converges to z^{β}, (z^{β}, z_0^{β}) is consistent with σ, the w_i^{β} are consistent with σ_i, and iii, v, and vi hold at β, and $f^{\beta}(i) \leq i$ for all $i \in \omega$.

Suppose first that $g(\beta) = 0$. iii, v, vi and $f^{\beta}(i) \leq i$ guarantee that the hypotheses of Lemma 2.1 are satisfied, with Y replaced by $X_{\beta+1}$, k replaced by k_{β}, with (the appropriate fragments of) the σ_i, and with $f(i) = f^{\beta}(i)$. Let $n, n_1, n_2, \ldots$ be as given by Lemma 2.1. Let n_0 be the move given by σ at $(z^{\beta\frown}n, z_0^{\beta})$. Let f^* be as given by Lemma 2.1. Let $f^{\beta+1}(i) = f^*(i)$. Let $w_i^{\beta+1}$ be w_i^{β} followed by the play given by Lemma 2.1. Let $z(\beta) = n$ and $z_i(\beta) = n_i$. Lemma 2.1 guarantees that $\langle z_i^{\beta+1} : i \in \omega\rangle$ converges to $z^{\beta+1}$, the $w_i^{\beta+1}$ are consistent with the σ_i, iii, v, and vi hold at $\beta+1$, $f^{\beta+1}(i) \leq f^{\beta}(i) \leq i$ for all i, and $f^{\beta+1}(i) \geq \min\{f^{\beta}(i), k_{\beta}\}$.

Next suppose that $g(\beta) = 1$. We proceed just as in the first case, using Lemma 3.1 in place of Lemma 2.1. We let n be the move given by σ at (z^{β}, z_0^{β}). We omit the details.

Finally suppose that $\lambda < \alpha$ is a limit ordinal, that z^{λ}, the z_i^{λ}, the w_i^{λ}, and the f^{β} for $\beta < \lambda$, are defined. Suppose $\langle z_i^{\beta} : i \in \omega\rangle$ converges to z^{β} for each $\beta < \lambda$, (z^{β}, z_0^{β}) is consistent with σ for $\beta < \lambda$, the w_i^{β} are consistent with σ_i for $\beta < \lambda$, iv holds, iii, v, vi, and vii hold for $\beta < \lambda$, ix holds for limit ordinals $< \lambda$, and each $f^{\beta}(i)$ is non-increasing as a function of $\beta < \lambda$.

It follows that $\langle z_i^{\lambda} : i \in \omega\rangle$ converges to z^{λ}, $\langle z^{\lambda}, z_0^{\lambda}\rangle$ is consistent with σ, and each w_i^{λ} is consistent with σ_i. Define f_i^{λ} by ix. It is readily seen that iii and v hold at λ. To see that vi holds at λ, let $n \in \omega$. We must show $f^{\lambda}(i) \geq n$ for some i. Let $\beta < \lambda$ be such that $k_{\delta} \geq n$ for every δ such that $\beta \leq \delta < \lambda$. Let i be such that $f^{\beta}(i) \geq n$. For $\beta \leq \delta < \lambda$, $f^{\delta+1}(i) \geq \min\{f^{\delta}(i), k_{\delta}\}$ by v and vii. Thus induction and ix give that $f^{\delta}(i) \geq n$ for all $\delta < \lambda$. Hence $f^{\lambda}(i) \geq n$.

We have given the construction and verified that it has the required properties. Now let us prove the lemma. Since $\langle z, z_0\rangle$ is a play consistent with σ, $x_0{}^\frown z_0 \in Y$ and either $x^\frown z \notin Y$ or $\varphi_0(x_0{}^\frown z_0) < \varphi_0(x^\frown z)$. Since each w_i is consistent with σ_i and the terminal condition of w_i is $\langle x_{i+1}{}^\frown z_{i+1}, x_i{}^\frown z_i, f^{\alpha}(i)\rangle$, it follows by induction that $x_i{}^\frown z_i \in Y$ and $\varphi_{f^{\alpha}(i)}(x_{i+1}{}^\frown z_{i+1} \leq \varphi_{f^{\alpha}(i)}(x_i{}^\frown z_i)$ for each $i \in \omega$. Since $\langle z_i : i \in \omega\rangle$ converges to z and $\langle x_i : i \in \omega\rangle$ converges to x, it follows, since $\vec{\varphi}$ is 0-lsc, that $x^\frown z \in Y$ and $\varphi_0(x^\frown z) \leq \lim_i \varphi_0(x_i{}^\frown z_i) \leq \varphi_0(x_0{}^\frown z_0)$. This is a contradiction. ⊣

THEOREM 5.5. If $\vec{\varphi}$ is a scale, so is $\vec{\psi}^0$.

PROOF. Let $\langle n, \beta \rangle$ be $\lhd$-minimal such that $\vec{\varphi}^\beta$ is not n-lsc. If $k_\beta < n$, it follows from Lemmas 2.2 and 3.2 that $\vec{\varphi}^{\beta+1}$ is not n-lsc. Since $\langle n, \beta + 1 \rangle \lhd \langle n, \beta \rangle$ this is a contradiction. If $k_\beta \geq n$, let $\gamma > \beta$ be the least ordinal such that $\gamma = \alpha$ or $k_\gamma < n$. Apply Lemma 5.4 with Y replaced by X_γ, g replaced by $g_{\beta,\gamma}$, and $\langle \varphi_i : i \in \omega \rangle$ replaced by $\langle \psi^\gamma_{n+i} : i \in \omega \rangle$. Since $\vec{\psi}^\beta$ is not n-lsc, this gives that $\vec{\psi}^\gamma$ is not n-lsc. Since $\langle n, \gamma \rangle \lhd \langle n, \beta \rangle$, we have a contradiction. ⊣

§6. Canonical winning strategies. Let g, Y, $\beta \mapsto k_\beta$, $\vec{\varphi}$, $\langle \vec{\psi}^\beta : \beta \leq \alpha \rangle$, and $\langle X_\beta : \beta \leq \alpha \rangle$ be as in §5, with $\vec{\varphi}$ a scale on Y. Let $x \in g(y)$. We define the **canonical strategy for White** as follows:

At position z^β with $g(\beta) = 0$, if $x^\frown z^\beta \notin X^\beta$, White plays $z(\beta) = 0$. If $x^\frown z^\beta \in X_\beta$, player I plays the smallest n such that $x^\frown z^{\beta\frown} n \in X_{\beta+1}$ and, for all m such that $x^\frown z^{\beta\frown} m \in X_{\beta+1}$, $\psi^{\beta+1}_{k_\beta}(x^\frown z^{\beta\frown} n) \leq \psi^{\beta+1}_{k_\beta}(x^\frown z^{\beta\frown} m)$.

THEOREM 6.1. Assume all games involved in the definition of the $\vec{\psi}^\beta$ are determined. The canonical strategy is a winning strategy.

PROOF. We prove the following stronger fact. Let z^β be any position consistent with the canonical strategy. Let $n \in \omega$ and $\gamma \geq \beta$ be such that, for all δ with $\beta \leq \delta < \gamma$, $k_\delta \geq n$. We define a game $\mathrm{G}(z^\beta, n, \gamma)$ whose length is 2ρ, where ρ is the least ordinal such that $\beta + \rho = \gamma$. Let $z^\beta_1 = z^\beta_2 = z^\beta$. For $\beta \leq \delta < \gamma$, if $g(\delta) = 1$, player I plays $z_1(\delta)$ and then player II plays $z_2(\delta)$. If $\beta \leq \delta < \gamma$, and $g(\delta) = 0$, player II must play $z_1(\delta) =$ the move given by the canonical strategy at z^δ_1; then player I plays $z_2(\delta)$. Player II wins just in case $x^\frown z^\gamma_1 \in X_\gamma$ and either $x^\frown z^\gamma_2 \notin X_\gamma$ or $\psi^\gamma_n(x^\frown z^\gamma_1) \leq \psi^\gamma_n(x^\frown z^\gamma_2)$.

We shall prove that player II has a winning strategy for $\mathrm{G}(z^\beta, n, \gamma)$. Suppose this is false for some $\langle \gamma, \beta, n \rangle$ and choose the lexocographically least $\langle \gamma, \beta, n \rangle$. Choose some z^β witnessing this fact.

We first show that γ is a limit ordinal $> \beta$. If $\gamma = \beta$, then $z^\beta \notin X_\beta$. Let $\gamma' = \beta$, let $n' = 0$, and let $\beta' = 0$. Player I can win $\mathrm{G}(z^0, n', \gamma')$ by playing $z_1(\delta) = z^\beta(\delta)$ and playing $z_2(\delta)$ according to some strategy witnessing $x \in X_0$. This contradicts the minimality of $\langle \gamma, \beta, n \rangle$ unless $\beta = 0$, which is clearly impossible. If $\beta < \gamma = \delta + 1$, let player II play a winning strategy for $\mathrm{G}(z^\beta, n, \delta)$. By the definition of the canonical strategy, the play must reach a winning position $\langle z^\delta_1, z^\delta_2 \rangle$ in $\mathrm{G}(z^\beta, n, \gamma)$.

Let us then consider the case γ is a limit ordinal $> \beta$. We choose a winning strategy τ for player I for $\mathrm{G}(z^\beta, n, \gamma)$ as follows. If some z^γ extending z^β and consistent with the canonical strategy is such that $x^\frown z^\gamma \notin X_\gamma$, let player I play $z_1(\delta) = z^\gamma(\delta)$. By the argument of the last paragraph, z^β is won for White, so let player I play a $z_2(\delta)$ according to a winning strategy for White. If every position z^γ extending z^β which is consistent with the canonical strategy satisfies $x^\frown z^\gamma \in X_\gamma$, let player I play an arbitrary winning strategy for $G(z^\beta, n, \gamma)$. Note that, in either case, every play $\langle z_1^\gamma, z_2^\gamma\rangle$ consistent with τ satisfies $x^\frown z_2^\gamma \in X_\gamma$ and either $x^\frown z_1^\gamma \notin X_\gamma$ or $\psi_n^\gamma(x^\frown z_1^\gamma) > \psi_n^\gamma(x^\frown z_2^\gamma)$.

Let $z_0^\beta = z^\beta$ and $\beta_0 = \beta$. Suppose inductively we have defined $\beta_0 < \cdots < \beta_i < \gamma$ and $z_0^{\beta_j}, z_1^{\beta_j}, \ldots, z_j^{\beta_j}, z^{\beta_j}$ for each $j \leq i$, and strategies σ_i for $j < i$ such that

a. $z_j^{\beta_j} \subseteq z_j^{\beta_{j+1}} \subseteq \cdots \subseteq z_j^{\beta_i}$;
b. $z^{\beta_0} \subseteq z^{\beta_1} \subseteq \cdots \subseteq z^{\beta_i}$;
c. $(z^{\beta_i}, z_0^{\beta_i})$ is consistent with τ;
d. $z_j^{\beta_j} \in X_{\beta_j}$ for all $j \leq i$;
e. For all $j < i$, $\psi_{n+j}^{\beta_{j+1}}(x^\frown z_{j+1}^{\beta_{j+1}}) \leq \psi_{n+j}^{\beta_{j+1}}(x^\frown z_j^{\beta_{j+1}})$, σ_j is a winning strategy for player II witnessing this fact, and $(z_{j+1}^{\beta_i}, z_j^{\beta_i})$ is consistent with σ_j. Furthermore, $k_\delta \geq n + j$ for all δ such that $\beta_j \leq \delta < \gamma$.
f. $x^\frown z^{\beta_i} \in X_{\beta_i}$ and $\psi_{n+1}^{\beta_i}(x^\frown z^{\beta_i}) \leq \psi_{n+i}^{\beta_i}(x^\frown z_i^{\beta_i})$.

Let $\beta_{i+1} > \beta_i$ be such that $\beta_{i+1} < \gamma$ and $k_\delta \geq n + i + 1$ for all δ such that $\beta_{i+1} \leq \delta < \gamma$. By the minimality of γ, let τ_i be a winning strategy for player II for $G(z^{\beta_i}, n + i, \beta_{i+1})$. We get $z_0^{\beta_{i+1}}, \ldots, z_{i+1}^{\beta_{i+1}}$ as follows: $(z_j^{\beta_{i+1}}, z_{j+1}^{\beta_{i+1}})$ will be consistent with σ_j, for all $j < i$. $(z_{i+1}^{\beta_{i+1}}, z_i^{\beta_{i+1}})$ will be consistent with some strategy for player II witnessing f. (In particular, $z_i^{\beta_{i+1}}$ will extend z^{β_i}.) $(z^{\beta_{i+1}}, z_{i+1}^{\beta_{i+1}})$ will be consistent with τ_i. $(z^{\beta_{i+1}}, z_0^{\beta_{i+1}})$ will be consistent with τ. Since τ is a strategy for player I and all other strategies are for player II, the reader will easily check that the $z_0^{\beta_{i+1}}, \ldots, z_{i+1}^{\beta_{i+1}}, z^{\beta_{i+1}}$ are determined uniquely, once the strategy witnessing f is chosen. Since $\psi_{n+i}^{\beta_{i+1}}(x^\frown z_{i+1}^{\beta_{i+1}}) \leq \psi_{n+i}^{\beta_{i+1}}(x^\frown z_i^{\beta_{i+1}})$, we may now complete the construction by choosing σ_i witnessing this fact.

Now let $z^\gamma \supseteq z^{\beta_i}$ for each i and let $z_j^\gamma \supseteq z_j^{\beta_i}$ for each $i \geq j$. Since the σ_j are winning strategies, we have $x^\frown z_{j+1}^\gamma \in X_\gamma$ and $\psi_{n+j}^\gamma(x^\frown z_{j+1}^\gamma) \leq \psi_{n+j}^\gamma(x^\frown z_j^\gamma)$ for each j. By the properties of τ, we have $x^\frown z_0^\gamma \in X_\gamma$ and $\psi_n^\gamma(x^\frown z_0^\gamma) < \psi_n^\gamma(x^\frown z^\gamma)$. Since $\langle x^\frown z_j^\gamma : j \in \omega\rangle$ converges to $x^\frown z^\gamma$ (note that $z_{i+1} \restriction \beta_i = z \restriction \beta_i$), we must have $x^\frown z^\gamma \in X^\gamma$ and $\psi_n^\gamma(x^\frown z^\gamma) \leq \lim_j \psi_n^\gamma(x^\frown z_j^\gamma) \leq \psi_n^\gamma(x^\frown z_0^\gamma)$. This is a contradiction. ⊣

§7. Definability. For pointclasses Γ and game types g, we wish to define a pointclass $g(\Gamma)$ and prove theorems such as Scale(Γ) $\Rightarrow$ Scale($g(\Gamma)$). Since Γ is to be a pointclass in the sense of [Mos80], we shall define $g(\Gamma)$ only under the following assumption:

(∗) There is a well-ordering R of a subset A of ω of order type α, where $g\colon \alpha \to \{0, 1\}$, such that, if $\beta \mapsto k_\beta$ is the isomorphism between $(\alpha, <)$ and (A, R), and $\hat{g}(k_\beta) = g(\beta)$, then Γ is closed under preimages by functions recursive in $(R, \hat{g})$.

Let $\beta \mapsto k_\beta$ be a one-one function from an ordinal α into ω. Let R be the induced well-ordering of a subset of ω. For $z \in {}^\omega\omega$, let $z^*_R \in {}^\alpha\omega$ be given by $z^*_R(\beta) = z(k_\beta)$.

Let $g\colon \alpha \to \{0, 1\}$. Let γ be a pointclass satisfying (*). For simplicity we shall assume that pointclasses are collections of sets each of which is a subset of ${}^\gamma\omega$, where $\gamma < \omega^2$ but γ is not fixed.

$X \subseteq {}^\gamma\omega$ belongs to $g(\Gamma)$ just in case there is an R witnessing that Γ satisfies (*) and a $Y \in \Gamma$ such that, for the associated $\beta \mapsto k_\beta$,

$$X = g(\{x^\frown z^*_R : x^\frown z \in Y\}).$$

THEOREM 7.1. Let α be a countable limit ordinal. Let $g\colon \alpha \to \{0, 1\}$ be such that, for every limit ordinal $\lambda < \alpha$ and every $n \in \omega$, $\{g(\lambda + m) : m \geq n\} = \{0, 1\}$. (In other words, White and Black both make infinitely many moves in every ω-block.) Suppose Γ satisfies (*) and that all games involved in defining $g(\Gamma)$ are determined. Then Scale(Γ) implies Scale($g(\Gamma)$).

PROOF. Let Y, R witness that $X \in g(\Gamma)$. Let $\beta \mapsto k_\beta$ be the associated function. Let $\vec{\varphi}$ be a scale on Y as given by Scale(Γ). Let $\varphi^*_i(x^\frown z^*_R) = \varphi_i(x^\frown z)$. $\vec{\varphi^*}$ is a scale on Y^* with $X = g(Y^*)$. Let $\vec{\psi^\beta}$, $\beta \leq \alpha$ be defined as in §5. We must show that the relations

a. $x_1 \in X \wedge (x_2 \notin X$ or $\psi^0_i(x_1) \leq \psi^0_i(x_2))$

b. $x_1 \in X \wedge (x_2 \notin X$ or $\psi^0_i(x_1) < \psi^0_i(x_2))$

are in Γ. (We take these relations to be subsets of ${}^{(\omega+\omega+1)}\omega$.) Since the two cases are similar, we consider only $\leq$.

By inserting dummy moves where necessary, we can make a hold just in case player II wins $\mathrm{G}(x_1, x_2, i)$, where $\mathrm{G}(x_1, x_2, i)$ is a game of type g. The winning conditions of a play of this game are of one of the forms

$$n_1 \leq n_2$$
$$n_1 < n_2$$
$$x_1{}^\frown z_1 \in Y \wedge (x_2{}^\frown z_2 \notin Y \text{ or } \varphi_j(x_1{}^\frown z_1) \leq \varphi_j(x_2{}^\frown z_2))$$
$$x_1{}^\frown z_1 \in Y \wedge (x_2{}^\frown z_2 \notin Y \text{ or } \varphi_j(x_1{}^\frown z_1) < \varphi_j(x_2{}^\frown z_2))$$

where z_1, z_2 and n_1, n_2 or j are determined from the play in the obvious way.

We must find an R' and a Y' which witnesses that the property that player II has a winning strategy for $\mathrm{G}(x_1, x_2, i)$ belongs to $g(\Gamma)$.

We do this as follows: If the βth move in the new game corresponds to picking $z^*_{1R}(\delta)$, we let $k'_\beta = 4k_\delta$. If it corresponds to $z^*_{2R}(\delta)$, we let $k'_\beta = 4k_\delta + 2$. We let the other non-dummy moves corresponding to δ be $p^1_{k_\delta}$, $p^2_{k_\delta}$, and $p^3_{k_\delta}$ and $p^4_{k_\delta}$ if needed, where p_i is the $i + 1$st prime. We let the dummy moves corresponding to β be $p^5_{k_\delta}$, $p^6_{k_\delta}$, etc. The corresponding R' and $\hat{g}'$ are recursive in $(R, \hat{g})$. If $z^*_{R'}$ is a play of the game, z_1 and z_2 are clearly recursive uniformly in $(R, \hat{g})$ and z. We need to be able to find which of the four winning conditions hold and the value of j or (n_1, n_2) as a function of (x_1, x_2, i, z) recursive in $(R, \hat{g})$. Note that, given $m = k_\beta$ and $k \in \omega$, we can find effectively from R the q such that $q = k_\delta$, where δ is the least ordinal $\geq \beta$ such that $k_\delta < k$ (if it exists, and we can determine whether it exists). Repeating this procedure at most i times, beginning with $k = i$ and $m = k_0$, we can find all the places in the game $\mathrm{G}(x_1, x_2, i)$ where the numbers j_1, j_2 as in §2 and §3 are played. This allows us to compute the desired information. We omit the details. ⊣

COROLLARY 7.2. Let g be the type of the real game, i.e., let $g : \omega^2 \to \{0, 1\}$ with $g(\beta) = 0 \Leftrightarrow \beta = \omega \cdot 2k + n$ for some $k, n \in \omega$. Assume that all integer games of length ω^2 in which White moves at exactly the even ordinals and whose payoffs are in Γ (in the obvious sense) are determined. If Scale(Γ) then Scale($g(\Gamma)$).

PROOF. Let Y witness $X \in g(\Gamma)$. As in the proof of Theorem 7.1, we get a scale on X which belongs to $g'(\Gamma)$, where $g' : \omega^2 \to \{0, 1\}$ and $g'(\beta) = 0 \Leftrightarrow \beta$ is even. But $g'(\Gamma) = g(\Gamma)$. To see this, replace the g' game by a g game as follows: Replace each ω-block by two ω-blocks. White plays a strategy for the next ω moves of the original game and then Black chooses a play consistent with the strategy. ⊣

We could prove move complicated definability theorems by letting the φ_i belong to different classes Γ_i. We could also prove a generalization of Corollary 7.2 for real games of arbitrary countable length. We could also prove definability results for our canonical strategies. Since there are no ideas involved beyond those already presented and those of [Mos80] and [Mos83], we shall do none of this.

REFERENCES

ALEXANDER S. KECHRIS, DONALD A. MARTIN, AND YIANNIS N. MOSCHOVAKIS

[CABAL iii] *Cabal seminar* 79–81, Lecture Notes in Mathematics, no. 1019, Berlin, Springer, 1983.

YIANNIS N. MOSCHOVAKIS

[Mos80] ***Descriptive set theory***, Studies in Logic and the Foundations of Mathematics, no. 100, North-Holland, Amsterdam, 1980.

[Mos83] *Scales on coinductive sets*, this volume, originally published in Kechris et al. [CABAL iii], pp. 77–85.

DEPARTMENT OF MATHEMATICS
UNIVERSITY OF CALIFORNIA
LOS ANGELES, CA 90095, USA
E-mail: dam@math.ucla.edu

LONG GAMES

JOHN R. STEEL

The hypothesis that definable games are determined has proven very powerful in its realm, the realm of reals and definable sets of reals. For example, ZFC + $\mathsf{AD}^{\mathbf{L}(\mathbb{R})}$ seems to yield a "complete" theory of $\mathbf{L}(\mathbb{R})$, in the same way that ZFC alone yields a "complete" theory of $\mathbf{L}$. This success makes it natural to investigate stronger forms of definable determinacy, and the universes larger than $\mathbf{L}(\mathbb{R})$ which these might civilize. One might hope to ultimately bring determinacy techniques to bear on questions involving quantification over arbitrary sets of reals, for example, the question of the prewellordering property for Π^2_n.

Various papers, in particular Becker [Bec85], Blass [Bla75], Martin [Mar83B], Solovay [Sol78B], and Woodin (unpublished), have contributed to this investigation. These papers have been concerned with games of length strictly less than ω_1, on ω or on $\mathbb{R}$. In this paper we shall go a bit further and consider certain clopen (i.e., decided after countably many moves) games of length ω_1.

In §1 we show that the game quantifiers associated to these clopen games propagate scales, and in §2 we show that the games have canonical winning strategies. Of course, both results require the determinacy of the games in question. Our methods here extend those of Moschovakis [Mos80, Chapter 6], who proved these "third periodicity" theorems for game of length ω on ω, and of Martin [Mar83B] who extended Moschovakis' proof to games of length less than ω_1 on ω or $\mathbb{R}$. In §3 we show that the determinacy of clopen games of length ω_1 with payoff $\utilde{\mathbf{\Pi}}^1_1$ in the codes implies the existence of a natural inner model in which all games of length less than ω_1 on $\mathbb{R}$ (with arbitrary payoff) are determined. This section makes use of the results of §1 and §2, but not of their proofs. It also makes heavy use of unpublished seminar notes of Woodin, which show how to construct such a model if one exists; our contribution is just to show that some form of definable determinacy implies its existence. We also indicate in §3 how to construct variants of Woodin's model satisfying stronger forms of determinacy for arbitrary payoff. In §4 we consider the problem of proving the determinacy hypotheses we have been using, and obtain a partial result using the methods of Blass [Bla75] and

The author was partially supported by National Science Foundation grant DMS-3802555.

The Cabal Seminar. Volume I: Games, Scales and Suslin Cardinals
Edited by A. S. Kechris, B. Löwe, J. R. Steel
Lecture Notes in Logic, 31

Martin [Mar75]. Along the way we show that the inner model of §3 satisfies the determinacy of certain games on $\wp(\mathbb{R})$. Finally, §5 is devoted to remarks and questions.

As an expository device we work in ZF + DC throughout, and state our additional hypotheses as we need them. By $\mathbb{R}$ we mean ${}^{\omega}\omega$, the Baire space. What concepts we take for granted, in particular that of a scale, are explained in [Mos80].

§1. Some game quantifiers which propagate scales. Let us call $T \subseteq \bigcup_{\alpha<\omega_1} {}^{\alpha}\omega$ a tree if $\forall p \in T\ \forall \beta \in \mathrm{dom}(p)\ (p \restriction \beta \in T)$. Associated to such a T is

$$E = \{p \in \bigcup_{\alpha<\omega_1} {}^{\alpha}\omega : p \notin T\ \&\ \forall \beta \in \mathrm{dom}(p)\ (p \restriction \beta \in T)\}.$$

Suppose we are given some such T with associated E, and suppose that some $A \subseteq E$ is given. We then have a game $\mathrm{G}(A, T)$: player I and player II alternate playing natural numbers, player I moving first at limit ordinals. The game is over when they reach a position $p \in E$, in which case player I wins iff $p \in A$. (It is convenient to agree that if p is a position, then for all $\alpha \in \mathrm{dom}(p)$, $p(\alpha) = \langle m, n\rangle$ where m is player I's αth move and n is player II's αth move.) Let

$$\Game^T A = \{p \in {}^{\omega}\omega\ :\ p \in T\ \&\ \text{player I has a winning strategy in } \mathrm{G}(A, T) \text{ starting from } p\}.$$

Of course, if $p \in E \Longrightarrow \mathrm{dom}(p) < \omega_1$, so that T has no ω_1-branches, then $\mathrm{G}(A, T)$ is just the general form for a clopen (that is, decided after countably many moves) game of length ω_1. We want to restrict this notion a bit. Let WO be the set of (codes of) wellorders of ω. Let

$$F\colon (T \cup E) \cap \{p\ :\ \mathrm{dom}(p) \geq \omega\} \to \mathrm{WO},$$

where

$$\mathrm{dom}(p) = |F(p)| = \text{order type of } F(p)$$

for all $p \in \mathrm{dom}(F)$. For $p \in T \cup E$ such that $\mathrm{dom}(p) \geq \omega$, let

$$p^* = \langle F(p), x\rangle$$

where

$$x(n) = p(|n|_{F(p)}) = p\big(\text{ordinal rank of } n \text{ in } F(p)\big).$$

(Here $\langle F(p), x\rangle \in {}^{\omega}\omega \times {}^{\omega}\omega \approx {}^{\omega}\omega$.) For any $S \subseteq \mathrm{dom}(F)$, we let

$$S^* = \{p^*\ :\ p \in S\}.$$

Now let

$$C(i,j,k,y) \text{ iff } \exists p \in \operatorname{dom}(F)\ \big(y = p^* \ \&\ |i|_{F(p)} = |j|_{F(p\restriction\alpha)},\\ \text{where } \alpha = |k|_{F(p)}\big).$$

We call F a **scaled coding of** T if both T^* and C admit $\mathbf{HOD}(\mathbb{R})$ scales.

EXAMPLE 1.1. Fix $n \geq 1$. Let

$$T = \Big\{p \in \bigcup_{\alpha<\omega_1} {}^{\alpha}\omega : \forall\beta \leq \operatorname{dom}(p)\big((L_\beta[p\restriction\beta], \in, p\restriction\beta) \text{ is not } \Sigma_n \text{ admissible}\big)\Big\}.$$

For $q \in T \cup E$, $\operatorname{dom}(q) \geq \omega$, let $F(q)$ = the first wellorder of ω of order type $\operatorname{dom}(q)$ constructed in $\mathbf{L}[q]$. Then T^* and C are Π^1_1, so that F is a scaled coding of T.

We shall now prove our general scale propagation theorem. We remark afterward on refinements of the theorem involving weaker determinacy hypotheses and better definability estimates on the scale produced.

THEOREM 1.2. *Assume that all clopen games of length ω_1 with $\mathbf{HOD}(\mathbb{R})$ payoff are determined. Let T admit scaled coding, and suppose that $A \subseteq E$ is such that A^* admits a $\mathbf{HOD}(\mathbb{R})$ scale. Then $\Game^T A$ admits a $\mathbf{HOD}(\mathbb{R})$ scale.*

PROOF. We assume that if $p \in T$ and $\alpha \in \operatorname{dom}(p)$, then $p(\alpha)_1 \in \{0,1\}$; that is, player II can only play 0's and 1's in $\mathrm{G}(A,T)$. This is no loss of generality.

Let F be a scaled coding of T, and $\vec{\rho}$ a $\mathbf{HOD}(\mathbb{R})$ scale on the associated C. This gives us some useful norms on A^* relating "global and local codes" of ordinals. For $p^* \in A^*$, let

$$\vartheta^0_i(p^*) = \big(n, \rho_{(i)0}(n,(i)_1,(i)_2,p^*)\big)$$

where n is unique such that $C(n,(i)_1,(i)_2,p^*)$.

REMARKS.

(a) We code elements of $\omega^{<\omega}$ by prime powers, so that $\langle n_0 \dots n_k\rangle = 2^{n_0+1} \cdot 3^{n_1+1} \cdots$, and $(i)_k$ = (exponent of p_k in i) $- 1$. Let $(0)_k = 0$.

(b) $\vartheta^0_i(p^*)$ is an ordinal gotten by using the lexicographic order of $\omega \times \operatorname{ran}\vec{\rho}$.

Let also, for $p^* \in A^*$,

$$\vartheta^1_i(p^*) = \big(n, \rho_{(i)_0}((i)_1, n, (i)_2, p^*)\big)$$

where n is unique such that $C((i)_1, n, (i)_2, p^*)$ if any such n exists (i.e., if $|(i)_1|_{F(p)} < |(i)_2|_{F(p)}$). Let $\vartheta^1_i(p^*) = 0$ if no such n exists.

Let $\vec{\sigma}$ be a $\mathbf{HOD}(\mathbb{R})$ scale on T^*, and for $p^* \in A^*$ let

$$\vartheta^2_i(p^*) = \sigma_{(i)_0}((p\restriction\alpha)^*)$$

where $\alpha = |(i)_1|_{F(p)}$, and

$$\vartheta_i^3(p^*) = |i|_{F(p)}$$

and

$$\vartheta_i^4(p^*) = n, \text{ where } |n|_{F(p)} = i.$$

Finally, let $\vec{\vartheta_i^5}$ be a very good scale on A^*, and set, for $p^* \in A^*$,

$$\psi_i(p^*) = \langle \mathrm{dom}(p), \vartheta_0^0(p^*), \dots, \vartheta_0^5(p^*), \vartheta_1^0(p^*), \dots, \vartheta_1^5(p^*), \dots, \vartheta_i^0(p^*), \dots, \vartheta_i^5(p^*)\rangle.$$

We now describe some games which lead to a scale on $\Game^T A$. As in Moschovakis' proof that the ordinary game quantifier propagates scales, we compare positions $p, q \in \Game^T A$ by playing out $\mathrm{G}(A, T)$ from each position simultaneously on two boards. This assigns an ordinal value to each such position. The new ingredient is that in these comparison games the players must now make additional moves. These moves reflect the ordinal value they assign to one-move variants of intermediate positions they reach during the game.

So let $p, q \in \Game^T A$, and let $k \in \omega$. We shall define a game $\mathrm{G}_k(p, q)$. The players in $\mathrm{G}_k(p, q)$ are F and S. They play on two boards, the p board and the q board, and make additional moves lying on neither board. On the p board S plays $\mathrm{G}(A, T)$ from p as player I while F plays as player II. On the q board F plays $\mathrm{G}(A, T)$ from q as player I while S plays as player II. Play is divided into rounds; we now describe the typical round.

Round α.

(a) F makes player I's αth move on q board, then S makes player I's move on the p board.
(b) F now proposes some i such that $0 \leq i \leq k$ and $(i)_0 \in \{0, 1\}$.
(c) S either accepts i or proposes some i', $0 \leq i' < i$ and $(i')_0 \in \{0, 1\}$.
(d) Let $t \leq k$ be the least proposal made during (b) and (c):
Case 1. $t \neq 0$. Then F and S must play $(t)_0$ as player II's αth move on the p and q boards respectively.
Case 2. $t = 0$. Then F plays any $m \in \{0, 1\}$ as player II's αth move on the p board, after which S plays any $m' \in \{0, 1\}$ as player II's αth move on the q board.

This completes round α.

Remarks.

(a) We call the 0-proposal "freedom".

(b) Our description of round α is valid only if neither board has reached a position in E. As soon as one board reaches a position in E, F and S start simply playing $\mathrm{G}(A, T)$ in their proper roles on the other board, until it too reaches a position in E. $\mathrm{G}_k(p,q)$ ends when both boards have reached a position in E.

(c) A position in $\mathrm{G}_k(p,q)$ is a function $u\colon \alpha \to \omega$, where $u(\beta)$ codes the (up to) 6 moves during round β.

Suppose now that u is a run of $\mathrm{G}_k(p,q)$ and that $r \supseteq p$ and $s \supseteq q$ are the runs of $\mathrm{G}(A,T)$ produced in u on the two boards. Let $e \in \omega$ be the least n such that, letting $\alpha = |(n)_0|_{F(r)}$, either $\alpha \geq \mathrm{dom}(s)$ or $(n)_1$ was the least proposal made during round α. Let $\alpha = |(e)_0|_{F(r)}$. If either $\alpha \geq \mathrm{dom}(s)$ or F proposed $(e)_1$ during round α, then

$$S \text{ wins } u \text{ iff } \psi_e(r^*) \leq \psi_e(s^*).$$

If S proposed $(e)_1$ during round α, then

$$S \text{ wins } u \text{ iff } \psi_e(r^*) < \psi_e(s^*).$$

REMARKS.

(a) We call e the critical number of u, and write $e = \mathrm{crit}(u)$. Notice that $\mathrm{crit}(u)$ exists, and in fact $\mathrm{crit}(u) \leq \langle 0, k\rangle$, for every run u of $\mathrm{G}_k(p,q)$.

(b) Our convention is that ψ_e takes value ∞ off of A^*, that $\infty \leq \infty$, and that $x < \infty$ iff $x \in \mathrm{Ord}$. So, for example, if S proposed e_1 then S loses unless $r \in A$. For p and q in $\Game^T A$, let

$$p \leq_k q \text{ iff } S \text{ has a winning strategy in } \mathrm{G}_k(p,q).$$

The next lemma implies that $\leq_k$ is a prewellorder.

LEMMA 1.3. Let $p_0 \in \Game^T A$, and suppose that for all $n \geq 0$, Σ_n is either a winning strategy for F in $\mathrm{G}_k(p_n, p_{n+1})$ or a winning strategy for S in $\mathrm{G}_k(p_{n+1}, p_n)$. Then only finitely many Σ_n's are for F.

PROOF. Fix a winning strategy τ for player I from p_0 in $\mathrm{G}(A,T)$. Assume toward a contradiction that infinitely many Σ_n's are for F. We shall construct runs u_n according to Σ_n such that u_n and u_{n+1} agree on a common play $r_{n+1} \subseteq p_{n+1}$ on the p_{n+1} board, and the play $r_0 \subseteq p_0$ on the p_0 board is according to τ. (Notice Σ_n always plays as player I on the p_{n+1} board and player II on the p_n board.) The definition is by induction on rounds. Suppose we have $u_n \restriction \alpha$ for all n, and consequently $r_n \restriction \omega + \alpha$ for all n. (Here $r_n \restriction \omega = p_n$.) Suppose also that $r_n \restriction \omega + \alpha \in T$ for all n; that is, no board has reached a position in E. We now define round α of the u_n's. Let

$$a_0 = \tau(r_0 \restriction \omega + \alpha)$$

and

$$a_{n+1} = \begin{cases} \Sigma_n(u_n \restriction \alpha^\frown a_n) & \text{if } \Sigma_n \text{ is for } S \\ \Sigma_n(u_n \restriction \alpha) & \text{if } \Sigma_n \text{ is for } F \end{cases}$$

be the αth moves for player I given by τ and the Σ_n's. We must now define the proposal phase of $u_n(\alpha)$, for all n. We represent a proposal-acceptance/counterproposal by a pair (b, c) of numbers.

CLAIM 1.4. There are a $t \leq k$, and n_0, and pairs (b_n, c_n) for $n \in \omega$, such that

(i) $u_n \restriction \alpha^\frown \langle a_n, a_{n+1}, b_n, c_n \rangle$ is according to Σ_n (if Σ_n is for S; otherwise $u_n \restriction \alpha^\frown \langle a_{n+1}, a_n, b_n, c_n \rangle$ is according to Σ_n), and

(ii) $\langle b, c \rangle$ settles on freedom for $n < n_0$ and on t for $n \geq n_0$, and

(iii) for infinitely many n, $\langle b_n, c_n \rangle$ involves Σ_n proposing t.

PROOF (SKETCH). Let $t_0 \leq k$ be least such that infinitely many Σ's which are F-strategies will now propose t_0. Suppose that t_n is given. If all but finitely many Σ's which are S strategies will accept a t_n proposal, stop the induction and set $t = t_n$. Otherwise let $t_{n+1} < t_n$ be such an infinitely many Σ's for S counterpropose t_{n+1} when their opponent proposes t_n.

Since $t_{n+1} < t_n$, we eventually get t. One can check that t works. (If $t_0 = t$, the Σ's verifying (iii) are for F. Otherwise, the Σ's verifying (iii) are for S.) ⊣

Now let $t, n_0, \langle (b_n, c_n) : n \in \omega \rangle$ satisfy the claim. We imagine (b_n, c_n) as the proposal phase of $u_n(\alpha)$. We want finally an αth move d_n for player II on the p_n board.

Case 1. $t > 0$.

Then let

$$d_n = (t)_0 \quad \text{for } n \geq n_0$$

and

$$d_n = \begin{cases} \Sigma_n(u_n \restriction \alpha^\frown \langle a_n, a_{n+1}, b_n, c_n, d_{n+1} \rangle) & \text{if } \Sigma_n \text{ is for } S \\ \Sigma_n(u_n \restriction \alpha^\frown \langle a_{n+1}, a_n, b_n, c_n \rangle) & \text{if } \Sigma_n \text{ is for } F \end{cases}$$

for $n < n_0$.

Case 2. $t = 0$.

Let

$$d_n = \begin{cases} \Sigma_n(u_n \restriction \alpha^\frown \langle a_n, a_{n+1}, b_n, c_n, d_{n+1} \rangle) & \text{if } \Sigma_n \text{ is for } S \\ \Sigma_n(u_n \restriction \alpha^\frown \langle a_{n+1}, a_n, b_n, c_n \rangle) & \text{if } \Sigma_n \text{ is for } F \end{cases}$$

Since infinitely many Σ_n's are for F, this definition makes sense.

Finally, we set

$$u_n(\alpha) = \begin{cases} \langle a_n, a_{n+1}, b_n, c_n, d_{n+1}, d_n \rangle & \text{if } \Sigma_n \text{ is for } S \\ \langle a_{n+1}, a_n, b_n, c_n, d_n, d_{n+1} \rangle & \text{if } \Sigma_n \text{ is for } F. \end{cases}$$

Thus in any case $u_n(\alpha)$ and $u_{n+1}(\alpha)$ agree that $r_{n+1}(\alpha) = \langle a_{n+1}, d_{n+1}\rangle$, while $r_0(\alpha) = \langle a_0, d_0\rangle$ according to τ.

We can continue to define $u_n(\alpha)$ this way until we reach an α such that $r_i{\restriction}\omega + \alpha \in E$ for some i. This must happen. Fix the first such α, and fix i. Now since τ is a winning strategy for player I and each Σ_n is a winning strategy for F or S in G_k, we see that $r_n{\restriction}\omega + \alpha \in \Game^T A$ for all n. (Cf. the payoff for G_k and our convention $x < \infty$ for $x \in \mathrm{Ord}$.) But then since the first component in any $\psi_e(r^*)$ is $\mathrm{dom}(r)$, $r_n{\restriction}\omega + \alpha \in E$ for all $n \geq i$.

Now let $e \leq \langle 0, k\rangle$ be least such that $e = \mathrm{crit}(u_n)$ for infinitely many $n \geq i$. So $e \leq \mathrm{crit}(u_n)$ for cofinitely many $n \geq i$, so that $\vartheta^3_{(e)_0}((r_n{\restriction}\omega + \alpha)^*)$ is eventually constant as $n \to \infty$. Write $r_n = r_n{\restriction}\omega + \alpha$, and let

$$\beta = \text{eventual value of } \vartheta^3_{(e)_0}(r_0^*) = |(e)_0|_{F(r_n)}.$$

Thus infinitely many, and hence by construction cofinitely many, u_n's settle on the $(e)_1$ proposal round at β. For infinitely many n, Σ_n is responsible for the proposal. Thus

$$\psi_e(r^*_{n+1}) \leq \psi_e(r^*_n)$$

for cofinitely many n, while

$$\psi_e(r^*_{n+1}) < \psi_e(r^*_n)$$

for infinitely many n. This is a contradiction. ⊣

Corollary 1.5. $\leq_k$ is a prewellorder of $\Game^T A$.

Proof. *Reflexive*: if $p \not\leq_k p$ then $p, p, p, \ldots$ violates Lemma 1.3. *Connected*: if $p \not\leq_k q$ and $q \not\leq_k p$ then $pqpqpq\ldots$ violates Lemma 1.3. *Transitive*: if $p \leq_k q \leq_k r$, and $p \not\leq_k r$, then $prqprqprq\ldots$ violates Lemma 1.3. *Wellfounded*: Clear from Lemma 1.3. ⊣

Actually, reflexivity, connectedness, and transitivity could be proved by more direct finite diagrams. Now for $p \in \Game^T A$, let

$$\varphi_k(p) = \text{ordinal of } p \text{ in } \leq_k.$$

Lemma 1.6. $\vec{\varphi}$ is a semiscale on $\Game^T A$.

Proof. Let $p_n \to p$ as $n \to \infty$, $p_n \in \Game^T A$ for all n, and $\forall i\ (\varphi_i(p_n)$ eventually constant as $n \to \infty)$. By thinning the sequence of p_n's we may assume we have a winning strategy Σ_n for S in $G_n(p_{n+1}, p_n)$ for all n. Let τ be a winning strategy for player I in $\mathrm{G}(A, T)$ from p_0, and, towards a contradiction, let σ be a winning strategy for player II in $\mathrm{G}(A, T)$ from p.

We need a definition. Suppose $r_n \in T \cup E$ for $n \in \omega$, and for all k

$$|k|_{F(r_n)} \text{ is eventually constant} = \alpha_k,$$

and

$$r_n(\alpha_k) \text{ is eventually constant}$$

as $n \to \infty$. Let $\pi\colon \beta \to \{\alpha_k : k \in \omega\}$ be the enumeration of $\{\alpha_k : k \in \omega\}$ in increasing order, and define $r\colon \beta \to \omega$ by

$$r(\gamma) = \text{eventual value of } r_n(\pi(\gamma)) \text{ as } n \to \infty.$$

We write then

$$r = \lim^*_{n\to\infty} r_n.$$

This notion of "convergence in the codes" is more useful than pointwise convergence in what follows. ($\lim^*_{n\to\infty} r_n = r$ means just that r^*_n converges to r^* in a certain scale.)

We now define by induction on rounds runs u_n of $G_n(p_{n+1}, p_n)$ according to Σ_n. We arrange that u_n and u_{n+1} agree on a common play $r_{n+1} \supseteq p_{n+1}$ for the p_{n+1} board, and that the play $r_0 \supseteq p_0$ on the p_0 board is by τ. So assume that $u_n \restriction \alpha$, hence $r_n \restriction \omega + \alpha$, is given for all n. (Let $r_n \restriction \omega = p_n$.) Let

$$a_0 = \tau(r_o \restriction \omega + \alpha)$$

and

$$a_{n+1} = \Sigma_n(u_n \restriction \alpha^\frown a_n)$$

be the moves for player I on the various boards generated by τ and the Σ_n's. If a_n is eventually constant, say $a_n = a$ eventually, and if $\lim^*_{n\to\infty} r_n \restriction \omega + \alpha = r$ for some r which is a play by σ, let

$$d = \sigma(r^\frown a).$$

Otherwise, let $d = 0$. Now for any n let i_n be the largest i such that $\langle d, i\rangle \leq n$ and $\Sigma_n(u_n \restriction a^\frown a_n{}^\frown\langle d, i\rangle) =$ "accept", if such an i exists. Let $i_n = 0$ otherwise.

Case 1. $i_n \to \infty$ as $n \to \infty$.

Pick n_0 such that $i_n > 0$ for $n \geq n_0$. The proposal pair in $u_n(\alpha)$ is (b_n, c_n) where $b_n = \langle d, i_n\rangle$ for $n \geq n_0$ and $b_n = 0 =$ freedom for $n < n_0$, and $c_n =$ accept for all n. Let

$$d_n = \begin{cases} d & \text{if } n \geq n_0 \\ \Sigma_n(u_n \restriction a^\frown\langle a_n, a_{n+1}, b_n, c_n, d_{n+1}\rangle) & \text{if } n < n_0 \end{cases}$$

and set

$$u_n(\alpha) = \langle a_n, a_{n+1}, b_n, c_n, d_{n+1}, d_n\rangle.$$

Case 2. Otherwise.

Then there is a $\langle d, i\rangle$ such that infinitely many Σ_n's reject $\langle d, i\rangle$. Just as in the claim in Lemma 1.3, we get a $t < \langle d, i\rangle$, an n_0, and pairs (b_n, c_n) such that

(i) $u_n \restriction \alpha^\frown \langle a_n, a_{n+1}, b_n, c_n \rangle$ is according to Σ_n,
(ii) $\langle b_n, c_n \rangle$ settles on freedom for $n < n_0$, and on t for $n \geq n_0$, and
(iii) for infinitely many n, (b_n, c_n) involves Σ_n proposing t.

Subcase A. $t > 0$. Then we set

$$d_n = \begin{cases} (t)_0 & \text{if } n \geq n_0 \\ \Sigma_n(u_n \restriction \alpha^\frown \langle a_n, a_{n+1}, b_n, c_n, d_{n+1} \rangle) & \text{if } n < n_0. \end{cases}$$

Subcase B. $t = 0$. Define

$$S = \{\langle d_0, \dots, d_k \rangle : \forall i \leq k\ (d_i \in \{0, 1\}) \text{ and } \\ \forall i < k\ d_i = \Sigma_i(u_i \restriction \alpha^\frown \langle a_i, a_{i+1}, b_i, c_i, d_{i+1} \rangle)\}.$$

Clearly S is an infinite, finitely branching tree. (This was the reason we restricted player II's plays in $\mathrm{G}(A, T)$ to $\{0, 1\}$.) Let f be an infinite branch of S, and set $d_i = f(i)$.

Finally, in Case 2 we set

$$u_n(\alpha) = \langle a_n, a_{n+1}, b_n, c_n, d_{n+1}, d_n \rangle.$$

We can continue to define $u_n(\alpha)$ this way until we reach, as we must, an α such that $r_i \restriction \omega + \alpha \in E$ for some i. Fix the least such α, and fix i. As before, $r_n \restriction \omega + \alpha \in \Game^T A$ for all n, so $r_n \restriction \omega + \alpha \in A$ for all $n \geq i$. Let $u_n = u_n \restriction \alpha$ and $r_n = r_n \restriction \omega + \alpha$.

We claim $\mathrm{crit}(u_n) \to \infty$ as $n \to \infty$. If not, let e be least such that $e = \mathrm{crit}(u_n)$ for infinitely many n. So $e \leq \mathrm{crit}(u_n)$ for cofinitely many n, and $\vartheta^3_{(e)_0}(r_n^*) = |(e)_0|_{F(r_n)}$ is eventually constant as $n \to \infty$. Let β be this constant value. Then infinitely many u_n's settle on the $(e)_1$ proposal at round β. Thus Case 2 must apply at round β, so by construction cofinitely many u_n's settle on $(e)_1$ at round β, and for infinitely many n, Σ_n is responsible for $(e)_1$. But then

$$\psi_e(r_{n+1})^* \leq \psi_e(r_n^*)$$

for cofinitely many n, while

$$\psi_e(r_{n+1})^* < \psi_e(r_n^*)$$

for infinitely many n, a contradiction.

Since $\mathrm{crit}(u_n) \to \infty$ as $n \to \infty$, $\psi_e(r_n^*)$ is eventually constant as $n \to \infty$, for all e. Thus r_n^* converges in $\vec{\vartheta}^3$, $\vec{\vartheta}^4$, and $\vec{\vartheta}^5$, so that

$$\lim^*_{n \to \infty} r_n = r$$

for some $r \supseteq p$ such that $r \in A$. We are done if we show that r is a play according to σ. If not, let β be least so that $r(\beta)$ is not according to σ. Let

$$\beta = |k|_{F(r)}$$

and let γ be the eventual value of $|k|_{F(r_n)}$ as $n \to \infty$.

We claim that

$$\lim^*_{n\to\infty} r_n \restriction \gamma = r \restriction \beta.$$

This follows from the convergence of r_n^* in $\vec{\vartheta^0}$ and $\vec{\vartheta^1}$: if $\delta < \gamma$, then the code for δ relative to $F(r_n \restriction \gamma)$ stabilizes as $n \to \infty$ iff the code for δ relative to $F(r_n)$ stabilizes as $n \to \infty$.

Since r^* converges in $\vec{\vartheta^2}$, we get that $r \restriction \beta \in T$.

Since r^* converges in $\vec{\vartheta^5}$, which is very good, the a_n's defined at round γ are eventually constant $= a$. But then at round γ, $d = \sigma(r \restriction \beta^\frown a)$. Moreover, Case 1 must apply at round γ, since otherwise $\text{crit}(u_n)$ has a finite lim inf. Thus $r_n(\gamma) = \langle a, d \rangle$ for all sufficiently large n. Since $\vec{\vartheta^5}$ is very good, $r(\beta) = \langle a, d \rangle$. But then $r \restriction \beta + 1$ is according to σ, a contradiction which completes the proof of Lemma 1.6. ⊣ (Lemma 1.6)

Lemma 1.6 completes the proof of Theorem 1.2, since a well known construction produces a $\mathbf{HOD}(\mathbb{R})$ scale on any set which carries a $\mathbf{HOD}(\mathbb{R})$ semiscale. ⊣ (Theorem 1.2)

Given a scaled coding F of T, and a pointclass Γ of a set of reals, let

$$\Game^{T,F}\Gamma = \{\Game^T A : A^* \in \Gamma\}.$$

The proof of Theorem 1.2 shows that if T and F are "reasonable" and Γ has the semiscale property, so does $\Game^{T,F}\Gamma$. For example, let

$$T_n = \{p \in \bigcup_{\alpha<\omega_1} {}^\alpha\omega : \forall \beta \in \text{dom}(p)\ ((L_\beta[p \restriction \beta], \in, p \restriction \beta) \text{ is not admissible})\},$$

and let F_n be the scaled coding of T_n described at the beginning of this section. Let us write

$$\Game^{\Sigma_n}\Gamma = \Game^{T_n, F_n}\Gamma,$$

and agree that

$$\Gamma\text{-AD}^{\Sigma_n} \text{ iff } G(A, T_n) \text{ is determined whenever } A^* \in \Gamma.$$

(For "$\wp(\mathbb{R})$-AD$^{\Sigma_n}$" we write simply "AD$^{\Sigma_n}$".

Corollary 1.7. Let $n \geq 1$, and assume Γ-AD$^{\sigma_n}$, where Γ is closed under $\forall^{\mathbb{R}}$. Suppose all Γ sets admit Γ semiscales. Then all $\Game^{\Sigma_n}\Gamma$ sets admit $\Game^{\Sigma_n}\Gamma$ scales, so that all $\Game^{\Sigma_n}\Gamma$ relations admit $\Game^{\Sigma_n}\Gamma$ uniformizations.

Proof. Suppose $A^* \in \Gamma$. Let us trace through the proof of Theorem 1.2. We can take $\vartheta_i^0 \dots \vartheta_i^5$ to be Γ norms. We must modify ψ_i slightly to get a Γ norm; let

$$\psi_i'(p^*) = \langle \vartheta_0^5(p^*), \psi_i(p^*) \rangle.$$

Finally, let $G'_k(p,q)$ be just like $G_k(p,q)$ except that S cannot win $G_k(p,q)$ unless he wins as player I on the p board. The proof of Theorem 1.2 goes through with these modifications (cf. [Mos80, Chapter 6]) and yields a $\Game^{\Sigma_n}\Gamma$ semiscale $\vec{\varphi}'$ on $\Game^{\Sigma_n}A$. One can easily check that, since $\Game^{\Sigma_n}\Gamma$ is closed under real quantification, the scale of the tree of $\vec{\varphi}'$ is in fact a $\Game^{\Sigma_n}\Gamma$ scale. ⊣

We would like to point out two curious features of the proof of Theorem 1.2. First, it handles directly only games where player II must play from $\{0,1\}$. Second, the verification that $p_\omega \in \Game^T A$ when $p_n \to p_\omega \mod \vec{\varphi}$ is indirect: we do not construct a strategy for player I in $G(A,T)$ from p_ω, but instead defeat a strategy for player II. Is there a more direct proof avoiding these devices?

§2. Canonical strategies. We shall construct definable winning strategies for the games of the form $G(A,T)$, where T admits a scaled coding and A^* a **HOD**$(\mathbb{R})$ scale. The strategies are in some sense "best", as in Moschovakis [Mos80, Chapter 6]. However, the games $G_k(p,q)$ are not adequate to evaluate what's best for player I; the problem is there is no satisfactory way to decide which k to use in evaluating a given position. So we use the game $G_\omega(p,q)$, which is like $G_k(p,q)$ except that no bound is put on the size of proposals F may make. Now $G_\omega(p,q)$ leads to a probably illfounded value order on player I's possible next moves, but we can avoid that problem by considering directly only games where player I must play from $\{0,1\}$. (Curiouser and curiouser!)

Once again, we prove our most general theorem first, then state its sharpened form from games ending at the first Σ_n admissible relative to the play as a corollary.

If Σ_n is a strategy for $G(A,T)$, where T admits a scaled coding, then $\Sigma^* = \{p^* : p \text{ is a position according to } \Sigma\}$.

THEOREM 2.1. Assume all clopen games of length ω_1 with **HOD**$(\mathbb{R})$ payoff are determined. Suppose T admits a scaled coding, and $A \subseteq E$ is such that A^* admits a **HOD**$(\mathbb{R})$ scale. Then if player I wins $G(A,T)$ he wins via a strategy Σ such that Σ^* admits a **HOD**$(\mathbb{R})$ scale.

PROOF. We assume without loss of generality that $(p \in T \;\&\; \delta \in \text{dom}(p)) \Rightarrow (p(\delta)_0 \in \{0,1\} \;\&\; p(\delta)_1 \in \{0,1\})$; that is both players must play 0 or 1 in $G(A,T)$. In order to conform to the notation of §1, we assume we have a fixed $p_0 \in T$ such that $\text{dom}(p_0) = \omega$ and player I wins $G(A,T)$ from p_0: the canonical Σ we are to construct must win from p_0. This is no loss of generality.

Let $\vec{\rho}, \vec{\vartheta}^0 \ldots \vec{\vartheta}^5$, and $\vec{\psi}$ be the families of norms on A^* defined from a scaled coding of T and a scale on A^* just as in the proof of Theorem 1.2. For $p,q \in T$ such that $\text{dom}(p) = \text{dom}(q)$ we define a game $G_\omega(p,q)$. The definition is the same as that of $G_k(p,q)$ for $k \in \omega$, except that: (a) we do not require $\text{dom}(p) = \omega$, or even that p or q are winning positions for player I in $G(A,T)$

(this assumption played no role in the definition of $\mathrm{G}_k(p,q)$ anyway), and (b) we allow F to propose any $t \in \omega$, not just $t \leq k$, when it's his turn to propose in $\mathrm{G}_\omega(p,q)$ (If S rejects t, he must still counterpropose a $t' < t$.) Now, let for $p \supseteq p_0$,

$$\Sigma(p) = \begin{cases} 0 & \text{if } \forall a \in \{0,1\}\, \exists b \in \{0,1\} \text{ such that } S \text{ has a winning strategy} \\ & \qquad \text{in } \mathrm{G}_\omega(p^\frown\langle 0,a\rangle, p^\frown\langle 1,b\rangle), \\ 1 & \text{otherwise.} \end{cases}$$

The determinacy of G_ω and its symmetry imply that if $\Sigma(p) = 1$, then

$$\forall a \in \{0,1\}\, \exists b \in \{0,1\}\, \big(S \text{ has a winning strategy in } \mathrm{G}_\omega(p^\frown\langle 1,a\rangle, p^\frown\langle 0,b\rangle)\big).$$

Thus

$$\Sigma(p) = i \Rightarrow S \text{ has a winning strategy in } H(p),$$

where $H(p)$ is the game in which F plays a, then S plays b, then F and S play out $\mathrm{G}_\omega(p^\frown\langle i,a\rangle, p^\frown\langle 1-i,b\rangle)$.

It is easy to see that Σ^* is of the form $\Game^S B$, for some S admitting a scaled coding and B such that B^* admits a $\mathbf{HOD}(\mathbb{R})$ scale. By Theorem 1.2, then, it is enough to show that Σ is a winning strategy for player I in $\mathrm{G}(A,T)$ from p_0. So let $q \in E$ be an arbitrary play according to Σ; we want to show $q \in A$.

Fix a winning strategy τ for player I in $\mathrm{G}(A,T)$ from p_0. For $\delta \in \mathrm{dom}(q)$, let Σ_δ be a winning strategy for S in $H(q\restriction\delta)$. We represent a position or completed run of $H(q\restriction\delta)$ by a function $u\colon \alpha \to \omega$, where $\alpha \geq \delta$ and $u\restriction\delta = q\restriction\delta$, $u(\delta) = \langle 1-q(\delta)_0, q(\delta)_0, a, b\rangle$ where a, b are the first two moves of $H(q\restriction\delta)$, and $u(\delta + 1 + \eta)$ codes the (up to) six moves of round η in $\mathrm{G}_\omega(q\restriction\delta^\frown\langle q(\delta)_0, a\rangle, q\restriction\delta^\frown\langle 1 - q(\delta)_0, b\rangle)$. If u is a position or run of $H(q\restriction\delta)$, then the lower board of u is r, where $r\restriction\delta = q\restriction\delta$, $r(\delta) = \langle q(\delta)_0, u(\delta)_2\rangle$, and $r(\delta+1+\eta) = \langle u(\delta+1+\eta)_1, u(\delta+1+\eta)_4\rangle$. Similarly, the upper board of u is s, where $s\restriction\delta = q\restriction\delta$, $s(\delta) = \langle 1-q(\delta)_0, u(\delta)_3\rangle$, and $s(\delta+1+\eta) = \langle u(\delta+1+\eta)_0, u(\delta+1+\eta)_5\rangle$.

Notice that from position u of $H(q\restriction\delta)$ we can recover δ; δ is the least $\alpha \in \mathrm{dom}(u)$ such that $u(\alpha)_0 \neq q(\alpha)_0$. Let us write $\delta = \delta(u)$.

Let us call a sequence $\langle u_\beta : \beta < \gamma\rangle$ a **diagram** if $\gamma \leq \mathrm{dom}(u_0)$ and

(a) for $\beta < \gamma$, u_β is a position in $H(q\restriction\delta)$ according to Σ_δ, where $\delta = \delta(u_\beta)$, and

(b) letting r_β be the upper board of u_β for $\beta < \gamma$, we have: r_0 according to τ, and $r_{\beta+1}$ is the lower board of u_β for $\beta + 1 < \gamma$.

Our plan is to construct a diagram $\langle u_\beta : \beta < \gamma\rangle$ such that for any limit $\lambda \leq \gamma$, $\mathrm{crit}(u_\beta) \to \omega$ as $\beta \to \lambda$. We also arrange that $r_0 \in E$, that $r_\lambda = \lim^*_{\beta\to\lambda} r_\beta$ if $\lambda < \gamma$ is a limit, and that $q = \lim^*_{\beta\to\gamma} r_\beta$ if γ is a limit, while q is the lower board of $u_{\gamma-1}$ if γ is a successor. The existence of such a diagram easily implies $q \in A$. [$r_0 \in A$ since r_0 is by τ. But then $r_\beta \in A$ for $\beta < \gamma$ by induction: since r_β and $r_{\beta+1}$ are the upper and lower boards of u_β, which is by some Σ_δ,

$r_\beta \in A \Rightarrow r_{\beta+1} \in A$. If $\lambda < \gamma$ is a limit, then since $\text{crit}(u_\beta) \to \omega$ as $\beta \to \lambda$, r^*_β converges in $\vec{\vartheta}^3, \vec{\vartheta}^4, \vec{\vartheta}^5$ to some r^* in A^*; moreover $r = r_\lambda$ since $\lim^*_{\beta\to\lambda} r_\beta = r_\lambda$. So $r_\beta \in A$ for $\beta < \gamma$; repeating the argument we get $q \in A$.]

We obtain the desired diagram by means of a sequence $D_\alpha = \langle u^\alpha_\beta : \beta < \gamma_\alpha\rangle$ of diagrams defined by induction on α. We maintain by induction that $\alpha < \sigma \Rightarrow u^\alpha_0 \subseteq u^\sigma_0$. In fact, if $n < \min(\omega, \gamma_\alpha)$ and $\alpha < \sigma$, then $n < \gamma_\sigma$ and $u^\alpha_n \subseteq u^\sigma_n$. (On the other hand, we may well have $\alpha < \sigma$ such that u^α_ω and u^σ_ω are defined and incompatible. We are really building a tree of approximations to the desired u_ω, "continuously" associating to each branch of this tree a tree of approximations to $u_{\omega+\omega}$, *etc.* This is because we do not know definitely even initial segments of the eventual r_ω, $r_{\omega+\omega}$, *etc.*, as we build our diagram. For notational simplicity, however, we shall suppress explicit mention of these trees.

If $\langle u_\beta : \beta < \gamma\rangle$ is any diagram, then $\langle (u_\beta, a_\beta) : \beta < \gamma\rangle$ is an **enlarged diagram** if $a_0 = \tau(r_0)$ and $a_{\beta+1} = \Sigma_\delta(u_\beta{}^\frown\langle a_\beta\rangle)$ for $\beta + 1 < \gamma$ and $\delta = \delta(u_\beta)$. Given such an enlarged diagram, let

$$i_\beta = \begin{cases} \text{the least } t \text{ such that } (t)_0 \in \{0,1\} \text{ and} \\ \Sigma_\delta(u_\beta{}^\frown\langle a_\beta, a_{\beta+1}, t\rangle) = \text{reject, where } \delta = \delta(u_\beta), \\ \omega, \text{ if no such } t \text{ exists.} \end{cases}$$

For $\lambda \leq \gamma$ a limit, we say $\langle (u_\beta, a_\beta) : \beta < \gamma\rangle$ **accepts readily at** λ iff $\lim_{\beta\to\lambda} i_\beta = \omega$.

Given an enlarged diagram $\langle (u_\beta, a_\beta) : \beta < \gamma\rangle$ with $\gamma = \beta + 1$, define a_γ by: $a_\gamma = \Sigma_\delta(u_\beta{}^\frown\langle a_\beta\rangle)$, where $\delta = \delta(u_\beta)$.

LEMMA 2.2. Suppose $\langle (u_\beta, a_\beta) : \beta < \gamma\rangle$ is as enlarged diagram which accepts readily at all limit $\lambda \leq \gamma$. Let $d \in \{0,1\}$. Then there are b_β, c_β, and d_β, for $\beta < \gamma$, such that, setting $d_\gamma = d$,

(a) $(u_\beta{}^\frown\langle a_\beta, a_{\beta+1}, b_\beta, c_\beta, d_{\beta+1}, d_\beta\rangle)$ is by Σ_δ, $\delta = \delta(u_\beta)$, for $\beta < \gamma$,
(b) $d_\beta \to d_\lambda$ as $\beta \to \lambda$, for $\lambda \leq \gamma$ a limit,
and
(c) $b_\beta \to \omega$ as $\beta \to \lambda$, for $\lambda \leq \gamma$ a limit
$c_\beta =$ accept, for all but finitely many β.

PROOF. By induction on γ. The successor step is easy, so let γ be a limit. Let $\pi : \gamma \hookrightarrow \omega$ be determined by r_0. Let i_β, for $\beta < \gamma$, be as in the definition of ready acceptance. Let $\eta < \gamma$ be such that $\langle d, 0\rangle \leq i_\beta$ for $\eta \leq \beta$ (so $\Sigma_{\delta(u_\beta)}$ accepts $\langle d, 0\rangle$ for $\eta \leq \beta$.) Let λ be the largest limit $\leq \eta$. For $\eta \leq \beta < \gamma$, let

$$b_\beta = \text{the largest } t \leq \max(\pi(\beta), i_\beta) \text{ such that } (t)_0 = d \text{ and}$$
$$\Sigma_\delta \text{ accepts } t \text{ after } u_\beta{}^\frown\langle a_\beta, a_{\beta+1}, t\rangle, \text{ where } \delta = \delta(u_\beta),$$

and

$$c_\beta = \text{accept}, \quad d_\beta = d.$$

For $\lambda \leq \beta < \eta$, let

$$b_\beta = 0 = \text{freedom}, \quad c_\beta = \text{accept}$$

and

$$d_\beta = \Sigma_\delta(u_\beta{}^\frown\langle a_\beta, a_{\beta+1}, b_\beta, c_\beta, d_{\beta+1}\rangle), \text{ for } \delta = \delta(u_\beta).$$

Finally, the induction hypothesis with $d = d_\lambda$ will give us the desired b_β, c_β and d_β for $\beta < \lambda$. ⊣ (Lemma 2.2)

We shall use tacitly the fact that there is a function which, given a sequence of satisfying the hypothesis of the lemma and a $d \in \{0, 1\}$, produces a sequence satisfying its conclusion. (We don't have AC!)

We are ready to define our approximations $D_\alpha = \langle u^\alpha_\beta : \beta < \gamma_\alpha\rangle$ to the desired diagram; the definition is by induction on α.

$\alpha = 0$: We may assume q is not by τ; otherwise $q \in A$ and we're done. So let δ be least such that $\tau(q{\restriction}\delta) \neq q(\delta)_0$. Set

$$\gamma_0 = 1,$$

and

$$u^0_0 = q{\restriction}\delta^\frown\langle 1 - q(\delta)_0, q(\delta)_0, q(\delta)_1, \Sigma_\delta(q{\restriction}\delta^\frown\langle 1 - q(\delta)_0, q(\delta)_0, q(\delta)_1\rangle)\rangle.$$

$\alpha > 0$: Set $v_0 = \bigcup_{\eta<\alpha} u^\eta_0$. We use s_β for the upper board of v_β. If $s_0 \in E$, then our induction on α stops; D_α is undefined. Otherwise, let $a_0 = \tau(s_0)$. We shall define v_β and a_β by induction on β. Suppose we have v_β and a_β for $\beta < \gamma$, and that $\langle(v_\beta, a_\beta) : \beta < \gamma\rangle$ is an enlarged diagram. Suppose also the following four conditions are met.

(1) γ limit $\Rightarrow \langle(v_\beta, a_\beta) : \beta < \gamma\rangle$ accepts readily at γ.

(2) If γ is a limit, then $\lim_{\beta\to\gamma} a_\beta$ and $\lim^*_{\beta\to\gamma} s_\beta$ exists, moreover $p_0 \supseteq \lim^*_{\beta\to\gamma} s_\beta$,

(3) Let $s = \lim^*_{\beta\to\gamma} s_\beta$ if γ is a limit, and s be the lower board of $v_{\gamma-1}$ otherwise. Then $\exists\delta \in \text{dom}(s) \cap \text{dom}(q)$ $(s(\delta) \neq q(\delta))$; moreover, if δ is the least such ordinal, then $s(\delta)_0 \neq q(\delta)_0$.

(4) For $\vartheta \in \text{dom}(s_0)$, define t_ϑ by

$$t_\vartheta(0) = \vartheta, \quad t_\vartheta(\beta + 1) = t_\vartheta(\beta) \text{ for } \beta + 1 \leq \gamma,$$

and

$$t_\vartheta(\lambda) = |e|_{F(r)}, \text{ where } r = \lim^*_{\beta\to\gamma} s_\beta \text{ and } |e|_{F(s_\beta)} = t_\vartheta(\beta)$$
$$\text{for all sufficiently large } \beta < \lambda,$$

for $\lambda \leq \gamma$ a limit; $t_\vartheta(\lambda)$ is undefined if no such $e \in \omega$ exists. (One should visualize t_ϑ as the ϑth "column" of the diagram $\langle v_\beta : \beta < \gamma \rangle$. Not all columns extend to γ, since not all contribute to $*$-limits all the way down.) Let

$$C_\gamma = \{\vartheta : t_\vartheta(\gamma) \text{ is defined}\}.$$

Let s, δ be as in (3). (Our assumptions imply $\mathrm{dom}(s) = \{t_\delta(\gamma) : \vartheta \in C_\gamma\}$.) Fix ϑ (unique) such that $\delta = t_\vartheta(\gamma)$. Then we require that for $\vartheta \leq \sigma \leq \rho$ and $\sigma, \rho \in C_\gamma$

$$\gamma < \min(\gamma_\sigma, \gamma_\rho) \text{ (i.e., } u_\gamma^\sigma, u_\gamma^\rho \text{ are defined)},$$
$$u_\gamma^\sigma \subseteq u_\gamma^\rho, \text{ and } s \restriction (t_\sigma(\gamma) + 1) \text{ is the upper board of } u_\gamma^\sigma.$$

If conditions (1)–(4) are met, then we set

$$a_\gamma = \lim_{\beta \to \gamma} a_\beta,$$

and

$$v_\gamma = \bigcup \{u_\gamma^\sigma : \sigma \in C_\gamma \ \&\ t_\sigma(\gamma) \geq \delta\}.$$

Notice that $\langle (v_\beta, a_\beta) : \beta < \gamma + 1 \rangle$ remains an enlarged diagram.

If one of (1)–(4) fails, then v_γ is undefined. This must happen at some $\gamma \leq \sup_{\sigma < \alpha} \gamma_\sigma$ since at least (4) will fail. So suppose γ is at least such that v_γ is undefined. We shall define D_α by taking cases on which of (1)–(4) fail at γ.

Case 1. (1) fails at γ.

So γ is a limit. Set $\gamma_\alpha = \gamma$. Now since (1) fails, we get a $t \in \omega$ and a cofinal $B \subseteq \gamma$ such that for $\beta \in B$, there is a b such that $(b)_0 \in \{0, 1\}$ and Σ_δ counterproposes t when F proposes b after $v_\beta{}^\frown\langle a_\beta, a_{\beta+1}\rangle$, where $\delta = \delta(v_\beta)$. Let t be least such that a cofinal $B \supseteq \gamma$ exists, and fix such a B of order type ω.

Subcase A. $t > 0$.

There is an $\eta < \gamma$ such that for $\eta \leq \beta < \gamma$ and $\delta = \delta(v_\beta)$, Σ_δ accepts t after $v_\beta{}^\frown\langle a_\beta, a_{\beta+1}\rangle$; this follows from the minimality of t. Let $\pi \colon \gamma \hookrightarrow \omega$, be determined by s_0. For $\eta \leq \beta \leq \gamma$, let

$$\begin{aligned} b_\beta &= \text{some } b \text{ witnessing } \beta \in B, \text{ if } \beta \in B, \\ c_\beta &= (\text{reject}, t), \text{ if } \beta \in B, \\ b_\beta &= \text{the largest } i \leq \max(t, \pi(\beta)) \text{ such that } (i)_0 = (t)_0 \text{ and} \\ &\qquad \Sigma_\delta \text{ accepts } i \text{ after } v_\beta{}^\frown\langle a_\beta, a_{\beta+1}\rangle, \text{ where } \delta = \delta(v_\beta), \text{ if } \beta \notin B, \\ c_\beta &= \text{accept, if } \beta \notin B, \end{aligned}$$

and

$$d_\beta = (t)_0.$$

Let λ be the largest limit ordinal $\leq \eta$. For $\lambda \leq \beta \leq \eta$, set

$$b_\beta = 0 = \text{freedom}, \quad c_\beta = \text{accept},$$

and

$$d_\beta = \Sigma_\delta(v_\beta{}^\frown\langle a_\beta, a_{\beta+1}, b_\beta, c_\beta, d_{\beta+1}\rangle), \text{ for } \delta = \delta(v_\beta).$$

Finally, since $\langle (v_\beta, a_\beta) : \beta < \gamma\rangle$ accepts readily at all limit $\lambda' \leq \lambda$, we may apply Lemma 2.2 with $d = d_\lambda$ to generate b_β, c_β and d_β for $\beta < \lambda$. Then let

$$u^\alpha_\beta = v_\beta{}^\frown\langle a_\beta, a_{\beta+1}, b_\beta, c_\beta, d_{\beta+1}, d_\beta\rangle$$

for $\beta < \gamma = \gamma_\alpha$.

Subcase B. $t = 0$.

Let i_β be as in the definition of ready acceptance, and let

$$S = \{\sigma < \gamma : \exists\beta[\sigma \leq \beta < \sigma + \omega \; \& \; (\beta \in B \text{ or } i_\beta \leq \max(\langle 0,0\rangle, \langle 1,0\rangle))]\}$$

Then S has order type ω, as otherwise (1) fails at some $\lambda < \gamma$. Let

$$b_\beta = \begin{cases} \text{some } b \text{ witnessing } \beta \in B, & \text{if } \beta \in B \\ 0, & \text{if } \beta \in S - B \end{cases}$$

and

$$c_\beta = \begin{cases} (\text{reject}, 0), & \text{if } \beta \in B \\ \text{accept}, & \text{if } \beta \in S - B. \end{cases}$$

Let $\{\sigma_i : i \in \omega\}$ be the increasing enumeration of S, and consider

$$U = \{\langle d_0 \ldots d_k\rangle \in 2^\omega : \forall i < k \; (\text{if } \sigma = \sigma_i \text{ and } \delta = \delta(v_\sigma)), \\ \text{then } d_i = \Sigma_\delta(v_\sigma{}^\frown\langle a_\sigma, a_{\sigma+1}, b_\sigma, b_\sigma, c_\sigma, d_{\sigma+1}\rangle)\}$$

U is an infinite tree on $\{0, 1\}$ so we have an $f \in {}^\omega 2$ such that $f \restriction k \in U$ for all k. Set

$$d_{\sigma_i} = f(i), \quad \text{for } i \in \omega.$$

Finally, suppose $\beta \in \gamma - S$. Let $\sigma \in S$ be least such that $\beta < \sigma$. (σ is a limit.) Set

$$b_\beta = \text{the largest } i \leq \max(i_\beta, \pi(\beta)) \text{ such that } (i)_0 = d_\sigma \text{ and} \\ \Sigma_\delta \text{ accepts } i \text{ after } v_\beta{}^\frown\langle a_\beta, a_{\beta+1}\rangle, \text{ where } \delta = \delta(v_\beta),$$

and

$$c_\beta = \text{accept}, \quad d_\beta = d_\sigma.$$

Now, we let, for $\beta < \gamma = \gamma_\alpha$

$$u^\alpha_\beta = v_\beta{}^\frown\langle a_\beta, a_{\beta+1}, b_\beta, c_\beta, d_{\beta+1}, d_\beta\rangle.$$

This completes the definition of D_α in Case 1.

Case 2. (1) holds and (2) fails at γ.

Since $\langle (v_\beta), a_\beta : \beta < \gamma \rangle$ accepts readily at all limit $\lambda \leq \gamma$, Lemma 2.2 applied with $d = 0$ yields b_β, c_β, and d_β for $\beta < \gamma$. Set $\gamma_\alpha = \gamma$, and $u^\alpha_\beta = v_\beta{}^\frown\langle a_\beta, a_{\beta+1}, b_\beta, c_\beta, d_{\beta+1}, d_\beta\rangle$ for $\beta < \gamma$.

Case 3. (1) and (2) hold but (3) fails at γ.

Let s be as in (3). Let $a = a_\gamma$ if γ is a successor, $a = \lim_{\beta\to\gamma} a_\beta$ otherwise.

Subcase A. $q \subseteq s$, or $\exists\delta \in \mathrm{dom}(s) \cap \mathrm{dom}(q)$ $(s(\delta) \neq q(\delta))$, but $s(\delta)_0 = q(\delta)_0$ for the least such δ.

In this case, proceed exactly as in Case 2.

Subcase B. $s \subsetneq q$ and $q(\mathrm{dom}(s))_0 = a$. In this case apply Lemma 2.2 with $d = q(\mathrm{dom}(s))_1$ to get b_β, c_β and d_β for $\beta < \gamma$. Set $\gamma_\sigma = \gamma$, and $u^\alpha_\beta = v_\beta{}^\frown\langle a_\beta, a_{\beta+1}, b_\beta, c_\beta, d_{\beta+1}, d_\beta\rangle$ for $\beta < \gamma$.

Subcase C. $s \subsetneq q$ and $q(\mathrm{dom}(s))_0 = 1 - a$.

This is the only case in which we set up a new board. Let $\gamma_\alpha = \gamma + 1$. Set

$$u^\alpha_\gamma = s^\frown\langle a, 1-a, q(\delta)_1, \Sigma_\delta(s^\frown\langle a, 1-a, q(\delta)_1\rangle)\rangle,$$

where $\delta = \mathrm{dom}(s)$. Now use the lemma with $d = \Sigma_\delta(s^\frown\langle a, 1-a, q(\delta)_1\rangle)$ to define u^α_β for $\beta < \gamma$.

Case 4. (1)–(3) hold, but (4) fails at γ.

Let $\gamma_\alpha = \gamma$, and apply Lemma 2.2 with $d = 0$ to get u^α_β for $\beta < \gamma$.

This completes the inductive definition of the D_α's. Let v^α_β, s^α_β be the v_β and s_β occurring in the definitions of D_α, $\alpha > 0$.

Let $M = \langle u_\beta : \beta < \gamma\rangle$ be a diagram with boards r_β, $\beta < \gamma$. We call M **good** iff $\lim^*_{\beta\to\lambda} r_\beta$ exists for all limit $\lambda \leq \gamma$, and $\lim^*_{\beta\to\lambda} r_\beta = r_\lambda$ for all limit $\lambda \leq \gamma$. If $M = \langle u_\beta : \beta < \gamma\rangle$ is good, we define t^M_ϑ for $\vartheta \in \mathrm{dom}(u_0)$ to be the ϑth "column" of M, as in condition (4) above: $t^M_\vartheta(0) = \vartheta$, $t^M_\vartheta(\beta + 1) = t^M_\vartheta(\beta)$ for $\beta + 1 \leq \lambda$, and $t^M_\vartheta(\lambda) = |e|_{F(r)}$, where $\lambda \leq \gamma$ is a limit, $r = \lim^*_{\beta\to\lambda} r_\beta$, and $|e|_{F(r_\beta)} = t^M_\vartheta(\beta)$ for all sufficiently large $\beta < \lambda$. ($\lambda \notin \mathrm{dom}(t^M_\vartheta)$ if no such e exists.) Let also $C^M_\gamma = \{\vartheta : t^M_\vartheta(\gamma) \text{ is defined}\}$.

We now construct the desired diagram. Let $u_0 = \bigcup\{u^\alpha_0 : u^\alpha_0 \text{ is defined}\}$, and note that the upper bound r_0 of u_0 is in A. Now suppose $M = \langle u_\beta : \beta < \gamma\rangle$ is given, and that

(a) $\mathrm{crit}(u_\beta) \to \omega$ as $\beta \to \lambda$, for any limit $\lambda \leq \gamma$,

(b) M is a good diagram, and

(c) for $\beta < \gamma$, if $\alpha \in C^M_\beta$ and $t^M_\alpha(\beta) \leq \delta(u_\beta)$, then $u^\alpha_\beta = u_\beta \upharpoonright (t^M_\alpha(\beta) + 1)$. (So u^α_β is defined.)

Let $r = \lim^*_{\beta\to\gamma} r_\beta$ if γ is a limit, and let r be the lower board of $u_{\gamma-1}$ otherwise. If $r = q$ then (a) and (b) guarantee that M is the desired diagram.

Otherwise, since $r, q \in E$, they are incompatible; let δ be least such that $r(\delta) \neq q(\delta)$. We define u_γ by

$$u_\gamma = \bigcup\{u_\gamma^\alpha : \alpha \in C_\gamma^M \ \& \ t_\alpha(\gamma) \geq \delta\};$$

we must see that (a), (b) and (c) remain true. The following claims insure this.

CLAIM 2.3. If $\alpha \in C_\gamma^M$, $\beta < \gamma$, and $t_\alpha^M(\beta) < \delta(u_\beta)$, then $t_\alpha^M(\gamma) < \delta$.

PROOF. By induction on $\eta \geq \beta$ we see that $t_\alpha^M(\eta) = t_\alpha^M(\beta)$ and $r_\eta \restriction t_\alpha^M(\eta) = q \restriction t_\alpha^M(\eta)$. If η is a successor, this follows from the definition of "diagram" and "column". If η is a limit, then since $\alpha \in C_\eta^M$, the r_ϑ-code of $t_\alpha^M(\vartheta)$ is eventually constant as $\vartheta \to \eta$. Since r_ϑ^* converges in $\vec{\vartheta}^1$ and $\vec{\vartheta}^2$, this means $r_\eta \restriction t_\alpha^M(\eta) = \lim^*_{\vartheta \to \eta} r_\vartheta \restriction t_\alpha^M(\vartheta) = q \restriction t_\alpha^M(\beta)$.

We can repeat the above argument with γ for η and r for r_η to get $r \restriction t_\alpha^M(\gamma) = q \restriction t_\alpha^M(\gamma)$, so $t_\alpha^M(\gamma) < \delta$. ⊣ (Claim 2.3)

CLAIM 2.4. If $\alpha \in C_\gamma^M$ and $t_\alpha^M(\gamma) = \delta$, then $\gamma_\alpha = \gamma + 1$ and Case 3C held in the definition of D_α. Moreover $r \restriction t_\alpha^M(\gamma) + 1$ is the upper board of u_γ^α.

PROOF. By Claim 2.3, $t_\alpha^M(\beta) \geq \delta(u_\beta)$ for $\beta < \gamma$. So by hypothesis (c) on M,

$$u_\beta^\alpha = u_\beta \restriction t_\alpha^M(\beta) + 1, \text{ for } \beta < \gamma \leq \gamma_\alpha.$$

Also $v_\beta^\alpha = u_\beta \restriction t_\alpha^M(\beta)$, and $s_\beta^\alpha = r_\beta \restriction t_\alpha^M(\beta)$ for $\beta < \gamma$. Now condition (1) holds at γ in the definition of the v_β^α's, as otherwise our construction in Case 1 guarantees that $\mathrm{crit}(u_\beta) \leq \langle e, t\rangle$ for infinitely many $\beta < \gamma$, where e is the eventual r_β code of $t_\alpha^M(\beta)$ and t is as in Case 1. Condition (2) holds: $\lim^*_{\beta \to \gamma} r_\beta \restriction t_\alpha^M(\beta) = r \restriction t_\alpha^M(\gamma)$ exists, so $\lim^*_{\beta \to \gamma} s_\beta^\alpha$ exists. Further, $a_\beta^\alpha = u_\beta^\alpha(t_\alpha^M(\beta))_0 = u_\beta(t_\alpha^M(\beta))_0 = r_\beta(t_\alpha^M(\beta))_0$ for $\beta < \gamma$, since r_β^* converges in $\vec{\vartheta}^5$, which is very good, $\lim_{\beta \to \gamma} a_\beta^\alpha = r(t_\alpha^M(\gamma))_0$.

Let s be as in condition (3); then $s = r \restriction (t_\alpha^M(\gamma)) = r \restriction \delta$.

Thus (3) fails at γ in the definition of D_α, and Case 3 holds. Since $s \supseteq r \supseteq q$, 3A cannot hold. Let a be as in Case 3. Then $a = r(\delta)_0$. If $a = q(\mathrm{dom}(s))_0 = q(s)_0$, then the construction in 3B guarantees $r(\delta)_1 = q(\delta)_1$, so $r(\delta) = q(\delta)$, contrary to the definition of δ. Thus 3B cannot hold, and 3C does. The rest of Claim 2.4 is obvious by now. ⊣ (Claim 2.4)

CLAIM 2.5. Let $\alpha \in C_\gamma^M$ and $t_\alpha^M(\gamma) > \delta$; then (1)–(4) hold at γ in the definition of D_α. Moreover, $\mathrm{dom}(u_\gamma)^\alpha = t_\alpha^M(\gamma) + 1$ is the upper board of u_γ^α. Finally, if $\eta < \alpha$, $\eta \in C_\gamma^M$, $t_\eta^M(\gamma) > \delta$, then $u_\gamma^\eta \subseteq u_\gamma^\alpha$.

PROOF. This is a tedious induction on α which we leave to the interested reader. The main point is that if $\alpha \in C_\gamma^M$, then letting $N = \langle v_\beta^\alpha : \beta < \gamma\rangle$, $C_\gamma^N = C_\gamma^M \cap \alpha$ and $t_\eta^N = t_\eta^M \restriction \gamma + 1$ for all $\eta < \alpha$. ⊣ (Claim 2.5)

It is clear from Claim 2.5 that if we define u_γ as above, then (a), (b), and (c) remain true for $\langle u_\beta : \beta < \gamma + 1\rangle$.

Lastly, we must see that (a), (b), and (c) are preserved at limit γ. Now (c) is trivial at limits, and for (b) we need only worry that $\lim^*_{\beta\to\gamma} r_\beta$ may not exist; however, this limit must exist granted (a). So it is enough to check (a).

If (a) fails, then we have $\langle e, t\rangle$ such that

$$\langle e, t\rangle = \liminf_{\beta\to\gamma} \operatorname{crit}(u_\beta).$$

Now $r_\beta \in A$ for $\beta < \gamma$ by (a)–(c) below γ, and so $\psi_k(r_\beta)$ converges as $\beta \to \gamma$ for all $k \leq \langle e, t\rangle$. Thus $|e|_{F(r_\eta)}$ is eventually constant, say for $\beta \neq \eta$. Let $t^M_\alpha(\eta) = |e|_{F(r_\eta)}$, so that $t^M_\alpha(\beta) = |e|_{F(r_\beta)}$ for all $\beta \geq \eta$, $\beta < \gamma$. [Though we haven't yet shown $M = \langle u_\beta : \beta < \gamma\rangle$ is good, we can define t^M_α and C^M_α by: $t^M_\alpha = \bigcup_{\beta<\gamma} t^{M\restriction\beta}_\alpha$ and $C^M_\alpha = \bigcup_{\beta<\gamma} C^{M\restriction\beta}_\alpha$.] Now by (c) below γ we have $u^\alpha_\beta = u_\beta\restriction(t^M_\alpha(\beta) + 1)$ for all $\beta < \gamma$. (In particular v^α_β is defined for $\beta < \gamma$.) If v^α_γ is defined, then $\gamma < \gamma_\alpha$, and then we see from the construction in Cases 1–4 that the least proposal in $u^\alpha_\beta(\operatorname{dom}(u^\alpha_\beta) - 1)$ goes to ω as $\beta \to \gamma$. (In this case, our current γ is less than the γ referred to in the case hypothesis.) But this proposal is t cofinally often. So v^α_γ is undefined. In Cases 2–4 apply in the definition of D_α, then again the least proposal in $u^\alpha_\beta(\operatorname{dom}(u^\alpha_\beta) - 1)$ goes to ω as $\beta \to \gamma$. So Case 1 applies. But then our construction guarantees that $\Sigma_{\delta(u_\beta)}$ is responsible for the t proposal cofinally often in $u^\alpha_\beta(\operatorname{dom}(u^\alpha_\beta) - 1)$. Setting $i = \langle e, t\rangle$, we get a $\vartheta < \gamma$ such that

$$\psi_i(r_\eta) \leq \psi_i(r_\beta)$$

for all $\eta \geq \beta \geq \vartheta$, and

$$\psi_i(r_{\eta+1}) < \psi_i(r_\eta)$$

for infinitely many $\eta \geq \vartheta$, a contradiction. (The first inequality requires that $\vec{\psi}$ be a scale, not just a semiscale.)

This completes the proof of Theorem 2.1. ⊣ (Theorem 2.1)

Recall that

$$T_n = \{p \in \bigcup_{\alpha<\omega_1} {}^\alpha\omega : \forall\beta \in \operatorname{dom}(p)((\boldsymbol{L}_\beta[p\restriction\beta], \in, p\restriction\beta) \text{ is not } \Sigma_n \text{ admissible})\}.$$

Corollary 2.6. Let $n \geq 1$, and assume Γ-AD^{Σ_n}, where Γ is closed under $\forall^{\mathbb{R}}$ and has the scale property. Then if player I has a winning strategy in $G(A, T_n)$, where $A^* \in \Gamma$, player I has a winning strategy Σ such that Σ^* is $\Game^{\Sigma_n}\Gamma$.

Proof. The proof of this is implicit in the proof of Theorem 2.1. ⊣

§3. An inner model of AD^{Σ_n}**.** One standard justification for deriving consequences of $\mathsf{ZF}+\mathsf{AD}+\mathsf{DC}$ is that, assuming definable determinacy, this theory has interesting inner models. In particular, if all games of length ω on ω with payoff in $\mathbf{L}(\mathbb{R})$ are determined, then $\mathbf{L}(\mathbb{R}) \models \mathsf{ZF}+\mathsf{AD}+\mathsf{DC}$. In this section we shall provide an analogous justification for deriving consequences of $\mathsf{ZF}+\mathsf{AD}^{\Sigma_n}+\mathsf{DC}$.

For $\alpha \leq \omega_1$, let Γ-AD^α be the assertion that all games of length α on ω whose payoff set is in Γ are determined. AD^α is the same assertion but with no restriction on payoffs. AD^{ω^2} is the weakest form of full determinacy stronger than that $\mathsf{AD}^\omega = \mathsf{AD}$. Blass and Mycielski [Bla75] have shown $\mathsf{AD}^{\omega^2} \iff \mathsf{AD}_\mathbb{R}$, while Solovay [Sol78B] has shown that $\mathsf{ZF}+\mathsf{AD}_\mathbb{R}+\mathsf{DC}$ proves $\mathrm{Con}(\mathsf{ZF}+\mathsf{AD}+\mathsf{DC})$. We shall give in detail only the construction of a model of $\mathsf{ZF}+\mathsf{AD}_\mathbb{R}+\mathsf{DC}$; even this seems to involved the machinery of long games and scales developed in §1 and §2. We indicate how to modify the construction in order to get models of $\mathsf{ZF}+\mathsf{AD}^{\Sigma_n}+\mathsf{DC}$ at the end of the section.

We shall rely heavily on work of Woodin (unpublished). Woodin constructs "from below" a class M, and shows that if there is any model of $\mathsf{ZF}+\mathsf{AD}_\mathbb{R}+\mathsf{DC}$ containing al reals and ordinals, then M is the smallest such model. So our task is to show that some amount of definable determinacy implies $M \models \mathsf{ZF}+\mathsf{AD}_\mathbb{R}+\mathsf{DC}$. By Woodin's work on M, it suffices for this to show that every set of reals in M admits a definable scale. We shall show using a *Friedman game* that every $(\Delta^2_1)^M$ set is $\Game^{\Sigma_2}\underset{\sim}{\Pi}^1_1$. Corollary 2.6 then implies every set of reals in M is $\Game^{\Sigma_2}\underset{\sim}{\Pi}^1_1$, and Corollary 1.7 implies that such sets admit definable scales.

In the end, we require $\underset{\sim}{\Delta}^1_2$-$\mathsf{AD}^{\Sigma_2}$ to show that $M \models \mathsf{ZF}+\mathsf{AD}_\mathbb{R}+\mathsf{DC}$.

In §4 we shall present some evidence that in fact $\mathsf{ZFC}+\mathbf{HOD}(\mathbb{R})$-$\mathsf{AD}^{\Sigma_1}$ is too weak to construct an inner model of $\mathsf{ZF}+\mathsf{AD}_\mathbb{R}+\mathsf{DC}$. We believe that $\underset{\sim}{\Delta}^1_2$-$\mathsf{AD}^{\Sigma_2}$ is very close to being the weakest definable determinacy hypothesis yielding an inner model of $\mathsf{AD}_\mathbb{R}$.

Some terminology: a filter $\mathcal{F}$ on $\wp_{\omega_1}(X) = \{A \subseteq X : A \text{ is countable}\}$ is **normal** iff $\mathcal{F}$ is closed under diagonal intersection ($A_x \in \mathcal{F}$ for all $x \in X \Rightarrow \{A \in \wp_{\omega_1}(X) : \forall x{\in}A\,(A \in A_x)\} \in \mathcal{F}$) and **fine** iff $\forall x{\in}X\,(\{A \in \wp_{\omega_1}(X) : x \in A\} \in \mathcal{F})$. A set $C \subseteq \wp_{\omega_1}(X)$ is a **club** iff C is closed under countable increasing unions and $\forall A{\in}\wp_{\omega_1}(X)\exists B{\in}C\,(A \subseteq B)$. The club filter on $\wp_{\omega_1}(X)$ consists of all $A \subseteq \wp_{\omega_1}(X)$ such that $C \subseteq A$ for some club C; it is normal and fine. (Normality uses AC.) We say ω_1 is X**-supercompact** iff there is a normal, fine ultrafilter on $\wp_{\omega_1}(X)$. One of the basic consequences of $\mathsf{AD}_\mathbb{R}$, due to Solovay, is that the club filter on $\wp_{\omega_1}(\mathbb{R})$ is a normal ultrafilter, so that ω_1 is $\mathbb{R}$-supercompact [Sol78B].

We proceed to the main result of this section. It is convenient at this point to add full AC to the metatheory $\mathsf{ZF}+\mathsf{DC}$ of this paper; this makes possible some simple manipulations of club sets in the proof to follow. (We doubt that AC is actually necessary.)

THEOREM 3.1 (ZFC). Assume $\underset{\sim}{\Delta}^1_2$-$\mathsf{AD}^{\Sigma_2}$. Then there is an inner model containing all reals and ordinals and satisfying $\mathsf{ZF} + \mathsf{AD}_{\mathbb{R}} + \mathsf{DC}$ + "ω_1 is $\wp(\mathbb{R})$-supercompact" + "every set of reals admits a scale".

PROOF. We shall define a slight variant of Woodin's model (in order to get $\wp(\mathbb{R})$-supercompactness.) Let

$$M_0 = \mathbf{V}_{\omega+1} = \text{the set of rank} \leq \omega$$

and

$$M_\lambda = \bigcup_{\beta<\lambda} M_\beta, \text{ for } \lambda \text{ a limit.}$$

Now suppose M_α is given. If $M_\alpha \not\models \mathsf{ZF}^-$ + "$\wp(\wp(\mathbb{R}))$ exists", set

$$M_{\alpha+1} = \{a \subseteq M_\alpha \,:\, a \text{ is 1st order definable over } (M_\alpha, \in) \text{ from parameters}\}.$$

If $M_\alpha \models \mathsf{ZF}^-$ + "$\wp(\wp(\mathbb{R}))$ exists" + $\neg\mathsf{AD}$, then set

$$M_{\alpha+1} = M_\alpha$$

(i.e., stop the construction). Finally, suppose $M_\alpha \models \mathsf{ZF}^-$+"$\wp(\wp(\mathbb{R}))$ exists"+ AD, and let $\gamma = \vartheta^{M_\alpha}$ = sup of lengths of prewellorders of $\mathbb{R}$ in M_α.

Case 1. $\mathrm{cf}(\gamma) = \omega$.

Pick any sequence $\langle A_n : n < \omega\rangle$ such that $\forall n(A_n \in \wp(\mathbb{R}) \cap M_\alpha)$, but $\langle A_n : n < \omega\rangle \notin M_\alpha$. (By Wadge, this means $\forall B \in (\wp(\mathbb{R}) \cap M_\alpha)\exists n\, (B \leq_{\mathrm{W}} A_n)$.) Set

$$M_{\alpha+1} = \{a \subseteq M_\alpha \,:\, a \text{ is first order definable from parameters over } (M_\alpha, \in, \langle A_n : n < \omega\rangle)\}.$$

(By Wadge, $M_{\alpha+1}$ is independent of the $\langle A_n : n < \omega\rangle$ chosen.)

Case 2. $\mathrm{cf}(\gamma) > \omega$.

Let $X = \wp(\mathbb{R}) \cap M_\alpha$, and let F_α be the club filter on $\wp_{\omega_1}(X)$.

Subcase (a). F is not an ultrafilter over $\wp(\wp_{\omega_1}(X)) \cap M_\alpha$. (That is, there is an $A \subseteq \wp_{\omega_1}(X)$, $A \in M_\alpha$, such that neither A nor $\wp_{\omega_1}(X) - A$ is in F.)

Then set

$$M_{\alpha+1} = M_\alpha.$$

Subcase (b). Otherwise.

Then set

$$M_{\alpha+1} = \{a \subseteq M_\alpha \,:\, a \text{ is first order definable from parameters over } (M_\alpha, \in, F_\alpha \cap M_\alpha)\}.$$

This completes the definition of the M_α's. Clearly they constitute an increasing sequence of transitive sets. Set

$$M = \bigcup_{a \in \text{Ord}} M_\alpha.$$

Let also

$$\mathcal{F}(\alpha, A) \text{ iff Case 2 applied at } \alpha \text{ and } A \in F_\alpha,$$

and $\mathcal{M}_\alpha = (M_\alpha, \in, \mathcal{F} \cap M_\alpha)$.

The point is that $\mathcal{M}_\alpha$ has the information it needs to define $\langle \mathcal{M}_\beta : \beta < \alpha \rangle$. Thus there is a fixed formula defining $\langle \mathcal{M}_\beta : \beta < \alpha \rangle$ over $\mathcal{M}_\alpha$ for all α. There is also a natural sentence expressing "I am an $\mathcal{M}_\alpha$".

LEMMA 3.2. *Every* $(\utilde{\Delta}^2_1)^M$, *set of reals is* $\Game^{\Sigma_2}\utilde{\Pi}^1_1$.

PROOF. Let S be $(\Delta^2_1)^M$, where we have dropped the real parameter for convenience. Say

$$S(x) \iff M \models \exists A \subseteq \mathbb{R}\, \varphi(A, x)$$

and

$$-S(x) \iff M \models \exists A \subseteq \mathbb{R}\, \psi(A, x),$$

where ψ and φ have real quantifiers only. We must recursively associate to any $x \in \mathbb{R}$ an AD^{Σ_2} type game G_x such that $S(x)$ iff player I wins G_x. For convenience, we shall make the individual moves of G_x reals rather than natural numbers; for games of the AD^{Σ_2} variety this affects nothing of importance.

Let x be given. We describe the payoff of G_x by specifying the rules of play. The rules governing round α are defined by induction on α.

Round 0. Player I must play a real coding transitive structure $\mathcal{A}_0$ such that $x \in |\mathcal{A}_0|$ and

$$\mathcal{A}_0 \models \text{"I am an } \mathcal{M}_\alpha\text{"} \;\&\; \exists A \subseteq \mathbb{R}\, \varphi(A, x) \;\&\; \forall A \subseteq \mathbb{R} \neg\psi(A, x)$$

Player II must then play a real coding in a transitive $\mathcal{B}_0$ with $x \in |\mathcal{B}_0|$ and

$$\mathcal{B}_0 \models \text{"I am an } \mathcal{M}_\alpha\text{"} \;\&\; \exists A \supseteq \mathbb{R}\, \psi(A, x) \;\&\; \forall A \supseteq \mathbb{R} \neg\varphi(A, x).$$

Failure by one of the players is violation of the rules at 0.

Round $\alpha + 1$. If neither player has violated the rules at or before α, then we shall have transitive structures $\mathcal{A}_\alpha$ and $\mathcal{B}_\alpha$ at the end of round α. Player I must now play the code of a transitive $\mathcal{A}_{\alpha+1}$ such that

$$\mathbb{R}^{\mathcal{B}_\alpha} \subseteq \mathcal{A}_{\alpha+1}$$

and an elementary

$$i_\alpha \colon \mathcal{A}_\alpha \to \mathcal{A}_{\alpha+1}$$

with the following properties: Suppose $\mathcal{A}_\alpha \models$ "case 2 occurred at stage δ"; that is, $\mathcal{A}_\alpha \models \exists A\, \mathcal{F}(\delta, A)$. Let

$$C = (\wp(\mathbb{R}) \cap M_\delta)^{\mathcal{A}_\alpha}.$$

Then for $A \subseteq (\wp_{\omega_1}(\wp(\mathbb{R} \cap \mathcal{M}_\delta))^{\mathcal{A}_\alpha}$, player I must arrange

$$\mathcal{A} \models \mathcal{F}(\delta, A) \text{ iff } i_\alpha[C] \in i_\alpha(A)$$

(in particular, $i_\alpha[C] \in |\mathcal{A}_{\alpha+1}|$, and is countable in $\mathcal{A}_{\alpha+1}$).

Further, suppose $A_n \in \wp(\mathbb{R})^{\mathcal{A}_\alpha}$ for all $n < \omega$, and $\langle A_n : n \in \omega\rangle \in |\mathcal{B}_\alpha|$. Then player I must arrange

$$\langle i_\alpha(A_n) : n < \omega\rangle \in |\mathcal{A}_{\alpha+1}|.$$

Failure to meet these requirements is a violation of the rules at $\alpha + 1$ by player I.

On his $\alpha + 1$st move player II must play a $\mathcal{B}_{\alpha+1}$ and a $j_\alpha \colon \mathcal{B}_\alpha \to \mathcal{B}_{\alpha+1}$ meeting requirements completely symmetric to those on $\mathcal{A}_{\alpha+1}$ and $i_{\alpha+1}$. If he doesn't, player II violates the rules at α.

Round λ, λ limit. If no one has violated the rules before λ we shall have direct limit systems

$$\langle \mathcal{A}_\alpha, i_{\alpha\beta} : \alpha < \beta < \lambda\rangle$$

and

$$\langle \mathcal{B}_\alpha, j_{\alpha\beta} : \alpha < \beta < \lambda\rangle.$$

Set

$$\mathcal{A}_\lambda = \text{direct limit of } \langle \mathcal{A}_\alpha, i_{\alpha\beta} : \alpha < \beta < \lambda\rangle$$
$$\mathcal{B}_\lambda = \text{direct limit of } \langle \mathcal{B}_\alpha, j_{\alpha\beta} : \alpha < \beta < \lambda\rangle.$$

Then player I violates the rules at λ if $\mathcal{A}_\lambda$ is illfounded, and player II violates the rules at λ if $\mathcal{B}_\lambda$ is illfounded. If there is no violation, we assume $\mathcal{A}_\lambda$ and $\mathcal{B}_\lambda$ are transitive.

The reals played during round λ are meaningless for G_x.

This completes the rules of G_x. The first player to violate these rules loses G_x. In case of a tie, player I loses. We shall now show that by the time we reach the first Σ_2 admissible relative to the play, someone must lose.

CLAIM 3.3. Let $p \colon \lambda \times \omega \to \omega$ be a partial play of G_x, where λ is a limit ordinal. Suppose in p neither player has violated the rules at or before λ. Then $L_\lambda[p]$ is not Σ_2 admissible.

PROOF. (Here $n \mapsto p(\alpha, n)$ codes the reals played by player I and player II at round α in p.) Let p be a counterexample to the claim. Let $\langle \mathcal{A}_\alpha, i_{\alpha\beta} : \alpha < \beta < \lambda\rangle$ and $\langle \mathcal{B}_\alpha, j_{\alpha\beta} : \alpha < \beta < \lambda\rangle$ be the systems produced in p by player I and player II respectively. Let $\mathcal{A}_\lambda$ and $\mathcal{B}_\lambda$ be their transitive direct limits, and $i_{\alpha\lambda}, j_{\alpha\lambda}$ (for $\alpha < \lambda$) the natural maps.

Now

$$\mathbb{R}^{\mathcal{A}_\lambda} = \bigcup_{\alpha<\lambda} \mathbb{R}^{\mathcal{A}_\alpha} = \bigcup_{\alpha<\lambda} \mathbb{R}^{\mathcal{B}_\alpha} = \mathbb{R}^{\mathcal{B}_\lambda}.$$

On the other hand, $\mathcal{A}_\lambda \models \exists A \subseteq \mathbb{R}\, \varphi(A, x)\ \&\ \forall A \subseteq \mathbb{R} \neg\psi(A, x)$ while $\mathcal{B}_\lambda \models \exists A \subseteq \mathbb{R}\, \psi(A, x)\ \&\ \forall A \subseteq \mathbb{R} \neg\varphi(A, x)$, so

$$\wp(\mathbb{R})^{\mathcal{A}_\lambda} \not\subseteq \mathcal{B}_\lambda$$

and

$$\wp(\mathbb{R})^{\mathcal{B}_\lambda} \not\subseteq \mathcal{A}_\lambda.$$

Thus we have a $\delta \in \mathrm{Ord}^{\mathcal{A}_\lambda} \cap \mathrm{Ord}^{\mathcal{B}_\lambda}$ such that

$$M_\delta^{\mathcal{A}_\lambda} = M_\delta^{\mathcal{B}_\lambda} \text{ and } M_{\delta+1}^{\mathcal{A}_\lambda} \neq M_{\delta+1}^{\mathcal{B}_\lambda}.$$

Clearly then, $M_\delta^{\mathcal{A}_\lambda} \models \mathsf{ZF}^- \ \&\ \mathsf{AD}$. Let

$$\vartheta = \vartheta^{(M_\delta^{\mathcal{A}_\lambda})} = \vartheta^{(M_\delta^{\mathcal{B}_\lambda})}.$$

We claim that both $\mathcal{A}_\lambda$ and $\mathcal{B}_\lambda$ satisfy "$\mathrm{cf}(\vartheta) > \omega$". For suppose e.g., $\mathcal{A}_\lambda$ satisfies "$\mathrm{cf}(\vartheta) = \omega$". Pick $\langle A_n : n < \omega\rangle$ in $|\mathcal{A}|$ such that $\mathcal{A}$ satisfies: $\forall B \in M_\delta \cap \wp(\mathbb{R})\, \exists n (B \leq_{\mathrm{W}} A_n)$. Then for $\alpha < \lambda$ sufficiently large, say $\alpha \geq \alpha_0$, we have a $\langle A_n^\alpha : n < \omega\rangle \in |\mathcal{A}_\alpha|$ such that

$$i_{\alpha\lambda}(\langle A_n^\alpha : n < \omega\rangle) = \langle A_n : n < \omega\rangle.$$

Now $A_n \in |\mathcal{B}_\lambda|$ for all n. So for fixed n, we have for all sufficiently large, say $\alpha \geq \alpha_0$, a $\mathcal{B}_n^\alpha \in |\mathcal{B}_\alpha|$ such that

$$j_{\alpha\lambda}(B_n^\alpha) = A_n.$$

Notice that if α is a limit then $\mathbb{R}^{\mathcal{A}_\alpha} = \mathbb{R}^{\mathcal{B}_\alpha}$ so if α is a limit such that both A_n^α and B_n^α are defined, then $A_n^\alpha = B_n^\alpha$. Let

$$S = \{\alpha < \lambda : L_\alpha[p{\restriction}\alpha] \prec_{\Sigma_1} L_\lambda[p]\}.$$

Then S is club in λ and $L_\lambda[p, S]$ is admissible. By our observation above, for each $n < \omega$ we can find a $\beta_n > \alpha_0$ and $A_n^{\beta_n} \in |\mathcal{B}_{\beta_n}|$. By the stability of β_n, this means $\beta_n \geq \alpha_0$ and $A_n^{\beta_n} = B_n^{\beta_n}$. Since the map $n \mapsto \beta_n$ is $\mathbf{\Delta}_1(L_\lambda, [p, S])$, we have a limit γ such that $\gamma > \beta_n$ for all n.

Now the requirements on player II at round $\gamma + 1$ imply that

$$\langle j_{\gamma,\gamma+1}(B_n^\gamma) : n < \omega\rangle \in |\mathcal{B}_{\gamma+1}|$$

and since $\gamma > \beta_n$ for all n,

$$j_{\gamma+1,\lambda}(\langle j_{\gamma,\gamma+1}(B_n^\gamma) : n < \omega\rangle) = \langle A_n : n \in \omega\rangle.$$

Thus $\langle A_n : n < \omega\rangle \in |\mathcal{B}_\lambda|$, and $\mathcal{B}_\lambda$ satisfies $\mathrm{cf}(\vartheta) = \omega$. Thus both $\mathcal{A}_\lambda$ and $\mathcal{B}_\lambda$ think $M_{\delta+1}$ comes from M_δ via Case 1 using $\langle A_n : n < \omega\rangle$, so $M_{\delta+1}^{\mathcal{A}_\lambda} = M_{\delta+1}^{\mathcal{B}_\lambda}$, a contradiction.

Thus $\mathrm{cf}(\vartheta) > \omega$ in both $\mathcal{A}_\lambda$ and $\mathcal{B}_\lambda$. Pick $A \in M_\delta^{\mathcal{A}_\lambda} = M_\delta^{\mathcal{B}_\lambda}$ such that

$$\mathcal{A}_\lambda \models \mathcal{F}(\delta, A) \text{ and } \mathcal{B}_\lambda \models \neg\mathcal{F}(\delta, A).$$

(Such an A must exist.) Pick α_0 large enough that $i_{\alpha_0,\lambda}^{-1}(A)$, $i_{\alpha_0,\lambda}^{-1}(\delta)$, $j_{\alpha_0,\lambda}^{-1}(A)$ and $j_{\alpha_0,\lambda}^{-1}(\delta)$ all exist. Let

$$\nu_\gamma = i_{\gamma\lambda}^{-1}(\delta)$$
$$\mu_\gamma = j_{\gamma\lambda}^{-1}(\delta)$$
$$B_\gamma = i_{\gamma\lambda}^{-1}(A)$$
$$C_\gamma = j_{\gamma\lambda}^{-1}(A)$$

for $\alpha_0 \leq \gamma < \lambda$. Then $\langle(\nu_\gamma, \mu_\gamma, B_\gamma, C_\gamma) : \gamma < \lambda\rangle$ is Δ_1 over $L_\lambda[p]$.

Now if $X \in \wp(\mathbb{R}) \cap M_{\nu_\gamma}^{\mathcal{A}_\gamma}$, then there is a $\beta > \gamma$, $\beta \in S$, and a Y in $\wp(\mathbb{R}) \cap M_{\mu_\beta}^{\mathcal{B}_\beta}$ such that $i_{\gamma\lambda}(X) = j_{\beta\lambda}(Y)$. Notice that for $\beta \in S$, $i_{\gamma\lambda}(X) = j_{\beta\lambda}(Y)$ iff $i_{\gamma\beta}(X) = Y$. By a closure argument then, we can find $\beta \in S$, $\beta > \alpha_0$, such that

$$\wp(\mathbb{R}) \cap M_{\nu_\beta}^{\mathcal{A}_\beta} = \wp(\mathbb{R}) \cap M_{\mu_\beta}^{\mathcal{B}_\beta}.$$

Call this set C. Then $i_{\beta,\beta+1}[C]$ is countable in $\mathcal{A}_{\beta+1}$, and $j_{\beta,\beta+1}[C]$ is countable in $\mathcal{B}_{\beta+1}$, so

$$i_{\beta+1,\lambda}(i_{\beta,\beta+1}[C]) = i_{\beta\lambda}[C]$$

and

$$j_{\beta+1,\lambda}(i_{\beta,\beta+1}[C]) = j_{\beta\lambda}[C].$$

But β is stable, so $j_{\beta\lambda}(X) = i_{\beta\lambda}(X)$ for all $X \in C$, so

$$i_{\beta\lambda}[C] = j_{\beta\lambda}[C].$$

But by our rules

$$i_{\beta,\beta+1}[C] \in B_{\beta+1}$$
$$j_{\beta,\beta+1}[C] \notin C_{\beta+1}.$$

Hence

$$i_{\beta\lambda}[C] \in A$$

and

$$j_{\beta\lambda}[C] \notin A$$

a contradiction. This proves the claim. ⊣ (Claim 3.3)

To finish the proof of Lemma 3.2, we must show that player I has a winning strategy in G_x iff $S(x)$. So suppose $S(x)$. Pick an α such that

$$M_\alpha \models \mathsf{ZF}^- \ \&\ \exists A \subseteq \mathbb{R}\, \varphi(A, x) \ \&\ \forall A \subseteq \mathbb{R} \neg\varphi(A, x).$$

Then we can find a club $\mathcal{C} \subseteq \wp_{\omega_1}(M_\alpha)$ such that if $P \in \mathcal{C}$, then

$$(P, \in, \mathcal{F} \cap P) \prec \mathcal{M}_\alpha$$

and

$$(A, \delta \in P \ \& \ \mathcal{F}(\delta, A)) \text{ implies } (\wp(\mathbb{R}) \cap M_\delta \cap P) \in A.$$

[*Proof.* For each δ, $A \in M_\alpha$ such that $\mathcal{F}(\delta, A)$, pick a club $\mathcal{C}_{\delta,A} \subseteq A$. Define

$$f(\delta, A, Y) = \text{some } P \in \mathcal{C}_{\delta,A} \text{ such that } Y \subseteq P$$

for $Y \in \wp_{\omega_1}(\wp(\mathbb{R}) \cap M_\delta)$. Let $\mathcal{C}$ be the set of all $Q \in \wp_{\omega_1}(M_\alpha)$ such that Q is closed under f and Skolem functions for M_α.]

Now for $P \in \mathcal{C}$, let

$$\pi_p \colon \mathcal{A}_p \cong (P, \in, \mathcal{F} \cap P)$$

be the inverse of the collapse. Player I should play in G_x so that for all β

$$\mathcal{A}_\beta = \mathcal{A}_{P_\beta}, \text{ for some } P_\beta \in \mathcal{C}$$

and

$$P_\beta \subseteq P_{\beta+1}, \quad P_\lambda = \bigcup_{\beta<\lambda} P_\beta \text{ for } \lambda \text{ limit}$$

and

$$i_{\beta,\beta+1} = \pi_{P_{\beta+1}}^{-1} \circ \pi_{P_\beta}.$$

It is clear that the properties of $\mathcal{C}$ guarantee that he can play this way forever and violate no rules in doing so.

The proof that if $\neg S(x)$ then player II has a winning strategy in G_x is entirely symmetric. ⊣ (Lemma 3.2)

By Lemma 3.2, there must be a game of the form $\mathrm{G}(A, T_2)$ where $A^* \in \utilde{\Pi}^1_1$, which has no winning strategy in M. (This just means $M \not\models \utilde{\Pi}^1_1\text{-AD}^{\Sigma_2}$.) For otherwise the universal $\Game^{\Sigma_2}\Pi^1_1$ set of reals is $(\Delta^2_1)^M$, and hence there is a fixed $(\Delta^2_1)^M$ set universal for $(\utilde{\Delta}^2_1)^M$ sets, an absurdity. By Theorem 2.1, then, there is a $\Game^{\Sigma_2}\utilde{\Delta}^1_2$ set S which is not in M.

Let μ be the club filter on $\wp_{\omega_1}(\mathbb{R})$. Let Q be a $\Game^{\Sigma_2}\utilde{\Delta}^1_2$ set which codes a scale on S.

Claim 3.4. $\mathbf{L}(Q, \mu, R) \models \mathsf{AD} + $ "μ is an ultrafilter".

Proof. Let

$$N_0 = V_{\omega+1}$$

and

$$N_\lambda = \bigcup_{\beta<\lambda} N_\beta, \text{ for } \lambda \text{ limit.}$$

Given N_α such that $N_\alpha \not\models \mathsf{ZF}^-$, let

$$N_{\alpha+1} = \{A \subseteq N_\alpha : A \text{ is first order definable over } (N_\alpha, \in, Q) \text{ from parameters}\}.$$

Given N_α such that $N_\alpha \models \mathsf{ZF}^-$, but $N_\alpha \not\models \mathsf{AD}$ or μ is not an ultrafilter over $\wp(\wp_{\omega_1}(\mathbb{R})) \cap N_\alpha$, then set $N_{\alpha+1} = N_\alpha$. Otherwise, let

$$N_{\alpha+1} = \{A \supseteq N_\alpha : A \text{ is first order definable over } (N_\alpha, \in, Q, \mu \cap N_\alpha) \text{ from parameters}\}.$$

Let $N = \bigcup_{\alpha \in \mathrm{Ord}} N_\alpha$. An argument similar to the proof of Lemma 3.2 but much simpler shows that every $(\underset{\sim}{\mathbf{\Delta}}^2_1)^N$ set of reals is $\Game^{\omega^2,\mathbb{R}}\underset{\sim}{\mathbf{\Pi}}^1_1(Q)$, where $\Game^{\alpha,\mathbb{R}}$ is the game quantifier corresponding to games of length α on $\mathbb{R}$, and $\underset{\sim}{\mathbf{\Pi}}^1_1(Q)$ is the least pointclass containing Q and closed under $\forall^{\mathbb{R}}$, $\cap$, $\cup$, and continuous substitution. Thus there is a $\underset{\sim}{\mathbf{\Pi}}^1_1(Q)$ game of length ω^2 on $\mathbb{R}$ which is not determined in N. Now Martin [Mar83B] shows that every such game has a winning strategy which is $\Game^{\omega^2,\mathbb{R}}\underset{\sim}{\mathbf{\Delta}}^1_2(P)$, where P is $\Game^{\Sigma_2}\underset{\sim}{\mathbf{\Delta}}^1_2$ and codes a scale on Q. (The relevant games of length ω^2 on $\mathbb{R}$ are determined by $\underset{\sim}{\mathbf{\Delta}}^1_2$-$\mathsf{AD}^{\Sigma_2}$: any game of length $\alpha < \omega_1$ on $\mathbb{R}$ with $\Game^{\Sigma_2}\underset{\sim}{\mathbf{\Delta}}^1_2$ payoff is determined.) Thus there is a $\Game^{\omega^2,\mathbb{R}}\underset{\sim}{\mathbf{\Delta}}^1_2(P)$, hence a $\Game^{\Sigma_2}\underset{\sim}{\mathbf{\Delta}}^1_2$, set not in N. Let B be such a set. Let $C \subseteq \mathbb{R}$, $B \leq_W C$; and C codes up winning strategies for all games of length ω on $\mathbb{R}$ with payoff $\leq_W B$; we can take C to be $\Game^{\Sigma_2}\underset{\sim}{\mathbf{\Delta}}^1_2$ by [Mar83B].

SUBCLAIM 3.5. $\mathbf{L}(C, \mathbb{R}) \models \mathsf{AD}$.

PROOF. It is enough to see $C^\#$ exists and is $\Game^{\Sigma_2}\underset{\sim}{\mathbf{\Delta}}^1_2$. Now every $(\Sigma^2_1)^{\mathbf{L}(C,\mathbb{R})}$ set is $\Game^{\omega,\mathbb{R}}\underset{\sim}{\mathbf{\Pi}}^1_1(C)$, so we have a $\Game^{\Sigma_2}\underset{\sim}{\mathbf{\Delta}}^1_2$ set D such that $D \notin \mathbf{L}(C, \mathbb{R})$. Let α be least such that $\mathbf{L}_\alpha(D, \mathbb{R}) \models \mathsf{KP}$; then $\mathbf{L}_\alpha(D, \mathbb{R}) \models \mathsf{AD}$ since $\Game^{\Sigma_2}\underset{\sim}{\mathbf{\Delta}}^1_2$ is closed under "inductive in". Work in $\mathbf{L}_\alpha(D, \mathbb{R})$. Then by Wadge, if $X \subseteq \mathbb{R}$ and $X \in \mathbf{L}(C, \mathbb{R})$, $X \leq_W D$. Thus there is a measurable cardinal greater than $\Theta^{\mathbf{L}(C,\mathbb{R})}$, so $C^\#$ exists. Since $\mathbf{L}_\alpha(D, \mathbb{R}) \models C^\#$ exists, $C^\#$ does exist and is in $\mathbf{L}_\alpha(D, \mathbb{R})$, hence $\Game^{\Sigma_2}\underset{\sim}{\mathbf{\Delta}}^1_2$. ⊣ (Subclaim 3.5)

To finish the proof of the claim, let

$$N'_\alpha = N_\alpha^{\mathbf{L}(C,\mathbb{R})}.$$

By induction on α we see that $N'_\alpha = N_\alpha$, $\wp(\mathbb{R}) \cap N \subseteq \{X : X \leq_W B\}$, $N_\alpha \models \mathsf{AD}$, and μ is an ultrafilter over $\wp(\wp_{\omega_1}(\mathbb{R})) \cap N_\alpha$. This yields the claim at once. ⊣ (Claim 3.4)

Now notice that there is a canonical $\pi\colon \mathbb{R} \twoheadrightarrow \{B : B \leq_W Q\}$, $\pi \in \mathbf{L}(Q, \mu, \mathbb{R})$. Suppose now

$$M_\alpha \in \mathbf{L}(Q, \mathbb{R}, \mu);$$

then by Wadge $\wp(\mathbb{R}) \cap M_\alpha \subseteq \{B : B \leq Q\}$ so that π includes a canonical

$$\pi_\alpha : \mathbb{R} \twoheadrightarrow \wp(\mathbb{R}) \cap M_\alpha = X_\alpha.$$

Suppose Case 2 occurs in the definition of $M_{\alpha+1}$. For any $\mathcal{A} \subseteq \wp_{\omega_1}(X_\alpha)$, let

$$\mathcal{A}^* = \{A \in \wp_{\omega_1}(\mathbb{R}) : \pi_\alpha[A] \in \mathcal{A}\};$$

then $\mathcal{A}^*$ is in $\mathbf{L}(Q, \mathbb{R}, \mu)$ if $\mathcal{A}$ is. Also, if $\mathcal{A}^* \in \mu$, then $\mathcal{A} \in F_\alpha$, while if $\wp_{\omega_1}(\mathbb{R}) - \mathcal{A}^* \in \mu$, then $\wp_{\omega_1}(X_\alpha) - \mathcal{A} \in F_\alpha$. [For this reason we seem to need AC in V. Pick $\sigma : X_\alpha \to \mathbb{R}$ such that $\sigma \circ \pi_\alpha = \text{id}$. Then if $\mathcal{C} \subseteq \wp_{\omega_1}(\mathbb{R})$ is club, then $\{A \in \wp_{\omega_1}(X_\alpha) : \sigma^{-1}(A) \in \mathcal{C}\}$ is club in $\wp_{\omega_1}(X_\alpha)$; moreover $A \in \mathcal{A}$ iff $\sigma^{-1}(A) \in \mathcal{A}^*$.]

It follows that F_α is an ultrafilter over $\wp(\wp_{\omega_1}(X_\alpha)) \cap M_\alpha$; moreover F_α is uniformly-in-M_α definable over $\mathbf{L}(Q, \mathbb{R}, \mu)$.

It is now easy to see by induction that $M_\alpha \in \mathbf{L}(Q, \mathbb{R}, \mu)$ for all α and that $M_\alpha \neq M_{\alpha+1}$ for all α. In fact, the function $\alpha \mapsto M_\alpha$ is definable over $\mathbf{L}(Q, \mathbb{R}, \mu)$. Thus M satisfies ZF + AD + "ν is a normal ultrafilter over $\wp(\wp_{\omega_1}(\mathbb{R}))$", where ν is the club filter on $\wp_{\omega_1}(\wp(\mathbb{R}) \cap M)$ restricted to M. Since $M \models \text{cf}(\vartheta) > \omega$ by construction, $M \models$ DC (cf. [Sol78B]). Finally, every set of reals in M is $\leq_W S$ since $M \subseteq \mathbf{L}(Q, \mathbb{R}, \mu)$. Thus each such set has a scale in $\mathbf{L}(Q, \mathbb{R}, \mu)$. Woodin (unpublished) shows that this implies each set in M has a scale in M. So $M \models$ "Every set of reals has a scale", and by Martin [MarB] and unpublished work of Woodin, $M \models \forall \alpha < \omega_1(\text{AD}^\alpha)$. ⊣ (Theorem 3.1)

Notice that we could have augmented the construction of M by throwing in canonical winning strategies for games of the form $\text{G}(A, T_1)$, $A \in M$. This would give an M^* having the properties of M and satisfying AD^{Σ_1} in addition. [The proof of Lemma 3.2 must be modified as follows: suppose we have an A in $\mathcal{A}_\alpha \cap \mathcal{B}_\alpha$ such that $\mathcal{A} \models$ player I wins $\text{G}(A, T_1)$ and $\mathcal{B}_\alpha \models$ player II wins $\text{G}(A, T_1)$. Then the players of G_x take time out to play $\text{G}(A, T_1)$ and produce thereby a $q \in E_{T_1}$. We then require $q \in i_{\alpha,\alpha+1}(A)$ and $q \notin j_{\alpha,\alpha+1}(A)$. With this modification, the proof of Theorem 3.1 goes through.] A similar argument shows

Theorem 3.6. Assume $\underset{\sim}{\boldsymbol{\Delta}}^1_2\text{-AD}^{\Sigma_{n+1}}$, where $n \leq 1$. Then there is an inner model containing all reals and ordinals, and satisfying $\text{ZF}+\text{AD}^{\Sigma_n}+\text{DC}+$"every set admits a scale" + "ω_1 is $\mathbf{V}_{\omega+n+1}$-supercompact".

In fact, it seems likely that the proof of Theorem 3.1 will adapt to show: if all clopen games of length ω_1 with $\mathbf{HOD}(\mathbb{R})$ (or even, "$\underset{\sim}{\boldsymbol{\Delta}}^1_2$ in the codes") payoff are determined , then there is an inner models satisfying ZF + DC + "Every set of reals admits a scale" + "$\text{G}(A, T)$ is determined whenever T admits a scaled coding". If so, then this is nearly as well as one can do in the direction of models of strong forms of full determinacy without modifying the notion of a winning strategy; cf. §5.

§4. **A determinacy proof.** The basic problem in this area is to prove that determinacy of the long games we have been considering from ZFC together with some large cardinal hypothesis. One method for doing this involves reducing the long games to games of length ω on higher type objects, as in the Blass-Mycielski proof that $\mathsf{AD}_{\mathbb{R}}$ implies AD^{ω^2}. In this section we shall present such that a reduction for games of the form $\mathrm{G}(A, T_1)$, that is, games ending at the first admissible relative to the play. We shall also prove the determinacy of the length of ω games to which we reduce within the theory ZF+AD+DC+"Every set admits a scale" + "ω_1 is $\wp(\mathbb{R})$-supercompact". Thus this theory proves AD^{Σ_1}. Now Woodin has recently shown, in ZFC + $\exists\kappa(\kappa$ is supercompact), that there is an inner model of ZF + AD + DC + "Every set admits a scale" containing all reals and ordinals. It seems likely that his techniques will/do produce a model satisfying "ω_1 is $\wp(\mathbb{R})$-supercompact" as well. If so, then we have a proof from ZFC + $\exists\kappa(\kappa$ is supercompact) of Γ-AD^{Σ_1}, where Γ is the class of sets in Woodin's model (and thus a proof of e.g., $\underset{\sim}{\Pi}^1_1$-AD^{Σ_1}).

Martin [MarB] and Woodin (unpublished) have independently shown that ZF + AD + DC + "Every set admits a scale" proves $\forall\alpha < \omega_1(\mathsf{AD}^\alpha)$. Thus $\underset{\sim}{\Pi}^1_1$-AD^α, for $\alpha < \omega_1$, follows from the existence of supercompact cardinals. Martin's idea seems likely to yield, with more work, a proof of AD^{Σ_1} from ZF + AD + DC + "Every set admits a scale"; if so, then $\underset{\sim}{\Pi}^1_1$-AD^{Σ_1} follows from ZFC + $\exists\kappa$ (κ is supercompact). However, the ideas presented in this section seem more promising than Martin's approach when it comes to proving $\underset{\sim}{\Pi}^1_1$-AD^{Σ_n}, $n \geq 2$, from ZFC + $\exists\kappa$ (κ is supercompact).

Suppose $T \subseteq \bigcup_{\alpha<\omega_1} {}^\alpha\omega$, and $\pi\colon T \to \omega$ is such that $\forall p, q \in T$ $(p \subseteq q \Rightarrow \pi(p) \neq \pi(q))$; then we shall call π a **continuous coding of** T. Our determinacy proof actually applies to game of the form $\mathrm{G}(A, T)$, where T admits a continuous coding. It is an easy exercise to show that T_1 admits a continuous coding, as do the trees for the games ending at the first recursively inaccessible, recursive Mahlo, ..., relative to play. It is also easy to see that T_2, the tree for games ending at the first Σ_2 admissible relative to the play, does not admit a continuous coding.

We begin with the higher type games to which we shall reduce our long games. Let X be any set, and $\mathcal{A} \subseteq {}^\omega X$. By $\mathrm{G}_{\mathcal{A}}$ we mean the game length ω on X with payoff set $\mathcal{A}$. We call $\mathrm{G}_{\mathcal{A}}$ determined if one of the players has a winning quasi-strategy. We call $\mathcal{A}$ **Suslin** if for some ordinal κ and tree T on $X \times \kappa$, $\mathcal{A} = \mathrm{p}[T] = \{f \in {}^\omega X : \exists g \in {}^\omega\kappa\, \forall n(f\restriction n, g\restriction n) \in T\}$. We call $\mathcal{A}$ co-Suslin if ${}^\omega X \setminus \mathcal{A}$ is Suslin. The author learned the proof of the following lemma from A.S. Kechris; it is due to him. By $\mathbf{OD}(X)$ we mean the class of sets ordinal definable from finitely many elements of X.

LEMMA 4.1. Assume ZF + AD + DC. Suppose ω_1 is X-supercompact, as witnessed by the ultrafilter μ. Suppose that $\mathcal{A} \subseteq {}^\omega X$ is Suslin and co-Suslin, as witnessed by the trees T and U. Then there is a $\mathbf{OD}(\{\mu, T, U\})$ winning quasi-strategy for $G_{\mathcal{A}}$.

PROOF. For $C \in P_\omega(X)$, let $G^C_{\mathcal{A}}$ be just like $G_{\mathcal{A}}$, except that the player's moves are restricted to lie in C. Then $G^C_{\mathcal{A}}$ is determined by AD. For any C such that player I wins $G^C_{\mathcal{A}}$, we can define a canonical "best" winning quasi-strategy Σ_C for player I in $G^C_{\mathcal{A}}$, as in Moschovakis [Mos80, Chapter 6]. (In order to make the strategy canonical, we must make it "quasi".) Σ_C is $\mathbf{OD}(\{T, C\})$ uniformly in C. Similarly, if player II wins $G^C_{\mathcal{A}}$ we get a canonical winning quasi-strategy for τ_C for player II which is $\mathbf{OD}(\{U, C\})$, uniformly in C.

Now suppose player I wins G^C_a for μ–a.e. C. Define Σ for player I in $G_{\mathcal{A}}$ by:

$$p \text{ is according to } \Sigma \text{ iff for } \mu\text{–a.e. } C,\ p \text{ is according to } \Sigma_C.$$

Σ is $\mathbf{OD}(\{T, \mu\})$, and it is easy to check that Σ is a winning quasi-strategy τ for player I in $G_{\mathcal{A}}$.

Similarly, if player II wins $G^C_{\mathcal{A}}$ for μ–a.e. C, we get an $\mathbf{OD}(\{U, \mu\})$ winning quasi-strategy τ for player II in $G_{\mathcal{A}}$. ⊣

The most natural reduction of continuously coded long games to games of length ω takes us to games on $\mathbb{R} \times \wp(\mathbb{R})$. Unfortunately, there is a very simple $\mathcal{A} \subseteq {}^\omega(\mathbb{R} \times \wp(\mathbb{R}))$ which is not Suslin in models of AD, namely $\mathcal{A} = \{(\vec{x}, \vec{A}) : \lambda_n \cdot x_n(0) \in A_0\}$. Thus we must take care to make Lemma 4.1 applicable. We do this by reducing to a game on $\mathbb{R} \times \mathcal{S}$, where

$$\mathcal{S} = \{(\vec{\varphi}, \vec{\psi}) : \vec{\varphi} \text{ and } \vec{\psi} \text{ are scales, and } \mathrm{p}(\vec{\varphi}) = \mathbb{R} - \mathrm{p}(\vec{\varphi})\}.$$

Here $\mathrm{p}(\vec{\varphi})$ is the projection of $\vec{\varphi}$, that is the common domain of the norms φ_i, $i \in \omega$. For $g \in {}^\omega\mathcal{S}$, let $\mathrm{p}(g) = h$, where $h(i) = \mathrm{p}(g(i)_0)$ for all $i \in \omega$. For $\mathcal{A} \subseteq {}^\omega(\mathbb{R} \times \mathcal{S})$, let

$$\mathrm{p}(\mathcal{A}) = \{(f, h) : \exists g((f, g) \in \mathcal{A} \;\&\; h = \mathrm{p}(g))\};$$

thus $\mathrm{p}(\mathcal{A}) \subseteq {}^\omega(\mathbb{R} \times \wp(\mathbb{R}))$. We say that $\mathcal{A} \subseteq {}^\omega(\mathbb{R} \times \mathcal{A})$ is **projection-invariant** iff whenever $\mathrm{p}(\mathcal{B}) = \mathrm{p}(\mathcal{A})$, then $\mathcal{B} \subseteq \mathcal{A}$. Now one can easily modify the example of the last paragraph to obtain an $\mathcal{A} \subseteq {}^\omega(\mathbb{R} \times \mathcal{S})$ which is not Suslin in models of AD; however there is no simple projection-invariant $\mathcal{A}$ with this property.

CONJECTURE 4.2. There is an inner model of ZF + AD + DC + "Every projection-invariant $\mathcal{A} \subseteq {}^\omega(\mathbb{R} \times \mathcal{S})$ is Suslin".

Such a model must of course satisfy "Every set of reals admits a scale". The model M of §3 may verify the conjecture; the author is not sufficiently in command of the relevant work of Woodin (unpublished) to decide.

Fortunately, we shall need the conjecture only for reasonable simple $\mathcal{A}$. Let $B \subseteq \mathbb{R}$ and $\mathcal{C} \subseteq {}^\omega(\mathbb{R} \times \wp(\mathbb{R}))$; we call $\mathcal{C}$ B**-projective** iff $\mathcal{C}$ is the smallest class of relations containing the basic relations

$$R(\vec{x}^1 \dots \vec{x}^n, \vec{A}) \text{ iff } x_i^m \in A_j,$$
$$S(\vec{x}^1 \dots \vec{x}^n, \vec{A}) \text{ iff } x_i^m \in B,$$

and

$$Q(\vec{x}^1 \dots \vec{x}^n, \vec{A}) \text{ iff } x_i^m(j) = \ell,$$

for all i, j, ℓ, m in ω, and closed under countable intersection, countable union, complement, and quantification over ${}^{\omega}\mathbb{R}$. The proof of the following lemma is implicit in that of Moschovakis' 2nd periodicity theorem, so we omit it.

LEMMA 4.3. *Assume* $\mathsf{ZF}+\mathsf{AD}+\mathsf{DC}$. *Suppose* $B \subseteq \mathbb{R}$ *is Suslin and co-Suslin via the trees* T *and* U. *Let* $\mathcal{A} \subseteq {}^{\omega}(\mathbb{R}\times\mathcal{S})$ *be projection-invariant, and suppose* $\mathrm{p}(\mathcal{A})$ *is* B*-projective. Then* $\mathcal{A}$ *is Suslin via a tree in* $\mathbf{OD}(\{T, U\})$.

THEOREM 4.4. *Assume* $\mathsf{ZF}+\mathsf{AD}+\mathsf{DC}+$"*Every set of reals admits a scale*"$+$ "ω_1 *is* $\wp(\mathbb{R})$ *supercompact*". *Then* $\mathrm{G}(A, T)$ *is determined whenever* T *admits a continuous coding. In particular,* AD^{Σ_1} *holds.*

PROOF. Let π be a continuous coding of T. There is a scaled coding F of T induced by π. Let $\hat{\pi} = \{(p^*, n) : \pi(p) = n\}$. Let $B \subseteq \mathbb{R}$ be such that $\hat{\pi}$, T^*, E_T^*, A^*, C, and their complements all have scales Wadge reducible to B. (Here $C = C(i, j, k, y)$ is as the definition of "scaled coding ".) Let μ be a normal, fine ultrafilter on $\wp_{\omega_1}(\mathbb{R}\times\mathcal{S})$; μ exists since we can map $\mathbb{R}\times\mathcal{S}$ into $\wp(\mathbb{R})$ in a 1-1 way.

Since we have AD + "Every set has a scale", $\wp(\mathbb{R}) \not\subseteq \mathbf{OD}(\{S\}) \cup \mathbb{R})$ for any set S. Thus for any set S there is an $A \subseteq \mathbb{R}$ such that any set of reals in $\mathbf{OD}(\{S\} \cup \mathbb{R})$ has a scale Wadge reducible to A. Let

$$H_0 = \mathbf{OD}(\{B\} \cup \{\mu\} \cup \mathbb{R})$$

and

$$H_{n+1} = \mathbf{OD}(\{\mu\} \cup \{A\} \cup \mathbb{R}), \text{ where } A \text{ Wadge-minimal such that every set of reals in } H_n \text{ has a scale } \leq_{\mathrm{W}} A.$$

We now describe a game G^* on $\mathbb{R}\times\mathcal{S}$ auxilliary to $\mathrm{G}(A, T)$. The game G^* is quite similar to the auxilliary game in Martin's proof of Borel determinacy in [Mar75]. A set $\Sigma \subseteq \bigcup_{\alpha<\omega_1} {}^{\alpha}\omega$ is a tree if $\forall p \in \Sigma\ \forall \beta \in \mathrm{dom}(p)\ (p\restriction\beta \in \Sigma)$. Play in G^* is divided into rounds; before beginning a round we have a G^*-position and an associated G position. For convenience we assume that if $p \in E$, then $\mathrm{dom}(p)$ is a limit.

Round 0. We begin with G^* and G-positions $\varnothing$.

(a) Player I plays $a \in \omega$, then player II plays $b \in \omega$. Let e be least in $\omega - \{\pi(\langle a, b\rangle)\}$; we call e critical for round 0.

(b) Player I now plays a $(\vec{\varphi}, \vec{\psi})$ in $\mathcal{S} \cap H_1$ such that $\mathrm{p}(\vec{\varphi}) = \Sigma$ is a subtree of T ($\Sigma = \varnothing$ is o.k.). Player II may either accept or reject Σ.

Player II accepts: Player I is now obliged to keep the G-position in Σ henceforth. Player II is obliged to insure that $e \neq \pi(p \restriction \beta)$ for any $\beta \leq \mathrm{dom}(p)$ and any G-position reached henceforth.

Player II rejects: Player II must play a $q \in \Sigma$ extending $\langle a, b\rangle$ and such that $e = \pi(q)$. No one incurs any obligations.

The new G position is $\langle a, b\rangle$ if player II accepts and q if player II rejects.

Round n, n > 0. We have a G^* position r with associated G position p.

(a) Player I plays $a \in \omega$, player II plays $b \in \omega$. Let e be least in $\omega - \{\pi((p^\frown\langle a, b\rangle)\restriction\beta) : \beta \leq \mathrm{dom}(p) + 1\}$; e is critical for round n.

(b) Player I now plays a $(\vec{\varphi}, \vec{\psi})$ in $\mathcal{S} \cap H_{n+1}$ such that $\mathrm{p}(\vec{\varphi}) = \Sigma$ is a (set of reals coding a) subtree of T. Player II may either accept or reject Σ.

Player II accepts: Player I is now obliged to keep the G position in Σ henceforth. Player II is obliged to insure that if q is a G position later reached, then $e \neq \pi(q \restriction \beta)$ for all $\beta < \mathrm{dom}(q)$.

Player II rejects: Player II must play a $q \in \Sigma$ extending $p^\frown\langle a, b\rangle$ and such that $e = \pi(q)$. No obligations are incurred.

The new G position is $p^\frown\langle a, b\rangle$ if player II accepts, and q if player II rejects.

The first player to fail to meet any one if his obligations loses G^*. If both players fail first at the same round, then the one violating the obligation incurred earliest loses, with player I's obligation incurred at round n coming after those incurred by player II at rounds $m < n$ and before those incurred by player II at rounds $m \geq n$.

Suppose both players meet all obligations incurred in G^*. After ω moves they have produced a G position p. If $p \in E$, then player I wins G^* iff $p \in A$. Otherwise, notice that $\pi(p)$ was critical at some round, and player II accepted at that round, so that p violates player II's obligation incurred at that round (and no earlier player II obligations). So if $p \notin E$, and $\pi(p)$ was critical at round n, then player I wins G^* iff p violates no player I obligation incurred at some round $m \leq n$.

Clearly we may regard G^* as a game on $\mathbb{R} \times \mathcal{S}$. Since the relation $y \in H_n$ is $\mathbf{OD}(\{B, \mu\})$, then there is a tree U_1 on $\mathbb{R} \times \mathcal{S}$ such that $U_1 \in \mathbf{OD}(\{B, \mu\})$ and

$$r \in [U_1] \text{ iff } r \text{ is a play of } \mathrm{G}^* \text{ in which player I loses at no finite stage.}$$

Similarly, there is an $\mathbf{OD}(\{B, \mu\})$ tree U_2 on $\mathbb{R} \times \mathcal{S} \times \omega$ such that

$$r \in p[U_2] \text{ iff } r \text{ is a play of } \mathrm{G}^* \text{ in which player II loses at some finite stage.}$$

Finally, there is a projection invariant $\mathcal{B} \subseteq {}^{\omega}(\mathbb{R} \times \mathcal{S})$ such that $\mathrm{p}(\mathcal{B})$ is $\langle \hat{\pi}, T^*, E^*, A^* C\rangle$-projective and whenever r is a play of G^* where no one loses at a finite stage,

$$r \in \mathcal{B} \text{ iff } r \text{ is a win for player I at stage } \omega.$$

It follows from Lemma 4.3 that $\mathcal{B}$ is Suslin via some $\mathbf{OD}(\{B, \mu\})$ tree U_3. But then

$$r \text{ is a win for player I in } \mathrm{G}^* \text{ iff } (r \in [U_1] \;\&\; (r \in p[U_2] \text{ or } r \in p[U_3]))$$

so that $\mathrm{G}^* = \mathrm{G}_{\mathcal{A}}$, where $\mathcal{A}$ is Suslin via an $\mathbf{OD}(\{B, \mu\})$ tree. Similarly, $\mathcal{A}$ is co-Suslin via an $\mathbf{OD}(\{B, \mu\})$ tree, so Lemma 4.1 gives a winning quasi-strategy for G^* which is $\mathbf{OD}(\{B, \mu\})$.

Case 1. Player I had a winning quasi-strategy in G^*.

We shall construct a winning strategy Γ for player I in $\mathrm{G}(A, T)$. Notice that there is a set of reals not in $\bigcup_n H_n$ by DC, and thus there is a map from $\mathbb{R}$ onto $\bigcup_n H_n \cap \mathcal{S}$. By uniformization we can thus convert player I's winning quasi-strategy in G^* into a full winning strategy; call it $\hat{\Gamma}$.

We define Γ from $\hat{\Gamma}$ by associating inductively to each p we reach following Γ a G^* position $\hat{p}$ which is according to $\hat{\Gamma}$ and whose associated G position is p. We arrange that in $\hat{p}$ player II has met all his obligations.

Suppose we have reached $p \in T$ by following Γ, and we have the associated $\hat{p}$ according to $\hat{\Gamma}$. Let $\hat{\Gamma}(\hat{p}) = a$; then set $\Gamma(p) = a$. Now suppose player II responds b in $\mathrm{G}(A, T)$.

Case A. $p^\frown\langle a, b\rangle$ meets all player II obligations incurred in $\hat{p}$.

In this case, let $(\vec{\varphi}, \vec{\psi}) = \hat{\Gamma}(\hat{p}^\frown\langle a, b\rangle)$, and set

$$\widehat{(p^\frown\langle a, b\rangle)} = \hat{p}^\frown\langle a, b, (\vec{\varphi}, \vec{\psi}), \text{accept}\rangle.$$

Case B. Otherwise.

Let the first player II obligation violated by $p^\frown\langle a, b\rangle$ be incurred at round n (i.e., in $\hat{p}(n)$). Let $\hat{p}(n) = \langle c, d, (\vec{\varphi}, \vec{\psi}), \text{accept}\rangle$. Set

$$\widehat{(p^\frown\langle a, b\rangle)} = \hat{p}\restriction n^\frown\langle c, d, (\vec{\varphi}, \vec{\psi}), \text{reject}, p^\frown\langle a, b\rangle\rangle.$$

(Notice $p^\frown\langle a, b\rangle \in \mathrm{p}(\vec{\varphi})$, as otherwise $\hat{p}^\frown\langle a, b\rangle$ is a loss for $\hat{\Gamma}$.)

In either case, the relation between p and $\hat{p}$ still holds between $p^\frown\langle a, b\rangle$ and $\widehat{(p^\frown\langle a, b\rangle)}$.

Finally, we must define $\hat{p}$ for p a position of limit length reached by following Γ. Let $\lambda = \mathrm{dom}(p)$. Notice that if n is least such that $\widehat{(p\restriction\beta)}(n) \neq \widehat{(p\restriction\beta + 1)}(n)$, then $\widehat{(p\restriction\beta)}(n)_3 = \text{accept}$ and $\widehat{(p\restriction\beta + 1)}(n)_3 = \text{reject}$. So $q(n) = \lim_{\beta\to\lambda}\widehat{(p\restriction\beta)}(n)$ exists for all n. If $p \in E$, set $\hat{p} = q$. In this case $\hat{p}$ is by $\hat{\Gamma}$, so $p \in A$, and we have verified that Γ wins. If $p \notin E$, let $e = \pi(p)$. Then e was critical at some round, say round n, of q. Let $q(n) = \langle c, d, (\vec{\varphi}, \vec{\psi}), \text{accept}\rangle$. Set

$$\hat{p} = q\restriction n^\frown\langle c, d, (\vec{\varphi}, \vec{\psi}), \text{reject}, p\rangle.$$

Since q was by $\hat{\Gamma}$, which wins for player I in G^*, $p \in \mathrm{p}(\vec{\varphi})$. Thus $\hat{p}$ is a position according to $\hat{\Gamma}$ in which player II meets all his obligations, with associated position p, as desired.

This defines Γ. We have also verified Γ wins $\mathrm{G}(A, T)$ for player I.

Case 2. Player II has a winning quasi-strategy in G^*.

We shall construct a winning strategy Γ for player II in $\mathrm{G}(A, T)$. Let $\hat{\Gamma}$ be a winning quasi-strategy for player II in G^* which is $\mathbf{OD}(\{B, \mu\})$.

Let $\sigma \colon \mathbb{R} \twoheadrightarrow \{q : q \text{ is a } \mathrm{G}^* \text{ position}\}$. If $\sigma(x) = q$ we call x an index of q. We shall define Γ by associating inductively to each G position p we reach the following Γ in G^* position $\hat{p}$ according to $\hat{\Gamma}$, together with a real indexing $\hat{p}$. We arrange that in $\hat{p}$ player I has met all his obligations and that p is the G-position associated to $\hat{p}$.

Suppose we have reached p by following Γ, and we have $\hat{p}$ and x such that $\sigma(x) = \hat{p}$. Suppose player I now plays $a \in \omega$ in $\mathrm{G}(A, T)$. Let $b \in \omega$ be least such that $\hat{p}^\frown\langle a, b\rangle$ is by $\hat{\Gamma}$; set $\Gamma(p^\frown\langle a\rangle) = b$. Now let $q = p^\frown\langle a, b\rangle$; we want to define $\hat{q}$ and an index for $\hat{q}$.

Case A. q violates no player I-obligation in $\hat{p}$.

In this case, let

$$\Sigma = \{r \in \bigcup_{\alpha<\omega_1} {}^{\alpha}\omega : \text{ if } \beta \leq \mathrm{dom}(r), \text{ then no position of the form}$$
$$\hat{p}^\frown\langle a, b, (\vec{\varphi}, \vec{\psi}), \text{reject}, r{\restriction}\beta\rangle, \text{ with } (\vec{\varphi}, \vec{\psi}) \in \mathcal{S} \cap H_{n+1},$$
$$n = \mathrm{dom}(\hat{p}), \text{ is in accord with } \hat{\Gamma}\}.$$

Then $\Sigma \in H_n$, where $n = \mathrm{dom}(\hat{p})$, since $\hat{p} \in H_n$ and $\hat{\Gamma} \in H_n$. So there is a $(\vec{\varphi}, \vec{\psi}) \in \mathcal{S} \cap H_{n+1}$ such that $\Sigma = \mathrm{p}(\vec{\varphi})$. Now $\hat{\Gamma}$ must accept Σ after $\hat{p}^\frown\langle a, b\rangle$ by the definition of Σ, so we set

$$\hat{q} = \hat{p}^\frown\langle a, b, (\vec{\varphi}, \vec{\psi}), \text{accept}\rangle,$$

where $(\vec{\varphi}, \vec{\psi}) \in \mathcal{S} \cap H_{n+1}$, $\Sigma = \mathrm{p}(\vec{\varphi})$, and the index of $\hat{q}$ is obtained from x by the appropriate uniformizing function. (We ought to have fixed at the outset a function $h \colon \mathbb{R} \to \mathbb{R}$ such that if $\sigma(x) = \hat{p}$ has the properties above, and Σ is defined as above, then $\sigma(h(x, a)) = \hat{q}$ is related to Σ and $\hat{p}$ as above.)

Case B. Otherwise.

Let n be such that the player I-obligation incurred in $\hat{p}(n)$ is violated by q. Then $\hat{p} = \langle c, d, (\vec{\varphi}, \vec{\psi}), \text{accept}\rangle$, where $\mathrm{p}(\vec{\varphi}) = \Sigma$ is as in Case A, and $q \notin \Sigma$ while $q{\restriction}\beta \in \Sigma$ for $\beta < \mathrm{dom}(q)$. By definition of Σ, there is a $(\vec{\vartheta}, \vec{\rho}) \in \mathcal{S} \cap H_{n+1}$, such that $\hat{p}{\restriction}n^\frown\langle c, d, (\vec{\vartheta}, \vec{\rho}), \text{reject}, q\rangle$ is in accord with $\hat{\Gamma}$. Using uniformization, we can pick such a $(\vec{\vartheta}, \vec{\rho})$ together with an index of

$$\hat{q} = \hat{p}{\restriction}n^\frown\langle c, d, (\vec{\vartheta}, \vec{\rho}), \text{reject}, q\rangle.$$

Finally, suppose we reach a position p of limit length following Γ; we must define $\hat{p}$ and an index of same. As in Case 1, $q(n) = \lim_{\beta \to \mathrm{dom}(p)} (\widehat{p \restriction \beta})(n)$ exists for all n, so that q is a completed run of G^* by $\hat{\Gamma}$. If $p \in E$, then since p is the run of $\mathrm{G}(A, T)$ associated to q, $p \notin A$, so that Γ has won. If $p \notin E$, then let $\pi(p) = e$, and let n be such that e is critical at round n in q. Since $\hat{\Gamma}$ wins for player II, q violates some player I obligation incurred in $q(m)$ for some $m \leq n$. Then let

$$\hat{p} = q \restriction m^\frown \langle c, d, (\vec{\vartheta}, \vec{p}), \text{reject}, q \rangle,$$

together with an index of $\hat{p}$, be obtained as in Case B above.

This completes the definition of Γ and the verification that it wins for player II. ⊣ (Theorem 4.4)

There is a natural strategy for iterating the reduction achieved in Theorem 4.4, and thereby proving $\undertilde{\Pi}^1_1$-AD^{Σ_n} (assuming, ultimately, $\mathsf{ZFC} + \exists\kappa(\kappa$ is supercompact$)$). It seems one must strengthen the hypothesis "Every set has a scale" of Theorem 4.4 (as well as requiring ω_1 be $\wp_{\omega_1}(\mathbf{V}_{\omega+\omega}$-supercompact$)$) in order to do this. The first step of the strengthening is given by the conjecture mentioned in this section.

The proof of the Kechris-Woodin theorem of §3 adapts at once to show

Corollary 4.5. $\mathsf{ZF} + \mathsf{AD} + \mathsf{DC} +$ "Every set has a scale" $+$ "ω_1 is $\wp(\mathbb{R})$-supercompact" provides the consistency of $\mathsf{ZFC} + \mathbf{HOD}(\mathbb{R})$-$\mathsf{AD}^{\Sigma_1}$.

Thus $\mathbf{HOD}(\mathbb{R})$-AD^{Σ_1} would not have sufficed to produce the model M of §3.

§5. **Questions.** The reader who has arrived at this section by slogging through the intermediate ones will probably have his own list of questions by now. We list here only a few salient ones.

Ad §1, the natural question is whether definable scales extend still further. For $p \in \bigcup_{\alpha<\omega_1} {}^\alpha\omega$, let

$$N_p = \{f \in {}^{\omega_1}\omega : p \subseteq f\},$$

and call a set $A \subseteq {}^{\omega_1}\omega$ open-$\undertilde{\Pi}^1_1$ if there is a $\undertilde{\Pi}^1_1$ set A^* of codes for elements of $\bigcup_{\alpha<\omega_1} {}^\alpha\omega$ such that

$$A = \bigcup\{N_p : \exists x \in A^* \ (x \text{ codes } p)\}.$$

If CH holds, then $\Game^{\omega_1}$ (closed-$\undertilde{\Pi}^1_1$) is just $\undertilde{\Sigma}^2_1$, and by Moschovakis' argument, $\undertilde{\Pi}^2_1$ has the prewellordering property. Does some form of definable determinacy, e.g., $\mathbf{HOD}(\mathbb{R})$-AD^{ω_1}, imply that all $\Game^{\omega_1}$ (open-$\undertilde{\Pi}^1_1$) sets of reals admit definable scales?

Ad §2, does every length ω_1 game with open-$\undertilde{\Pi}^1_1$ payoff won by player I have a definable winning strategy (assuming e.g., $\mathbf{HOD}(\mathbb{R})$-AD^{ω_1}?). There is a simple such game won by player II for which player II has no definable winning strategy. It is a part of the folklore: player I and player II take a nap until, at some move α, player I awakens with a start and asks player II to produce $x \in \mathrm{WO}$ such that $|x| = \alpha$. If player I never awakens, he loses, otherwise player II must comply or lose. For all we know, however, every length ω_1 game with $\mathbf{HOD}(\mathbb{R})$ payoff has a definable **pseudo-strategy,** where S is a pseudo-strategy winning G_A for (say) player I if

$$S\colon \bigcup_{\alpha<\omega_1} ({}^{\alpha}\omega \times \mathrm{WO}_\alpha) \to \omega,$$

where $\mathrm{WO}_\alpha = \{x \in \mathrm{WO} : |x| = \alpha\}$, and whenever $g(\alpha) \in \mathrm{WO}_\alpha$ for all $\alpha < \omega_1$,

$$\forall\alpha(f(\alpha)_0 = S(f \restriction \alpha, g(\alpha))) \Rightarrow f \in A,$$

for all $f \in {}^{\omega_1}\omega$. Does $\mathbf{HOD}(\mathbb{R})$-AD^{ω_1} guarantee $\mathbf{HOD}(\mathbb{R})$ pseudo-strategies for length ω_1 with $\mathbf{HOD}(\mathbb{R})$ payoff? (The notion of a pseudo-strategy is due independently to Woodin.)

Ad §3, notice that the folklore example above shows $\mathsf{ZF} + \mathsf{AD} + (\text{open-}\underset{\sim}{\mathbf{\Pi}}^1_1)$-$\mathsf{AD}_{\omega_1}$ is inconsistent. A slight modification shows that $\mathsf{ZF} + \mathsf{AD}$ implies the existence of a non-determined "clopen-$\underset{\sim}{\mathbf{\Pi}}^1_1$" game of length ω_1. [At any move β before he awakes, player I must pay for continuing to nap by playing an $x \in \mathrm{WO}_\beta$. Notice that, in contrast to the games admitting scaled codings, the game is not clopen in $\mathbf{V}$, only in models of AD.] Thus §3 goes nearly as far as possible in producing inner models with full winning strategies for games not limited by definability of the payoff. Can we produce stronger inner models using pseudo-strategies?

Another question suggested by §3: what is the consistency strength of $\mathsf{AD}_{\mathbb{R}}$ + "Θ is regular", or even $\mathsf{AD}_{\mathbb{R}} + \mathrm{cf}(\Theta) > \omega_1$, vis-a-vis the long game hierarchy? Does $\underset{\sim}{\mathbf{\Delta}}^1_2$-$\mathsf{AD}^{\Sigma_2}$ give us a model of $\mathsf{AD}_{\mathbb{R}} + \mathrm{cf}(\Theta) > \omega_1$?

Finally, §4 suggests the obvious question: does $\mathsf{ZFC}+\exists\kappa(\kappa$ is supercompact) prove $\underset{\sim}{\mathbf{\Pi}}^1_1$-$\mathsf{AD}^{\Sigma_n}$ for all $n < \omega$? Woodin has shown in $\mathsf{ZFC}+\exists\kappa(\kappa$ is supercompact) that all $\Game^{\omega_1}$ (closed-$\underset{\sim}{\mathbf{\Pi}}^1_1$) sets are Lebesgue measurable. This suggests that in fact $(\text{open-}\underset{\sim}{\mathbf{\Pi}}^1_1)$-$\mathsf{AD}^{\omega_1}$ might follow from $\mathsf{ZFC}+\exists\kappa(\kappa$ is supercompact).

We conclude by mentioning the strongest form of determinacy we know (other than the inconsistent forms). There are three parameters entering into the description of a class of games: complexity of payoff, complexity of individual moves, and length. The known limitations in each direction are:

(a) *payoff*: there is a non-determined game on $\{0, 1\}$ of length ω. (Gale-Stewart)

The natural response is to limit the payoffs considered to $\mathbf{OD}(\mathbb{R})$.

(b) *moves*: there is a non-determined game on $\wp(\mathbb{R})$ of length ω whose payoff is $\mathbf{OD}$. [player I plays a non-determined $A \subseteq {}^{\omega}\omega$, then player I and player II play out G_A.] (folklore)

The natural response is to limit the moves to be $\mathbf{OD}(\mathbb{R})$.

(c) *length*: there is a non-determined game on $\{0, 1\}$ with **OD** payoff of length $\omega_1 + \omega$. [In first ω_1 moves player I must describe an uncountable $A \subseteq {}^{\omega}\omega$ with no perfect subset. In the next ω moves, player I and player II play the perfect set game for A.] (Galvin, Laver)

The natural (most generous) response is to limit the intermediate positions to be $\mathbf{OD}(\mathbb{R})$. (This subsumes our response to (b).) This leads to a "maximum determinacy" principle, or MD.

MD. Let G be a game on $\mathbf{OD}(\mathbb{R})$ which ends as soon as the players reach a position which is not $\mathbf{OD}(\mathbb{R})$, with the winner declared according to some $\mathbf{OD}(\mathbb{R})$ payoff condition. Then G is determined.

Superficially, anyway, MD allows games of any ordinal length. Is MD consistent? Is it good for anything?

REFERENCES

HOWARD S. BECKER

[Bec85] *A property equivalent to the existence of scales*, ***Transactions of the AMS***, vol. 287 (1985), pp. 591–612.

ANDREAS BLASS

[Bla75] *Equivalence of two strong forms of determinacy*, ***Proceedings of the American Mathematical Society***, vol. 52 (1975), pp. 373–376.

ALEXANDER S. KECHRIS, DONALD A. MARTIN, AND YIANNIS N. MOSCHOVAKIS

[CABAL iii] *Cabal seminar* 79–81, Lecture Notes in Mathematics, no. 1019, Berlin, Springer, 1983.

ALEXANDER S. KECHRIS AND YIANNIS N. MOSCHOVAKIS

[CABAL i] *Cabal seminar* 76–77, Lecture Notes in Mathematics, no. 689, Berlin, Springer, 1978.

DONALD A. MARTIN

[MarB] *Games of countable length*, to appear.

[Mar75] *Borel determinacy*, ***Annals of Mathematics***, vol. 102 (1975), no. 2, pp. 363–371.

[Mar83B] *The real game quantifier propagates scales*, this volume, originally published in Kechris et al. [CABAL iii], pp. 157–171.

YIANNIS N. MOSCHOVAKIS

[Mos80] ***Descriptive set theory***, Studies in Logic and the Foundations of Mathematics, no. 100, North-Holland, Amsterdam, 1980.

ROBERT M. SOLOVAY

[Sol78B] *The independence of* DC *from* AD, In Kechris and Moschovakis [CABAL i], pp. 171–184.

DEPARTMENT OF MATHEMATICS
UNIVERSITY OF CALIFORNIA
BERKELEY, CA 94720, USA
E-mail: steel@math.berkeley.edu

THE LENGTH-ω_1 OPEN GAME QUANTIFIER PROPAGATES SCALES

JOHN R. STEEL

§1. Introduction. We shall call a set T an ω_1**-tree** if an only if

$$T \subseteq \bigcup_{\alpha<\omega_1} {}^{\alpha}\omega,$$

and whenever $p \in T$, then $p\restriction\beta \in T$ for all $\beta < \mathrm{dom}(p)$. (This diverges a bit from standard usage.) The set of ω_1-branches of T is

$$[T] = \{f \in {}^{\omega_1}\omega : \forall\alpha < \omega_1(f\restriction\alpha \in T)\}.$$

$[T]$ is just a typical closed set in the topology on ${}^{\omega_1}\omega$ whose basic neighborhoods are the sets of the form $N_p = \{f \in {}^{\omega_1}\omega : p \subseteq f\}$, where $p\colon \alpha \to \omega$ for some countable α. Associated to T is a closed game $\mathcal{G}(T)$ on ω of length ω_1: at round α in $\mathcal{G}(T)$, player I plays m_α, then player II plays n_α. Letting $f(\alpha) = \langle m_\alpha, n_\alpha\rangle$ for all $\alpha < \omega_1$, we say that player II wins the run of $\mathcal{G}(T)$ determined by f iff $f \in [T]$. (We code sequences from ω by prime powers, so that $\langle n_0, \dots, n_k\rangle = \prod_{i\leq k} p_i^{n_i+1}$, and the decoding is given by setting $(n)_i = k$ if the exponent of p_i in n is $k+1$ where p_i is the ith prime.)

Of course, not all such closed games are determined; indeed, every game of length ω can be regarded as a clopen game of length ω_1. However, the determinacy of $\mathcal{G}(T)$ for definable T does follow from a large cardinal hypothesis, in virtue of the following beautiful theorem of Itay Neeman:

THEOREM 1.1 (Neeman, [Nee, Nee04]). *Suppose that for any real x, there is a countable, ω_1+1-iterable mouse M such that $x \in M$ and $M \models \mathsf{ZFC}^- +$ "there is a measurable Woodin cardinal"; then $\mathcal{G}(T)$ is determined whenever T is a ω_1-tree such that T is definable over $\langle \mathbf{H}_{\omega_1}, \in\rangle$ from parameters.*

The definability restriction on T is equivalent to requiring that T be coded by a projective set of reals. For that reason, we shall call the conclusion of Neeman's theorem ω_1**-open-projective determinacy**. Neeman's proof works, under a natural strengthening of its large cardinal hypothesis, whenever T is

The Cabal Seminar. Volume I: Games, Scales and Suslin Cardinals
Edited by A. S. Kechris, B. Löwe, J. R. Steel
Lecture Notes in Logic, 31

coded by a Hom_∞ set of reals. In this form, Neeman's result obtains as much determinacy from large-cardinal-like hypotheses as has been proved to date.[1]

Associated to any game we have a game quantifier:

DEFINITION 1.2. Let T be an ω_1-tree; then

$$\Game^{\mathrm{I}}(T) = \{p \in {}^\omega\omega \,:\, p \text{ is a winning position for player I in } \mathcal{G}(T)\},$$

and

$$\Game^{\mathrm{II}}(T) = \{p \in {}^\omega\omega \,:\, p \text{ is a winning position for player II in } \mathcal{G}(T)\}.$$

DEFINITION 1.3. We define

$$\Game^{\omega_1}(\text{open-analytical})$$

to be the class $\{\Game^{\mathrm{I}}(T) \,:\, T$ is an ω_1-tree which is definable over $\langle \mathbf{H}_{\omega_1}, \in\rangle\}$ and

$$\Game^{\omega_1}(\text{closed-analytical})$$

to be the class $\{\Game^{\mathrm{II}}(T) \,:\, T$ is an ω_1-tree which is definable over $\langle \mathbf{H}_{\omega_1}, \in\rangle\}$.

Clearly, the determinacy of the games in question implies that the pointclasses $\Game^{\omega_1}$(open-analytical) and $\Game^{\omega_1}$(closed-analytical) are duals. The following observations relate these pointclasses to more familiar ones.

PROPOSITION 1.4. For any $A \subseteq \omega^\omega$, the following are equivalent:

(1) A is $\Game^{\omega_1}$(closed-analytical),
(2) there is a Σ^2_1 formula $\vartheta(v)$ such that for all $x \in \omega^\omega$,

$$x \in A \text{ iff } \mathrm{Col}(\omega_1, \mathbb{R}) \Vdash \vartheta(\check{x}),$$

where $\mathrm{Col}(\omega_1, \mathbb{R})$ is the partial order for adding a map from ω_1 onto $\mathbb{R}$ with countable conditions.

PROPOSITION 1.5. If CH holds, then $\Game^{\omega_1}(\text{closed-analytical}) = \Sigma^2_1$.

COROLLARY 1.6. If CH and ω_1-open-analytical determinacy hold, then $\Game^{\omega_1}(\text{open-analytical}) = \Pi^2_1$.

We shall omit the easy proofs of these results.

It is natural to ask whether ω_1-open-projective determinacy implies the pointclasses $\Game^{\omega_1}$(open-analytical) and $\Game^{\omega_1}$(closed-analytical) are well-behaved from the point of view of descriptive set theory. In this note we shall extend Moschovakis' periodicity theorems to this context, and show thereby that ω_1-open-projective determinacy implies that $\Game^{\omega_1}$(open-analytical) has the Scale

[1] Literally speaking, Neeman's theorem uses a mouse existence hypothesis, rather than a large cardinal hypothesis. It is not known how to derive this mouse existence hypothesis from any true large cardinal hypothesis, because it is not known how to prove the required iterability. Presumably, if there is a measurable Woodin cardinal, then there are mice of the sort needed in Neeman's theorem.

Property, and that there are canonical winning strategies for the ω_1-open-projective games won by player player I. By Corollary 1.6, adding CH to our hypotheses gives the Scale Property for Π^2_1.

Our arguments are quite close to those of [Ste88, §§ 1& 2], which prove the same results for the quantifiers associated to certain *clopen* games of length ω_1. Therefore, in this paper we shall only describe the simple changes to [Ste88] which are needed for the full theorems on open games, and refer the reader to [Ste88] for the long stretches of the proofs which agree in every detail.

§2. The Prewellordering Property. Moschovakis' argument easily yields the prewellordering property for $\Game^{\omega_1}$(open-analytical).

THEOREM 2.1. *If ω_1-open-projective determinacy holds, then the pointclass $\Game^{\omega_1}$(open-analytical) has the prewellordering property.*

PROOF. Let T be an ω_1-tree which is definable over $\langle \mathbf{H}_{\omega_1}, \in \rangle$, and let

$$B = \Game^{\mathrm{I}}(T).$$

We define a prewellordering on B by means of a game $\mathrm{G}(p,q)$ which compares the values for player I of positions $p, q \in B$. In $\mathrm{G}(p,q)$ there are two players, F and S, and two boards, the p-board and the q-board. The play takes place in rounds, the first being round ω. In round $\alpha \geq \omega$ of $\mathrm{G}(p,q)$:

(a) F plays as player I in round α of $\mathcal{G}(T)$ on the q-board, then
(b) S plays as player I in round α of $\mathcal{G}(T)$ on the p-board, then
(c) F plays as player II in round α of $\mathcal{G}(T)$ on the p-board, then
(d) S plays as player II in round α of $\mathcal{G}(T)$ on the q-board.

This play produces $f \supseteq p$ on the p-board and $g \supseteq q$ on the q-board, with $f, g \in {}^{\omega_1}\omega$. Player S wins this run of $\mathrm{G}(p,q)$ iff for some $\alpha < \omega_1$, $f \restriction \alpha \notin T$ but for all $\beta < \alpha$, $g \restriction \beta \in T$. In other words, S must win $\mathcal{G}(T)$ as player I on the p-board, and not strictly after F wins as player I (if he does) on the q-board. Put

$$p \leq^* q \Leftrightarrow p, q \in A \text{ and } S \text{ has a winning strategy in } \mathrm{G}(p,q).$$

Let $\mathrm{G}^0(p,q)$ be the same as $\mathrm{G}(p,q)$, except that a run $\langle f, g \rangle$ such that neither player has won as player I (i.e., such that $f \in [T]$ and $g \in [T]$) is a win for S, rather than F.

LEMMA 2.2. *Let $p_0 \in \Game^{\mathrm{I}}(T)$, and suppose that for all $n \geq 0$, Σ_n is either a winning strategy for F in $\mathrm{G}(p_n, p_{n+1})$, or a winning strategy for S in $\mathrm{G}^0(p_{n+1}, p_n)$; then only finitely many Σ_n are for F.*

PROOF. Let $\langle p_n : n < \omega \rangle$ and $\langle \Sigma_n : n < \omega \rangle$ be as in the hypothesis. Let τ be winning for player I in $\mathcal{G}(T)$ from p_0.

Note that in any case, Σ_n plays for player II on the p_n board, and for player I on the p_{n+1} board. Playing τ and the Σ_n's together in the standard game diagram, we get $\langle u_n : n < \omega \rangle$ such that for all n,

(1) u_n is a run according to Σ_n of the appropriate game ($G(p_n, p_{n+1})$ if Σ_n is for F, and $G^0(p_{n+1}, p_n)$ if Σ_n is for S), and
(2) u_n and u_{n+1} have a common play r_{n+1} on their p_{n+1}-boards,
(3) the play r_0 on the p_0-board of u_0 is according to τ.

As usual, $u_n(\alpha)$ is determined by induction on α, simultaneously for all n. Because τ was winning for player I, we have $r_0 \notin [T]$. It follows by induction that $r_n \notin [T]$ for all n. Let α_n be least such that $r_n \restriction \alpha_n \notin [T]$. Since Σ_n won its game, we have that $\alpha_{n+1} \leq \alpha_n$ if Σ_n is for S, and $\alpha_{n+1} < \alpha_n$ if Σ_n is for F. Thus only finitely many Σ_n are for F. ⊣ (Lemma 2.2)

COROLLARY 2.3. $\leq^*$ is a prewellorder of B.

PROOF. (1) *Reflexive*: if $\neg p \leq^* p$, then $\langle p, p, p, \ldots \rangle$ violates Lemma 2.2.
(2) *Transitive*: if $p \leq^* q \leq^* r$, but $\neg p \leq^* r$, then $\langle r, q, p, r, q, p, r, \ldots \rangle$ violates Lemma 2.2.
(3) *Connected*: otherwise $\langle p, q, p, q, \ldots \rangle$ violates Lemma 2.2.
(4) *Wellfounded*: clear from Lemma 2.2. ⊣

COROLLARY 2.4. If $q \in B$, then for any p, $p \leq^* q \Leftrightarrow$ player II has a winning strategy in $G^0(p, q)$.

PROOF. Otherwise, let $p_n = q$ if n is even, and $p_n = p$ if n is odd. Let Σ_n be winning for S in $G^0(p, q)$ if n is even, and let Σ_n be winning for F in $G(p, q)$ if n is odd. We obtain a contradiction to Lemma 2.2. ⊣

So if $q \in B$, then we have

$$p \in B \wedge p \leq^* q \Leftrightarrow S \text{ has a winning strategy in } G(p, q)$$
$$\Leftrightarrow S \text{ has a winning strategy in } G^0(p, q).$$

The first equivalence shows $\{p : p \leq^* q\}$ is uniformly $\Game^{\omega_1}$(open-analytical) in q, the second shows that it is uniformly $\Game^{\omega_1}$(closed-analytical) in q. Thus $\leq^*$ determines a $\Game^{\omega_1}$(open-analytical) norm on B. ⊣ (Theorem 2.1)

§3. The Scale Property. In order to prove the scale property for $\Game^{\omega_1}$(open-analytical), we need to add to the comparison games of Theorem 2.1 additional moves. As in [Ste88], these additional moves reflect the value which the players assign to one-move variants of positions they reach during a run of the game. One needs that such positions are canonically coded by reals in order to make sense of this. For this reason, in [Ste88] we demanded that there be a function F assigning to possible positions $p \in T$ a wellorder $F(p) \in \mathrm{WO}$ such that $F(p)$ has order type $\mathrm{dom}(p)$. What is new here is just the realization that one can relax this a bit, by demanding only that our game tree T incorporates the rule that player player I must provide a code of dom(p) as soon as p has been reached, before player II is required to make any further moves.

THEOREM 3.1. If ω_1-open-projective determinacy holds, then the pointclass $\Game^{\omega_1}$ (open-analytical) has the scale property.

PROOF. Let $\hat{T}$ be an ω_1-tree which is definable over $\langle \mathbf{H}_{\omega_1}, \in \rangle$, and let $B = \Game^{\mathrm{I}}(\hat{T})$. We assume without loss of generality that in $\mathcal{G}(\hat{T})$, player II plays only from $\{0, 1\}$. We may as well also assume that player I and player II only move in $\mathcal{G}(\hat{T})$ at finite rounds, or at rounds of the form $\omega\eta + \omega$ where $\eta \geq 1$. Finally, we assume that for $\eta \geq 1$, player player I must produce a code of $\omega\eta + \omega$ in rounds $\omega\eta + 1$ through $\omega\eta + \omega$. Formally, for any $r \in {}^{<\omega_1}\omega$, let

$$q_r(n) = r(n) \text{ if } \eta < \omega$$

and

$$q_r(\omega + \eta) = r(\omega + \omega\eta + \omega)$$

for all $\eta \geq 0$ such that $\omega + \omega\eta + \omega \in \mathrm{dom}(r)$. Set

$$w_r^\eta(n) = r(\omega\eta + n + 1)_0,$$

for η such that $\omega + \omega\eta + \omega \leq \mathrm{dom}(r)$. For $r \in {}^{<\omega_1}\omega$, we put $r \in A$ iff

(a) $\mathrm{dom}(r) = \omega + \omega\xi + \omega$, for some ξ,
(b) $\forall \eta \in \mathrm{dom}(q_r)(q_r{\restriction}\eta \in \hat{T}$, but $q_r \notin \hat{T})$, and
(c) if $\omega\eta + \omega \leq \mathrm{dom}(r)$, then $w_r^\eta \in \mathrm{WO}^*$ and $|w_r^\eta| = \omega\eta + \omega$.

Here WO* is the set of reals x coding wellorders of ω of length $|x| > \omega$, and such that $|0|_x = \omega$. (This last very technical stipulation simplifies something later.) For $r \in {}^{<\omega_1}\omega$, put

$$r \in T \Leftrightarrow \forall \xi \leq \mathrm{dom}(r)(r{\restriction}\xi \notin A).$$

Thus in $\mathcal{G}(T)$, player I is just trying to reach a position in A.

CLAIM 3.2. $B = \Game^{\mathrm{I}}(T)$.

PROOF. This is clear. It is worth noting that we use the Axiom of Choice at this point, however. ⊣ (Claim 3.2)

As we said, our only new idea here is to work with T, rather than $\hat{T}$, in defining the comparison games producing a scale on B.

We shall use the notation of the proof of [Ste88, Theorem 1.2] as much as possible. The counterpart of F there is given by: dom(F) consists of those $r \in {}^{<\omega_1}\omega$ which satisfy (a) and (c) above, and for such r

$$F(r) = w_r^\eta, \text{ where } \mathrm{dom}(r) = \omega\eta + \omega.$$
$$r^* = \langle F(r), x \rangle,$$

where $x(n) = r(|n|_{F(r)})$. For $D \subseteq \mathrm{dom}(F)$, we let $D^* = \{r^* : r \in D\}$. Let

$$C(i, j, k, y) \Leftrightarrow \exists r \in \mathrm{dom}(F)(y = r^* \wedge |i|_{F(r)} = |j|_{F(r \restriction \alpha)})$$
$$\text{where } |k|_{F(r)} = \alpha = \omega\xi + \omega, \text{ for some } \xi.$$

Note that C and F are Δ^1_2. Let $\vec{\rho}$ be a Δ^1_2 scale on C, and let $\vec{\sigma}$ be an analytical scale on T^*. Let ϑ^k_i, for $0 \leq k \leq 4$ and $i < \omega$, be the norms on A^* defined from $\vec{\rho}$ and $\vec{\sigma}$ exactly as in the proof of [Ste88, Theorem 1.2]. Let

$$D(k, y) \Leftrightarrow \exists r \in \mathrm{dom}(F)(y = r^* \wedge \exists \xi \geq 1(|k|_{F(r)} = \omega\xi + \omega)),$$

let $\vec{\nu}^0, \vec{\nu}^1$ be an analytical scales on D and $\neg D$, and set

$$\vartheta^5_{\langle k,i \rangle}(y) = \begin{cases} \nu^0_i(k, y), & \text{when } D(k, y), \\ \nu^1_i(k, y), & \text{when } \neg D(k, y). \end{cases}$$

Let $\vec{\vartheta}^6$ be a very good scale on A^*, and set, for $p^* \in A^*$,

$$\psi_i(p^*) = \langle \mathrm{dom}(p), \vartheta^0_0(p^*), \dots, \vartheta^6_0(p^*), \dots, \vartheta^0_i(p^*), \dots, \vartheta^6_i(p^*) \rangle,$$

or rather, let $\psi_i(p^*)$ be the ordinal of this tuple in the lexicographic order. Just as in [Ste88], the norms ψ_i determine the values we assign to runs of the comparison games which yield a scale on B.

Let $p, q \in \Game^{\mathrm{I}}(T)$, and let $k \in \omega$. We shall define a game $\mathrm{G}_k(p, q)$. The players in $\mathrm{G}_k(p, q)$ are F and S. They play on two boards, the p board and the q board, and make additional moves lying on neither board. On the p board, S plays $\mathcal{G}(T)$ from p as I, while F plays as player II. On the q board, F plays $\mathcal{G}(T)$ from q as player I, while S plays as player II. Play is divided into rounds, the first being round ω. We now describe the typical round.

We shall require that the moves in $\mathrm{G}_k(p, q)$ on the p and q boards which are meaningless for $\mathcal{G}(T)$ be 0. So in round α of $\mathrm{G}_k(p, q)$, for $\alpha = \omega$ or α a limit of limit ordinals, F and S must each play 0's on both boards. We make the technical stipulation that F always proposes k at round ω. (This will make it more convenient to describe the payoff condition of $\mathrm{G}_k(p, q)$.) No further moves are made in round α, for $\alpha = \omega$ or α a limit of limits.

Round α, for α a successor ordinal:

(a) F plays as player I in round α of $\mathcal{G}(T)$ on the q-board, then
(b) S plays as player I in round α of $\mathcal{G}(T)$ on the p-board, then
(c) F, S play 0 as player II in round α of $\mathcal{G}(T)$ on the p, q-boards, respectively.

Round α, when $\exists \xi \geq 1(\alpha = \omega\xi + \omega)$:

(a) F makes player I's α-th move on the q board, then S makes player I's α-th move on the p-board.
(b) F now proposes some i such that $0 \leq i \leq k$ and $(i)_0 \in \{0, 1\}$.
(c) S either accepts i, or proposes some j such that $0 \leq j < i$ and $(j)_0 \in \{0, 1\}$.

(d) Let $t \leq k$ be the least proposal made during (b) and (c):
Case 1: $t \neq 0$. Then F and S must play $(t)_0$ as player II's $\omega + \alpha$-th move on the p and q boards respectively.
Case 2: $t = 0$. Then F plays any $m \in \{0, 1\}$ as player II's $\omega + \alpha$-th move on the p board, after which S plays any $n \in \{0, 1\}$ as player II's $\omega + \alpha$-th move on the q board.

This completes our description of round α.

Play in $\mathrm{G}_k(p, q)$ continues until one of the two boards reaches a position in A. If this never happens, then F wins $\mathrm{G}_k(p, q)$. If one of the two boards reaches a position in A strictly before the other board does, then the player playing as player I on that board wins $\mathrm{G}_k(p, q)$. We are left with the case that we have a run u of $\mathrm{G}_k(p, q)$ and r, s with $p \subseteq r$ and $q \subseteq s$ the runs of $\mathcal{G}(T)$ on the two boards, and $r, s \in A$. Letting $\beta = \mathrm{dom}(r) = \mathrm{dom}(s)$, we must have $\beta = \omega\xi + \omega$ for some $\xi \geq 1$. Let $e \in \omega$ be the least n such that for some $\alpha \geq \omega$, $|(n)_0|_{F(r)}| = \alpha$, and $(n)_1$ was the least proposal made during round α. (Our technical stipulations guarantee that $\langle 0, k \rangle$ is such an n.) We call e the **critical number** of u, and write $e = \mathrm{crit}(u)$; thus $\mathrm{crit}(u) \leq \langle 0, k \rangle$. Let $\alpha = |(e)_0|_{F(r)}$. If F proposed $(e)_1$ during round α, then

$$S \text{ wins } u \text{ iff } \psi_e(r^*) \leq \psi_e(s^*).$$

If S proposed $(e)_1$ during round α, then

$$S \text{ wins } u \text{ iff } \psi_e(r^*) < \psi_e(s^*).$$

This completes our description of $\mathrm{G}_k(p, q)$.

Let $\mathrm{G}^0_k(p, q)$ be just like $\mathrm{G}_k(p, q)$, except that if neither board reaches a position in A after ω_1 moves, then it is S who wins, rather than F.

LEMMA 3.3. Let $p_0 \in \Game^{\mathrm{I}}(T)$, and suppose that for all $n \geq 0$, Σ_n is either a winning strategy for F in $\mathrm{G}_k(p_n, p_{n+1})$, or a winning strategy for S in $\mathrm{G}^0(p_{n+1}, p_n)$; then only finitely many Σ_n are for F.

We omit the proof of Lemma 3.3, as it is a direct transcription of the corresponding lemma in [Ste88].

For $p, q \in \Game^{\mathrm{I}}(T)$, we put

$$p \leq_k q \text{ iff player II has a winning strategy in } \mathrm{G}_k(p, q).$$

COROLLARY 3.4.

(a) $\leq_k$ is a prewellorder of $\Game^{\mathrm{I}}(T)$.
(b) For $q \in \Game^{\mathrm{I}}(T)$, S has a winning strategy in $\mathrm{G}_k(p, q)$ iff S has a winning strategy in $\mathrm{G}^0_k(p, q)$.
(c) $\leq_k$ determines a $\Game^{\omega_1}$(open-analytical) norm on $\Game^{\mathrm{I}}(T)$.

PROOF. Just as in the proofs of Corollaries 2.3 and 2.4. ⊣

Now for $p \in \Game^{\mathrm{I}}(T)$, let

$$\varphi_k(p) = \text{ ordinal of } p \text{ in } \leq_k.$$

LEMMA 3.5. $\vec{\varphi}$ is a semiscale on $\Game^{\mathrm{I}}(T)$.

PROOF. The proof is very close to that of [Ste88, Lemma 1.6], so we just indicate the small changes needed.

Let $p_n \to p$ as $n \to \infty$, with $p_n \in \Game^{\mathrm{I}}(T)$ for all n. Let τ be a winning strategy for player I in $\mathcal{G}(T)$. Suppose $p_{n+1} \leq_n p_n$, as witnessed by the winning strategy Σ_n for S in $\mathrm{G}_n(p_{n+1}, p_n)$, for all n. We must show that $p \in \Game^{\mathrm{I}}(T)$. Suppose toward contradiction that σ is a winning strategy for player II in $\mathcal{G}(T)$.[2]

If $r_n \in \mathrm{dom}(F)$ for all n, then we write

$$r = \lim^*_{n\to\infty} r_n$$

for the same notion of "convergence in the codes" as defined in [Ste88].

We define by induction on rounds runs u_n of $\mathrm{G}_n(p_{n+1}, p_n)$ according to Σ_n. We arrange that u_n and u_{n+1} agree on a common play r_{n+1} extending p_{n+1} on the p_{n+1} board, and that the play r_0 extending p_0 on the p_0 board is by τ. So assume that $u_n{\restriction}\alpha$ and $r_n{\restriction}\alpha$ are given for all n.

If $\alpha = \omega$ or α is a limit of limit ordinals, we set $r_n(\alpha) = \langle 0, 0\rangle$, and thus $u_n(\alpha) = \langle 0, 0, n, 0, 0\rangle$, for all n. None of these moves count for anything, of course.

If α is a successor ordinal, then only player I's move in $\mathcal{G}(T)$ counts. Let

$$a_0 = \tau(r_0{\restriction}(\alpha)),$$

and

$$a_{n+1} = \Sigma_n((u_n{\restriction}\alpha)^\frown\langle a_n\rangle)$$

fill in the α-th column of player I's plays in our diagram, and set $r_n(\alpha) = \langle a_n, 0\rangle$, and thus $u_n(\alpha) = \langle a_n, a_{n+1}, 0, 0\rangle$, for all n.

Finally, suppose $\alpha = \omega\xi + \omega$ for some $\xi \geq 1$. Again, let $a_0 = \tau(r_0{\restriction}(\alpha))$, and $a_{n+1} = \Sigma_n((u_n{\restriction}\alpha)^\frown\langle a_n\rangle)$ for all n. If a_n is eventually equal to some fixed a, and if $\lim^*_{n\to\infty} r_n{\restriction}(\alpha) = r$, where r is a play by σ of length $\omega\xi + \omega$ for some $\xi \geq 1$, then we set

$$d = \sigma(r^\frown\langle a\rangle).$$

If not, then we set $d = 0$.

We proceed now exactly as in [Ste88]. For any n, let i_n be the largest i such that $\langle d, i\rangle \leq n$ and $\Sigma_n((u_n{\restriction}\alpha)^\frown\langle a_n, \langle d, i\rangle\rangle) =$ "accept", if such an i exists. Let $i_n = 0$ otherwise.

[2]Unfortunately, there are some typos in the proof of [Ste88, Lemma 1.6] which confuse τ with σ at various points.

Case 1: $i_n \to \infty$ as $n \to \infty$.

Pick n_0 such that $i_n > 0$ for $n \geq n_0$. The proposal pair in $u_n(\alpha)$ is $\langle b_n, c_n \rangle$, where $b_n = \langle d, i_n \rangle$ for $n \geq n_0$ and $b_n = 0 =$ freedom for $n < n_0$, and $c_n =$ accept for all n. Let $d_n = d$ if $n \geq n_0$, and

$$d_n = \Sigma_n((u_n \restriction \alpha)^\frown \langle a_n, a_{n+1}, b_n, c_n, d_{n+1} \rangle)$$

if $n < n_0$.

Case 2: Otherwise.

Again, we define the $u_n(\alpha)$ exactly as in [Ste88]. We shall not repeat the definition here in this case.

This completes the definition of the u_n and r_n. Since r_0 is a play by τ, $r_0 \restriction \alpha \in A$ for some α. Letting α_0 be the least α such that $r_n \restriction \alpha \in A$ for some n, we have $r_n \restriction \alpha_0 \in A$ for all sufficiently large n because the Σ_n's won for S. To save notation, let us assume $r_n \restriction \alpha_0 \in A$ for all n. Note that $\alpha_0 = \omega\xi + \omega$ for some $\xi \geq 1$. Let us write $u_n = u_n \restriction \alpha_0$ and $r_n = r_n \restriction \alpha_0$ for all n.

Claim 3.6. $\operatorname{crit}(u_n) \to \infty$ as $n \to \infty$.

Proof. See [Ste88]. The only additional point here is that if $e = \operatorname{crit}(u_n)$ for infinitely many n, then $(e)_0 \neq 0$. ⊣ (Claim 3.6)

It follows from the claim that for all e, $\psi_e(r_n^*)$ is eventually constant as $n \to \infty$. From this we get that $\lim^*_{n\to\infty} r_n = r$, for some $r \in A$ such that $p \subseteq r$. We shall show that r is a play by σ, so that σ was not winning for player II in $\mathcal{G}(T)$, a contradiction.

Let β be least such that $r \restriction (\beta + 1)$ is not by σ. Fix η a limit ordinal such that $\beta = \eta + \omega$; such an η must exist because σ is for player II, who only moves at stages of the form $\eta + \omega$. Let

$$\beta = |k|_{F(r)}$$

and

$$\alpha = \text{ eventual value of } |k|_{F(r_n)} \text{ as } n \to \infty.$$

Since the r_n^* converge in the scales $\vec{\vartheta}^5$, we have that $\alpha = \mu + \omega$ for some limit ordinal μ. We now look at how column α of our diagram was constructed. Since r_n^* converged in $\vec{\vartheta}^0$ and $\vec{\vartheta}^1$, we have

$$\lim^*_{n\to\infty} r_n \restriction \alpha = r \restriction \beta.$$

Since the r_n^* converge in $\vec{\vartheta}^6$, which is very good, the a_n's defined at round α are eventually constant, with value $a = r(\beta)_0$. But then at round α in the construction we set $d = \sigma((r \restriction \beta)^\frown \langle a \rangle)$. Moreover, Case 1 must have applied at α, as otherwise $\operatorname{crit}(u_n)$ would have a finite lim inf. Thus $r_n(\gamma) = \langle a, d \rangle$ for all sufficiently large n. Since $\vec{\vartheta}^6$ is very good, $r(\beta) = \langle a, d \rangle$, so that $r \restriction (\beta + 1)$ is by σ, a contradiction. ⊣ (Lemma 3.5)

Let U be the tree of the semiscale $\vec{\varphi}$ given by Lemma 3.5, and let $\vec{\vartheta}$ be the scale of U. One can easily check that, since $\Game^{\omega_1}$(open-analytical) is closed under real quantification, the ϑ_i are $\Game^{\omega_1}$(open-analytical) norms, uniformly in i. ⊣ (Theorem 3.1)

There are some awkward features of our proof of Theorem 3.1. First, it only applies directly to games in which player II is restricted to playing from $\{0, 1\}$. Second, our comparison games seem to only yield a semiscale directly, and not a scale. This is connected to the fact that we only show that if $p_n \to p$ modulo our semiscale, then player II has no winning strategy in $\mathcal{G}(T)$ from p; we do not construct a winning strategy for player I from p. One could probably obtain a more direct proof by bringing in the construction of definable winning strategies for player I. In this connection, one has

THEOREM 3.7. Assume that ω_1-open-projective determinacy holds, and that player I has a winning strategy in $\mathcal{G}(T)$, where T is an ω_1-tree which is definable over $\langle \mathbf{H}_{\omega_1}, \in \rangle$; then player I has a $\Game^{\omega_1}$(open-analytical) winning strategy in $\mathcal{G}(T)$.

One can prove Theorem 3.7 by modifying the proof of [Ste88, Theorem 2.1], in the same way that we modified the proof of [Ste88, Theorem 1.2] in order to prove Theorem 3.1.

REFERENCES

ALEXANDER S. KECHRIS, DONALD A. MARTIN, AND JOHN R. STEEL

[CABAL iv] *Cabal seminar* 81–85, Lecture Notes in Mathematics, no. 1333, Berlin, Springer, 1988.

ITAY NEEMAN

[Nee] *Games of length* ω_1, ***Journal of Mathematical Logic***, to appear.

[Nee04] ***The determinacy of long games***, de Gruyter Series in Logic and its Applications, vol. 7, Walter de Gruyter, Berlin, 2004.

JOHN R. STEEL

[Ste88] *Long games*, this volume, originally published in Kechris et al. [CABAL iv], pp. 56–97.

DEPARTMENT OF MATHEMATICS
UNIVERSITY OF CALIFORNIA
BERKELEY, CA 94720, USA
E-mail: steel@math.berkeley.edu

PART II: SUSLIN CARDINALS, PARTITION PROPERTIES, HOMOGENEITY

SUSLIN CARDINALS, PARTITION PROPERTIES, HOMOGENEITY
INTRODUCTION TO PART II

STEVE JACKSON

§1. Introduction. In this paper we survey the basic concepts and results pertaining to Suslin cardinals, partition properties, and homogeneous trees. This paper in part serves as an introduction to the papers in this section. We explain the basic material, and also survey some of the more recent developments as they pertain to these concepts.

Throughout, by "real" we mean an element of the Baire space ${}^\omega\omega$. We fix recursive bijections $(a_0, \dots, a_{n-1}) \mapsto \langle a_0, \dots, a_{n-1} \rangle$ from ${}^n\omega$ to ω which are reasonable, say increasing in each argument. We let $t \mapsto ((t)_0, \dots, (t)_n)$ denote the recursive inverse map. These maps induce recursive bijections between ${}^n({}^\omega\omega)$ and ${}^\omega\omega$, and we use the same notation $\langle \alpha_0, \dots, \alpha_{n-1} \rangle$ and $(\alpha)_i$ to denote these bijections. There is also a recursive bijection between ${}^\omega({}^\omega\omega)$ and ${}^\omega\omega$, and we again use the same notation $\langle \alpha_0, \alpha_1, \dots \rangle$, $(\alpha)_i$ to denote it; which bijection we are referring to will be clear from the context.

Our base theory throughout is $\mathsf{ZF} + \mathsf{DC}$, although we will frequently be assuming some form of determinacy as well. We let AD abbreviate the axiom of determinacy—the statement that every two player integer game is determined. We remind the reader that the basic properties of ordinals and cardinals remain valid assuming just ZF with the notable exception that successor cardinals no longer need be regular.

By a **tree** on set X we always mean a tree in the descriptive set theoretic sense, that is, $T \subseteq {}^{<\omega}X$ and if $\vec{x} = (x_0, \dots, x_{n-1}) \in T$ then $\vec{x} \restriction m = (x_0, \dots, x_{m-1}) \in T$ for all $m \leq n$. We write $\mathrm{lh}(\vec{x})$ to denote the length of the sequence $\vec{x}$. We say $\vec{y}$ extends $\vec{x}$ if $\mathrm{lh}(\vec{y}) \geq \mathrm{lh}(\vec{x})$ and $y \restriction (\mathrm{lh}(x)) = \vec{x}$. As is customary, we abuse notation slightly and consider elements of a tree on $X \times Y$ as pairs $(\vec{x}, \vec{y})$ where $\mathrm{lh}(\vec{x}) = \mathrm{lh}(\vec{y})$. By an infinite **branch** through T we mean a $\vec{x} = (x_0, x_1, \dots) \in {}^\omega X$ such that $\vec{x} \restriction n \in T$ for all n. We say T is ill-founded if it has an infinite branch, and otherwise say T is wellfounded. We let $[T]$ denote the set of infinite branches through T. If T is a tree on $X_1 \times X_2 \times \cdots \times X_n$, we let $\mathrm{p}_i[T]$ denote the projection of $[T]$ onto the ith coordinate, that is, $\vec{x} \in \mathrm{p}_i[T]$ iff $\exists \vec{y}_1, \dots, \vec{y}_{n-1}\ (\vec{y}_1, \dots, \vec{y}_{i-1}, \vec{x}, \vec{y}_{i+1}, \dots, \vec{y}_{n-1}) \in [T]$. We write

The Cabal Seminar. Volume I: Games, Scales and Suslin Cardinals
Edited by A. S. Kechris, B. Löwe, J. R. Steel
Lecture Notes in Logic, 31

$p[T]$ for $p_1[T]$. If T is a tree on $X \times Y$ and $s = (x_0, \dots, x_{n-1}) \in {}^{<\omega}X$, we define $T_s = \{t \in {}^{\leq n}Y : (s \restriction \mathrm{lh}(t), t) \in T\}$. We also use this notation for trees on general products $X_1 \times \cdots \times X_n$ and allow $s \in {}^{<\omega}(X_i)$, with the obvious meaning.

If T is an illfounded tree on X, and $\prec$ is a wellordering of X, it makes sense to speak of the **left-most** branch ℓ of T. This is simply the infinite branch of T which is $\prec$-lexicographically less than any other infinite branch.

If $\prec$ is a wellordering of X and T is a tree on X, then $\prec$ induces a natural linear order on T called the **Kleene-Brouwer order**. It is defined by $\vec{x} <_{\mathrm{BK}} \vec{y}$ iff $\vec{y}$ extends $\vec{x}$ or there is an $n < \min\{\mathrm{lh}(\vec{x}), \mathrm{lh}(\vec{y})\}$ such that $x_n \neq y_n$ and for the least such n we have $x_n \prec y_n$. It is easy to check that T is wellfounded iff the relation $<_{\mathrm{BK}}$ on T is a wellordering.

By a **measure** on a set X we mean a countably additive ultrafilter on X. It is a theorem of AD that every ultrafilter on a set X is a measure.

If $A, B \subseteq {}^{\omega}\omega$, we write $A \leq_{\mathrm{W}} B$ (A is **Wadge reducible** to B) if there is a continuous function $f\colon {}^{\omega}\omega \to {}^{\omega}\omega$ with $A = f^{-1}[B]$. We say $A \equiv_{\mathrm{W}} B$ if $A \leq_{\mathrm{W}} B$ and $B \leq_{\mathrm{W}} A$. By a (boldface) pointclass $\boldsymbol{\Gamma}$ we mean a collection $\boldsymbol{\Gamma} \subseteq \wp({}^{\omega}\omega)$ closed under Wadge reduction, i.e., if $A \in \boldsymbol{\Gamma}$ and $B \leq_{\mathrm{W}} A$, then $B \in \boldsymbol{\Gamma}$. For $\boldsymbol{\Gamma}$ a pointclass we let $\check{\boldsymbol{\Gamma}}$ denote the **dual pointclass** $\check{\boldsymbol{\Gamma}} = \{A : {}^{\omega}\omega - A \in \boldsymbol{\Gamma}\}$. For $A \subseteq {}^{\omega}\omega$ we sometimes write $\check{A}$ for ${}^{\omega}\omega - A$. If $\boldsymbol{\Gamma}_1, \boldsymbol{\Gamma}_2$ are pointclasses, we let $\boldsymbol{\Gamma}_1 \wedge \boldsymbol{\Gamma}_2 = \{A \cap B : A \in \boldsymbol{\Gamma}_1 \wedge B \in \boldsymbol{\Gamma}_2\}$, and similarly define $\boldsymbol{\Gamma}_1 \vee \boldsymbol{\Gamma}_2$. We let $\bigcup_\alpha \boldsymbol{\Gamma}$ denote the sets which can be written as α-length unions of $\boldsymbol{\Gamma}$ sets.

We let $\Game\boldsymbol{\Gamma}$ denote the integer game quantifier applied to sets in $\boldsymbol{\Gamma}$, where for $A \subseteq {}^{\omega}\omega \times {}^{\omega}\omega$ in $\boldsymbol{\Gamma}$, $x \in \Game A$ iff player II has a winning strategy in the game where player I and player II alternate playing integers $y(i)$ to produce a real y, and player II wins the run iff $A(x, y)$. We likewise define the real game quantifier $\Game^{\mathbb{R}}$ (letting $y = \langle y_i \rangle$ where y_i are the reals they play).

We say a pointclass $\boldsymbol{\Gamma}$ has the **reduction** property, $\mathrm{Red}(\boldsymbol{\Gamma})$, if whenever A, $B \in \boldsymbol{\Gamma}$ then there is a pair $A', B' \in \boldsymbol{\Gamma}$ which reduces A, B, that is, $A' \cap B' = \varnothing$, $A' \subseteq A$, $B' \subseteq B$, and $A' \cup B' = A \cup B$. We say $\boldsymbol{\Gamma}$ has the **separation** property, $\mathrm{Sep}(\boldsymbol{\Gamma})$, if any disjoint $\boldsymbol{\Gamma}$ sets A and B can be separated by a $\boldsymbol{\Delta}$ set D (where $\boldsymbol{\Delta} = \boldsymbol{\Gamma} \cap \check{\boldsymbol{\Gamma}}$), that is, $A \subseteq D$ and $D \cap B = \varnothing$. We say $\boldsymbol{\Gamma}$ has the **prewellordering property**, $\mathrm{PWO}(\boldsymbol{\Gamma})$, if every $\boldsymbol{\Gamma}$ set admits a $\boldsymbol{\Gamma}$-norm where:

DEFINITION 1.1. A **norm** φ on $A \subseteq {}^{\omega}\omega$ is a map $\varphi\colon A \to \mathrm{Ord}$. The norm is said to be **regular** if it is onto an ordinal. The norm is a $\boldsymbol{\Gamma}$-norm if the relations $<^*_\varphi$, $\leq^*_\varphi$ are both in $\boldsymbol{\Gamma}$, where:

$$x <^*_\varphi y \Leftrightarrow x \in A \wedge (y \notin A \vee (y \in A \wedge \varphi(x) < \varphi(y))),$$
$$x \leq^*_\varphi y \Leftrightarrow x \in A \wedge (y \notin A \vee (y \in A \wedge \varphi(x) \leq \varphi(y))).$$

Another definition of φ being a $\boldsymbol{\Gamma}$-norm that is frequently used is to say that there are binary relations $\leq_{\boldsymbol{\Gamma}}$, $\leq_{\check{\boldsymbol{\Gamma}}}$ in $\boldsymbol{\Gamma}$ and $\check{\boldsymbol{\Gamma}}$ respectively such that for all

$y \in A$ we have

$$(x \in A \wedge \varphi(x) \leq \varphi(y)) \Leftrightarrow x \leq_{\Gamma} y \Leftrightarrow x \leq_{\check{\Gamma}} y.$$

It is easy to see that Definition 1.1 implies this alternate definition, and the alternate definition implies that of Definition 1.1 if Γ is closed under $\wedge, \vee$ (see [Mos80]). The prewellordering property of Γ is a strong way of saying that every Γ set is a union of Δ sets.

A regular norm on a set A can be identified with a **prewellordering** $\preceq$ of A, that is a reflexive, transitive, connected (i.e., $\forall x, y \in A\ (x \preceq y \vee y \preceq x)$) binary relation on A whose strict part $\prec$ is wellfounded, where $x \prec y \Leftrightarrow (x \preceq y \wedge \neg y \preceq x)$. We call A the domain of the prewellordering.

We let Θ denote the supremum of the lengths of the prewellorderings of $\mathbb{R}$.

It is a standard fact (see [Mos80]) that $\mathrm{PWO}(\Gamma) \Rightarrow \mathrm{Red}(\Gamma)$ if Γ is closed under disjunction. Also, $\mathrm{Red}(\Gamma) \Rightarrow \mathrm{Sep}(\check{\Gamma})$, and if Γ has a universal set, then $\mathrm{Red}(\Gamma) \Rightarrow \neg\ \mathrm{Sep}(\Gamma)$ (these are all ZF results).

We refer the reader to [Van78B] for the basic facts about Wadge degrees assuming AD. We mention a few of the results of significance for us. Assume AD for the remainder of this discussion. Wadge's Lemma asserts that for any $A, B \subseteq {}^{\omega}\omega$ that either $A \leq_{\mathrm{W}} B$ or $B \leq_{\mathrm{W}} {}^{\omega}\omega - A$. One consequence is that a pointclass has a universal set iff it is non-selfdual. In view of Wadge's Lemma, It is natural to consider the equivalence class $[A]_{\mathrm{W}}$ of sets Wadge equivalent to A if A is **self-dual**, that is $A \equiv_{\mathrm{W}} {}^{\omega}\omega - A$, and pairs of equivalence classes $([A]_{\mathrm{W}}, [{}^{\omega}\omega - A]_{\mathrm{W}})$ when A is non-selfdual. Defined this way, Martin showed that the Wadge degrees are wellfounded. Moreover, in [Van78B] it is shown that the dual and non-selfdual degrees alternate, and at limit ordinals of cofinality ω there is a self-dual degree, and a non-selfdual degree at ordinals of uncountable cofinality. For $A \subseteq {}^{\omega}\omega$, let $|A|_{\mathrm{W}}$ denote the Wadge rank of A.

By a **Lévy** pointclass we mean a non-selfdual pointclass Γ closed under either $\exists^{\mathbb{R}}$ or $\forall^{\mathbb{R}}$ (or both). Note that if Γ is non-selfdual and closed under $\exists^{\mathbb{R}}$, then Γ is closed under countable unions, and likewise for $\forall^{\mathbb{R}}$ and countable intersections. For suppose each $A_n \in \Gamma$. Let $B \in \Gamma - \check{\Gamma}$. By Wadge's Lemma, each $A_n \leq_{\mathrm{W}} B$. It follows that $\bigoplus_n A_n \leq_{\mathrm{W}} B$, where the join is defined by $\bigoplus_n A_n = \{n^\frown x : x \in A_n\}$. But then $x \in \bigcup_n A_n$ iff $\exists y\ (y(0)^\frown x \in \bigoplus_n A_n)$ which suffices as $\{(x, y) : y(0)^\frown x \in \bigoplus_n A_n\} \in \Gamma$.

DEFINITION 1.2. For Γ a (possibly selfdual) pointclass, let

$$o(\Gamma) = \sup\{|A|_{\mathrm{W}} : A \in \Gamma\}.$$

Let $\delta(\Gamma)$ be the supremum of the lengths of the Δ prewellorderings of ${}^{\omega}\omega$, where $\Delta = \Gamma \cap \check{\Gamma}$.

If Δ is closed under $\vee$ then $\delta(\Delta)$ is a limit ordinal, and if Δ is closed under countable unions then $\mathrm{cf}(\delta(\Delta)) > \omega$. If Γ is any pointclass and φ is a regular

$\underset{\sim}{\Gamma}$-norm on $A \in \underset{\sim}{\Gamma}$, then $|\varphi| \leq \delta(\underset{\sim}{\Gamma})$, as any initial segment of the norm corresponds to a $\underset{\sim}{\Delta}$ prewellordering. The following frequently used lemma (see 4C.14 of [Mos80]) is a partial converse to this.

LEMMA 1.3 (Moschovakis). Assume $\underset{\sim}{\Gamma}$ has a universal set, and $\underset{\sim}{\Gamma}$ is closed under $\forall^{\mathbb{R}}$, $\vee$. If φ is a $\underset{\sim}{\Gamma}$-norm on a $\underset{\sim}{\Gamma}$-universal set A, then $|\varphi| = \delta(\underset{\sim}{\Gamma})$.

So, assuming AD, this holds for any non-selfdual pointclass $\underset{\sim}{\Gamma}$ and any $\underset{\sim}{\Gamma}$-norm on a set $A \in \underset{\sim}{\Gamma} - \check{\underset{\sim}{\Gamma}}$.

In [KSS81] the following is shown. It shows that for sufficiently closed pointclasses $\underset{\sim}{\Delta}$, $o(\underset{\sim}{\Delta})$ and $\delta(\underset{\sim}{\Delta})$ are the same.

THEOREM 1.4 (AD). Let $\underset{\sim}{\Delta}$ be selfdual and closed under $\exists^{\mathbb{R}}$, $\wedge$. Then $o(\underset{\sim}{\Delta}) = \delta(\underset{\sim}{\Delta}) =$ the supremum of the lengths of the $\underset{\sim}{\Delta}$ wellfounded relations on ${}^{\omega}\omega$.

We note that the hypothesis that $\underset{\sim}{\Delta}$ is closed under $\wedge$ is almost redundant in Theorem 1.4; it only rules out the case where $\underset{\sim}{\Delta} = \underset{\sim}{\Gamma} \cup \check{\underset{\sim}{\Gamma}}$ for some non-selfdual $\underset{\sim}{\Gamma}$ closed under quantifiers.

Steel [Ste81B] showed that for a non-selfdual $\underset{\sim}{\Gamma}$ one of Sep($\underset{\sim}{\Gamma}$), Sep($\check{\underset{\sim}{\Gamma}}$) holds. Van Wesep [Van78A] showed that Sep($\underset{\sim}{\Gamma}$) and Sep($\check{\underset{\sim}{\Gamma}}$) cannot both hold. Thus we have:

THEOREM 1.5 (AD). For any non-selfdual pointclass $\underset{\sim}{\Gamma}$ exactly one of Sep($\underset{\sim}{\Gamma}$), Sep($\check{\underset{\sim}{\Gamma}}$) holds.

For $\underset{\sim}{\Gamma}$ a Lévy class, Kechris, Solovay, Steel [KSS81] and Steel [Ste81A] show the following.

THEOREM 1.6 (AD). Let $\underset{\sim}{\Gamma}$ be non-selfdual, closed under $\exists^{\mathbb{R}}$ or $\forall^{\mathbb{R}}$, and assume $\underset{\sim}{\Delta} = \underset{\sim}{\Gamma} \cap \check{\underset{\sim}{\Gamma}}$ is not closed under wellordered unions. Then exactly one of PWO($\underset{\sim}{\Gamma}$), PWO($\check{\underset{\sim}{\Gamma}}$) holds.

The technical hypothesis that $\underset{\sim}{\Delta}$ is not closed under wellordered unions holds, for example, if every $\underset{\sim}{\Gamma}$ set is ∞-Borel (see Definition 2.11 below). It follows therefore from the slight strengthening AD^+ of AD introduced by Woodin (see [Woo99]).

In fact, [KSS81] and [Ste81A] completely analyze the prewellordering property for Lévy classes. If $\underset{\sim}{\Gamma}$ is a Lévy class, following [KSS81] and [Ste81A] consider

$$\delta_0 = \sup\{o(\underset{\sim}{\Delta}) : \underset{\sim}{\Delta} \text{ is closed under } \exists^{\mathbb{R}}, \wedge, \neg, \text{ and } \underset{\sim}{\Delta} \subseteq \underset{\sim}{\Gamma}\}.$$

Let $\underset{\sim}{\Delta}_0$ be the selfdual class of sets of Wadge rank less than δ_0. It is easily checked that δ_0 is a limit ordinal. If $\mathrm{cf}(\delta_0) > \omega$, then there is a non-selfdual $\underset{\sim}{\Gamma}$ closed under $\forall^{\mathbb{R}}$ with $\underset{\sim}{\Delta}_0 = \underset{\sim}{\Gamma}_0 \cap \check{\underset{\sim}{\Gamma}}_0$ (see [Ste81A] for details). $\underset{\sim}{\Gamma}$ is then in the **projective hierarchy** over $\underset{\sim}{\Gamma}_0$. If $\underset{\sim}{\Gamma}_0$ is not closed under $\exists^{\mathbb{R}}$, then we generate this hierarchy by applying quantifiers to $\underset{\sim}{\Gamma}_0$ and/or $\check{\underset{\sim}{\Gamma}}_0$. If $\underset{\sim}{\Gamma}_0$ is closed under quantifiers, we must apply quantifiers to $\boldsymbol{\Gamma}_0 \wedge \check{\boldsymbol{\Gamma}}_0$. We call the class $\boldsymbol{\Gamma}_0$ at the base of this hierarchy a **Steel pointclass**. In the case where $\mathrm{cf}(\delta_0) > \omega$, we have $\mathrm{PWO}(\boldsymbol{\Gamma}_0)$, and the prewellordering property propagates up the hierarchy by periodicity.

[If $\boldsymbol{\Gamma}_0$ is not closed under quantifiers then $\mathrm{PWO}(\exists^{\mathbb{R}}\boldsymbol{\Gamma}_0)$, $\mathrm{PWO}(\forall^{\mathbb{R}}\exists^{\mathbb{R}}\boldsymbol{\Gamma}_0)$, etc. If $\boldsymbol{\Gamma}_0$ is closed under quantifiers, then $\mathrm{PWO}(\exists^{\mathbb{R}}(\boldsymbol{\Gamma}_0 \wedge \check{\boldsymbol{\Gamma}}_0)$, $\mathrm{PWO}(\forall^{\mathbb{R}}\exists^{\mathbb{R}}(\boldsymbol{\Gamma}_0 \wedge \check{\boldsymbol{\Gamma}}_0)$, etc. See [KSS81] and [Ste81A] for details.]

If $\mathrm{cf}(\delta_0) = \omega$, then we let $\boldsymbol{\Gamma}_0 = \bigcup_\omega \boldsymbol{\Delta}_0$, the collection of countable unions of sets in $\boldsymbol{\Delta}_0$. Then $\mathrm{PWO}(\boldsymbol{\Gamma}_0)$, and prewellordering propagates up by periodicity. Namely, we have $\mathrm{PWO}(\forall^{\mathbb{R}}\boldsymbol{\Gamma}_0)$, $\mathrm{PWO}(\exists^{\mathbb{R}}\forall^{\mathbb{R}}\boldsymbol{\Gamma}_0)$, etc.

We call the pointclass $\boldsymbol{\Gamma}_0$ as above the **base** of a projective hierarchy.

Concerning the reduction property we have the following. Steel shows in [Ste81B] that for $\boldsymbol{\Gamma}$ non-selfdual with $\boldsymbol{\Delta}$ closed under $\wedge$, that $\mathrm{Red}(\boldsymbol{\Gamma})$ or $\mathrm{Red}(\check{\boldsymbol{\Gamma}})$ holds. This includes all $\boldsymbol{\Gamma}_0$ at the base of a projective hierarchy. Inspecting the possible projective hierarchies shows that this also includes all Lévy classes except the case where $\boldsymbol{\Gamma} = \boldsymbol{\Gamma}_0 \wedge \check{\boldsymbol{\Gamma}}_0$ (or the dual of this class) and $\boldsymbol{\Gamma}_0$ is closed under real quantification. In this case we can argue directly that $\mathrm{Red}(\boldsymbol{\Gamma})$ holds assuming again the technical hypothesis that $\boldsymbol{\Delta}_0$ is not closed under wellordered unions. For in this case we may assume without loss of generality that $\mathrm{PWO}(\boldsymbol{\Gamma}_0)$. Let $\kappa = o(\boldsymbol{\Delta}_0)$.

From the Coding Lemma (see Theorem 1.8 below) it follows that $\boldsymbol{\Gamma}_0 = \bigcup_\kappa \boldsymbol{\Delta}_0 = \bigcup_\kappa \boldsymbol{\Gamma}_0$. Suppose $A = A_0 \cap A_1$, $B = B_0 \cap B_1$ are in $\boldsymbol{\Gamma}$, with $A_0, B_0 \in \boldsymbol{\Gamma}_0$ and $A_1, B_1 \in \check{\boldsymbol{\Gamma}}_0$. Write $A_0 = \bigcup_{\alpha<\kappa} C_\alpha$, $B_0 = \bigcup_{\alpha<\kappa} D_\alpha$, with $C_\alpha, D_\alpha \in \boldsymbol{\Delta}_0$. Let

$$A' = A_1 \cap \bigcup_{\alpha<\kappa}(C_\alpha \cap ({}^\omega\omega - \bigcup_{\beta<\alpha}(D_\beta \cap B_1))),$$

$$B' = B_1 \cap \bigcup_{\alpha<\kappa}(D_\alpha \cap ({}^\omega\omega - \bigcup_{\beta\leq\alpha}(C_\beta \cap A_1))).$$

From the Coding Lemma, $\check{\boldsymbol{\Gamma}}_0$ is closed under $< \kappa$ unions, and it follows that A', B' are in $\boldsymbol{\Gamma}$, and they are easily seen to reduce A, B.

Corollary 1.7 (AD). Let $\boldsymbol{\Gamma}$ be non-selfdual, closed under $\exists^{\mathbb{R}}$ or $\forall^{\mathbb{R}}$, and assume $\boldsymbol{\Delta} = \boldsymbol{\Gamma} \cap \check{\boldsymbol{\Gamma}}$ is not closed under wellordered unions. Then exactly one of $\mathrm{Red}(\boldsymbol{\Gamma})$, $\mathrm{Red}(\check{\boldsymbol{\Gamma}})$ holds.

The Moschovakis Coding Lemma is a basic tool in determinacy theory. It requires full AD, even for prewellorderings of short length. It can be stated in several different forms, one of which is the following.

Theorem 1.8 (Coding Lemma). Assume AD. Let $\boldsymbol{\Gamma}$ be a non-self-dual pointclass closed under $\exists^{\mathbb{R}}$, $\wedge$. Suppose $\prec$ is a $\boldsymbol{\Gamma}$ wellfounded relation on

${}^{\omega}\omega$. Then for any $R \subseteq \mathrm{dom}(\prec) \times {}^{\omega}\omega$ such that $\forall x \in \mathrm{dom}(\prec)\ \exists y\ R(x, y)$, there is an $A \subseteq \mathrm{dom}(\prec) \times {}^{\omega}\omega$, $A \in \boldsymbol{\Gamma}$, which is a choice set for R. That is,

1. $\forall \alpha < |\prec|\ \exists x \in \mathrm{dom}(\prec)\ \exists y\ [|x|_{\prec} = \alpha \wedge A(x, y)]$.
2. $\forall x, y\ [A(x, y) \Rightarrow R(x, y)]$.

Proof (sketch). Let $U \subseteq ({}^{\omega}\omega)^3$ be in $\boldsymbol{\Gamma}$ and universal for the $\boldsymbol{\Gamma}$ subsets of ${}^{\omega}\omega \times {}^{\omega}\omega$. We may assume that $\prec$ is chosen of minimal length so that the theorem fails. It follows that δ is a limit. For $\delta < |\prec|$, we say $u \in {}^{\omega}\omega$ is a δ choice code if (1) above holds for all $\alpha < \delta$ using U_u in place of A, and in place of (2) we use $\forall x, y\ [U_u(x, y) \Rightarrow |x|_{\prec} < \delta \wedge R(x, y)]$. By minimality of $|\prec|$ it follows that for all $\delta < |\prec|$ that there is a δ choice code. Consider the integer game where player I plays out $u \in {}^{\omega}\omega$ and player II plays out v, and player II wins iff whenever u codes a δ choice set for some δ, then v codes a δ' choice set for some $\delta' > \delta$. If player I had a winning strategy then we would get a $\boldsymbol{\Sigma}^1_1$ set $S \subseteq {}^{\omega}\omega$ such that each $u \in S$ is a δ choice code for some $\delta < |\prec|$, and $\forall \delta < |\prec|\ \exists \delta' > \delta\ \exists u \in S$ (u is a δ' choice set). We can then form the "union" of S by: $A(x, y) \Leftrightarrow \exists u \in S\ U_u(x, y)$. A then easily satisfies (1) and (2) above. Suppose player II had a winning strategy τ. We attempt to define a $\boldsymbol{\Gamma}$ relation $U_\varepsilon(x, y)$ (by defining its code ε) such that $\mathrm{dom}(U_\varepsilon) = \mathrm{dom}(\prec)$ and if $U_\varepsilon(x, y)$ then y is a δ choice code for some $\delta \geq |x|_{\prec}$. Using the Recursion Theorem (which holds for any $\boldsymbol{\Gamma}$ with a universal set, hence from AD for any non-self-dual $\boldsymbol{\Gamma}$), let $\varepsilon \in {}^{\omega}\omega$ be such that $U_\varepsilon(x, u) \Leftrightarrow u = \tau(f(\varepsilon, x))$, where f is a continuous function from the s-m-n Theorem such that $U_{f(\varepsilon,x)}(y, z) \Leftrightarrow \exists x' \prec x\ \exists u\ [U_\varepsilon(x', u) \wedge U_u(y, z)]$. We clearly have $\forall x \in \mathrm{dom}(\prec)\ \exists y\ U_\varepsilon(x, y)$. Induction on $|x|_{\prec}$ also shows that $U_\varepsilon(x, y)$ implies that y is a δ choice code for some $\delta \geq |x|_{\prec}$. Finally, define A by $A(x, y) \Leftrightarrow \exists x'\ \exists u\ [U_\varepsilon(x', u) \wedge U_u(x, y)]$. Then A satisfies (1), (2). ⊣

One frequently arising case is when the relation $\prec$ is the strict part of a prewellordering $\preceq$ with both $\prec$ and $\preceq$ in $\boldsymbol{\Gamma}$. This will be the case if $\preceq$ is the prewellordering from a $\boldsymbol{\Gamma}$-norm on a $\boldsymbol{\Gamma}$ set. Also, it frequently occurs that the R of Theorem 1.8 is invariant, that is, if $R(x, y)$, $x \preceq x'$, and $x' \preceq x$, then $R(x', y)$. In this case we may view R as assigning a non-empty set to each ordinal $\alpha < |\prec|$. Note that the A of Theorem 1.8 may also be taken to be invariant, as given A from Theorem 1.8 we may define $A'(x, y) \Leftrightarrow \exists x'\ [(x' \preceq x) \wedge (x \preceq x') \wedge A(x', y)]$.

Definition 1.9. If $\preceq$ is a prewellordering (or equivalently a norm), and $A \subseteq |\preceq|$, we say A is $\boldsymbol{\Delta}$ in the codes (for some pointclass $\boldsymbol{\Gamma}$) if there are $\boldsymbol{\Gamma}$, $\check{\boldsymbol{\Gamma}}$ (respectively) sets C and D such that $\forall x \in \mathrm{dom}(\preceq)\ (|x|_{\preceq} \in A \Leftrightarrow C(x) \Leftrightarrow D(x))$.

If $\boldsymbol{\Gamma}$ is as in Theorem 1.8, $\preceq$ is a prewellordering with $\preceq$ and $\prec$ in $\boldsymbol{\Gamma}$, and $A \subseteq \lambda = |\preceq|$, then A is $\boldsymbol{\Delta}$ in the codes. This follows by applying Theorem

1.8 to the characteristic function of A. If $\underset{\sim}{\Gamma}$ is non-self-dual, closed under $\forall^{\mathbb{R}}$, and PWO($\underset{\sim}{\Gamma}$), then we can improve this to say that any $A \subseteq \delta(\underset{\sim}{\Gamma})$ is $\underset{\sim}{\Delta}$ in the codes (as opposed to $\underset{\sim}{\Delta}(\exists^{\mathbb{R}}\underset{\sim}{\Gamma})$). For example, any subset of $\underset{\sim}{\delta}^1_{2n+1}$ is $\underset{\sim}{\Delta}^1_{2n+1}$ in the codes (see Definition 2.6 below). To see this, note that if $\underset{\sim}{\Gamma}$ is closed under $\exists^{\mathbb{R}}$ then we are done by Theorem 1.8. Otherwise, let φ be a $\underset{\sim}{\Gamma}$ norm on a $\underset{\sim}{\Gamma}$-complete set P. Assume for the moment $\underset{\sim}{\Gamma}$ is also closed under $\vee$. Then by Lemma 1.3, φ is onto $\delta(\underset{\sim}{\Gamma})$. Let $U \subseteq {}^{\omega}\omega \times {}^{\omega}\omega \times \omega$ be universal for $\check{\underset{\sim}{\Gamma}}$. Given $A \subseteq \delta(\underset{\sim}{\Gamma})$, say $y \in {}^{\omega}\omega$ codes $A \restriction \beta$ (for $\beta < \delta(\underset{\sim}{\Gamma})$) if

$$\begin{aligned} U_y(z, a) \Leftrightarrow (z \in P \wedge \varphi(z) < \beta \wedge z \in A \wedge a = 1) \\ \vee (z \in P \wedge \varphi(z) < \beta \wedge z \notin A \wedge a = 0). \end{aligned}$$

From the Coding Lemma, for all $\beta < \delta(\underset{\sim}{\Gamma})$ there is a y coding $A \restriction \beta$. Play the integer game where player I plays x, player II plays y, and player II wins iff $[x \in P \Rightarrow \exists \beta > \varphi(x)\ (y \text{ codes } A \restriction \beta)]$. player II wins by boundedness as a winning strategy for player I would give a $\underset{\sim}{\Sigma}^1_1$ set $S \subseteq P$ coding cofinally in $\delta(\underset{\sim}{\Gamma})$ many ordinals, from which we would compute $P \in \check{\underset{\sim}{\Gamma}}$ by $x \in P \Leftrightarrow \exists y \in S\ x \leq_{\underset{\sim}{\Gamma}} y$. If τ is winning for player II and $x \in P$, then $\varphi(x) \in A$ iff $U_{\tau(x)}(x, 1)$ iff $\neg U_{\tau(x)}(x, 0)$. If $\underset{\sim}{\Gamma}$ is closed under $\forall^{\mathbb{R}}$ but not $\exists^{\mathbb{R}}$, PWO($\underset{\sim}{\Gamma}$), and $\underset{\sim}{\Gamma}$ is not closed under $\vee$, then $\underset{\sim}{\Gamma}$ is a Steel pointclass at the base of a projective hierarchy. The analysis of [Ste81A] shows that there is a $\underset{\sim}{\Gamma}$ prewellordering on a $\underset{\sim}{\Gamma}$ complete set of length $\delta(\underset{\sim}{\Gamma})$ which is $\underset{\sim}{\Sigma}^1_1$ bounded, and the above argument then applies, using a $\underset{\sim}{\Gamma}$ universal set to do the coding. We summarize this in the following corollary.

Corollary 1.10 (AD). If $\underset{\sim}{\Gamma}$ is non-self-dual, closed under $\forall^{\mathbb{R}}$, and PWO($\underset{\sim}{\Gamma}$), then every $A \subseteq \delta(\underset{\sim}{\Gamma})$ is $\underset{\sim}{\Delta}$ in the codes with respect to a $\underset{\sim}{\Gamma}$-norm on a $\underset{\sim}{\Gamma}$-complete set.

§2. Suslin Cardinals. We first introduce the notion of a Suslin cardinal.

Definition 2.1. For $\alpha \in \text{Ord}$, X a set, and $A \subseteq {}^{\omega}X$, we say A is α-Suslin if there is a tree T on $X \times \alpha$ such that $A = \text{p}[T]$. We say $\kappa \in \text{Ord}$ is a **Suslin cardinal** if there is an $A \subseteq {}^{\omega}\omega$ which is α-Suslin but not β-Suslin for any $\beta < \alpha$.

Being α-Suslin only depends on $|\alpha|$, the cardinality of α, so it is immediate that a Suslin cardinal is a cardinal. We say $A \subseteq {}^{\omega}X$ is co-α-Suslin if the complement ${}^{\omega}X - A$ is α-Suslin. We say A is Suslin if it is κ-Suslin for some κ and likewise for co-Suslin.

We let Ξ denote the supremum of the Suslin cardinals. We might have $\Xi = \Theta$, in which case every set has a scale, or might have $\Xi < \Theta$. In the latter case it is not known assuming AD whether Ξ must be a Suslin cardinal, although the Suslin cardinals are closed below Ξ.

From AD^+ it follows that Ξ is a Suslin cardinal if $\Xi < \Theta$ (see Theorem 2.13 below; see also Lemma 2.20 for more about Ξ).

DEFINITION 2.2. For $\kappa \in \mathrm{Ord}$ we let $\mathbf{S}(\kappa)$ be the pointclass of κ-Suslin sets.

It is easy to see that $\mathbf{S}(\kappa)$ is a (boldface) pointclass (i.e., is closed under Wadge reduction) and is closed under $\exists^{\mathbb{R}}$, countable unions and intersections. For the case of interest when κ is a Suslin cardinal, Kechris [Kec81B] shows, assuming AD, that $\mathbf{S}(\kappa)$ is non-selfdual. In fact, Kechris shows that for any κ, $\mathbf{S}(\kappa)$ is non-selfdual provided the Suslin cardinals are closed below κ.

From Theorem 2.13 below, this always holds except perhaps in the case when $\kappa \geq \Xi$ and Ξ is not a Suslin cardinal (in which case $\mathbf{S}(\kappa)$ is the collection of all Suslin sets).

Suslin representations are closely related to scales. The notion of a scale, and the scale property, was introduced by Moschovakis. It was introduced originally as a distillation of the key ingredients in the Novikov-Kondô proof of $\underset{\sim}{\mathbf{\Pi}}^1_1$ uniformization, but it quickly evolved into one of the main structural notions in descriptive set theory. We recall the definition. The reader can also consult [KM78B] for further discussion.

DEFINITION 2.3. A **semi-scale** $\langle \varphi_n : n \in \omega \rangle$ on a set $A \subseteq {}^{\omega}X$ (X a set) is a collection of norms φ_n on A such that if $\langle x_m : m \in \omega \rangle \subseteq A$ is a sequence of points in A converging coordinate-wise (i.e., in the product of the discrete topology on X) to $x \in {}^{\omega}X$, and for all n, $\varphi_n(x_m)$ is eventually constant, then $x \in A$. We say $\langle \varphi_n \rangle$ is a **scale** if it in addition satisfies the lower semi-continuity property: $\forall n\ \varphi_n(x) \leq \lim_{m\to\infty} \varphi_n(x_m)$.

A (semi)-scale $\vec{\varphi}$ on A is a **good** (semi)-scale if whenever $x_m \in A$ and for all n $\varphi_n(x_m)$ is eventually constant, then $x = \lim_{m\to\infty} x_m$ exists (and thus $x \in A$). A (semi)-scale is called **very good** if it is good and whenever $x, y \in A$ and $\varphi_n(x) \leq \varphi_n(y)$, then $\varphi_i(x) \leq \varphi_i(y)$ for all $i < n$.

We say $\vec{\varphi}$ is an α semi-scale, etc., if all norms map into α.

The existence of an α-semiscale on $A \subseteq {}^{\omega}X$ is easily equivalent to X being α-Suslin. For if $A = \mathrm{p}[T]$, T a tree on $X \times \alpha$, then we can define a semiscale $\vec{\varphi}$ on A by $\varphi_n(\vec{x})$ = the nth coordinate of the leftmost branch $\ell_{\vec{x}}$ of $T_{\vec{x}}$. Conversely, given an α-semiscale $\vec{\varphi}$ on A, define the **tree of the semi-scale** T_φ by: $((x_0, \dots, x_{n-1}), (\beta_0, \dots, \beta_{n-1})) \in T_\varphi$ iff $\exists \vec{x} \in {}^{\omega}X$ extending $(x_0, \dots, x_{n-1})$ with $\varphi_0(\vec{x}) = \beta_0, \dots, \varphi_{n-1}(\vec{x}) = \beta_{n-1}$. Using the definition of semi-scale we easily get that $A = \mathrm{p}[T]$.

In the main case of interest, when $X = \omega$, we can start from a α-Suslin representation for $A \subseteq {}^{\omega}\omega$ and produce a very-good α'-scale on A, where α' has the same cardinality as α. This is done by letting, for $x \in A$, $\varphi_n(x) = \langle \ell_x(0), x(0), \dots, \ell_x(n-1), x(n-1) \rangle$ where ℓ_x again denotes the leftmost branch of T_x and $\langle \eta_0, \dots, \eta_{k-1} \rangle$ denotes the rank of the tuple in the lexicographic ordering on α^k. It is easy to check that the φ_n form a very-good scale. With a little extra effort we can in fact take $\alpha' = \alpha$ as the next lemma shows (see also [KM78B]).

LEMMA 2.4. For any $\alpha \geq \omega$ and $A \subseteq {}^\omega\omega$, the following are equivalent.

1. A is α-Suslin.
2. A admits an α-very-good scale.

PROOF. Suppose A is α-Suslin. Then A is κ-Suslin where $\kappa = |\alpha|$, so we may assume that $\alpha = \kappa$. Let T be a tree on $\omega \times \kappa$ with $A = \mathrm{p}[T]$. First assume $\mathrm{cf}(\kappa) > \omega$. Define T' on $\omega \times \kappa$ by: $((x(0), \dots, x(n-1)), (\beta_0, \dots, \beta_{n-1})) \in T'$ iff $\beta_0 \geq \max\{\beta_1, \dots, \beta_{n-1}\}$ and $((x(0), \dots, x(n-2)), (\beta_1, \dots, \beta_{n-1})) \in T$. Clearly $A = \mathrm{p}[T']$ as well. Let $\vec{\varphi}$ be the very-good scale from T' as above where now $\langle \beta_0, x(0), \dots, \beta_{n-1}, x(n-1) \rangle$ denotes rank in the lexicographic order on tuples from κ^{2n} whose first entry is the maximum of the entries. Clearly $\varphi_n(x) < \kappa$, and this defines a very-good scale.

Suppose $\mathrm{cf}(\kappa) = \omega$. Let $\kappa = \sup_n \kappa_n$, each $\kappa_n \geq \omega$. Define T' to consist of all pairs $((x(0), \dots, x(n-1)), (\beta_0, \dots, \beta_{n-1}))$ where $\beta_i < \kappa_i$, and $(\beta_0, \dots, \beta_{n-1}))$ is an initial segment of a sequence of the form

$$(i_0, 0, 0, \dots, 0, \gamma_0, i_1, 0, 0, \dots, 0, \gamma_1, \dots)$$

where $i_0, i_1, \dots \in \omega$, there are exactly i_k zeros after i_k, and $(\vec{x}, \vec{\gamma}) \in T$. Clearly $A = \mathrm{p}[T']$. Using lexicographic order on those $(\beta_0, \dots, \beta_n)$ satisfying $\forall i\ \beta_i < \kappa_i$ now produces a very-good κ-scale on A. ⊣

We recall now the definition of a $\underset{\sim}{\boldsymbol{\Gamma}}$-scale, a fundamental notion in descriptive set theory introduced by Moschovakis.

DEFINITION 2.5. A scale $\vec{\varphi}$ on A is said to be a $\underset{\sim}{\boldsymbol{\Gamma}}$-scale (or $\underset{\sim}{\boldsymbol{\Gamma}}$-very good scale, etc.) if all the norms φ_n are $\underset{\sim}{\boldsymbol{\Gamma}}$-norms. We say $\underset{\sim}{\boldsymbol{\Gamma}}$ has the scale property if every $A \in \underset{\sim}{\boldsymbol{\Gamma}}$ admits a $\underset{\sim}{\boldsymbol{\Gamma}}$-scale.

Note that if $\vec{\varphi}$ is a $\underset{\sim}{\boldsymbol{\Gamma}}$-scale on A, then its regularization $\vec{\varphi}'$ is also a $\underset{\sim}{\boldsymbol{\Gamma}}$-scale on A (it is easily still a scale and the relations $<^*_n$, $\leq^*_n$ are the same). Also, if $\underset{\sim}{\boldsymbol{\Gamma}}$ has the scale property and is closed under $\wedge$, $\vee$, then every $A \in \underset{\sim}{\boldsymbol{\Gamma}}$ admits a $\underset{\sim}{\boldsymbol{\Gamma}}$-very good scale since if $\vec{\varphi}$ is a $\underset{\sim}{\boldsymbol{\Gamma}}$-scale on A, then we can define a very good scale by $\psi_n(x) = \langle \varphi_0(x), x(0), \dots, \varphi_n(x), x(n) \rangle$ (it is important to order by the ordinal first).

The original significance of the scale property was that it provides definable uniformizations. For example, if $\underset{\sim}{\boldsymbol{\Gamma}}$ has the scale property and $\underset{\sim}{\boldsymbol{\Gamma}}$ is closed under $\wedge$, $\vee$, and $\forall^{\mathbb{R}}$, then every $A \subseteq {}^\omega\omega \times {}^\omega\omega$ in $\underset{\sim}{\boldsymbol{\Gamma}}$ has a $\underset{\sim}{\boldsymbol{\Gamma}}$ uniformization (i.e., a $\underset{\sim}{\boldsymbol{\Gamma}}$ set $A' \subseteq A$ such that $\forall x\ (\exists y A(x, y) \Leftrightarrow \exists! y A'(x, y))$). We refer the reader to [Mos80] for the proof. Also, if A admits a very good scale all of whose norms are in a pointclass $\underset{\sim}{\boldsymbol{\Gamma}}$, then A has a uniformizing function f

which is $\Game\underset{\sim}{\Gamma}$-measurable, that is, the relation $R(x, n, m) \Leftrightarrow f(x)(n) = m$ is in $\Game\underset{\sim}{\Gamma}$.

Another significance of $\underset{\sim}{\Gamma}$-scales is that they give extra information about the Suslin representation of the set A. If A admits a $\underset{\sim}{\Gamma}$-scale $\vec{\varphi}$ and all the φ_n are regular (without loss of generality), then the tree T_φ of the scale is a tree on $\omega \times \delta(\underset{\sim}{\Gamma})$ (and possibly on a smaller ordinal as well). This is because each φ_n has length $\leq \delta$ by definition of δ and the fact that all initial segments of the φ_n prewellordering lie in $\underset{\sim}{\Delta}$. An important special case of this concerns the **projective ordinals**.

Definition 2.6. Let $\underset{\sim}{\delta}^1_n = \delta(\underset{\sim}{\Sigma}^1_n) =$ the supremum of the lengths of the $\underset{\sim}{\Delta}^1_n$ prewellorderings of ${}^\omega\omega$.

In particular, every regular $\underset{\sim}{\Pi}^1_{2n+1}$ scale on a $\underset{\sim}{\Pi}^1_{2n+1}$ set maps into $\underset{\sim}{\delta}^1_{2n+1}$. On the other hand, from Lemma 1.3 each norm of the scale must map onto $\underset{\sim}{\delta}^1_{2n+1}$. As an immediate corollary we have:

Corollary 2.7. If $\vec{\varphi}$ is a regular $\underset{\sim}{\Pi}^1_{2n+1}$ scale on a $\underset{\sim}{\Pi}^1_{2n+1}$ set, then all of the norms map onto $\underset{\sim}{\delta}^1_{2n+1}$.

We review now briefly the basic facts about transferring scales and the scale property to higher pointclasses. As we mentioned above, $\mathbf{S}(\kappa)$ is closed under $\exists^{\mathbb{R}}$, so if every set in $\underset{\sim}{\Gamma}$ admits a scale, the same is true for $\exists^{\mathbb{R}}\underset{\sim}{\Gamma}$. The Second Periodicity Theorem of Moschovkis [Mos80] shows assuming AD that if $A \subseteq {}^\omega\omega \times {}^\omega\omega$ admits a scale then so does $\forall^{\mathbb{R}} A$. In fact, the proof also works for $A \subseteq {}^\omega X \times {}^\omega\omega$, X any set. Concerning the scale property we have the following. If $\underset{\sim}{\Gamma}$ is closed under $\forall^{\mathbb{R}}$ and every $\underset{\sim}{\Gamma}$ set admits a $\underset{\sim}{\Gamma}$-very good scale, then $\exists^{\mathbb{R}}\underset{\sim}{\Gamma}$ has the scale property. In particular, if $\underset{\sim}{\Gamma}$ is closed under $\forall^{\mathbb{R}}$, $\wedge$, $\vee$ and has the scale property, then $\exists^{\mathbb{R}}\underset{\sim}{\Gamma}$ has the scale property (and is closed under $\wedge$, $\vee$). The scale on $A \in \exists^{\mathbb{R}} B$, $B \in \underset{\sim}{\Gamma}$, is obtained by taking the "infimum" of the very good scale on B. We refer the reader to [Mos80] for details. One version of the Second Periodicity Theorem says that if $\underset{\sim}{\Gamma}$ is closed under $\exists^{\mathbb{R}}$, $\wedge$ $\vee$ and has the scale property, then assuming $\underset{\sim}{\Delta}$ determinacy $\forall^{\mathbb{R}}\underset{\sim}{\Gamma}$ has the scale property. Both of these scale transfer results are contained in the following somewhat more general form of the Second Periodicity Theorem (see [Mos80]).

Theorem 2.8 (Moschovakis). Assume every set in $\underset{\sim}{\Gamma}$ admits a very good scale $\vec{\varphi}$, with the norm relations $\leq^{*,\varphi}_n$, $<^{*,\varphi}_n \in \underset{\sim}{\Gamma}_n$. Assume every game in $\Lambda = \bigcup_n \underset{\sim}{\Gamma}_n$ is determined. Then every set in $\Game\underset{\sim}{\Gamma}$ admits a very good scale $\vec{\psi}$ where the norm relations $\leq^{*,\psi}_n$, $<^{*,\psi}_n$ are in $\Game\underset{\sim}{\Gamma}_n$.

The proof of the previous theorem also shows that if $A \subseteq ({}^\omega X \times {}^\omega\omega)$ admits a scale, then assuming AD, $\Game A$ also admits a scale.

Starting with the classical fact that $\underset{\sim}{\Pi}^1_1$ has the scale property (in ZF), it follows from second periodicity that, assuming projective determinacy, the pointclasses $\underset{\sim}{\Pi}^1_{2n+1}$, $\underset{\sim}{\Sigma}^1_{2n+2}$ have the scale property. We will use this in a moment to analyze the Suslin cardinals below the supremum of the $\underset{\sim}{\delta}^1_n$. Steel [Ste83B] shows that every $\underset{\sim}{\Sigma}^1_1$ set admits a very-good scale $\vec{\varphi}$ with $\varphi_n \in \omega \cdot (n+1)$-$\underset{\sim}{\Pi}^1_1$. It follows from Theorem 2.8 that every $\underset{\sim}{\Pi}^1_{2n}$ set admits a scale all of whose norms lie in $\Game^{2n-1}\omega \cdot k$-$\underset{\sim}{\Pi}^1_1$ for some k (here $\Game^{2n-1}$ is the $2n-1$ iterate of the integer game quantifier).

We mention two other scale transfer results. Martin [Mar83B] shows that under a suitable determinacy hypothesis (the determinacy of certain games on ω of countable length), the integer game quantifier $\Game_\alpha$ of countable length α propagates scales and the scale property. That is, if $\underset{\sim}{\Gamma}$ is closed under $\wedge$, $\vee$ and has the scale property and if $A \subseteq ({}^\omega\omega \times \omega^\alpha)$ is in $\underset{\sim}{\Gamma}$, then $\Game_\alpha A$ also has the scale property. We refer the reader to [Mar83B] for the precise definitions and statements. In particular, assuming $\mathsf{AD}_{\mathbb{R}}$ if $\underset{\sim}{\Gamma}$ (as above) has the scale property then so does $\Game^{\mathbb{R}}\underset{\sim}{\Gamma}$. Martin's proof is purely game theoretic, generalizing the methods of the periodicity theorems. If one doesn't care about the definability of the scales, Martin and Steel [MS83] by completely different methods showed that just assuming AD, if A admits a scale then $\Game^{\mathbb{R}}A$ also admits a scale. Their method is related to the homogeneous tree construction.

Two important consequences of a κ-Suslin representation are the Kunen-Martin Theorem and a κ^+-Borel representation for the set. First we recall the Kunen-Martin Theorem.

Theorem 2.9 (Kunen, Martin). Let $\prec$ be a κ-Suslin wellfounded relation on ${}^\omega\omega$. Then $|\prec| < \kappa^+$, where $|\prec|$ denotes the rank of $\prec$.

Proof. Let T be a tree on $\omega \times \omega \times \kappa$ with $\prec\; = \mathrm{p}[T]$. Let U be the wellfounded tree consisting of finite $\prec$-decreasing sequences $(x_0, \dots, x_n)$, that is, $x_n \prec \cdots \prec x_1 \prec x_0$. Easily $\prec$ and U have the same rank. To each $\vec{x} = (x_0, \dots, x_n) \in U$ assign $\pi(\vec{x}) = (x_0 \restriction (n+1), \dots, x_n \restriction (n+1), \ell(x_1, x_0) \restriction (n+1), \dots, \ell(x_n, x_{n-1}) \restriction (n+1))$, where $\ell(y, z) \in {}^\omega\kappa$ is the leftmost branch of $T_{y,z}$. If $\vec{y}$ extends $\vec{x}$, we view $\pi(\vec{y})$ as extending $\pi(\vec{x})$ in a natural way. π gives an order-preserving map of U into a wellfounded relation on (essentially) κ. Thus $|\prec| = |U| < \kappa^+$. ⊣

We recall the notion of a κ-Borel set (see also [KM78B]). We must be careful working in ZF to distinguish between the "effective" and "non-effective" versions of the definition. The next definition is the non-effective version.

Definition 2.10. $\boldsymbol{B}_\kappa$ is the smallest collection containing the open sets and closed under complements and wellordered unions (equivalently intersections) of length $< \kappa$. We let $\boldsymbol{B}_\infty = \bigcup_\kappa \boldsymbol{B}_\kappa$.

By an effective κ-Borel code $S = \langle T, \varphi \rangle$ we mean a wellfounded tree T on a $\lambda < \kappa$ and a map φ which assigns to each terminal node s of T a basic open set $\varphi(s)$ in ${}^\omega\omega$. Such an S builds up an element of $\boldsymbol{B}_\kappa$ in a natural manner.

Namely, extend φ to all of S by $\varphi(s) = {}^{\omega}\omega - \bigcup_{t \prec s} \varphi(t)$, the union ranging over t extending s with $\text{lh}(t) = \text{lh}(s) + 1$. Then set $\varphi(S) = \varphi(\varnothing)$ ($\varnothing$ is the maximal node of T). Sometimes we use minor variations of this definition, like allowing each node of the tree to be either a "union" or an "intersection" node, in the obvious sense. It is easy to see that the relation $x \in \varphi(S)$ is absolute between transitive models containing x, S.

DEFINITION 2.11. We say $A \subseteq {}^{\omega}\omega$ is effectively κ-Borel if there is an effective κ-Borel code S for A (i.e., $A = \varphi(S)$). We say A is ∞-Borel if it is effectively κ-Borel for some κ.

Woodin has given another characterization of ∞-Borel sets. Namely, A is ∞-Borel iff there is an $S \subseteq \text{Ord}$ and a formula φ such that $\forall x\ (x \in A \Leftrightarrow \text{L}[S, x] \models \varphi(S, x))$.

Of course, any effectively κ-Borel set is in $\boldsymbol{B}_{\kappa}$, but the converse is not clear in ZF (or ZF + AD). Suslin representations give effective Borel codes according to the following lemma (see also [KM78B]).

LEMMA 2.12. If A is κ-Suslin or co-κ-Suslin then A is effectively κ^{++}-Borel. In fact, A is (effectively) a κ^{+} union of sets which are κ^{+}-Borel. If $\text{cf}(\kappa) > \omega$ then every κ-Suslin set is an effective κ union of κ-Borel sets.

PROOF. Let $A = \text{p}[T]$, T a tree on $\omega \times \kappa$. For $s \in {}^{<\omega}\kappa$ and $\alpha < \kappa^{+}$, let $B(s, \alpha) = \{x : |s|_{T_x} \leq \alpha\}$, where $|s|_{T_x}$ denotes the rank of s in the tree T_x if s is in the wellfounded part of T_x, is undefined if s is in the illfounded part of T_x, and for convenience we set $|s|_{T_x} = 0$ if $s \notin T_x$. So, terminal nodes of the tree T_x have rank 0 as do all further extensions. Clearly every $B(s, 0)$ is clopen. For $\alpha > 0$ we have:

$$B(s, \alpha) = \bigcap_{\eta < \kappa} \bigcup_{\beta < \alpha} B(s^\frown \eta, \beta).$$

It follows by induction that all $B(s, \alpha)$ are (effectively) κ^{+}-Borel. Since ${}^{\omega}\omega - A = \bigcup_{\alpha < \kappa^{+}} B(\varnothing, \alpha)$, we have that ${}^{\omega}\omega - A$ is a κ^{+} union of κ^{+}-Borel sets.

Note that if $x \in A$ and α is greater than the supremum of the $|s|_{T_x}$ for s in the wellfounded part of T_x, then (1) the root node $\varnothing$ of T_x is not wellfounded of rank $\leq \alpha$ and (2) for all $s \in T_x$ if s does not have rank $\leq \alpha$ in T_x then there is an immediate extension $s^\frown \eta$ which does not have rank $\leq \alpha$ in T_x. Conversely, (1) and (2) imply T_x is illfounded. Thus,

$$A = \bigcup_{\alpha < \kappa^{+}} [\breve{B}(\varnothing, \alpha) \cap \bigcap_{s \in {}^{<\omega}\kappa} (B(s, \alpha) \cup \bigcup_{\eta < \kappa} \breve{B}(s^\frown \eta, \alpha))]$$

This shows A is also a κ^{+} union of κ^{+}-Borel sets.

If $\text{cf}(\kappa) > \omega$ then $\text{p}[T] = \bigcup_{\alpha < \kappa} \text{p}[T \restriction \alpha]$, and the last statement of the lemma follows. ⊣

The converse to Lemma 2.12 is not in general true. Assuming AD+$\mathbf{V}$=$\mathbf{L}(\mathbb{R})$, there is a largest Suslin class, $\utilde{\Sigma}^2_1$, but every set of reals is ∞-Borel (we discuss this in more detail below).

The following result is a fundamental results about the Suslin cardinals. It was first proved by Steel from the assumption AD + $\mathbf{V}$=$\mathbf{L}(\mathbb{R})$ (cf. [Ste83A]), and then proved by Woodin from AD. Woodin's proof introduced several new notions and methods including the concept of a **strong ∞-Borel code**.

THEOREM 2.13 (Steel, Woodin). Assume AD. Then the Suslin cardinals are closed below their supremum.

Woodin has isolated a natural strengthening of AD called AD$^+$ which implies that the Suslin cardinals are closed below Θ (see [Woo99]). Thus, AD$^+$ implies $S(\kappa)$ is always non-self-dual (this answers one of the questions of [Kec81B]). In fact Woodin shows that assuming AD, AD$^+$ is equivalent to the Suslin cardinals being closed below Θ.

Though we do not prove the Steel-Woodin Theorem here, we define the notion of a strong ∞-Borel code, which is crucial for their analysis. If $S = \langle T, \varphi \rangle$ is an ∞-Borel code and $C \subseteq \mathrm{Ord}$, by $S \restriction C$ we mean the ∞-Borel code obtained by restricting to the subtree of T generated by C (i.e., the set of nodes $s \in {}^{<\omega}C$). If s is not terminal in T but is terminal in $T \restriction C$, we set $\varphi(s) = \varnothing$ in computing φ for $T \restriction C$ (in practice we may assume C has the property that if $s \in {}^{<\omega}C$ is non-terminal in T then s is non-terminal in $T \restriction C$ so this minor annoyance does not arise).

DEFINITION 2.14 (Woodin). A **strong ∞-Borel code** is an ∞-Borel code S with the property that player II wins the following ordinal game G_S: player I and player II alternate playing ordinals α_i. Let $C = \langle \alpha_i : i \in \omega \rangle$. Player II wins the run iff $\varphi(S \restriction C) \subseteq \varphi(S)$.

Note that for C countable, $S \restriction C$ is just an ordinary Borel code (actually, isomorphic to one in an obvious sense).

LEMMA 2.15 (Woodin). $A \subseteq {}^{\omega}\omega$ is Suslin iff A has a strong ∞-Borel code. In fact, if A has a strong ∞-Borel code of size κ, then A is κ-Suslin.

PROOF. Suppose first that $A = \mathrm{p}[T]$ where T is a tree on $\omega \times \kappa$. Let $S \subseteq \kappa^+$ ($S \subseteq \kappa$ if $\mathrm{cf}(\kappa) > \omega$) be the ∞-Borel code for A produced by Lemma 2.12. We show that S is actually a strong ∞-Borel code. In fact, we show that for all $C \subseteq \mathrm{Ord}$ that $\varphi(S \restriction C) \subseteq \varphi(S)$. Suppose $x \in \varphi(S \restriction C)$. We produce an $f \in {}^{\omega}\mathrm{Ord}$ with $(x, f) \in [T]$. The top node $\varnothing$ is a union node of S which writes S as a κ^+ union of codes S_α. Thus, for some $\alpha \in C$ we have $x \in \varphi(S_\alpha \restriction C)$. The code S_α is the conjunction of a code D for $\breve{B}(\varnothing, \alpha)$ and a code E which is a conjunction over $s \in {}^{<\omega}\kappa$ of codes E_s for $B(s, \alpha) \cup \bigcup_\eta \breve{B}(s^\frown\eta, \alpha)$. Let F_s denote the code for $\bigcup_\eta \breve{B}(s^\frown\eta, \alpha)$. F_s is a κ disjunction of codes $G_{s^\frown\eta}$ for $\breve{B}(s^\frown\eta, \alpha)$. Now $x \in \varphi(D \restriction C)$ and taking

$s = \varnothing$ we have $x \in ({}^{\omega}\omega - \varphi(D \restriction C)) \cup \varphi(F_{\varnothing} \restriction C)$. Thus, $x \in \varphi(F_{\varnothing} \restriction C)$. Thus there is an $f(0) \in C$ such that $x \in \varphi(G_{s_0} \restriction C)$, where $s_0 = \langle f(0) \rangle$. Since $x \in \varphi(E \restriction C)$ and $f(0) \in C$, we have $x \in \varphi(E_{s_0} \restriction C)$. Since $x \in \varphi(G_{s_0} \restriction C)$, we also have $x \in \varphi(F_{s_0} \restriction C)$. Thus there is an $f(1) \in C$ such that letting $s_1 = \langle f(0), f(1) \rangle$ we have $x \in \varphi(G_{s_1} \restriction C)$. Since $s_1 \in {}^{<\omega}C$ and $x \in \varphi(E \restriction C)$, we have $x \in \varphi(E_{s_1} \restriction C)$. Continuing, we produce f as claimed.

Suppose next that $S \subseteq \kappa$ is a strong ∞-Borel code for A. Fix a winning strategy τ for player II in the game G_S. It is straightforward to construct a tree T on $\omega \times \kappa$ such that if $(x, f) \in [T]$ then $C := \{f(0), f(1), \dots\}$ is closed under τ and $x \in \varphi(S \restriction C)$, and furthermore, for any C closed under τ and $x \in \varphi(S \restriction C)$ there is an f enumerating C with $(x, f) \in [T]$. [The definition of T dovetails the requirements that $\operatorname{ran}(f)$ be closed under τ and $x \in \varphi(S \restriction \operatorname{ran}(f))$. The latter construction is similar to the proof that every Borel set is ω-Suslin.] If $x \in [T]$ then for some $f \in {}^{\omega}\kappa$, $(x, f) \in [T]$ and so $x \in \varphi(S \restriction C) \subseteq \varphi(S) = A$ since $C = \{f(0), f(1), \dots\}$ is closed under τ and hence can be enumerated by a run of the game G_s following τ. If $x \in \varphi(S)$, then there is c.u.b. set of $C \in \wp_{\omega_1}(\kappa)$ for which $x \in \varphi(S \restriction C)$. If we choose C to also be closed under τ, then we can find an f enumerating C with $(x, f) \in [T]$, so $x \in \mathrm{p}[T]$. ⊣

The theory of Suslin cardinals and scales is closely related to the theory of wellordered unions of pointclasses. We recall a few of the results of this theory. In [JM83] it is shown that if $\underset{\sim}{\Gamma}$ is closed under $\exists^{\mathbb{R}}$, $\forall^{\mathbb{R}}$ and $\mathrm{PWO}(\underset{\sim}{\Gamma})$, then $\underset{\sim}{\Gamma}$ is closed under wellordered unions. In [KSS81] it is show that if $\underset{\sim}{\Gamma}$ is closed under $\exists^{\mathbb{R}}$ but not $\forall^{\mathbb{R}}$, closed under countable unions and intersections, and $\mathrm{PWO}(\underset{\sim}{\Gamma})$, then $\underset{\sim}{\Gamma}$ is closed under wellordered unions. In [Jac07A] it is noted that this last result generalizes to the case $\underset{\sim}{\Gamma}$ not closed under countable unions and intersections. Thus we have:

Theorem 2.16 (AD). If $\underset{\sim}{\Gamma}$ is a non-selfdual pointclass closed under $\exists^{\mathbb{R}}$ and $\mathrm{PWO}(\underset{\sim}{\Gamma})$, then $\underset{\sim}{\Gamma}$ is closed under wellordered unions.

When the prewellordering property falls on the side closed under $\forall^{\mathbb{R}}$, we have the following result of Martin.

Theorem 2.17 (Martin). Assume AD. Let $\underset{\sim}{\Gamma}$ be non-selfdual, closed under $\forall^{\mathbb{R}}$, $\wedge$, $\vee$, and $\mathrm{PWO}(\underset{\sim}{\Gamma})$. Then $\underset{\sim}{\Delta}$ is closed under $< \delta(\underset{\sim}{\Gamma})$ length unions and intersections.

Proof. Suppose $\alpha < \delta(\underset{\sim}{\Gamma})$ be least such that $\underset{\sim}{\Delta}$ is not closed under α length unions. By the Coding Lemma, every α union of $\underset{\sim}{\Delta}$ sets is in $\check{\underset{\sim}{\Gamma}}$. By Wadge's Lemma it follows that some $\check{\underset{\sim}{\Gamma}}$-complete set, and hence every $\check{\underset{\sim}{\Gamma}}$ set is an α-union of $\underset{\sim}{\Delta}$ sets. If $A \in \check{\underset{\sim}{\Gamma}}$, write $A = \bigcup_{\beta<\alpha} A_\beta$, where each $A_\beta \in \underset{\sim}{\Delta}$. For $x \in A$, let $\varphi(x) = \mu\beta\ (x \in A_\beta)$. Since $\check{\underset{\sim}{\Gamma}} = \bigcup_\alpha \underset{\sim}{\Delta}$, we get that the norm relations $<^*_\varphi$, $\leq^*_\varphi$ are in $\check{\underset{\sim}{\Gamma}}$ [for example, $<^*_\varphi = \bigcup_{\beta<\alpha} B_\beta$ where

$B_\beta = \{(x, y) : x \in A_\beta \wedge y \notin A_\beta\}$]. This shows PWO($\check{\Gamma}$), a contradiction since not both Γ, $\check{\Gamma}$ can have the prewellordering property. ⊣

In some cases we can show that a stronger principle than Theorem 2.16 applies, namely, there are no long increasing sequences of Γ sets. In [JM83] it is shown that any increasing or decreasing sequence of Σ^1_{2n} sets must have length $< \delta^1_{2n}$ (in particular, any Π^1_{2n} prewellordering has length $< \delta^1_{2n}$). More generally, the following was shown:

THEOREM 2.18 (Jackson, Martin). Assume AD. Let κ be a Suslin cardinal with $\mathrm{cf}(\kappa) > \omega$. Then there is no increasing or decreasing sequence of $\mathbf{S}(\kappa)$ sets of length κ^+.

Interestingly, Hjorth has obtained improvements to some of these results using completely different methods, namely inner model theory. In [Hjo96], Hjorth shows that there is no ω_2 sequence of distinct Σ^1_2 sets. In [Hjo01], Hjorth generalizes this to show that there is no ω_{n+2} sequence of distinct $\Game \omega \cdot n$-Π^1_1 sets (and this is shown from Π^1_2-determinacy).

When κ is a Suslin cardinal with $\mathrm{cf}(\kappa) > \omega$ then in fact $\mathbf{S}(\kappa)$ has the scale property (cf. [JM83]). However, there are cases where we can establish the non-existence of long sequences without scales. This is the content of the next result, due to Chuang, which uses also the methods of [JM83] (cf. [Chu82], [Jac07A]).

THEOREM 2.19 (Chuang). Assume AD. Let Γ be non-selfdual, closed under $\forall^{\mathbb{R}}$, countable unions and intersections and PWO(Γ). Then there is no $\delta(\Gamma)^+$ increasing or decreasing sequence of Γ sets.

In particular, if Γ is non-selfdual and closed under quantifiers then there are no $o(\Delta)^+$ increasing or decreasing sequences of Γ sets.

Recall Ξ denotes the supremum of the Suslin cardinals. The next lemma shows that the largest Suslin cardinal, if it exists, must be a limit of Suslin cardinals.

LEMMA 2.20 (AD). If Ξ is a Suslin cardinal, then the Suslin cardinals are unbounded below Ξ, and Ξ is a regular cardinal. Furthermore, $\mathbf{S}(\Xi)$ (the pointclass of Ξ-Suslin sets) is closed under $\exists^{\mathbb{R}}$, $\forall^{\mathbb{R}}$, countable unions and intersections and has the scale property with norms onto Ξ.

PROOF. Let $\Gamma = \mathbf{S}(\Xi)$. By Kechris [Kec81B] Γ is non-selfdual, and closed under $\exists^{\mathbb{R}}$, and countable unions and intersections. Γ is also closed under $\forall^{\mathbb{R}}$ as otherwise by second periodicity $\forall^{\mathbb{R}}\Gamma$ would be a strictly larger pointclass admitting scales. Let $\delta = \delta(\Gamma) = o(\Delta)$ by Theorem 1.4. From the Coding Lemma, $\delta \leq \Xi$ (if $\Xi < \delta$, then we could code any tree T on $\omega \times \Xi$ via a pointclass in Δ, which would then compute $\mathrm{p}[T] \in \Delta$, a contradiction). One of Γ, $\check{\Gamma}$ has the prewellordering property by Theorem 1.6, and so Theorem 2.19

applies to either $\boldsymbol{\Gamma}$ or $\check{\boldsymbol{\Gamma}}$. In either case, there is no δ^+ increasing sequence of $\boldsymbol{\Gamma}$ sets. Suppose $\delta < \Xi$. Let $A \in \boldsymbol{\Gamma} - \check{\boldsymbol{\Gamma}}$ and let $A = \mathrm{p}[T]$ with T a tree on $\omega \times \Xi$. Let $\vec{\varphi}$ be the corresponding regular scale, with $\kappa_n = \sup \varphi_n$. Since A is not α-Suslin for any $\alpha < \Xi$, we have that $\sup_n \kappa_n = \Xi$. Using $\varphi_0, \dots, \varphi_n$ we easily get a κ_n increasing sequence of $\boldsymbol{\Gamma}$ sets, and so $\kappa_n < \delta^+$. Since $\mathrm{cf}(\delta^+) > \omega$, $\Xi = \sup_n \kappa_n < \delta^+$. This contradicts Ξ being a cardinal. So, $\Xi = \delta$. Easily, $\mathrm{cf}(\delta) > \omega$. Again letting A, T be as above, we have that A is a δ union of $\boldsymbol{\Delta}$ sets (since $\mathrm{p}[T] = \bigcup_{\alpha<\delta} \mathrm{p}[T \restriction \alpha]$ as $\mathrm{cf}(\delta) > \omega$). This shows PWO($\boldsymbol{\Gamma}$) since if PWO($\check{\boldsymbol{\Gamma}}$), then $\check{\boldsymbol{\Gamma}}$ would be closed under wellordered unions by 2.16, and then $A \in \check{\boldsymbol{\Gamma}}$. So $\boldsymbol{\Gamma}$ is closed under wellordered unions. It follows that δ is regular, as otherwise by the coding lemma we could compute any Ξ union of $\boldsymbol{\Delta}$ sets to be in $\boldsymbol{\Delta}$, using the closure properties of $\boldsymbol{\Gamma}$; however $\boldsymbol{\Delta}$ is not closed under δ unions from PWO($\boldsymbol{\Gamma}$). To see Scale($\boldsymbol{\Gamma}$), let $A \in \boldsymbol{\Gamma}$ with $A = \mathrm{p}[T]$ as above. For $x \in A$, let $\psi_0(x)$ be the least $\alpha < \Xi$ such that $x \in \mathrm{p}[T \restriction \alpha]$. Let $\psi_{i+1}(x) = \langle \psi_0(x), \ell_0^{\psi_0(x)}(x), \dots, \ell_i^{\psi_0(x)}(x) \rangle$, where $\ell_j^\alpha(x)$ is the jth coordinate of the leftmost branch of $T_x \restriction \alpha$. It is straightforward to check that $\langle \psi_i : i \in \omega \rangle$ is a scale on A. Moreover, all of the norm relations $<^*_{\psi_n}$, $\leq^*_{\psi_n}$ are easily expressible as δ unions of $\boldsymbol{\Delta}$ sets, hence they are $\boldsymbol{\Gamma}$ relations.

Let λ be the supremum of the Suslin cardinals which are less than Ξ. Suppose towards a contradiction that $\lambda < \Xi$. First suppose that λ is not a Suslin cardinal (this case cannot actually arise by Theorem 2.13, but we argue directly here). Again let $A = \mathrm{p}[T]$, where $A \in \boldsymbol{\Gamma} - \check{\boldsymbol{\Gamma}}$ and T a tree on $\omega \times \Xi$. This writes $A = \bigcup_{\alpha<\Xi} A_\alpha$, where each A_α is λ'-Suslin for some $\lambda' < \lambda$ (letting $A_\alpha = \mathrm{p}[T \restriction \alpha]$, which is α-Suslin, and hence λ'-Suslin for some $\lambda' < \lambda$ by definition of λ). Since Ξ is regular, for some fixed $\lambda_1 < \lambda$ we have that A is a Ξ union of λ_1-Suslin sets. If $\boldsymbol{\Gamma}_1 = \mathbf{S}(\lambda_1)$, then either PWO($\boldsymbol{\Gamma}_1$) or PWO($\check{\boldsymbol{\Gamma}}_1$), and so either $\boldsymbol{\Gamma}_1$ or $\exists^{\mathbb{R}} \check{\boldsymbol{\Gamma}}_1$ is closed under wellordered unions. In either case this shows $A \in \boldsymbol{\Delta}$, a contradiction. If λ is a Suslin cardinal, the argument is almost identical. We have A is a Ξ union of λ-Suslin sets now, and either $\mathbf{S}(\lambda)$ or $\exists^{\mathbb{R}} \check{S}(\lambda)$ is closed under wellordered unions which gives the same contradiction. ⊣

We are now ready to classify the Suslin cardinals within the supremum of the $\boldsymbol{\delta}^1_n$. Various parts of the following theorem are due to Kechris, Martin, and Moschovakis; we refer the reader to [Mos80] or [Kec78A] for a more detailed accounting.

Theorem 2.21 (Kechris, Martin, Moschovakis). *Assume* AD. *All of the $\boldsymbol{\delta}^1_n$ are regular. $\boldsymbol{\delta}^1_{2n+2} = (\boldsymbol{\delta}^1_{2n+1})^+$, and $\boldsymbol{\delta}^1_{2n+1} = (\boldsymbol{\lambda}_{2n+1})^+$, where $\boldsymbol{\lambda}_{2n+1}$ is a cardinal of cofinality ω. The Suslin cardinals below $\sup_n \boldsymbol{\delta}^1_n$ are exactly the $\boldsymbol{\lambda}_{2n+1}$, $\boldsymbol{\delta}^1_{2n+1}$. Also $\mathbf{S}(\boldsymbol{\lambda}_{2n+1}) = \boldsymbol{\Sigma}^1_{2n+1}$, $\mathbf{S}(\boldsymbol{\delta}^1_{2n+1}) = \boldsymbol{\Sigma}^1_{2n+2}$.*

PROOF (SKETCH). Let φ be a $\underset{\sim}{\Pi}^1_{2n+1}$ norm on a $\underset{\sim}{\Pi}^1_{2n+1}$-complete set P. Thus φ is onto $\underset{\sim}{\delta}^1_{2n+1}$. If $f : \alpha \to \underset{\sim}{\delta}^1_{2n+1}$ were cofinal and $\alpha < \underset{\sim}{\delta}^1_{2n+1}$, then the Coding Lemma applied to f and a $\underset{\sim}{\Delta}^1_{2n+1}$ prewellordering of length α would show that there is a $\underset{\sim}{\Sigma}^1_{2n+1}$ set $S \subseteq P$ which coded cofinally many ordinals. This would compute $P \in \underset{\sim}{\Sigma}^1_{2n+1}$, a contradiction ($x \in P \Leftrightarrow \exists y \in S\ (x \leq_{\Sigma^1_{2n+1}} y)$). Thus, $\underset{\sim}{\delta}^1_{2n+1}$ is regular.

Every $\underset{\sim}{\Pi}^1_{2n+1}$ set, and hence every $\underset{\sim}{\Sigma}^1_{2n+2}$ set is $\underset{\sim}{\delta}^1_{2n+1}$-Suslin from the scale property for $\underset{\sim}{\Pi}^1_{2n+1}$. From the Kunen-Martin Theorem it follows that every $\underset{\sim}{\Sigma}^1_{2n+2}$ wellfounded relation has length $< (\underset{\sim}{\delta}^1_{2n+1})^+$. On the other hand, Corollary 1.10 shows that every wellorder of $\underset{\sim}{\delta}^1_{2n+1}$ is $\underset{\sim}{\Pi}^1_{2n+1}$ in the codes, and thus there are $\underset{\sim}{\Pi}^1_{2n+1}$ prewellorderings of any length $< (\underset{\sim}{\delta}^1_{2n+1})^+$. Hence $\underset{\sim}{\delta}^1_{2n+2} =$ the supremum of the length of the $\underset{\sim}{\Sigma}^1_{2n+2}$ wellfounded relations $=$ the supremum of the lengths of the $\underset{\sim}{\Pi}^1_{2n+1}$ prewellorderings $= (\underset{\sim}{\delta}^1_{2n+1})^+$.

If $f : \alpha \to \underset{\sim}{\delta}^1_{2n+2}$ were cofinal with $\alpha < \underset{\sim}{\delta}^1_{2n+2}$, then the Coding Lemma would give a $\underset{\sim}{\Sigma}^1_{2n+2}$ set S of codes for $\underset{\sim}{\Sigma}^1_{2n+2}$ wellfounded relations (say via a $\underset{\sim}{\Sigma}^1_{2n+2}$ universal set U). We could put these together into a single $\underset{\sim}{\Sigma}^1_{2n+2}$ wellfounded relation: $(x, y) \prec (x', y')$ iff $x = x' \in S \wedge U_x(y, y')$. This contradicts the Kunen-Martin Theorem. Thus, $\underset{\sim}{\delta}^1_{2n+2}$ is regular.

A $\underset{\sim}{\Pi}^1_{2n+1}$-complete set P cannot be $< \underset{\sim}{\delta}^1_{2n+1}$-Suslin as an easy computation from the Coding Lemma would then show that $P \in \underset{\sim}{\Sigma}^1_{2n+1}$. Thus, $\underset{\sim}{\delta}^1_{2n+1}$ is a Suslin cardinal. The Coding Lemma shows that every $\underset{\sim}{\delta}^1_{2n+1}$-Suslin set is $\underset{\sim}{\Sigma}^1_{2n+2}$, and thus $\mathrm{S}(\underset{\sim}{\delta}^1_{2n+1}) = \underset{\sim}{\Sigma}^1_{2n+2}$.

Define $\underset{\sim}{\lambda}_{2n+1}$ to be the least cardinal κ such that some $\underset{\sim}{\Pi}^1_{2n}$-complete set is κ-Suslin. So, $\underset{\sim}{\Sigma}^1_{2n+1} \subseteq \mathrm{S}(\underset{\sim}{\lambda}_{2n+1})$. From the scale property for $\underset{\sim}{\Pi}^1_{2n+1}$ it follows that every $\underset{\sim}{\Pi}^1_{2n}$ admits a (regular) scale all of whose norms are in $\underset{\sim}{\Delta}^1_{2n+1}$. Each of these norms must therefore have length $< \underset{\sim}{\delta}^1_{2n+1}$, and thus so does their supremum. This shows $\underset{\sim}{\lambda}_{2n+1} < \underset{\sim}{\delta}^1_{2n+1}$. By definition, $\underset{\sim}{\delta}^1_{2n+1}$ is the supremum of the lengths of the $\underset{\sim}{\Delta}^1_{2n+1}$ prewellorderings, and from the Kunen-Martin Theorem each of these has length $< \underset{\sim}{\lambda}^+_{2n+1}$. Thus, $\underset{\sim}{\delta}^1_{2n+1} \leq \underset{\sim}{\lambda}^+_{2n+1}$, and hence $\underset{\sim}{\delta}^1_{2n+1} = \underset{\sim}{\lambda}^+_{2n+1}$.

Finally, we show that $\mathrm{cf}(\underset{\sim}{\lambda}_{2n+1}) = \omega$. If $\mathrm{cf}(\underset{\sim}{\lambda}_{2n+1}) > \omega$, then every $\underset{\sim}{\Sigma}^1_{2n+1}$ set would be a $\underset{\sim}{\lambda}_{2n+1}$ union of sets each of which was $< \underset{\sim}{\lambda}_{2n+1}$-Suslin. By definition of $\underset{\sim}{\lambda}_{2n+1}$, this would show that every $\underset{\sim}{\Sigma}^1_{2n+1}$ set was a $\underset{\sim}{\lambda}_{2n+1}$ union of $\underset{\sim}{\Sigma}^1_{2n}$ sets, and hence by Theorem 2.17 would be $\underset{\sim}{\Delta}^1_{2n+1}$, a contradiction. ⊣

The previous theorem can be generalized to obtain a classification theorem for the Suslin cardinals from AD. We refer the reader to [Jac07A] for a precise statement and proof.

Assuming AD, the values of the projective ordinals $\underset{\sim}{\delta}^1_n$ have been determined. The answer is stated in the next theorem (the values of $\underset{\sim}{\delta}^1_1$ through $\underset{\sim}{\delta}^1_4$ were known earlier; see [Kec78A] for details and history). The proof uses a theorem of Martin (see Theorem 4.23 below) and a certain algebra of "descriptions." We refer the reader to [Jac07A] for an introduction to this theory and [Jac88], [Jac99] for more detailed proofs.

THEOREM 2.22 (Jackson). Assume AD. Then $\underset{\sim}{\delta}^1_1 = \omega_1$, $\underset{\sim}{\delta}^1_2 = \omega_2$, and for $n \geq 1$, $\underset{\sim}{\delta}^1_{2n+1} = \omega_{w(2n-1)+1}$, $\underset{\sim}{\delta}^1_{2n+2} = (\underset{\sim}{\delta}^1_{2n+1})^+$ where $w(1) = \omega$ and $w(m+1) = \omega^{w(m)}$ (ordinal exponentiation).

§3. **Partition Properties.** Partition properties play an important role in determinacy theory. Establishing the strong partition relation on the odd projective ordinals $\underset{\sim}{\delta}^1_{2n+1}$ is a critical step in the inductive analysis of the projective sets (and similarly for a ways beyond). The fact that from AD there are arbitrarily large below Θ cardinals with the strong partition property has useful consequences, for example, it gives a proof that every Suslin, co-Suslin ordinal game is determined. Kechris and Woodin [KW83] have shown, working over the base theory ZF + DC + $\mathbf{V}$=$\mathbf{L}(\mathbb{R})$, the equivalence of AD with the existence of arbitrarily large below Θ cardinals with the strong partition property.

We first recall the basic definitions. We let S^α (for $S \subseteq$ Ord) denote the set of increasing functions from α to S. Recall the Erdős-Rado partition notation: we write $\kappa \to (\lambda)^\rho_\mu$ to mean that for every partition $F : \kappa^\rho \to \mu$ of the increasing functions from ρ to κ into μ pieces, there is a set $A \subseteq \kappa$ of size λ which is **homogeneous** for the partition, that is, there is an $\alpha < \mu$ such that for all $f \in A^\rho$, $F(f) = \alpha$. We call ρ the **exponent** on the partition relation. If we omit the subscript, we mean $\mu = 2$. Assuming choice, no cardinal can satisfy a partition relation with infinite exponent (see [Kan94, Proposition 7.1]), however assuming AD very strong partition relations are possible. Since even $\kappa \to (\kappa)^2$ implies (by an easy argument) that κ is regular, we henceforth assume that κ denotes a regular cardinal.

DEFINITION 3.1. We say κ has the **weak partition property** if for all $\lambda < \kappa$ we have $\kappa \to (\kappa)^\lambda$. We say κ has the **strong partition property** if $\kappa \to (\kappa)^\kappa$.

For proving partition relations from AD and also in deriving consequences from them, it is more more convenient to adopt an alternative form of the partition relations. The alternate form allows us to get homogeneous sets which are c.u.b. in κ, but we must fix the "type" of the function $f : \lambda \to \kappa$. For (everywhere) discontinuous functions (i.e., for all limit α, $f(\alpha) > \sup_{\beta<\alpha} f(\beta)$) this amounts to specifying the **uniform cofinality** of f, which we make precise in the following definition.

DEFINITION 3.2. Let $g : \lambda \to \kappa$. We say $f : \lambda \to \kappa$ has **uniform cofinality** g if there is a function f' with domain $\{(\alpha, \beta) : \beta < g(\alpha)\}$ which is strictly increasing in the second argument such that for all $\alpha < \kappa$ we have $f(\alpha) = \sup_{\beta<g(\alpha)} f'(\alpha, \beta)$.

If g is a constant function with constant value ρ, then we abbreviate the above by saying f has uniform cofinality ρ. If μ is a measure on λ, we say f has uniform cofinality g μ-almost everywhere if there is an f' with domain $\{(\alpha, \beta) : \alpha \in S \wedge \beta < g(\alpha)\}$ for some S of μ measure one, and for all $\alpha \in S$, $f(\alpha) = \sup_{\beta<g(\alpha)} f'(\alpha, \beta)$. Assuming choice, the possible uniform cofinalities simply correspond to the functions g from λ to the regular cardinals below κ. Assuming AD, however, there are many more possible uniform cofinalities, and their consideration is important for many arguments.

Perhaps the simplest uniform cofinality is uniform cofinality ω. This occurs sufficiently often to warrant a special name. If $f : \lambda \to \kappa$ is increasing, everywhere discontinuous, and of uniform cofinality ω, then we say f is of the **correct type**.

Lemma 3.3. For any regular cardinal κ, the following are equivalent.

1. κ has the weak (respectively, strong) partition property.
2. For any $\lambda < \kappa$ (respectively, for $\lambda = \kappa$) and any partition F of the functions $f : \lambda \to \kappa$ of the correct type, there is a c.u.b. $C \subseteq \kappa$ which is homogeneous for F.
3. For any $\lambda < \kappa$ (respectively, for $\lambda = \kappa$), any uniform cofinality $g : \lambda \to \kappa$, and any partition F of the (increasing, everywhere discontinuous) functions $f : \lambda \to \kappa$ of uniform cofinality g, there is a c.u.b. $C \subseteq \kappa$ which is homogeneous for F.

Proof. We first show (1) implies (3). Fix a partition F of the functions $f : \lambda \to \kappa$ of uniform cofinality g. Consider the partition F' of functions $f' : \Omega \to \kappa$, where $\Omega = \{(\alpha, \beta) : \beta < g(\alpha)\}$, which are increasing with respect to the lexicographic ordering on Ω. We partition them by $F'(f') = i$ iff $F(f) = i$ where $f(\alpha) = \sup_{\beta<g(\alpha)} f'(\alpha, \beta)$. Let $A \subseteq \kappa$ be homogeneous of size κ. Let C_1 be the c.u.b. set of limit points of A, and let $C \subseteq C_1$ be c.u.b. and consist of closure points of C (i.e., $\alpha \in C$ is the αth element of C_1) and also closed under g (i.e., if $\gamma < \alpha$ then $g(\gamma) < \alpha$). If $f : \lambda \to C$ is increasing, discontinuous, and of uniform cofinality g, then there is an $f' : \Omega \to C_1$ which is increasing and induces f. It follows that C is homogeneous for F.

We next show (2) implies (1). Let F be a partition of the functions $f \in \kappa^\lambda$. Let F' denote F restricted to functions f of the correct type. Let $C \subseteq \kappa$ be c.u.b. and homogeneous for F'. Let $A \subseteq C$ be the range of ℓ, where $\ell(\alpha)$ is the ωth element of C greater than $\sup_{\beta<\alpha} \ell(\beta)$. Any increasing $f : \lambda \to A$ is necessarily of the correct type. Thus, A is homogeneous for F. ⊣

The basic tool for establishing partition properties from AD is the following theorem of Martin. In fact, all known proofs of partition properties from AD use this result. We state the result in a general form.

DEFINITION 3.4. A (λ, κ) coding map is a function $\varphi\colon {}^{\omega}\omega \to \wp(\lambda \times \kappa)$ such that for every function $f : \lambda \to \kappa$, there is an $x \in {}^{\omega}\omega$ with $\varphi(x) = f$.

The intention is to code functions $f : \lambda \to \kappa$ by reals, but in general $\varphi(x)$ is not required to be a function or to have domain λ.

If φ is a (λ, κ) coding map, we define for each $\alpha < \lambda$ and $\beta < \kappa$ the sets $R_{\alpha,\beta}$ and R_α. We define

$$x \in R_{\alpha,\beta} \Leftrightarrow \varphi(x)(\alpha, \beta) \wedge \forall \beta' < \kappa\, (\varphi(x)(\alpha, \beta') \Rightarrow \beta' = \beta).$$

Also,

$$x \in R_\alpha \Leftrightarrow \exists \beta < \kappa\, (x \in R_{\alpha,\beta}).$$

DEFINITION 3.5. Let κ be a regular cardinal, $\lambda \leq \kappa$. We say κ is λ-reasonable if there is a non-selfdual pointclass $\boldsymbol{\Gamma}$ closed under $\exists^{\mathbb{R}}$, and a (λ, κ) coding map φ satisfying (where $\boldsymbol{\Delta} = \boldsymbol{\Gamma} \cap \check{\boldsymbol{\Gamma}}$):

1. $\forall \alpha < \lambda\, \forall \beta < \kappa\; R_{\alpha,\beta} \in \boldsymbol{\Delta}$.
2. Suppose $\alpha < \lambda$, $A \in \exists^{\mathbb{R}}\boldsymbol{\Delta}$, and $A \subseteq R_\alpha$. Then $\exists \beta_0 < \kappa\, \forall x \in A\, \exists \beta < \beta_0\; R_{\alpha,\beta}(x)$.

We say κ is **reasonable** if it is κ-reasonable. If there is a unique $\beta < \kappa$ such that $\varphi(x)(\alpha, \beta)$, then we write $\varphi(x)(\alpha)$ for this β.

THEOREM 3.6 (Martin). Assume AD. If κ is $\omega\cdot\lambda$-reasonable, then $\kappa \to (\kappa)^\lambda$.

PROOF. We will show below that $\boldsymbol{\Delta}$ is closed under $< \kappa$ unions and intersections; we assume this for now. In most applications this will also follow from the argument showing reasonableness. We refer below to the sets R_α, $R_{\alpha,\beta}$ of Definition 3.5.

Fix a partition $\mathcal{P}\colon \kappa^\lambda \to \{0, 1\}$. Play the integer game where player I plays out $x \in {}^{\omega}\omega$, player II plays out $y \in {}^{\omega}\omega$. If there is a least ordinal $\alpha < \omega \cdot \lambda$ such that $x \notin R_\alpha$ or $y \notin R_\alpha$, then player II wins provided $x \notin R_\alpha$. Otherwise, let $f_x, f_y\colon \omega\cdot\lambda \to \kappa$ be the functions they determine (*e.g.*, $f_x(\alpha) = \varphi(x)(\alpha)$). Define in this case $f_{x,y}\colon \lambda \to \kappa$ by

$$f_{x,y}(\alpha) = \sup_{\alpha' < \omega\cdot(\alpha+1)} \max(f_x(\alpha'), f_y(\alpha')).$$

Player II then wins iff $\mathcal{P}(f_{x,y}) = 1$.

Assume without loss of generality that player II has a winning strategy τ. For $\alpha < \omega \cdot \lambda$ and $\beta < \kappa$, define $x \in S_{\alpha,\beta} \Leftrightarrow \forall \alpha' \leq \alpha\, \exists \beta' \leq \beta\, (x \in R_{\alpha',\beta'})$. Thus, $S_{\alpha,\beta} \in \boldsymbol{\Delta}$ by (1) of Definition 3.5 and the closure of $\boldsymbol{\Delta}$ under $< \kappa$ unions and intersections. Hence, for all $\alpha < \omega \cdot \lambda$ and $\beta < \kappa$, $\tau[S_{\alpha,\beta}] \in \exists^{\mathbb{R}}\boldsymbol{\Delta}$. Now, $\tau[S_{\alpha,\beta}] \subseteq R_\alpha$ as τ is winning for player II. Thus, $\theta(\alpha, \beta) := \sup\{\varphi(x)(\alpha) : x \in \tau[S_{\alpha,\beta}]\} < \kappa$, from (2) of Definition 3.5. Let $C \subseteq \kappa$ be the set of points closed under θ, and $C' \subseteq C$ the set of limit points of C.

Suppose $F\colon \lambda \to C'$ is of the correct type; we show that $\mathcal{P}(F) = 1$. Let x be such that $\varphi(x)$ determines a function $f_x\colon \omega \cdot \lambda \to C$ such that

$F(\beta) = \sup_{\beta' < \omega \cdot (\beta+1)} f_x(\beta')$. We may assume $f_x(\beta) \geq \beta$ for all β. Let $y = \tau(x)$. Since for all $\beta < \omega \cdot \lambda$ we have $x \in S_{\beta, f_x(\beta)}$, it follows from the the definition of C that $\varphi(y)$ determines a function $f_y : \omega \cdot \lambda \to \kappa$ such that $f_y(\beta) \leq N_C(f_x(\beta)) \leq f_x(\beta+1)$ for all β, where $N_C(\gamma)$ denotes the least element of C greater than γ. Thus, $F = f_{x,y}$, so $\mathcal{P}(F) = 1$.

We sketch the argument that $\underset{\sim}{\Delta}$ is closed under $< \kappa$ unions. Suppose not, and let $\delta < \kappa$ be least such that some union $A = \bigcup_{\alpha<\delta} A_\alpha$ is not in $\underset{\sim}{\Delta}$. Note that $R_0 = \bigcup_{\gamma<\kappa} R_{0,\gamma}$ is a κ union of $\underset{\sim}{\Delta}$ sets, and $R_0 \notin \exists^{\mathbb{R}} \underset{\sim}{\Delta}$. Suppose first $\text{PWO}(\underset{\sim}{\Gamma})$. Then $\underset{\sim}{\Gamma}$ is closed under well-ordered unions by Theorem 2.16. Thus $A \in \underset{\sim}{\Gamma}$, and in fact by Wadge's Lemma $\underset{\sim}{\Gamma} = \bigcup_\delta \underset{\sim}{\Delta}$ (the pointclass of δ-unions of sets in $\underset{\sim}{\Delta}$). Also, $R_0 \in \underset{\sim}{\Gamma}$ and so $R_0 = \bigcup_{\alpha<\delta} S_\alpha$ for some $S_\alpha \in \underset{\sim}{\Delta}$. Since κ is regular, one of the $S_\alpha \subseteq R_0$ must be "unbounded" in κ, a contradiction to $\omega \cdot \lambda$-reasonableness. So assume $\text{PWO}(\check{\underset{\sim}{\Gamma}})$, and thus from periodicity $\text{PWO}(\underset{\sim}{\Gamma}_1)$, where $\underset{\sim}{\Gamma}_1 = \exists^{\mathbb{R}} \check{\underset{\sim}{\Gamma}}$. From Theorem 2.16, $\underset{\sim}{\Gamma}_1$ is closed under well-ordered unions, and so $R_0 \in \underset{\sim}{\Gamma}_1$. We cannot have $\bigcup_\delta \underset{\sim}{\Delta} = \underset{\sim}{\Gamma}$, as otherwise Martin's argument (Theorem 2.17) shows $\text{PWO}(\underset{\sim}{\Gamma})$. It follows that $\bigcup_\delta \underset{\sim}{\Delta} \supseteq \check{\underset{\sim}{\Gamma}}$, and so $\bigcup_\delta \exists^{\mathbb{R}} \underset{\sim}{\Delta} \supseteq \underset{\sim}{\Gamma}_1$ (and hence actually $\underset{\sim}{\Gamma}_1 = \bigcup_\delta \exists^{\mathbb{R}} \underset{\sim}{\Delta}$). Thus, $R_0 = \bigcup_{\alpha<\delta} S_\alpha$, with each $S_\alpha \in \exists^{\mathbb{R}} \underset{\sim}{\Delta}$. As before, this contradicts reasonableness. ⊣

Perhaps the most fundamental partition result from AD is the strong partition relation on ω_1 (recall $\underset{\sim}{\delta}^1_1 = \omega_1$), a result due to Martin which we record next.

Theorem 3.7 (Martin). Assume AD. Then $\omega_1 \to (\omega_1)^{\omega_1}$.

First note that the weak partition relation on ω_1 follows fairly immediately from Theorem 3.6. For suppose $\lambda < \omega_1$. Taking a bijection $\pi : \lambda \to \omega$, we define the (λ, ω_1) coding map φ by $\varphi(x)(\alpha, \beta)$ iff $(x)_{\pi(\alpha)} \in \text{WO}$ and $|(x)_{\pi(\alpha)}| = \beta$. It is straightforward to check that this coding is reasonable, and the weak partition relation on ω_1 follows. An immediate consequence (considering partitions of functions $f \in (\omega_1)^\lambda$, with $\lambda = 1$) is that the c.u.b. filter on ω_1 is a countably additive ultrafilter (i.e., a **measure**) on ω_1. Another easy partition argument shows that this measure is normal (note one cannot directly appeal to Fodor's Lemma as this requires choice). The n-fold product of this normal measure is the measure induced from the weak partition property of ω_1 and functions $f : n \to \omega_1$. We let W^1_1 denote the normal measure on ω_1, and W^n_1 the n-fold product.

To get the strong partition relation on ω_1 requires a coding of the functions from ω_1 to ω_1. A first attempt might be to use the coding of functions (viewed as subsets of $(\omega_1)^2$) given by the Coding Lemma, Theorem 1.8. This, however, is not sufficient to satisfy the conditions of Definition 3.5. Martin's original proof (see [Kec78A]) used the fact that from AD every subset of ω_1 lies in $\mathbf{L}[x]$ for some real x, and a c.u.b. set of (Silver) indiscernibles exists for each $\mathbf{L}[x]$. Since Martin's proof, several other proofs of the strong partition relation on ω_1 have been found. Kechris [Kec77] gives a proof using generic codes for countable ordinals. Using their theory of generic codes for uncountable ordinals, Kechris and Woodin [KW07] have shown that $\underset{\sim}{\boldsymbol{\delta}}^1_{2n+1} \to (\underset{\sim}{\boldsymbol{\delta}}^1_{2n+1})^{\underset{\sim}{\boldsymbol{\delta}}^1_{2n-1}}$ for all $n \geq 1$. In [Jac90] a more combinatorial proof is given (which builds on some of the ideas in Kunen's proof of the weak partition relation on $\underset{\sim}{\boldsymbol{\delta}}^1_3$). This proof, which involves analyzing the measures on ω_1, is currently the only one known to generalize to the higher odd projective ordinals. Finally, in [JM04] a variation of the latter proof is given which avoids the complete analysis of measures. Whether this argument can be generalized to higher cardinals is not known.

The inductive analysis of the projective hierarchy does allow the analysis of measures on ω_1 to be generalized to the higher odd projective ordinals, and thus to get a coding a coding map for the functions $f : \underset{\sim}{\boldsymbol{\delta}}^1_{2n+1} \to \underset{\sim}{\boldsymbol{\delta}}^1_{2n+1}$ which is reasonable (using $\underset{\sim}{\boldsymbol{\Gamma}} = \underset{\sim}{\boldsymbol{\Sigma}}^1_{2n+1}$). Thus we have (cf. [Jac88], [Jac99]):

THEOREM 3.8 (Jackson). Assume AD. For all n, $\underset{\sim}{\boldsymbol{\delta}}^1_{2n+1} \to (\underset{\sim}{\boldsymbol{\delta}}^1_{2n+1})^{\underset{\sim}{\boldsymbol{\delta}}^1_{2n+1}}$.

This analysis extends a ways beyond the projective hierarchy. For example if $\underset{\sim}{\boldsymbol{\Sigma}}^1_\alpha$, $\alpha < \omega_1$ enumerates the first ω_1 non-self-dual pointclasses closed under $\exists^{\mathbb{R}}$, and if $\underset{\sim}{\boldsymbol{\delta}}^1_\alpha$ denotes the supremum of the lengths of the $\underset{\sim}{\boldsymbol{\Delta}}^1_\alpha$ prewellorderings, then for all odd α (where limit ordinals are considered even as usual), $\underset{\sim}{\boldsymbol{\delta}}^1_\alpha$ has the strong partition property. The analysis can be propagated further (and the analogs of the odd $\underset{\sim}{\boldsymbol{\delta}}^1_\alpha$ continue to have the strong partition property), but somewhere around the first inaccessible cardinal the induction breaks down. In [Jac91] results are shown which imply that an inductive "from below" analysis of the measures cannot succeed through $\kappa^{\mathbb{R}}$, the ordinal of the inductive sets (the Wadge rank $o(\underset{\sim}{\boldsymbol{\Gamma}})$ of the smallest non-selfdual pointclass $\underset{\sim}{\boldsymbol{\Gamma}}$ closed under real quantification). Thus, the following remains open.

CONJECTURE 3.9 (AD + DC). Let $\underset{\sim}{\boldsymbol{\Gamma}}$ be non-self-dual, closed under $\forall^{\mathbb{R}}$, countable unions and intersections, and PWO($\underset{\sim}{\boldsymbol{\Gamma}}$). Let $\boldsymbol{\delta}$ be the supremum of the lengths of the $\underset{\sim}{\boldsymbol{\Delta}}$ prewellorderings. Then $\boldsymbol{\delta} \to (\boldsymbol{\delta})^{\boldsymbol{\delta}}$.

In fact, this conjecture is open in general even assuming $\underset{\sim}{\boldsymbol{\Gamma}}$ has the scale property, in which case $\boldsymbol{\delta}$ is a regular Suslin cardinal.

By a result of Martin and Paris (see [Kec78A]), all of the even projective ordinals, $\underset{\sim}{\boldsymbol{\delta}}^1_{2n}$ have the weak partition property, but not the strong. There are other regular cardinals below the projective ordinals.

From the inductive analysis, there are $2^{n+1} - 1$ regular cardinals strictly between $\underset{\sim}{\boldsymbol{\delta}}^1_{2n+1}$ and $\underset{\sim}{\boldsymbol{\delta}}^1_{2n+3}$. The smallest of these is $\underset{\sim}{\boldsymbol{\delta}}^1_{2n+2} = (\underset{\sim}{\boldsymbol{\delta}}^1_{2n+1})^+$. The regular cardinals strictly between $\underset{\sim}{\delta}^1_{2n+1}$ and $\underset{\sim}{\delta}^1_{2n+3}$ are precisely the ultrapowers of $\underset{\sim}{\delta}^1_{2n+1}$ by the normal measures on $\underset{\sim}{\delta}^1_{2n+1}$, which in turn correspond to the regular cardinals below $\underset{\sim}{\delta}^1_{2n+1}$. For example, between $\underset{\sim}{\delta}^1_3$ and $\underset{\sim}{\delta}^1_5$ they are $\underset{\sim}{\delta}^1_4 = \aleph_{\omega+2}$, $\aleph_{\omega\cdot 2+1}$, and $\aleph_{\omega^\omega+1}$. All of these regular cardinals satisfy the same infinite exponent partition relations, that is, they satisfy $\kappa \to (\kappa)^{\underset{\sim}{\delta}^1_{2n+1}}$ but $\kappa \not\to (\kappa)^{\underset{\sim}{\delta}^1_{2n+2}}$ (see [Jac]).

Although the detailed inductive analysis mentioned above breaks down fairly early in the $\mathbf{L}(\mathbb{R})$ hierarchy, it is still possible to obtain partition results higher up. The Kechris-Woodin generic coding arguments mentioned earlier [KW07] show that if $\underset{\sim}{\Gamma}$ is non-self-dual, closed under $\forall^{\mathbb{R}}$, countable unions and intersections, and PWO($\underset{\sim}{\Gamma}$), and $\boldsymbol{\delta}$ is the supremum of the lengths of the $\underset{\sim}{\Delta}$ prewellorderings, and if $\lambda < \boldsymbol{\delta}$ is a Suslin cardinal with $\forall^{\mathbb{R}} S_\lambda \subseteq \underset{\sim}{\Delta}$, then $\boldsymbol{\delta} \to (\boldsymbol{\delta})^\lambda$. Also, if $\boldsymbol{\delta}$ corresponds to a "sufficiently closed" pointclass, then $\boldsymbol{\delta}$ will have the strong partition property. Results of this form are proved in [KKMW81]. We mention a result along these lines.

THEOREM 3.10 (AD). Let $\underset{\sim}{\Gamma}$ be non-self-dual, closed under $\forall^{\mathbb{R}}$, PWO($\underset{\sim}{\Gamma}$), $\exists^{\mathbb{R}}\underset{\sim}{\Delta} \subseteq \underset{\sim}{\Delta}$, and assume also $\underset{\sim}{\Gamma}$ is closed under $\vee$. Then $\delta(\underset{\sim}{\Delta})$ has the strong partition property.

The proof of Theorem 3.10 uses the "Uniform Coding Lemma," a version of the Coding Lemma which says roughly that given a prewellordering $\preceq$ and a relation $R \subseteq \mathrm{dom}(\preceq) \times {}^\omega\omega$, there is a choice set $A \subseteq R$ (in the sense of Theorem 1.8) such that the initial segments A_α of A are uniformly $\underset{\sim}{\Sigma}^1_1$ over the initial segments $\preceq_\alpha$ of $\preceq$. We refer the reader to [KKMW81] for a precise statement and proof.

We make some observations which put Theorem 3.10 in context. Following Steel [Ste81A], let

$$C = \{o(\underset{\sim}{\Delta}) \,:\, \underset{\sim}{\Delta} \text{ is self-dual} \wedge \exists^{\mathbb{R}}\underset{\sim}{\Delta} \subseteq \underset{\sim}{\Delta}\}.$$

So, C is c.u.b. in Θ and consists of places in the Wadge hierarchy where we are at the base of a projective-like hierarchy. By an earlier remark, it doesn't matter in the definition of C if we add that $\underset{\sim}{\Delta}$ is closed under finite unions and intersections. Also, an argument using the Coding Lemma shows that every $\kappa \in C$ is a cardinal.

Steel shows in [Ste81A] that if $\kappa \in C$ and $\mathrm{cf}(\kappa) > \omega$, then there is a non-selfdual pointclass $\underset{\sim}{\Gamma}$ closed under $\forall^{\mathbb{R}}$ with PWO($\underset{\sim}{\Gamma}$) such that $o(\underset{\sim}{\Delta}) = \kappa$ (where $\underset{\sim}{\Delta} = \underset{\sim}{\Gamma} \cap \check{\underset{\sim}{\Gamma}}$). This Steel pointclass satisfies all of the hypotheses of Theorem 3.10 except perhaps the assumption that $\underset{\sim}{\Gamma}$ is closed under $\vee$. [Ste81A] also shows that a Steel pointclass $\underset{\sim}{\Gamma}$ is closed under disjunction with $\underset{\sim}{\Delta}$ sets iff $\underset{\sim}{\Gamma}$ is closed under disjunction iff $\underset{\sim}{\Gamma}$ is closed under countable unions. We may thus rephrase Theorem 3.10 as follows.

Corollary 3.11. If $\kappa \in C$ is regular and the corresponding Steel pointclass $\boldsymbol{\Gamma}$ (i.e., $\kappa = o(\boldsymbol{\Delta})$ where $\boldsymbol{\Delta} = \boldsymbol{\Gamma} \cap \check{\boldsymbol{\Gamma}}$) is closed under $\vee$, then κ has the strong partition property.

From [Ste81A, Theorem 2.1] it follows that if κ is regular and a limit of Suslin cardinals, then the corresponding Steel pointclass $\boldsymbol{\Gamma}$ is closed under $\vee$. Also, $\kappa \in C$ and $\kappa = \delta(\boldsymbol{\Delta})$. Thus we have the following.

Corollary 3.12. If κ is a regular limit of Suslin cardinals, then κ has the strong partition property.

Steel conjectures in [Ste81A] that if $\kappa \in C$ is regular then the corresponding Steel pointclass $\boldsymbol{\Gamma}$ is closed under disjunctions. We state this explicitly as:

Conjecture 3.13. For $\kappa \in C$, κ is regular iff κ has the strong partition property.

It is conceivable that this conjecture could hold even if some of the Steel pointclasses for regular κ are not closed under disjunction. Although Steel's conjecture is still open, it does hold for a fairly large initial segment of κ's. Kechris [Kec81B] showed that for $A \subseteq {}^{\omega}\omega$ contained within the inductive sets, A is λ-Suslin, where λ is the length of some projective over A prewellordering. Thus for $\kappa \leq \kappa^{\mathbb{R}}$ (the Wadge rank of the inductive sets) in C, with Steel pointclass $\boldsymbol{\Gamma}$, every $\boldsymbol{\Delta}$ set is λ Suslin for some $\lambda < \kappa$. From [Ste81A] again it follows that $\boldsymbol{\Gamma}$ is closed under $\vee$. Thus, the above conjecture is true for $\kappa \leq \kappa^{\mathbb{R}}$. In fact, from Steel's analysis of scales in $\mathbf{L}(\mathbb{R})$ [Ste83A] Kechris' result can be extended to all κ up to the least non-trivial Σ_1 gap, in the terminology of [Ste83A] (the least such gap is strictly larger than $\kappa^{\mathbb{R}}$). Thus we have:

Corollary 3.14. If $\kappa \in C$ and κ is less than or equal to the least α beginning a non-trivial Σ_1 gap, then κ is regular iff κ has the strong partition property.

Finally, as discussed in [KKMW81], the pointclasses $\boldsymbol{\Gamma}$ satisfying the hypotheses of Theorem 3.10 are in abundance below Θ.

We mention one more application of the strong partition property. If κ has the strong partition property we can, following Steel, introduce a generalized notion of Mahloness on κ. We consider stationary sets $S \subset \kappa$, and we assume for convenience that S contains only ordinals of uncountable cofinality. Recall that S is said to be **thin** if for every $\alpha \in S$, $S \cap \alpha$ is not stationary in α. For every stationary S, its set of thin points S' is also stationary. We define an equivalence relation $\sim$ on the thin stationary sets by $S \sim T$ iff there is a c.u.b. $C \subseteq \kappa$ such that $C \cap S = C \cap T$. We define $S \prec T$ iff there is a c.u.b. C such that for all $\alpha \in C \cap T$, S is stationary in α.

Fact 3.15 (Steel). If $\kappa \to (\kappa)^{\kappa}$, then $\prec$ is a well-order of the $\sim$-equivalence classes of the thin stationary subsets of κ.

Proof (sketch). Given thin stationary sets S, T, consider the partition $\mathcal{P}$ of functions $f : \kappa \to \kappa$ of the correct type according to whether $\alpha_f(S) < \alpha_f(T)$, $\alpha_f(S) = \alpha_f(T)$, or $\alpha_f(S) > \alpha_f(T)$, where $\alpha_f(S)$ is the least limit point of f in S. It is easy to check that a homogeneous C would give respectively $S \prec T$, $S \sim T$, or $T \prec S$. ⊣

It is shown in [Jac91] that for $\underset{\sim}{\Gamma}$ non-selfdual and closed under real quantification, $\kappa = o(\underset{\sim}{\Gamma})$ has a generalized Mahlo order greatly exceeding κ.

§4. Homogeneous Trees. The notion of a homogeneous tree arose independently in the work of Kunen and Martin. Kechris and Martin independently (cf. [Kec81A]) then formulated the general notion of a homogeneous tree. This concept, and the related notion of a weakly homogeneous tree, plays an important role in some of the proofs of determinacy from large cardinals as well as in various theorems in the AD world. Thus, the notion is important in both the ZFC and ZF + AD contexts.

If $f : X \to Y$ is a map from the set X to the set Y, and μ is a measure (or filter, ultrafilter, etc.) on X, then we define $f(\mu)$ to be the measure on Y given by $f(\mu)(A) = \mu(f^{-1}[A])$. For X a set and $m \le n$, let $\pi^n_m : X^n \to X^m$ denote the projection map $\pi^n_m(s) = s{\restriction}m$.

Definition 4.1. We say a tree T on $\omega \times X$ is **homogeneous** if there is a family of measures $\langle \mu_s : s \in {}^{<\omega}\omega \rangle$ satisfying:

1. Each μ_s is a measure on T_s, that is, $\mu_s(T_s) = 1$.
2. If $s \le t$ (that is, t extends s), then $\pi^n_m(\mu_t) = \mu_s$ where $n = \mathrm{lh}(t)$, $m = \mathrm{lh}(s)$.
3. For every $x \in {}^\omega\omega$, if T_x is illfounded then for any sequence $A_1, A_2, \ldots$ with $\mu_{x\restriction n}(A_n) = 1$, there is an $f \in {}^\omega X$ such that for all n, $(x{\restriction}n, f{\restriction}n) \in A_n$.

We say T is δ-homogeneous if the measures μ_s may be taken to be δ-complete. We say T is $<\delta$-homogeneous if T is λ homogeneous for all $\lambda < \delta$. We say $A \subseteq {}^\omega\omega$ is homogeneously Suslin (or δ-homogeneously Suslin, etc.) if $A = \mathrm{p}[T]$ for some homogeneous T.

The last property (3) says that T appears "homogeneous," at least as far as restricting to measure one sets goes. This property is also equivalent to saying that the direct limit M_x of the ultrapowers by the measures $\mu_{x\restriction n}$ is wellfounded (using the natural embeddings from the μ_s ultrapower to the μ_t ultrapower when $s \le t$). To see this, first suppose that (3) holds for the $\mu_{x\restriction n}$. If M_x were illfounded, then there would be functions $g_n : T_{x\restriction n} \to \mathrm{Ord}$ and measure one sets A_n with respect to μ_n such that $\forall \vec{\alpha} \in A_{n+1}\ g_{n+1}(\vec{\alpha}) < g_n(\vec{\alpha}{\restriction}n)$. Let $f \in {}^\omega X$ be as in (3) for these A_n. Then $g_1(f{\restriction}1) > g_2(f{\restriction}2) > \ldots$, a contradiction. Suppose next that the direct limit M_x is wellfounded. If (3) failed, then then let $A_1, A_2, \ldots$ be measure one sets with respect to the

$\mu_{x\restriction n}$ with no $f \in {}^{\omega}X$ as in (3). Let V be the tree of all $\vec{\alpha} \in {}^{\omega}X$ such that $\forall n \leq \text{lh}(\vec{\alpha})\ \vec{\alpha}\restriction n \in A_n$. So, V is wellfounded. Note that the set of $\vec{\alpha} \in X^n$ such that $\vec{\alpha} \in V$ has $\mu_{x\restriction n}$ measure one by (2). For such $\vec{\alpha}$ define $g_n(\vec{\alpha}) = |\alpha|_T$, the rank of $\vec{\alpha}$ in T. We then have $[g_1]_{\mu_{x\restriction 1}} > [g_2]_{\mu_{x\restriction 2}} > \dots$, a contradiction to the wellfoundedness of M_x.

It is easy to check that if A is homogeneously Suslin and $B \leq_{\mathrm{W}} A$, then B is also homogeneously Suslin [if the function reducing B to A is Lipschitz, this is clear, if it is just continuous, one must pad using dummy principal measures on the integers; details are left to the reader].

One use of homogeneous trees is that they provide a means of giving determinacy. The basic method used is "integrating out" one of the players ordinal moves. This method goes back to Martin's proof of $\underset{\sim}{\Pi}^1_1$ determinacy from a measurable cardinal.

LEMMA 4.2 (Martin). *Suppose $A = \mathrm{p}[T]$ where T is a homogeneous tree on $\omega \times \lambda$ for some $\lambda \in \mathrm{Ord}$. Then A is determined.*

PROOF. Consider the auxiliary game G^* played as follows:

I	$(x(0), \alpha_0)$		$(x(2), \alpha_1, \alpha_2)$		$(x(4), \alpha_3, \alpha_4)$ $\dots$	
II		$x(1)$		$x(3)$		$x(5)$ $\dots$

Here $x(i) \in \omega$ and $\alpha_i \in \mathrm{Ord}$. Player I must play so that for all i,

$$((x(0), \dots, x(2i)), (\alpha_0, \dots, \alpha_{2i})) \in T$$

otherwise player II wins. If player I meets this requirement, then player I wins. G^* is a closed game for I, so is determined. If player I wins G^*, then clearly player I wins the game G for the set A by simply ignoring the extra ordinal moves that the strategy gives. So suppose τ^* is a winning strategy for player II in G^*. We define a strategy τ for player II in G as follows. Given a position $p = (x(0), \dots, x(2i))$ with player II to move, let $\tau(p)$ be the integer a such that for $\mu_{x\restriction(2i+1)}$ almost all $(\alpha_0, \dots, \alpha_{2i})$, τ^* will respond with a when player I plays $(x(0), x(2), \dots, x(2i))$ and $(\alpha_0, \dots, \alpha_{2i})$. This is well-defined by the countable completeness of the measure $\mu_{x\restriction(2i+1)}$.

Suppose $x \in {}^{\omega}\omega$ is a run according to τ. We must show that $x \notin A$. Suppose $x \in A = \mathrm{p}[T]$. For each i, let A_{2i+1} be a $\mu_{x\restriction(2i+1)}$ measure one set such that if player I plays $(x(0), \dots, x(2i))$ and $(\alpha_0, \dots, \alpha_{2i})$ in G^*, then τ^* responds with $x(2i+1)$. By homogeneity, let $f \in {}^{\omega}\mathrm{Ord}$ be such that $f\restriction(2i+1) \in A_{2i+1}$ for all i. Then x together with f gives an infinite run following τ^*, a contradiction to τ^* being a winning strategy for player II. ⊣

Variations of the above argument are also frequently used. For example, we have the following.

LEMMA 4.3. *Let G be the ordinal game where player I plays integers $x(2i)$ and ordinals $\alpha(i)$, and player II plays integers $x(2i+1)$, and ordinals $\beta(i) < \delta$ producing $x \in {}^{\omega}\omega$, $\vec{\alpha} \in {}^{\omega}\mathrm{Ord}$, and $\vec{\beta} \in {}^{\omega}\delta$:*

I	$(x(0), \alpha_0)$		$(x(2), \alpha_1)$		$(x(4), \alpha_2)$ ...	
II		$(x(1), \beta_0)$		$(x(3), \beta_2)$		$(x(5), \beta_3)$ ...

Assume the payoff set for player I is of the form

$$F(x, \vec{\alpha}, \vec{\beta}) \wedge A(x)$$

where $F \subseteq {}^{\omega}\omega \times {}^{\omega}\text{Ord} \times {}^{\omega}\delta$ is closed and $A = \text{p}[T]$ where T is homogeneous with δ^+-complete measures. Then G is determined.

Proof. Let U be a tree on $\omega \times \text{Ord} \times \delta$ such that $F = [U]$. Let G^* be the auxiliary closed game for player I where player I also makes moves γ_0, (γ_1, γ_2), $(\gamma_3, \gamma_4), \dots$ and player I must play so that $(x{\restriction}i, \vec{\alpha}{\restriction}i, \vec{\beta}{\restriction}i) \in U$ and $(x{\restriction}i, \vec{\gamma}{\restriction}i) \in T$ for all i. Again, if player I wins G^* then player I wins G. If τ^* is a winning strategy for player II in G^*, we again integrate τ^* to define a strategy τ for player II in G. This is well-defined using the δ^+-completeness of the measures $\mu_{x\restriction(2i+1)}$. The proof that τ is winning for player II in G is exactly as before. ⊣

For the purposes of propagating Suslin representations, and for other purposes, a weaker notion, that of a **weakly homogeneous** tree is useful. We will have that the weakly homogeneously Suslin sets are existential quantifications of the homogeneously Suslin sets.

There are several different definitions of a tree being weakly homogeneous. These turn out to be almost equivalent assuming AC (see Lemma 4.7 below), but not necessarily just in ZF. However, assuming AD essentially every tree turns out to be weakly homogeneous by the stronger definition anyway (see Theorem 4.11 below).

The first version we give is the stronger form of the definition. The following definition, which we take as our official one, essentially says that a tree T on $\omega \times X$ is weakly homogeneous if there is a homogeneous tree T' on $(\omega \times \omega) \times X$ which is isomorphic to T via a tree isomorphism which is the identity on the first coordinate. Thus, in weak homogeneity we have the option of splitting the tree at each node into countably many pieces, and using a different measure for each. In the actual definition we allow the option of a node in T' to split only finitely often on the second coordinate.

Definition 4.4. A tree T on $\omega \times X$ is **weakly homogeneous** if there are measures $\mu_{s,t}$ on ${}^{\text{lh}(s)}X$ and non-empty sets $A_{s,t} \subseteq {}^{\text{lh}(s)}X$ defined for all $s \in {}^{<\omega}\omega$ and some $t \in {}^{<\omega}\omega$ with $\text{lh}(s) = \text{lh}(t)$ and satisfying:

1. The set of (s, t) for which $\mu_{s,t}$, $A_{s,t}$ are defined is a tree. Furthermore, $\mu_{s,t}$ is a measure on T_s, and $A_{s,t} \subseteq T_s$.
2. If (s', t') extends (s, t) then $\mu_{s',t'}$ projects to $\mu_{s,t}$ and $A_{s',t'} \subseteq A_{s,t} \times {}^{\text{lh}(s')-\text{lh}(s)}X$.
3. For every (s, t) for which $\mu_{s,t}$, $A_{s,t}$ are defined, and every s' extending s with $\text{lh}(s') = \text{lh}(s) + 1$, the sets $A_{s',t'}$ partition $T_{s'} \cap (A_{s,t} \times X)$.

4. For all $x, y \in {}^{\omega}\omega$ such that for all n $\mu_{x\restriction n, y\restriction n}$ is defined, if there is an $f \in {}^{\omega}X$ such that $\forall n\ f\restriction n \in A_{x\restriction n, y\restriction n}$, then the direct limit $M_{x,y}$ of the ultrapowers by the $\mu_{x\restriction n, y\restriction n}$ is wellfounded.

Note that by (3) it follows that if T_x is illfounded, say $(x, f) \in [T]$, then there is a $y \in {}^{\omega}\omega$ such that for all n $f\restriction n \in A_{x\restriction n, y\restriction n}$. Then by (4) the direct limit of the $\mu_{x\restriction n, y\restriction n}$ ultrapowers is wellfounded. This shows that the above definition implies the following, which is also frequently taken to be the definition. In this version we don't split the tree, just the measures.

DEFINITION 4.5. A tree T on $\omega \times X$ is **weakly homogeneous** if there is a family of measures $\mu_{s,t}$, defined for all $s \in {}^{<\omega}\omega$ and some $t \in {}^{<\omega}\omega$ with $s, t \in {}^{<\omega}\omega$ with $\mathrm{lh}(s) = \mathrm{lh}(t)$ satisfying:

1. The set of (s, t) for which $\mu_{s,t}$ is defined is a tree. Each $\mu_{s,t}$ is a measure on ${}^{\mathrm{lh}(s)}X$ and $\mu_{s,t}(T_s) = 1$.
2. If (s', t') extends (s, t) and $\mu_{s',t'}$ is defined, then $\mu_{s',t'}$ projects to $\mu_{s,t}$.
3. For every $x \in {}^{\omega}\omega$, if T_x is illfounded then there is a $y \in {}^{\omega}\omega$ such that the measures $\mu_{x\restriction n, y\restriction n}$ are all defined and the direct limit $M_{x,y}$ of the ultrapowers by the measures $\mu_{x\restriction n, y\restriction n}$ is wellfounded.

Again, the conclusion of (3) is equivalent to saying that there is a $y \in {}^{\omega}\omega$ such that whenever $B_{x\restriction n, y\restriction n} \subseteq T_{x\restriction n}$ have $\mu_{x\restriction n, y\restriction n}$ measure one, then there is an $f \in {}^{\omega}X$ such that $\forall n\ f\restriction n \in B_{x\restriction n, y\restriction n}$.

We say $A \subseteq {}^{\omega}\omega$ is weakly homogeneously Suslin if $A = \mathrm{p}[T]$ for some weakly homogeneous tree T. The following is essentially immediate from the strong form of the definition of weakly homogeneous tree.

LEMMA 4.6. $A \subseteq {}^{\omega}\omega$ is weakly homogeneously Suslin iff we can write $A(x) \Leftrightarrow \exists y\ B(x, y)$ where B is homogeneously Suslin. In fact, for every ordinal α, A is the projection of a weakly homogeneous tree on $\omega \times \alpha$ iff we can write $A(x) \Leftrightarrow \exists y\ B(x, y)$ where B is the projection of a homogeneous tree on $(\omega \times \omega) \times \alpha$.

The conclusion of Lemma 4.6 is still true using the weaker definition of weakly homogeneous tree, Definition 4.5, provided we assume ZFC according to the following lemma, due to Woodin.

LEMMA 4.7 (ZFC). Let T be a tree on $\omega \times X$ which is weakly homogeneous according to the weaker Definition 4.5. Then there is a subtree $T' \subseteq T$ which is weakly homogeneous according to the stronger Definition 4.4 and such that $\mathrm{p}[T'] = \mathrm{p}[T]$. Furthermore T and T' use the same set of measures in their homogeneity systems.

PROOF. Let T be a tree on $\omega \times X$ satisfying Definition 4.5. Recall that the completeness of any measure is a measurable cardinal, and in ZFC these must be inaccessibles. Thus the measures $\mu_{s,t}$ (as in Definition 4.5) are all $(2^{\omega})^{+}$-complete. We need to define the sets $A_{s,t}$ so that the $\mu_{s,t}$ together with the $A_{s,t}$

satisfy Definition 4.4 for a subtree T'. Fix $s, t \in {}^{<\omega}\omega$ with $\mathrm{lh}(s) = \mathrm{lh}(t)$. For every $x, y \in {}^{\omega}\omega$ extending s, t, if the direct limit $M_{x,y}$ of the ultrapowers by the measures $\mu_{x\restriction n, y\restriction n}$ is illfounded, let $A_n^{x,y} \subseteq T_{x\restriction n}$ be $\mu_{x\restriction n, y\restriction n}$ measure one sets such that for no $f \in {}^{\omega}X$ do we have $\forall n\, f\restriction n \in A_n^{x,y}$ (we use AC here as well). Let

$$A_{s,t} = \bigcap_{x,y} A^{x,y}_{\mathrm{lh}(s)},$$

the intersection taken over all x, y extending s, t with $M_{x,y}$ illfounded. So, $\mu_{s,t}(A_{s,t}) = 1$. We may further shrink the $A_{s,t}$ (keeping them measure one) so that: (1) if t_1, t_2 are of the same length and incompatible then $A_{s,t_1} \cap A_{s,t_2} = \varnothing$ and (2) if (s', t') immediately extends (s, t), then $A_{s',t'} \subseteq A_{s,t} \times X$. [By re-indexing the measures $\mu_{s,t}$ we may assume that whenever $t_1 \perp t_2$ then $\mu_{s,t_1} \neq \mu_{s,t_2}$. For each such pair there is a $B = B_{s,t_1,t_2}$ with $\mu_{s,t_1}(B) = 1$, $\mu_{s,t_2}(B) = 0$. We intersect A_{s,t_1} with all such B_{s,t_1,t_2}. This ensures (1). Then by induction on $\mathrm{lh}(s) = \mathrm{lh}(t)$, intersect $A_{s,t}$ with $A_{\bar{s},\bar{t}} \times \kappa$, where (s, t) is an immediate extension of $(\bar{s}, \bar{t})$. This ensures (2).] Let $T' \subseteq T$ consist of those $(s, \vec{\alpha})$ such that for some $t \in {}^{<\omega}\omega$ with $\mathrm{lh}(s) = \mathrm{lh}(t)$ we have $\vec{\alpha} \in A_{s,t}$.

To see $\mathrm{p}[T'] = \mathrm{p}[T]$, let $x \in \mathrm{p}[T]$. From Definition 4.5, let $y \in {}^{\omega}\omega$ be such that the direct limit $M_{x,y}$ by the $\mu_{x\restriction n, y\restriction n}$ measures is wellfounded. Since the $A_{x\restriction n, y\restriction n}$ have $\mu_{x\restriction n, y\restriction n}$ measure one, it follows that there is an $f \in {}^{\omega}X$ such that $\forall n\, f\restriction n \in A_{x\restriction n, y\restriction n}$. Thus $x \in \mathrm{p}[T']$.

To see that T' is weakly homogeneous according to Definition 4.4, fix $x, y \in {}^{\omega}\omega$ such that for all n $\mu_{x\restriction n, y\restriction n}$ is defined, and suppose $f \in {}^{\omega}X$ with $\forall n\, f\restriction n \in A_{x\restriction n, y\restriction n}$. We must show that the direct limit $M_{x,y}$ of the $\mu_{x\restriction n, y\restriction n}$ ultrapowers is wellfounded. If not, then the measure one sets $A_n^{x,y}$ are defined and $A_{x\restriction n, y\restriction n} \subseteq A_n^{x,y}$ for all n. By definition of the $A_n^{x,y}$ there is no $g \in {}^{\omega}X$ such that $\forall n\, g\restriction n \in A_n^{x,y}$. But this contradicts the assumption on f. ⊣

In fact, in ZFC we can simplify the definition of weakly homogeneous tree a bit further in the following, due also to Woodin.

Lemma 4.8 (ZFC). A tree T on $\omega \times X$ is weakly homogeneous iff there is a countable set $\mathcal{M}$ of measures $\langle \mu_i : i \in \omega \rangle$, with each measure on ${}^{j}X$ for some j, such that whenever T_x is illfounded then there is a sequence $\mu_{n_1}, \mu_{n_2}, \ldots$ from $\mathcal{M}$ with each $\mu_{n_i}(T_{x\restriction i}) = 1$, each $\mu_{n_{i+1}}$ projects to μ_{n_i}, and with the direct limit of the ultrapowers by the μ_{n_i} wellfounded.

Proof. It is clear that Definition 4.5 implies the statement from the lemma. To see the other direction, let $\mathcal{M}$ be the countable family of measures for the tree T as in the statement of the lemma. Define the measures $\mu_{s,t} \in \mathcal{M}$ inductively as follows. If $\mu_{s,t}$ is defined, $\mu_{s,t}(T_s) = 1$, and s' is an immediate extension of s, then we let $\mu_{s',t'}$ enumerate all the measures $\mu \in \mathcal{M}$ which are measures on ${}^{\mathrm{lh}(s')}X$ which project to $\mu_{s,t}$ and such that $\mu(T_{s'}) = 1$. Suppose now that T_x is illfounded. Let $\mu_{n_1}, \mu_{n_2}, \ldots$ be as in the statement of the lemma.

Then there is a $y \in {}^{\omega}\omega$ such that for all i, $\mu_{n_i} = \mu_{x\restriction n, y\restriction n}$. By assumption the direct limit of the ultrapowers by the measures $\mu_{x\restriction n, y\restriction n}$ is defined and wellfounded. ⊣

We next show how weak homogeneity propagates Suslin representations. This is the construction of the Martin-Solovay tree, a useful construction in many contexts. Given a weakly homogeneous tree T on $\omega \times \kappa$, the Martin-Solovay construction produces a tree T' on $\omega \times \lambda$, for some $\lambda \in \mathrm{Ord}$, with $\mathrm{p}[T'] = {}^{\omega}\omega - \mathrm{p}[T]$. It suffices to assume T is weakly homogeneous according to the weaker Definition 4.5.

Definition 4.9 (Martin-Solovay Tree). Let T be a tree on $\omega \times \kappa$ which is weakly homogeneous according to Definition 4.5 with measures $\mu_{s,t}$, each of which we may assume is a measure on ${}^{\mathrm{lh}(s)}\kappa$. We define the **Martin-Solovay** tree T' as follows. Let (s_i, t_i) be an enumeration of all $(s, t) \in ({}^{<\omega}\omega)^2$ with $\mathrm{lh}(s) = \mathrm{lh}(t)$ such that any proper extension of any (s_i, t_i) is enumerated at a later stage. Define $(s, \vec{\alpha}) \in T'$ iff there is an $f : T_s \to \kappa^+$ which is order-preserving with respect to the Kleene-Brouwer ordering $<^s_{\mathrm{BK}}$ on T_s such that for all $i < \mathrm{lh}(s)$, $\alpha_i = [f^i]_{\mu_{s_i,t_i}}$. Here f^i is defined by

$$f^i = f \restriction \{\vec{\gamma} : (s_i, \vec{\gamma}) \in T\}$$

if s_i is an initial segment of s and μ_{s_i,t_i} is defined; otherwise $\alpha_i = 0$.

Variations on this definition are frequently used. For example, we can consider functions $f : T_s \to \delta$ for δ other than κ^+, and we can also impose restrictions on the "type" of the functions f used. We might, for example, require the functions to be of the correct type (Definition 3.2). We let $\mathrm{ms}(T, \delta)$ denote the Martin-Solovay tree constructed from T using function into δ of the correct type. We write $\mathrm{ms}(T, \delta, \vec{\mu})$ if we want to display the homogeneity measures being used.

Theorem 4.10 (ZF + DC). Let T be a weakly homogeneous tree (according to Definition 4.5), and let T' be the corresponding Martin-Solovay tree. Then $\mathrm{p}[T'] = {}^{\omega}\omega - \mathrm{p}[T]$.

Proof. First suppose $x \in {}^{\omega}\omega$ is such that $x \notin \mathrm{p}[T]$, that is, T_x is wellfounded. Thus the Kleene-Brouwer $<^x_{\mathrm{BK}}$ on T_x is a wellordering of length $< \kappa^+$. Let $f : <^x_{\mathrm{BK}} \to \kappa^+$ be order-preserving. For each i such that s_i is an initial segment of x and μ_{s_i,t_i} is defined, let again $f^i = f \restriction \{\vec{\gamma} : (s_i, \vec{\gamma}) \in T\}$. For such i let $\alpha_i = [f^i]_{\mu_{s_i,t_i}}$, and for other i let $\alpha_i = 0$. Then by definition of T', $(x\restriction n, \vec{\alpha}\restriction n) \in T'$ for all n, so $x \in \mathrm{p}[T']$.

Suppose next that $x \in \mathrm{p}[T]$. We must show that T'_x is wellfounded. Suppose T'_x were illfounded, and let $(x, \vec{\alpha}) \in [T']$. For each $j \in \omega$, let $f_j : T_{x\restriction j} \to \kappa^+$ be order-preserving with respect to $<^{x\restriction j}_{\mathrm{BK}}$ such that for all $i < j$ such that s_i is an initial segment of x and μ_{s_i,t_i} is defined we have $\alpha_i = [f^i_j]_{\mu_{s_i,t_i}}$. For each such i let B_i have μ_{s_i,t_i} measure one such that for all $j_1, j_2 > i$ and all $\vec{\gamma} \in B_i$, $f^i_{j_1}(\vec{\gamma}) = f^i_{j_2}(\vec{\gamma})$. Since T_x is illfounded, let $y \in {}^\omega\omega$ be such that all $\mu_{x\restriction n, y\restriction n}$ are defined and the direct limit $M_{x,y}$ of these ultrapowers is wellfounded. Consider $B_{i_1}, B_{i_2}, \dots$, where $(s_{i_k}, t_{i_k}) = (x\restriction k, y\restriction k)$. Since $M_{x,y}$ is wellfounded, there is a $g \in {}^\omega\kappa$ such that $\forall n\, g\restriction n \in B_{i_n}$.

From the definition of the B_i we now have that

$$f^{i_0}(\langle g(0)\rangle) > f^{i_1}(\langle g(0), g(1)\rangle) > f^{i_2}(\langle g(0), g(1), g(2)\rangle) > \dots,$$

a contradiction. Here $f^{i_k}(\vec{\gamma})$ denotes the common value of $f^{i_k}_j(\vec{\gamma})$ for $j > i_k$, which is well-defined for $\vec{\gamma} \in B_{i_k}$. For example, $f^{i_0}(\langle g(0)\rangle) = f^{i_0}_{i_0+1}(\langle g(0)\rangle) = f^{i_0}_{i_1+1}(\langle g(0)\rangle) > f^{i_1}_{i_1+1}(\langle g(0), g(1)\rangle)$. The first two equalities follow from $g(0) \in B_{i_0}$, and the last from f_{i_1+1} being order-preserving. ⊣

Martin and Woodin [MW07] have shown the remarkable result that from AD every tree T on $\omega \times \kappa$ for κ less than the supremum of the Suslin cardinals is weakly homogeneous, even according to the stronger Definition 4.4. We record this in the following theorem.

THEOREM 4.11 (Martin, Woodin). *Assume AD. Let κ be less than the supremum of the Suslin cardinals. Then every tree on $\omega \times \kappa$ is weakly homogeneous in the strong sense of Definition 4.4.*

Martin originally proved this theorem from $\mathsf{AD}_{\mathbb{R}}$ [MarD], and then Woodin modified and extended the argument to work from AD. The assumption that κ is below the largest Suslin cardinal (if there is one) is necessary if one is assuming just AD [If κ is the largest Suslin cardinal, and then $\mathbf{S}(\kappa)$ is non-selfdual by [Kec81B] and thus a tree on $\omega \times \kappa$ projecting to a $\mathbf{S}(\kappa)$-complete set A cannot be weakly homogeneous as otherwise by Lemma 4.10 ${}^\omega\omega - A$ would be Suslin]. $\mathsf{AD}_{\mathbb{R}}$ is equivalent over AD to "every set of reals is Suslin" by a theorem of Woodin, and so this case does not arise from $\mathsf{AD}_{\mathbb{R}}$.

Woodin has proved a "ZFC-analog" of Theorem 4.11. This theorem, or variations on it, play an important role in some of the proofs of determinacy from large cardinals.

DEFINITION 4.12. δ is a Woodin cardinal if for every $f\colon \delta \to \delta$ there is a $\kappa < \delta$ closed under f and an elementary embedding $j\colon \mathbf{V} \to M$ with $\mathbf{V}_{j(f)(\kappa)} \subseteq M$.

An equivalent definition (cf. [MS89]) is that for every $A \subseteq \mathbf{V}_\delta$ there is a $\kappa < \delta$ which is "A-strong," that is, for every $\lambda < \delta$ there is an embedding $j\colon \mathbf{V} \to M$ with $\mathbf{V}_\lambda \subseteq M$ and such that $j(A) \cap \mathbf{V}_\lambda = A \cap \mathbf{V}_\lambda$.

THEOREM 4.13 (Woodin). *Let δ be a Woodin cardinal. Let T be a tree on $\omega \times \alpha$, for $\alpha \in$ Ord. Then there is a $\kappa < \delta$ such that in the generic extension $\mathbf{V}[G]$, where G is generic for $\mathrm{coll}(\omega, \kappa)$, T is $< \delta$ weakly homogeneously Suslin.*

We refer the reader to [Lar04] or [SteA] for a proof, which is vaguely similar to that of Theorem 4.11.

Homogeneous trees are important for propagating Suslin representations according to Theorem 4.10. An important question is whether the resulting Martin-Solovay tree is itself homogeneous. Again, this question is important in both the AD and ZFC contexts.

In the AD context, partition properties are used to establish the homogeneity of the Martin-Solovay tree. This is the content of the next theorem.

THEOREM 4.14. Let T be a weakly homogeneous tree on $\omega \times \kappa$ in the (weaker) sense of Definition 4.5. Assume $\delta > \kappa$ and $\delta \to (\delta)^\kappa$. Then the Martin-Solovay tree $\text{ms}(T, \delta)$ is homogeneous via δ-complete measures.

The following variation is also useful.

THEOREM 4.15. Let T be a weakly homogeneous tree on $\omega \times \kappa$ in the sense of Definition 4.5, and assume T has the property that whenever $((a_0, \dots, a_n), (\alpha_0, \dots, \alpha_n)) \in T$ then $\alpha_0 \geq \max\{\alpha_1, \dots, \alpha_n\}$. Assume $\kappa \to (\kappa)^\kappa$. Then the Martin-Solovay tree $\text{ms}(T, \kappa)$ is homogeneous. Furthermore the measures may be taken to be κ^+ complete provided the following holds: All of the homogeneity measures μ for T give bounded subsets of κ measure 0 and there is a non-selfdual pointclass $\underset{\sim}{\Gamma}$ closed under $\exists^{\mathbb{R}}$ such that (1) The supremum of the length of the $\exists^{\mathbb{R}}\underset{\sim}{\Delta}$ wellfounded relations is κ and (2) There is a prewellordering of length κ which is $\exists^{\mathbb{R}}\underset{\sim}{\Delta}$-bounded.

Remark. The extra pointclass hypotheses to guarantee κ^+-completeness of the measures follows from κ being reasonable (according to Definition 3.5). Since the reasonableness of κ is the only known hypothesis to give the strong partition relation on κ, in practice this hypothesis is always satisfied. For assuming κ is reasonable, (1) follows by considering $\langle R_{0,\beta} : \beta < \delta \rangle$ for $\delta < \kappa$. Since $\underset{\sim}{\Delta}$ (using $\underset{\sim}{\Gamma}$ from reasonableness) is closed under $< \kappa$ unions (this was shown in the proof of Theorem 3.6), this gives a δ-length prewellordering in $\underset{\sim}{\Delta}$. (2) follows by considering $\langle R_{0,\beta} : \beta < \kappa \rangle$ which gives an $\exists^{\mathbb{R}}\underset{\sim}{\Delta}$-bounded prewellordering from the definition of reasonableness. The author does not know if the κ^+-completeness of the measures follows from just the strong partition property at κ.

The two proofs are almost identical, so we just prove the latter.

PROOF. Note that whenever T_x is wellfounded the rank of any node in the Kleene-Brouwer ordering on T_x is less than κ. Thus the proof of Theorem 4.10 goes through and shows that $\text{p}[T] = {}^\omega\omega - \text{p}[T']$ where $T' = \text{ms}(T, \kappa)$ is the Martin-Solovay tree.

Recall $(s, \vec{\alpha}) \in T'$ iff there is a $f : <^s_{\text{BK}} \to \kappa$ of the correct type which represents $\vec{\alpha}$ in the sense that for all $i < \text{lh}(s)$ $\alpha_i = [f^i]_{\mu_{s_i,t_i}}$ (if s_i is an initial segment of s and μ_{s_i,t_i} is defined). The strong partition property of κ gives a natural measure ν_s then on T'_s: we define $A \subseteq T'_s$ to have ν_s measure one if there is a c.u.b. $C \subseteq \kappa$ such that for any $f\colon \overset{s}{<_{\mathrm{BK}}}\to C$ of the correct type, $\vec{\alpha} \in A$ where $\alpha_i = [f^i]_{\mu_{s_i,t_i}}$. It is clear that if s' extends s then $\nu_{s'}$ projects to ν_s (since a subfunction of a function of correct type is also of correct type). By definition of T' we have $\nu_s(T'_s) = 1$. To verify homogeneity, fix $x \in {}^\omega\omega$ with T'_x illfounded, so T_x is wellfounded, and let A_n be measure one sets with respect to $\upsilon_{x\restriction n}$. Let $C_n \subseteq \kappa$ be c.u.b. witnessing A_n has measure one. Let $C = \bigcap_n C_n$, so C is c.u.b. in κ. Since T_x is wellfounded and every node in T_x has rank $< \kappa$ in the Kleene-Brouwer order on T_x, there is an order-preserving map $f\colon T_x \to C$ which is of the correct type. If we let $\alpha_i = [f^i]_{\mu_{s_i,t_i}}$ (if s_i is an initial segment of s and μ_{s_i,t_i} is defined; $\alpha_i = 0$ otherwise), then $(x, \vec{\alpha}) \in [T']$ and $\forall n\, (x\restriction n, \vec{\alpha}\restriction n) \in A_n$. The κ^+-completeness of the measures ν_s follows from the next general lemma. ⊣

Lemma 4.16. Assume $\kappa \to (\kappa)^\kappa$, and let μ be a measure on κ which gives every bounded subset of κ measure 0. Let ν be the measure on $j_\mu(\kappa)$ defined by: $\nu(A) = 1$ iff there is a c.u.b. $C \subseteq \kappa$ such that for all $f\colon \kappa \to C$ of the correct type, $[f]_\mu \in A$. Then assuming there is a pointclass $\underset{\sim}{\Gamma}$ as in Theorem 4.15, ν is a κ^+-complete measure.

Proof. The strong partition relation on κ easily gives that ν is a measure. To show κ^+-completeness, suppose $\delta \leq \kappa$ is least such that there is a sequence $\langle A_\alpha : \alpha < \delta\rangle$ with $\nu(A_\alpha) = 0$ for all $\alpha < \delta$ but $\nu(\bigcup_{\alpha<\delta} A_\alpha) = 1$. We first show that $\delta < \kappa$. Suppose $\delta = \kappa$. Fix a c.u.b. $C_0 \subseteq \kappa$ witnessing $\bigcup_{\alpha<\kappa} A_\alpha$ has ν measure one. For $f\colon \kappa \to C_0$ of the correct type let $\alpha(f)$ be the least $\alpha < \kappa$ such that $[f]_\mu \in A_\alpha$.

Consider the partition $\mathcal{P}_1$: we partition pairs (α, f) where $\alpha < \kappa$ and $f\colon \kappa \to \kappa$ of the correct type with $f(0) > \alpha$ according to whether $\alpha \geq \alpha(f)$ (more formally we are considering functions from the order type $1 \oplus \kappa$ into κ; of course $1 \oplus \kappa \cong \kappa$). We claim that on the homogeneous side the stated property holds. Suppose not, and let C be homogeneous for the contrary side. We may assume $C \subseteq C_0$. Let $f\colon \kappa \to C$ be of the correct type, and let $\alpha \geq \alpha(f)$ with $\alpha \in C$. It is easy to see that there is an $f'\colon \kappa \to C$ such that (1) $f'(0) > \alpha$, (2) f' is of the correct type, and (3) $f'(\beta) = f(\beta)$ for all $\beta > \beta_0$ for some fixed $\beta_0 < \kappa$. [Let $\beta_0 > \alpha$ be a closure point of $\mathrm{ran}(f)$, that is, β_0 is the β_0th element of $\mathrm{ran}(f)$. Let $f'(\beta) = f(\beta)$ for $\beta \geq \beta_0$, and for $\beta < \beta_0$ define $f'(\beta) = f(\alpha + 1 + \beta)$.] Since μ concentrates on unbounded sets, $[f']_\mu = [f]_\mu$. Thus, $\alpha \geq [f']_\mu$, contradicting the homogeneity of C for the contrary side.

Fix now $C_1 \subseteq C_0$ homogeneous for the stated side of $\mathcal{P}_1$. Let $\alpha_0 = \min(C_1)$. The argument of the previous paragraph now shows that for any $f\colon \kappa \to C_1$ of the correct type, $\alpha(f) < \alpha_0$. So, $\nu(\bigcup_{\alpha<\alpha_0} A_\alpha) = 1$, a contradiction.

Thus, $\delta < \kappa$. It suffices therefore to show that ν is κ-complete. Again suppose $\langle A_\alpha : \alpha < \delta\rangle$ were a counterexample. Fix a prewellordering $\preceq$ of length κ which is $\exists^{\mathbb{R}}\underset{\sim}{\Delta}$-bounded. Let $D = \text{fld}(\preceq)$, and for $x \in D$ let $|x|$ denote the rank of x in $\preceq$. Say $\sigma \in {}^{\omega}\omega$ is a **code** if σ codes a continuous function from ${}^{\omega}\omega$ to ${}^{\omega}\omega$ (which we also denote by σ) such that $\forall x \in D\ (\sigma(x) \in D)$. For σ a code define $f_\sigma \colon \kappa \to \kappa$ by

$$f_\sigma(\alpha) = \inf\{|\sigma(x)| : x \in D \wedge |x| = \alpha\}.$$

For σ a code let $C_\sigma \subseteq \kappa$ be the c.u.b. set of points closed under f_σ.

For any c.u.b. $C \subseteq \kappa$ there is a code σ such that $C_\sigma \subseteq C$. To see this, play the Solovay game where player I plays x, player II plays y, and player II wins iff

$$x \in D \Rightarrow (y \in D \wedge |y| > N_C(|x|)),$$

where $N_C(\alpha) =$ the least element of C greater than α. Since $\underset{\sim}{\Sigma}^1_1$ subsets of D are bounded in the prewellordering, player II has a winning strategy σ, and we have $C_\sigma \subseteq C$.

By assumption, there is an $\exists^{\mathbb{R}}\underset{\sim}{\Delta}$ wellfounded relation of length δ. From the Coding Lemma it follows that there is an $\exists^{\mathbb{R}}\underset{\sim}{\Delta}$ set $S \subseteq {}^{\omega}\omega$ consisting of codes and such that for all $\alpha < \delta$ there is a $\sigma \in S$ such that C_σ defines a ν measure one set contained in A_α. Define $f \colon \kappa \to \kappa$ as follows. For $\alpha < \kappa$, let

$$f(\alpha) = \sup\{f_\sigma(\alpha) : \sigma \in S\}.$$

To see that $f(\alpha) < \kappa$, fix $x \in D$ with $|x| = \alpha$. For any $\sigma \in S$, $f_\sigma(\alpha) \leq |\sigma(x)|$. Thus, $f(\alpha) \leq \sup_{\sigma \in S} |\sigma(x)|$. However $\{\sigma(x) : \sigma \in S\}$ is a continuous image of S, and thus is in $\exists^{\mathbb{R}}\underset{\sim}{\Delta}$ (note that $\exists^{\mathbb{R}}\underset{\sim}{\Delta}$ is closed under $\wedge$, $\vee$. This follows by considering cases as to whether $\underset{\sim}{\Delta}$ is closed under quantifiers, that is $\underset{\sim}{\Gamma}$ is at the base of a projective hierarchy, or not). So by boundedness, $f(\alpha) < \kappa$.

Let C be the c.u.b. set of points closed under f, and A the ν measure one set determined by C. For every $\alpha < \delta$ there is a $\sigma \in S$ with C_σ determining a measure one set $A_\sigma \subseteq A_\alpha$. Since $f > f_\sigma$ everywhere, $C \subseteq C_\sigma$. Thus $A \subseteq A_\alpha$. So, $\nu(\bigcap_{\alpha<\delta} A_\alpha) = 1$, a contradiction. ⊣

We note that the ultrapowers $j_\mu(\kappa)$ of Lemma 4.16 are all cardinals according to the following result of Martin. We refer the reader to [Jac07A] for a proof.

LEMMA 4.17. Assume $\kappa \to (\kappa)^\kappa$. Then for any measure μ on κ, the ultrapower $j_\mu(\kappa)$ is a cardinal.

A related result is the following (see also [Jac07A] for a proof). A measure is said to be semi-normal if it gives every c.u.b. set measure one (every normal measure is semi-normal).

LEMMA 4.18. Assume $\kappa \to (\kappa)^\kappa$. Then for any semi-normal measure μ on κ, the ultrapower $j_\mu(\kappa)$ is a regular cardinal.

From Theorems 4.14, 4.15 we get the following result on the projective ordinals. We assume as hypotheses two facts which end up being true, and in the complete analysis are proved by induction along with the following. Recall $\underset{\sim}{\delta}^1_{2n+1} = (\underset{\sim}{\lambda}_{2n+1})^+$ where $\mathrm{cf}(\underset{\sim}{\lambda}_{2n+1}) = \omega$.

THEOREM 4.19. Assume $\forall n\ \underset{\sim}{\delta}^1_{2n+1} \to (\underset{\sim}{\delta}^1_{2n+1})^{\underset{\sim}{\delta}^1_{2n+1}}$, and $\underset{\sim}{\delta}^1_{2n+1}$ is closed under ultrapowers, that is, for any $\alpha, \beta < \underset{\sim}{\delta}^1_{2n+1}$ and any measure μ on α, $j_\mu(\beta) < \underset{\sim}{\delta}^1_{2n+1}$. Then every $\underset{\sim}{\Pi}^1_{2n}$ set is the projection of a homogeneous tree on $\omega \times \underset{\sim}{\lambda}_{2n+1}$ with $(\underset{\sim}{\delta}^1_{2n-1})^+$-complete measures. Every $\underset{\sim}{\Pi}^1_{2n+1}$ set is the projection of a homogeneous tree on $\omega \times \underset{\sim}{\delta}^1_{2n+1}$ with $\underset{\sim}{\delta}^1_{2n+1}$-complete measures. Furthermore, $\underset{\sim}{\lambda}_{2n+3} = \sup_\mu j_\mu(\underset{\sim}{\delta}^1_{2n+1})$ where μ ranges over the measures appearing in the homogeneous tree for a $\underset{\sim}{\Pi}^1_{2n+1}$-complete set.

PROOF. If we assume inductively that every $\underset{\sim}{\Pi}^1_{2n}$ set is the projection of a homogeneous tree on $\omega \times \underset{\sim}{\lambda}_{2n+1}$, then every $\underset{\sim}{\Sigma}^1_{2n+1}$ set is the projection of a weakly homogeneous tree on $\omega \times \underset{\sim}{\lambda}_{2n+1}$.

From Theorem 4.14 and the closure of $\underset{\sim}{\delta}^1_{2n+1}$ under ultrapowers, every $\underset{\sim}{\Pi}^1_{2n+1}$ set is the projection of a homogeneous tree on $\omega \times \underset{\sim}{\delta}^1_{2n+1}$ with $\underset{\sim}{\delta}^1_{2n+1}$-complete measures. Moreover, we may easily arrange it so this tree satisfies the hypothesis of Theorem 4.15. Likewise, starting from such homogeneous trees on $\underset{\sim}{\Pi}^1_{2n+1}$ sets, we get that every $\underset{\sim}{\Sigma}^1_{2n+2}$ set is the projection of a weakly homogeneous tree on $\omega \times \underset{\sim}{\delta}^1_{2n+1}$ and from Theorem 4.15 it follows that every $\underset{\sim}{\Pi}^1_{2n+2}$ set is the projection of a homogeneous tree on $\lambda := \sup_\mu j_\mu(\underset{\sim}{\delta}^1_{2n+1})$, where the supremum ranges over the measures μ in the weakly homogeneous tree for a $\underset{\sim}{\Sigma}^1_{2n+2}$ set (equivalently the measures in a homogeneous tree for a $\underset{\sim}{\Pi}^1_{2n+1}$ set). Furthermore, the measures in this homogeneous tree for the $\underset{\sim}{\Pi}^1_{2n+2}$ set will be $(\underset{\sim}{\delta}^1_{2n+1})^+$-complete (the pointclass hypotheses of Theorem 4.15 are satisfied by $\underset{\sim}{\Gamma} = \underset{\sim}{\Sigma}^1_{2n+1}$). In particular, every $\underset{\sim}{\Sigma}^1_{2n+3}$ set is λ-Suslin. This shows $\lambda \geq \underset{\sim}{\lambda}_{2n+3}$. On the other hand, λ is a countable supremum of cardinals (by Theorem 4.17) each of which is less than $\underset{\sim}{\delta}^1_{2n+3}$ (by closure under ultrapowers). Thus, $\lambda \leq \underset{\sim}{\lambda}_{2n+3}$. ⊣

In fact, the families of measures necessary to compute $\underset{\sim}{\lambda}_{2n+3}$ can be simplified considerably. Assuming the strong partition relation on the $\underset{\sim}{\delta}^1_{2n+1}$ and the closure of the $\underset{\sim}{\delta}^1_{2n+1}$ under ultrapowers, we inductively define the following families of measures. Let W^i_1 denote the i-fold product of the normal measure on ω_1. We identify $\mathrm{dom}(W^i_1)$ with ω_1 by the ordering

$$(\alpha_1, \alpha_2, \ldots, \alpha_i) <_i (\beta_1, \beta_2, \ldots, \beta_i) \Leftrightarrow (\alpha_i, \alpha_1, \ldots, \alpha_{i-1}) <_{\text{lex}} (\beta_i, \beta_1, \ldots, \beta_{i-1}).$$

Let $S^{1,i}_1$ be the measure on $\underset{\sim}{\lambda}_3$ (actually on ω_{i+1}) defined by the strong partition relation on ω_1, functions $f : \omega_1 \to \omega_1$ of the correct type, and the measure W^i_1 on ω_1.

In general, assume that for $m < n$ we have defined the measures W^i_{2m+1} and $S^{j,i}_{2m+1}$ for $j = 1, \ldots, 2^{m+1} - 1$. We let W^i_{2n+1} be the measure on $\underset{\sim}{\delta}^1_{2n+1}$

induced by the weak partition relation on $\underset{\sim}{\delta}^1_{2n+1}$, functions $f\colon \operatorname{dom}(S^{2^n-1,i}_{2n-1}) \to \underset{\sim}{\delta}^1_{2n+1}$, and the measure $S^{2^n-1,i}_{2n-1}$. Let $S^{1,i}_{2n+1}$ be the measure defined just as $S^{1,i}_1$, replacing $\underset{\sim}{\delta}^1_1$ by $\underset{\sim}{\delta}^1_{2n+1}$. Let $S^{j,i}_{2n+1}$, for $2 \leq j \leq 2^{n+1}-1$ be the family of measures on $\underset{\sim}{\lambda}_{2n+3}$ defined using the strong partition relation on $\underset{\sim}{\delta}^1_{2n+1}$, functions $f : \underset{\sim}{\delta}^1_{2n+1} \to \underset{\sim}{\delta}^1_{2n+1}$ of the correct type and the following measure μ on $\underset{\sim}{\delta}^1_{2n+1}$: let ν denote the $(j-1)$th measure in the list W^i_1, $S^{1,i}_1$, W^i_3, $S^{1,i}_3$, $S^{2,i}_3$, $S^{3,i}_3$, $W^i_5, \dots$. Then μ is the measure induced by the weak partition relation on $\underset{\sim}{\delta}^1_{2n+1}$, functions $f\colon \operatorname{dom}(\nu) \to \underset{\sim}{\delta}^1_{2n+1}$, and the measure ν. If $j = 2$, then we regard W^i_1 as a measure on ω_1 by the ordering

$$(\alpha_1, \alpha_2, \dots, \alpha_i) <^i (\beta_1, \beta_2, \dots, \beta_i) \Leftrightarrow (\alpha_i, \alpha_{i-1}, \dots, \alpha_1) <_{\text{lex}} (\beta_i, \beta_{i-1}, \dots, \beta_1).$$

We then have the following facts (proofs are given in [Jac88]).

LEMMA 4.20. For any measure μ on $\underset{\sim}{\delta}^1_{2n+1}$ occurring in a homogeneous tree on a $\underset{\sim}{\Pi}^1_{2n+1}$ set, there is an i such that $j_\mu(\underset{\sim}{\delta}^1_{2n+1}) \leq j_{W^i_{2n+1}}(\underset{\sim}{\delta}^1_{2n+1})$. Thus, $\underset{\sim}{\lambda}_{2n+3} = \sup_i j_{W^i_{2n+1}}(\underset{\sim}{\delta}^1_{2n+1})$.

We also have the following "local version".

LEMMA 4.21. For any measure μ occurring in a homogeneous trees for a $\underset{\sim}{\Pi}^1_m$ set, for $m < 2n+1$, there is a measure ν in one of the families W^i_{2k+1}, or $S^{j,i}_{2k+1}$ for $k < n$ such that for a c.u.b. set of $\alpha < \underset{\sim}{\delta}^1_{2n+1}$ we have $j_\mu(\alpha) \leq j_\nu(\alpha)$.

In fact both of the above lemmas hold for all measures μ on $\underset{\sim}{\delta}^1_{2n+1}$ or $\underset{\sim}{\lambda}_{2n+1}$ respectively. The above lemmas reduce the computation of $\underset{\sim}{\lambda}_{2n+3}$ to the computation of certain iterated ultrapowers by the canonical measures W^i_{2m+1}, $S^{j,i}_{2m+1}$. These are analyzed through the use of "descriptions," finitary objects which describe how to build equivalence classes of functions with respect to these measures. The reader can consult [Jac07A] for an introduction to this theory, [Jac99] for a complete account of the first inductive step of this analysis (including the computation of $\underset{\sim}{\delta}^1_5$ and the proof of the strong partition relation on $\underset{\sim}{\delta}^1_3$), and [Jac88] for details on the general case.

We mention one more important application of homogeneous trees in determinacy theory, particularly in the theory of the projective ordinals. This is constructing the "Martin tree," a generalization of the Kunen tree. We recall first the Kunen tree in the following (see [Kec78A] or [Jac07A] for a proof).

THEOREM 4.22 (Kunen). Assume AD. There is a tree T on $\omega \times \omega_1$ such that for all $f : \omega_1 \to \omega_1$ there is a $x \in {}^\omega\omega$ with T_x wellfounded and such that for all infinite $\alpha < \omega_1$, $f(\alpha) < |T_x \restriction \alpha|$.

The Kunen tree can be used as the basis for developing the first level of the projective hierarchy analysis, that is, computing $\underset{\sim}{\delta}^1_3$, proving the strong partition relation on $\underset{\sim}{\delta}^1_1$, and the weak partition relation on $\underset{\sim}{\delta}^1_3$. The same proof gives the Kunen tree on the higher $\underset{\sim}{\delta}^1_{2n+1}$. That is, there is a tree T on $\omega \times \underset{\sim}{\delta}^1_{2n+1}$ such that for all $f: \underset{\sim}{\delta}^1_{2n+1} \to \underset{\sim}{\delta}^1_{2n+1}$ there is a $x \in {}^\omega\omega$ with T_x wellfounded and a c.u.b. $C \subseteq \underset{\sim}{\delta}^1_{2n+1}$ such that for all $\alpha \in C$ with $\mathrm{cf}(\alpha) = \omega$ we have $f(\alpha) < |T_x \restriction \alpha|$. The Kunen tree, however, only provides almost everywhere domination with respect to points of cofinality ω.

The higher levels of the analysis require the generalization of the Kunen tree which works at general cofinalities; this is the Martin tree.

THEOREM 4.23 (Martin). Assume AD. There is a tree T on $\omega \times \underset{\sim}{\delta}^1_{2n+1}$ such that for any $f: \underset{\sim}{\delta}^1_{2n+1} \to \underset{\sim}{\delta}^1_{2n+1}$ there is a $x \in {}^\omega\omega$ with T_x wellfounded and a c.u.b. $C \subseteq \underset{\sim}{\delta}^1_{2n+1}$ such that for all $\alpha \in C$, $f(\alpha) < |T_x \restriction (\sup_\mu j_\mu(\alpha))|$. Here the supremum ranges over the measures occurring in homogeneous trees on $\underset{\sim}{\Pi}^1_m$ sets for $m < 2n+1$.

We refer the reader to [Jac07A] or [Jac99] for a proof.

The ZFC analog of Theorem 4.14 for propagating homogeneity is the Martin-Steel Theorem [MS89]. We state their theorem as follows.

THEOREM 4.24 (Martin, Steel). Let T be a weakly homogeneous tree with δ complete measures, where δ is a Woodin cardinal. Then the Martin-Solovay tree T' is $< \delta$ homogeneous.

Weakly homogeneous trees are closely related to absoluteness. To see this, first recall the following standard, easily checked fact.

FACT 4.25. Let μ be a κ-complete measure on X, and let $\mathbf{V}[G]$ be a generic extension of $\mathbf{V}$ by a poset $\mathbb{P}$ of size $< \kappa$. Then in $\mathbf{V}[G]$,

$$\mu' := \{A \subseteq X : \exists B \in \mathbf{V}\,(B \subseteq A \wedge \mu(B) = 1)\}$$

is a κ-complete measure. Furthermore, if $Y \in \mathbf{V}$ then for every $f: X \to Y$ in $\mathbf{V}[G]$ there is a $g: X \to Y$ in $\mathbf{V}$ which agrees with f on a μ measure one set.

The following lemma is an immediate consequence of the previous fact, the definition of the Martin-Solovay tree and Theorem 4.10.

LEMMA 4.26. Let T be a weakly homogeneous tree with κ-complete measures and $T' = \mathrm{ms}(T)$. If $\mathbb{P}$ is a poset with $|\mathbb{P}| < \kappa$ then in $\mathbf{V}[G]$ we also have that $T' = \mathrm{ms}(T)$. In particular, $\mathbf{V}[G] \models (\mathrm{p}[T'] = {}^\omega\omega - \mathrm{p}[T])$.

Woodin has shown a converse of Lemma 4.26. Specifically he has shown the following (see [Lar04] or [SteA] for a proof).

THEOREM 4.27 (Woodin). Assume ZFC. Suppose T, U are trees, δ is a Woodin cardinal, and for all posets $\mathbb{P}$ of size $\leq \delta$ we have $\mathbf{V}[G] \models (\mathrm{p}[T] = {}^\omega\omega - \mathrm{p}[U])$. Then T is $<\delta$-weakly homogeneous.

The absolutely complementing property of the previous theorem is frequently made into the following definition (cf. [FMW92]).

Definition 4.28. $A \subseteq {}^\omega\omega$ is **κ-universally Baire** if there are trees T, U with $A = \mathrm{p}[T]$ such that for any poset $\mathbb{P}$ with $|\mathbb{P}| < \kappa$ we that $\mathbf{V}[G] \models (\mathrm{p}[T] = {}^\omega\omega - \mathrm{p}[U])$.

One significance of universal Baireness is that the usual forcing arguments for showing regularity properties of analytic set can be made to work from the assumption of universal Baireness. More precisely, in [FMW92] it is shown that if A is $(2^{\aleph_0})^+$-universally Baire then A has the standard regularity properties such as the Baire property, Lebesgue measurability, and being completely Ramsey. In particular, assuming ZFC, every weakly homogeneously Suslin set has these properties. Kechris, however, has shown [Kec88B] that weak homogeneity has consequences beyond these usual regularity properties. Specifically he shows:

Theorem 4.29 (Kechris). Suppose $A \subseteq \omega_1$ has a code set $A^* \subseteq \mathrm{WO}$ which is weakly homogeneously Suslin. Then $A \in \mathbf{L}[x]$ for some $x \in {}^\omega\omega$.

Theorem 4.29 was improved in [FMW92] to work from the assumption that A is universally Baire (assuming the existence of a measurable cardinal).

A weakly homogeneously Suslin representation for $A \subseteq {}^\omega\omega$ gives an unambiguous way of interpreting A in (small) generic extensions according to the following lemma.

Lemma 4.30. Let $A = \mathrm{p}[T] = \mathrm{p}[U]$ where T, U are weakly homogeneously Suslin with κ-complete measures. Then for any poset $\mathbb{P}$ with $|\mathbb{P}| < \kappa$ we have that $\mathbf{V}[G] \models (\mathrm{p}[T] = \mathrm{p}[U])$.

Proof. Let T', U' be the Martin-Solovay trees for T, U respectively. In $\mathbf{V}$, $\mathrm{p}[T] \cap \mathrm{p}[T'] = \varnothing$ and $\mathrm{p}[U] \cap \mathrm{p}[U'] = \varnothing$, and by absoluteness these intersections remain empty in $\mathbf{V}[G]$. Suppose $x \in \mathbf{V}[G]$ and $x \in \mathrm{p}[T] - \mathrm{p}[U]$. Then, in $\mathbf{V}[G]$, $x \in \mathrm{p}[T] \cap \mathrm{p}[U']$ as U, U' project to complements in $\mathbf{V}[G]$ by Lemma 4.26. By absoluteness there is an $x \in \mathbf{V}$ with $x \in \mathrm{p}[T] \cap \mathrm{p}[U']$ which is impossible since in $\mathbf{V}$, $\mathrm{p}[U'] = {}^\omega\omega - \mathrm{p}[U] = {}^\omega\omega - A = {}^\omega\omega - \mathrm{p}[T]$. ⊣

Finally along these lines we mention that weak homogeneity allows an extension of Lemma 4.26 from complements to "projective in" (and a ways beyond).

Lemma 4.31. Let $A \subseteq {}^\omega\omega$ and B be projective in A, say $B(x) \Leftrightarrow \varphi(x, A)$ where φ is a $\undertilde{\Sigma}^1_n$ formula over A. Suppose $A = \mathrm{p}[T]$, $B = \mathrm{p}[U]$ where T, U are weakly homogeneous trees with κ-complete measures. Assume also that every set projective in A is κ-weakly homogeneously Suslin. Let $\mathbb{P}$ be a poset with $|\mathbb{P}| < \kappa$. Let $\bar{A} = (\mathrm{p}[T])^{\mathbf{V}[G]}$, $\bar{B} = (\mathrm{p}[U])^{\mathbf{V}[G]}$. Then in $\mathbf{V}[G] \models (\forall x\, (\bar{B}(x) \Leftrightarrow \varphi(x, \bar{A}))$.

Proof. Let T_1 be weakly homogeneous (with κ-complete measures) and $\mathrm{p}[T_1] = {}^\omega\omega - \mathrm{p}[T] = {}^\omega\omega - A = \mathrm{p}[T']$, where $T' = \mathrm{ms}(T)$. We claim that in $\mathbf{V}[G]$ we also have $\mathrm{p}[T_1] = \mathrm{p}[T']$. Let $x \in \mathbf{V}[G]$ and suppose first that $x \in \mathrm{p}[T_1] - \mathrm{p}[T']$. Then by Lemma 4.26 $x \in \mathrm{p}[T_1] \cap \mathrm{p}[T]$. By absoluteness there is a $y \in \mathbf{V}$ with $y \in \mathrm{p}[T_1] \cap \mathrm{p}[T]$. But in $\mathbf{V}$, $\mathrm{p}[T_1] = {}^{\omega}\omega - \mathrm{p}[T]$, a contradiction. Likewise if $x \in \mathrm{p}[T'] - \mathrm{p}[T_1]$ then $x \in \mathrm{p}[T'] \cap \mathrm{p}[T_1']$ where $T_1' = \mathrm{ms}(T_1)$. By absoluteness again, there is a $y \in \mathbf{V}$ with $y \in \mathrm{p}[T'] \cap \mathrm{p}[T_1']$. However, in $\mathbf{V}$ $\mathrm{p}[T_1'] = {}^{\omega}\omega - \mathrm{p}[T_1] = \mathrm{p}[T] = {}^{\omega}\omega - \mathrm{p}[T']$, a contradiction.

Let A_2 be defined by $A_2(x) \Leftrightarrow \exists y\ \neg A(\langle x, y\rangle)$. In ZF there is a simple operation which takes a tree W projecting to a set B and produces a tree $\tilde{U}$ projecting to the existential quantification $B_2(x) \Leftrightarrow \exists y\ B(\langle x, y\rangle)$. Let $\tilde{T}'$, $\tilde{T}_1$ be the resulting trees constructed from T' and T_1. So in both $\mathbf{V}$ and $\mathbf{V}[G]$ we have $\mathrm{p}[\tilde{T}'] = \exists^{\mathbb{R}}\mathrm{p}[T'] = \exists^{\mathbb{R}}\mathrm{p}[T_1] = \mathrm{p}[\tilde{T}_1]$. Also, $U_2 := \tilde{T}_1$ will be weakly homogeneous since T_1 is. We have thus produced a weakly homogeneous U_2 such that $\mathrm{p}[U_2] = \exists^{\mathbb{R}}\neg\,\mathrm{p}[T]$ in both $\mathbf{V}$ and $\mathbf{V}[G]$. From Lemma 4.30 it doesn't matter which weakly homogeneous U_2 projecting to $\exists^{\mathbb{R}}\neg A$ in $\mathbf{V}$ we take.

This then proves the lemma in the case where $\varphi(x, A) = \exists y \neg A(\langle x, y\rangle)$. The general case now follows by repeating the argument. ⊣

REFERENCES

Howard S. Becker

[Bec81] *Determinacy implies that $\aleph_2$ is supercompact*, ***Israel Journal of Mathematics***, vol. 40 (1981), no. 3–4, pp. 229–234.

Chen-Lian Chuang

[Chu82] ***The propagation of scales by game quantifiers***, Ph.D. thesis, UCLA, 1982.

Qi Feng, Menachem Magidor, and W. Hugh Woodin

[FMW92] *Universally Baire sets of reals*, In Judah et al. [JJW92], pp. 203–242.

Gregory Hjorth

[Hjo96] *Two applications of inner model theory to the study of $\mathbf{\Sigma}^1_2$ sets*, ***The Bulletin of Symbolic Logic***, vol. 2 (1996), no. 1, pp. 94–107.

[Hjo97] *Some applications of coarse inner model theory*, ***The Journal of Symbolic Logic***, vol. 62 (1997), no. 2, pp. 337–365.

[Hjo01] *A boundedness lemma for iterations*, ***The Journal of Symbolic Logic***, vol. 66 (2001), no. 3, pp. 1058–1072.

Stephen Jackson

[Jac] *Non-partition results in the projective hierarchy*, to appear.

[Jac88] AD *and the projective ordinals*, In Kechris et al. [Cabal iv], pp. 117–220.

[Jac90] *A new proof of the strong partition relation on ω_1*, ***Transactions of the American Mathematical Society***, vol. 320 (1990), no. 2, pp. 737–745.

[Jac91] *Admissible Suslin cardinals in* $\mathrm{L}(\mathbb{R})$, ***The Journal of Symbolic Logic***, vol. 56 (1991), no. 1, pp. 260–275.

[Jac99] ***A computation of $\mathbf{\delta}^1_5$***, vol. 140, Memoirs of the AMS, no. 670, American Mathematical Society, July 1999.

[Jac07A] *Structural consequences of* AD, In Kanamori and Foreman [KF07].

STEPHEN JACKSON AND DONALD A. MARTIN

[JM83] *Pointclasses and wellordered unions*, In Kechris et al. [CABAL iii], pp. 55–66.

STEPHEN JACKSON AND RUSSELL MAY

[JM04] *The strong partition relation on ω_1 revisited*, ***Mathematical Logic Quarterly***, vol. 50 (2004), no. 1, pp. 33–40.

H. JUDAH, W. JUST, AND W. HUGH WOODIN

[JJW92] ***Set theory of the continuum***, MSRI publications, vol. 26, Springer-Verlag, 1992.

AKIHIRO KANAMORI

[Kan94] ***The higher infinite***, Springer-Verlag, Berlin, 1994.

AKIHIRO KANAMORI AND MATTHEW FOREMAN

[KF07] ***Handbook of set theory into the 21st century***, Springer, 2007.

ALEXANDER S. KECHRIS

[Kec77] AD *and infinite exponent partition relations*, Circulated manuscript, 1977.

[Kec78A] AD *and projective ordinals*, In Kechris and Moschovakis [CABAL i], pp. 91–132.

[Kec81A] *Homogeneous trees and projective scales*, In Kechris et al. [CABAL ii], pp. 33–74.

[Kec81B] *Suslin cardinals, κ-Suslin sets, and the scale property in the hyperprojective hierarchy*, this volume, originally published in Kechris et al. [CABAL ii], pp. 127–146.

[Kec88B] *A coding theorem for measures*, this volume, originally published in Kechris et al. [CABAL iv], pp. 103–109.

ALEXANDER S. KECHRIS, EUGENE M. KLEINBERG, YIANNIS N. MOSCHOVAKIS, AND W. HUGH WOODIN

[KKMW81] *The axiom of determinacy, strong partition properties and nonsingular measures*, this volume, originally published in Kechris et al. [CABAL ii], pp. 75–100.

ALEXANDER S. KECHRIS, DONALD A. MARTIN, AND YIANNIS N. MOSCHOVAKIS

[CABAL ii] *Cabal seminar* 77–79, Lecture Notes in Mathematics, no. 839, Berlin, Springer, 1981.

[CABAL iii] *Cabal seminar* 79–81, Lecture Notes in Mathematics, no. 1019, Berlin, Springer, 1983.

ALEXANDER S. KECHRIS, DONALD A. MARTIN, AND JOHN R. STEEL

[CABAL iv] *Cabal seminar* 81–85, Lecture Notes in Mathematics, no. 1333, Berlin, Springer, 1988.

ALEXANDER S. KECHRIS AND YIANNIS N. MOSCHOVAKIS

[CABAL i] *Cabal seminar* 76–77, Lecture Notes in Mathematics, no. 689, Berlin, Springer, 1978.

[KM78B] *Notes on the theory of scales*, this volume, originally published in *Cabal Seminar* 76–77 [CABAL i], pp. 1–53.

ALEXANDER S. KECHRIS, ROBERT M. SOLOVAY, AND JOHN R. STEEL

[KSS81] *The axiom of determinacy and the prewellordering property*, In Kechris et al. [CABAL ii], pp. 101–125.

ALEXANDER S. KECHRIS AND W. HUGH WOODIN

[KW83] *Equivalence of determinacy and partition properties*, ***Proceedings of the National Academy of Sciences of the United States of America***, vol. 80 (1983), no. 6 i., pp. 1783–1786.

[KW07] *Generic codes for uncountable ordinals*, this volume, originally circulated manuscript, 2007.

PAUL B. LARSON

[Lar04] ***The stationary tower: Notes on a course by W. Hugh Woodin***, University Lecture Series (AMS), vol. 32, American Mathematical Society, Providence, RI, 2004.

Donald A. Martin

[MarD] *Weakly homogeneous trees*, Circulated manuscript.

[Mar83B] *The real game quantifier propagates scales*, this volume, originally published in Kechris et al. [Cabal iii], pp. 157–171.

Donald A. Martin and John R. Steel

[MS83] *The extent of scales in* $\mathbf{L}(\mathbb{R})$, this volume, originally published in Kechris et al. [Cabal iii], pp. 86–96.

[MS89] *A proof of projective determinacy*, ***Journal of the American Mathematical Society***, vol. 2 (1989), pp. 71–125.

Donald A. Martin and W. Hugh Woodin

[MW07] *Weakly homogeneous trees*, this volume, 2007.

Yiannis N. Moschovakis

[Mos80] ***Descriptive set theory***, Studies in Logic and the Foundations of Mathematics, no. 100, North-Holland, Amsterdam, 1980.

John R. Steel

[SteA] *The derived model theorem*, Available at `http://www.math.berkeley.edu/~steel`.

[Ste81A] *Closure properties of pointclasses*, In Kechris et al. [Cabal ii], pp. 147–163.

[Ste81B] *Determinateness and the separation property*, ***The Journal of Symbolic Logic***, vol. 46 (1981), no. 1, pp. 41–44.

[Ste83A] *Scales in* $\mathbf{L}(\mathbb{R})$, this volume, originally published in Kechris et al. [Cabal iii], pp. 107–156.

[Ste83B] *Scales on* Σ^1_1*-sets*, this volume, originally published in Kechris et al. [Cabal iii], pp. 72–76.

Robert Van Wesep

[Van78A] *Separation principles and the axiom of determinateness*, ***The Journal of Symbolic Logic***, vol. 43 (1978), pp. 77–81.

[Van78B] *Wadge degrees and descriptive set theory*, In Kechris and Moschovakis [Cabal i], pp. 151–170.

W. Hugh Woodin

[Woo99] ***The axiom of determinacy, forcing axioms, and the nonstationary ideal***, De Gruyter Series in Logic and its Applications, Walter de Gruyter, Berlin, 1999.

DEPARTMENT OF MATHEMATICS
P.O. BOX 311430
UNIVERSITY OF NORTH TEXAS
DENTON, TX 76203-1430, USA
E-mail: jackson@unt.edu

SUSLIN CARDINALS, κ-SUSLIN SETS AND THE SCALE PROPERTY IN THE HYPERPROJECTIVE HIERARCHY

ALEXANDER S. KECHRIS

Let $\Theta = \sup\{\xi : \xi$ is the length of a prewellordering of the set of reals $\mathbb{R}$ $(= {}^{\omega}\omega)\}$. Let $\kappa < \Theta$ be an infinite cardinal. The class $\boldsymbol{S}(\kappa)$ of κ-Suslin sets has some well-known closure properties, i.e., it is closed under continuous substitutions, countable intersections and unions, and existential quantification over $\mathbb{R}$. We investigate in §§1 and 2 the question whether $\boldsymbol{S}(\kappa)$ is $\mathbb{R}$-parametrized (i.e., whether $\boldsymbol{S}(\kappa)$ admits universal sets), assuming AD. Let us call a cardinal κ **Suslin** iff there is a new κ**-Suslin** set i.e., $\boldsymbol{S}(\kappa) \setminus \bigcup_{\lambda<\kappa} \boldsymbol{S}(\lambda) \neq \varnothing$. Let $\kappa_0, \kappa_1, \kappa_2, \ldots, \kappa_\xi, \ldots$ be the increasing enumeration of the Suslin cardinals (the first few of them are $\kappa_0 = \omega, \kappa_1 = \omega_1, \kappa_2 = \omega_\omega, \kappa_3 = \omega_{\omega+1} \ldots$). We show in §1 that $\boldsymbol{S}(\kappa)$ is $\mathbb{R}$-parametrized iff there is a largest Suslin cardinal $\leq \kappa$. Thus $\boldsymbol{S}(\kappa)$ is $\mathbb{R}$-parametrized for all $\kappa < \Theta$ iff the sequence $\langle \kappa_\xi \rangle$ is normal (and has a largest element if bounded below Θ). We conjecture that this is indeed the case, and in §2 we verify this conjecture at least below $\kappa^{\mathbb{R}} =$ the first non-hyperprojective ordinal, so that $\boldsymbol{S}(\kappa)$ is always $\mathbb{R}$-parametrized for $\kappa \leq \kappa^{\mathbb{R}}$.

In §3 we study the hyperprojective hierarchy $(\utilde{\boldsymbol{\Sigma}}^1_\xi, \utilde{\boldsymbol{\Pi}}^1_\xi, \utilde{\boldsymbol{\Delta}}^1_\xi\}_{\xi<\kappa^{\mathbb{R}}}$ and, using AD again, we establish that the scale property propagates throughout this hierarchy following the pattern established for the prewellordering property in [KSS81]. In particular, it follows that if $\utilde{\Gamma}$ is a projective-like pointclass, contained in the inductive sets, then either $\utilde{\Gamma}$ or $\check{\utilde{\Gamma}}$ has the scale property. This smooth propagation of scales breaks down immediately past the inductive sets and this "gap phenomenon" is discussed briefly in §4. Finally §5 contains a number of open problems, conjectures and remarks related to the preceding work.

This paper draws heavily on the terminology, notation and results established in [KSS81]. We refer also to [Mos80] for the basic results from descriptive set theory that we use.

Our underlying theory is ZF+DC, and any further assumptions (mainly AD) are explicitly indicated. Of course those of our results which are also absolute

Research partially supported by NSF Grant MCS79-20465 and an A. P. Sloan Foundation fellowship.

The Cabal Seminar. Volume I: Games, Scales and Suslin Cardinals
Edited by A. S. Kechris, B. Löwe, J. R. Steel
Lecture Notes in Logic, 31

for the inner model $\mathbf{L}(\mathbb{R})$, although stated as proved from AD, need only $\mathrm{AD}^{\mathbf{L}(\mathbb{R})}$ and in fact (in most cases) much weaker forms of definable determinacy, like Hyperprojective Determinacy, etc. An interested and patient reader should have no trouble figuring out how much determinacy is needed in each case.

§1. Suslin cardinals and κ-Suslin sets. Let κ be an infinite ordinal. A set $A \subseteq \mathbb{R}$ is **κ-Suslin** if there is a tree T on $\omega \times \kappa$ such that $A = \mathrm{p}[T] \equiv \{\alpha \in \mathbb{R} : \exists f \in {}^{\omega}\kappa(\langle \alpha, f \rangle \in [T])\}$. Similar definitions apply to pointsets $A \subseteq \mathcal{X}$ contained in arbitrary product spaces. Denote by $\boldsymbol{S}(\kappa)$ the pointclass of κ-Suslin pointsets. It is immediate that $\boldsymbol{S}(\kappa) = \boldsymbol{S}(\kappa')$, where κ' is the cardinality of κ, so that it is enough to consider only $\boldsymbol{S}(\kappa)$, when κ is a cardinal.

The class $\boldsymbol{S}(\kappa)$ has some well-known closure properties which we summarize below.

Proposition 1.1. The pointclass $\boldsymbol{S}(\kappa)$ is closed under continuous substitutions, countable unions and intersections and $\exists^{\mathbb{R}}$.

The obvious question now is whether $\boldsymbol{S}(\kappa)$ is $\mathbb{R}$-parametrized, i.e., has universal sets. We provide below a sufficient condition for this to be true. For this we need to introduce first the following notion.

Definition 1.2. A cardinal is called **Suslin** iff $\boldsymbol{S}(\kappa) \setminus \bigcup_{\lambda<\kappa} \boldsymbol{S}(\lambda) \neq \varnothing$ i.e., there is a κ-Suslin set which is not λ-Sousin for any $\lambda < \kappa$.

Assuming AD, let $\kappa_0, \kappa_1, \kappa_2, \dots, \kappa_\xi$, $\xi < \Theta$, enumerate the Suslin cardinals. Thus

$$\kappa_0 = \omega, \kappa_1 = \omega_1 = \underset{\sim}{\delta}^1_1, \kappa_2 = \omega_\omega, \kappa_3 = \omega_{\omega+1} = \underset{\sim}{\delta}^1_3, \dots.$$

The only Suslin cardinals below $\sup_{n<\omega}\underset{\sim}{\delta}^1_n$ are $\underset{\sim}{\lambda}_{2n+1}, \underset{\sim}{\delta}^1_{2n+1}$, where $(\underset{\sim}{\lambda}_{2n+1})^+ = \underset{\sim}{\delta}^1_{2n+1}$. We have now the following main result.

Theorem 1.3 (AD). Let κ be a Suslin cardinal. Then $\boldsymbol{S}(\kappa)$ is $\mathbb{R}$-parametrized.

Proof. Assume not, towards a contradiction. Then, by Wadge, $\boldsymbol{S}(\kappa) \equiv \Lambda$ is closed under complements, so also under $\forall^{\mathbb{R}}$, i.e., it is **strongly closed** in the sense of [KSS81, §2.3]. By the results of that paper (especially 2.4.1) if $M \subsetneqq \Lambda$ then there is a pointclass Γ, with $M \subsetneqq \Gamma \subsetneqq \Lambda$ such that Γ is closed under wellordered unions. We shall use this fact repeatedly below.

Note first that it is sufficient to show that $\mathrm{cf}(\kappa) > \omega$. Because then letting $M = \bigcup_{\xi<\kappa} \boldsymbol{S}(\xi)$, if $A \in \boldsymbol{S}(\kappa)$, then A is a wellordered union of sets in M (if T is a tree on $\omega \times \kappa$, then $\mathrm{p}[T] = \bigcup_{\xi<\kappa} \mathrm{p}[T \restriction \xi]$). Let $M \subsetneqq \Gamma \subsetneqq \Lambda$ be closed under wellordered unions. Then $\boldsymbol{S}(\kappa) = \Lambda \subseteq \Gamma$, a contradiction.

Note now that if $\lambda = \sup\{\xi : \xi$ is the length of a Λ prewellordering of $\mathbb{R}\}$, then λ has cofinality $> \omega$, since Λ is closed under countable unions. Thus it is enough to prove the lemma below to obtain the desired contradiction.

LEMMA 1.4. $\lambda = \kappa$.

PROOF OF LEMMA 1.4. If $\kappa < \lambda$, then there is a Λ-norm $\varphi\colon \mathbb{R} \twoheadrightarrow \kappa$. By the Moschovakis Coding Lemma we can code trees on $\omega \times \kappa$ within $\undertilde{\Sigma}^1_1(\leq_\varphi, <_\varphi)$, where $x \leq_\varphi y \Leftrightarrow \varphi(x) \leq \varphi(y)$ and similarly for $<_\varphi$. Thus

$$\boldsymbol{S}(\kappa) \subseteq \undertilde{\Sigma}^1_1(\leq_\varphi, <_\varphi) \subsetneq \Lambda,$$

a contradiction.

So $\kappa \geq \lambda$. Pick now $A \in \boldsymbol{S}(\kappa) \setminus \bigcup_{\xi<\kappa} \boldsymbol{S}(\xi)$. Since A is κ-Suslin, A carries a κ'-scale $\langle \varphi_n : n \in \omega \rangle$, where κ' has the same cardinality as κ. We can also assume that $\langle \varphi_n \rangle$ is **regular** i.e., each φ_n maps A onto an initial segment of the ordinals. For $n \in \omega$, let μ_n = length of φ_n. Clearly $\kappa \leq \sup_{n\in\omega} \mu_n$, so it will be enough to show that for each n,

$$\text{card}(\mu_n) \leq \lambda.$$

So fix $n \in \omega$. For any $\xi < \mu_n$, let

$$A_\xi = \{\alpha \in A : \varphi_n(\alpha) = \xi\}.$$

Then if T is the tree associated with the scale $\langle \varphi_n \rangle$, i.e.,

$$T = \{\langle \alpha(0), \varphi_0(\alpha), \ldots, \alpha(m-1), \varphi_{m-1}(\alpha) \rangle : m \in \omega, \alpha \in A\}$$

and for $\xi \leq \mu_n$ we let

$$T_{\leq\xi} = \{\langle a_0, \xi_0, \ldots, a_{m-1}, \xi_{m-1} \rangle \in T : \xi_n \leq \xi\}$$

and similarly for $T_{<\xi}$, we have by the semicontinuity property of scales that

$$A_\xi = \text{p}[T_{\leq\xi}] \setminus \text{p}[T_{<\xi}].$$

Thus $A_\xi \in \Lambda$, since $\text{p}[T_{\leq\xi}], \text{p}[T_{<\xi}]$ are κ'-Suslin and Λ is closed under complements. Moreover

$$\xi \neq \eta < \mu_n \text{ implies } A_\xi \cap A_\eta = \varnothing.$$

Put for $\xi < \mu_n$,

$$g(\xi) = \|A_\xi\|_{\text{W}} \equiv \text{ the } \textbf{Wadge ordinal} \text{ of } A_\xi.$$

Since by [KSS81, Lemma 2.3.1], we have $\lambda = \sup\{\|B\|_{\text{W}} : B \in \Lambda\}$, it follows that $g\colon \mu_n \to \lambda$.

We claim now that for each $\rho < \lambda$, the order type of

$$\{\xi < \lambda : g(\xi) < \rho\} = g^{-1}[\rho]$$

is less than λ. Granting this, let for $\rho < \lambda$, $f_\rho\colon \lambda \twoheadrightarrow \varphi^{-1}[\rho]$. Then if $f(\rho, \xi) = f_\rho(\xi)$, $f : \lambda \times \lambda \twoheadrightarrow \bigcup_{\rho<\lambda} g^{-1}[\rho] = \mu_n$, thus $\text{card}(\mu_n) \leq \lambda$ and we are done.

We prove the claim now: Let for each $\rho < \lambda$,

$$N = \{A : \|A\|_{\mathrm{W}} < \rho\}.$$

Then $N \subsetneq \Lambda$, so find Γ closed under wellordered unions and such that $N \subsetneq \Gamma \subsetneq \Lambda$. Define

$$x \prec y \Leftrightarrow \exists \xi \exists \xi' [\xi, \xi' \in g^{-1}[\rho] \wedge \xi < \xi' \wedge x \in A_\xi \wedge y \in A_{\xi'}].$$

Clearly $\prec$ is wellfounded, has rank the order type of $g^{-1}[\rho]$ and belongs to Γ, thus the order type of $g^{-1}[\rho]$ is less than λ, and we are done. $\dashv$Lemma 1.4 and Theorem 1.3

By the preceding result the only case when $\boldsymbol{S}(\kappa)$ is not $\mathbb{R}$-parametrized is when

$$\boldsymbol{S}(\kappa) = \boldsymbol{S}(\sigma),$$

where σ is a limit of Suslin cardinals but is itself not Suslin. In this case $\boldsymbol{S}(\sigma) = \bigcup_{\kappa<\sigma} \boldsymbol{S}(\kappa)$ is closed under complements. We conjecture that this never happens, i.e.,

Conjecture 1.5 (AD + anything reasonable). If $\kappa < \Theta$ is a limit of Suslin cardinals, then κ is a Suslin cardinal.

Thus if $\langle \kappa_\xi : \xi < \Xi \rangle$ is the enumeration of the Suslin cardinals, our conjecture means that for $\nu < \Xi$ limit

$$\kappa_\nu = \sup_{\xi<\nu} \kappa_\xi,$$

and moreover that either $\sup_{\xi<\Xi} \kappa_\xi = \Theta$, i.e., if $\boldsymbol{S}_\infty = \bigcup_\lambda \boldsymbol{S}(\lambda)$, then

$$\boldsymbol{S}_\infty = \wp(\mathbb{R})$$

or else $\Xi = \xi + 1$ is successor and $\boldsymbol{S}_\infty = \boldsymbol{S}(\kappa_\xi)$. Thus in particular we have the following

Conjecture 1.6 (AD + anything reasonable). The pointclass $\boldsymbol{S}_\infty$ is $\mathbb{R}$-parametrized, unless $\boldsymbol{S}_\infty = \wp(\mathbb{R})$.

It has been proved by Martin and Steel that

$$\mathsf{AD} + \mathbf{V} = \mathbf{L}(\mathbb{R}) \Rightarrow \boldsymbol{S}_\infty = \utilde{\Sigma}^2_1 \wedge (\Xi = \utilde{\delta}^2_1 + 1, \kappa_{\utilde{\delta}^2_1} = \utilde{\delta}^2_1).$$

On the other hand [Sol78B] raises the question whether

$$\mathsf{AD}_{\mathbb{R}} + \Theta \text{ is regular } \Rightarrow \boldsymbol{S}_\infty = \wp(\mathbb{R})?$$

We shall see now in the next section that our first conjecture is verified for $\kappa \leq \kappa^{\mathbb{R}}$ (and in fact for a while beyond that, by §4).

§2. Suslin cardinals below $\kappa^{\mathbb{R}}$. Our main purpose here is to prove the following.

THEOREM 2.1 (AD). The class of Suslin cardinals is closed unbounded below $\kappa^{\mathbb{R}}$.

As an immediate corollary we have

THEOREM 2.2 (AD). For each $\kappa \leq \kappa^{\mathbb{R}}$, the pointclass $\boldsymbol{S}(\kappa)$ is $\mathbb{R}$-parametrized.

It will be convenient at this stage to introduce the **hyperprojective hierarchy** to facilitate the presentation of the proof of Theorem 2.1. In the next section we shall take up a detailed study of the structural properties of the hyperprojective hierarchy itself.

First let $\underset{\sim}{\mathbf{IND}}$ denote the class of **inductive** over the structure of analysis $\mathbb{R}$ pointsets and $\underset{\sim}{\mathbf{HYP}} \equiv \underset{\sim}{\mathbf{IND}} \cap \check{\underset{\sim}{\mathbf{IND}}}$ the pointclass of **hyperprojective** sets. We define below a hierarchy on $\underset{\sim}{\mathbf{HYP}}$.

DEFINITION 2.3 (AD). For each $1 \leq \xi < \kappa^{\mathbb{R}}$ let the pointclasses $\underset{\sim}{\mathbf{\Sigma}}^1_\xi, \underset{\sim}{\mathbf{\Pi}}^1_\xi, \underset{\sim}{\mathbf{\Delta}}^1_\xi$ be defined as follows:

(i) For $\xi < \omega$, these are the usual projective pointclasses.

(ii) $\underset{\sim}{\mathbf{\Sigma}}^1_{\xi+1} = \exists^{\mathbb{R}} \underset{\sim}{\mathbf{\Pi}}^1_\xi$, $\underset{\sim}{\mathbf{\Pi}}^1_{\xi+1} = \check{\underset{\sim}{\mathbf{\Sigma}}}^1_{\xi+1}$.

(iii) For limit λ, let $\underset{\sim}{\mathbf{\Sigma}}^1_\lambda$ be the smallest projective-like pointclass closed under $\exists^{\mathbb{R}}$ which contains $\bigcup_{\xi<\lambda} \underset{\sim}{\mathbf{\Sigma}}^1_\xi$, and let $\underset{\sim}{\mathbf{\Pi}}^1_\lambda = \check{\underset{\sim}{\mathbf{\Sigma}}}^1_\lambda$.

(iv) $\underset{\sim}{\mathbf{\Delta}}^1_\xi = \underset{\sim}{\mathbf{\Sigma}}^1_\xi \cap \underset{\sim}{\mathbf{\Pi}}^1_\xi$.

Put also for limit λ,

$$\underset{\sim}{\mathbf{\Delta}}^1_\lambda = \bigcup_{\xi<\lambda} \underset{\sim}{\mathbf{\Delta}}^1_\xi.$$

We have of course

$$\underset{\sim}{\mathbf{HYP}} = \bigcup_{\xi<\kappa^{\mathbb{R}}} \underset{\sim}{\mathbf{\Sigma}}^1_\xi \ (= \bigcup_{\xi<\kappa^{\mathbb{R}}} \underset{\sim}{\mathbf{\Delta}}^1_\xi).$$

Finally define

$$\underset{\sim}{\boldsymbol{\delta}}^1_\xi = \sup\{\eta : \eta \text{ is the rank of a } \underset{\sim}{\mathbf{\Sigma}}^1_\xi \text{ well founded relation}\}$$

and for limit ν

$$\underset{\sim}{\boldsymbol{\lambda}}^1_\nu = \sup_{\xi<\nu} \underset{\sim}{\boldsymbol{\delta}}^1_\xi.$$

(It follows from the results in §3 that also $\underset{\sim}{\boldsymbol{\delta}}^1_\xi = \sup\{\eta : \eta$ is the rank of a $\underset{\sim}{\mathbf{\Delta}}^1_\xi$ prewellordering$\}$, so that this notation is justified.)

We shall need of course crucially below some facts about positive elementary inductive definability on the structure of analysis $\mathbb{R}$; see [Mos74A] here.

For some monotone operator $\Phi(x, A)$, where x varies over a product space $\mathcal{X}$ and A over $\wp(\mathcal{X})$, let Φ^ξ be its ξth iterate, defined by

$$\Phi^\xi(x) \Leftrightarrow \Phi(x, \Phi^{<\xi}), \Phi^{<\xi} = \bigcup_{\eta<\xi} \Phi^\eta.$$

DEFINITION 2.4. For each limit ordinal $\lambda \leq \kappa^{\mathbb{R}}$ define the following pointclass

$$\underset{\sim}{\mathbf{IND}}_\lambda = \{A : A \text{ is Wadge reducible to some } \Phi^{<\lambda}, \text{ where}$$
$$\Phi \text{ is a positive elementary operator on } \mathbb{R}\}.$$

(Recall that A is **Wadge reducible** to B ($A \leq_{\mathrm{W}} B$) iff there is continuous f with $A = f^{-1}[B]$). Thus $\underset{\sim}{\mathbf{IND}}_{\kappa^{\mathbb{R}}} = \underset{\sim}{\mathbf{IND}}$.

We will need the following two lemmas of which the first one is implicit in [Mos74A]. We will sketch their proofs after deriving from them a proof of Theorem 2.1.

LEMMA 2.5. For each limit $\lambda \leq \kappa^{\mathbb{R}}$, $\underset{\sim}{\mathbf{IND}}$ is closed under continuous substitutions, $\cap, \cup, \cup^\omega, \exists^{\mathbb{R}}$ and is $\mathbb{R}$-parametrized. In fact there is a "universal" positive elementary over $\mathbb{R}$ operator $\Phi(\alpha, A)$ such that $\Phi^{<\lambda}$ is $\underset{\sim}{\mathbf{IND}}_\lambda$-complete for all limit $\lambda \leq \kappa^{\mathbb{R}}$ (i.e., each member of $\underset{\sim}{\mathbf{IND}}_\lambda$ is Wadge reducible to it). If $\mathrm{cf}(\lambda) > \omega$, then $\underset{\sim}{\mathbf{IND}}_\lambda$ is also closed under $\cap^\omega$ as well. Finally $\underset{\sim}{\mathbf{IND}}_\lambda$ has the prewellordering property.

(We have actually put a lot more information in Lemma 2.5 than we need for the proof of Theorem 2.1. It will be used in later sections.)

LEMMA 2.6 (AD). For each limit $\lambda < \kappa^{\mathbb{R}}$,

$$\underset{\sim}{\mathbf{\Delta}}^1_\lambda \subsetneq \underset{\sim}{\mathbf{IND}}_\lambda \subsetneq \underset{\sim}{\mathbf{\Delta}}^1_{\lambda+\omega}$$

(The class $\underset{\sim}{\mathbf{IND}}_\lambda$ can be actually computed precisely in the hyperprojective hierarchy by the results in the next section.)

Granting these two lemmas let us proceed to give the

PROOF OF THEOREM 2.1. It is clear that the sequence $\langle \kappa_\xi : \xi < \kappa^{\mathbb{R}} \rangle$ is unbounded below $\kappa^{\mathbb{R}}$. We have to show that it is closed. So let $\vartheta < \kappa^{\mathbb{R}}$ be limit. Put $\kappa = \sup_{\eta<\vartheta} \kappa_\eta$. We have to show that κ is a Suslin cardinal i.e., $\boldsymbol{S}(\kappa) \not\subseteq \bigcup_{\eta<\vartheta} \boldsymbol{S}(\kappa_\eta) \equiv \Lambda$. Since Λ is strongly closed and $\Lambda \subsetneq \underset{\sim}{\mathbf{HYP}}$, there is a limit ordinal $\sigma < \kappa^{\mathbb{R}}$ with

$$\Lambda = \underset{\sim}{\mathbf{\Delta}}^1_\sigma.$$

Note that $\sigma \leq \underset{\sim}{\lambda}^1_\sigma \equiv \sup_{\xi<\sigma} \underset{\sim}{\delta}^1_\xi \leq \kappa$, where the last bound comes from the Kunen-Martin Theorem.

Let now Φ be the universal operator asserted to exist in Lemma 2.5. For each ordinal $\xi \leq \kappa^{\mathbb{R}}$, let $\langle \varphi^\xi_n : n \in \omega \rangle$ be the canonical scale on Φ^ξ defined by the procedure of [Mos78]. Let $\rho_\xi = \sup_n(\text{lh}(\varphi_n^\xi))$. By the definition of $\langle\varphi_n^\xi\rangle$ it is easy to check that for each limit ordinal ν and each $m \in \omega$

$$\rho_{\nu+m} < \underset{\sim}{\lambda}^1_{\nu+\omega}.$$

Moreover again by the procedure of [Mos78], for each limit ordinal ϑ, $\Phi^{<\vartheta}$ admits a scale on

$$\max\{\vartheta, \sup_{\xi<\vartheta}\rho_\xi\},$$

thus $\Phi^{<\sigma}$ admits a scale on

$$\max\{\sigma, \sup_{\xi<\sigma}\rho_\xi\} \leq \underset{\sim}{\lambda}^1_\sigma \leq \kappa$$

i.e., a κ-scale. Since $\Phi^{<\sigma} \in \underset{\sim}{\mathbf{IND}}_\sigma \setminus \underset{\sim}{\check{\mathbf{IND}}}_\sigma$, we have by Lemma 2.6 that $\Phi^{<\sigma} \notin \underset{\sim}{\mathbf{\Delta}}^1_\sigma = \Lambda$. Thus $\Phi^{<\sigma} \in \boldsymbol{S}(\kappa) \setminus \Lambda$ and we are done. ⊣

We conclude this section by giving the proofs of Lemmas 2.5 and 2.6.

Proof of Lemma 2.5. The proof of the closure properties follows that of the closure properties of $\underset{\sim}{\mathbf{IND}}$ as in [Mos74A, Chapter 1]. As an example let us prove closure under $\exists^{\mathbb{R}}$: Let $A \in \underset{\sim}{\mathbf{IND}}_\lambda$ and let Ψ be a positive elementary on $\mathbb{R}$ operator such that for some continuous f

$$(x, \alpha) \in A \Leftrightarrow f(x, \alpha) \in \Psi^{<\lambda}.$$

Let

$$x \in B \Leftrightarrow \exists\alpha(x, \alpha) \in A.$$

Then

$$\begin{aligned} x \in B &\Leftrightarrow \exists\alpha\exists\xi < \lambda[f(x, \alpha) \in \Psi^\xi] \\ &\Leftrightarrow \exists\xi < \lambda\exists\alpha[f(x, \alpha) \in \Psi^\xi]. \end{aligned}$$

Consider now the following simultaneous induction

$$\begin{aligned} &\Psi_1^\xi(\beta) \Leftrightarrow \Psi(\beta, \Psi_1^{<\xi}) \\ &\Psi_2^\xi(x) \Leftrightarrow \exists\alpha[f(x, \alpha) \in \Psi_1^{<\xi}]. \end{aligned}$$

Then

$$\begin{aligned} &\Psi_1^\xi = \Psi^\xi \text{ and} \\ &\Psi_2^\xi(x) \Leftrightarrow \exists\alpha[f(x, \alpha) \in \Psi^{<\xi}], \end{aligned}$$

so for limit λ,

$$\begin{aligned} \Psi_2^{<\lambda}(x) &\Leftrightarrow \exists\xi < \lambda\exists\alpha[f(x, \alpha) \in \Psi^\xi] \\ &\Leftrightarrow x \in B. \end{aligned}$$

Now, by the Simultaneous Induction Lemma [Mos74A, 1C.1], there is a positive elementary on $\mathbb{R}$ operator Ω and constants x_0 such that

$$\Psi_2^\xi(x) = \Omega^\xi(x_0, x),$$

thus

$$x \in B \Leftrightarrow \Omega^{<\lambda}(x_0, x),$$

so $B \in \underset{\sim}{\mathbf{IND}}_\lambda$ and we are done.

For the prewellordering property we just use again the proof that $\underset{\sim}{\mathbf{IND}}$ has the prewellordering property; see [Mos74A, Chapter 2]. According to it, to each positive elementary on $\mathbb{R}$ operator Ψ we can assign two other positive elementary on $\mathbb{R}$ operators Ψ_1, Ψ_2 such that if ψ is the canonical norm on Ψ^∞ associated with the induction, and $\leq_\psi^*, <_\psi^*$ its corresponding relations, then for each ξ we have

$$\psi(x) \leq \xi \wedge x \leq_\psi^* y \Leftrightarrow \Psi_1^\xi(x, y)$$
$$\psi(x) \leq \xi \wedge x <_\psi^* y \Leftrightarrow \Psi_2^\xi(x, y),$$

so that for limit λ,

$$x \in \Psi^{<\lambda} \wedge x \leq_\psi^* y \Leftrightarrow \Psi_1^{<\lambda}(x, y)$$
$$x \in \Psi^{<\lambda} \wedge x <_\psi^* y \Leftrightarrow \Psi_2^{<\lambda}(x, y).$$

thus $\psi \restriction \Psi^{<\lambda}$ is a $\underset{\sim}{\mathbf{IND}}_\lambda$-norm and this establishes PWO($\underset{\sim}{\mathbf{IND}}_\lambda$).

Finally we prove the statement about parametrization:

First recall that by [Mos70A] every positive $\underset{\sim}{\mathbf{\Pi}}_1^1$ relation $\Psi(x, A)$, where $x \in \mathcal{X}, A \subseteq \mathbb{R}$ can be brought in the form

$$\Psi(x, A) \Leftrightarrow \forall \beta[R(x, \beta) \vee \exists n((\beta)_n \in A)] \wedge \bar{R}(x),$$

with $R, \bar{R} \in \underset{\sim}{\mathbf{\Pi}}_1^1$. Similarly every positive $\underset{\sim}{\mathbf{\Sigma}}_1^1$ relation $\Psi'(x, A)$ is equivalent to one in the form

$$\Psi'(x, A) \Leftrightarrow \exists \beta[S(x, \beta) \wedge \forall n((\beta)_n \in A)] \vee \bar{S}(x),$$

with $S, \bar{S} \in \underset{\sim}{\mathbf{\Sigma}}_1^1$. Similar normal forms are valid for positive $\underset{\sim}{\mathbf{\Pi}}_1^1$ or $\underset{\sim}{\mathbf{\Sigma}}_1^1$ relations $\Psi(x, A)$ for $A \subseteq \mathcal{Y}$ an arbitrary product space. As a result, for each fixed $\mathcal{X}, \mathcal{Y}$ there is a positive $\underset{\sim}{\mathbf{\Pi}}_1^1$ (resp. $\underset{\sim}{\mathbf{\Sigma}}_1^1$) relation of the form $\Psi(\varepsilon, x, A)$ ($\varepsilon \in \mathbb{R}$, $x \in \mathcal{X}$, $A \subseteq \mathcal{Y}$) which is universal for the positive $\underset{\sim}{\mathbf{\Pi}}_1^1$ (resp. $\underset{\sim}{\mathbf{\Sigma}}_1^1$) relations of the form $\Psi(x, A)$.

According to [Mos74A, Chapter 1, Ex. 1.15] one can find for each positive elementary on $\mathbb{R}$ operator $\Psi(x, A)$ an operator $\Psi'(y, x, A)$ which is a disjunction of a $\underset{\sim}{\mathbf{\Pi}}_1^1$ positive operator and a $\underset{\sim}{\mathbf{\Sigma}}_1^1$ positive operator, and constants y_0 such that for all limit λ,

$$\Psi^{<\lambda}(x) \Leftrightarrow \Psi'^{<\lambda}(y_0, x).$$

So it is clear that it is enough to find an operator $\Phi(\varepsilon, \alpha, A)$ ($\varepsilon \in \mathbb{R}, \alpha \in \mathbb{R}, A \subseteq \mathbb{R} \times \mathbb{R}$) which is the disjunction of a positive $\underset{\sim}{\mathbf{\Pi}}_1^1$ and a positive $\underset{\sim}{\mathbf{\Sigma}}_1^1$ operator such that for all such $\Psi(\alpha, X)$ ($\alpha \in \mathbb{R}, X \subseteq \mathbb{R}$) there is $\varepsilon_0 \in \mathbb{R}$ (depending on Ψ) so that:

$$\Psi^{<\lambda}(\alpha) = \Phi^{<\lambda}(\varepsilon_0, \alpha), \text{ for all limit } \lambda.$$

Note that to achieve this it is sufficient to find Φ as above such that for all Ψ as above there is ε_0 (depending on Ψ) such that

$$\Phi(\varepsilon_0, \alpha, A) \Leftrightarrow \Psi(\alpha, \{\alpha' : A(\varepsilon_0, \alpha')\})$$

(Then by a simple induction on ξ,

$$\Phi^{\xi}(\varepsilon_0, \alpha) \Leftrightarrow \Psi^{\xi}(\alpha).)$$

For that, let $\Xi(\delta, \alpha, X)$, where $\delta, \alpha \in \mathbb{R}, X \subseteq \mathbb{R}$, be universal in the class of disjunctions of positive $\underset{\sim}{\mathbf{\Pi}}^1_1$ and $\underset{\sim}{\mathbf{\Sigma}}^1_1$ formulas. Put then

$$\Phi(\varepsilon, \alpha, A) \Leftrightarrow \Xi(\varepsilon, \alpha, \{\alpha' : A(\varepsilon, \alpha')\}).$$

To verify that this works, fix an appropriate Ψ and let ε_0 be such that

$$\Phi(\alpha, X) \Leftrightarrow \Xi(\varepsilon_0, \alpha, X).$$

Then

$$\begin{aligned}\Phi(\varepsilon_0, \alpha, A) &\Leftrightarrow \Xi(\varepsilon_0, \alpha, \{\alpha' : A(\varepsilon_0, \alpha')\} \\ &\Leftrightarrow \Psi(\alpha, \{\alpha' : A(\varepsilon_0, \alpha')\}).\end{aligned}$$ ⊣

Proof of Lemma 2.6. By induction on λ. Obvious for $\lambda = \omega$. If $\lambda = \lambda' + \omega$ is a successor limit ordinal, then by induction hypothesis,

$$\underset{\sim}{\mathbf{\Delta}}^1_{\lambda'} \subsetneq \underset{\sim}{\mathbf{IND}}_{\lambda'} \subsetneq \underset{\sim}{\mathbf{\Delta}}^1_{\lambda'+\omega} = \underset{\sim}{\mathbf{\Delta}}^1_{\lambda}.$$

Since $\underset{\sim}{\mathbf{\Delta}}^1_\lambda$ is strongly closed, it is easy to check that for each positive elementary on $\mathbb{R}$ operator Ψ one has $\Psi^{\lambda'+n} \in \underset{\sim}{\mathbf{\Delta}}^1_\lambda$ for each $n \in \omega$, thus $\Psi^{<\lambda} \in \underset{\sim}{\mathbf{\Sigma}}^1_{\lambda+1} \subsetneq \underset{\sim}{\mathbf{\Delta}}^1_{\lambda+\omega}$. So $\underset{\sim}{\mathbf{IND}}_\lambda \subsetneq \underset{\sim}{\mathbf{\Delta}}^1_{\lambda+\omega}$. To see that $\underset{\sim}{\mathbf{\Delta}}^1_\lambda \subsetneq \underset{\sim}{\mathbf{IND}}_\lambda$, notice that for any $n \in \omega$, if $A \in \underset{\sim}{\mathbf{\Sigma}}^1_{\lambda'+n}$, then A is gotten from some $B \in \underset{\sim}{\mathbf{IND}}_{\lambda'}$ by repeatedly applying a finite number of real quantifications. This is because $\underset{\sim}{\mathbf{\Delta}}^1_{\lambda'} \subsetneq \underset{\sim}{\mathbf{IND}}_{\lambda'}$. Then by the Simultaneous Induction Lemma [Mos74A, 1C.1] it follows that for some positive elementary on $\mathbb{R}$ operator $\Psi_1, A \leq_{\mathrm{W}} \Psi_1^{\lambda'}$. But by the proof of the prewellordering property for $\underset{\sim}{\mathbf{IND}}_\lambda$, clearly $\Psi_1^{\lambda'} \in \underset{\sim}{\mathbf{IND}}_\lambda$, so $A \in \underset{\sim}{\mathbf{IND}}_{\lambda_1}$. Thus $\underset{\sim}{\mathbf{\Delta}}^1_\lambda = \bigcup_n \underset{\sim}{\mathbf{\Sigma}}^1_{\lambda'+n} \subseteq \underset{\sim}{\mathbf{IND}}_\lambda$ and since $\underset{\sim}{\mathbf{\Delta}}^1_\lambda$ is closed under complements $\underset{\sim}{\mathbf{\Delta}}^1_\lambda \subsetneq \underset{\sim}{\mathbf{IND}}_\lambda$.

For the case when λ is the limit of limit ordinals, it immediately follows by induction hypothesis that $\underset{\sim}{\mathbf{\Delta}}^1_\lambda \subsetneq \underset{\sim}{\mathbf{IND}}_\lambda$. For the other inclusion, notice that by our induction hypothesis each member of $\underset{\sim}{\mathbf{IND}}_\lambda$ is the wellordered union of sets in $\underset{\sim}{\mathbf{\Delta}}^1_\lambda$ and thus belongs to $\underset{\sim}{\mathbf{\Sigma}}^1_{\lambda+1}$ (since by [KSS81] either $\underset{\sim}{\mathbf{\Sigma}}^1_\lambda$ or $\underset{\sim}{\mathbf{\Sigma}}^1_{\lambda+1}$ has the prewellordering property and thus is closed under wellordered unions). So $\underset{\sim}{\mathbf{IND}}_\lambda \subsetneq \underset{\sim}{\mathbf{\Delta}}^1_{\lambda+\omega}$. ⊣

§3. The scale property in the hyperprojective hierarchy. As a special case of the results in [KSS81] one can determine the prewellordering pattern in the hyperprojective hierarchy. We do the same thing for the scale property. Moreover we identify the pointclasses $\boldsymbol{S}(\kappa)$, for κ a Suslin cardinal $< \kappa^{\mathbb{R}}$, within the hyperprojective hierarchy. It will be convenient to introduce first the following terminology, motivated by the classification of projective-like hierarchies in [KSS81, §4]:

For ϑ a limit ordinal $< \kappa^{\mathbb{R}}$, we shall say that,

(i) ϑ is type I iff $\mathrm{cf}(\vartheta) = \omega$, or equivalently iff the projective-like hierarchy $\{\underset{\sim}{\boldsymbol{\Pi}}^1_\vartheta, \underset{\sim}{\boldsymbol{\Sigma}}^1_{\vartheta+1}, \underset{\sim}{\boldsymbol{\Pi}}^1_{\vartheta+2}, \underset{\sim}{\boldsymbol{\Sigma}}^1_{\vartheta+3}, \dots\}$ is of type I.

(ii) ϑ is of type II iff $\{\underset{\sim}{\boldsymbol{\Pi}}^1_\vartheta, \underset{\sim}{\boldsymbol{\Sigma}}^1_{\vartheta+1}, \dots\}$ is of type II,

(iii) ϑ is of type III iff $\{\underset{\sim}{\boldsymbol{\Pi}}^1_\vartheta, \underset{\sim}{\boldsymbol{\Sigma}}^1_{\vartheta+1}, \dots\}$ is of type III.

John Steel has proved a beautiful result (see [Ste81A]) which allows us to characterize types II and III above in terms of ordinal invariants. The characterization is as follows.

THEOREM 3.1 (AD, [Ste81A]). *A limit cardinal $\vartheta < \kappa^{\mathbb{R}}$ is of type III iff $\underset{\sim}{\lambda}^1_\vartheta (= \kappa_\vartheta)$ is regular (iff $\vartheta = \underset{\sim}{\lambda}^1_\vartheta (= \kappa_\vartheta)$ and ϑ is regular). (Thus $\vartheta < \kappa^{\mathbb{R}}$ is of type II iff $\mathrm{cf}(\vartheta) > \omega$ and $\underset{\sim}{\lambda}^1_\vartheta (= \kappa_\vartheta)$ is singular.)*

We now have the following

THEOREM 3.2 (AD). *Let ϑ be a limit ordinal $< \kappa^{\mathbb{R}}$. Then we have,*

(i) *If ϑ is of type I, the following classes have the scale property*

$$\underset{\sim}{\boldsymbol{\Pi}}^1_\vartheta, \underset{\sim}{\boldsymbol{\Sigma}}^1_{\vartheta+1}, \underset{\sim}{\boldsymbol{\Pi}}^1_{\vartheta+2}, \underset{\sim}{\boldsymbol{\Sigma}}^1_{\vartheta+3}, \dots,$$

Moreover, $\boldsymbol{S}(\kappa_{\vartheta+m}) = \underset{\sim}{\boldsymbol{\Sigma}}^1_{\vartheta+m}$, for all $m \geq 0$. Finally, for all $m \geq 0$,

$$(\kappa_{\vartheta+m})^+ = \underset{\sim}{\boldsymbol{\delta}}^1_{\vartheta+m} \text{ and is measurable,}$$
$$\underset{\sim}{\boldsymbol{\delta}}^1_{\vartheta+2m+1} = (\underset{\sim}{\boldsymbol{\delta}}^1_{\vartheta+2m})^+, \text{ so } \kappa_{\vartheta+2m+1} = \underset{\sim}{\boldsymbol{\delta}}^1_{\vartheta+2m}, \text{ and}$$
$$\kappa_{\vartheta+2m} \text{ has cofinality } \omega.$$

(ii) *If ϑ is of type II, the following classes have the scale property*

$$\underset{\sim}{\boldsymbol{\Sigma}}^1_\vartheta, \underset{\sim}{\boldsymbol{\Pi}}^1_{\vartheta+1}, \underset{\sim}{\boldsymbol{\Sigma}}^1_{\vartheta+2}, \underset{\sim}{\boldsymbol{\Pi}}^1_{\vartheta+3}, \dots.$$

Moreover,

$$\boldsymbol{S}(\kappa_{\vartheta+m}) = \underset{\sim}{\boldsymbol{\Sigma}}^1_{\vartheta+m}, \text{ for all } m \geq 0.$$

Finally, for all $m \geq 0$,

$$(\kappa_{\vartheta+m})^+ = \underset{\sim}{\boldsymbol{\delta}}^1_{\vartheta+m} \text{ and is measurable,}$$
$$\underset{\sim}{\boldsymbol{\delta}}^1_{\vartheta+2m+2} = (\underset{\sim}{\boldsymbol{\delta}}^1_{\vartheta+2m+1})^+, \text{ so } \kappa_{\vartheta+2m+2} = \underset{\sim}{\boldsymbol{\delta}}^1_{\vartheta+2m+1}, \text{ and}$$
$$\kappa_{\vartheta+2m+1} \text{ has cofinality } \omega.$$

(iii) *If ϑ is of type III, the following classes have the scale property*

$$\underset{\sim}{\boldsymbol{\Pi}}^1_\vartheta, \underset{\sim}{\boldsymbol{\Sigma}}^1_{\vartheta+1}, \underset{\sim}{\boldsymbol{\Sigma}}^1_{\vartheta+2}, \underset{\sim}{\boldsymbol{\Sigma}}^1_{\vartheta+3}, \dots.$$

Moreover

$$S(\kappa_{\vartheta+m}) = \underset{\sim}{\Sigma}^1_{\vartheta+m+1}, \text{ for all } m \geq 0.$$

Finally, for all $m \geq 0$,

$$\begin{gathered}(\kappa_{\vartheta+m})^+ = \underset{\sim}{\delta}^1_{\vartheta+m+1} \text{ is measurable} \\ \underset{\sim}{\delta}^1_{\vartheta+2m+1} = (\underset{\sim}{\delta}^1_{\vartheta+2m})^+, \text{ so } \kappa_{\vartheta+2m} = \underset{\sim}{\delta}^1_{\vartheta+2m}, \\ \kappa_{\vartheta+2m+1} \text{ has cofinality } \omega, \text{ and} \\ \kappa_\vartheta = \underset{\sim}{\delta}^1_\vartheta \text{ and is measurable.}\end{gathered}$$

PROOF. By induction on ϑ.

(i) By induction hypothesis (or by standard facts about the projective hierarchy when $\vartheta = \omega$) $\underset{\sim}{\Delta}^1_\vartheta = \bigcup_{\eta<\vartheta} S(\kappa_\eta)$. Moreover $\kappa_\vartheta = \lim_{\eta<\vartheta} \kappa_\eta = \underset{\sim}{\lambda}^1_\vartheta$ has cofinality ω. Let $\Sigma = \{\bigcup_n A_n : \text{For all } n, A_n \in \underset{\sim}{\Delta}^1_\vartheta\}$. Then, by our induction hypothesis, Σ has the scale property and is closed under $\exists^{\mathbb{R}}$, so since $\underset{\sim}{\Pi}^1_\vartheta = \forall^{\mathbb{R}}\Sigma$ we have, by the Second Periodicity Theorem of [Mos80], that $\underset{\sim}{\Pi}^1_\vartheta, \underset{\sim}{\Sigma}^1_{\vartheta+1}, \underset{\sim}{\Pi}^1_{\vartheta+2}, \dots$ all have the scale property.

Since $\kappa_\vartheta = \underset{\sim}{\lambda}^1_\vartheta$, there is a $\underset{\sim}{\Delta}^1_\vartheta$ prewellordering of $\mathbb{R}$ of length κ_ϑ, thus $S(\kappa_\vartheta) \subseteq \underset{\sim}{\Sigma}^1_\vartheta$. That $\underset{\sim}{\Sigma}^1_\vartheta \subseteq S(\kappa_\vartheta)$ follows from the fact that $\underset{\sim}{\Delta}^1_\vartheta \subsetneq S(\kappa_\vartheta)$ and the closure properties of $S(\kappa_\vartheta)$. Thus $\underset{\sim}{\Sigma}^1_\vartheta = S(\kappa_\vartheta)$. The equalities $S(\kappa_{\vartheta+m}) = \underset{\sim}{\Sigma}^1_{\vartheta+m}$ follow now by the usual arguments as for the projective hierarchy. The same holds for the proofs of the fact about $\underset{\sim}{\delta}^1_{\vartheta+m}$ and $\kappa_{\vartheta+m}$. (A quicker proof that $\kappa_{\vartheta+2m+2}$ has cofinality ω makes use of the fact that $S(\kappa_{\vartheta+2m+1}) = \underset{\sim}{\Sigma}^1_{\vartheta+2m+1}$ is closed under wellordered unions.)

(ii) First notice that if the cofinality of ϑ is bigger than ω, then

$$S(\kappa_\vartheta) = \underset{\sim}{\mathbf{IND}}_\vartheta. \qquad (*)$$

To prove $(*)$ first use the fact that each element of $S(\kappa_\vartheta)$ is a wellordered union of sets in $\bigcup_{\eta<\vartheta} S(\kappa_\eta)$, since $\mathrm{cf}(\kappa_\vartheta) > \omega$. But $\bigcup_{\eta<\vartheta} S(\kappa_\eta) \subseteq \underset{\sim}{\Delta}^1_\vartheta$, by induction hypothesis, and $\underset{\sim}{\Delta}^1_\vartheta \subsetneq \underset{\sim}{\mathbf{IND}}_\vartheta$ by Lemma 2.6. Since $\underset{\sim}{\mathbf{IND}}_\vartheta$ is projective-like closed under $\exists^{\mathbb{R}}$ and has the prewellordering property by Lemma 2.5, it is closed under wellordered unions, thus $S(\kappa_\vartheta) \subseteq \underset{\sim}{\mathbf{IND}}_\vartheta$. For the inclusion $\underset{\sim}{\mathbf{IND}}_\vartheta \subseteq S(\kappa_\vartheta)$, just look at the proof of Theorem 2.1 and note that in the notation there, $\Lambda = \underset{\sim}{\Delta}^1_\vartheta$ by our induction hypothesis, thus $\vartheta = \sigma$.

Since in case ϑ is of type II, $\underset{\sim}{\Sigma}^1_\vartheta$ consists (by the analysis of type II hierarchies in [KSS81]) of all wellordered unions in $\underset{\sim}{\Delta}^1_\vartheta$, we also have in this case that

$$S(\kappa_\vartheta) = \underset{\sim}{\mathbf{IND}}_\vartheta = \underset{\sim}{\Sigma}^1_\vartheta.$$

The rest of the conclusions of (ii) will follow routinely once we can show that each set in $\underset{\sim}{\mathbf{IND}}_\vartheta$ admits an $\underset{\sim}{\mathbf{IND}}_\vartheta$-scale, each norm of which has length $\leq \kappa_\vartheta$.

For that let Φ be the universal positive elementary on $\mathbb{R}$ operator of Lemma 2.5. Let $\langle \varphi^\xi_n \rangle$ be the canonical scale on Φ^ξ as defined in [Mos78].

Let also $\langle \varphi_n \rangle$ be the scale on $\Phi^{<\vartheta}$ defined there, i.e.,

$$\varphi_0(x) = \text{least } \xi < \vartheta \text{ such that } x \in \Phi^{\xi}$$
$$\varphi_{n+1}(x) = \langle \varphi_0(x), \varphi_n^{\varphi_0(x)} \rangle.$$

Clearly φ_0 is a $\underset{\sim}{\mathbf{IND}}_\vartheta$-norm. Now we have

$$x \leq^*_{\varphi_{n+1}} y \text{ iff } x <^*_{\varphi_0} y \vee \exists \xi < \vartheta[\varphi_0(x) = \varphi_0(y) = \xi \wedge \varphi_n^{\xi}(x) \leq \varphi_n^{\xi}(y)].$$

Now it is easy to verify that for each fixed $\xi < \vartheta$ the relation

$$\varphi_0(x) = \varphi_0(y) = \xi \wedge \varphi_n^{\xi}(x) \leq \varphi_n^{\xi}(y)$$

is in $\underset{\sim}{\mathbf{\Delta}}^1_\vartheta$, thus $\leq^*_{\varphi_{n+1}}$ is the wellordered union for $\underset{\sim}{\mathbf{\Delta}}^1_\vartheta$ relations, thus belongs to $\underset{\sim}{\mathbf{IND}}_\vartheta$. Similarly for $<^*_{\varphi_{n+1}}$ and we are done. The fact that all these norms have length $\leq \kappa_\vartheta$ is included in the proof of Theorem 2.1.

(iii) By $(*)$ of case (ii) we know that

$$S(\kappa_\vartheta) = \underset{\sim}{\mathbf{IND}}_\vartheta = \underset{\sim}{\mathbf{\Sigma}}^1_{\vartheta+1},$$

so all the assertions of (iii) follow except that $\underset{\sim}{\mathbf{\Pi}}^1_\vartheta$ has the scale property. We prove this now.

First recall from [KSS81, 2.4] that $\underset{\sim}{\mathbf{\Pi}}^1_\vartheta$ is equal to the class of all $\underset{\sim}{\mathbf{\Delta}}^1_\vartheta$ $(= \underset{\sim}{\mathbf{\Delta}}^1_\vartheta)$-bounded wellordered unions of $\underset{\sim}{\mathbf{\Delta}}^1_\vartheta$ sets.

Fix now $A \in \underset{\sim}{\mathbf{\Pi}}^1_\vartheta$ and let Φ be the universal positive elementary on $\mathbb{R}$ operator considered before. For $x \in \Phi^\infty$, let $|x| \equiv \varphi_0(x) =$ least ξ such that $x \in \Phi^\xi$. Put then

$$(\varepsilon, x) \in C \Leftrightarrow x \in \Phi^{<\vartheta} \wedge \varepsilon \text{ codes a continuous function } (\equiv f_\varepsilon),$$

and for $(\varepsilon, x) \in C$, let

$$Y_{\varepsilon,x} = f_\varepsilon^{-1}[\Phi^{<|x|}].$$

Then put

$$B = \{(\varepsilon, x) \in C : Y_{\varepsilon,x} \subseteq A\},$$

so that $B \in \underset{\sim}{\mathbf{\Sigma}}^1_{\vartheta+1} = \underset{\sim}{\mathbf{IND}}_\vartheta$. Then there is continuous g such that

$$B = g^{-1}[\Phi^{<\vartheta}].$$

Put finally for $\xi < \vartheta$,

$$A_\xi = \bigcup \{Y_{\varepsilon,x} : g(\varepsilon, x) \in \Phi^{<\xi} \wedge x \in \Phi^{<\xi}\}.$$

Clearly $A = \bigcup_{\xi<\vartheta} A_\xi$, $A_\xi \in \underset{\sim}{\mathbf{\Delta}}^1_\vartheta$ and $\{A_\xi\}_{\xi<\vartheta}$ is a $\underset{\sim}{\mathbf{\Delta}}^1_\vartheta$-bounded union of sets in $\underset{\sim}{\mathbf{\Delta}}^1_\vartheta$. Now

$$\alpha \in A_\xi \Leftrightarrow \exists \varepsilon \exists x[g(\varepsilon, x) \in \Phi^{<\xi} \wedge x \in \Phi^{<\xi} \wedge f_\varepsilon(\alpha) \in \Phi^{<|x|}]$$
$$\Leftrightarrow \exists \varepsilon \exists x[g(\varepsilon, x) \in \Phi^{<\xi} \wedge x \in \Phi^{<\xi} \wedge f_\varepsilon(\alpha) <^*_{\varphi_0} x].$$

Let Φ_1 be a positive elementary on $\mathbb{R}$ operator such that

$$z \in \Phi^\eta \wedge z <^*_{\varphi_0} y \Leftrightarrow \Phi_1^\eta(z, y).$$

Then

$$\alpha \in A_\xi \Leftrightarrow \exists \varepsilon \exists x[g(\varepsilon, x) \in \Phi^{<\xi} \wedge x \in \Phi^{<\xi} \wedge \Phi_1^{<\xi}(f_\varepsilon(\alpha), x)].$$

From this it is clear that we can define a map $\xi \mapsto \langle \psi_n^\xi \rangle$, where for each $\xi < \vartheta$, the sequence $\langle \psi_n^\xi \rangle$ is a $\underset{\sim}{\mathbf{\Delta}}^1_\vartheta$-scale on A_ξ, making use of the canonical scales being put, as in [Mos78], on $\Phi^{<\xi}$ and $\Phi_1^{<\xi}$. This in turn defines the following scale on $A = \bigcup_{\xi<\vartheta} A_\xi$:

$$\psi_0(\alpha) = \text{ least } \xi \text{ such that } \alpha \in A_\xi,$$

$$\psi_{n+1}(\alpha) = \langle \psi_0(\alpha), \psi_n^{\psi_0(\alpha)}(\alpha) \rangle.$$

We show that this is a $\underset{\sim}{\mathbf{\Pi}}^1_\vartheta$-scale. Clearly $\leq^*_{\psi_\vartheta}, <^*_{\psi_0}$ are in $\underset{\sim}{\mathbf{\Pi}}^1_\vartheta$ (see [KSS81, 2.5]). Consider now $\leq^*_{\psi_{n+1}}$ (the argument for $<^*_{\psi_{n+1}}$ being similar). We have

$$\alpha \leq^*_{\psi_{n+1}} \beta \text{ iff } \alpha <^*_{\psi_0} \beta \vee \exists \xi < \vartheta[\exists \xi' \leq \xi(\psi_0(\alpha) = \psi_0(\beta) = \xi' \\ \wedge \psi_n^{\xi'}(\alpha) \leq \psi_n^{\xi'}(\beta))],$$

thus

$$\leq^*_{\psi_{n+1}} = <^*_{\psi_0} \cup \bigcup_{\xi<\vartheta} D_\xi,$$

where

$$D_\xi = \bigcup_{\xi' \leq \xi} \{(\alpha, \beta) : \psi_0(\alpha) = \psi_0(\beta) = \xi' \wedge \psi_n^{\xi'}(\alpha) \leq \psi_n^{\xi'}(\beta)\} \in \underset{\sim}{\mathbf{\Delta}}^1_\vartheta.$$

So it is enough to check that $\langle D_\xi : \xi < \vartheta \rangle$ is $\underset{\sim}{\mathbf{\Delta}}^1_\vartheta$-bounded. For that let $X \in \underset{\sim}{\mathbf{\Delta}}^1_\vartheta$, $X \subseteq \bigcup_{\xi<\vartheta} D_\xi$ be given. Then

$$\{\alpha : \exists \beta(\alpha, \beta) \in X\} = Y$$

is in $\underset{\sim}{\mathbf{\Delta}}^1_\vartheta$ and $Y \subseteq \bigcup_{\xi<\vartheta} A_\xi$, so for some $\xi < \vartheta$, $Y \subseteq A_\xi$. But then $X \subseteq D_\xi$. Indeed, if $(\alpha, \beta) \in X$, then $\alpha \in Y$, thus $\psi_0(\alpha) = \xi' \leq \xi$. So also $\psi_0(\beta) = \xi'$ and since $(\alpha, \beta) \in \bigcup_{\xi<\vartheta} D_\xi$ we have that $\psi_n^{\xi'}(\alpha) \leq \psi_n^{\xi'}(\beta)$, thus $(\alpha, \beta) \in D$. ⊣

We have now the following immediate corollaries:

Corollary 3.3 (AD). Let $\underset{\sim}{\Gamma}$ be any projective-like pointclass contained in $\underset{\sim}{\mathbf{IND}}$. Then one of $\underset{\sim}{\Gamma}$ or $\check{\underset{\sim}{\Gamma}}$ has the scale property.

Corollary 3.4 (AD). Let κ be a Suslin cardinal $\leq \kappa^{\mathbb{R}}$. Then

(i) $\boldsymbol{S}(\kappa)$ has the scale property iff $\mathrm{cf}(\kappa) > \omega$.
(ii) $\check{\boldsymbol{S}}(\kappa)$ has the scale property iff $\mathrm{cf}(\kappa) = \omega$.
(iii) κ^+ is always measurable.

(iv) If $\kappa = \kappa_\eta$ and $\kappa^* = \kappa_{\eta+1}$ is the next Suslin cardinal, then
(a) If κ has cofinality ω, then $\kappa^* = \kappa^+$ is a measurable cardinal, while
(b) If κ has cofinality $> \omega$, then κ^* has cofinality ω.

COROLLARY 3.5 (AD). Let $\kappa < \kappa^{\mathbb{R}}$ be a Suslin cardinal. Then there is a prewellordering of $\mathbb{R}$ in $\boldsymbol{S}(\kappa) \cap \breve{\boldsymbol{S}}(\kappa)$ of length κ.

Finally, define for each $A \in \boldsymbol{S}_\infty$, the cardinal $\kappa(A)$, the **Suslin cardinal of** A, to be the smallest κ such that $A \in \boldsymbol{S}(\kappa)$. Recall that $\|A\|_{\mathrm{W}}$ denotes the Wadge ordinal of A. We have now the following estimate.

COROLLARY 3.6 (AD). For each $A \in \underset{\sim}{\mathbf{IND}}$, $\kappa(A) \leq \|A\|_{\mathrm{W}}$.

PROOF. Let $\kappa(A) = \kappa_\eta, \eta < \kappa^{\mathbb{R}}$ (the case $\kappa(A) = \kappa^{\mathbb{R}}$ is obvious). If $\eta = \vartheta$ is limit and $\|A\|_{\mathrm{W}} < \kappa_\vartheta = \underset{\sim}{\lambda}^1_\vartheta$, then $A \in \underset{\sim}{\Delta}^1_\vartheta$, so $\kappa(A) < \underset{\sim}{\lambda}^1_\vartheta$, a contradiction. If $\eta = \vartheta + m$, ϑ limit, $m > 0$, consider cases according to the type of ϑ. ⊣

§4. Gaps in the propagation of scales. Assume AD for the discussion in this section.

The uninterrupted propagation of the scale property throughout the hyperprojective hierarchy is disturbed as one goes past the class of inductive sets. If $\{\underset{\sim}{\Gamma}_i\}$ is a projective-like hierarchy of type IV (see [KSS81]), then Kechris has shown that neither $\underset{\sim}{\Gamma}_1$ nor $\breve{\underset{\sim}{\Gamma}}_1$ can have the scale property and Martin substantially extended this to show that in fact no $\underset{\sim}{\Gamma}_i$ or $\breve{\underset{\sim}{\Gamma}}_i$ can have the scale property. In particular, $\{\underset{\sim}{\Pi}^*_1, \underset{\sim}{\Sigma}^*_2, \underset{\sim}{\Pi}^*_3, \underset{\sim}{\Sigma}^*_4, \dots\}$ is the least projective-like hierarchy of type IV (so that $\underset{\sim}{\Pi}^*_1 = \forall^{\mathbb{R}}(\underset{\sim}{\mathbf{IND}} \vee \breve{\underset{\sim}{\mathbf{IND}}})$), no $\underset{\sim}{\Sigma}^*_n$ or $\underset{\sim}{\Pi}^*_n$ can have the scale property. Thus Corollary 3.3 cannot be extended past $\underset{\sim}{\mathbf{IND}}$. Compare this with the corresponding result about the prewellordering property (see [KSS81]), which extends to all $\Gamma \subseteq \mathbf{L}(\mathbb{R})$ and beyond. (In particular $\underset{\sim}{\Sigma}^*_1, \underset{\sim}{\Pi}^*_2, \underset{\sim}{\Sigma}^*_3, \dots$ all have the prewellordering property.)

Now if $\underset{\sim}{\Sigma}^*_\omega$ denotes the smallest projective-like pointclass closed under $\exists^{\mathbb{R}}$ and containing $\bigcup_n \underset{\sim}{\Sigma}^*_n$, then [Mos83] has shown that Π^*_ω has the scale property and thus so do $\underset{\sim}{\Sigma}^*_{\omega+1}, \Pi^*_{\omega+2}, \dots$, and the familiar pattern for the propagation of scales resumes again for a while, after this gap of length ω. But later on wider and wider gaps occur, reflecting eventually the "unbounded" gap occurring in $\mathbf{L}(\mathbb{R})$ beyond $\boldsymbol{S}_\infty = \underset{\sim}{\Sigma}^2_1$. We will not however pursue this matter any further here.

Beyond this first occurrence of gaps, other phenomena happen for the first time at the level of $\kappa^{\mathbb{R}}$. For example, $\kappa^{\mathbb{R}}$ is the least Suslin cardinal κ for which the conclusion of Corollary 3.5 fails and also the least ordinal κ for which there is an A with $\|A\|_{\mathrm{W}} = \kappa$ but $\kappa(A) > \|A\|_{\mathrm{W}}$. (Take A to be a complete co-inductive set.)

§5. Miscellaneous remarks, questions, and conjectures. We assume again AD throughout this section.

5.1. Closure properties of $\boldsymbol{S}(\kappa)$. Let κ be a Suslin cardinal. Of course $\boldsymbol{S}(\kappa)$ is closed under $\exists^{\mathbb{R}}$, but when is it also closed under $\forall^{\mathbb{R}}$? The following result gives a necessary and sufficient condition when $\mathrm{cf}(\kappa) > \omega$.

We need first to define a notion of indescribability of ordinals. For any limit ordinal λ let $B_\lambda = \{x : \exists \xi < \lambda, x \subseteq \mathbf{L}_\xi\}$. Thus $\mathbf{L}_\lambda \subseteq B_\lambda$ and B_λ is transitive. Recall that a formula in the language of ZF augmented by extra predicates is Π_2 if it has the form $\forall x \exists y \varphi$, where φ is bounded. We call now an ordinal λ **${}^b\Pi^1_2$-indescribable** if for each $X \subseteq \mathbf{L}_\lambda$ and each Π_2 formula φ of the appropriate language, with parameters from B_λ, we have

$$(B_\lambda, \in, X) \models \varphi \Rightarrow \exists \vartheta < \lambda, (B_\vartheta, \in, X \cap \mathbf{L}_\vartheta) \models \varphi.$$

It is easy to check that such λ's are regular cardinals. We have now

THEOREM 5.1 (AD). *Let κ be a Suslin cardinal of cofinality greater than ω. Then the following are equivalent:*

(i) *$\boldsymbol{S}(\kappa)$ is closed under $\forall^{\mathbb{R}}$,*

(ii) *κ is ${}^b\Pi^1_2$-indescribable.*

PROOF. (i) $\Rightarrow$ (ii). Let $\Gamma = \boldsymbol{S}(\kappa)$. Then $\underset{\sim}{\Gamma}$ is a Spector class on the structure of analysis $\mathbb{R}$, so if $\underset{\sim}{\delta} = \sup\{\xi : \xi$ is the length of a $\underset{\sim}{\Delta}$ prewellordering of $\mathbb{R}\}$, by the Companion Theorem of [Mos74A], $\underset{\sim}{\delta}$ is the ordinal of its companion admissible set M above $\mathbb{R}$. As every admissible set is Π_2 reflecting and every set $A \subseteq \mathbf{L}_{\underset{\sim}{\delta}}$ is Δ_1 in M, by the Moschovakis Coding Lemma (see [Mos70A]), it is easy to verify that $\underset{\sim}{\delta}$ is ${}^b\Pi^1_2$-indescribable (this argument is due to Moschovakis).

So it is enough to show that $\underset{\sim}{\delta} = \kappa$. By Kunen-Martin, $\underset{\sim}{\delta} \leq \kappa^+$ and, since $\underset{\sim}{\delta}$ is a limit cardinal, $\underset{\sim}{\delta} \leq \kappa$. Let now $A \in \boldsymbol{S}(\kappa) \setminus \bigcup_{\eta<\kappa} \boldsymbol{S}(\eta)$. Let $\langle \varphi_n \rangle$ be a regular κ-scale on A. We can assume moreover that $\langle \varphi_n \rangle$ is good, i.e.,

$$\varphi_i(x) \leq \varphi_i(y) \Rightarrow \forall j \leq i(\varphi_j(x) \leq \varphi_j(y)).$$

Fix now n. For $\xi < \mathrm{length}(\varphi_n)$, let

$$K_n(\xi) = \sup\{\varphi_j(x) : x \in A \wedge \varphi_n(x) = \xi\}.$$

Then $K_n(\xi) < \kappa$, since if $\varphi_n(x_0) > \xi$, then for any x with $\varphi_n(x) = \xi < \varphi_n(x_0)$ we must have $\varphi_j(x) \leq \varphi_j(x_0)$, $\forall j \leq n$, by goodness, but also $\varphi_j(x) < \varphi_j(x_0)$, $\forall j \geq n$, by goodness again, so $K_n(\xi) \leq \sup \varphi_j(x_0) < \kappa$.

Let now T be the tree on $\omega \times \kappa$ associated with $\langle \varphi_i \rangle$. Let

$$T_n(\xi) = \{\langle a_0, u_0, \dots, a_{m-1}, u_{m-1} \rangle \in T : u_n \leq \xi \wedge \forall i < m, u_i \leq K_n(\xi)\}$$

and similarly for $T'_n(\xi)$, replacing $u_n \leq \xi$ by $u_n < \xi$. Then

$$A_n^\xi = \{x : \varphi_n(x) \leq \xi\} = \mathrm{p}[T_n(\xi)],$$

$$B_n^\xi = \{x : \varphi_n(x) < \xi\} = \mathrm{p}[T_n'(\xi)],$$

so since $T_n(\xi)$, $T_n'(\xi)$ are trees on $K_n(\xi) < \kappa$, clearly $A_n^\xi, B_\eta^\xi \in \underset{\sim}{\mathbf{\Delta}}$. Thus

$$C_n^\xi = \{x : \varphi_n(x) = \xi\} \in \underset{\sim}{\mathbf{\Delta}}.$$

The rest of the argument is as in the proof of the Lemma 1.4.

(ii) $\Rightarrow$ (i) Let T be a tree on $\omega \times \kappa$. Let $(x, \alpha) \in B \Leftrightarrow T(x, \alpha)$ is not wellfounded. Let

$$\begin{aligned} x \in A &\Leftrightarrow \forall \alpha (x, \alpha) \in B \\ &\Leftrightarrow \forall \alpha, T(x, \alpha) \text{ is not wellfounded} \\ &\Leftrightarrow (B_\kappa, \in, x, T) \models \forall \alpha \exists \xi (T \restriction \xi(x, \alpha) \text{ is not wellfounded}) \\ &\quad \text{(since } \mathrm{cf}(\kappa) > \omega), \\ &\Leftrightarrow (B_\kappa, \in, x, T\} \models \forall \alpha \exists \xi \exists f \in \xi^\omega \forall n (x \restriction n, \alpha \restriction n, f \restriction n) \in T \\ &\Rightarrow \exists \kappa' < \kappa, (B_{\kappa'}, \in, x, T \restriction \kappa') \models (*) \end{aligned}$$

(where $(*)$ is the Π_2 formula $\forall \alpha \exists \xi \exists f \in \xi^\omega (x \restriction n, \alpha \restriction n, f \restriction n) \in T)$)

$$\Rightarrow \exists \kappa' < \kappa, \forall \alpha [T \restriction \kappa'(x, \alpha) \text{ is not wellfounded}]).$$

Thus

$$x \in A \Leftrightarrow \forall \alpha [S(x, \alpha) \text{ is not wellfounded}],$$

where S is a tree on $\omega^2 \times \kappa'$. So

$$A \in \forall^{\mathbb{R}} \boldsymbol{S}(\kappa').$$

Since B is an arbitrary element of $\boldsymbol{S}(\kappa)$, we have that $\boldsymbol{S}(\kappa) \subseteq \forall^{\mathbb{R}} \boldsymbol{S}(\kappa')$, so if $\forall^{\mathbb{R}} \boldsymbol{S}(\kappa') \subseteq \boldsymbol{S}(\kappa)$ we are done. Otherwise, by Wadge, $\boldsymbol{S}(\kappa) \subseteq \exists^{\mathbb{R}} \check{\boldsymbol{S}}(\kappa')$, so since $\boldsymbol{S}(\kappa) \supseteq \boldsymbol{S}(\kappa')$, we have $\boldsymbol{S}(\kappa) = \exists^{\mathbb{R}} \check{\boldsymbol{S}}(\kappa')$. Let $\bar{\kappa}$ be the largest Suslin cardinal $\leq \kappa'$ (which must exist since $\boldsymbol{S}(\kappa)$ is $\mathbb{R}$-parametrized). Then $\boldsymbol{S}(\kappa) = \exists^{\mathbb{R}} \check{\boldsymbol{S}}(\bar{\kappa})$ and $\boldsymbol{S}(\bar{\kappa})$ is a projective-like pointclass closed under $\exists^{\mathbb{R}}$, thus $\exists^{\mathbb{R}} \check{\boldsymbol{S}}(\bar{\kappa}) = \boldsymbol{S}(\kappa) \not\subseteq \forall^{\mathbb{R}} \boldsymbol{S}(\bar{\kappa})$, a contradiction. ⊣

Note that from the preceding argument and the fact that for every Spector class $\underset{\sim}{\Gamma}$ on $\mathbb{R}$ its associated ordinal $\underset{\sim}{\delta}$ is Mahlo (see [KKMW81]) we have that if either of the equivalent conditions (i), (ii) above hold then actually κ is also Mahlo.

As a corollary of Theorem 5.1, we see that $\kappa^{\mathbb{R}}$ is the least ${}^b\Pi^1_2$-indescribable Suslin cardinal.

Conjecture 5.2. The assumption $\mathrm{cf}(\kappa) > \omega$ is not needed in Theorem 5.1.

Question 5.3. Is $\kappa^{\mathbb{R}}$ the least Mahlo ${}^b\Pi^1_2$-indescribable cardinal?

5.2. The Prewellordering and scale properties for $S(\kappa)$. By [KSS81], for each Suslin cardinal κ, either $S(\kappa)$ or $\check{S}(\kappa)$ has the prewellordering property. From Corollary 3.4 we see that, when $\kappa \leq \kappa^{\mathbb{R}}$, what distingushes the first case from the second is whether $\mathrm{cf}(\kappa) > \omega$ or not.

QUESTION 5.4. Does this hold also for arbitrary κ?

Similarly about the scale property.

One fact that we have noticed is that if $S(\kappa)$ is closed under $\forall^{\mathbb{R}}$, then $S(\kappa)$ has the prewellordering property.

5.3. Properties of the Suslin cardinals.

QUESTION 5.5. Do the properties of Corollary 3.4 (iii), (iv) hold for arbitrary Suslin cardinals?

5.4. About S_∞. According to our conjecture in §2 we expect S_∞ either to be all of $\wp(\mathbb{R})$ or else to be $\mathbb{R}$-parametrized. In the latter case $S_\infty = S(\kappa)$, where κ is the largest Suslin cardinal, so, by our remarks in Section 5.2, S_∞ (being closed under $\forall^{\mathbb{R}}$) also has the prewellordering property, i.e., is a Spector class on the structure of analysis $\mathbb{R}$. We mentioned in §1 that assuming $\mathsf{AD} + \mathbf{V} = \mathbf{L}(\mathbb{R})$, $S_\infty = \utilde{\Sigma}^2_1$. It is conceivable that in some reasonable theory extending AD one can prove that if S_∞ is $\mathbb{R}$-parametrized, then $S_\infty \subseteq \utilde{\Sigma}^2_1$. It is already known (in AD only) that there are some forbidden values for S_∞ when we go past $\utilde{\Sigma}^2_1$. For example, S_∞ cannot be $\utilde{\Sigma}^2_n$ for $n > 1$ or $\utilde{\Pi}^2_n$ for $n \geq 1$, and similarly for $\utilde{\Sigma}^k_n, \utilde{\Pi}^k_n$ for all $k \geq 3, n \geq 1$. Also it cannot be $\utilde{\Delta}^k_n$ for any $k + n > 3$. These observations are based on the fact that

$$\{A : A \in S_\infty\} \text{ is } \utilde{\Sigma}^2_1,$$

(as a collection of sets of reals).

5.5. Reliable cardinals. According to [Bec79] an ordinal λ is called **reliable** if there is a regular scale $\langle \varphi_i \rangle$ on a set $A \subseteq \mathbb{R}$ such that $\{\varphi_i(\alpha) : i \in \omega, \alpha \in A\} = \lambda$. Clearly every Suslin cardinal is reliable.

CONJECTURE 5.6 (AD). At least for cardinals $\leq \kappa^{\mathbb{R}}$, the notions of being Suslin and reliable coincide.

If one goes back to our analysis of the hyperprojective hierarchy in §3 it is easy to check that the following conjecture (which is motivated by some conjectures in [Kec78C]) implies the preceding one.

CONJECTURE 5.7 (AD). If κ is a Suslin cardinal and $S(\kappa)$ has the scale property, then any strictly increasing wellordered sequence of sets in $S(\kappa)$ has length $< \kappa^+$.

This is not even known for $\kappa = \aleph_1$, i.e., for $S(\kappa) = \utilde{\Sigma}^1_2$. It is known by a different argument that there are no reliable cardinals between $\aleph_1$ and $\aleph_\omega$. But it is not known if there are any reliable cardinals between $\underset{\sim}{\delta}^1_3 = \aleph_{\omega+1}$ and the predecessor of $\underset{\sim}{\delta}^1_5$.

The proof that there are no reliable cardinals between $\aleph_1$ and $\aleph_\omega$ is based on the following more general fact.

PROPOSITION 5.8 (AD). Assume λ is reliable and there is a tree V on $\omega \times \lambda$, with $\sup\{\text{rank}(V(\alpha)) : V(\alpha) \text{ is wellfounded}\} = \lambda^+$. Then λ^+ is regular.

REFERENCES

HOWARD S. BECKER

[Bec79] ***Some applications of ordinal games***, Ph.D. thesis, UCLA, 1979.

ALEXANDER S. KECHRIS

[Kec78C] *On transfinite sequences of projective sets with an application to $\underset{\sim}{\Sigma}^1_2$ equivalence relations*, ***Logic colloquium '77*** (A. Macintyre, L. Pacholski, and J. Paris, editors), North-Holland, 1978, pp. 155–160.

ALEXANDER S. KECHRIS, EUGENE M. KLEINBERG, YIANNIS N. MOSCHOVAKIS, AND W. HUGH WOODIN

[KKMW81] *The axiom of determinacy, strong partition properties and nonsingular measures*, this volume, originally published in Kechris et al. [CABAL ii], pp. 75–100.

ALEXANDER S. KECHRIS, DONALD A. MARTIN, AND YIANNIS N. MOSCHOVAKIS

[CABAL ii] *Cabal seminar* 77–79, Lecture Notes in Mathematics, no. 839, Berlin, Springer, 1981.

[CABAL iii] *Cabal seminar* 79–81, Lecture Notes in Mathematics, no. 1019, Berlin, Springer, 1983.

ALEXANDER S. KECHRIS AND YIANNIS N. MOSCHOVAKIS

[CABAL i] *Cabal seminar* 76–77, Lecture Notes in Mathematics, no. 689, Berlin, Springer, 1978.

ALEXANDER S. KECHRIS, ROBERT M. SOLOVAY, AND JOHN R. STEEL

[KSS81] *The axiom of determinacy and the prewellordering property*, In Kechris et al. [CABAL ii], pp. 101–125.

YIANNIS N. MOSCHOVAKIS

[Mos70A] *Determinacy and prewellorderings of the continuum*, ***Mathematical logic and foundations of set theory. Proceedings of an international colloquium held under the auspices of the Israel Academy of Sciences and Humanities, Jerusalem, 11–14 November 1968*** (Y. Bar-Hillel, editor), Studies in Logic and the Foundations of Mathematics, North-Holland, Amsterdam-London, 1970, pp. 24–62.

[Mos74A] ***Elementary induction on abstract structures***, North-Holland, 1974.

[Mos78] *Inductive scales on inductive sets*, this volume, originally published in Kechris and Moschovakis [CABAL i], pp. 185–192.

[Mos80] ***Descriptive set theory***, Studies in Logic and the Foundations of Mathematics, no. 100, North-Holland, Amsterdam, 1980.

[Mos83] *Scales on coinductive sets*, this volume, originally published in Kechris et al. [CABAL iii], pp. 77–85.

ROBERT M. SOLOVAY

[Sol78B] *The independence of* DC *from* AD, In Kechris and Moschovakis [CABAL i], pp. 171–184.

John R. Steel
[Ste81A] *Closure properties of pointclasses*, In Kechris et al. [Cabal ii], pp. 147–163.

DEPARTMENT OF MATHEMATICS
CALIFORNIA INSTITUTE OF TECHNOLOGY
PASADENA, CA 91125, USA
E-mail: kechris@caltech.edu

THE AXIOM OF DETERMINACY, STRONG PARTITION PROPERTIES AND NONSINGULAR MEASURES

ALEXANDER S. KECHRIS, EUGENE M. KLEINBERG, YIANNIS N. MOSCHOVAKIS, AND W. HUGH WOODIN

In this paper we study the relationship between AD and strong partition properties of cardinals as well as some consequences of these properties themselves.

Let us say that an uncountable cardinal κ has **the strong partition property** if

$$(\forall \mu < \kappa)[\kappa \rightarrow (\kappa)^{\kappa}_{\mu}],$$

i.e., if for every partition of all increasing κ-term sequences into fewer than κ parts, there is a set $C \subseteq \kappa$ of cardinality κ such that all the increasing sequences into C lie in the same part of the partition. We will show in §1 that the axiom of determinacy (AD) implies the existence of unboundedly many cardinals with the strong partition property below Θ, where Θ is the supremum of the ranks of the prewellorderings of the continuum.

In §2, we will show that conversely, if there is a cardinal κ with the strong partition property above an ordinal λ, then every λ-Suslin set is determined. Combining these two results we obtain an elegant purely set-theoretic characterization of AD within $\mathbb{R}^+$, the smallest admissible set containing the continuum; namely, in $\mathbb{R}^+$, AD holds if and only if the power set of every ordinal exists, and there are arbitrarily large cardinals with the strong partition property.

In §3 we will strengthen the main result of §1 to obtain from AD unboundedly many cardinals κ below ϑ, such that not only κ has the strong partition property but also $\{\lambda < \kappa : \lambda \text{ has the strong partition property}\}$ is stationary in κ (thus κ is also Mahlo). Finally, in §4 we will show that every Mahlo cardinal with the strong partition property carries a normal measure concentrating on regular cardinals. Up until now all the normal measures on cardinals κ, produced by AD, were of the "singular" type, i.e., for some regular $\lambda < \kappa$ they concentrated on the ordinals of cofinality λ.

A.S.K. was partially supported by NSF Grant MCS79-20465 and an A.P. Sloan Foundation Fellowship. E.M.K. was partially supported by NSF Grant MCS78-03744. Y.N.M. was partially supported by NSF Grant MCS78-02989.

In matters of descriptive set theory, we will follow in general the terminology and notation of [Mos80]. We refer to [Kle77] for information concerning partition properties.

Our work in this paper takes place in ZF + DC with all other hypothesis (mainly AD) stated explicitly.

§1. A partition theorem. In the main result of this section, we will establish that AD implies the existence of many cardinals with the strong partition property.

As usual, a **space** will be any product

$$\mathfrak{X} = X_1 \times \cdots \times X_n$$

of copies of ω and $\mathbb{R} = {}^{\omega}\omega$, a **pointset** is any subset of a space and a **pointclass** is any collection Γ of pointsets. A **Γ-norm** on a pointset $P \subseteq \mathfrak{X}$ is any function

$$\varphi : P \to \mathrm{Ord}$$

such that both $\leq^*_\varphi$ and $<^*_\varphi$ are in Γ, where

$$x \leq^*_\varphi y \Leftrightarrow x \in P \wedge [y \notin P \vee \varphi(x) \leq \varphi(y)],$$
$$x <^*_\varphi y \Leftrightarrow x \in P \wedge [y \notin P \vee \varphi(x) < \varphi(y)].$$

We say the Γ has the **prewellordering property** if every pointset in Γ admits a Γ-norm.

Recall from [Mos80] that a **Spector pointclass closed under** $\forall^{\mathbb{R}}$ is a pointclass Γ which is closed under recursive substitutions, $\wedge, \vee, \exists^\omega, \forall^\omega$ and $\forall^{\mathbb{R}}$, which is ω-parametrized and which has the prewellordering property. We will need here a slightly stronger notion.

A partial function

$$f : \mathfrak{X} \to \omega$$

is **Γ-recursive** (or in Γ), if

$$\mathrm{Graph}(f) = \{(x, i) : f(x) = i\}$$

is in Γ. We say that Γ is **closed under Kleene's** ${}^3E$ (deterministic quantification on $\mathbb{R}$), if whenever $f : \mathfrak{X} \times \mathbb{R} \to \omega$ is a Γ-recursive partial function, then the relation

$$P(x) \Leftrightarrow (\forall\alpha)[f(x, \alpha)\downarrow] \wedge (\exists\alpha)[f(x, \alpha) = 0]$$

is in Γ.

If Γ is closed under Kleene's ${}^3E$, then (trivially) Γ is closed under $\forall^{\mathbb{R}}$ and Δ, $\underset{\sim}{\Delta}$ are both closed under $\forall^{\mathbb{R}}$ and $\exists^{\mathbb{R}}$; and if Γ is closed under both $\forall^{\mathbb{R}}$ and $\exists^{\mathbb{R}}$, then (again trivially) Γ is closed under Kleene's ${}^3E$. By [Mos67], if A is any pointset, then the **envelope** $\mathrm{Env}(A, {}^3E)$ is defined to be the class of all pointsets which are Kleene-semirecursive in ${}^3E$ and (the characteristic function of) A. This is a Spector pointclass closed under ${}^3E$, which furthermore contains both A and its complement and is *not closed* under $\exists^{\mathbb{R}}$. These Kleene envelopes are very important for the applications of our present results, although one need not know any recursion in type-3 in order to follow the proofs in §1.

We can now state our main result in this section.

THEOREM 1.1. Assume AD, let Γ be Spector pointclass closed under ${}^3E$ and let

$$\begin{aligned}\kappa &= \mathrm{o}(\underset{\sim}{\mathbf{\Delta}})\\ &= \text{supremum of the ranks of prewellorderings of } \mathbb{R} \text{ in } \underset{\sim}{\mathbf{\Delta}}\end{aligned}$$

be the ordinal associated with Γ; then κ has the strong partition property.

This result implies that (granting AD),

$$\begin{aligned}\kappa({}^3E) &= \text{the Kleene ordinal of the continuum}\\ &= \text{the ordinal associated with } \mathrm{Env}({}^3E)\end{aligned}$$

has the strong partition property, as does the closure ordinal of the continuum

$$\kappa^{\mathbb{R}} = \sup\{\xi : \xi \text{ in the rank of a hyperprojective prewellordering of } \mathbb{R}\}.$$

It also implies that (under AD again), there are arbitrarily large cardinals with the strong partition property below Θ.

One very important problem is whether the projective ordinals $\underset{\sim}{\delta}^1_{2n+1}, n \geq 1$ (in particular $\underset{\sim}{\delta}^1_3$), have the strong partition property, granting AD. Clearly our Theorem 1.1 does not tell us anything in this case, since the Spector pointclasses $\Gamma = \Pi^1_{2n+1}$ are not closed under ${}^3E$.

Using a method of Martin (see [Kec78A, Lemma 11.1]), the proof of Theorem 1.1 is reduced to the problem of finding an appropriate coding (by elements of $\mathbb{R}$) of functions $f : \kappa \to \kappa$. Let us first reformulate Martin's Lemma in a form convenient for our purposes here.

Suppose Γ is a Spector pointclass closed under $\forall^{\mathbb{R}}$ and with associated ordinal κ, let

$$\varphi : S \to \kappa$$

be a Γ-norm which maps some set $S \subseteq \mathbb{R}$ in Γ onto κ and suppose that for each $\varepsilon \in \mathbb{R}$ we have a partial function

$$f_\varepsilon : \kappa \to \kappa;$$

we will say that φ and $\{f_\varepsilon : \varepsilon \in \mathbb{R}\}$ define a **good coding in** Γ of the functions on κ to κ if conditions (i)–(iii) below hold, where for ordinals $\xi, \vartheta < \kappa$, we let

$$\begin{aligned}C_\xi(\varepsilon) &\Leftrightarrow f_\varepsilon(\xi)\downarrow,\\ C_{\xi,\vartheta}(\varepsilon) &\Leftrightarrow (\forall \xi' \leq \xi)[f_\varepsilon(\xi')\downarrow \wedge f_\varepsilon(\xi') \leq \vartheta],\end{aligned}$$

(i) For each fixed $\xi, \vartheta < \kappa$, the relation $C_{\xi,\vartheta}$ is in $\underset{\sim}{\Delta}$.

(ii) There is some relation $V(\varepsilon, \alpha, \beta)$ in $\check{\underset{\sim}{\Gamma}}$ which computes the values of each f_ε relative to the norm φ in the following sense:

$$\alpha \in S \wedge f_\varepsilon(\varphi(\alpha))\downarrow \Rightarrow (\exists \beta) V(\varepsilon, \alpha, \beta) \\ \wedge (\forall \beta)\{V(\varepsilon, \alpha, \beta) \Rightarrow [\beta \in S \wedge f_\varepsilon(\varphi(\alpha)) = \varphi(\beta)]\}.$$

(iii) For every total function $f : \kappa \to \kappa$, there is some $\varepsilon \in \mathbb{R}$ such that $f = f_\varepsilon$.

LEMMA 1.2. Assume AD and let Γ be a Spector pointclass closed under $\forall^{\mathbb{R}}$ with ordinal κ, which admits a good coding of the functions on κ to κ; then κ has the strong partition property.

PROOF. Notice first that a good coding in Γ of the functions on κ to κ has the following additional boundedness property.

(ii′) If $A \in \check{\underset{\sim}{\Gamma}}$ and for some $\xi < \kappa$, $A \subseteq C_\xi$, then

$$\sup\{f_\varepsilon(\xi) : \varepsilon \in A\} < \kappa.$$

This is because, by the definitions, if $\alpha \in S$ and $\varphi(\alpha) = \xi$, then

$$\sup\{f_\varepsilon(\xi) : \varepsilon \in A\} \leq \sup\{\varphi(\beta) : (\exists \varepsilon)[\varepsilon \in A \wedge V(\varepsilon, \alpha, \beta)]\}$$

and the set $\{\beta : (\exists \varepsilon)[\varepsilon \in A \wedge V(\varepsilon, \alpha, \beta)]\}$ is a subset of S which lies in $\check{\underset{\sim}{\Gamma}}$, so that the standard boundedness argument for Spector pointclasses closed under $\forall^{\mathbb{R}}$ applies (see [Mos80, 4C.11]).

Using this observation, we can prove the lemma by a small modification of the argument given in [Kec78A, Lemma 11.1], whose notation we will use.

Let $\langle A_\eta : \eta < \mu \rangle$, where $\mu < \kappa$, be a partition of ${}^{\kappa}[\kappa]$ into μ many pieces. For each $\eta < \mu$ consider the game presented there with A replaced by $\neg A_\eta$. Call it G_η. If for some $\eta < \mu$ player I has a winning strategy in G_η, then clearly there is a homogeneous set landing in A_η, so we are done. So assume player II has a winning strategy in each G_η, $\eta < \mu$, towards a contradiction. By the Coding Lemma [Mos80, 7D.5] let $\langle S_\eta : \eta < \mu \rangle$ be a sequence of sets such that

(a) $S_\eta \neq 0$,
(b) $\sigma \in S_\eta \Rightarrow \sigma$ is a winning strategy in G_η,
(c) $\bigcup_{\eta<\mu} S_\eta \in \check{\underset{\sim}{\Gamma}}$,

and put

$$G(\xi, \vartheta) = \sup\{f_{\sigma(\varepsilon)}(\xi) + 1 : \sigma \in \bigcup_{\eta<\mu} S_\eta \wedge \varepsilon \in C_{\xi,\vartheta}\}.$$

By boundedness again $G(\xi, \vartheta) < \kappa$, and from this it is easy to find a closed unbounded set $D \subseteq \kappa$ such that

$$^{\kappa}D\!\uparrow\, = D\!\uparrow\, \subseteq \neg A_\eta, \quad \forall \eta < \mu;$$

thus $D\!\uparrow\, \subseteq \bigcap_{\eta<\mu} \neg A_\eta = \varnothing$, a contradiction. Here $^{\kappa}D\!\uparrow$ denotes the set of all increasing maps $h\colon \kappa \to D$ such that $h(x) > \sup\{h(y) : y < x\}$ and there is $\langle \xi_n^x : n \in \omega\rangle_{x\in\kappa}$ such that $\xi_0^x < \xi_1^x < \cdots < \xi_n^x < \cdots \to h(x)$. ⊣

Thus to prove Theorem 1.1 it will be sufficient to show that if Γ is a Spector pointclass closed under Kleene's ^3E, then Γ admits a good coding of the functions on its ordinal κ.

The key to this proof is a strong (uniform) version of the Coding Lemma (I), [Mos80, 7D.5], which is implicit in the proof of that result, particularly as that was described in the original paper [Mos70A].

Consider then the customary **language of analysis** (or second-order number theory), where we have variables $n, k, i, \ldots$ over ω and $\alpha, \beta, \gamma, \ldots$ over $\mathbb{R}$, symbols for $0, 1, =, +$ and $\cdot$ on ω and application "$\alpha(t)$" of variables over $\mathbb{R}$ on terms. We obtain an extension $\mathcal{L}(\chi)$ of this language by adding new prime formulas of the form

$$\chi(\alpha, \beta) \simeq m;$$

in the intended interpretation, χ will denote a partial function from $\mathbb{R} \times \mathbb{R}$ into ω in the obvious way. We will often denote a formula of $\mathcal{L}(\chi)$ by a symbol such as

$$\varphi(\chi, x_1, \ldots, x_n),$$

which shows explicitly the occurrence of χ; we will then use (ambiguously) the same symbol $\varphi(\chi, x_1, \ldots, x_n)$ to denote the relation on $\chi, x_1, \ldots, x_n$ defined by the formula.

Let $\Sigma^1_1(\chi)$ be the smallest collection of formulas of $\mathcal{L}(\chi)$ which contains all ordinary Σ^1_1 formulas (with no χ) and the prime formula $\chi(\alpha, \beta) \simeq m$ and which is closed under the positive operations $\wedge$, $\vee$, $\exists^\omega$, $\forall^\omega$, and $\exists^{\mathbb{R}}$. It is important that we do not allow the negative formula $\neg(\chi(\alpha, \beta) \simeq m)$ in $\Sigma^1_1(\chi)$.

Lemma 1.3. For each $\Sigma^1_1(\chi)$ formula $\varphi(\chi, x_1, \ldots, x_n)$, there are (ordinary) Σ^1_1 formulas $\rho(x_1, \ldots, x_n)$ and $\sigma(x_1, \ldots, x_n, \alpha, \beta, \gamma)$ such that for all $\chi, x_1, \ldots, x_n$,

$$\begin{aligned}\varphi(\chi, x_1, \ldots, x_n) \Leftrightarrow \rho(x_1, \ldots, x_n) \\ \vee\, \exists\alpha\exists\beta\exists\gamma\big(\forall t[\chi((\alpha)_t, (\beta)_t) \simeq \gamma(t)] \\ \wedge\, \sigma(x_1, \ldots, x_n, \alpha, \beta, \gamma)\big).\end{aligned}$$

Proof. The proof of this is very easy, as in [Mos80, 7D.7] or [Mos70A, Lemma 1]—where the argument is not completely correct, as it does not work for the empty partial function χ. ⊣

From Lemma 1.3 and the known parametrization theorems for Σ^1_1, we can obtain easily a uniform (in χ) parametrization theorem for $\Sigma^1_1(\chi)$.

LEMMA 1.4. For each product

$$\mathfrak{X} = X_1 \times \cdots \times X_n$$

of copies of ω and $\mathbb{R}$, there is a fixed $\Sigma^1_1(\chi)$ formula

$$G^{\mathfrak{X}}(\chi, \varepsilon, x) \Leftrightarrow G(\chi, \varepsilon, x),$$

where ε varies over η, x varies over $\mathfrak{X}$ and the following are true:

(i) If $\varphi(\chi, x)$ is any $\Sigma^1_1(\chi)$ formula with x varying over $\mathfrak{X}$, then for some recursive ε and all χ and x,

$$\varphi(\chi, x) \Leftrightarrow G(\chi, \varepsilon, x).$$

(ii) For each pair of spaces $\mathfrak{Y}, \mathfrak{X}$, there is a recursive function

$$S^{\mathfrak{X},\mathfrak{Y}} = S : \mathbb{R} \times \mathfrak{Y} \to \mathbb{R},$$

so that for all χ, x, y (omitting superscripts)

$$G(\chi, \varepsilon, y, x) \Leftrightarrow G(\chi, S(\varepsilon, y), x).$$

After these preliminary lemmas, fix a prewellordering $\leq$ on a subset S of $\mathbb{R}$ with rank function

$$\rho : S \to \lambda,$$

and let

$$f : \lambda^n \to \wp(\mathfrak{Y})$$

be any function which assigns to n-tuples from λ subsets (possibly empty) of a space $\mathfrak{Y}$. A **choice set** for f is any subset

$$C \subseteq \mathbb{R}^n \times \mathfrak{Y},$$

such that

(i) $(\alpha_1, \ldots, \alpha_n, y) \in C \Rightarrow \alpha_1, \ldots, \alpha_n \in S \wedge y \in f(\rho(\alpha_1), \ldots, \rho(\alpha_n))$,

(ii) $f(\xi_1, \ldots, \xi_n) \neq \varnothing \Rightarrow$ for each $\alpha_1, \ldots, \alpha_n \in S$ with $\rho(\alpha_1) = \xi_1, \ldots, \rho(\alpha_n) = \xi_n$, there is some y such that $(\alpha_1, \ldots, \alpha_n, y) \in C$.

In effect, C assigns to each $\xi_1, \ldots, \xi_n < \lambda$ such that $f(\xi_1, \ldots, \xi_n) \neq \varnothing$ a non-empty subset of $f(\xi_1, \ldots, \xi_n)$.

If $\alpha \notin S$, put by convention

$$\rho(\alpha) = \infty > \lambda$$

and consider the partial function $\chi = \chi(\leq)$ encoding $\leq$:

$$\chi(\alpha, \beta) \simeq \begin{cases} 1, & \text{if } \alpha \in S \wedge \rho(\alpha) \leq \rho(\beta) \\ 0, & \text{if } \beta \in S \wedge \rho(\beta) < \rho(\alpha). \end{cases}$$

For each $\mu < \lambda$ consider also the approximations

$$\chi_\mu(\alpha, \beta) \simeq \begin{cases} 1, & \text{if } \alpha \in S \wedge \rho(\alpha) \leq \mu \wedge \rho(\alpha) \leq \rho(\beta), \\ 0, & \text{if } \beta \in S \wedge \rho(\beta) \leq \mu \wedge \rho(\beta) < \rho(\alpha). \end{cases}$$

Thus $\chi_\mu \subseteq \chi_{\mu'}$ if $\mu \leq \mu' < \lambda$ and

$$\bigcup_{\mu<\lambda} \chi_\mu = \chi.$$

If $f : \lambda^n \to \wp(\mathfrak{Y})$ as above and $\mu < \lambda$, we let f_μ be the restriction of f to μ,

$$f_\mu(\xi_1, \ldots, \xi_n) = \begin{cases} f(\xi_1, \ldots, \xi_n), & \text{if } \xi_1, \ldots, \xi_n \leq \mu, \\ \varnothing, & \text{otherwise.} \end{cases}$$

Lemma 1.5 (The Uniform Coding Lemma). Assume AD, let $\leq$ be a prewellordering on a subset S of $\mathbb{R}$ with rank function

$$\rho : S \twoheadrightarrow \lambda$$

and associated partial functions χ_μ ($\mu < \lambda$), let

$$f : \lambda^n \to \wp(\mathfrak{Y})$$

be a function and let $G(\chi, \varepsilon, \alpha_1, \ldots, \alpha_n, y)$ be the universal $\Sigma^1_1(\chi)$ formula of Lemma 1.4. There is a fixed $\varepsilon^* \in \mathbb{R}$ such that for all $\mu < \lambda$ the relation

$$C_\mu(\alpha_1, \ldots, \alpha_n, y) \Leftrightarrow G(\chi_\mu, \varepsilon^*, \alpha_1, \ldots, \alpha_n, y)$$

is a choice set for f_μ.

Proof. Proof of this is a modification of the proof of [Mos80, 7D.5].

Take $n = 1$ for convenience, and in the notation of Lemma 1.5, call ε^* a **uniform** code of a choice set for f (relative to $\leq$). Assume, towards a contradiction, that this lemma fails and pick the least λ such that for some $\leq$ and some f as above there is no uniform code ε^* for f. It is easy to check that λ is limit. Fix such a counterexample $\leq$ and f for λ now. Consider then the game where player I plays α and player II plays β and

Player II wins iff [α is not a uniform code for any $f_\xi, \xi < \lambda$
(relative to $\leq\restriction\xi = \{(\alpha, \beta) : \alpha \leq \beta \wedge \rho(\beta) \leq \xi\}$)]
$\vee$ [α is a uniform code for some f_ξ and there is $\eta > \xi$
such that β is a uniform code for f_η].

The proof can be now completed as in [Mos80, 7D.5]. ⊣

We can finally produce the desired coding of functions $f : \kappa \to \kappa$ by elements of $\mathbb{R}$.

LEMMA 1.6. Assume AD, let Γ be a Spector pointclass closed under ${}^3E$ and let κ be the ordinal of Γ. Then Γ admits a good coding of the functions on κ to κ.

PROOF. Choose a set $S \subseteq \mathbb{R}$ in Γ and a Γ-norm

$$\varphi : S \twoheadrightarrow \kappa$$

and let $\leq = \leq_\varphi$ be the associated prewellordering, so that in the notation we have established, for $\mu < \kappa$,

$$\chi_\mu(\alpha, \beta) \simeq \begin{cases} 1, & \text{if } \alpha \in S \wedge \varphi(\alpha) \leq \mu \wedge \varphi(\alpha) < \varphi(\beta), \\ 0, & \text{if } \beta \in S \wedge \varphi(\beta) \leq \mu \wedge \varphi(\beta) < \varphi(\alpha). \end{cases}$$

(Again $\varphi(\alpha) = \infty$, if $\alpha \notin S$.) Let $G(\chi, \varepsilon, \alpha, \beta)$ be the universal $\Sigma^1_1(\chi)$ formula of Lemma 1.4, and put for $\xi < \kappa$:

$$\begin{aligned} f_\varepsilon(\xi)\downarrow \Leftrightarrow \forall \alpha\{[\alpha \in S \wedge \varphi(\alpha) = \xi] \Rightarrow \exists \beta G(\chi_\xi, \varepsilon, \alpha, \beta)\} \\ \wedge \forall \alpha \forall \alpha' \forall \beta \forall \beta' \{[\alpha, \alpha' \in S \wedge \varphi(\alpha) = \varphi(\alpha') = \xi \\ \wedge G(\chi_\xi, \varepsilon, \alpha, \beta) \wedge G(\chi_\xi, \varepsilon, \alpha', \beta')] \\ \Rightarrow [\beta \in S \wedge \beta' \in S \wedge \varphi(\beta) = \varphi(\beta')]\}; \end{aligned} \tag{$*$}$$

If $f_\varepsilon(\xi)\downarrow$ put

$$\begin{aligned} f_\varepsilon(\xi) = {} & \text{the unique } \zeta \text{ such that for some } \alpha, \beta \in S \text{ with } \varphi(\alpha) = \xi, \\ & \varphi(\beta) = \zeta, \text{ we have } G(\chi_\xi, \varepsilon, \alpha, \beta). \end{aligned}$$

We now verify for this φ and $\{f_\varepsilon : \varepsilon \in \mathbb{R}\}$ the conditions (i)–(iii) in the definition of a good coding.

Condition (iii) is a direct consequence of the Uniform Coding Lemma, applied to

$$f^*(\xi) = \{\beta : \beta \in S \wedge f(\xi) = \varphi(\beta)\}.$$

To prove (i) and (ii) notice first that for each fixed $\xi < \kappa$, the relation

$$\chi_\xi(\alpha, \beta) \simeq m$$

is in $\underset{\sim}{\Delta}$, and in fact the following stronger assertion is true: there is a Γ-recursive partial function $F(\alpha, \beta, m, \gamma)$ such that

$$\begin{aligned} \gamma \in S \Rightarrow F(\alpha, \beta, m, \gamma)\downarrow \wedge \\ [F(\alpha, \beta, m, \gamma) \simeq 1 \Leftrightarrow \chi_{\varphi(\gamma)}(\alpha, \beta) \simeq m]; \end{aligned}$$

this follows directly from the fact that φ is a Γ-norm. Now using the hypothesis that Γ is closed under ${}^3E$, it follows immediately that there are relations $P(\varepsilon, \alpha, \beta, \gamma)$ and $\check{P}(\varepsilon, \alpha, \beta, \gamma)$ in Γ and $\check{\Gamma}$ respectively, such that

$$\begin{aligned} \gamma \in S \Rightarrow [P(\varepsilon, \alpha, \beta, \gamma) \Leftrightarrow \check{P}(\varepsilon, \alpha, \beta, \gamma) \\ \Leftrightarrow G(\chi_{\varphi(\gamma)}, \varepsilon, \alpha, \beta)], \end{aligned}$$

i.e., the key relation $G(\chi_\xi, \varepsilon, \alpha, \beta)$ is *in* $\underset{\sim}{\Delta}$, *for each* ξ, *uniformly in* ξ.

We can establish (ii) immediately by setting

$$V(\varepsilon, \alpha, \beta) \Leftrightarrow \check{P}(\varepsilon, \alpha, \beta, \alpha).$$

To check (i), notice that the relation

$$f_\varepsilon(\xi)\downarrow \wedge f_\varepsilon(\xi) \leq \vartheta$$

is defined by replacing in $(*)$ above the last clause

$$\beta \in S \ \wedge \ \beta' \in S \ \wedge \ \varphi(\beta) = \varphi(\beta')$$

by

$$\beta \in S \ \wedge \ \beta' \in S \ \wedge \ \varphi(\beta) = \varphi(\beta') \leq \vartheta;$$

for this too we can find some $Q, \check{Q}$ in $\Gamma, \check{\Gamma}$ respectively such that

$$\begin{aligned} \gamma \in S \Rightarrow & [Q(\beta, \beta', \gamma) \Leftrightarrow \check{Q}(\beta, \beta', \gamma) \\ & \Leftrightarrow \beta \in S \ \wedge \ \beta' \in S \ \wedge \ \varphi(\beta) = \varphi(\beta') \leq \varphi(\gamma)]. \end{aligned}$$

Putting these equivalences together then and using the closure properties of Γ, we find some $R, \check{R}$ in $\Gamma, \check{\Gamma}$ respectively so that

$$\begin{aligned} \gamma, \gamma' \in S \Rightarrow & [R(\varepsilon, \gamma, \gamma') \Leftrightarrow \check{R}(\varepsilon, \gamma, \gamma') \\ & \Leftrightarrow f_\varepsilon(\varphi(\gamma))\downarrow \wedge f_\varepsilon(\varphi(\gamma)) \leq \varphi(\gamma')], \end{aligned}$$

from which (i) follows directly. ⊣

§2. **Partition properties imply determinacy.** Our main goal now will be to show that partition properties of cardinals imply the determinacy of Suslin sets. Combining this and the result of §1 we obtain an elegant set theoretical equivalent of AD within the smallest admissible set above the continuum.

What actually comes up in the proof is a relatively weak consequence of the strong partition property which we establish first.

LEMMA 2.1. If $\kappa \rightarrow (\kappa)^\mu$ for each $\mu < \kappa$, then we can associate with each wellordering W of rank $\leq \kappa$ a countably additive measure μ_W on the set ${}^W[\kappa]$ of increasing W-term sequences in κ, so that the following coherence property holds: if $W' \subseteq W$ is a subordering of W and $A \subseteq {}^W[\kappa]$, then

$$\mu_W(A) = 1 \Rightarrow \mu_{W'}\{f \restriction W' : f \in A\} = 1.$$

PROOF. As usual, let ${}^W A\!\uparrow$ $(A \subseteq \kappa)$ be the set of all increasing maps h from W into A with the property that $h(x) > \sup\{h(y) : y < x\}$ and there is $\langle \xi_n^x : n \in \omega \rangle_{x \in W}$ such that for each $x \in W$, $\xi_0^x < \xi_1^x < \cdots < \xi_n^x < \cdots \rightarrow h(x)$. Then the partition property $\kappa \rightarrow (\kappa)^\mu, \forall \mu < \kappa$, easily implies that the following is a (countably additive) measure on ${}^W[\kappa]$:

$$\mu_W(A) = 1 \text{ iff there is a closed unbounded } C \subseteq \kappa \text{ with } {}^W C\!\uparrow \subseteq A.$$

The coherence property for these measures follows immediately. ⊣

Recall that a set $A \subseteq \mathbb{R}$ is called **λ-Suslin** if there is a tree T on $\omega \times \lambda$ such that

$$A = \mathrm{p}[T] = \{\alpha \in \mathbb{R} : \exists f \in {}^{\omega}\lambda (\alpha, f) \in [T]\}.$$

THEOREM 2.2. Let λ be an ordinal and assume that there is a cardinal $\kappa > \lambda$ with the property that

$$\kappa \to (\kappa)^{\mu}, \quad \forall \mu < \kappa.$$

Then every λ-Suslin subset of $\mathbb{R}$ is determined.

PROOF. Let $A = \mathrm{p}[T]$, where T is a tree on $\omega \times \lambda$. For each $\varnothing \neq \sigma \in {}^{<\omega}\omega =$ the set of all finite sequences from ω, let $T'(\sigma) = \{u : \mathrm{lh}(u)(= m) \leq \mathrm{length}(\sigma) \wedge (\sigma{\restriction}m, u) \in T\}$. Let $<_{\mathrm{KB}}$ be the Kleene-Brouwer ordering on ${}^{<\omega}\lambda$ and let W_σ be $<_{\mathrm{KB}}$ restricted to $T'(\sigma)$, so that W_σ is a wellordering. For convenience, we shall identify many times W_σ with its domain $T'(\sigma)$. If $T(\alpha) = \bigcup_{n\geq 1} T'(\alpha{\restriction}n)$ is wellfounded, let W_α be again $<_{\mathrm{KB}}$ restricted on $T(\alpha)$, so that $W_\alpha = \bigcup_{n\geq 1} W_{\alpha{\restriction}n}$. Note of course that

$$\varnothing \neq \sigma \subseteq \sigma' = W_\sigma \subseteq W_{\sigma'}.$$

Using Lemma 2.1, let $\{\mu_\sigma : \varnothing \neq \sigma \in {}^{<\omega}\omega\}$ be an assignment of countably additive measures to the function sets ${}^{W_\sigma}[\kappa]$ so that the coherence property of the lemma holds and let μ_α be the analogous measure on ${}^{W_\alpha}[\kappa]$, whenever W_α is a wellordering. We will use these measures to reduce the game (associated with) A to a closed game A^*.

The game A is played as follows:

	I	$\alpha(0)$		$\alpha(2)$		$\dots$
A						
	II		$\alpha(1)$		$\alpha(3)$	

$$\text{Player II wins} \Leftrightarrow \alpha \notin A$$
$$\Leftrightarrow T(\alpha) \text{ is wellfounded.}$$

In the auxiliary game A^*, player II makes additional moves as follows:

	I	$\alpha(0)$		$\alpha(2)$		$\dots$
A^*						
	II		$\alpha(1), f_1$		$\alpha(3), f_3$	

To win, player II must insure that for each n and for each $i \leq n$, f_{2i+1} is an order preserving map from $W_{\alpha\restriction 2i+1}$ into κ, and $f_1 \subseteq f_3 \subseteq \cdots \subseteq f_{2n+1}$.

Clearly this is a closed game for player II, so it is determined. Actually, since the auxiliary moves by player II come from a set which is not necessarily wellorderable, without the axiom of choice, one can only assert that either player I has a winning strategy or else player II has a "multiple-valued" winning strategy or **quasistrategy**. The easiest way to visualize a winning quasistrategy for player II is as a set Q of sequences $\langle\alpha(0);\alpha(1), f_1;\dots;\alpha(2n+1), f_{2n+1}\rangle$ closed under subsequences such that the winning conditions for player II are satisfied and such that

(a) $\forall\alpha(0)\exists\alpha(1), f_1\,\langle\alpha(0);\alpha(1), f_1\rangle \in Q$, and
(b) for every $\langle\alpha(0);\alpha(1), f_1;\dots;\alpha(2n+1), f_{2n+1}\rangle \in Q$ and every $\alpha(2n+2)$ there is some $\alpha(2n+3), f_{2n+3}$ such that $\langle\alpha(0);\alpha(1), f_1;\dots;\alpha(2n+1), f_{2n+1};\alpha(2n+2);\alpha(2n+3), f_{2n+3}\rangle \in Q$.

Since however player I plays only natural numbers, an easy application of DC shows that if player II has a winning quasistrategy he actually has a winning strategy.

If now indeed player II has a winning strategy in the auxiliary game A^*, he clearly has a winning strategy in the original game. So assume player I has a winning strategy τ^* in the auxiliary game. We will define a strategy τ for player I in A by using the measures μ_σ to "integrate out" player II's auxiliary moves in the usual way.

For any

$$\sigma = \langle\alpha(0), \alpha(1), \dots, \alpha(2n+1)\rangle$$

and any order preserving

$$f : T'(\sigma) \to \kappa,$$

let

$$f_1, f_3, \dots, f_{2n+1}$$

be the auxiliary moves induced by f, i.e.,

$$f_{2i+1} = f \restriction T'(\alpha(0), \dots, \alpha(2i+1)) \quad (i \le n).$$

At position σ of the game A then, have player I play by

$$\tau(\sigma) = m \Leftrightarrow \text{for } \mu_\sigma\text{-almost all } f,$$
$$\tau^*(\alpha(0), \alpha(1), f_1, \dots, \alpha(2n+1), f_{2n+1}) = m.$$

Assume towards a contradiction that player I follows τ but loses, in a run of A which produces the play

$$\alpha = \alpha(0), \alpha(1), \alpha(2), \dots,$$

so that $T(\alpha)$ is wellfounded, i.e., W_α is a wellordering. By the construction, for each n we have a set

$$B_n \subseteq {}^{W_{\alpha\restriction(2n+1)}}[\kappa]$$

of measure 1 in the canonical measure on this space, so that player I's play by τ in A together with any member of B_n is a partial play by player I in A^*

which follows τ^*. Now by the coherence property of the measures, for each n, the set

$$\{f \in {}^{W_\alpha}[\kappa] : f \restriction T'(\alpha(0), \dots, \alpha(2n+1)) \in B_n\}$$

has measure 1, so the intersection of all these sets has measure 1 and is not empty; if f is in this set, then

$$\alpha(0), \alpha(1), f \restriction T'(\alpha(0), \alpha(1)), \dots, \alpha(2n+1), f \restriction T'(\alpha(0), \dots, \alpha(2n+1)), \dots$$

is a play in A^*, where player I plays by τ^* and loses, contrary to our assumptions. ⊣

Let $\mathbb{R}^+$ be the smallest admissible set containing $\mathbb{R}$. By combining Theorems 1.1 and 2.2 we now have

THEOREM 2.3. The following are equivalent

(i) $\mathbb{R}^+ \models \mathsf{AD}$.
(ii) $\mathbb{R}^+ \models \forall\kappa(\wp(\kappa)$ exists) and $\forall\lambda\exists\kappa > \lambda(\kappa$ has the strong partition property).

PROOF. Assume $\mathbb{R}^+ \models \mathsf{AD}$. That $\mathbb{R}^+ \models \forall\kappa(\wp(\kappa)$ exists), follows immediately from the Coding Lemma [Mos80, 7D.5]. That $\mathbb{R}^+ \models \forall\lambda\exists\kappa > \lambda(\kappa$ has the strong partition property), follows from Theorem 1.1.

Conversely assume (ii). By the proof of Theorem 2.2 we have that

$$\mathbb{R}^+ \models \forall A \subseteq \mathbb{R}(A \text{ is } \lambda\text{-Suslin for some } \lambda \Rightarrow A \text{ is determined}) \qquad (*)$$

(To adapt the proof of Theorem 2.2 in this context one needs to recall that for any admissible set $\mathcal{A}$ and any open game on a set $x \in \mathcal{A}$, if the player that tries to win the open side has a winning quasistrategy, he has one which is in $\mathcal{A}$.) Recall now ([Mos80, 7C] and [Mos74A, Chapter 9]) that the pointsets $A \subseteq \mathbb{R}$ in $\mathbb{R}^+$ are exactly the hyperprojective ones. So it is enough to show that every hyperprojective pointset A carries a hyperprojective scale. As a warmup let us see how to do this for the projective sets.

Since every $\utilde{\Sigma}^1_2$ set carries a $\utilde{\Sigma}^1_2$-scale we immediately have that it is determined by $(*)$. So we have $\mathsf{Det}(\utilde{\Sigma}^1_2)$, and therefore by the Second Periodicity Theorem [Mos80, 6C.3 and 6C.1] every $\utilde{\Sigma}^1_4$ set carries a $\utilde{\Sigma}^1_4$-scale. Again by $(*)$ we have $\mathsf{Det}(\utilde{\Sigma}^1_4)$, thus by another use of the Second Periodicity Theorem, every $\utilde{\Sigma}^1_6$ set carries a $\utilde{\Sigma}^1_6$-scale, etc.

We will prove the general result about hyperprojective sets now, by defining an appropriate hierarchy on these sets and extending the above argument through the transfinite.

In the rest of this proof we will follow the terminology and notation of [Mos80, 7C]. Recall that $\utilde{\mathbf{IND}}$ is the pointclass of all inductive pointsets. For each ordinal ξ and each positive analytical operator $\phi(x, A)$, let ϕ^ξ be its ξth iterate, i.e.,

$$\phi^{\xi}(x) \Leftrightarrow \phi(x, \phi^{<\xi}),$$

where $\phi^{<\xi}(x) \Leftrightarrow \exists\eta < \xi\phi^{\eta}(x)$. For each limit ordinal $\lambda < \kappa^{\mathbb{R}}$, let

$$\underset{\sim}{\mathbf{IND}}_{\lambda} = \{A : A \text{ is the continuous preimage of } \phi^{<\lambda}, \quad \text{for some positive analytical operator } \phi\}.$$

We summarize some basic structural properties of $\underset{\sim}{\mathbf{IND}}_{\lambda}$.

LEMMA 2.4. For each limit $\lambda < \kappa^{\mathbb{R}}$, the pointclass $\underset{\sim}{\mathbf{IND}}_{\lambda}$ is closed under continuous substitutions, $\wedge$, $\vee$, $\cup^{\omega}$, $\exists^{\mathbb{R}}$, has the prewellordering property and is $\mathbb{R}$-parametrized. In fact there is a single "universal" positive analytical operator $\phi_0(\varepsilon, \alpha, A)$, A varying over subsets of $\mathbb{R} \times \mathbb{R}$, such that for each limit $\lambda < \kappa^{\mathbb{R}}$, $\phi_0^{<\lambda}$ is universal for the $\underset{\sim}{\mathbf{IND}}_{\lambda}$ subsets of $\mathbb{R}$, and moreover for each positive analytical $\phi(\alpha, B)$ (with B varying over subsets of $\mathbb{R}$) there is a recursive ε such that for all limit $\lambda < \kappa^{\mathbb{R}}$, $\phi^{<\lambda}(\alpha) \Leftrightarrow \phi_0^{<\lambda}(\varepsilon, \alpha)$.

For a proof see for example [Kec81B].

We define also the pointclass $\underset{\sim}{\mathbf{IND}}_{\lambda+2n}$ for each $n \geq 0$ by the induction

$$\underset{\sim}{\mathbf{IND}}_{\lambda+2n+2} = \exists^{\mathbb{R}}\forall^{\mathbb{R}}\underset{\sim}{\mathbf{IND}}_{\lambda+2n}.$$

Again these are closed under $\wedge$, $\vee$, $\cup^{\omega}$, $\cap^{\omega}$, $\exists^{\mathbb{R}}$ and are $\mathbb{R}$-parametrized. Let us in fact fix canonical universal sets $U_{\lambda+2n}$ for each $\underset{\sim}{\mathbf{IND}}_{\lambda+2n}$ as follows:

$$U_{\lambda} = \phi_0^{<\lambda},$$
$$U_{\lambda+2n+2} = \{(\varepsilon, \alpha) : \exists\beta\forall\gamma U_{\lambda+2n}(\varepsilon, \langle\alpha, \beta, \gamma\rangle)\}.$$

These are universal sets in $\underset{\sim}{\mathbf{IND}}_{\lambda+2n}$ for the $\underset{\sim}{\mathbf{IND}}_{\lambda+2n}$ subsets of $\mathbb{R}$, but as in [Mos80, 3H.1] one can easily build upon them a good universal system $\{U^{\mathfrak{X}}_{\lambda+2n}\}$, where $U^{\mathfrak{X}}_{\lambda+2n}$ is universal for the subsets of a product space $\mathfrak{X}$ in $\underset{\sim}{\mathbf{IND}}_{\lambda+2n}$. If $A \subseteq \mathfrak{X}$ and $x \in A \Leftrightarrow (\varepsilon, x) \in U^{\mathfrak{X}}_{\lambda+2n}$, we call ε an $\underset{\sim}{\mathbf{IND}}_{\lambda+2n}$**-code** of A. One of the basic properties of a good universal system is that the closure properties of $\underset{\sim}{\mathbf{IND}}_{\lambda+2n}$ are uniform in the $\underset{\sim}{\mathbf{IND}}_{\lambda+2n}$-codes; see [Mos80, 3H.2].

In order to show that every hyperprojective set carries a hyperprojective scale, we will prove that if $\phi(\alpha, A)$ is a positive Σ^1_2 operator, then for each $\lambda < \kappa^{\mathbb{R}}$ limit, $\phi^{<\lambda}$ carries an $\underset{\sim}{\mathbf{IND}}_{\lambda}$-scale and $\phi^{\lambda+n}$ carries a $\underset{\sim}{\mathbf{IND}}_{\lambda+2n+2}$-scale for each $n \geq 0$. Since every hyperprojective pointset is the continuous preimage of ϕ^{ξ} for some such ϕ and some $\xi < \kappa^{\mathbb{R}}$, this will complete our proof.

The proof will be by effective transfinite induction, on a suitable coding system for ordinals $< \kappa^{\mathbb{R}}$, using the Recursion Theorem.

Let first $\psi(w, A)$, $w \in \mathbb{R}$, $A \subseteq \mathbb{R}$, be positive analytical such that if

$$|w| = \text{least } \xi \text{ such that } \psi^{\xi}(w),$$

and $W = \psi^\infty$, then

$$|\cdot|: W \twoheadrightarrow \kappa^{\mathbb{R}}$$

and there are recursive pointsets $\mathrm{Lim} \subseteq \mathbb{R}$, $\mathrm{Succ} \subseteq \mathbb{R}$ and a total recursive function pd such that

$$\begin{aligned} w \in W \Rightarrow & [\mathrm{Lim}(w) \Leftrightarrow |w| \text{ is limit}] \wedge \\ & [\mathrm{Succ}(w) \Leftrightarrow |w| \text{ is successor}] \wedge \\ & [\mathrm{Succ}(w) \Rightarrow |\mathrm{pd}(w)| = |w| - 1]. \end{aligned}$$

The following lemma will now complete the proof. ⊣

Lemma 2.5. Given $\phi(\alpha, A)$ a Σ^1_2 positive operator, there are scales $\langle \sigma_i^{<\lambda} \rangle$, $\langle \sigma_i^{\lambda+n} \rangle$ on $\phi^{<\lambda}$, $\phi^{\lambda+n}$ respectively, for each limit $\lambda < \kappa^{\mathbb{R}}$, and partial recursive functions f_0, g_0, f_1, g_1 such that, letting for each scale $\langle \sigma_i \rangle$ on a pointset P,

$$\begin{aligned} S(i, x, y) &\Leftrightarrow x \leq^*_{\sigma_i} y, \\ T(i, x, y) &\Leftrightarrow x <^*_{\sigma_i} y, \end{aligned}$$

be its associated relations, we have

$$\begin{aligned} w \in W \Rightarrow & \text{ if } |w| = \lambda + n,\ \lambda \text{ limit},\ n \geq 0, \text{ then for } n = 0 \text{ we have} \\ & \text{that } f_0(w), g_0(w) \text{ are } \underset{\sim}{\mathbf{IND}}_\lambda\text{-codes of the relations} \\ & \text{associated with } \langle \sigma_i^{<\lambda} \rangle, \text{ while for all } n, f_1(w), g_1(w) \text{ are} \\ & \underset{\sim}{\mathbf{IND}}_{\lambda+2n+2}\text{-codes of the relations associated with } \langle \sigma_i^{\lambda+n} \rangle. \end{aligned}$$

Proof. The scales are defined as in [Mos78] inductively:

$$\begin{aligned} \sigma_0^{<\lambda}(\alpha) &= |\alpha|_\phi = \text{ least } \xi \text{ such that } \alpha \in \phi^\xi, \\ \sigma_{i+1}^{<\lambda}(\alpha) &= \langle |\alpha|_\phi, \sigma_i^{|\alpha|_\phi}(\alpha) \rangle; \end{aligned}$$

while $\langle \sigma_i^{\lambda+n} \rangle$ is defined by the Second Periodicity Theorem inductively on n starting from $\langle \sigma_i^{<\lambda} \rangle$. For this to be a legitimate scale, it is sufficient to know that $\underset{\sim}{\mathbf{IND}}_\lambda$ has the scale property, by an argument similar to that given for the projective sets in the early stages of the present proof. In any case, however, the associated relations of $\langle \sigma_i^{<\lambda} \rangle$, $\langle \sigma_i^{\lambda+n} \rangle$ are defined independently of this, so if we can prove the second assertion of the lemma about these associated relations, then by induction on λ we have immediately that $\underset{\sim}{\mathbf{IND}}_\lambda$ has the scale property (recall that every $A \in \underset{\sim}{\mathbf{IND}}_\lambda$ is the continuous preimage of $\phi^{<\lambda}$, where ϕ can be taken to be Σ^1_2 positive analytical), and thus the proof is complete.

In order to construct f_0, f_1, g_0, g_1, we shall use effective transfinite induction. It is rather routine to define $f_1(w), g_1(w)$ once all $f_0(v), g_0(v), f_1(v), g_1(v)$ are known for $|v| < |w|$, thus we can concentrate on $f_0(w), g_0(w)$. So assume $|w| = \lambda$ and that all $f_0(v), g_0(v), f_1(v), g_1(v)$ are known for $|v| < \lambda$.

We write the first relation associated with $\langle \sigma_i^{<\lambda} \rangle$ (the calculation for the other one is similar):

$$\begin{aligned} S(i, \alpha, \beta) \Leftrightarrow & [i = 0 \wedge |\alpha|_\phi \leq |\beta|_\phi < \lambda] \\ & \vee \{i > 0 \wedge [(|\alpha|_\phi < |\beta|_\phi < \lambda) \\ & \qquad \vee (|\alpha|_\phi = |\beta|_\phi < \lambda \wedge \sigma_i^{|\alpha|_\phi}(\alpha) \leq \sigma_i^{|\alpha|_\phi}(\beta))]\}. \end{aligned}$$

By the Stage Comparison Theorem (see [Mos74A]) there are positive analytical $\phi_1, \phi_2, \phi_3, \phi_4$ such that

$$\begin{aligned} |\alpha|_\phi \leq |\beta|_\phi < \lambda &\Leftrightarrow \phi_1^{<\lambda}(\alpha, \beta) \wedge \phi^{<\lambda}(\beta) \\ |\alpha|_\phi < |\beta|_\phi < \lambda &\Leftrightarrow \phi_2^{<\lambda}(\alpha, \beta) \wedge \phi^{<\lambda}(\beta) \\ |\alpha|_\phi = |\beta|_\phi = |v|^{<\lambda} &\Leftrightarrow \phi_3^{<\lambda}(\alpha, v) \wedge \phi_3^{<\lambda}(\beta, v) \\ & \qquad \wedge \phi_4^{<\lambda}(v, \alpha) \wedge \phi_4^{<\lambda}(v, \beta) \wedge \psi^{<\lambda}(v). \end{aligned}$$

Thus

$$\begin{aligned} S(i, \alpha, \beta) \Leftrightarrow & [i = 0 \wedge \phi_1^{<\lambda}(\alpha, \beta) \wedge \phi^{<\lambda}(\beta)] \\ & \vee \{i > 0 \wedge [\phi_2^{<\lambda}(\alpha, \beta) \wedge \phi^{<\lambda}(\beta)] \\ & \qquad \vee (\exists v)[\phi_3^{<\lambda}(\alpha, v) \wedge \phi_3^{<\lambda}(\beta, v) \wedge \phi_4^{<\lambda}(v, \alpha) \wedge \\ & \qquad\qquad \phi_4^{<\lambda}(v, \beta) \wedge \psi^{<\lambda}(v) \wedge \sigma_i^{|v|}(\alpha) \leq \sigma_i^{|v|}(\beta)]\}. \end{aligned}$$

By induction hypothesis, if $|v| < \lambda$, say $|v| = \lambda' + n'$, we have that $f_1(v)$ is a $\underset{\sim}{\mathbf{IND}}_{\lambda'+2n'+2}$ code of the first relation associated with $\langle \sigma_i^{\lambda'+n'} \rangle = \langle \sigma_i^{|v|} \rangle$. It is therefore enough to establish the following fact:

There is a positive analytical $\vartheta(\varepsilon, \alpha, w, A)$, $A \subseteq \mathbb{R}^3$, and ε_0 recursive, such that for each ξ,

$$|w| \leq \xi \wedge (\varepsilon, \alpha) \in U_{|w|'} \Leftrightarrow (\varepsilon_0, \varepsilon, \alpha, w) \in \vartheta^\xi,$$

where $|w|' = \lambda + 2n + 2$, if $|w| = \lambda + n$.

By the Simultaneous Induction Lemma ([Mos74A, 1C.1] or [Mos80, 7C.11]), it is actually enough to construct a system $\vartheta_1, \vartheta_2, \vartheta_3$ such that

$$\vartheta_3^\xi(\varepsilon, \alpha, w) \Leftrightarrow |w| \leq \xi \wedge (\varepsilon, \alpha) \in U_{|w'|}.$$

Let first, by the Stage Comparison Theorem, $\chi(\varepsilon, \alpha, w, A)$ be a positive analytical operator such that letting

$$|\varepsilon, \alpha|_{\phi_0} = \text{ least } \xi \text{ such that } (\varepsilon, \alpha) \in \phi_0^\xi,$$

we have

$$|\varepsilon, \alpha|_{\phi_0} \leq \xi \wedge (w \notin W \vee |\varepsilon, \alpha|_{\phi_0} < |w|) \Leftrightarrow \chi^\xi(\varepsilon, \alpha, w).$$

The system is now as follows:

$$\begin{aligned}
\vartheta_1^{\xi}(w) &\Leftrightarrow \psi(w, \vartheta_1^{<\xi}) \\
\vartheta_2^{\xi}(\varepsilon, \alpha, w) &\Leftrightarrow \chi(\varepsilon, \alpha, w, \vartheta_2^{<\xi}) \\
\vartheta_3^{\xi}(\varepsilon, \alpha, w) &\Leftrightarrow \psi(w, \vartheta_1^{<\xi}) \wedge \\
&\qquad \{[\mathrm{Lim}(w) \wedge \exists\beta\forall\gamma\vartheta_2^{<\xi}(\varepsilon, \langle\alpha, \beta, \gamma\rangle, w)] \\
&\qquad \vee [\mathrm{Succ}(w) \wedge \exists\beta\forall\gamma\vartheta_3^{<\xi}(\varepsilon, \langle\alpha, \beta, \gamma\rangle, \mathrm{pd}(w))]\}. \qquad \dashv
\end{aligned}$$

In view of the preceding result, we are naturally led to pose the following question.

OPEN PROBLEM 2.6. Is it true that in $\mathbf{L}(\mathbb{R})$, AD is equivalent to the existence of arbitrarily large below Θ cardinals with the strong partition property?

And we conclude this section by proving an extension of Theorem 2.2 for games on ordinals. If $A \subseteq {}^{\omega}\lambda$ is a pointset on λ, we say that A is **ν-Suslin** if there is a tree T on $\lambda \times \nu$ such that

$$f \in A \Leftrightarrow \exists g \in {}^{\omega}\nu\big((f, g) \in [T]\big).$$

We now have

THEOREM 2.7. *Let λ, ν be ordinals and assume that there is a cardinal $\kappa > \lambda, \nu$ such that*

$$\kappa \to (\kappa)^{\mu}_{\xi}, \quad \forall \mu, \xi < \kappa.$$

If $A \subseteq {}^{\omega}\lambda$ is such that both A and $\neg A$ are ν-Suslin, then A is determined.

PROOF. We try to imitate the argument in the proof of Theorem 2.2. Consider the game $\mathrm{G}(A)$ associated with A, and, picking a tree T with $\mathrm{p}[T] = A$, consider the auxiliary game $\mathrm{G}^*(A, T)$ defined there. If player I has a winning strategy in $\mathrm{G}^*(A, T)$ we are done. So assume player II has a winning quasistrategy in $\mathrm{G}^*(A, T)$. This we cannot necessarily convert into a winning strategy however, since λ will be in general uncountable and we do not have AC_{λ}. So we cannot immediately conclude that player I will have a winning strategy for $\mathrm{G}(A)$.

Here we have to use the fact that $\neg A$ is also ν-Suslin. We fix a tree S so that $\mathrm{p}[S] = \neg A$ and we consider an auxiliary game $\mathrm{G}'(A, S)$, which is as before except that it is now player I's responsibility to make the extra moves. If player II has a winning strategy in $\mathrm{G}'(A, S)$, we easily again conclude that player II has a winning strategy in $\mathrm{G}(A)$. Else player I has a winning quasistrategy in $\mathrm{G}'(A, S)$. Using DC now it is easy to see that there are runs in $\mathrm{G}^*(A, T)$ and $\mathrm{G}'(A, S)$ with *the same real part* $\alpha(0), \alpha(1), \ldots$ in which player II (resp. player I) has followed his winning quasistrategy in $\mathrm{G}^*(A, T)$ (resp. $\mathrm{G}'(A, S)$). This is clearly a contradiction. $\dashv$

From this result and Theorem 1.1 we immediately obtain the following, where we call $A \subseteq {}^{\omega}\lambda$ **Suslin** if it is μ-Suslin for some μ.

THEOREM 2.8. Assume AD. Let $\lambda < \Theta$ and let $A \subseteq {}^{\omega}\lambda$ be such that both A and $\neg A$ are Suslin. Then A is determined.

This strengthens the last conclusion of [Mos81, Theorem 2.2], by removing the restriction that $\lambda \leq \kappa^{\mathbb{R}}$. It also provides an alternative proof of that result, which however does not give the key definability estimates of the original argument that are needed in the rest of that paper. (Note that a set $A \subseteq {}^{\omega}\lambda$ is Suslin iff it admits uniform semiscales in the terminology of [Mos81].)

§3. **Mahlo cardinals from determinacy.** In this section we prove a strengthening of Theorem 1.1 for Spector pointclasses Γ which are additionally closed under both $\exists^{\mathbb{R}}$ and $\forall^{\mathbb{R}}$. The result is as follows.

THEOREM 3.1. Assume AD, let Γ be a Spector pointclass closed under both $\exists^{\mathbb{R}}$ and $\forall^{\mathbb{R}}$ and let $\kappa = \text{o}(\utilde{\mathbf{\Delta}})$ be the ordinal associated with Γ. Then κ has the strong partition property, and moreover $\{\lambda : \lambda < \kappa \wedge \lambda$ has the strong partition property$\}$ is stationary in κ. In particular, κ is Mahlo.

Among other things, this implies that there are arbitrarily large cardinals κ with the above properties below Θ. This is because for each pointset A the pointclass

$$\text{IND}(A) = \text{ all pointsets which are inductive in } A,$$

is a Spector pointclass containing both A and its complement and is closed under $\exists^{\mathbb{R}}$ and $\forall^{\mathbb{R}}$.

Also it follows that $\kappa^{\mathbb{R}}$ is Mahlo and in fact $\{\lambda : \lambda < \kappa^{\mathbb{R}} \wedge \lambda$ has the strong partition property$\}$ is stationary in $\kappa^{\mathbb{R}}$ (since $\kappa^{\mathbb{R}} = \text{o}(\utilde{\mathbf{\Delta}})$, where $\Gamma = \text{IND} = \text{IND}(\varnothing)$). Note here that by [Ste81A], $\kappa({}^3E)$ is not Mahlo. In fact Moschovakis has conjectured that $\kappa({}^3E)$ is the first (weakly) inaccessible cardinal (granting AD of course). It can be seen that $\kappa^{\mathbb{R}}$ is not the first Mahlo cardinal. It is conjectured that the first Mahlo cardinal is the ordinal of ${}^4S$ (= the type 4 superjump), granting AD again (see [Har73] for results about the superjump).

We now give the proof of Theorem 3.1.

PROOF. It is enough by Theorem 1.1 to show that

$$\{\text{o}(\utilde{\mathbf{\Delta}}^*) : \Gamma^* \text{ is a Spector pointclass closed under } {}^3E \text{ and contained in } \utilde{\mathbf{\Delta}}\}$$

is stationary in $\kappa = \text{o}(\utilde{\mathbf{\Delta}})$.

Let $f : \kappa \to \kappa$. We shall find Γ^* as above, such that $\text{o}(\utilde{\mathbf{\Delta}}^*)$ is closed under f, i.e.,

$$\xi < \text{o}(\utilde{\mathbf{\Delta}}^*) \Rightarrow f(\xi) < \text{o}(\utilde{\mathbf{\Delta}}^*).$$

Fix $S \subseteq \mathbb{R}$ and a Γ-norm

$$\phi \colon S \twoheadrightarrow \kappa.$$

Then in the notation of the proof of Lemma 1.6 find ε such that $f = f_\varepsilon$. From the definition of f_ε it is obvious that there are two relations $R, \breve{R}$ in $\underset{\sim}{\Gamma}, \underset{\sim}{\breve{\Gamma}}$ respectively such that

$$\begin{aligned} \alpha \in S \Rightarrow [R(\alpha, \beta) &\Leftrightarrow \breve{R}(\alpha, \beta) \\ &\Leftrightarrow \beta \in S \wedge f(\phi(a)) = \phi(\beta)]. \end{aligned} \tag{$*$}$$

To simplify the notation, assume that actually $R, \breve{R}$ are in $\Gamma, \breve{\Gamma}$. (Otherwise replace everywhere below Γ by $\Gamma(\alpha_0)$ for some appropriate parameter α_0.)

A **type 3 object** is a function

$${}^3F \colon \omega^{\mathfrak{Z}} \times \mathfrak{Y} \to \omega,$$

where $\mathfrak{Z}, \mathfrak{Y}$ are product spaces. For example Kleene's ${}^3E$ is the type 3 object

$${}^3E \colon \omega^{\mathbb{R}} \to \omega$$

given by

$${}^3E(h) = \begin{cases} 0, & \text{if } \exists \alpha[h(\alpha) = 0] \\ 1, & \text{if } \forall \alpha[h(\alpha) \neq 0]. \end{cases}$$

We say that a Spector pointclass Γ is **closed under** ${}^3F$ if for each $h \colon \mathfrak{X} \times \mathfrak{Z} \to \omega$, a Γ-recursive partial function, the relation

$$P(i, x, y) \Leftrightarrow \forall z[h(x, z)\downarrow] \wedge {}^3F(\lambda z h(x, z), y) = i$$

is in Γ. To each such ${}^3F$ we associate the envelop

$$\text{Env}({}^3E, {}^3F) = \text{all pointsets which are Kleene-semirecursive in } {}^3E, {}^3F.$$

This is a Spector pointclass closed under ${}^3E$, ${}^3F$ and in fact by [Mos74B] it is the smallest one with these properties, thus if Γ is a Spector pointclass closed under ${}^3E$, ${}^3F$, then $\text{Env}({}^3E, {}^3F) \subseteq \Gamma$. Moreover by [Mos67] if $A \in \text{Env}({}^3E, {}^3F)$, there is $B \in \text{Env}({}^3E, {}^3F)$ with

$$x \notin A \Leftrightarrow \exists \alpha (x, \alpha) \in B.$$

This immediately implies that if Γ is a Spector pointclass closed under both $\exists^{\mathbb{R}}, \forall^{\mathbb{R}}$ and if ${}^3F$ is a type 3 object, so that Γ is closed under ${}^3F$, then

$$\text{Env}({}^3E, {}^3F) \subseteq \Delta.$$

This is the key fact that we will need below.

Consider now the following type 3 object ${}^3F$ associated with $f \colon \kappa \to \kappa$:

$${}^3F \colon \omega^{\mathbb{R}\times\mathbb{R}} \times \mathbb{R}^2 \to \omega,$$

and for each $h \in \omega^{\mathbb{R}\times\mathbb{R}}, \alpha \in \mathbb{R}$

$$^3F(h,\alpha,\beta) = \begin{cases} 0, & \text{if } h \text{ is the characteristic function of a prewellordering} \\ & \quad \leq \text{ on } \mathbb{R} \text{ such that } |\leq| < \kappa, \alpha, \beta \in S \\ & \quad \wedge \phi(\alpha) < \phi(\beta) < f(|\leq|); \\ 1, & \text{otherwise.} \end{cases}$$

LEMMA 3.2. Γ is closed under ^3F.

Granting the lemma consider

$$\Gamma^* = \mathrm{Env}(^3E, {}^3F).$$

By our preceding remarks, it is enough to show that $\kappa^* = o(\underset{\sim}{\mathbf{\Delta}}^*)$ is closed under f. For that let $S^* \in \Gamma^*$ and $\phi^* \colon S^* \twoheadrightarrow \kappa^*$ be a Γ^*-norm. If $\xi < \kappa^*$, let $h\colon \mathbb{R}\times\mathbb{R} \to \omega$ be the characteristic function of $\leq = \{(\alpha,\beta) : \phi^*(\alpha) \leq \phi^*(\beta) < \xi\}$. Clearly h is in $\underset{\sim}{\mathbf{\Delta}}^*$. Since Γ^* is closed under ^3F, this implies that

$$\prec = \{(\alpha,\beta) : {}^3F(h,\alpha,\beta) = 0\}$$

is also in $\underset{\sim}{\mathbf{\Delta}}^*$. But

$$\prec = \{(\alpha,\beta) : \phi(\alpha) < \phi(\beta) < f(\xi)\}.$$

Thus $f(\xi)$ is the length of a $\underset{\sim}{\mathbf{\Delta}}^*$ prewellordering and so $f(\xi) < \kappa^* = o(\underset{\sim}{\mathbf{\Delta}}^*)$, which is what we wanted to prove.

To verify the lemma note that if $h\colon \mathfrak{X}\times\mathbb{R}\times\mathbb{R} \to \omega$ has graph in Γ and for some $x \in \mathfrak{X}, \forall z[h(x,z)\downarrow]$, then one can check in a $\Delta(x)$ way, uniformly in x, whether or not $\lambda z h(x,z)$ is a characteristic function of a prewellordering of $\mathbb{R}$. If this is the case and we denote by $\leq_x = \leq$ this prewellordering, then by [Mos80, 4C.14] we can find, effectively in x, a $\gamma_x = \gamma \in S$ with $\phi(\gamma) > |\leq|$, and thus using the Coding Lemma [Mos80, 7D.5] we can see that $\{\delta : \phi(\delta) = |\leq|\}$ is in $\Delta(x)$, uniformly in x again. From this and $(*)$ it is immediate that we can check in a $\Delta(x)$ way, uniformly in x, whether $\phi(\alpha) < \phi(\beta) < f(|\leq|)$ is true or not, thus completing the proof that Γ is closed under ^3F. ⊣

§4. **Nonsingular measures from** AD. It has been known for quite some time that AD implies the existence of many measurable cardinals below Θ (see for example [Kle77] or [Kec78A]). All the normal measures on cardinals κ that were produced, however, were of the "singular" type, i.e., they concentrated on the ordinals of cofinality λ, for some regular $\lambda < \kappa$. It has thus been open whether one could obtain from AD measurable cardinals below Θ that carry measures which concentrate on regular cardinals. Our result below provides (when combined with Theorem 3.1) many such examples below Θ.

THEOREM 4.1. Assume that κ is a Mahlo cardinal satisfying the strong partition property. Then there exists a nontrivial κ-additive normal measure on κ, giving the regular cardinals less than κ measure 1.

Proof. Let Q be *any* stationary set of regular cardinals less than κ. We will construct a normal measure giving Q measure 1.

First let us define $\hat{Q}$ to be the set of those ξ in Q such that some closed unbounded subset of ξ is disjoint from Q. Note that $\hat{Q}$ consists of the difference between Q and the result of applying the Mahlo operation to Q.

We can now define a function $\mu_Q\colon {}^{\kappa}2 \to 2$ by

$$\mu_Q(X) = 1 \text{ iff for some closed unbounded subset } C \text{ of } \kappa,\ X \supseteq C \cap \hat{Q}.$$

Lemma 4.2. μ_Q is a non-trivial κ-additive measure on κ.

Proof of Lemma 4.2. Let $\mathcal{U}_Q = \{A \subseteq K : \mu_Q(A) = 1\}$. We will show that $\mathcal{U}_Q$ is a nonprincipal κ-additive ultrafilter on κ.

It is first important to note that $\hat{Q}$ is stationary. For given any closed unbounded set C, the least limit point of C in Q is a member of $\hat{Q} \cap C$. Since, now, $\hat{Q}$ is stationary, $\mathcal{U}_Q$ must be nonprincipal. Also, clearly, $\mathcal{U}_Q$ is a filter.

Suppose $X \subseteq \kappa$ is given. We must show that $X \in \mathcal{U}_Q$ or $\neg X \in \mathcal{U}_Q$, and so let us define a partition $F\colon {}^{\kappa}[\kappa] \to 2$ by

$$F(Y) = 1 \text{ iff the least limit point of } Y \text{ which is a member of } Q \text{ is also a member of } X.$$

(Identify here ${}^{\kappa}[\kappa]$ with the set of subsets of κ of cardinality κ.) Let D be a set of cardinality κ homogeneous for F and let us suppose that $F[{}^{\kappa}[D]] = \{0\}$.

Claim. $X \supseteq D_{\text{l.p.}} \cap \hat{Q}$, where $D_{\text{l.p.}}$ denotes the set of limit points of D.

Proof of Claim. Suppose ξ is a limit point of D in $\hat{Q}$. Let C_ξ be a closed unbounded subset of ξ disjoint from Q. By an interlocking argument, we can thin out $D \cap \xi$ and C_ξ simultaneously to $(D \cap \xi)'$ and C'_ξ respectively such that $(D \cap \xi)'$ and C'_ξ have the same limit points and such that each is unbounded in ξ. Since $C_\xi \cap Q = \varnothing$, there is no limit point of $(D \cap \xi)'$ less than ξ in Q, and so ξ is the least limit point of $(D \cap \xi)'$ in Q. Since $F((D \cap \xi)' \cup (D - \xi)) = 0$, $\xi \in X$. ⊣ (Claim)

By a similar claim, if we had that $F[{}^{\kappa}[D]] = \{1\}$, then we would have that $\neg X \supseteq D_{\text{l.p.}} \cap \hat{Q}$. Thus either X or $\neg X$ is in $\mathcal{U}_Q$, and so $\mathcal{U}_Q$ is an ultrafilter.

Suppose now that $X_\xi \in \mathcal{U}_Q$ for each $\xi < \vartheta < \kappa$. We must show that $\bigcap_{\xi<\vartheta} X_\xi \in \mathcal{U}_Q$. Let us define a partition $G\colon {}^{\kappa}[\kappa] \to \vartheta + 1$ as follows: given $Y \in {}^{\kappa}[\kappa]$, let η be the least limit point of Y which is a member of Q. Then if $\eta \in \bigcap_{\xi<\vartheta} X_\xi$ we define $G(Y)$ to be ϑ. Otherwise, $G(Y)$ is the least $\xi < \vartheta$ such that $\eta \notin X_\xi$.

Let, now, E be a set of cardinality κ homogeneous for G. Then if $G[{}^{\kappa}[E]] = \{\vartheta\}$, an argument similar to one used above would show that $\bigcap_{\xi<\vartheta} X_\xi \supseteq E_{\text{l.p.}} \cap \hat{Q}$, and hence that $\bigcap_{\xi<\vartheta} X_\xi \in \mathcal{U}_Q$. Otherwise, our argument above would show that $\neg X_\xi \supseteq E_{\text{l.p.}} \cap \hat{Q}$ for some $\xi < \vartheta$. Since for some closed unbounded C, $X_\xi \supseteq C \cap \hat{Q}$, we would have $E_{\text{l.p.}} \cap C \cap \hat{Q} \subseteq X_\xi \cap \neg X_\xi = \varnothing$. Since $E_{\text{l.p.}} \cap C \cap \hat{Q}$ is stationary, this is impossible. Thus $\bigcap_{\xi<\vartheta} X_\xi \in \mathcal{U}_Q$. ⊣

LEMMA 4.3. μ_Q is normal.

PROOF OF LEMMA 4.3. Suppose $f : \kappa \to \kappa$ and $\mu_Q(\{\xi : f(\xi) < \xi\}) = 1$. Let C be a closed unbounded set such that $\{\xi : f(\xi) < \xi\} \supseteq C \cap \hat{Q}$, and let us define a partition $F : {}^\kappa[C \cap \hat{Q}] \to 2$ by

$F(Y) = 0$ iff the value of f on the least limit point of Y which is a member of Q is less than the least member of Y.

Let D be homogeneous for F and of cardinality κ.

Claim. $F[{}^\kappa[D]] = \{0\}$.

PROOF OF CLAIM. Since $D \supseteq C \cap \hat{Q}$, the least limit point of D which is a member of Q, η, is sent by f to some ordinal less than η. Clearly η is still the least limit point of $D \setminus (f(\eta) + 1)$ which is a member of Q, and so $F(D \setminus (f(\eta) + 1)) = 0$. Thus $F[{}^\kappa[D]] = \{0\}$. ⊣ (Claim)

By an argument similar to one used earlier, the above claim yields that $f[D_{\text{l.p.}} \cap \hat{Q}] \subseteq \bigcap D$. Since μ_Q is κ-additive, $\mu_Q(f^{-1}\{\xi_0\}) = 1$ for some $\xi_0 < \bigcap D$. ⊣

By Lemmas 4.2 and 4.3, μ_Q is a κ-additive nontrivial normal measure on κ. Since $Q \supseteq \kappa \cap \hat{Q}$, $\mu_Q(Q) = 1$, and our proof is complete. ⊣

With a bit of extra effort, we could carry out the above proof starting with any stationary set Q. Thus for uncountable κ satisfying the strong partition property, each stationary set gets measure 1 under some normal measure. For these and further results we refer to [Kle82].

REFERENCES

LEO A. HARRINGTON

[Har73] ***Contributions to recursion theory in higher types***, Ph.D. thesis, MIT.

ALEXANDER S. KECHRIS

[Kec78A] AD *and projective ordinals*, In Kechris and Moschovakis [CABAL i], pp. 91–132.

[Kec81B] *Suslin cardinals, κ-Suslin sets, and the scale property in the hyperprojective hierarchy*, this volume, originally published in Kechris et al. [CABAL ii], pp. 127–146.

ALEXANDER S. KECHRIS, DONALD A. MARTIN, AND YIANNIS N. MOSCHOVAKIS

[CABAL ii] *Cabal seminar* 77–79, Lecture Notes in Mathematics, no. 839, Berlin, Springer.

ALEXANDER S. KECHRIS AND YIANNIS N. MOSCHOVAKIS

[CABAL i] *Cabal seminar* 76–77, Lecture Notes in Mathematics, no. 689, Berlin, Springer.

EUGENE M. KLEINBERG

[Kle77] ***Infinitary combinatorics and the axiom of determinacy***, Lecture Notes in Mathematics, vol. 612, Springer-Verlag.

[Kle82] *A measure representation theorem for strong partition cardinals*, ***The Journal of Symbolic Logic***, vol. 47, no. 1, pp. 161–168.

YIANNIS N. MOSCHOVAKIS

[Mos67] *Hyperanalytic predicates*, ***Transactions of the American Mathematical Society***, vol. 129, pp. 249–282.

[Mos70A] *Determinacy and prewellorderings of the continuum*, ***Mathematical logic and foundations of set theory. Proceedings of an international colloquium held under the auspices of the Israel Academy of Sciences and Humanities, Jerusalem, 11–14 November 1968*** (Y. Bar-Hillel, editor), Studies in Logic and the Foundations of Mathematics, North-Holland, Amsterdam-London, pp. 24–62.

[Mos74A] ***Elementary induction on abstract structures***, North-Holland.

[Mos74B] *Structural characterizations of classes of relations*, ***Generalized recursion theory. Proceedings of the 1972 Oslo symposium*** (Jens Erik Fenstad and Peter G. Hinman, editors), Studies in Logic and the Foundations of Mathematics, North-Holland, pp. 53–79.

[Mos78] *Inductive scales on inductive sets*, this volume, originally published in Kechris and Moschovakis [CABAL i], pp. 185–192.

[Mos80] ***Descriptive set theory***, Studies in Logic and the Foundations of Mathematics, no. 100, North-Holland, Amsterdam.

[Mos81] *Ordinal games and playful models*, In Kechris et al. [CABAL ii], pp. 169–201.

JOHN R. STEEL

[Ste81A] *Closure properties of pointclasses*, In Kechris et al. [CABAL ii], pp. 147–163.

DEPARTMENT OF MATHEMATICS
CALIFORNIA INSTITUTE OF TECHNOLOGY
PASADENA, CA 91125, USA
E-mail: kechris@caltech.edu

DEPARTMENT OF MATHEMATICS
STATE UNIVERSITY OF NEW YORK
BUFFALO, NY 14260, USA
E-mail: kleinbrg@cse.buffalo.edu

DEPARTMENT OF MATHEMATICS
UNIVERSITY OF CALIFORNIA
LOS ANGELES, CA 90095, USA
E-mail: ynm@math.ucla.edu

DEPARTMENT OF MATHEMATICS
UNIVERSITY OF CALIFORNIA
BERKELEY, CA 94720, USA
E-mail: woodin@math.berkeley.edu

THE EQUIVALENCE OF PARTITION PROPERTIES AND DETERMINACY

ALEXANDER S. KECHRIS AND W. HUGH WOODIN

Historical Remark. This paper was circulated in handwritten form in March 1982 and contained Sections 1-4 below. There is an additional Section 5 containing information about the solution of a problem mentioned in the last paragraph of Section 1.

§1. Statements of results.

1.1. Let $\mathbf{L}(\mathbb{R})$ be the smallest inner model of ZF containing the set of reals $\mathbb{R}$. Let Θ be the sup of the ordinals which are subjective images of $\mathbb{R}$.

THEOREM 1.1. Assume ZF+DC. Then the following are equivalent:

(1) $\mathbf{L}(\mathbb{R}) \models \mathsf{AD}$;
(2) $\mathbf{L}(\mathbb{R}) \models \forall\lambda < \Theta\exists\kappa(\lambda < \kappa \wedge \kappa \rightarrow (\kappa)^{\kappa})$;
(3) $\mathbf{L}(\mathbb{R}) \models \forall\lambda < \Theta\exists\kappa(\lambda < \kappa \wedge \kappa \rightarrow (\kappa)^{\lambda})$.

It has been already known, see [KKMW81], that

$$\mathsf{ZF}+\mathsf{DC}+\mathsf{AD} \Rightarrow \forall\lambda < \Theta\exists\kappa(\lambda < \kappa \wedge \kappa \rightarrow (\kappa)^{\kappa}).$$

Recall that $A \subseteq \mathbb{R}$ is called **λ-Suslin** if there is a tree T on $\omega \times \lambda$ such that $A = \mathrm{p}[T] = \{\alpha \in \mathbb{R} : \exists f \in {}^{\omega}\lambda\forall n(\langle\alpha\restriction n, f\restriction n\rangle \in T)\}$, and A is called **Suslin** if it is λ-Suslin for some λ. Again in [KKMW81] it is proved that (in ZF+DC):

$$\forall\lambda < \Theta\exists\kappa(\lambda < \kappa \wedge \kappa \rightarrow (\kappa)^{\lambda}) \Rightarrow \text{“Every Suslin set of reals is determined”}.$$

Thus Theorem 1.1 is an immediate consequence of

THEOREM 1.2. Assume ZF+DC. Then the following are equivalent:

(1) $\mathbf{L}(\mathbb{R}) \models$ Every Suslin set of reals is determined;
(2) $\mathbf{L}(\mathbb{R}) \models \mathsf{AD}$.

Preparation of this paper is partially supported by NSF Grants DMS-0455285 (A.S.K.) and DMS-0355334 (W.H.W.).

The Cabal Seminar. Volume I: Games, Scales and Suslin Cardinals
Edited by A. S. Kechris, B. Löwe, J. R. Steel
Lecture Notes in Logic, 31

The proof of Theorem 1.2 relies heavily on work of [Ste83A], which analyzes the propagation of the scale property in $\mathbf{L}(\mathbb{R})$, using the fine structure of this inner model. Also in the proof of Theorem 1.2, as well as Theorems 1.8 and 1.10 below, essential use is made of a technique of [Mar83A] for handling finite strings of alternating quantifiers over $\mathbb{R}$ via iterated products of the Martin measure on the Turing degrees.

There are several strengthenings and corollaries of the preceding theorems. For instance in Theorem 1.2 we can weaken (1) as follows: Call a pointclass $\underset{\sim}{\Gamma}$ **reasonable** if it is closed under $\vee, \wedge$, bounded number quantification, continuous substitutions and $\exists^{\mathbb{R}}$ *or* $\forall^{\mathbb{R}}$ and is $\mathbb{R}$-parametrized. We say that $\underset{\sim}{\Gamma}$ **is determined** if every $A \in \underset{\sim}{\Gamma}$, $A \subseteq \mathbb{R}$ is determined. Then we have that, in ZF+DC, the following are equivalent:

(1) $\mathbf{L}(\mathbb{R}) \models$ Every reasonable $\underset{\sim}{\Gamma}$ with the scale property is determined;
(2) $\mathbf{L}(\mathbb{R}) \models$ AD.

The following is also an immediate corollary of Theorem 1.2.

COROLLARY 1.3. Assume ZF+DC. If there is a cardinal κ with $\kappa \geq \Theta^{\mathbf{L}(\mathbb{R})}$ and $\kappa \to (\kappa)^\kappa$ (or even $\kappa \to (\kappa)^\lambda, \forall \lambda < \Theta^{\mathbf{L}(\mathbb{R})}$), then $\mathbf{L}(\mathbb{R}) \models$ AD.

In particular

$$\mathrm{Con}(\mathsf{ZF{+}DC}{+}\exists\kappa(\kappa \geq \Theta^{\mathbf{L}(\mathbb{R})} \wedge \kappa \to (\kappa)^\kappa)) \text{ implies } \mathrm{Con}(\mathsf{ZF{+}DC{+}AD}).$$

Note that in Corollary 1.3, κ is assumed to have the partition property in the universe, not necessarily in $\mathbf{L}(\mathbb{R})$. Conceivably this result could be used to demonstrate the consistency of ZF+DC+AD from appropriate large cardinal assumptions.

The following is a related open problem.

QUESTION 1.4. Is $\mathsf{ZF{+}DC}{+}\exists\kappa(\kappa \geq \Theta \wedge \kappa \to (\kappa)^\kappa)$ consistent? If so, what is its strength, in particular does it imply $\mathrm{Con}(\mathsf{ZF{+}DC{+}AD}_{\mathbb{R}})$?

We describe next some corollaries about the relativization of partition properties to $\mathbf{L}(\mathbb{R})$.

COROLLARY 1.5. Let σ be any of the statements $\forall\lambda < \Theta\exists\kappa(\lambda < \kappa \wedge \kappa \to (\kappa)^\kappa)$ or $\forall\lambda < \Theta\exists\kappa(\lambda < \kappa \wedge \kappa \to (\kappa)^\lambda)$. Then, in ZF+DC, σ relativizes to $\mathbf{L}(\mathbb{R})$ (i.e., σ implies $\sigma^{\mathbf{L}(\mathbb{R})}$). Also, in ZF+DC, σ implies $\mathbf{L}(\mathbb{R}) \models \omega_1 \to (\omega_1)^{\omega_1}$.

The following is a rather curious reflection property of $\mathbf{L}(\mathbb{R})$.

COROLLARY 1.6. Assume ZF+DC+$\mathbf{V} = \mathbf{L}(\mathbb{R})$. Let $\kappa = \underset{\sim}{\delta}^2_1$ (i.e., the least ξ such that $L_\xi(\mathbb{R}) \prec_1 \mathbf{L}(\mathbb{R})$). Then, if $\kappa \to (\kappa)^\kappa$, we have that $\lambda \to (\lambda)^\lambda$ for many small $\lambda < \kappa$, e.g., $\lambda = \omega_1$, and also $\lambda \to (\lambda)^\lambda$ for cofinally in Θ many λ's.

Finally we examine the relation between partition properties and partition measures in $\mathbf{L}(\mathbb{R})$. Let us say that a cardinal κ has a **partition measure** if for each wellordering W of order type $\leq \kappa$ there is a countably additive measure μ_W on the set ${}^W[\kappa]$ of increasing W-term sequences from κ with the following coherence property: If $W' \subseteq W$ and $A \subseteq {}^W[\kappa]$ then $\mu_W(A) = 1$ implies $\mu_{W'}(\{f \restriction W' : f \in A\}) = 1$. Clearly $\kappa \to (\kappa)^\kappa$ implies that κ has a partition measure, but the existence of a partition measure on κ does not imply that $\kappa \to (\kappa)^\kappa$. Steel [Ste80] shows that $\mathsf{ZF}+\mathsf{DC}+\mathsf{AD} \Rightarrow \forall\kappa < \Theta(\kappa$ has a partition measure). Also in [KKMW81] it is shown that, in $\mathsf{ZF}+\mathsf{DC}$, if $\forall\lambda < \Theta\exists\kappa(\kappa > \lambda \wedge \kappa$ has a partition measure), then every Suslin set is determined. So we have the following

Corollary 1.7. The following are equivalent, in $\mathsf{ZF}+\mathsf{DC}$,

(1) $\mathbf{L}(\mathbb{R}) \models \forall\lambda < \Theta\exists\kappa(\lambda < \kappa \wedge \kappa$ has a partition measure);
(2) $\mathbf{L}(\mathbb{R}) \models \forall\lambda < \Theta\exists\kappa(\lambda < \kappa \wedge \kappa \to (\kappa)^\kappa)$;
(3) $\mathbf{L}(\mathbb{R}) \models \mathsf{AD}$.

Thus *globally in* $\mathbf{L}(\mathbb{R})$ partition measures are equivalent to partition properties.

1.2. The key new tool in the proof of Theorem 1.2 is a "transfer" theorem of the form (in $\mathsf{ZF}+\mathsf{DC}$):

$$\mathsf{Det}(\boldsymbol{\Gamma}_1) \Rightarrow \mathsf{Det}(\boldsymbol{\Gamma}_2),$$

where $\boldsymbol{\Gamma}_1, \boldsymbol{\Gamma}_2$ are pointclasses with appropriate properties and interrelationships, and $\boldsymbol{\Gamma}_2$ is "much bigger" than $\boldsymbol{\Gamma}_1$. A basic instance of this type of theorem can be stated as follows:

For each pointclass $\boldsymbol{\Gamma}$ closed under continuous substitutions, $\wedge, \vee$, let $\boldsymbol{\Sigma}^*_n(\boldsymbol{\Gamma})$ be defined as follows,

$$\boldsymbol{\Sigma}^*_1(\boldsymbol{\Gamma}) = \Big\{A \subseteq \mathfrak{X} : \text{For some } B \in \boldsymbol{\Gamma}, C \in \check{\boldsymbol{\Gamma}} \quad (x \in A \Leftrightarrow \exists y[B(x,y) \wedge C(x,y)])\Big\},$$

$$\boldsymbol{\Pi}^*_n(\boldsymbol{\Gamma}) = \{\neg A : A \in \boldsymbol{\Sigma}^*_n(\boldsymbol{\Gamma})\},$$
$$\boldsymbol{\Sigma}^*_{n+1}(\boldsymbol{\Gamma}) = \{\exists y A(x,y) : A \in \boldsymbol{\Pi}^*_n(\boldsymbol{\Gamma})\}.$$

A typical, for our purposes, example of such a $\boldsymbol{\Gamma}$ is $\boldsymbol{\Gamma} = \mathbf{IND}$, the pointclass of inductive sets of reals, in which case we just write

$$\boldsymbol{\Sigma}^*_n \equiv \boldsymbol{\Sigma}^*_n(\mathbf{IND}), \boldsymbol{\Pi}^*_n \equiv \boldsymbol{\Pi}^*_n(\mathbf{IND}).$$

We have now:

Theorem 1.8. Assume $\mathsf{ZF}+\mathsf{DC}$. Let $\boldsymbol{\Gamma}$ be a pointclass closed under continuous substitutions, $\wedge, \vee, \exists^{\mathbb{R}}, \forall^{\mathbb{R}}$. If $\boldsymbol{\Gamma}$ has the prewellordering and uniformization properties, then $\mathsf{Det}(\boldsymbol{\Delta})$ implies that for all $n < \omega$, we have $\mathsf{Det}(\boldsymbol{\Sigma}^*_n(\boldsymbol{\Gamma}))$, where as usual $\boldsymbol{\Delta} = \boldsymbol{\Gamma} \cap \check{\boldsymbol{\Gamma}}$.

COROLLARY 1.9. Assume ZF+DC. Then Det($\underset{\sim}{\mathbf{HYP}}$) implies that for all $n < \omega$, we have Det($\underset{\sim}{\Sigma}^*_n$) (where $\underset{\sim}{\mathbf{HYP}} = \underset{\sim}{\mathbf{IND}} \cap \underset{\sim}{\check{\mathbf{IND}}}$).

[Har78] and [MarA] have shown that, in ZF+DC, Det($\underset{\sim}{\Pi}^1_1$) is equivalent to Det($< \omega^2$-$\underset{\sim}{\Pi}^1_1$). Abbreviate for convenience

$$\mathbf{M}_n \equiv \omega \cdot n\text{-}\underset{\sim}{\Pi}^1_1.$$

Let $\Game$ be the game quantifier and for any pointclass $\underset{\sim}{\Gamma}$ let

$$\Game^m \underset{\sim}{\Gamma} = \underbrace{\Game\Game\Game \dots \Game}_{m} \underset{\sim}{\Gamma},$$

where $\Game \underset{\sim}{\Gamma} = \{\Game \alpha A(x, \alpha) : A \in \underset{\sim}{\Gamma}\}$. Another instance of the "transfer" theorem provides a generalization of the Harrington-Martin result to all even levels of the projective hierarchy.

THEOREM 1.10. Assume ZF+DC. Then for $k \geq 1$,

$$\mathsf{Det}(\underset{\sim}{\Delta}^1_{2k}) \Leftrightarrow \forall n < \omega(\mathsf{Det}(\Game^{2k-1}\mathbf{M}_n)).$$

For instance, if $k = 1$ this says that, in ZF+DC,

$$\mathsf{Det}(\underset{\sim}{\Delta}^1_2) \Leftrightarrow \forall n < \omega(\mathsf{Det}(\Game\mathbf{M}_n)).$$

In particular, combining this with Martin's result in [Mar78], one sees that, in ZFC,

if there is an iterable $j : V_\lambda \to V_\lambda$, then for all $n < \omega$, Det($\Game\mathbf{M}_n$) holds.

The classes $\Game\mathbf{M}_n$ are substantially bigger than $\underset{\sim}{\Sigma}^1_2$, but still well within $\underset{\sim}{\Delta}^1_3$.

Because it turns out that (Det($\omega^2 + 1$-$\underset{\sim}{\Pi}^1_1$) proves the consistency of Det($\underset{\sim}{\Pi}^1_1$) (see [MarA]), the Martin-Harrington result is basically best possible. Similarly Theorem 1.10 is basically best possible, since from [KS85] it follows that Det($\Game^{2k-1}(\omega^2 + 1$-$\underset{\sim}{\Pi}^1_1)$) proves the consistency of $\forall n < \omega$(Det($\Game^{2k-1}(\omega \cdot n$-$\underset{\sim}{\Pi}^1_1)$))).

It should be also the case that Theorem 1.10 holds at odd levels as well, i.e., that for $k \geq 1$, we have

$$\mathsf{Det}(\underset{\sim}{\Pi}^1_{2k+1}) \Leftrightarrow \forall n < \omega(\mathsf{Det}(\Game^{2k}\mathbf{M}_n))$$

but at this writing we do not know a proof of that. It is also interesting to see whether the techniques used here can be employed to give a different proof of the Harrington-Martin Theorem (the only known proof of it goes through the theory of sharps, while our methods are direct and purely analytical).

§2. Proof of Theorem 1.8. We start from the proof of Theorem 1.8, since it will be used in the proof of Theorem 1.2.

Assume $\mathsf{Det}(\boldsymbol{\Delta})$.

Let $A \subseteq \mathbb{R} \times \mathbb{R}$ be a $\boldsymbol{\Sigma}^*_n(\boldsymbol{\Gamma})$ game, say $n = 2$ for notational simplicity. Thus

$$A(\alpha, \beta) \Leftrightarrow \exists\gamma\forall\delta[R(\alpha, \beta, \gamma, \delta) \wedge S(\alpha, \beta, \gamma, \delta)],$$

where $R \in \boldsymbol{\Gamma}$, $S \in \check{\boldsymbol{\Gamma}}$. Let

$$P(i, x) \Leftrightarrow [i = 0 \wedge R(x)] \vee [i \neq 0 \wedge \neg S(x)].$$

So $P \in \boldsymbol{\Gamma}$. Let $\varphi \colon P \twoheadrightarrow \kappa$ be a $\boldsymbol{\Gamma}$-norm on P. Note that κ is limit and

$$\mathrm{cf}(\kappa) > \omega,$$

since otherwise P and therefore R, S and A are in $\boldsymbol{\Delta}$ and there is nothing to prove.

For $\xi < \kappa$, let

$$R_\xi(x) \Leftrightarrow \varphi(0, x) < \xi,$$
$$S_\xi(x) \Leftrightarrow \neg[\varphi(1, x) < \xi].$$

Thus

$$R = \bigcup_{\xi<\kappa} R_\xi,\ S = \bigcap_{\xi<\kappa} S_\xi,$$
$$\xi < \eta \Rightarrow R_\xi \subseteq R_\eta \wedge S_\xi \supseteq S_\eta$$
$$\lambda < \kappa \text{ limit } \Rightarrow R_\lambda = \bigcup_{\xi<\lambda} S_\xi,\ S_\lambda = \bigcap_{\xi<\lambda} S_\xi.$$

Thus

$$A(\alpha, \beta) \Leftrightarrow \exists\gamma\forall\delta[\exists\xi R_\xi(\alpha, \beta, \gamma, \delta) \wedge \forall\xi S_\xi(\alpha, \beta, \gamma, \delta)],$$

where from now on, in order to simplify the notation, we agree that ordinal variables ξ, η, λ vary over ordinals $< \kappa$.

Also we agree that:

σ denotes a strategy for player I,

τ denotes a strategy for player II,

$\sigma * \beta$ denotes player I's answer if player II plays β,

$\tau * \alpha$ denotes player II's answer if player I plays α.

Assume now A is not determined, towards a contradiction. Then

$$\forall\sigma, \tau\exists\alpha, \beta[A(\alpha, \tau * \alpha) \wedge \neg A(\sigma * \beta, \beta)],$$

or explicitly

$$\begin{aligned}\forall\sigma, \tau\exists\alpha, \beta\{&\exists\gamma\forall\delta[\exists\xi R_\xi(\alpha, \tau * \alpha, \gamma, \delta) \vee \forall\xi S_\xi(\alpha, \tau * \alpha, \gamma, \delta)] \wedge \\ &\forall\gamma\exists\delta[\forall\xi\neg R_\xi(\sigma * \beta, \beta, \gamma, \delta) \wedge \exists\xi\neg S_\xi(\sigma * \beta, \beta, \gamma, \delta)]\}.\end{aligned} \tag{1}$$

We use below letters $\boldsymbol{c}, \boldsymbol{d}, \boldsymbol{e}, \ldots$ to denote Turing degrees. Abbreviate:

$$\forall^* \boldsymbol{c} P(\boldsymbol{c}) \Leftrightarrow \exists \boldsymbol{c}_0 \forall \boldsymbol{c} \geq \boldsymbol{c}_0 P(\boldsymbol{c}),$$

where $\boldsymbol{c} \leq \boldsymbol{d}$ is the usual partial order on the Turing degrees. If γ is a real, let also $\gamma \leq \boldsymbol{d}$ mean that $\gamma \leq_{\mathrm{T}} \delta$ for any δ of Turing degree $\boldsymbol{d}$.

Note now the following simple implications:

$$\exists \gamma \forall \delta P(\gamma, \delta) \Rightarrow \forall^* \boldsymbol{c} \forall^* \boldsymbol{d} \exists \gamma \leq \boldsymbol{c} \forall \delta \leq \boldsymbol{d} P(\gamma, \delta).$$

(*Proof.*
$$\begin{aligned} \exists \gamma \forall \delta P(\gamma, \delta) &\Rightarrow \forall^* \boldsymbol{c} \exists \gamma \leq \boldsymbol{c} \forall^* \boldsymbol{d} \forall \delta \leq \boldsymbol{d} P(\gamma, \delta) \\ &\Rightarrow \forall^* \boldsymbol{c} \forall^* \boldsymbol{d} \exists \gamma \leq \boldsymbol{c} \forall \delta \leq \boldsymbol{d} P(\gamma, \delta).) \end{aligned}$$

$$\forall \gamma \exists \delta P(\gamma, \delta) \Rightarrow \forall^* \boldsymbol{c} \forall^* \boldsymbol{d} \forall \gamma \leq \boldsymbol{c} \exists \delta \leq \boldsymbol{d} P(\gamma, \delta).$$

(*Proof.*
$$\begin{aligned} \forall \gamma \exists \delta P(\gamma, \delta) &\Rightarrow \forall^* \boldsymbol{c} \forall \gamma \leq \boldsymbol{c} \forall^* \boldsymbol{d} \exists \delta \leq \boldsymbol{d} P(\gamma, \delta) \\ &\Rightarrow \forall^* \boldsymbol{c} \forall^* \boldsymbol{d} \forall \gamma \leq \boldsymbol{c} \exists \delta \leq \boldsymbol{d} P(\gamma, \delta), \end{aligned}$$
since $\forall n[\forall^* \boldsymbol{e} P_n(\boldsymbol{e})] \Rightarrow \forall^* \boldsymbol{e}[\forall n P_n(\boldsymbol{e})]$.)

Thus we have from (1), using these manipulations:

$$\begin{aligned} \forall \sigma, \tau \exists \alpha, \beta \{ \forall^* \boldsymbol{c} \forall^* \boldsymbol{d} [\exists \gamma \leq \boldsymbol{c} \forall \delta \leq \boldsymbol{d} (\exists \xi R_\xi(\alpha, \tau * \alpha, \gamma, \delta) \\ \vee \forall \xi S_\xi(\alpha, \tau * \alpha, \gamma, \delta))] \\ \wedge \forall^* \boldsymbol{c} \forall^* \boldsymbol{d} [\forall \gamma \leq \boldsymbol{c} \exists \delta \leq \boldsymbol{d} (\forall \xi \neg R_\xi(\sigma * \beta, \beta, \gamma, \delta) \wedge \\ \exists \xi \neg S_\xi(\sigma * \beta, \beta, \gamma, \delta))] \}. \end{aligned}$$

Note now that

$$\forall^* \boldsymbol{c} \forall^* \boldsymbol{d} P_1(\boldsymbol{c}, \boldsymbol{d}) \wedge \forall^* \boldsymbol{c} \forall^* \boldsymbol{d} P_2(\boldsymbol{c}, \boldsymbol{d}) \Rightarrow \forall^* \boldsymbol{c} \forall^* \boldsymbol{d} [P_1(\boldsymbol{c}, \boldsymbol{d}) \wedge P_2(\boldsymbol{c}, \boldsymbol{d})],$$

thus

$$\begin{aligned} \forall \sigma, \tau \exists \alpha, \beta \{ \forall^* \boldsymbol{c} \forall^* \boldsymbol{d} [\exists \gamma \leq \boldsymbol{c} \forall \delta \leq \boldsymbol{d} (\exists \xi R_\xi(\alpha, \tau * \boldsymbol{d}, \gamma, \delta) \\ \vee \forall \xi S_\xi(\alpha, \tau * \alpha, \gamma, \delta)) \\ \wedge \forall \gamma \leq \boldsymbol{c} \exists \delta \leq \boldsymbol{d} (\forall \xi \neg R_\xi(\sigma * \beta, \beta, \gamma, \delta) \\ \wedge \exists \xi \neg S_\xi(\sigma * \beta, \beta, \gamma, \delta))] \}. \end{aligned} \tag{2}$$

Now fix a real $x \in \mathbb{R}$. Then by a simple Skolem-Löwenheim argument, we can find a countable set of reals $M \subseteq \mathbb{R}$ containing x, closed under pairing and also downward closed under $\leq_{\mathrm{T}}$ (i.e., $y \in M \wedge z \leq_{\mathrm{T}} y \Rightarrow z \in M$), such that

$$\begin{aligned} \forall \sigma, \tau \in M \exists \alpha, \beta \in M \{ \forall^* \boldsymbol{c} \forall^* \boldsymbol{d} [\exists \gamma \leq \boldsymbol{c} \forall \delta \leq \boldsymbol{d} (\exists \xi R_\xi(\alpha, \tau * \boldsymbol{d}, \gamma, \delta) \\ \vee \forall \xi S_\xi(\alpha, \tau * \alpha, \gamma, \delta)) \\ \wedge \forall \gamma \leq \boldsymbol{c} \exists \delta \leq \boldsymbol{d} (\forall \xi \neg R_\xi(\sigma * \beta, \beta, \gamma, \delta) \\ \wedge \exists \xi \neg S_\xi(\sigma * \beta, \beta, \gamma, \delta))] \}. \end{aligned} \tag{3}$$

But since M is countable we have again

$$\forall\sigma,\tau\in M\exists\alpha,\beta\in M\forall^*\boldsymbol{c}\forall^*\boldsymbol{d}R(\alpha,\beta,\sigma,\tau,\boldsymbol{c},\boldsymbol{d}) \Rightarrow \forall^*\boldsymbol{c}\forall^*\boldsymbol{d}\forall\sigma,\tau\in M\exists\alpha,\beta\in MR(\alpha,\beta,\sigma,\tau,\boldsymbol{c},\boldsymbol{d}).$$

Thus finally we conclude that

$$\begin{aligned}\forall^*\boldsymbol{c}\forall^*\boldsymbol{d}\forall\sigma,\tau\in M\exists\alpha,\beta\in M[&\exists\gamma\le\boldsymbol{c}\forall\delta\le\boldsymbol{d}(\exists\xi R_\xi(\alpha,\tau*\alpha,\gamma,\delta)\\&\qquad\vee\forall\xi S_\xi(\alpha,\tau*\alpha,\gamma,\delta))\\&\wedge\forall\gamma\le\boldsymbol{c}\exists\delta\le\boldsymbol{d}(\forall\xi\neg R_\xi(\sigma*\beta,\beta,\gamma,\delta)\\&\qquad\wedge\exists\xi\neg S_\xi(\sigma*\beta,\beta,\gamma,\delta))].\end{aligned}\tag{4}$$

Note now that since the R_ξ's, $\neg S_\xi$'s are increasing and ξ varies over the ordinal κ of cofinality $>\omega$, we must have

$$\begin{aligned}&\exists\gamma\le\boldsymbol{c}\forall\delta\le\boldsymbol{d}(\exists\xi R_\xi(\alpha,\tau*\alpha,\gamma,\delta)\vee\forall\xi S_\xi(\alpha,\tau*\alpha,\gamma,\delta))\\&\quad\Rightarrow\exists\eta\exists\gamma\le\boldsymbol{c}\forall\delta\le\boldsymbol{d}(R_\eta(\alpha,\tau*\alpha,\gamma,\delta)\vee\forall\xi S_\xi(\alpha,\tau*\alpha,\gamma,\delta)),\end{aligned}\tag{5}$$

and similarly

$$\begin{aligned}&\forall\gamma\le\boldsymbol{c}\exists\delta\le\boldsymbol{d}(\forall\xi\neg R_\xi(\sigma*\beta,\beta,\gamma,\delta)\wedge\exists\xi\neg S_\xi(\sigma*\beta,\beta,\gamma,\delta))\\&\quad\Rightarrow\exists\eta\forall\gamma\le\boldsymbol{c}\exists\delta\le\boldsymbol{d}(\forall\xi\neg R_\xi(\sigma*\beta,\beta,\gamma,\delta)\vee\neg S_\eta(\sigma*\beta,\beta,\gamma,\delta)).\end{aligned}\tag{6}$$

Applying the same procedure once again, we see that

$$\begin{aligned}\forall\sigma,\tau\in M\exists\alpha,\beta\in M[&\exists\gamma\le\boldsymbol{c}\forall\delta\le\boldsymbol{d}(\exists\xi R_\xi(\alpha,\tau*\alpha,\gamma,\delta)\\&\qquad\vee\forall\xi S_\xi(\alpha,\tau*\alpha,\gamma,\delta))\\&\wedge\forall\gamma\le\boldsymbol{c}\exists\delta\le\boldsymbol{d}(\forall\xi\neg R_\xi(\sigma*\beta,\beta,\gamma,\delta)\\&\qquad\wedge\exists\xi\neg S_\xi(\sigma*\beta,\beta,\gamma,\delta))].\\&\Rightarrow\\\exists\eta\forall\sigma,\tau\in M\exists\alpha,\beta\in M[&\exists\gamma\le\boldsymbol{c}\forall\delta\le\boldsymbol{d}(R_\eta(\alpha,\tau*\alpha,\gamma,\delta)\\&\qquad\vee\forall\xi S_\xi(\alpha,\tau*\alpha,\gamma,\delta))\\&\wedge\forall\gamma\le\boldsymbol{c}\exists\delta\le\boldsymbol{d}(\forall\xi\neg R_\xi(\sigma*\beta,\beta,\gamma,\delta)\\&\qquad\wedge\neg S_\eta(\sigma*\beta,\beta,\gamma,\delta))].\end{aligned}$$

Thus, in particular,

$$\begin{aligned}\forall\sigma,\tau\in M\exists\alpha,\beta\in M[&\exists\gamma\le\boldsymbol{c}\forall\delta\le\boldsymbol{d}(\exists\xi R_\xi(\alpha,\tau*\alpha,\gamma,\delta)\\&\qquad\vee\forall\xi S_\xi(\alpha,\tau*\alpha,\gamma,\delta))\\&\wedge\forall\gamma\le\boldsymbol{c}\exists\delta\le\boldsymbol{d}(\forall\xi\neg R_\xi(\sigma*\beta,\beta,\gamma,\delta)\\&\qquad\wedge\exists\xi\neg S_\xi(\sigma*\beta,\beta,\gamma,\delta))].\\&\Rightarrow\end{aligned}$$

$$\begin{aligned}\exists\eta\forall\sigma,\tau\in M\exists\alpha,\beta\in M[\exists\gamma\leq \boldsymbol{c}\forall\delta\leq \boldsymbol{d}(R_\eta(\alpha,\tau*\alpha,\gamma,\delta)\\ \vee S_\eta(\alpha,\tau*\alpha,\gamma,\delta))\\ \wedge\forall\gamma\leq \boldsymbol{c}\exists\delta\leq \boldsymbol{d}(\neg R_\eta(\sigma*\beta,\beta,\gamma,\delta)\\ \wedge\neg S_\eta(\sigma*\beta,\beta,\gamma,\delta))].\end{aligned}\tag{7}$$

Abbreviate

$$\begin{aligned}Q_\eta(\sigma,\tau,\alpha,\beta,\boldsymbol{c},\boldsymbol{d})\Leftrightarrow\exists\gamma\leq \boldsymbol{c}\forall\delta\leq \boldsymbol{d}(R_\eta(\alpha,\tau*\alpha,\gamma,\delta)\\ \vee S_\eta(\alpha,\tau*\alpha,\gamma,\delta))\\ \wedge\forall\gamma\leq \boldsymbol{c}\exists\delta\leq \boldsymbol{d}(\neg R_\eta(\sigma*\beta,\beta,\gamma,\delta)\\ \wedge\neg S_\eta(\sigma*\beta,\beta,\gamma,\delta)).\end{aligned}$$

So from (4) and (7) we have finally

$$\forall^*\boldsymbol{c}\forall^*\boldsymbol{d}\exists\eta\forall\sigma,\tau\in M\exists\alpha,\beta\in M(Q_\eta(\sigma,\tau,\alpha,\beta,\boldsymbol{c},\boldsymbol{d})).$$

Call a countable $M\subseteq\mathbb{R}$ x-**good** if it contains $x\in\mathbb{R}$, is closed under pairing, and downward closed under $\leq_{\mathrm{T}}$. We think of M as a real via some appropriate coding. So we have shown that

$$\forall x\exists M(M\text{ is }x\text{-good }\wedge\forall^*\boldsymbol{c}\forall^*\boldsymbol{d}\exists\eta\forall\sigma,\tau\in M\exists\alpha,\beta\in M(Q_\eta(\sigma,\tau,\alpha,\beta,\boldsymbol{c},\boldsymbol{d}))).$$

Now an easy calculation shows that the relation

$$P(x,M)\Leftrightarrow M\text{ is }x\text{-good}\wedge\forall^*\boldsymbol{c}\forall^*\boldsymbol{d}\exists\eta\forall\sigma,\tau\in M\exists\alpha,\beta\in M(Q_\eta(\sigma,\tau,\alpha,\beta,\boldsymbol{c},\boldsymbol{d}))$$

is in $\utilde{\boldsymbol{\Gamma}}$. Since $\utilde{\boldsymbol{\Gamma}}$ has the uniformization property, there is a function $F:\mathbb{R}\to\mathbb{R}$ in $\utilde{\boldsymbol{\Delta}}$ such that for each $x\in\mathbb{R}$, if

$$F(x)=M_x,$$

we have $P(x,M_x)$. Put

$$\mathcal{M}(x)=\{M_y\ :\ y\leq_{\mathrm{T}}x\wedge M_y\text{ is }x\text{-good}\}.$$

Then clearly

$$\forall x\forall^*\boldsymbol{c}\forall^*\boldsymbol{d}\exists\eta\exists M\in\mathcal{M}(x)\forall\sigma,\tau\in M\exists\alpha,\beta\in M(Q_\eta(\sigma,\tau,\alpha,\beta,\boldsymbol{c},\boldsymbol{d})).$$

Let $\mathcal{D}$ be the set of Turing degrees, and define for each real $x\in\mathbb{R}$ a partial function $f_x\colon\mathcal{D}\times\mathcal{D}\to\kappa$ by letting $f_x(\boldsymbol{c},\boldsymbol{d})$ be the least η such that

$$\exists M\in\mathcal{M}(x)\forall\sigma,\tau\in M\exists\alpha,\beta\in M(Q_\eta(\sigma,\tau,\alpha,\beta,\boldsymbol{c},\boldsymbol{d}))$$

if there is one. Thus

$$\forall x\forall^*\boldsymbol{c}\forall^*\boldsymbol{d}(f_x(\boldsymbol{c},\boldsymbol{d})\text{ is defined}).$$

Consider now the following game:

$$\begin{array}{cc}\mathrm{I} & \mathrm{II}\\ \alpha,x_0 & x_1,\beta\end{array}\tag{8}$$

Player I plays $\alpha, x_0 \in \mathbb{R}$; player II plays $x_1, \beta \in \mathbb{R}$. Let $x = \langle x_0, x_1 \rangle$. Then player I wins iff $\forall^* c \forall^* d \exists \gamma \leq c \forall \delta \leq d [R_{f_x(c,d)}(\alpha, \beta, \gamma, \delta) \vee S_{f_x(c,d)}(\alpha, \beta, \gamma, \delta)]$.

An easy calculation shows that this game is in $\underset{\sim}{\Delta}$, so it is determined. We will derive a contradiction from this fact. Say player I has a winning strategy $\bar{\sigma}$. (The argument is entirely similar if player II has a winning strategy.)

We need first the following

LEMMA 2.1.

$$\forall v \exists w \geq_T v \forall w' \geq_T w \forall^* c \forall^* d (f_{w'}(c,d) \geq f_w(c,d)).$$

PROOF. Assume not, and pick v such that

$$\forall w \geq_T v \exists w' \geq_T w \neg \forall^* c \forall^* d (f_{w'}(c,d) \geq f_w(c,d)).$$

Note that if $c_0 \in \mathcal{D}$ is such that

$$\forall c \geq c_0 \forall^* d (f_w(c,d) \text{ is defined } \wedge f_{w'}(c,d) \text{ is defined}),$$

then the relation

$$c \geq c_0 \wedge \forall^* d (f_w(c,d) \geq f_w(c,d))$$

is in $\underset{\sim}{\Delta}$, so by using $\underset{\sim}{\Delta}$-Turing Determinacy, we have

$$\neg \forall^* c \forall^* d (f_{w'}(c,d) \geq f_w(c,d)) \Rightarrow \forall^* c \neg \forall^* d (f_{w'}(c,d) \geq f_w(c,d)).$$

Similarly

$$\begin{aligned} &\forall^* c \neg \forall^* d (f_{w'}(c,d) \geq f_w(c,d)) \\ &\quad \Rightarrow \forall^* c \forall^* d \neg (f_{w'}(c,d) \geq f_w(c,d)) \\ &\quad \Leftrightarrow \forall^* c \forall^* d (f_{w'}(c,d) < f_w(c,d)). \end{aligned}$$

So we have

$$\forall w \geq_T v \exists w' \geq_T w \forall^* c \forall^* d (f_{w'}(c,d) < f_w(c,d)).$$

Thus we can choose $v \leq_T w_0 \leq_T w_1 \leq_T w_2 \leq_T \ldots$ such that

$$\forall n \forall^* c \forall^* d (f_{w_{n+1}}(c,d) < f_{w_n}(c,d)),$$

so that

$$\forall^* c \forall^* d \forall n (f_{w_{n+1}}(c,d) < f_{w_n}(c,d)),$$

a contradiction. ⊣

By Lemma 2.1, find a real $w \geq_T \bar{\sigma}$ such that

$$\forall w' \geq_T w \forall^* c \forall^* d (f_{w'}(c,d) \geq f_w(c,d)).$$

Consider now the strategy σ_0 for player I in a game of the form

I	II
α	β

given by

$$\sigma_0 * \beta = (\bar{\sigma} * \langle w, \beta \rangle)_0.$$

Recall that $\bar{\sigma} * \langle w, \beta \rangle = \langle \alpha, x_0 \rangle$ for some α, x_0 and thus $\alpha = (\bar{\sigma} * \langle w, \beta \rangle)_0$. Thus, if in the game (8) player II plays w, β, player I answers by $\sigma_0 * \beta, x_0$ (for some x_0). Clearly

$$\sigma_0 \leq_{\mathrm{T}} w.$$

By the definition of $f_w(\boldsymbol{c}, \boldsymbol{d})$, we have

$$\forall^* \boldsymbol{c} \forall^* \boldsymbol{d} \exists M \in \mathcal{M}(w) \forall \sigma, \tau \in M \exists \alpha, \beta \in M(Q_{f_w(\boldsymbol{c},\boldsymbol{d})}(\sigma, \tau, \alpha, \beta, \boldsymbol{c}, \boldsymbol{d})).$$

Since $\mathcal{M}(w)$ is a countable set it follows easily from $\underset{\sim}{\boldsymbol{\Delta}}$-Turing Determinacy that

$$\exists M \in \mathcal{M}(w) \forall^* \boldsymbol{c} \forall^* \boldsymbol{d} \forall \sigma, \tau \in M \exists \alpha, \beta \in M(Q_{f_w(\boldsymbol{c},\boldsymbol{d})}(\sigma, \tau, \alpha, \beta, \boldsymbol{c}, \boldsymbol{d})).$$

(We are using here that

$$\forall^* \boldsymbol{c} \forall^* \boldsymbol{d} \exists m R(m, \boldsymbol{c}, \boldsymbol{d}) \Rightarrow \exists m \forall^* \boldsymbol{c} \forall^* \boldsymbol{d} R(m, \boldsymbol{c}, \boldsymbol{d}),$$

if R is in $\underset{\sim}{\boldsymbol{\Delta}}$, a fact which follows from $\underset{\sim}{\boldsymbol{\Delta}}$-Turing Determinacy.) So fix $M_0 \in \mathcal{M}(w)$ such that

$$\forall^* \boldsymbol{c} \forall^* \boldsymbol{d} \forall \sigma, \tau \in M_0 \exists \alpha, \beta \in M_0(Q_{f_w(\boldsymbol{c},\boldsymbol{d})}(\sigma, \tau, \alpha, \beta, \boldsymbol{c}, \boldsymbol{d})). \tag{9}$$

Since $M_0 \in \mathcal{M}(w)$, M_0 is w-good, so in particular, as $\sigma_0 \leq_{\mathrm{T}} w$, $\sigma_0 \in M_0$.

Recalling the definition of Q_η, we have

$$\forall^* \boldsymbol{c} \forall^* \boldsymbol{d} \forall \sigma \in M_0 \exists \beta \in M_0 \forall \gamma \leq \boldsymbol{c} \exists \delta \leq \boldsymbol{d}$$
$$[\neg R_{f_w(\boldsymbol{c},\boldsymbol{d})}(\sigma * \beta, \beta, \gamma, \delta) \wedge \neg S_{f_w(\boldsymbol{c},\boldsymbol{d})}(\sigma * \beta, \beta, \gamma, \delta)],$$

so also, by taking $\sigma = \sigma_0$,

$$\forall^* \boldsymbol{c} \forall^* \boldsymbol{d} \exists \beta \in M_0 \forall \gamma \leq \boldsymbol{c} \exists \delta \leq \boldsymbol{d}$$
$$[\neg R_{f_w(\boldsymbol{c},\boldsymbol{d})}(\sigma_0 * \beta, \beta, \gamma, \delta) \wedge \neg S_{f_w(\boldsymbol{c},\boldsymbol{d})}(\sigma_0 * \beta, \beta, \gamma, \delta)].$$

But M_0 is countable, so, exactly as before, we conclude that

$$\exists \beta \in M_0 \forall^* \boldsymbol{c} \forall^* \boldsymbol{d} \forall \gamma \leq \boldsymbol{c} \exists \delta \leq \boldsymbol{d}$$
$$[\neg R_{f_w(\boldsymbol{c},\boldsymbol{d})}(\sigma_0 * \beta, \beta, \gamma, \delta) \wedge \neg S_{f_w(\boldsymbol{c},\boldsymbol{d})}(\sigma_0 * \beta, \beta, \gamma, \delta)].$$

Thus fix $\beta_0 \in M_0$ such that

$$\forall^* \boldsymbol{c} \forall^* \boldsymbol{d} \forall \gamma \leq \boldsymbol{c} \exists \delta \leq \boldsymbol{d} [\neg R_{f_w(\boldsymbol{c},\boldsymbol{d})}(\sigma_0 * \beta_0, \beta_0, \gamma, \delta) \wedge \neg S_{f_w(\boldsymbol{c},\boldsymbol{d})}(\sigma_0 * \beta_0, \beta_0, \gamma, \delta)].$$

Put

$$\alpha_0 = \sigma_0 * \beta_0.$$

Then, for some x_0,

I	II
α_0, x_0	w, β_0

is a run of the game (8) in which player I followed $\bar{\sigma}$. Thus, if

$$x = \langle x_0, w \rangle,$$

we have

$$\forall^* \boldsymbol{c} \forall^* \boldsymbol{d} \exists \gamma \leq \boldsymbol{c} \forall \delta \leq \boldsymbol{d}[R_{f_x(\boldsymbol{c},\boldsymbol{d})}(\alpha_0, \beta_0, \gamma, \delta) \vee S_{f_x(\boldsymbol{c},\boldsymbol{d})}(\alpha_0, \beta_0, \gamma, \delta)].$$

This contradicts immediately the above, if we can only show that

$$\forall^* \boldsymbol{c} \forall^* \boldsymbol{d}(f_x(\boldsymbol{c}, \boldsymbol{d}) = f_w(\boldsymbol{c}, \boldsymbol{d})).$$

This can be proved as follows: First, since $w \leq_{\mathrm{T}} x$, we have from the lemma that

$$\forall^* \boldsymbol{c} \forall^* \boldsymbol{d}(f_w(\boldsymbol{c}, \boldsymbol{d}) \leq f_x(\boldsymbol{c}, \boldsymbol{d})),$$

so it is enough to check that

$$\forall^* \boldsymbol{c} \forall^* \boldsymbol{d}(f_w(\boldsymbol{c}, \boldsymbol{d}) \geq f_x(\boldsymbol{c}, \boldsymbol{d})).$$

Since M_0 is w-good and $\beta_0 \in M_0$, $\sigma_0 \in M_0$, we also have $\alpha_0, x_0 \in M_0$ thus $x \in M_0$. So M_0 is x-good. Since $M_0 \in \mathcal{M}(w)$, $M_0 = M_y$ for some $y \leq_{\mathrm{T}} w \leq_{\mathrm{T}} x$, thus also $M_0 \in \mathcal{M}(x)$. But from (9)

$$\forall^* \boldsymbol{c} \forall^* \boldsymbol{d} \forall \sigma, \tau \in M_0 \exists \alpha, \beta \in M(Q_{f_w(\boldsymbol{c},\boldsymbol{d})}(\sigma, \tau, \alpha, \beta, \boldsymbol{c}, \boldsymbol{d})),$$

therefore, in particular,

$$\forall^* \boldsymbol{c} \forall^* \boldsymbol{d} \exists M \in \mathcal{M}(x) \forall \sigma, \tau \in M_0 \exists \alpha, \beta \in M(Q_{f_w(\boldsymbol{c},\boldsymbol{d})}(\sigma, \tau, \alpha, \beta, \boldsymbol{c}, \boldsymbol{d})).$$

But by definition, for almost all $\boldsymbol{c}$ and $\boldsymbol{d}$, $f_x(\boldsymbol{c}, \boldsymbol{d})$ is the least η such that

$$(\exists M \in \mathcal{M}(x) \forall \sigma, \tau \in M \exists \alpha, \beta \in M(Q_\eta(\sigma, \tau, \alpha, \beta, \boldsymbol{c}, \boldsymbol{d}))),$$

so for almost all $\boldsymbol{c}$ and $\boldsymbol{d}$, we have $f_x(\boldsymbol{c}, d) \leq f_w(\boldsymbol{c}, \boldsymbol{d})$, and the proof of Theorem 1.8 is complete.

Now consider the case $\Gamma = \mathrm{IND}$. The preceding proof shows that if IND has the uniformization property, then Det(HYP) implies $\forall n < \omega(\mathsf{Det}(\Sigma^*_n))$, i.e., we have a lightface version. However one needs Det($\underset{\sim}{\mathbf{HYP}}$) to prove that IND has the uniformization property. So we can ask whether this lightface implication is true outright. At this moment we can only establish this for $n = 1$, and we do not know if it is true for higher n. The argument is as follows:

Assume Det(HYP). Let

$$A(\alpha, \beta) \Leftrightarrow \exists \gamma[\exists \xi R_\xi(\alpha, \beta, \gamma) \wedge \forall \xi S_\xi(\alpha, \beta, \gamma)],$$

be a typical Σ^*_1 set, where R_ξ, S_ξ are as in the preceding proof. Assume again A is not determined towards a contradiction. Thus

$$\forall \sigma, \tau \exists \alpha, \beta[A(\alpha, \tau * \alpha) \wedge \neg A(\sigma * \beta, \beta)].$$

Fix a real $x \in \mathbb{R}$. Then there is an x-good M such that

$$\forall \sigma, \tau \in M \exists \alpha, \beta \in M[A(\alpha, \tau * \alpha) \wedge \neg A(\sigma * \beta, \beta)].$$

Thus,

$$\forall \tau \in M \exists \alpha \in M \exists \gamma[\exists \xi R_\xi(\alpha, \tau * \alpha, \gamma) \wedge \forall \xi S_\xi(\alpha, \tau * \alpha, \beta)]$$
$$\wedge \forall \sigma \in M \exists \beta \in M \forall \gamma[\forall \xi \neg R_\xi(\sigma * \beta, \beta, \gamma) \vee \exists \xi \neg S_\xi(\sigma * \beta, \beta, \gamma)].$$

From the first conjunct we have again some ξ_0 such that

$$\forall \tau \in M \exists \alpha \in M \exists \gamma[R_{\xi_0}(\alpha, \tau * \alpha, \gamma) \wedge \forall \xi S_\xi(\alpha, \tau * \alpha, \gamma)]. \qquad (*)$$

Then from the second conjunct we have

$$\forall \sigma \in M \exists \beta \in M \forall \gamma[\neg R_{\xi_0}(\sigma * \beta, \beta, \gamma) \vee \exists \xi \neg S_\xi(\sigma * \beta, \beta, \gamma)],$$

so that by standard reflection properties of IND (recall that IND is the class of Σ_1 over the next admissible of $\mathbb{R}$ sets of reals), we have some ξ_1 with

$$\forall \sigma \in M \exists \beta \in M \forall \gamma[\neg R_{\xi_0}(\sigma * \beta, \beta, \gamma) \vee \neg S_{\xi_1}(\sigma * \beta, \beta, \gamma)].$$

thus going back to $(*)$ again we have also

$$\forall \tau \in \mathcal{M} \exists \alpha \in \mathcal{M} \exists \gamma[R_{\xi_0}(\alpha, \tau * \alpha, \gamma) \wedge S_{\xi_1}(\sigma * \beta, \beta, \gamma)].$$

So we have proved that

$$\exists \xi_0, \xi_1 \forall \sigma, \tau \in M \exists \alpha, \beta \in M[\forall \gamma(\neg R_{\xi_0}(\sigma * \beta, \beta, \gamma) \vee \neg S_{\xi_1}(\sigma * \beta, \beta, \gamma))$$
$$\wedge \exists \gamma(R_{\xi_0}(\alpha, \tau * \alpha, \gamma) \wedge S_{\xi_1}(\alpha, \tau * \alpha, \gamma))].$$

So

$$\forall x \exists \xi_0, \xi_1 \exists M[M \text{ is } x\text{-good} \wedge$$
$$\forall \sigma, \tau \in M \exists \alpha, \beta \in M(Q_{\xi_0, \xi_1}(\sigma, \tau, \alpha, \beta))],$$

where

$$Q_{\xi_0, \xi_1}(\sigma, \tau, \alpha, \beta) \Leftrightarrow \forall \gamma[\neg R_{\xi_0}(\sigma * \beta, \beta, \gamma) \vee \neg S_{\xi_1}(\sigma * \beta, \beta, \gamma)] \wedge$$
$$\exists \gamma[R_{\xi_0}(\alpha, \tau * \alpha, \gamma) \wedge S_{\xi_1}(\alpha, \tau * \alpha, \gamma)].$$

Define now $f : \mathbb{R} \to \kappa$ by

$$f(x) = \text{ least } \langle \xi_0, \xi_1 \rangle \exists M[M \text{ is } x\text{-good } \wedge$$
$$\forall \sigma, \tau \in M \exists \alpha, \beta \in M(Q_{\xi_0, \xi_1}(\sigma, \tau, \alpha, \beta))],$$

and write $f(x) = \langle f(x)_0, f)x_1 \rangle$. Then $f \in$ HYP, and if we consider the game

I	II
α, x_0	x_1, β

where $x = \langle x_0, x_1 \rangle$; player I wins iff $\exists \gamma[R_{f(x)_0}(\alpha, \beta, \gamma) \wedge S_{f(x)_1}(\alpha, \beta, \gamma)]$, we obtain a contradiction, exactly as before, noticing that this is now a HYP game.

§3. Proof of Theorem 1.10. This is a variant of the proof in §2. We take $k = 2, n = 2$ for notational simplicity. We thus assume below $\mathrm{Det}(\underset{\sim}{\mathbf{\Delta}}^1_4)$.

First note that, using $\forall\alpha(\alpha^\# \text{ exists})$, we have that every $\Game\mathcal{M}_2$ set A can be represented in the form

$$x \in A \Leftrightarrow \mathbf{L}[x] \models \varphi(x, u_1, u_2),$$

where φ is a formula of set theory and $u_1, u_2, \dots$ the uniform indiscernibles; see [Mar83A]. Thus a typical $\Game^3\mathcal{M}_2$ game has the form

$$A(\alpha, \beta) \Leftrightarrow \Game\gamma\Game\delta(\mathbf{L}[\alpha, \beta, \gamma, \delta] \models \varphi(\alpha, \beta, \gamma, \delta, u_1, u_2)).$$

Assume now, towards a contradiction, that this game is not determined. Thus we have

$$\forall\sigma, \tau\exists\alpha, \beta[\Game\gamma\Game\delta(\mathbf{L}[\alpha, \beta, \gamma, \delta] \models \varphi(\alpha, \tau * \alpha, \gamma, \delta, u_1, u_2))$$
$$\wedge \neg\Game\gamma\Game\delta(\mathbf{L}[\sigma * \beta, \beta, \gamma, \delta] \models \varphi(\sigma * \beta, \beta, \gamma, \delta, u_1, u_2))].$$

Now

$$\Game\gamma P(\gamma) \Leftrightarrow \exists\rho\forall\varepsilon P(\rho \cdot \varepsilon),$$

where if ρ is a strategy for player I and ε is what player II plays, $\rho \cdot \varepsilon$ is the run of the game with player I following ρ and player II playing ε. Thus

$$\Game\gamma P(\gamma) \Rightarrow \forall^* \boldsymbol{c}\exists\rho \leq \boldsymbol{c}\forall\varepsilon \leq \boldsymbol{c}P(\rho \cdot \varepsilon).$$

Abbreviate:

$$\Game\gamma \leq \boldsymbol{c}(R(\gamma)) \Leftrightarrow \exists\rho \leq \boldsymbol{c}\forall\varepsilon \leq \boldsymbol{c}R(\rho \cdot \varepsilon). \tag{10}$$

Thus we have

$$\Game\gamma\Game\delta(\mathbf{L}[\alpha, \tau * \alpha, \gamma\delta] \models \varphi(\alpha, \tau * \alpha, \gamma, \delta, u_1, u_2)) \Rightarrow$$
$$\forall^* \boldsymbol{c}\forall^* \boldsymbol{d}\Game\gamma \leq \boldsymbol{c}\Game\delta \leq \boldsymbol{d}(\mathbf{L}[\alpha, \tau * \alpha, \gamma, \delta] \models \varphi(\alpha, \tau * \alpha, \gamma, \delta, u_1, u_2)).$$

Now note that

$$S(\gamma) \Leftrightarrow \Game\delta(\mathbf{L}[\sigma * \beta, \beta, \gamma, \delta] \models \varphi(\sigma * \beta, \beta, \gamma, \delta, u_1, u_2))$$

is a $\Game^2\mathcal{M}_2$ game, thus in particular $\underset{\sim}{\mathbf{\Delta}}^1_4$, so it is determined. So

$$\neg\Game\gamma\Game\delta(\mathbf{L}[\sigma * \beta, \beta, \gamma, \delta] \models \varphi(\sigma * \beta, \beta, \gamma, \delta, u_1, u_2))$$
$$\Rightarrow \Game'\gamma\neg\Game\delta(\mathbf{L}[\sigma * \beta, \beta, \gamma, \delta] \models \varphi(\sigma * \beta, \beta, \gamma, \delta, u_1, u_2)),$$

where

$$\Game'\gamma\neg P(\gamma) \Leftrightarrow \text{player II has a winning strategy in } \{\gamma : P(\gamma)\}.$$

Applying this once more we have

$$\neg\Game\gamma\Game\delta(\mathbf{L}[\sigma * \beta, \beta, \gamma, \delta] \models \varphi(\sigma * \beta, \beta, \gamma, \delta, u_1, u_2))$$
$$\Rightarrow \Game'\gamma\Game'\delta(\mathbf{L}[\sigma * \beta, \beta, \gamma, \delta] \models \neg\varphi(\sigma * \beta, \beta, \gamma, \delta, u_1, u_2)).$$

Thus, exactly as before,

$$\neg\Game\gamma\Game\delta(\mathbf{L}[\sigma * \beta, \beta, \gamma, \delta] \models \varphi(\sigma * \beta, \beta, \gamma, \delta, u_1, u_2))$$
$$\Rightarrow \forall^* \boldsymbol{c} \forall^* \boldsymbol{d} \Game'\gamma \leq \boldsymbol{c} \Game'\delta \leq d(\mathbf{L}[\sigma * \beta, \beta, \gamma, \delta] \models \neg\varphi(\sigma * \beta, \beta, \gamma, \delta, u_1, u_2)),$$

where $\Game'\gamma \leq \boldsymbol{c}$ has the obvious meaning as in (10), by interchanging the roles of players player I and player II. So finally we have

$$\forall\sigma, \tau\exists\alpha, \beta\forall^* \boldsymbol{c}\forall^* \boldsymbol{d}[\Game\gamma \leq \boldsymbol{c}\Game\delta \leq \boldsymbol{d}(\mathbf{L}[\alpha, \tau * \alpha, \gamma, \delta] \models \varphi(\alpha, \tau * \alpha, \gamma, \delta, u_1, u_2))$$
$$\wedge \Game'\gamma \leq \boldsymbol{c}\Game'\delta(\mathbf{L}[\sigma * \beta, \beta, \gamma, \delta] \models \neg\varphi(\sigma * \beta, \beta, \gamma, \delta, u_1, u_2))].$$

Now fix $x \in \mathbb{R}$ and find M which is x-good such that

$$\forall^* \boldsymbol{c}\forall^* \boldsymbol{d}\forall\sigma, \tau \in M\exists\alpha, \beta \in M$$
$$[\Game\gamma \leq \boldsymbol{c}\Game\delta \leq \boldsymbol{d}(\mathbf{L}[\alpha, \tau * \alpha, \gamma, \delta] \models \varphi(\alpha, \tau * \alpha, \gamma, \delta, u_1, u_2))$$
$$\wedge \Game'\gamma \leq \boldsymbol{c}\Game'\delta \leq \boldsymbol{d}(\mathbf{L}[\sigma * \beta, \beta, \gamma, \delta] \models \neg\varphi(\sigma * \beta, \beta, \gamma, \delta, u_1, u_2))].$$

By a simple indiscernibility argument, this implies that

$$\forall^* \boldsymbol{c}\forall^* \boldsymbol{d}\exists\xi_0, \xi_1, \xi_2\, \forall\sigma, \tau \in M\exists\alpha, \beta \in M$$
$$[\Game\gamma \leq \boldsymbol{c}\Game\delta \leq \boldsymbol{d}(\mathbf{L}_{\xi_0}[\alpha, \tau * \alpha, \gamma, \delta] \models \varphi(\alpha, \tau * \alpha, \gamma, \delta, \xi_1, \xi_2))$$
$$\wedge \Game'\gamma \leq \boldsymbol{c}\Game'\delta \leq \boldsymbol{d}(\mathbf{L}_{\xi_0}[\sigma * \beta, \beta, \gamma, \delta] \models \neg\varphi(\sigma * \beta, \beta, \gamma, \delta, \xi_1, \xi_2))].$$

Here and below ordinal variables ξ_0, ξ_1, ξ_2 are supposed to vary over ω_1 and it is implicit in the above notation that $\xi_1 < \xi_2 < \xi_0$. Abbreviate

$$Q_{\xi_0,\xi_1,\xi_2}(\sigma, \tau, \alpha, \beta, \boldsymbol{c}, \boldsymbol{d}) \Leftrightarrow$$
$$\Game\gamma \leq \boldsymbol{c}\Game\delta \leq \boldsymbol{d}(\mathbf{L}_{\xi_0}[\alpha, \tau * \alpha, \gamma, \delta) \models \varphi(\alpha, \tau * \alpha, \gamma, \delta, \xi_1, \xi_2) \wedge$$
$$\Game'\gamma \leq \boldsymbol{c}\Game'\delta \leq \boldsymbol{d}(\mathbf{L}_{\xi_0}[\sigma * \beta, \beta, \gamma, \delta] \models \neg\varphi(\sigma * \beta, \beta, \gamma, \delta, \xi_1, \xi_2)).$$

Thus we have shown that

$$\forall x\exists M(M \text{ is } x\text{-good } \wedge$$
$$\forall^* \boldsymbol{c}\forall^* \boldsymbol{d}\exists\xi_0, \xi_1, \xi_2\forall\sigma, \tau \in M\exists\alpha, \beta \in M(Q_{\xi_0,\xi_1,\xi_2}(\sigma, \tau, \alpha, \beta, \boldsymbol{c}, \boldsymbol{d})))$$

Since the relation $P(x, \mathcal{M})$ following $\forall x\exists\mathcal{M}$ is Σ^1_4, we have, by the uniformization theorem for Σ^1_4, a function $F : \mathbb{R} \to \mathbb{R}$ in Δ^1_4 such that if $F(x) = M_x$, then $P(x, M_x)$.

(That $P(x, M)$ is Σ^1_4 follows from the fact that the expression $R(\boldsymbol{c}, \boldsymbol{d})$ following $\forall^* \boldsymbol{c}\forall^* \boldsymbol{d}$ is $\utilde{\mathbf{\Sigma}}^1_2$, thus

$$\neg\forall^* \boldsymbol{d}R(\boldsymbol{c}, \boldsymbol{d}) \Leftrightarrow \forall^* \boldsymbol{d}\neg R(\boldsymbol{c}, \boldsymbol{d}),$$

so $\forall^* \boldsymbol{d}R(\boldsymbol{c}, \boldsymbol{d})$ is $\utilde{\mathbf{\Pi}}^1_3$, therefore

$$\forall^* \boldsymbol{c}\forall^* \boldsymbol{d}R(\boldsymbol{c}, \boldsymbol{d})$$

is $\utilde{\mathbf{\Sigma}}^1_4$, uniformly in all the parameters involved, i.e. R is Σ^1_4).

Let again

$$\mathcal{M}(x) = \{M_y : y \leq_T x \wedge M_y \text{ is } x\text{-good}\}.$$

Thus

$$\forall x \forall^* \boldsymbol{c} \forall^* \boldsymbol{d} \exists \xi_0, \xi_1, \xi_2 \exists M \in \mathcal{M}(x) \forall \sigma, \tau \in M \exists \alpha, \beta \in M (Q_{\xi_0, \xi_1, \xi_2}(\sigma, \tau, \alpha, \beta, \boldsymbol{c}, \boldsymbol{d})).$$

Define as before a partial function by letting $f_x(\boldsymbol{c}, \boldsymbol{d})$ be the least $\langle \xi_0, \xi_1, \xi_2 \rangle$ such that

$$\exists M \in \mathcal{M}(x) \forall \sigma, \tau \in M \exists \alpha, \beta \in M (Q_{\xi_0, \xi_1, \xi_2}(\sigma, \tau, \alpha, \beta, \boldsymbol{c}, \boldsymbol{d})),$$

and let

$$f_x(\boldsymbol{c}, \boldsymbol{d}) = \langle f_x(\boldsymbol{c}, \boldsymbol{d})_0, f_x(\boldsymbol{c}, \boldsymbol{d})_1, f_x(\boldsymbol{c}, \boldsymbol{d})_2 \rangle.$$

Then consider the game

I	II
α, x_0	x_1, β

where $x = \langle x_0, x_1 \rangle$; player I wins iff

$$\forall^* \boldsymbol{c} \forall^* \boldsymbol{d} \Game \gamma \leq \boldsymbol{c} \Game \delta \leq \boldsymbol{d} (\mathbf{L}_{f_x(\boldsymbol{c}, \boldsymbol{d})_0}[\alpha, \beta, \gamma, \delta] \models \varphi(\alpha, \beta, \gamma, \delta, f_x(\boldsymbol{c}, \boldsymbol{d})_1, f_x(\boldsymbol{c}, \boldsymbol{d})_2)).$$

This is a Δ^1_4 game, so it is determined, and we proceed as in §2 to derive a contradiction.

NOTE. In the proof of the analog of Lemma 2.1, we make use of $\mathsf{Det}(\boldsymbol{\Pi}^1_3)$ and similarly for the arguments following it.

Let us remark that from the proof we just gave, we have the following lightface version of Theorem 1.10:

For all $k \geq 1$,

$$\forall m (\mathsf{Det}(\Game^{2k-2} \mathbf{M}_m)) \Rightarrow [\mathsf{Det}(\Delta^1_{2k}) \Leftrightarrow \mathsf{Det}(\Game^{2k-1} \mathbf{M}_n)], \ \forall n < \omega.$$

In particular, for $k = 1$, we have

$$\mathsf{Det}(\Delta^1_{2k}) \Leftrightarrow \mathsf{Det}(\Game \mathbf{M}_n), \ \forall n < \omega$$

granting $\forall \alpha (\alpha^\# \text{ exists})$. If our conjecture at the end of § 1 holds, then the above hypothesis should be just equivalent to $\mathsf{Det}(\boldsymbol{\Pi}^1_{2k-1})$.

§4. Proof of Theorem 1.2. Assume $\mathsf{AD}+\mathsf{DC}+\mathbf{V}=\mathbf{L}(\mathbb{R})$ and every Suslin set of reals is determined. We shall prove by induction on $\xi \in \mathrm{Ord}$, that every set in $\mathbf{L}_\xi(\mathbb{R})$ is determined. Abbreviate this by

$$\mathsf{Det}(\mathbf{L}_\xi(\mathbb{R})).$$

We can assume $\xi \geq \omega + \omega$. (In [KKMW81], it is actually proved that, under our hypotheses, we have $\mathsf{Det}(\mathbf{L}_{\kappa^\mathbb{R}}(\mathbb{R}))$, where $\kappa^\mathbb{R}$ is the first $\mathbb{R}$-admissible ordinal.)

So assume $\mathsf{Det}(\mathbf{L}_\xi(\mathbb{R}))$. We want to prove $\mathsf{Det}(\mathbf{L}_{\xi+1}(\mathbb{R}))$ We can assume that there is a new set of reals in $\mathbf{L}_{\xi+1}(\mathbb{R})$, otherwise there is nothing to prove.

Following the analysis of [Ste83A], this will be done by considering two main cases on ξ.

From now on our blanket assumptions on ξ are

$$\mathsf{Det}(\mathbf{L}_\xi(\mathbb{R})) \text{ and } \exists A \subseteq \mathbb{R}(A \in \mathbf{L}_{\xi+1}(\mathbb{R}) \setminus \mathbf{L}_\xi(\mathbb{R})).$$

CASE I. There is no ordinal $\xi' < \xi$ such that $\mathbf{L}_{\xi'}(\mathbb{R}) \prec_1^{\mathbb{R}} \mathbf{L}_\xi(\mathbb{R})$, where this notation means that for every Σ_1 formula φ and every real parameter r,

$$\mathbf{L}_{\xi'}(\mathbb{R}) \models \varphi(r) \Leftrightarrow \mathbf{L}_\xi(\mathbb{R}) \models \varphi(r).$$

LEMMA 4.1. There is a partial map

$$p : \mathbb{R} \twoheadrightarrow \mathbf{L}_\xi(\mathbb{R}),$$

whose graph is $\underset{\sim}{\Sigma}_1$ over $\mathbf{L}_\xi(\mathbb{R})$ with only real parameters.

PROOF. First note that there is a total function $f : \xi \times \xi \times \mathbb{R} \twoheadrightarrow \mathbf{L}_\xi(\mathbb{R})$ which is $\underset{\sim}{\Sigma}_1$ over $\mathbf{L}_\xi(\mathbb{R})$ with only parameter the largest limit ordinal $\lambda \leq \xi - 1$, if such exists, i.e., if ξ is successor. In this case, since $\lambda < \xi$, we have $\mathbf{L}_\lambda(\mathbb{R}) \not\prec_1^{\mathbb{R}} \mathbf{L}_\xi(\mathbb{R})$, so there is a Σ_0 formula $\varphi(r, x)$ and a real parameter r_0 such that $\mathbf{L}_\xi(\mathbb{R}) \models \exists x \varphi(r_0, x)$ but $\mathbf{L}_\lambda(\mathbb{R}) \models \forall x \neg\varphi(r_0, x)$. Say $x_0 \in \mathbf{L}_{\lambda+n+1}(\mathbb{R})$, $\lambda + n + 1 \leq \xi$, be such that $\mathbf{L}_\xi(\mathbb{R}) \models \varphi(r_0, x_0)$.

Let ϑ be a formula such that

$$x_0 = \{y \in \mathbf{L}_{\lambda+n}(\mathbb{R}) : \mathbf{L}_{\lambda+n}(\mathbb{R}) \models \vartheta(y, z_0)\},$$

where $z_0 \in \mathbf{L}_{\lambda+n}(\mathbb{R})$. Then we have in $\mathbf{L}_\xi(\mathbb{R})$:

$$\begin{aligned} \lambda' = \lambda \Leftrightarrow\ & \lambda' \text{ is limit} \\ & \wedge \exists M \exists z \exists x (M = \mathbf{L}_{\lambda'+n}(\mathbb{R}) \wedge z \in M \\ & \qquad \wedge x = \{y \in M : M \models \vartheta(y, z)\} \wedge \varphi(r_0, z)). \end{aligned}$$

So $\{\lambda\}$ is $\underset{\sim}{\Sigma}_1$ in $\mathbf{L}_\xi(\mathbb{R})$ with only real parameters. Thus we have that there is total $f : \xi \times \xi \times \mathbb{R} \twoheadrightarrow \mathbf{L}_\xi(\mathbb{R})$ which is $\underset{\sim}{\Sigma}_1$ over $\mathbf{L}_\xi(\mathbb{R})$ with only real parameters.

Using f we can define a Σ_1-Skolem function for $\mathbf{L}_\xi(\mathbb{R})$ as follows: Let $\psi(i, x, y)$ be a universal Σ_1 formula. Say

$$\psi(i, x, y) \Leftrightarrow \exists z \vartheta_0(i, x, y, z),\ \vartheta_0 \in \Sigma_0.$$

Define in $\mathbf{L}_\xi(\mathbb{R})$

$$\begin{aligned} g_1(i, x) = \eta_1 \Leftrightarrow\ & \exists \eta_2 \exists \zeta_1 \exists \zeta_2 [\exists r \exists s \vartheta_0(i, f(\eta_1, \eta_2, r), f(\zeta_1, \zeta_2, s)) \\ & \wedge \forall (\eta_1', \eta_2', \zeta_1', \zeta_2') <_{\text{Gödel}} (\eta_1, \eta_2, \zeta_1, \zeta_2) \\ & \forall r' \forall s' \neg \vartheta_0(i, x, f(\eta_1', \eta_2', r'), f(\zeta_1', \zeta_2', s'))], \end{aligned}$$

$$g_2(i.x) = \eta_2 \Leftrightarrow \exists \eta_1 \exists \zeta_1, \exists \zeta_2 [\ldots \text{as before} \ldots],$$

where r, s vary over $\mathbb{R}$. Then

$$\exists y \psi(i, x, y) \text{ implies } \exists r \psi(i, x, f(g_1(i, x), g_2(i, x), r)).$$

Put

$$S(i,z) = \begin{cases} f(g_1(i,(z)_0), g_2(i,(z)_0), (z)_1) & \text{if } (z)_1 \in \mathbb{R}, \\ 0 & \text{otherwise.} \end{cases}$$

Here $z = \langle (z)_0, (z)_1, (z)_2 \rangle$ is some Σ_1 coding function in $\mathbf{L}_\xi(\mathbb{R})$. Thus S is a partial function with graph $\underset{\sim}{\Sigma}_1$ over $\mathbf{L}_\xi(\mathbb{R})$ involving only real parameters, such that

$$\exists y \psi(i,x,y) \text{ implies } \exists r \psi(i,x,S(i,\langle x,r\rangle)).$$

Thus we have

$$\mathbb{R} \subseteq S(\omega \times \mathbb{R}) = M \prec_1 \mathbf{L}_\xi(\mathbb{R}),$$

where $M \prec_1 N$ means that M is a Σ_1 substructure of N, i.e., for every Σ_1 formula φ and every $x_0 \in M$, $M \models \varphi(x_0) \Leftrightarrow N \models \varphi(x_0)$. Collapse now M to some $\mathbf{L}_\eta(\mathbb{R})$ for $\eta \leq \xi$. Clearly $\mathbf{L}_\eta(\mathbb{R}) \prec_1^{\mathbb{R}} \mathbf{L}_\xi(\mathbb{R})$, so $\eta = \xi$. Since $S' = S{\restriction}M$ is $\underset{\sim}{\Sigma}_1$ over M with only real parameters, the image of S', say S^*, under the collapse is $\underset{\sim}{\Sigma}_1$ over $\mathbf{L}_\xi(\mathbb{R})$ with only real parameters and $S(\omega \times \mathbb{R}) = \mathbf{L}_\xi(\mathbb{R})$. Let

$$p(r) = S^*(r(0), \lambda t r(t+1)).$$

Then $p \colon \mathbb{R} \twoheadrightarrow \mathbf{L}_\xi(\mathbb{R})$ is a partial map with $\underset{\sim}{\Sigma}_1$ over $\mathbf{L}_\xi(\mathbb{R})$ graph involving only real parameters. ⊣

We shall also need the following basic result of [Ste83A]:

THEOREM 4.2. [Ste83A] For any infinite cardinal ξ, if $\mathsf{Det}(\mathbf{L}_\xi(\mathbb{R}))$ holds, the pointclass of sets of reals which are $\underset{\sim}{\Sigma}_1$ over $\mathbf{L}_\xi(\mathbb{R})$ with *only real parameters* has the scale property.

Let now $A \subseteq \mathbb{R}$ belong to $\mathbf{L}_{\xi+1}(\mathbb{R})$. Say, for notational convenience, A is $\underset{\sim}{\Pi}_3$ over $\mathbf{L}_\xi(\mathbb{R})$. Then for some $\varphi \in \Sigma_0$ and some $z_0 \in \mathbf{L}_\xi(\mathbb{R})$:

$$r \in A \Leftrightarrow \mathbf{L}_\xi(\mathbb{R}) \models \forall x_0 \exists x_1 \forall x_2 \varphi(r, x_0, x_1, x_2, z_0).$$

Here and below letters r, s, t will be reserved for reals only. Using the projection map $p \colon \mathbb{R} \twoheadrightarrow \mathbf{L}_\xi(\mathbb{R})$, this can be written as

$$r \in A \Leftrightarrow \mathbf{L}_\xi(\mathbb{R}) \models \forall r_0 \exists r_1 \forall r_2 \bigvee_{i=1}^{k} [\psi_i(r, r_0, r_1, r_2, s_0) \wedge \chi_i(r, r_0, r_1, r_2, s_0)],$$

where $\psi_i \in \Sigma_1$, $\chi_1 \in \Pi_1$ and $s_0 \in \mathbb{R}$.

Let us call $\underset{\sim}{\Gamma}$ the class of all sets of reals which are $\underset{\sim}{\Sigma}_1$ over $\mathbf{L}_\xi(\mathbb{R})$ with real parameters, so that $\underset{\sim}{\Gamma}$ has the scale property by Steel's Theorem. Let us consider now the following subcases on ξ.

SUBCASE 1. ξ is limit and for every Σ_0 formula φ and every $s \in \mathbb{R}$

$$\mathbf{L}_\xi(\mathbb{R}) \models \forall r \exists x \varphi(r,x,s) \Rightarrow \exists \eta < \xi (\mathbf{L}_\eta(\mathbb{R}) \models \forall r \exists x \varphi(r,x,s)).$$

Then clearly $\underset{\sim}{\Gamma}$ is closed under $\wedge, \vee, \forall^{\mathbb{R}}, \exists^{\mathbb{R}}$, continuous substitutions and has the scale property, so in particular the prewellordering and uniformization properties. Thus by Theorem 1.2

$$\mathsf{Det}(\underset{\sim}{\Gamma}) \Rightarrow \mathsf{Det}(\underset{\sim}{\Sigma}^*_n(\underset{\sim}{\Gamma})),\ \forall n < \omega.$$

But since every set in $\underset{\sim}{\Gamma}$ is Suslin, we have $\mathsf{Det}(\underset{\sim}{\Gamma})$. So we have $\mathsf{Det}(\underset{\sim}{\Sigma}^*_n(\underset{\sim}{\Gamma}))$. But from the preceding it follows that the typical set $A \in \mathbf{L}_{\xi+1}(\mathbb{R})$ is in $\underset{\sim}{\Sigma}^*_n(\underset{\sim}{\Gamma})$ for some n, thus we have $\mathsf{Det}(\mathbf{L}_{\xi+1}(\mathbb{R}))$, and we are done.

Subcase 2. ξ is limit but Subcase 1 fails.

Then there is some Σ_0 formula φ and some $s_0 \in \mathbb{R}$ such that $\mathbf{L}_\xi(\mathbb{R}) \models \forall r \exists x \varphi(r, x, s_0)$, but for all $\eta < \xi$, we have $\mathbf{L}_\eta(\mathbb{R}) \not\models \forall r \exists x \varphi(r, x, s_0)$. For each real $r \in \mathbb{R}$, let $h(r)$ be the least $\eta < \xi$ such that

$$\mathbf{L}_\eta(\mathbb{R}) \models \exists x \varphi(r, s, x_0).$$

Then $h \colon \mathbb{R} \to \xi$ is total with $\underset{\sim}{\Sigma}_1$ over $\mathbf{L}_\xi(\mathbb{R})$ graph involving only real parameters and h is cofinal in ξ. Thus if $B \subseteq \mathbb{R}$ is $\check{\underset{\sim}{\Gamma}}$ we have some $\psi \in \Sigma_0$ and some $r_0 \in \mathbb{R}$ such that

$$\begin{aligned} r \in B &\Leftrightarrow \mathbf{L}_\xi(\mathbb{R}) \models \forall x \psi(r, x, r_0) \\ &\Leftrightarrow \forall s \mathbf{L}_{h(s)}(\mathbb{R}) \models \forall x \psi(r, x, r_0) \\ &\Leftrightarrow \mathbf{L}_\xi(\mathbb{R}) \models \forall s \psi'(r, s, r_0), \end{aligned}$$

where $\psi' \in \Sigma_1$. This shows that

$$\check{\underset{\sim}{\Gamma}} \subseteq \forall^{\mathbb{R}} \underset{\sim}{\Gamma} = \{\forall s P(r, s) \ :\ P \in \underset{\sim}{\Gamma}\}.$$

So we see that the set of reals A is in the class $\forall^{\mathbb{R}}\exists^{\mathbb{R}}\forall^{\mathbb{R}}\underset{\sim}{\Gamma}$. Now $\underset{\sim}{\Gamma}$ is closed under continuous substitutions, $\wedge, \vee, \exists^{\mathbb{R}}$ and has the scale property, so by the Second Periodicity Theorem (see [Mos80]) $\forall^{\mathbb{R}}\underset{\sim}{\Gamma}$ has the scale property, granting $\mathsf{Det}(\underset{\sim}{\Gamma})$, which is given to us by the fact that every set in $\underset{\sim}{\Gamma}$ is Suslin. By the Second Periodicity Theorem again, $\exists^{\mathbb{R}}\forall^{\mathbb{R}}\Gamma$ has the scale property, so every set in this class is determined and by one last application of Second Periodicity $\forall^{\mathbb{R}}\exists^{\mathbb{R}}\forall^{\mathbb{R}}\Gamma$ has the scale property, so A is Suslin, thus determined. So every set in $\mathbf{L}_{\xi+1}(\mathbb{R})$ is determined and we are done.

Subcase 3. ξ is successor.

Say $\xi = \eta + 1$. As in Subcase 2 it will be enough to show that $\check{\underset{\sim}{\Gamma}} \subseteq \forall^{\mathbb{R}}\underset{\sim}{\Gamma}$. In fact we shall prove here that every set in $\check{\underset{\sim}{\Gamma}}$ is a countable intersection of sets in $\underset{\sim}{\Gamma}$.

Indeed let $B \in \check{\underset{\sim}{\Gamma}}$. Then for some Σ_0 formula φ and some real parameter s_0, we have

$$\begin{aligned} r \in B &\Leftrightarrow \mathbf{L}_{\eta+1}(\mathbb{R}) \models \forall x \varphi(r, x, s_0) \\ &\Leftrightarrow \forall m \forall z[z \subseteq \mathbf{L}_\eta(\mathbb{R}) \wedge z \text{ is } \underset{\sim}{\Sigma}_m \text{ over } \mathbf{L}_\eta(\mathbb{R}) \Rightarrow \mathbf{L}_{\eta+1}(\mathbb{R}) \models \varphi(r, z, s_0)]. \end{aligned}$$

Let

$$B_m = \{r \in \mathbb{R} : \forall z[z \subseteq \mathbf{L}_\eta(\mathbb{R}) \wedge z \text{ is } \underset{\sim}{\mathbf{\Sigma}}_m \text{ over } \mathbf{L}_\eta(\mathbb{R}) \Rightarrow \mathbf{L}_{\eta+1}(\mathbb{R}) \models \varphi(r, z, s_0)]\}.$$

Then $B = \bigcap_m B_m$ and each B_m is Σ_1 over $\mathbf{L}_{\eta+1}(\mathbb{R})$ with real parameters only, since if $\psi(i, x, y)$ is universal Σ_n we have

$$r \in B_m \Leftrightarrow \mathbf{L}_{\eta+1}(\mathbb{R}) \models \exists M \exists \zeta (M = \mathbf{L}_\zeta(\mathbb{R}) \wedge \\ \wedge \forall x \in M \forall i \varphi(r, \{y \in M : \psi^M(i, x, y)\}, s_0)).$$

So our proof in Case I is complete. ⊣ CASE I

CASE II. There is an ordinal $\xi' < \xi$ such that $\mathbf{L}_{\xi'}(\mathbb{R}) \prec_1^{\mathbb{R}} \mathbf{L}_\xi(\mathbb{R})$.

Let $\xi^* < \xi$ be least such that $\mathbf{L}_{\xi^*}(\mathbb{R}) \prec_1^{\mathbb{R}} \mathbf{L}_\xi(\mathbb{R})$. Let $\underset{\sim}{\Gamma}$ be the pointclass of all sets of reals which are $\underset{\sim}{\Sigma}_1$ over $\mathbf{L}_\xi(\mathbb{R})$ with only real parameters. Clearly these are the same as the sets of reals which are $\underset{\sim}{\Sigma}_1$ over $\mathbf{L}_{\xi^*}(\mathbb{R})$ with only real parameters. We note first that $\underset{\sim}{\Gamma}$ is closed under $\forall^{\mathbb{R}}$:

Indeed let ψ be Σ_1. Then

$$\begin{aligned} \mathbf{L}_{\xi^*}(\mathbb{R}) \models \forall r \psi(r, s) &\Rightarrow \mathbf{L}_\xi(\mathbb{R}) \models \exists \eta (\mathbf{L}_\eta(\mathbb{R}) \models \forall r \psi(r, s)) \\ &\Rightarrow \mathbf{L}_{\xi^*}(\mathbb{R}) \models \exists \eta (\mathbf{L}_\eta(\mathbb{R}) \models \forall r \psi(r, s)) \\ &\Rightarrow \exists \eta < \xi^* (\mathbf{L}_\eta(\mathbb{R}) \models \forall r \psi(r, s)). \end{aligned}$$

Thus

$$\mathbf{L}_{\xi^*} \models \forall r \psi(r, s) \Leftrightarrow \mathbf{L}_{\xi^*}(\mathbb{R}) \models \exists \eta (\mathbf{L}_\eta(\mathbb{R}) \models \forall r \psi(r, s))).$$

So $\underset{\sim}{\Gamma}$ is closed under continuous substitutions, $\wedge, \vee, \exists^{\mathbb{R}}, \forall^{\mathbb{R}}$ and has the scale property by Steel's Theorem 4.2.

Let now $n \geq 1$ be the least integer such that there is a new $\underset{\sim}{\Sigma}_n$ over $\mathbf{L}_\xi(\mathbb{R})$ set of reals in $\mathbf{L}_{\xi+1}(\mathbb{R})$. We shall need below the second basic result of [Ste83A].

THEOREM 4.3. ([Ste83A]) Let ξ be an infinite ordinal and assume $\exists A \subseteq \mathbb{R}(A \in \mathbf{L}_{\xi+1}(\mathbb{R}) \setminus \mathbf{L}_\xi(\mathbb{R}))$. Let $n \geq 1$ be least such that there is $A \subseteq \mathbb{R}$ which is $\underset{\sim}{\Sigma}_n$ over $\mathbf{L}_\xi(\mathbb{R})$ but $A \notin \mathbf{L}_\xi(\mathbb{R})$. Assume finally that there is a countable sequence of Σ_n formulas $\sigma_0, \sigma_1, \ldots$, ordinals $\eta_0, \vartheta_0 < \xi$ and a real $r_0 \in \mathbb{R}$ such that for no ordinals $\xi' < \xi, \eta', \vartheta' < \xi'$ we have

$$\forall i [\mathbf{L}_\xi(\mathbb{R}) \models \sigma_i(r_0, \eta_0, \vartheta_0) \Leftrightarrow \mathbf{L}_{\xi'}(\mathbb{R}) \models \sigma_i(r_0, \eta', \vartheta')].$$

Then, if $\mathsf{Det}(\mathbf{L}_\xi(\mathbb{R}))$ holds, the pointclass $\underset{\sim}{\Sigma}$ of sets of reals which are $\underset{\sim}{\Sigma}_n$ over $\mathbf{L}_\xi(\mathbb{R})$ (with arbitrary parameters) has the scale property and the complement of every set in $\underset{\sim}{\Sigma}$ is a countable intersection of sets of reals in $\mathbf{L}_\xi(\mathbb{R})$, so in particular in $\forall^{\mathbb{R}} \underset{\sim}{\Sigma}$.

Since there is a new $\underset{\sim}{\Sigma}_n$ over $\mathbf{L}_\xi(\mathbb{R})$ subset of $\mathbb{R}$, there is a partial map $p \colon \mathbb{R} \twoheadrightarrow \mathbf{L}_\xi(\mathbb{R})$ with $\underset{\sim}{\Sigma}_n$ over $\mathbf{L}_\xi(\mathbb{R})$ graph.

Assume now, towards a contradiction, that there exists an undetermined game in $\mathbf{L}_{\xi+1}(\mathbb{R})$. Say, for notational simplicity, it is $\mathbf{\Sigma}_{n+2}$, so it has the form

$$A(\alpha,\beta) \Leftrightarrow \mathbf{L}_\xi(\mathbb{R}) \models \exists x \forall y \psi(\alpha,\beta,x,y,z_1),$$

where $\psi \in \Sigma_n$, $z_1 \in \mathbf{L}_\xi(\mathbb{R})$. Using the projection map p, we can write this as follows

$$A(\alpha,\beta) \Leftrightarrow \mathbf{L}_\xi(\mathbb{R}) \models \exists\gamma\forall\delta \bigvee_{i=1}^{k_0}[\psi_i(\alpha,\beta,\gamma,r_1,\eta_0,\vartheta_0) \wedge \chi_i(\alpha,\beta,\gamma,\delta,r_1,\eta_0,\vartheta_0)],$$

for some real $r_1 \in \mathbb{R}$, $\eta_0, \vartheta_0 < \xi$, where $\psi_i \in \Sigma_n$, $\chi_i \in \Pi_n$ (here η_0, ϑ_0 are such that z_1 is definable in $\mathbf{L}_{\vartheta_0}(\mathbb{R})$ from η_0 and reals).

Since A is not determined, we have as usual that

$$\forall\sigma,\tau\exists\alpha,\beta\{\exists\gamma\forall\delta \mathbf{L}_\xi(\mathbb{R}) \models \bigvee_{i=1}^{k_0}[\psi_i(\alpha,\tau*\alpha,\gamma,\delta,r_1,\eta_0,\vartheta_0) \wedge \chi_i(\alpha,\tau*\alpha,\gamma,\delta,r_1,\eta_0,\vartheta_0)]$$

$$\wedge \forall\gamma\exists\delta \mathbf{L}_\xi(\mathbb{R}) \models \bigwedge_{i=1}^{k_0}[\neg\psi_i(\sigma*\beta,\beta,\gamma,\delta,r_1,\eta_0,\vartheta_0) \vee \neg\chi_i(\sigma*\beta,\beta,\gamma,\delta,r_1,\eta_0,\vartheta_0)]\}.$$

Then, by the manipulations we used in § 2, we also have

$$\forall x\exists M\{M \text{ is } x\text{-good} \wedge \forall^* \boldsymbol{c}\forall^* \boldsymbol{d}\forall\sigma,\tau \in \mathcal{M}\exists\alpha,\beta \in M :$$

$$\exists\gamma \le \boldsymbol{c}\forall\delta \le \boldsymbol{d}(\mathbf{L}_\xi(\mathbb{R}) \models \bigvee_{i=1}^{k_0}[\psi_i(\alpha,\tau*\alpha,\gamma,\delta,r_1,\eta_0,\vartheta_0) \wedge \chi_i(\alpha,\tau*\alpha,\gamma,\delta,r_1,\eta_0,\vartheta_0)])$$

$$\wedge \forall\gamma \le \boldsymbol{c}\exists\delta \le \boldsymbol{d}(\mathbf{L}_\xi(\mathbb{R}) \models \bigwedge_{i=1}^{k_0}[\neg\psi_i(\sigma*\beta,\beta,\gamma,\delta,r_1,\eta_0,\vartheta_0) \vee \neg\chi_i(\sigma*\beta,\beta,\gamma,\delta,r_1,\eta_0,\vartheta_0)]).$$

We consider now two subcases:

SUBCASE 1.

$$\forall x\exists M\{M \text{ is } x\text{-good} \wedge \forall^*\boldsymbol{c}\forall^*\boldsymbol{d}\exists\xi' < \xi\exists\eta',\vartheta' < \xi'\forall\gamma,\tau \in M\exists\alpha,\beta \in M :$$

$$\left.\begin{aligned}&\exists\gamma \le \boldsymbol{c}\forall\delta \le \boldsymbol{d}(\mathbf{L}_{\xi'}(\mathbb{R}) \models \textstyle\bigvee_{i=1}^{k_0}[\psi_i(\alpha,\tau*\alpha,\gamma,\delta,r_1,\eta',\vartheta') \wedge \\ &\qquad \chi_i(\alpha,\tau*\alpha,\gamma,\delta,r_1,\eta',\vartheta')]) \\ &\wedge\forall\gamma \le \boldsymbol{c}\exists\delta \le \boldsymbol{d}(\mathbf{L}_{\xi'}(\mathbb{R}) \models \textstyle\bigwedge_{i=1}^{k_0}[\neg\psi_i(\sigma*\beta,\beta,\gamma,\delta,r_1,\eta',\vartheta') \vee \\ &\qquad \neg\chi_i(\sigma*\beta,\beta,\gamma,\delta,r_1,\eta',\vartheta')]).\end{aligned}\right\} \quad (**)$$

Let then $P(x, M)$ be relation following $\forall x\exists M$ above. It is clearly in the class $\underset{\sim}{\Gamma}$, since it is $\mathbf{\Sigma}_1$ over $\mathbf{L}_\xi(\mathbb{R})$ with real parameters only. Thus since

$\forall x \exists M P(x, M)$ and Γ has the uniformization property, we can find $F : \mathbb{R} \to \mathbb{R}$ in Δ such that if $F(x) = M_x$, then $P(x, M_x)$ holds. Put, as usual,

$$\mathcal{M}(x) = \{M_y : y \leq_{\mathrm{T}} x \wedge M_y \text{ is } x\text{-good}\},$$

and abbreviate by

$$Q_{\xi', \eta', \vartheta'}(\sigma, \tau, \alpha, \beta, \boldsymbol{c}, \boldsymbol{d})$$

the expression $(**)$ above, assuming that $\eta', \vartheta' < \xi'$. Thus we have

$$\forall x \forall^* \boldsymbol{c} \forall^* \boldsymbol{d} \exists \xi' \exists \eta', \vartheta' < \xi' \exists M \in \mathcal{M}(x) \forall \sigma, \tau \in M \exists \alpha, \beta \in M :$$

$$Q_{\xi', \eta', \vartheta'}(\sigma, \tau, \alpha, \beta, \boldsymbol{c}, \boldsymbol{d}).$$

But, since $\mathbf{L}_{\xi^*}(\mathbb{R}) \prec_1^{\mathbb{R}} \mathbf{L}_{\xi}(\mathbb{R})$, we also have

$$\forall x \forall^* \boldsymbol{c} \forall^* \boldsymbol{d} \exists \xi' < \xi^* \exists \eta', \vartheta' < \xi' \exists M \in \mathcal{M}(x) \forall \sigma, \tau \in M \exists \alpha, \beta \in M :$$
$$Q_{\xi', \eta', \vartheta'}(\sigma, \tau, \alpha, \beta, \boldsymbol{c}, \boldsymbol{d}).$$

Let now $f_x^0(\boldsymbol{c}, \boldsymbol{d})$ be the least $\xi' < \xi^*$ such that

$$\exists \eta', \vartheta' < \xi' \exists M \in \mathcal{M}(x) \forall \sigma, \tau \in M \exists \alpha, \beta \in M(Q_{\xi', \eta', \vartheta'}(\sigma, \tau, \alpha, \beta, \boldsymbol{c}, \boldsymbol{d})),$$

let $f_x^1(\boldsymbol{c}, \boldsymbol{d})$ be the least $\eta' < f_x^0(\boldsymbol{c}, \boldsymbol{d})$ such that

$$\exists \vartheta' < f_x^0(\boldsymbol{c}, \boldsymbol{d}) \exists M \in \mathcal{M}(x) \forall \sigma, \tau \in M \exists \alpha, \beta \in M(Q_{f_x^0(\boldsymbol{c}, \boldsymbol{d}), \eta', \vartheta'}(\sigma, \tau, \alpha, \beta, \boldsymbol{c}, \boldsymbol{d})),$$

and let $f_x^2(\boldsymbol{c}, \boldsymbol{d})$ be the least $\vartheta' < f_x^0(\boldsymbol{c}, \boldsymbol{d})$ such that

$$\exists M \in \mathcal{M}(x) \forall \sigma, \tau \in M \exists \alpha, \beta \in M(Q_{f_x^0(\boldsymbol{c}, \boldsymbol{d}), f_x^1(\boldsymbol{c}, \boldsymbol{d}), \vartheta'}(\sigma, \tau, \alpha, \beta, \boldsymbol{c}, \boldsymbol{d})).$$

Again $\forall^* \boldsymbol{c} \forall^* \boldsymbol{d}(f_x^i(\boldsymbol{c}, \boldsymbol{d})$ is defined$)$, $i = 0, 1, 2$.

Consider now the game

I	II
α, x_0	x_1, β

where $x = \langle x_0, x_1 \rangle$; player I wins iff

$$\forall^* \boldsymbol{c} \forall^* \boldsymbol{d} \exists \gamma \leq \boldsymbol{c} \forall \delta \leq \boldsymbol{d}(\mathbf{L}_{f_x^0(\boldsymbol{c}, \boldsymbol{d})}(\mathbb{R}) \models \bigvee_{i=1}^{k_0} [\psi_i(\alpha, \beta, \gamma, \delta, r_0, f_x^1(\boldsymbol{c}, \boldsymbol{d}), f_x^2(\boldsymbol{c}, \boldsymbol{d}))$$
$$\wedge \chi_i(\alpha, \beta, \gamma, \delta, r_0, f_x^1(\boldsymbol{c}, \boldsymbol{d}), f_x^2(\boldsymbol{c}, \boldsymbol{d}))]).$$

This is a game in Δ, so it is determined (since every set in Γ is Suslin). Then we obtain a contradiction exactly as in § 2.

Subcase 2. Subcase 1 fails.

Then find $x_0 \in \mathbb{R}$ such that for all M which are x_0-good we have

$$\neg \forall^* \boldsymbol{c} \forall^* \boldsymbol{d} \exists \xi' < \xi \exists \eta', \vartheta' < \xi' \forall \sigma, \tau \in M \exists \alpha, \beta \in M(Q_{\xi', \eta', \vartheta'}(\sigma, \tau, \alpha, \beta, \boldsymbol{c}, \boldsymbol{d})).$$

Then, by the formula just before SUBCASE 1, find M_0 such that M_0 is x_0-good and

$$\forall^* \boldsymbol{c} \forall^* \boldsymbol{d} \forall \sigma, \tau \in M_0 \exists \alpha, \beta \in M_0 (Q_{\xi,\eta,\vartheta_0}(\sigma, \tau, \alpha, \beta, \boldsymbol{c}, \boldsymbol{d})).$$

Then

$$\neg \forall^* \boldsymbol{c} \forall^* \boldsymbol{d} \exists \xi' < \xi \exists \eta', \vartheta' < \xi' \forall \sigma, \tau \in M_0 (Q_{\xi',\eta',\vartheta'}(\sigma, \tau, \alpha, \beta, \boldsymbol{c}, \boldsymbol{d})).$$

Since the expression following $\neg \forall^* \boldsymbol{c} \forall^* \boldsymbol{d}$ is $\underset{\sim}{\boldsymbol{\Sigma}}_1$ over $\mathbf{L}_\xi(\mathbb{R})$ with only real parameters, it is in $\underset{\sim}{\boldsymbol{\Gamma}}$, and we have $\underset{\sim}{\boldsymbol{\Gamma}}$-Turing Determinacy, since every set in $\underset{\sim}{\boldsymbol{\Gamma}}$ is Suslin, so we conclude that

$$\forall^* \boldsymbol{c} \forall^* \boldsymbol{d} \neg \exists \xi' < \xi \exists \eta', \vartheta' < \xi' \forall \sigma, \tau \in \mathcal{M}_0 \exists \alpha, \beta \in \mathcal{M}_0 : \\ Q_{\xi',\eta',\vartheta'}(\sigma, \tau, \alpha, \beta, \boldsymbol{c}, \boldsymbol{d}),$$

thus

$$\forall^* \boldsymbol{c} \forall^* \boldsymbol{d} \forall \xi' < \xi \forall \eta', \vartheta' < \xi' \exists \sigma, \tau \in M_0 \forall \alpha, \beta \in M_0 : \\ \neg Q_{\xi',\eta',\vartheta'}(\sigma, \tau, \alpha, \beta, \boldsymbol{c}, \boldsymbol{d}).$$

Fix then $\boldsymbol{c_0}, \boldsymbol{d_0} \in \mathcal{D}$ such that

$$\forall \sigma, \tau \in M_0 \exists \alpha, \beta \in M_0 (Q_{\xi,\eta_0,\vartheta_0}(\sigma, \tau, \alpha, \beta, \boldsymbol{c_0}, \boldsymbol{d_0}))$$

but

$$\forall \xi' < \xi \forall \eta', \vartheta' < \xi' \exists \sigma, \tau \in M_0 \forall \alpha, \beta \in M_0 \neg Q_{\xi',\eta',\vartheta'}(\sigma, \tau, \alpha, \beta, \boldsymbol{c_0}, \boldsymbol{d_0}).$$

Recalling the definition of $Q_{\xi,\eta,\vartheta}(\sigma, \tau, \alpha, \beta, \boldsymbol{c}, \boldsymbol{d})$, we see that if $\{\sigma_j\}, \{\tau_k\}$ enumerate the strategies in M_0, $\{\alpha_\ell\}, \{\beta_m\}$ the reals in M_0, $\{\gamma_n\}, \{\delta_p\}$ the reals $\leq \boldsymbol{c_0}, \boldsymbol{d_0}$, resp., and we define

$$\rho_i^1(\alpha, \beta, \sigma, \tau, \gamma, \delta, r, \eta, \vartheta) \Leftrightarrow \psi_i(\alpha, \tau * \alpha, \gamma, \delta, r, \eta, \vartheta)$$
$$\rho_1^2(\alpha, \beta, \sigma, \tau, \gamma, \delta, r, \eta, \vartheta) \Leftrightarrow \neg \chi_i(\alpha, \tau * \alpha, \gamma, \delta, r, \eta, \vartheta)$$
$$\rho_i^3(\alpha, \beta, \sigma, \tau, \gamma, \delta, r, \eta, \vartheta) \Leftrightarrow \psi_i(\sigma * \beta, \beta, \gamma, \delta, r, \eta, \vartheta)$$
$$\rho_i^4(\alpha, \beta, \sigma, \tau, \gamma, \delta, r, \eta, \vartheta) \Leftrightarrow \neg \chi_i(\sigma * \beta, \beta, \gamma, \delta, r, \eta, \vartheta),$$

then *there is no* $\xi' < \xi, \eta', \vartheta' < \xi'$ such that: *for all* $1 \leq t \leq 4, j, k, \ell, m, n, p \in \omega$ and $i \leq k_0$, we have

$$\mathbf{L}_\xi(\mathbb{R}) \models \rho_i^t(\alpha_\ell, \beta_m, \sigma_j, \tau_k, \gamma_n, \delta_p, r_1, \eta_0, \vartheta_0) \\ \Leftrightarrow \mathbf{L}_{\xi'}(\mathbb{R}) \models \rho_i^t(\alpha_\ell, \beta_m, \sigma_j, \tau_k, \gamma_n, \delta_p, r_1, \eta', \vartheta').$$

By suitably renumbering and coding, we conclude that there is a sequence $\sigma_0, \sigma_1, \ldots$ of Σ_n formulas, there exist ordinals $\eta_0, \vartheta_0 < \xi$ and a real $r_0 \in \mathbb{R}$ such that for no ordinals $\xi' < \xi, \eta', \vartheta' < \xi'$ we have

$$\forall i [\mathbf{L}_\xi(\mathbb{R}) \models \sigma_i(r_0, \eta_0, \vartheta_0) \Leftrightarrow \mathbf{L}_{\xi'}(\mathbb{R}) \models \sigma_i(r_0, \eta', \vartheta')].$$

Thus by Steel's Theorem 4.3 the pointclass $\underset{\sim}{\boldsymbol{\Sigma}}$ of sets of reals which are $\underset{\sim}{\boldsymbol{\Sigma}}_n$ over $\mathbf{L}_\xi(\mathbb{R})$ has the scale property, and moreover every set of reals which is

$\underset{\sim}{\Pi}_n$ over $\mathbf{L}_\xi(\mathbb{R})$ is in the class $\forall^{\mathbb{R}}\underset{\sim}{\Sigma}$. So we have that our undetermined set A is in $\exists^{\mathbb{R}}\forall^{\mathbb{R}}\underset{\sim}{\Sigma}$. Now $\underset{\sim}{\Sigma}$ is closed under continuous substitutions, $\wedge, \vee, \exists^{\mathbb{R}}$, so by the Second Periodicity Theorem, since $\underset{\sim}{\Sigma}$ has the scale property, so does $\forall^{\mathbb{R}}\underset{\sim}{\Sigma}$, granting $\mathsf{Det}(\underset{\sim}{\Sigma})$, which is given to us by the fact that every set in $\underset{\sim}{\Sigma}$ is Suslin. So, by Second Periodicity again, $\exists^{\mathbb{R}}\forall^{\mathbb{R}}\underset{\sim}{\Sigma}$ has the scale property, so A is Suslin, so determined, a contradiction. Thus the proof of SUBCASE 2 and Theorem 1.2 is complete.

§5. **Addendum: Update on an open problem.** The conjecture mentioned in the last paragraph of Section 1 has now been proved by Neeman and Woodin. It is a matter of combining two theorems. Fix $k > 0$. Neeman [Nee95] proved that if for each real x, $\mathbf{M}^{\#}_{2k}(x)$ exists and is countably iterable, where $\mathbf{M}^{\#}_{2k}$ is the sharp of the Mitchell-Steel inner model with $2k$ Woodin cardinals, then for each n, all games in $⅁^{2k}(\mathbf{M}_n)$ are determined. Woodin (unpublished) proved that if all $\underset{\sim}{\Pi}^1_{2k+1}$ games are determined, then for all reals x, $\mathbf{M}^{\#}_{2k}(x)$ exists and is countably iterable.

REFERENCES

LEO A. HARRINGTON

[Har78] *Analytic determinacy and* $0^{\#}$, ***The Journal of Symbolic Logic***, vol. 43, pp. 685–693.

ALEXANDER S. KECHRIS, EUGENE M. KLEINBERG, YIANNIS N. MOSCHOVAKIS, AND W. HUGH WOODIN

[KKMW81] *The axiom of determinacy, strong partition properties and nonsingular measures*, this volume, originally published in Kechris et al. [CABAL ii], pp. 75–100.

ALEXANDER S. KECHRIS, DONALD A. MARTIN, AND YIANNIS N. MOSCHOVAKIS

[CABAL ii] *Cabal seminar* 77–79, Lecture Notes in Mathematics, no. 839, Berlin, Springer.

[CABAL iii] *Cabal seminar* 79–81, Lecture Notes in Mathematics, no. 1019, Berlin, Springer.

ALEXANDER S. KECHRIS AND ROBERT M. SOLOVAY

[KS85] *On the relative consistency strength of determinacy hypotheses*, ***Transactions of the American Mathematical Society***, vol. 290, no. 1, pp. 179–211.

DONALD A. MARTIN

[MarA] *Borel and projective games*, to appear.

[Mar78] *Infinite games*, ***Proceedings of the international congress of mathematicatians, Helsinki 1978*** (Olli Lehto, editor), Finnish Academy of Sciences, pp. 269–273.

[Mar83A] *The largest countable this, that, and the other*, this volume, originally published in Kechris et al. [CABAL iii], pp. 97–106.

YIANNIS N. MOSCHOVAKIS

[Mos80] ***Descriptive set theory***, Studies in Logic and the Foundations of Mathematics, no. 100, North-Holland, Amsterdam.

ITAY NEEMAN

[Nee95] *Optimal proofs of determinacy*, ***The Bulletin of Symbolic Logic***, vol. 1, pp. 327–339.

John R. Steel
[Ste80] *More measures from* AD, mimeographed notes.
[Ste83A] *Scales in* $\mathbf{L}(\mathbb{R})$, this volume, originally published in Kechris et al. [Cabal iii], pp. 107–156.

DEPARTMENT OF MATHEMATICS
CALIFORNIA INSTITUTE OF TECHNOLOGY
PASADENA, CA 91125, USA
E-mail: kechris@caltech.edu

DEPARTMENT OF MATHEMATICS
UNIVERSITY OF CALIFORNIA
BERKELEY, CA 94720, USA
E-mail: woodin@math.berkeley.edu

GENERIC CODES FOR UNCOUNTABLE ORDINALS, PARTITION PROPERTIES, AND ELEMENTARY EMBEDDINGS

ALEXANDER S. KECHRIS AND W. HUGH WOODIN

Historical Remark. This paper was circulated in handwritten form in December 1980 and contained Sections 1–7 below. There are two additional Sections 8 and 9 here that contain further material and comments.

§1. A simple lemma. Let $\mathbb{R} = {}^{\omega}\omega$, the Baire space. Adopting (a variant of) a definition of [Bec79], let us call an ordinal λ **reliable** if there is a scale $\langle \varphi_i \rangle$ on a set $W \subseteq \mathbb{R}$ such that $\varphi_0 \colon W \twoheadrightarrow \lambda, \varphi_i \colon W \to \lambda$ and the two relations

$$x, y \in W \wedge \varphi_0(x) \leq \varphi_0(y)$$
$$x, y \in W \wedge \varphi_0(x) < \varphi_0(y)$$

admit scales. [Note that if AD+V=L($\mathbb{R}$) holds, then every reliable ordinal is $\leq \underset{\sim}{\delta}^2_1$, and in that case λ is reliable iff $\exists \langle \varphi_i \rangle$, W as above such that $\varphi_0 \colon W \twoheadrightarrow \lambda \wedge \varphi_i \colon W \to \lambda$ (so that last condition is not required).] If $\langle \varphi_i \rangle$, W are as above, we will say that $\langle \varphi_i \rangle$, W **witness the reliability** of λ or simply are **witnesses** for λ. We call a countable subset of λ, S, ξ**-honest**, where $\xi \in S$, if there is a code $w \in W$ of ξ, i.e., a $w \in W$ such that $\varphi_0(w) = \xi$, with $\varphi_i(w) \in S$, $\forall i$. We call such an S **honest** if it is ξ-honest for all $\xi \in S$ (this is of course all relative to $\langle \varphi_i \rangle$, W). Note that for each $\xi < \lambda$ there is some countable S_0 containing ξ such that $S \supseteq S_0 \Rightarrow S$ is ξ-honest (however such an S_0 cannot in general be canonically attached to each ξ). Note also that $\{S \in \wp_{\omega_1}(\lambda) : S \text{ is honest}\}$ is a strongly closed unbounded, in short **scub**, subset of $\wp_{\omega_1}(\lambda)$ [Bec79].

LEMMA 1.1 (AD). Let λ be reliable with witnesses $\langle \varphi_i \rangle$, W.

(i) There is a Lipschitz function $F_0 \colon {}^{\omega}\lambda \to \mathbb{R}$ (i.e., $F_0(f)\restriction n$ depends only on $f\restriction n$) such that $\operatorname{ran}(F_0) \subseteq W$ and for any $f \in {}^{\omega}\lambda$:

$$\{f(0), f(1), \dots\} \text{ is } f(0)\text{-honest} \Rightarrow \varphi_0(F_0(f)) = f(0).$$

Preparation of this paper is partially supported by NSF Grants DMS-0455285 (A.S.K.) and DMS-0355334 (W.H.W.).

The Cabal Seminar. Volume I: Games, Scales and Suslin Cardinals
Edited by A. S. Kechris, B. Löwe, J. R. Steel
Lecture Notes in Logic, 31

(ii) There is a Lipschitz function $F\colon {}^{\omega}\lambda \to \mathbb{R}$ such that $\operatorname{ran}(F) \subseteq \{w : \forall n((w)_n \in W)\}$ and for any $f \in {}^{\omega}\lambda$:

$$\{f(0), f(1), \dots\} \text{ is honest } \Rightarrow \forall n, \varphi_0((F(f))_n) = f(n).$$

Proof. We prove (i), the proof of (ii) being similar.

Let T be the tree on $\omega \times \lambda$ coming from the scale $\langle \varphi_i \rangle$ on W. For each $\xi < \lambda$, let

$$T_{(\xi)} = \{(s, u) \in T : u(0) = \xi\},$$

be the subtree of T whose ordinal sequences start with ξ.

Now consider the following game on λ:

$$\begin{array}{ll} \text{I} & \text{II} \\ f(0) & \\ & w(0), h(0) \\ f(1) & \\ & w(1), h(1) \\ & \vdots \\ f & w\ h \end{array}$$

where $f(i) < \lambda$; $w(i) \in \omega$; $h(i) < \lambda$; player II wins iff $(w, h) \in [T_{(f(0))}] \wedge \forall v[v \in \mathrm{p}[T_{(f(0))} \restriction \{f(0), f(1), \dots\}] \Rightarrow \varphi_0(v) \le \varphi_0(w)]$. (For any tree J on $\omega \times \lambda$, and any $S \subseteq \lambda$, $J \restriction S$ is the restriction of J to $\omega \times S$.)

It is enough to show that this game is determined. [If so, then since player I clearly can't have a winning strategy (when player II sees $f(0)$, he plays a w such that $\varphi_0(x) = f(0)$ and $h(i) = \varphi_i(w)$), player II must have a winning strategy τ. Let then $F(f) = w$ iff for some h, (f, w, h) is a run of the game in which player II follows τ.]

To prove now the determinacy of the game, it is enough, by [Mos81], to verify that the λ-pointset

$$R(f, w, h) \Leftrightarrow \text{ the run } f, w, h \text{ is a win for player I}$$

and its negation, admit scales. This is clear from the closure properties of λ-pointsets carrying scales (see [Mos81]), noting that

$$v \in \mathrm{p}[T_{f(0)} \restriction \{f(0), f(1), \dots\}] \Leftrightarrow \exists \alpha \in {}^{\omega}\omega((v, f \circ \alpha) \in [T_{f(0)}]). \qquad \dashv$$

Remark. (1) Note that the game

$$\begin{array}{ll} \text{I} & \text{II} \\ \xi_0 & \\ & \xi_1 \\ \xi_2 & \\ & \xi_3 \\ \vdots & \end{array} \qquad \text{player II wins iff } \{\xi_0, \xi_1, \dots\} \text{ is honest}$$

is not determined, since otherwise for each $\xi < \lambda$ we could explicitly find a ξ-honest S_ξ, violating a result of [Bec79] (or, by the preceding result, to each $\xi = \lambda$ we could explicitly assign w_ξ, a code of ξ).

(2) For $\lambda = \omega_1$, there are witnesses $\langle \varphi_i \rangle$, W such that the honest S are precisely the proper initial segments of ω_1. In the general case of arbitrary λ the honest S play in many respects a role similar to that of the proper initial segments in ω_1.

§2. **An ordinal determinacy result.** For each $P \subseteq {}^\omega\lambda$ (λ reliable as before) consider the associated game, also denoted by P:

$$\begin{array}{ll} \text{I} & \text{II} \\ f(0) & \\ & f(1) \\ f(2) & \\ & f(3) \\ & \vdots \end{array} \qquad f(i) < \lambda;\ \text{player II wins iff } P(f).$$

Let, also assuming AD, $\mathcal{U}$ be a supercompactness measure on $\wp_{\omega_1}(\lambda)$. Put

$$\forall^*_{\mathcal{U}} S(\dots S \dots) \Leftrightarrow \forall^* S(\dots S \dots)$$
$$\stackrel{\text{def}}{\Longleftrightarrow} \{S : \dots S \dots\} \in \mathcal{U}.$$

For each $S \in \wp_{\omega_1}(\lambda)$, let $P \restriction S$ be the game P restricted to S, i.e.

$$\begin{array}{ll} \text{I} & \text{II} \\ f(0) & \\ & f(1) \\ & \vdots \end{array} \qquad f(i) \in S;\ \text{player II wins iff } P(f).$$

We now have

player I (player II) has a winning strategy in P

$\Leftrightarrow \forall^* S$(player I (player II) has a winning strategy in $P \restriction S$)

$\Leftrightarrow \exists F : \wp_{\omega_1}(S) \to$ strategies for games on S such that

$\forall^* S(F(S)$ is a winning strategy for player I (player II) in $P \restriction S$).

Clearly, by AD, for any $S \in \wp_{\omega_1}(\lambda)$ player I or player II has a winning strategy in $P \restriction S$ and thus either $\forall^* S$ (player I has a winning strategy in $P \restriction S$) or $\forall^* S$ (player II has a winning strategy in $P \restriction S$); we describe this situation by saying that P is **weakly determined**. If however P itself is not determined, then by the above equivalences, even if say $\forall^* S$ (player I has a winning strategy in $P \restriction S$), we can't find a strategy for player I in $P \restriction S$ explicitly from S (for $\forall^* S$). We show that one can do the next best thing, i.e., have such a strategy depend explicitly (and in fact continuously) on any enumeration of S (for $\forall^* S$).

THEOREM 2.1 (AD+V=L($\mathbb{R}$)). Let $\lambda < \underset{\sim}{\delta}_1^2$ be reliable. Let $P \subseteq {}^\omega\lambda$. Then

(i) $\exists H : {}^{<\omega}\lambda \times {}^{<\omega}\lambda \to \lambda$ such that for some scub subset C of $\wp_{\omega_1}(\lambda)$, if $f \in {}^\omega\lambda$ and $S = \{f(0), f(1), \dots) \in C$, then H_f is a winning strategy for player I in $P\restriction S$, where

$$H_f(s) = H(f\restriction \mathrm{lh}(s), s).$$

or

(ii) Similarly for player II.

PROOF. Let $\langle\varphi_i\rangle$, W in $\underset{\sim}{\Delta}_1^2$ witness the reliability of λ. Let

$$P^*(w) \Leftrightarrow \forall n((w)_n \in W) \wedge \langle\varphi_0((w)_n)\rangle_{n=1}^\infty \in P,$$

be the coded version of P.

Now consider the following game, where F is as in 1.1:

I	II
ξ_0, η_0	
	η_1, ξ_1
ξ_2, η_2	
	η_3, ξ_3
$\vdots$	

where $\eta_i < \lambda_i, \xi_i < \lambda$; player II wins iff $\exists\alpha\{\forall n(\eta_n = \xi_{\alpha(n)}) \wedge \exists w[\forall n(\varphi_0((w)_n) = \varphi_0((F(\vec{\xi}))_{\alpha(n)}) \wedge P^*(w)]\}$.

Assume that this game is determined. To find out who wins this game, note that if $\mathcal{U}$ is a supercompactness measure on $\wp_{\omega_1}(\lambda)$, then $\forall^* S$ (player I has a winning strategy in $P\restriction S$) or $\forall^* S$ (player II has a winning strategy in $P\restriction S$). Say the second case occurs. We shall see then that player II has a winning strategy in the above game. Indeed if player I won by σ, fix S honest, closed under σ and such that player II has a winning strategy in $P\restriction S$. Let $\langle\xi_1, \xi_3, \dots\rangle$ enumerate S. Let $\xi_0, \eta_0, \xi_2, \eta_2, \dots$ be determined by σ and $\eta_1, \eta_3, \dots$ by following player II's strategy in $P\restriction S$ against $\eta_0, \eta_2, \dots$ Then $\{\eta_0, \eta_1, \dots\} \subseteq \{\xi_0, \xi_1, \dots\} = S$, so let α be such that $\eta_n = \xi_{\alpha(n)}$. As $\{\xi_0, \xi_1, \dots\}$ enumerates an honest S, $\varphi_0(F(\langle\xi_n\rangle_{n=1}^\infty)_k) = \xi_k$ for all k, so if $v = F(\langle\xi_n\rangle)$, we have $\varphi_0((v)_{\alpha(n)}) = \xi_{\alpha(n)} = \eta_n$. Let w be such that $(w)_n = (v)_{\alpha(n)}$. Then $\varphi_0((w)_n) = \eta_n$, and as $\langle\eta_n\rangle \in P$, we have $w \in P^*$, so player II won against σ, a contradiction.

So player II has a winning strategy, say τ. Let H be τ, forgetting about the $\xi_1, \xi_3, \dots$ If S is honest and closed under τ, then, if f enumerates S, we claim that H_f is a winning strategy for player II in $P\restriction S$. Indeed assume player I played $\eta_0, \eta_2, \dots$ in $P\restriction S$ and player II produced $\eta_1, \eta_3, \dots$ Let $\xi_1, \xi_3, \dots$ be such that if $\xi_{2i} = f(i)$, then $\xi_0, \eta_0, \eta_1, \xi_1, \xi_2, \eta_2, \dots$ is a run of the above game according to τ. Then $\{\eta_0, \eta_1, \dots\} \subseteq \{\xi_0, \xi_1 \dots\}$ and for some α, w, $\eta_n = \xi_{\alpha(n)}$ and $\varphi_0((w)_n) = \varphi_0(F(\vec{\xi})_{\alpha(n)})$, and $P^*(w)$. But $\{\xi_0, \xi_1, \dots\}$ enumerates also S, therefore

$$\varphi_0(F(\vec{\xi})_{\alpha(n)}) = \xi_{\alpha(n)} = \eta_n,$$

so $P(\langle\eta_n\rangle$, i.e., player II won $P\restriction S$ and we are done.

Now clearly the statement "The above game is determined" is a $\utilde{\mathbf{\Delta}}^2_1$ property of P^*. So if it fails for some P^*, then by Solovay's Basis Theorem it fails for some $P^* \in \utilde{\mathbf{\Delta}}^2_1$. But, by the Martin-Moschovakis-Steel Theorem, if $P^* \in \utilde{\mathbf{\Delta}}^2_1$ then both P^* and $\neg P^*$ carry scales, therefore the above game carries scales, so is determined, and we are done. ⊣

NOTE. It is not immediately clear that the above argument works for $\lambda = \utilde{\delta}^2_1$, although the result ought to be true in that case as well.

§3. Generic codes for ordinals. Given $P \subseteq {}^\omega\lambda$ consider the Banach-Mazur game:

I	II	
s_0		where $s_i \in {}^{<\omega}\lambda \setminus \{\varnothing\}$;
	s_1	player II wins iff $P(s_0{}^\frown s_1{}^\frown s_2{}^\frown \dots)$.
s_2		
	s_3	
$\vdots$		

Write $\forall^* f(P(f)) \overset{\text{def}}{\Longleftrightarrow} \forall s_0 \exists s_1 \forall s_2 \exists s_3 \dots P(s_0{}^\frown s_1{}^\frown \dots)$. Also, for any given $s \in {}^{<\omega}\lambda$,

$$\forall^* f \supseteq s P(f) \Leftrightarrow \forall s_0 \supseteq s \exists s_1 \forall s_2 \dots P(s_0{}^\frown s_1{}^\frown \dots),$$

where the formula on the right asserts as usual the existence of a winning strategy for player II.

FACT 3.1 (AD). *Let λ be reliable, with witnesses $\langle\varphi_i\rangle$, W and let F_0 be the function of 1.1. Then for all $\xi < \lambda$:*

$$\forall^* f \supseteq (\xi)[\varphi_0(F_0(f)) = \xi].$$

PROOF. In the Banach-Mazur game

I	II	
s_0		player II wins iff $\varphi_0(F_0(s_0{}^\frown s_1{}^\frown \dots)) = \xi$,
	s_1	
s_2		
	s_3	
$\vdots$		

if player I starts with $s_0 \supseteq (\xi)$, then player II just enumerates any ξ-honest S_0. So player I cannot have a winning strategy By [Mos81] this game is determined, so player II has a winning strategy ⊣

For convenience, put for $w \in W$

$$|w| = \varphi_0(w).$$

The preceding fact allows us to define category notions on each

$$W_\xi = \{w \in W : |w| = \xi\}.$$

Indeed, for each $A \subseteq W_\xi$, say that

$$A \text{ is } \mathbf{comeager} \Leftrightarrow \forall^* f \supseteq (\xi)(F_0(f) \in A), \text{ and}$$
$$A \text{ is } \mathbf{meager} \Leftrightarrow C_\xi \setminus A \text{ is comeager.}$$

Then note that:

(1) W_ξ is comeager.
(2) If A_n is comeager for each n, $\bigcap_n A_n$ is comeager.
(3) If A is comeager, $A \neq \varnothing$.
(4) For each $s \supseteq (\xi)$, $s \in {}^{<\omega}\lambda$ say that A **is comeager** on s iff

$$\forall^* f \supseteq s(F_0(f) \in A).$$

Then, if $\lambda < \underset{\sim}{\delta}^2_1$ and $\mathbf{V}{=}\mathbf{L}(\mathbb{R})$ holds, we have that every $A \subseteq W_\xi$ has the property of Baire, i.e., either A is comeager or $W_\xi \setminus A$ is comeager on some s, i.e., $\forall^* f \supseteq (\xi)(F_0(f) \in A)$ or $\exists s \supseteq (\xi)\forall^* f \supseteq (s)(F_0(f) \notin A)$. This is because the Banach-Mazur game "$F_0(f) \in A$" is determined by an argument similar to that in § 2 (recall that F_0 is continuous).

It is now convenient to introduce the following notations (where $A \subseteq \mathbb{R}$, $s \in {}^{<\omega}\lambda$, $s \supseteq (\xi)$):

$$s \models_\xi A \Leftrightarrow \forall^*_s |w| = \xi A(w) \overset{\text{def}}{\Longleftrightarrow} \forall^* f \supseteq s(F_0(f) \in A)$$
$$\Leftrightarrow A \text{ is comeager on } s,$$

$$0 \models_\xi A \Leftrightarrow (\xi) \models_\xi A \Leftrightarrow \forall^* |w| = \xi A(w) \overset{\text{def}}{\Longleftrightarrow} \forall^* f \supseteq (\xi) F_0(f) \in A)$$
$$\Leftrightarrow A \text{ is comeager.}$$

Then again for all $\lambda < \underset{\sim}{\delta}^2_1$, assuming AD+$\mathbf{V}{=}\mathbf{L}(\mathbb{R})$, we have the usual rules

1. $s \models_\xi A \Rightarrow A \neq \varnothing$,
2. $s \models_\xi \bigcap_n A_n \Leftrightarrow \forall n, s \models_\xi A_n$,
3. $s \models_\xi \neg A \Leftrightarrow \forall t \supseteq s(t \not\models_\xi A)$,
4. Moreover we also have the following unfolding formula, where $A \subseteq \mathbb{R} \times \mathbb{R}$,

$$s \models_\xi \forall\alpha(0)\exists\alpha(1)\forall\alpha(2)\exists\alpha(3)\ldots A(\cdot, \alpha)$$

if and only if

$$\forall s_0 \supseteq s \exists s_1 \forall s_2 \exists s_3 \ldots \forall\alpha(0)\exists\alpha(1)\forall\alpha(2)\ldots A(F_0(s_0{}^\frown s_1{}^\frown\ldots), \alpha)$$

if and only if

$$\forall s_0 \supseteq s \forall \alpha(0) \exists s_1 \exists \alpha(1) \forall s_2 \forall \alpha(2) \exists s_3 \exists \alpha(3) \ldots A(F_0(s_0{}^\frown s_1{}^\frown \ldots), \alpha). \quad (*)$$

PROOF. Let $\mathcal{U}$ be a supercompactness measure on $\wp_{\omega_1}(\lambda)$. As in §2, we can assume that the game in $(*)$ is determined, so we have

$$\forall s_0 \supseteq s \forall \alpha(0) \exists s_1 \exists \alpha(1) \ldots A(F_0(s_0{}^\frown s_1{}^\frown \ldots), \alpha)$$
$$\Leftrightarrow \forall^* S \forall s_0 \supseteq s, s_0 \in {}^{<\omega}S \forall \alpha(0) \exists s_1 \in {}^{<\omega}S \exists \alpha(1) \ldots A(F_0(s_0{}^\frown s_1{}^\frown \ldots), \alpha)$$

so (by the standard unfolding formula, see [Kec78B])

$$\Leftrightarrow \forall^* S \forall s_0 \supseteq s, s_0 \in {}^{<\omega}S \exists s_1 \in {}^{<\omega}S \ldots \forall \alpha(0) \exists \alpha(1) \ldots A(F_0(s_0{}^\frown s_1{}^\frown \ldots), \alpha)$$
$$\Leftrightarrow \forall s_0 \supseteq s \exists s_1 \forall s_2 \ldots \forall \alpha(0) \exists \alpha(1) \ldots A(F_0(s_0{}^\frown s_1 \ldots), \alpha). \quad \dashv$$

§4. Some new partition properties.

THEOREM 4.1 (AD). For each $n \geq 1$,

$$\underset{\sim}{\delta}^1_{2n+1} \to (\underset{\sim}{\delta}^1_{2n+1})^{\lambda}_{\eta}; \forall \lambda < \underset{\sim}{\delta}^1_{2n}, \forall \eta < \underset{\sim}{\delta}^1_{2n+1}.$$

COROLLARY 4.2 (AD). For each $n \geq 1$,

$$\underset{\sim}{\delta}^1_{2n+3} \geq \underset{\sim}{\delta}^1_{2n+1} + \aleph_{\omega^3+1},$$
$$\text{so } \underset{\sim}{\delta}^1_{2n+3} \geq \aleph_{\omega^3 \cdot n+1}.$$

PROOF. By imitating [MarC]. ⊣

COROLLARY 4.3 (AD). Let $\underset{\sim}{\delta}^1_{\omega} = \sup \underset{\sim}{\delta}^1_n$. For each projective $P \subseteq \mathbb{R}$ and each $\kappa < \underset{\sim}{\delta}^1_{\omega}$, there is a homogeneous tree T on $\omega \times \lambda$, for some $\lambda < \underset{\sim}{\delta}^1_{\omega}$, such that $P = \mathrm{p}[T]$ and the measures witnessing the homogeneity of P are κ-additive.

PROOF. As in [Kec81A]. ⊣

[Ste80] has first established the above result, without the extra additivity property.

PROOF OF THEOREM 4.1. We verify the Martin Criterion as formulated in [Kec78A, Lemma 11.1]:

From [HK81] it follows that if $\langle \varphi_i \rangle$ is a Π^1_{2n-1}-scale on $W \in \Pi^1_{2n-1}$, $\varphi_i : W \twoheadrightarrow \underset{\sim}{\delta}^1_{2n-1}$, and $\psi : V \twoheadrightarrow \underset{\sim}{\delta}^1_{2n+1}$ a Π^1_{2n+1}-norm, then there is a total $\underset{\sim}{\Delta}^1_{2n}$-function $G : \mathbb{R}^2 \to \mathbb{R}$ such that for every $f : \underset{\sim}{\delta}^1_{2n-1} \to \underset{\sim}{\delta}^1_{2n+1}$ there is $\varepsilon \in \mathbb{R}$ such that

(i) $w \in W \Rightarrow G(\varepsilon, w) \in V$,

(ii) $\forall w \in W[\psi(G(\varepsilon, w)) = f(\varphi_0(w))]$.

Fix now $t : \omega \cdot \lambda \twoheadrightarrow \underset{\sim}{\delta}^1_{2n-1}$ and define for $\xi < \omega \cdot \lambda$, letting $|w| \equiv \varphi_0(w)$,

$$C_{\xi} = \{\varepsilon : \forall^* |w| = t(\xi)(G(\varepsilon, w) \in V)\};$$

then, for $\varepsilon \in C_\xi$, let

$$f^\xi(\varepsilon) = \min\{\psi(G(\varepsilon, w) : |w| = t(\xi)\}.$$

We now verify 1), 2), 3), of [Kec78A, Lemma 11.1]. 1) is obvious by the above remarks.

Let for $\xi < \omega \cdot \lambda$, $\vartheta < \underset{\sim}{\delta}^1_{2n+1}$:

$$C_{\xi,\vartheta} = \{\varepsilon : \forall \xi' \leq \xi \forall^* |w| = t(\xi') \\ [G(\varepsilon, w) \in V \wedge \psi(G(\varepsilon, w)) \leq \vartheta]\}.$$

Clearly $C_{\xi,\vartheta} \subseteq \bigcap_{\xi' \leq \xi} C_\xi$ and 3) is trivially satisfied. So it is enough to verify 2). Thus let σ be continuous such that $\sigma[\bigcap_{\xi' \subseteq \xi} C_{\xi'}] \subseteq C_\xi$, so that in particular $A = \sigma[C_{\xi,\vartheta}] \subseteq C_\xi$.

We need now the following key lemma:

LEMMA 4.4. If $R(w, x)$ is $\underset{\sim}{\Sigma}^1_{2n+1}$, so is $P(x) \Leftrightarrow \forall^* |w| = \rho R(w, x)$, for each $\rho < \underset{\sim}{\delta}^1_{2n-1}$.

Granting this lemma, we have, using also [HK81], that $C_{\xi,\vartheta} \in \underset{\sim}{\Sigma}^1_{2n+1}$, thus also $A \in \underset{\sim}{\Sigma}^1_{2n+1}$. Put

$$G^\sigma_{(\xi,\vartheta)} = \sup\{f^\xi(\sigma(\varepsilon)) + 1 : \varepsilon \in C_{\xi,\vartheta}\} \\ = \sup\{f^\xi(\varepsilon') + 1 : \varepsilon' \in A\}.$$

We have to show $G^\sigma_{(\xi,\vartheta)} < \underset{\sim}{\delta}^1_{2n+1}$. Assume not, i.e., $G^\sigma_{(\xi,\vartheta)} = \underset{\sim}{\delta}^1_{2n+1}$, towards a contradiction. Then we have

$$y \in V \Leftrightarrow \exists \varepsilon' \in A \forall^* |w| = t(\xi)(y \leq_{\Sigma^1_{2n+1}} G(\varepsilon', w)), \tag{$**$}$$

where

$$\leq_{\Sigma^1_{2n+1}} \in \Sigma^1_{2n+1}$$

and

$$y \in V \Rightarrow [x \in V \wedge \psi(x) \leq \psi(y) \Leftrightarrow x \leq_{\Sigma^1_{2n+1}} y].$$

PROOF OF $(**)$. $\Rightarrow$: If $y \in V$, let $\varepsilon' \in A$ be such that $\psi(y) \leq f^\xi(\varepsilon')$. Then, since $\forall^* |w| = t(\xi)(G(\varepsilon', w) \in V)$ and $\forall^* |w| = t(\xi)(\psi(y) \leq \psi(G(\varepsilon', w)))$, we have $\forall^* |w| = t(\xi)(\psi(y) \leq \psi(G(\varepsilon', w)) \wedge G(\varepsilon', w) \in V)$, so $\forall^* |w| = t(\xi)(y \leq_{\Sigma^1_{2n+1}} G(\varepsilon', w))$.

$\Leftarrow$: Let ε' satisfy the right-hand side of $(**)$. Then $\forall^* |w| = t(\xi)(y \leq_{\Sigma^1_{2n+1}} G(\varepsilon', w))$. But also $\forall^* |w| = t(\xi)(G(\varepsilon', w) \in V)$, so $\forall^* |w| = t(\xi)(y \leq_{\Sigma^1_{2n+1}} G(\varepsilon', w) \wedge G(\varepsilon', w) \in V$, thus $y \in V$. $\dashv (**)$

Now $(**)$ and Lemma 4.4 imply that $V \in \underset{\sim}{\Sigma}^1_{2n+1}$, a contradiction.

Proof of Lemma 4.4. Let $R(w,x) \Leftrightarrow \exists\alpha Q(w,x,\alpha)$, where $Q \in \Pi^1_{2n}$. Then

$$P(x) \Leftrightarrow \forall^*|w| = p\exists\alpha Q(w,x,\alpha)$$
$$\Leftrightarrow \forall s_0 \supseteq (p)\exists\alpha(0)\exists s_1\forall s_2\exists\alpha(1)\exists s_3 \dots Q(F_0(s_0{}^\frown s_1{}^\frown\dots),x,\alpha),$$

where $s_i \in {}^{<\omega}(\underset{\sim}{\delta}^1_{2n-1})$, by the formula 4. in § 3. Now

$$Q(w,x,\alpha) \Leftrightarrow \forall f \in {}^{\omega}(\underset{\sim}{\delta}^1_{2n-1})\exists n(w\restriction n, x\restriction n, \alpha\restriction n, f\restriction n) \notin S,$$

for some tree S on $\omega \times \omega \times \omega \times \underset{\sim}{\delta}^1_{2n-1}$, so

$$P(x) \Leftrightarrow \forall s_0 \supseteq (p)\exists\alpha(0)\exists s_1\forall s_2\exists\alpha(1)\dots\forall f \in {}^{\omega}(\underset{\sim}{\delta}^1_{2n-1})\exists n$$
$$(F_0(s_0{}^\frown s_1{}^\frown\dots)\restriction n, x\restriction n, \alpha\restriction n, f\restriction n) \notin S.$$

Since $F_0\colon {}^{\omega}(\underset{\sim}{\delta}^1_{2n-1}) \to {}^{\omega}\omega$ is Lipschitz, it follows easily that $P(x)$ is $\underset{\sim}{\Sigma}^1_1$ in the structure $\langle \underset{\sim}{\delta}^1_{2n-1}, <, A\rangle$, for some $A \subseteq \underset{\sim}{\delta}^1_{2n-1}$. (Here A encodes S and F_0.) But then by [HK81], $P(x)$ is $\underset{\sim}{\Sigma}^1_{2n+1}$ and we are done. ⊣ (Lemma 4.4)

⊣ (Theorem 4.1)

Actually a direct application of [Kec78A, Lemma 11.1] only shows that $\underset{\sim}{\delta}^1_{2n+1} \to (\underset{\sim}{\delta}^1_{2n+1})^{\lambda}, \forall\lambda < \underset{\sim}{\delta}^1_{2n}$. To prove that $\underset{\sim}{\delta}^1_{2n+1} \to (\underset{\sim}{\delta}^1_{2n+1})^{\lambda}_{\eta}, \forall\eta < \underset{\sim}{\delta}^1_{2n+1}$, we need a minor modification as in [KKMW81, Lemma 1.2].

§5. The weak Baire theory for ${}^{\omega}\lambda$.

For each set X, ${}^{\omega}X$ is the space of infinite sequences from X with the product topology, X taken to be discrete. The category notions for ${}^{\omega}X$ are defined in the standard way. Abbreviate

$$\forall^* f \in {}^{\omega}XP(f) \Leftrightarrow \{f \in {}^{\omega}X : P(f)\} \text{ is comeager.}$$

Call now $P \subseteq {}^{\omega}X$ **weakly comeager** if there is $C \subseteq \wp_{\omega_1}(X)$ strongly closed unbounded such that

$$\forall S \in C\forall^* f \in {}^{\omega}SP(f).$$

(Clearly comeager sets are weakly comeager.) The notion of weakly meager is defined in the obvious way, as the complement of a weakly comeager set. We now show that the weak Baire notions on ${}^{\omega}\lambda$ work as the Baire notions on ${}^{\omega}\omega$.

Theorem 5.1 (AD + V=L(ℝ)). *Assume $\lambda < \underset{\sim}{\delta}^2_1$ is reliable.*

(i) *If $A \subseteq {}^{\omega}\lambda$, then either A is weakly meager or else there is $s \in {}^{<\omega}\lambda$ such that $N^{(\lambda)}_s \setminus A$ is weakly meager, where $N^{(\lambda)}_s = \{f \in {}^{\omega}\lambda : f \supseteq s\}$ (i.e, A is weakly comeager on $N^{(\lambda)}_s$).*

(ii) *Let $R \subseteq {}^{\omega}\lambda \times V$ be a relation such that $\forall f \in {}^{\omega}\lambda\exists xR(f,x)$. Then there is $G\colon {}^{\omega}\lambda \to V$ such that $R(f,G(f))$ for a weakly comeager set of f's.*

PROOF. (i) Consider the Banach-Mazur game on ${}^{\omega}\lambda$ defined as follows, where we fix $\langle\varphi_i\rangle$, W witnesses for λ, F_0 the function in 1.1, and we let $A^*(w) \Leftrightarrow \forall n((w)_n \in W) \wedge \langle\varphi_0(w)_n\rangle \in A$:

$$\begin{array}{ll} \text{I} & \text{II} \\ s_0 & \\ & s_1 \\ s_2 & \\ & s_3 \\ \vdots & \end{array} \qquad \begin{array}{l} s_i \in {}^{<\omega}\lambda;\ \text{player II wins iff} \\ F_0(s_0{}^\frown s_1{}^\frown \dots) \notin A^*. \end{array}$$

By the usual arguments we have that this game is determined. If player II has a winning strategy, clearly A is weakly meager, and if player I has a winning strategy A is comeager on some $N_s^{(\lambda)}$.

(ii) Since $\mathbf{V}=\mathbf{L}(\mathbb{R})$, we can assume that $R \subseteq {}^{\omega}\lambda \times \mathbb{R}$. Let $\langle\varphi_i\rangle$, W witness the reliability of λ, and let F_0 be as in Lemma 1.1. Consider now the following game:

$$\begin{array}{ll} \text{I} & \text{II} \\ s_0 & \\ & s_1, \alpha(0) \\ s_2 & \\ & s_3, \alpha(1) \\ & \vdots \end{array} \qquad \begin{array}{l} s_i \in {}^{<\omega}\lambda;\ s_0{}^\frown s_1{}^\frown \dots = f; \\ \text{player II wins iff} \\ R^*(F_0(f), \alpha), \end{array}$$

where $R^*(w, \alpha) \Leftrightarrow \forall n((w)_n \in W) \wedge (\langle\varphi_0(w_n)\rangle, \alpha) \in R$. By the usual argument, we can assume that this game is determined.

CLAIM 5.2. Player I can't have a winning strategy Because, if he had one, say σ, and $S \in \wp_{\omega_1}(\lambda)$ is honest and closed under σ, we would have

$$\exists s_0 \in {}^{<\omega}S \forall s_1 \in {}^{<\omega}S \dots \forall\alpha \neg R(s_0{}^\frown s_1{}^\frown \dots, \alpha),$$

so by the usual game formula,

$$\exists s_0 \in {}^{<\omega}S \forall s_1 \in {}^{<\omega}S \dots \forall\alpha \neg R(s_0{}^\frown s_1{}^\frown \dots, \alpha),$$

thus $\exists f \in {}^{\omega}\lambda \forall\alpha \neg R(f, \alpha)$, a contradiction.

So player II has a winning strategy. Call $f \in {}^{\omega}\lambda$ **consistent** with τ if there is a run of the game $s_0, s_1, \alpha(0), s_2, s_3, \alpha(1), \dots$ in which player II follows τ and $s_0{}^\frown s_1{}^\frown \dots = f$. If f is consistent with τ, define a canonical run $\bar{s}_0, \bar{s}_1, \bar{\alpha}(0), \dots$ with $\bar{s}_0{}^\frown \bar{s}_1{}^\frown \dots = f$ and player II following τ, as follows:

- $\bar{s}_0$ is the shortest initial segment s_0 of f such that there is a run $s_0, s_1, \alpha(0), \dots$ with $s_0{}^\frown s_1{}^\frown \dots = f$ and player II following τ,
- $\bar{s}_1, \bar{\alpha}(0) = \tau(\bar{s}_0)$ (thus $\bar{s}_0{}^\frown \bar{s}_1 \subseteq f$),

- $\bar{s}_2$ is the shortest sequence s_2 such that $\bar{s}_0{}^\frown\bar{s}_1{}^\frown s_2 \subseteq f$ and there is a run

$$\bar{s}_0, \bar{s}_1, \bar{\alpha}(0), s_2, s, \alpha(1), \ldots$$

with $\bar{s}_0{}^\frown\bar{s}_1{}^\frown s_2{}^\frown s^\frown \cdots = f$ and player II following τ,
- etc.

Now define $G\colon {}^\omega\lambda \to \mathbb{R}$ by

$$G(f) = \begin{cases} \text{the } \bar{\alpha} \text{ produced as above,} & \text{if } f \text{ is consistent with } \tau, \\ \lambda t \cdot 0, & \text{otherwise.} \end{cases}$$

Note now that if $S \in \wp_{\omega_1}(\lambda)$ is closed under τ and honest, then since $\forall^* f \in {}^\omega S$ (f is onto) and $\forall^* f \in {}^\omega S$ (f is consistent with τ), we have $\forall^* f \in {}^\omega S (f, G(f)) \in R$, and we are done. ⊣

§6. Generic elementary embeddings for the Lévy collapse. We shall now use the results in § 5 to translate Woodin's theory of generic elementary embeddings generated by the Lévy collapse (i.e., generic enumeration) of $\mathbb{R}$ in the framework of $\mathsf{AD}_{\mathbb{R}}$ (see [WooB]), to the context of the Lévy collapse of a reliable $\lambda < \underset{\sim}{\delta}^2_1$ in the framework of AD+$\mathbf{V}$=$\mathbf{L}(\mathbb{R})$.

From now on, in this section, assume $\lambda < \underset{\sim}{\delta}^2_1$ is reliable and AD+$\mathbf{V}$=$\mathbf{L}(\mathbb{R})$ *holds.*

Denote by $\mathcal{I}$ the σ-ideal of weakly meager subsets of ${}^\omega\lambda$. For $A \subseteq {}^\omega\lambda$, let $A/\mathcal{I}$ be the equivalence class of A modulo $\mathcal{I}$. Let also $\wp({}^\omega\lambda)/\mathcal{I}$ be the quotient Boolean algebra. We view this as a notion of forcing $\mathbb{C} = \langle \wp({}^\omega\lambda)/\mathcal{I}, \leq \rangle$, where

$$A/\mathcal{I} \leq B/\mathcal{I} \Leftrightarrow A \subseteq B \text{ (modulo } \mathcal{I}).$$

Note now that if

$$\mathbb{C}_\lambda = \langle {}^{<\omega}\lambda, \leq \rangle, \text{ where } t \leq s \Leftrightarrow t \supseteq s,$$

is the notion of forcing for collapsing λ to ω, then $\mathbb{C}_\lambda$ is canonically isomorphic to a dense subset of $\mathbb{C}$. Indeed, for each $s \in {}^{<\omega}\lambda$, let

$$N_s^{(\lambda)} = \{f \in {}^\omega\lambda \colon f \supseteq s\}.$$

Then if $\pi(s) = N_s^{(\lambda)}/\mathcal{I}$, π is an embedding of $\mathbb{C}_\lambda$ onto a dense subset $\pi[\mathbb{C}_\lambda]$ of $\mathbb{C}$. This follows immediately from Theorem 5.1 (i). Then if G is a generic (over $\mathbf{V}$) subset of $\mathbb{C}_\lambda$, G gives rise canonically to an ultrafilter $H \in \mathbf{V}[G]$ for the Boolean algebra $\mathbb{C}$ and thus to an ultrafilter $\mathcal{U} \in \mathbf{V}[G]$ on $(\wp({}^\omega\lambda))^{\mathbf{V}}$ such that

$$\{A \in \mathbf{V} : A \text{ is a weakly comeager subset of } ({}^\omega\lambda)^{\mathbf{V}}\} \subseteq \mathcal{U}.$$

Let now $J = ({}^\omega\lambda)^{\mathbf{V}}$ and consider the ultrapower (taken in $\mathbf{V}[G]$):

$$\mathbf{V}^J \cap \mathbf{V}/\mathcal{U};$$

we want first to check that the Łos Theorem goes through in this case. As usual this comes down to verifying (working in $\mathbf{V}$), that if $R(f,x)$ is a relation, $A \subseteq {}^{\omega}\lambda$ is in $\mathcal{U}$ and $\forall f \in A \exists x R(f,x)$, then there is $G\colon {}^{\omega}\lambda \to \mathbf{V}$ such that for some $B \in \mathcal{U}$, $(f, G(f)) \in R$, for all $f \in B$. This however is obvious from Theorem 5.1 (ii) and the fact that $\mathcal{U}$ contains all weakly comeager sets.

To show the wellfoundedness of this ultrapower and study some of its further properties, we digress briefly to discuss the supercompactness measure on $\wp_{\omega_1}(\lambda)$, working until further notice in $\mathbf{V}$. By [HK81], there is a supercompactness measure on $\wp_{\omega_1}(\lambda)$, and by [Bec79] it is unique, and coincides with the strongly closed unbounded filter on $\wp_{\omega_1}(\lambda)$. Denote it by $\mathcal{U}^{\lambda}$.

Clearly each $F\colon \wp_{\omega_1}(\lambda) \to \mathbf{V}$ gives rise to a map $F^*\colon {}^{\omega}\lambda \to \mathbf{V}$ defined by

$$F^*(f) = F(\operatorname{ran}(f)).$$

LEMMA 6.1. The map (in $\mathbf{V}[G]$)

$$[F]_{\mathcal{U}^{\lambda}} \mapsto [F^*]_{\mathcal{U}},$$

is an isomorphism between

$$\mathrm{Ord}^{\wp_{\omega_1}(\lambda)}/\mathcal{U}_{\lambda} \text{ and } \mathrm{Ord}^J \cap \mathbf{V}/\mathcal{U}.$$

In particular, $\mathbf{V}^J \cap \mathbf{V}/\mathcal{U}$ is wellfounded.

PROOF. It is routine to verify that this is indeed an embedding. For example, if $[F]_{\mathcal{U}^{\lambda}} \in [G]_{\mathcal{U}^{\lambda}}$, then for some $C \subseteq \wp_{\omega_1}(\lambda)$ scub we have $S \in C \Rightarrow F(S) \in G(S)$, so for $S \in C$,

$$\begin{aligned} &\forall^* f \in {}^{\omega}S(F(\operatorname{ran}(f)) \in G(\operatorname{ran}(f)), \\ &\text{so } \forall S \in C \forall^* f \in {}^{\omega}S(F^*(f) \in G^*(f)), \\ &\text{so } \{f : F^*(f) \in G^*(f)\} \text{ is weakly comeager,} \\ &\text{thus } [F^*]_{\mathcal{U}} \in [G^*]_{\mathcal{U}}. \end{aligned}$$

We now show that this map is onto $\mathrm{Ord}^J \cap \mathbf{V}/\mathcal{U}$. So let $H\colon {}^{\omega}\lambda \to \mathbf{V}$ be in $\mathbf{V}$. We shall show that: (∗∗∗) For any $A \subseteq {}^{\omega}\lambda$ not weakly meager, there is $B \subseteq {}^{\omega}\lambda$ not weakly meager, $B \subseteq A$ such that for $f, g \in B$:

$$\operatorname{ran}(f) = \operatorname{ran}(g) \Rightarrow H(f) = H(g).$$

By genericity this implies that there is $C \in \mathcal{U}$ such that

$$f, g \in C \wedge \operatorname{ran}(f) = \operatorname{ran}(g) \Rightarrow H(f) = H(g).$$

Define then $F\colon \wp_{\omega_1}(\lambda) \to \mathbf{V}$ in $\mathbf{V}$ by

$$F(S) = \begin{cases} H(f), \text{ where } f \in C \text{ and } S = \operatorname{ran}(f), & \text{if such exists} \\ 0, & \text{otherwise.} \end{cases}$$

Clearly $[F^*]_{\mathcal{U}} = [H]_{\mathcal{U}}$ and we are done.

PROOF OF $(***)$. Fix $A \subseteq {}^{\omega}\lambda$ not weakly meager. Then A is weakly comeager on some $N_{s_0}^{(\lambda)}$; say $\mathcal{A}$ is scub such that $S \in \mathcal{A} \wedge s_0 \in {}^{<\omega}S \Rightarrow \forall^* f \in N_{s_0}^{(\lambda)} \cap {}^{\omega}S(f \in A)$. For each such S, we have by Theorem 5.1 that there is some $s \in {}^{<\omega}S$ so that $s \supseteq s_0$ and $H(f)$ is constant on a comeager in $N_s^{(\lambda)} \cap {}^{\omega}S$ set of f's. By normality this s is independent of S for $S \in \mathcal{D}$, where $\mathcal{D}$ is scub in $\wp_{\omega_1}(\lambda)$, call it s_1 (clearly $s_1 \supseteq s_0$). Thus for $S \in \mathcal{D}$, $s_1 \in {}^{<\omega}S$ and for comeager many $f \in N_{s_0}^{(\lambda)} \cap {}^{\omega}S, H(f) = \text{fixed} = \alpha_S$. Let

$$B = \{f \in {}^{\omega}\lambda \colon f \in A \wedge \operatorname{ran}(f) \in \mathcal{D} \wedge f \supseteq s_1 \wedge H(f) = \alpha_{\operatorname{ran}(f)}\}.$$

Clearly for $S \in \mathcal{D} \wedge s_1 \in {}^{<\omega}S, \forall^* f \in N_{s_1}^{(\lambda)} \cap {}^{\omega}S(f \in B)$, thus B is not weakly meager. Moreover if $f, g \in B$ and $\operatorname{range}(f) = \operatorname{range}(g) = S \in \mathcal{D}$, then $H(f) = H(g) = \alpha_s$ and we are done. ⊣ $(***)$

⊣ (Lemma 6.1)

Let us now denote by $M \subseteq \mathbf{V}[G]$ the transitive collapse of $\mathbf{V}^J \cap \mathbf{V}/\mathcal{U}$, and by

$$j \equiv j_G \colon \mathbf{V} \to M \cong \mathbf{V}^J \cap \mathbf{V}/\mathcal{U}$$

the associated elementary embedding. We identify also each $[H]_{\mathcal{U}}$ with its image under the isomorphism of $\mathbf{V}^J \cap \mathbf{V}/\mathcal{U}$ with M.

Working now in $\mathbf{V}$, let

$$\text{HWO} = \{a \colon \operatorname{tcl}(a) \text{ is wellorderable}\}$$

and consider the ultrapower

$$\operatorname{Ult}(\text{HWO}, \mathcal{U}) = \{F \colon \wp_{\omega_1}(\lambda) \to \mathbf{V} : \operatorname{ran}(F) \in \text{HWO}\}/\mathcal{U}^{\lambda}$$

It is easy to verify that the Łos Theorem goes through for Δ_0 formulas, and thus this ultrapower is wellfounded and extensional, so it can be identified with its transitive collapse N. Let then j_λ be the associated Δ_0-elementary embedding

$$j_\lambda \colon \text{HWO} \to N \cong \operatorname{Ult}(\text{HWO}, \mathcal{U}^{\lambda}).$$

The argument used in Lemma 6.1 clearly also establishes that the map

$$\rho([F]_{\mathcal{U}_\lambda}) = [F^*]_{\mathcal{U}}$$

is an isomorphism between

$$\operatorname{Ult}(\text{HWO}, \mathcal{U}^{\lambda})$$

and

$$\{H \colon {}^{\omega}\lambda \to \mathbf{V} : \operatorname{ran}(H) \in \text{HWO}\}/\mathcal{U}$$

and clearly we have $j_G \restriction \text{HWO} = j_\lambda \circ \rho$. In particular, for $\vartheta \in \text{Ord}$,

$$j_\lambda(\vartheta) = j_G(\vartheta)$$

and also for $X \subseteq \mathrm{Ord}$

$$j_\lambda(X) = j_G(X).$$

As $j_G : \mathbf{V} \to M$ is elementary and $\mathbf{V} \models \mathbf{V}{=}\mathbf{L}(\mathbb{R})$, clearly $M \models \mathbf{V}{=}\mathbf{L}(\mathbb{R})$. We actually show now that

$$M = \mathbf{L}(\mathbb{R})^{\mathbf{V}[G]}.$$

For that we have to show that, if $x \in \mathbb{R}^{\mathbf{V}[G]}$, then $x \in M$. Let τ be a term in the forcing language denoting x. Let $p_0 \in G$ be such that $p_0 \models \tau \in \mathbb{R}$. Note then that $N_{p_0}^{(\lambda)} \in \mathcal{U}$. We shall find some $F : {}^{\omega}\lambda \to \mathbf{V}$ in $\mathbf{V}$ such that $[F]_{\mathcal{U}} = x$. We have $p_0 \models \forall n \exists m(\tau(n) = m)$, so $\forall n \forall p \leq p_0 \exists q \leq p \exists m(q \models \tau(n) = m)$ (conditions are in $\mathbb{C}_\lambda = {}^{<\omega}\lambda$). Thus $\forall^* f \supseteq p_0$, there is a unique $\alpha \in {}^{\omega}\omega$ such that for all n there is k with $f \restriction k \models \tau(n) = \alpha(n)$. Denote this α by $\tau(f)$. This is only defined for $f \in C \subseteq {}^{\omega}\lambda$, where $\forall^* f \supseteq p_0(f \in C)$. Put

$$F(f) = \begin{cases} \tau(f), & \text{if } f \in C, \\ \lambda t \cdot 0, & \text{otherwise.} \end{cases}$$

Clearly $[F]_{\mathcal{U}} \in \mathbb{R}^{\mathbf{V}[G]}$. Fix n. We show that $[F]_{\mathcal{U}}(n) = x(n)$. Let $[F]_{\mathcal{U}}(n) = m$. Then for some $A \in \mathcal{U}, f \in A \Rightarrow F(f)(n) = m$. If now $x(n) \neq m$, there is $p_1 \leq p_0, p_1 \in G$ such that $p_1 \models \tau(n) \neq m$. Thus $\forall^* f \supseteq p_1(F(f)(n) \neq m)$. But $N_{p_1}^{(\lambda)} \in \mathcal{U}$, so $\{f \supseteq p_1 : F(f)(n) \neq m\} \cap A \neq \varnothing$, which is a contradiction.

We can summarize what we have proved until now as follows:

Theorem 6.2 (AD+$\mathbf{V}{=}\mathbf{L}(\mathbb{R})$). *Let $\lambda < \underset{\sim}{\delta}{}_1^2$ be reliable. Let $\mathbb{C}_\lambda = {}^{<\omega}\lambda$ be the notion of forcing for collapsing λ to ω. If G is a generic subset of $\mathbb{C}_\lambda$ (over $\mathbf{V}$), then in $\mathbf{V}[G]$ we can define an elementary embedding*

$$j_G : \mathbf{V} \to M = L[\mathbb{R}]^{\mathbf{V}[G]}$$

Moreover, if $\mathrm{HWO} = \{a \in \mathbf{V} : \mathrm{tcl}(a) \text{ is wellorderable}\}$, $\mathcal{U}^\lambda$ is the scub measure on $p_{\omega_1}(\lambda)$ and

$$j_\lambda : \mathrm{HWO} \to N \cong \{F : p_{\omega_1}(\lambda) \to \mathbf{V} : \mathrm{ran}(F) \in \mathrm{HWO}\}/\mathcal{U}^\lambda$$

is the Δ_0-embedding generated by $\mathcal{U}^\lambda$, then $j_G = j_\lambda \circ p$, where p is the inclusion map from N to M, in particular $j_\lambda = j_G \restriction \mathrm{HWO}$.

We conclude this section by pointing out an equivalence between forcing and Banach-Mazur games in the preceding context.

Let τ be a term (always for forcing with $\mathbb{C}_\lambda$) and $p \in \mathbb{C}_\lambda$ be a condition such that $p \models \tau \in \mathbb{R}$. Let φ be a formula, $x \in \mathbf{V}$. We claim that the following equivalence holds: For any $q \leq p$,

$$q \models \varphi^{\mathbf{L}(\mathbb{R})}(\tau, j_G(x)) \Leftrightarrow \forall^* f \supseteq q(\varphi(\tau(f), x)).$$

Indeed, if $\forall^* f \supseteq q(\varphi(\tau(f)), x)$ and G is generic containing q, therefore $N_q^{(\lambda)} \in \mathcal{U}$, we have that $\{f \in {}^{\omega}\lambda : \varphi(\tau(f), x)\} \in \mathcal{U}$, as this set is the intersection of a comeager (therefore weakly comeager) set in ${}^{\omega}\lambda$ and $N_q^{(\lambda)}$. So, since $[f \mapsto \tau(f)]_{\mathcal{U}}$ = the real represented by τ in $\mathbf{V}[G]$ (= say, a) we have, by Łos, $M = \mathbf{L}(\mathbb{R})^{\mathbf{V}[G]} \models \varphi(a, j_G(x))$, i.e., $\varphi^{\mathbf{L}(\mathbb{R})}(a, j_G(x))$. Assume now $\neg\forall^* f \supseteq q(\varphi(\tau(f), x))$. As in §2, we can assume that $\{f \supseteq q : \varphi(\tau(f), x))$ is a determined Banach-Mazur game, therefore $\exists r \leq q$ such that $\forall^* f \supseteq r \neg(\varphi(\tau(f), x))$, so as before $r \models \neg\varphi^{\mathbf{L}(\mathbb{R})}(\tau, j_G(x))$, thus $q \not\models \varphi^{\mathbf{L}(\mathbb{R})}(\tau, j_G(x))$ and we are done.

§7. Some applications of the generic elementary embeddings.

THEOREM 7.1 (AD). For each $n \geq 1$ and $\alpha < \omega_2$:

$$\underset{\sim}{\delta}^1_{2n} \leq \omega_{\alpha+1} \Rightarrow \mathrm{cf}(\omega_{\alpha+1}) \geq \underset{\sim}{\delta}^1_{2n}.$$

In particular, $\mathrm{cf}(\omega_{\alpha+1}) \geq \omega_{\omega+2}$ $(= \underset{\sim}{\delta}^1_4)$, for all $\alpha \geq \omega + 1$.

PROOF. Assume $\underset{\sim}{\delta}^1_{2n} \leq \omega_{\alpha+1}$, with $\alpha < \omega_1$. Let $\lambda = \underset{\sim}{\delta}^1_{2n-1}$; λ is reliable. Let j_λ, j_G be the two embeddings as in § 6. Let λ' be the predecessor of λ (λ' is semireliable), let $j_{\lambda'}$ be the embedding generated by the supercompactness measure on $\wp_{\omega_1}(\lambda')$. By [Bec79], $j_{\lambda'}(\lambda) < \lambda^+ = \underset{\sim}{\delta}^1_{2n}$. But if μ_ω is the ω-closed unbounded measure on λ, then it is easy to check that $j_\lambda(\vartheta) \leq j_{\mu_\omega}(j_{\lambda'}(\vartheta))$, so $j_\lambda(\lambda) < \lambda^{++}$. Similarly, $j_{\lambda'}(\lambda^+) = \lambda^+$, thus $j_\lambda(\lambda') \leq \lambda^{++}$, so, since $j_\lambda(\vartheta) \geq j_{\mu_\omega}(\vartheta)$ and $j_{\mu_\omega}(\lambda^+) = \lambda^{++}$, we have $j_\lambda(\lambda^+) = \lambda^{++}$, and more generally $j_\lambda(\lambda^{+(n)}) = \lambda^{+(n+1)}$, for $n \geq 1$. Since however $j_\lambda = j_G$, and $\mathbf{V} \to \mathbf{V}[G]$ does not collapse cardinals $\geq \lambda^+$, we also have $j_\lambda(\omega_\beta) = \omega_\beta$, for all $\beta < \omega_1$, with $\omega_\beta \geq \lambda^{+(\omega)}$. Consider now $\omega_{\alpha+1}$ as above. If $\omega_{\alpha+1} = \lambda^{+(k)}$ for some $k \geq 1$, we are done. Also, by the above, $j_\lambda(\omega_{\alpha+1}) = \omega_{\alpha+1}$, in particular $j_\lambda(\omega_{\alpha+1}) = \sup\{j_\lambda(\xi) : \xi < \omega_{\alpha+1}\}$ is a continuity point of j_λ. Now assume, towards a contradiction, that $\mathrm{cf}(\omega_{\alpha+1}) = \kappa \leq \lambda$. Let $\pi : \wp_{\omega_1}(\lambda) \to \kappa$ be given by $\pi(S) = \sup(S \cap \kappa)$. Let $\pi_*(\mathcal{U}^\lambda) = \mathcal{V}$ be the induced measure on κ. Clearly $\mathcal{V}$ is uniform and $j_{\mathcal{V}}(\vartheta) \leq j_\lambda(\vartheta)$, so $\omega_{\alpha+1}$ is a continuity point of $j_{\mathcal{V}}$ as well, a contradiction, since if $f : \kappa \to \omega_{\alpha+1}$ is cofinal and nondecreasing, then $\sup\{j_{\mathcal{V}}(\xi) : \xi < \omega_{\alpha+1}\} < [f]_v < j_{\mathcal{V}}(\omega_{\alpha+1})$. ⊣

In [Bec79], he calculates that if $\lambda < \underset{\sim}{\delta}^2_1$ is reliable (and AD+V=L($\mathbb{R}$) holds), then $j_\lambda(w_1) = \lambda^+$.

THEOREM 7.2 (AD+V=L($\mathbb{R}$)). Let $\lambda \leq \underset{\sim}{\delta}^2_1$ be reliable and let $u_\alpha^{(\lambda)} = \alpha$th uniform indiscernible for subsets of λ ($u_1^{(\lambda)} = \lambda^+$). Then

$$j_\lambda(\omega_n) = u_n^{(\lambda)}.$$

PROOF. Since $j_\lambda = j_G$, $j_\lambda(\omega_n) = j_G(\omega_n) = j_G(u_n)$ (where $u_n \equiv u_n^{(\omega)}$) $= (u_n)^{\mathbf{V}[G]}$. But subsets of λ in $\mathbf{V}$ are reals in $\mathbf{V}[G]$, while reals in $\mathbf{V}[G]$ are represented by terms which are essentially subsets of λ in $\mathbf{V}$, so are generic over subsets of λ in $\mathbf{V}$. Thus $(u_n)^{\mathbf{V}(G)} = u_n^{(\lambda)}$. ⊣

It is clear that (granting AD again), $u_n^{(\omega_m)} = u_{m+n+1}$. For ω_ω we have however $u_n^{(\omega_\omega)} < \omega_{\omega+2}$ and $\mathrm{cf}(u_n^{(\omega_\omega)}) = \omega_{\omega+1}$ $(= \underset{\sim}{\delta}_3^1)$.

PROOF. From [Bec79], $j_{\omega_\omega}(\omega_n) < \omega_{\omega+2}$, so $u_n^{(\omega_\omega)} = j_{\omega_\omega}(\omega_n) < \omega_{\omega+2}$. Now assume $\mathrm{cf}(u_2^{(\omega_\omega)}) < \omega_{\omega+1}$, towards a contradiction. Then $u_2^{(\omega_\omega)}$ is a continuity point for j_{μ_ω}, where μ_ω is the ω-cub measure on $\omega_{\omega+1}$. Thus

$$\begin{aligned} j_{\mu_\omega}(u_2^{(\omega_\omega)}) &= \sup\{j_{\mu_\omega}(\alpha) : \alpha < u_2^{(\omega_\omega)}\} \\ &\leq u_2^{(\omega_{\omega+1})} \\ &= j_{\omega_{\omega+1}}(\omega_2) \\ &\leq j_{\mu_\omega}(j_{\omega_\omega}(\omega_2)) \quad = \quad j_{\mu_\omega}(u_2^{(\omega_\omega)}), \end{aligned}$$

therefore $u_2^{(\omega_{\omega+1})}$ has also cofinality $< \omega_{\omega+1}$, so it can't be a continuity point for j_{ω_ω}, therefore it can't be a fixed point for $j_{\omega_\omega} = j_G$, where G is generic for $\mathbb{C}_{\omega_\omega}$. Let $M = \mathbf{L}(\mathbb{R})^{\mathbf{V}[G]}$. Then $(u_2^{(\underset{\sim}{\delta}_3^1)})^M \leq (u_2^{(\underset{\sim}{\delta}_3^1)})^{\mathbf{V}}$, since every $A \subseteq (\underset{\sim}{\delta}_3^1)^M$, $A \in M$, is a subset of $(\underset{\sim}{\delta}_3^1)^M = j_G((\underset{\sim}{\delta}_3^1)^{\mathbf{V}}) < (\underset{\sim}{\delta}_4^1)^{\mathbf{V}}$, so is generic over a subset of $(\underset{\sim}{\delta}_3^1)^{\mathbf{V}}$ in $\mathbf{V}$. But also

$$(u_2^{(\underset{\sim}{\delta}_3^1)})^{\mathbf{V}} \leq j_G((u_2^{(\underset{\sim}{\delta}_3^1)})^{\mathbf{V}}) \leq (u_2^{(\underset{\sim}{\delta}_3^1)})^M,$$

thus $u_2^{(\underset{\sim}{\delta}_3^1)}$ is a fixed point of j_G, a contradiction. ⊣

Similarly $u_n^{(\omega_{\omega+1})} < \omega_{\omega+3}$, $\forall n$, although the obvious conjecture that

$$\mathrm{cf}(u_n^{(\omega_{\omega+1})}) = \omega_{\omega+2} = \underset{\sim}{\delta}_4^1$$

has not been proven yet. Similar results hold about the higher $\underset{\sim}{\delta}_{2n+1}^1$ and their predecessors.

A final observation: Granting AD, it is known that the only reliable cardinals above $\omega_{\omega+1}$ and below ω_{ω^3} can be $\omega_{\omega\cdot 2}$ or ω_{ω^2}, although it ought to be true that neither of them actually is. We can eliminate $\omega_{\omega\cdot 2}$ as follows: If $\lambda = \omega_{\omega\cdot 2}$ was reliable, let j_λ be the associated supercompactness measure and j_G be the associated generic embedding. As no cardinal above λ^+ is collapsed in the extension $\mathbf{V} \to \mathbf{V}[G]$, it is easy to see that for $\alpha < \omega_1$, $j_G(\omega_\alpha) \leq \omega_{\omega\cdot 2+\alpha}$, so if $\alpha \geq \omega^2$, $j_G(\omega_\alpha) = \omega_\alpha$, thus as before $\alpha \geq \omega^2 \Rightarrow \mathrm{cf}(\omega_{\alpha+1}) \geq \lambda = \omega_{\omega\cdot 2}$. But by [MarC], there are unboundedly many cardinals in ω_{ω^3} of cofinality $\omega_{\omega+2}$, a contradiction.

§8. Addendum; non-existence of strong codes. (The nonexistence of strong codes was proved by Woodin and a simplified argument was given afterwards by Kechris but the proofs were not included in the original note. We would like to thank John Steel for writing up this addendum.)

We shall show that, assuming AD, there is no coding of the ordinals below ω_2 which admits a notion of category with the properties described in §3, and such that the ideals of meager sets are ω_2-additive, and have quotients with uniformly wellordered dense sets.

More precisely, suppose $\langle C_\alpha : \alpha < \omega_2 \rangle$ is a sequence of non-empty, pairwise disjoint sets of reals. Suppose also we have a sequence $\langle (I_\alpha, \mathbb{P}_\alpha, <_\alpha) : \alpha < \omega_2 \rangle$ such that

(a) I_α is a nontrivial ideal on C_α which is closed under wellordered unions of length $\leq \omega_1$, that is, $I_\alpha \subseteq P(C_\alpha)$, $C_\alpha \notin I_\alpha$, $A \subseteq B \in I_\alpha \Rightarrow A \in I_\alpha$, and $\forall \xi < \omega_1 (A_\xi \in I_\alpha)) \Rightarrow \bigcup_{\xi < \omega_1} A_\xi \in I_\alpha$.,

(b) P_α is dense in the I_α-positive sets under almost inclusion, that is, letting $I_\alpha^+ = P(C_\alpha) \setminus I_\alpha$, we have $P_\alpha \subseteq I_\alpha^+$, and for all $X \in I_\alpha^+$, there is a $p \in P_\alpha$ such that $p \setminus X \in I_\alpha$, and

(c) $<_\alpha$ is a wellorder of P_α.

We then call $\langle (C_\alpha, I_\alpha, P_\alpha, <_\alpha) : \alpha < \omega_2 \rangle$ a *strong coding system for* ω_2. It doesn't matter too much how apt this term is, because we can show

THEOREM 8.1. Assume AD; then there is no strong coding system for ω_2.

PROOF. Assume that $\langle (C_\alpha, I_\alpha, P_\alpha, <_\alpha) : \alpha < \omega_2 \rangle$ is a strong coding system for ω_2. Let $C = \bigcup_{\omega_1 \leq \alpha < \omega_2} C_\alpha$, and for $w \in C$, let $|w|$ be the unique α such that $w \in C_\alpha$. By the Coding Lemma, there is a function $w \mapsto f^w$ with domain C such that

$$f^w \text{ is a bijection from } \omega_1 \text{ onto } |w|,$$

for all $w \in C$.

CLAIM 8.2. For any $\alpha \in [\omega_1, \omega_2)$, there are club many $\sigma \in \wp_{\omega_1}(\alpha)$ such that $\exists p \in P_\alpha \exists \xi < \omega_1 \forall^* w \in p(f^w[\xi] = \sigma)$.

PROOF. Here and below, whenever $Y \subseteq I_\beta^+$, then $\forall^* w \in Y \varphi(w)$ means that $\{w \in Y : \neg\varphi(w)\} \in I_\beta$.

Fix α. It is enough to show there are stationary many such σ, since the club filter on $\wp_{\omega_1}(\alpha)$ is an ultrafilter. For that, it is enough to show that whenever $g \colon \omega_1 \to \alpha$ is a bijection, then $\exists \xi < \omega_1 \exists p \in P_\alpha \forall^* w \in p(f^w[\xi] = g[\xi])$. So fix a g.

Clearly $\forall^* w \in C_\alpha \exists \xi < \omega_1 (f^w[\xi] = g[\xi])$. But I_α is ω_2-additive, so $\exists \xi (\{w : f^w[\xi] = g[\xi]\} \in I_\alpha^+)$. Fixing such a ξ, by clause (b) in the definition of strong coding we have a $p \in P_\alpha$ such that $\forall^* w \in p(f^w[\xi] = g[\xi])$, as desired. ⊣ (Claim 8.2)

If $p \in P_\alpha$, $\xi < \omega_1$, and $\sigma \in \wp_{\omega_1}(\alpha)$ are such that $\forall^* w \in p(f^w[\xi] = \sigma)$, then we set $B_{p,\xi} = \sigma$. So Claim 1 tells us that club many $\sigma \in pow_{\omega_1}(\alpha)$ are of the form $B_{p,\xi}$. Note that, because of the $<_\alpha$'s, the family of all $B_{p,\xi}$ is wellordered.

Claim 8.3. For ω-club many $\gamma < \omega_2$, $\exists p \exists \xi (B_{p,\xi}$ is cofinal in $\gamma)$.

Proof. Otherwise, we have an ω-club $D \subseteq \omega_2$ such that for no $\gamma \in D$ are there such p, ξ. However, let α be a limit point of D of cofinality ω_1, and let $U = \{\sigma \in pow_{\omega_1}(\alpha) : \sup(\sigma) \in D\}$. Since U is club in $pow_{\omega_1}(\alpha)$, we have by Claim 1 some p, ξ such that $B_{p,\xi} \in U$. But then $\sup(B_{p,\xi}) = \gamma \in D$, a contradiction. ⊣ (Claim 8.3)

Now set

$$h(\gamma) = \text{ least } B_{p,\xi} \subset \gamma \text{ which is cofinal in } \gamma,$$

whenever there is one. Here "least" refers to some fixed wellorder of the family of all $B_{p,\xi}$. By Claim 2, h is defined on an ω-club in ω_2. It is easy to see using the ω_2-additivity of the ω-club ultrafilter on ω_2 that h is constant on an ω-club. This is impossible, as $h(\gamma)$ is cofinal in γ. ⊣

§9. Addendum; Ordinal games and reliable cardinals. The ordinal games used in the proofs of both Lemma 1.1 and Theorem 2.1 are special cases of the following class of ordinal games: Suppose $\lambda < \Theta$ and that

$$\pi : \lambda^\omega \to \omega^\omega$$

is a continuous function, where the topology on λ^ω is the product topology induced by the discrete topology. Suppose $A \subseteq \omega^\omega$. Then associated to the pair (π, A) is the ordinal game on λ given by $\pi^{-1}[A]$, the preimage of A under π.

By [KKMW81, Theorem 2.7], assuming AD, if A is Suslin and co-Suslin, then this ordinal game is determined.

As corollary, by the Solovay Basis Theorem and the Martin-Steel Theorem for Scale(Σ^2_1) in $\mathbf{L}(\mathbb{R})$, assuming AD holds in $\mathbf{L}(\mathbb{R})$, then all the ordinal games given by a pair (π, A) as above, are determined. This is *ordinal determinacy*. Ordinal determinacy in this form has emerged has a fundamental concept; it is one of the axioms of AD^+ which is a structural generalization of AD. A major open question is whether AD *implies* AD^+. There are a number of partial results, for example, assuming ZF+DC, $\mathsf{AD}_\mathbb{R}$ implies AD^+.

In analyzing the Suslin cardinals of $\mathbf{L}(\mathbb{R})$, Steel [Ste83A] obtained as a corollary (assuming AD+$\mathbf{V}$=$\mathbf{L}(\mathbb{R})$) that every reliable cardinal is a Suslin cardinal. For this equivalence one need only assume that the cardinal κ is *weakly* reliable. This is the assertion that there exists a scale on a set B with norms ρ_i such that for each i,

$$\rho_i : B \to \kappa$$

and such that ρ_0 is a surjection.

Applying the theory of AD^+, one obtains this equivalence of weakly reliable cardinals and Suslin cardinals just assuming AD^+ (so for example, as a consequence of ZF+DC+$\mathsf{AD}_\mathbb{R}$).

REFERENCES

Howard S. Becker

[Bec79] ***Some applications of ordinal games***, Ph.D. thesis, UCLA, 1979.

Leo A. Harrington and Alexander S. Kechris

[HK81] *On the determinacy of games on ordinals*, ***Annals of Mathematical Logic***, vol. 20 (1981), pp. 109–154.

Alexander S. Kechris

[Kec78A] AD *and projective ordinals*, In Kechris and Moschovakis [Cabal i], pp. 91–132.

[Kec78B] *Forcing in analysis*, ***Higher set theory. Proceedings of a conference held at the Mathematisches Forschungsinstitut, Oberwolfach, April 13–23, 1977*** (Gert H. Müller and Dana Scott, editors), Lecture Notes in Mathematics, vol. 669, Springer, 1978, pp. 277–302.

[Kec81A] *Homogeneous trees and projective scales*, In Kechris et al. [Cabal ii], pp. 33–74.

Alexander S. Kechris, Eugene M. Kleinberg, Yiannis N. Moschovakis, and W. Hugh Woodin

[KKMW81] *The axiom of determinacy, strong partition properties and nonsingular measures*, this volume, originally published in Kechris et al. [Cabal ii], pp. 75–100.

Alexander S. Kechris, Donald A. Martin, and Yiannis N. Moschovakis

[Cabal ii] *Cabal seminar* 77–79, Lecture Notes in Mathematics, no. 839, Berlin, Springer, 1981.

[Cabal iii] *Cabal seminar* 79–81, Lecture Notes in Mathematics, no. 1019, Berlin, Springer, 1983.

Alexander S. Kechris and Yiannis N. Moschovakis

[Cabal i] *Cabal seminar* 76–77, Lecture Notes in Mathematics, no. 689, Berlin, Springer, 1978.

Donald A. Martin

[MarC] *On Victoria Delfino problem number* 1, notes.

Yiannis N. Moschovakis

[Mos81] *Ordinal games and playful models*, In Kechris et al. [Cabal ii], pp. 169–201.

John R. Steel

[Ste80] *More measures from* AD, mimeographed notes, 1980.

[Ste83A] *Scales in* $\mathbf{L}(\mathbb{R})$, this volume, originally published in Kechris et al. [Cabal iii], pp. 107–156.

W. Hugh Woodin

[WooB] $\aleph_1$*-dense ideals*, to appear.

DEPARTMENT OF MATHEMATICS
CALIFORNIA INSTITUTE OF TECHNOLOGY
PASADENA, CA 91125, USA
E-mail: kechris@caltech.edu

DEPARTMENT OF MATHEMATICS
UNIVERSITY OF CALIFORNIA
BERKELEY, CA 94720, USA
E-mail: woodin@math.berkeley.edu

A CODING THEOREM FOR MEASURES

ALEXANDER S. KECHRIS

§1. Introduction. Assuming ZF+DC+AD, Moschovakis (see [Mos80, §7D]) has shown that if there is a surjection $\pi\colon \mathbb{R} \to \lambda$ from the reals ($\mathbb{R} = {}^{\omega}\omega$ in this paper) onto an ordinal λ, then there is a surjection $\pi^*\colon \mathbb{R} \to \wp(\lambda)$ from the reals onto the power set of λ. Let us denote by $\beta(\lambda)$ the set of ultrafilters on λ. The question was raised whether there is an analog of Moschovakis' Theorem for $\beta(\lambda)$, i.e., if there is a surjection from $\mathbb{R}$ onto λ, is there one from $\mathbb{R}$ onto $\beta(\lambda)$? Martin showed that this cannot be proved in ZF+DC+AD alone, because if $\mathbf{V} = \mathbf{L}(\mathbb{R})$ and $\lambda = \underset{\sim}{\delta}^2_1$, there is no surjection of $\mathbb{R}$ onto $\beta(\lambda)$. We prove in this paper that if one strengthens the determinacy hypothesis from AD to $\mathsf{AD}^{1/2}_{\mathbb{R}}$ (see [Kec88A]), then this question has a positive answer. Our main theorem is then:

Theorem 1.1. Assume $\mathsf{ZF+DC+AD}^{1/2}_{\mathbb{R}}$. If there is a surjection from $\mathbb{R}$ onto an ordinal λ, then there is a surjection from $\mathbb{R}$ onto $\beta(\lambda)$.

This result can be rephrased as follows. By a result of Kunen (for a proof see [Kec85]) $\beta(\lambda)$ is wellorderable, so it has a definite cardinality, which we denote also by $\beta(\lambda)$. Since the sup of the ordinals onto which we can map the continuum is denoted by Θ, the Theorem 1.1 says:

Corollary 1.2. Assuming $\mathsf{ZF+DC+AD}^{1/2}_{\mathbb{R}}$, if $\lambda < \Theta$, then $\beta(\lambda) < \Theta$.

Combining the above result with Moschovakis' Theorem, one can actually obtain a stronger statement which implies both. Recall first the standard fact that in ZF+DC+AD every ultrafilter is countably complete. Denote then, for each ordinal λ, by $\mathrm{q}(\lambda)$ the set of countably complete filters on λ. This contains $\beta(\lambda)$ but also contains $\wp(\lambda)$ by the natural identification of $A \subseteq \lambda$ with the filter $\hat{A} = \{X \subseteq \lambda : A \subseteq X\}$. We now have the following:

Corollary 1.3. Assume $\mathsf{ZF+DC+AD}^{1/2}_{\mathbb{R}}$. Then if $\lambda < \Theta$, there is a surjection of $\mathbb{R}$ onto $\mathrm{q}(\lambda)$.

Research partially supported by NSF Grants MCS-8117804 and DMS-8416349.

The Cabal Seminar. Volume I: Games, Scales and Suslin Cardinals
Edited by A. S. Kechris, B. Löwe, J. R. Steel
Lecture Notes in Logic, 31

Actually in Moschovakis' Theorem one obtains, for each $\pi\colon \mathbb{R} \to \lambda$, estimates of the complexity of $A^\pi = \{x \in \mathbb{R} : \pi(x) \in A\}$, for $A \subseteq \lambda$, in terms of $\leq_\pi$, where $x \leq_\pi y \Leftrightarrow \pi(x) \leq \pi(y)$, which are very useful in various applications. Similarly we obtain definability estimates for the complexity of ultrafilters. As an immediate consequence we have various local versions of the main theorem, of which we mention as an example the following:

Corollary 1.4. Assume ZF+DC+AD (this is all we need here). Then

(i) If $\lambda < \underset{\sim}{\delta}^1_\omega = \sup\{\underset{\sim}{\delta}^1_n : n < \omega\}$, then $\beta(\lambda) < \underset{\sim}{\delta}^1_\omega$.

(ii) If $\lambda < \kappa^{\mathrm{KL}}$ (the Kleene ordinal of the continuum), then $\beta(\lambda) < \kappa^{\mathrm{KL}}$. Similarly for $\kappa^{\mathbb{R}}$ (for the definition of $\kappa^{\mathrm{KL}}, \kappa^{\mathbb{R}}$, see, for instance [Kec85]).

(iii) If $\lambda < (\underset{\sim}{\delta}^2_1)^{\mathbf{L}(\mathbb{R})}$, $\beta(\lambda) < (\underset{\sim}{\delta}^1_1)^{\mathbf{L}(\mathbb{R})}$.

Finally let us mention another corollary concerning the absoluteness of ultrafilters for certain inner models of AD.

Corollary 1.5. Assume ZF+DC+AD and let M be an inner model of ZF+DC+$\mathsf{AD}^{1/2}_{\mathbb{R}}$ containing $\mathbb{R}$. Let $\lambda < \Theta^M$. Then every ultrafilter on λ is in M. In particular, if λ is measurable, then $M \models$ "λ is measurable".

Actually a finer version of this result is possible, which implies the following (note that $\mathbf{L}(\mathbb{R}) \models \neg\mathsf{AD}^{1/2}_{\mathbb{R}}$).

Corollary 1.6. Assume ZF+DC+AD. If $\lambda < \Theta^{\mathbf{L}(\mathbb{R})}$, then every ultrafilter on λ belongs to $\mathbf{L}(\mathbb{R})$, thus if λ is measurable, $\mathbf{L}(\mathbb{R}) \models$ "λ is measurable".

§2. A game for coding ultrafilters. Let $C \subseteq \mathbb{R}$, $\pi\colon C \to \wp(\lambda)$, λ an ordinal, and write for simplicity $X_\alpha \equiv \pi(\alpha)$. Consider then the following game $\mathrm{G} \equiv \mathrm{G}_\pi$, which is a "coded" version of a "cut-and-choose" game on λ:

$$\begin{array}{lcccccc} \mathrm{I} & \alpha_0 & & \alpha_1 & & & \alpha_j \in \mathbb{R} \\ & & & & & \dots & \\ \mathrm{II} & & i_0 & & i_1 & & i_j \in \{0,1\} \end{array}$$

Player I wins iff: (i) $\forall n(\alpha_n \in C)$ and (ii)

$$\bigcap\{X_{\alpha_n} : i_n = 0\} \cap \bigcap\{\neg X_{\alpha_n} : i_n = 1\} \neq \varnothing.$$

We claim first that, assuming ZF+DC+AD, player I has no winning strategy in G. Indeed if τ was such a strategy, towards a contradiction, let, for every $x \in {}^\omega 2, \alpha^x_0, \alpha^x_1, \dots$ be player I's moves following τ, when player II plays $x(0), x(1), \dots$. Put

$$f(x) = \min\Bigl\{\xi : \xi \in \bigcap\{X_{\alpha^x_n} : x(n) = 0\} \cap \bigcap\{\neg X_{\alpha^x_n} : x(n) = 1\}\Bigr\}.$$

Then $f\colon {}^{\omega}2 \to \lambda$ is an injection, i.e., there is a wellordering of ${}^{\omega}2$, which violates AD.

A strategy F for player II in this game is a map $F\colon {}^{<\omega}\mathbb{R} \to (0,1)$. We now have the key lemma.

LEMMA 2.1. If F is a winning strategy for player II in G, and $\mathcal{U} \subseteq \wp(\lambda)$ is an ultrafilter, then there are $\alpha_0, \alpha_1, \dots, \alpha_{n-1} \in C$, such that for all $\alpha \in C$,

$$X_\alpha \in \mathcal{U} \Leftrightarrow F(\alpha_0, \dots, \alpha_{n-1}, \alpha) = 1.$$

In particular each such $\mathcal{U}$ is completely determined by $F \upharpoonright \mathbb{R}^{n+1}$.

PROOF. Call a sequence $(\alpha_0, i_0, \alpha_1, i_1, \dots, a_{m-1}, i_{m-1})$ **$\mathcal{U}$-good** if it is a finite run of the game G in which player II follows F, $\alpha_0, \dots, \alpha_{m-1} \in C$ and $\forall j \leq m-1 (X_{\alpha_j} \in \mathcal{U} \Leftrightarrow i_j = 0)$. The empty sequence is $\mathcal{U}$-good by convention. If every $\mathcal{U}$-good sequence has a $\mathcal{U}$-good proper extension, we can obtain an infinite run $(\alpha_0, i_0, \alpha_1, \dots)$ of G in which player II followed F, $\alpha_0, \alpha_1, \dots \in C$ and $X_{\alpha_j} \in \mathcal{U} \Leftrightarrow i_j = 0$, for all j. Then clearly $\bigcap\{X_{\alpha_n} : i_n = 0\} \cap \bigcap\{\neg X_{\alpha_n} : i_n = 1\} \in \mathcal{U}$ (recall that $\mathcal{U}$ is countably complete), so this intersection is non-$\varnothing$, and player I won, a contradiction.

So let $(\alpha_0, i_0, \dots, \alpha_{n-1}, i_{n-1})$ be a maximal $\mathcal{U}$-good sequence. Let $\alpha \in C$. Then $(\alpha_0, i_0, \dots, \alpha_{n-1}, \alpha, F(\alpha_0, \dots, \alpha_{n-1}, \alpha))$ is not $\mathcal{U}$-good, i.e.,

$$X_\alpha \in \mathcal{U} \Leftrightarrow F(\alpha_0, \dots, \alpha_{n-1}, \alpha) = 1$$

and we are done. ⊣

§3. On real-integer games. Note that the preceding game, G_π, is a game in which one player plays reals and the other integers, so it is not necessarily determined using AD alone. We denote the determinacy of such games by $\mathsf{AD}_{\mathbb{R}}^{1/2}$. (In [Kec88A] it is shown that in ZF+DC, $\mathsf{AD}_{\mathbb{R}}^{1/2}$ is equivalent to Unif.) So we have immediately

THEOREM 3.1. Assume ZF+DC+$\mathsf{AD}_{\mathbb{R}}^{1/2}$. If $\lambda < \Theta$, then $\beta(\lambda) < \Theta$.

PROOF. Using the notation of §2, with $C = \mathbb{R}$, $\pi\colon \mathbb{R} \to \lambda$, map each $(n+1)$-tuple $\vec{\alpha} = (\alpha_0, \dots, \alpha_{n-1}, \alpha_n)$ in $\mathbb{R}^{n+1}$ to $V_{\vec{\alpha}} = \{X_{\alpha_n} : F(\vec{\alpha}) = 1\}$. Then

$$\beta(\lambda) \subseteq \{V_{\vec{\alpha}} : \vec{\alpha} \in C^{n+1},\ n = 0, 1, 2, \dots\}.$$

So there is a surjection of $\mathbb{R}$ onto $\beta(\lambda)$. ⊣

We also have easily the result about countably complete filters.

COROLLARY 3.2. Assume ZF+DC+$\mathsf{AD}_{\mathbb{R}}^{1/2}$. If $\lambda < \Theta$, then there is a surjection from $\mathbb{R}$ onto $\mathrm{q}(\lambda)$, the set of countably complete filters on λ.

PROOF. By a result of Kunen, if $\mathcal{I}$ is a proper countably complete filter on λ, then there is an ultrafilter $\mathcal{U}$ on λ extending $\mathcal{I}$. (For a proof see [Sol78A, p. 148]) Thus if $\mathcal{I}$ is a proper countably complete filter, then $\mathcal{I} = \bigcap\{\mathcal{U} : \mathcal{U}$ in an ultrafilter on λ containing $\mathcal{I}\}$. Thus $\wp(\beta(\lambda))$ can be mapped onto $\mathrm{q}(\lambda)$ and, since $\beta(\lambda) < \Theta, \mathbb{R}$ can be mapped onto $\mathrm{q}(\lambda)$. ⊣

§4. The complexity of ultrafilters. In the notation of §2 again, let for each ultrafilter $\mathcal{U}$ on λ:

$$\mathcal{U}^* \equiv \mathcal{U}^*_\pi = \{\alpha \in C : X_\alpha \in \mathcal{U}\}.$$

It follows that for some $\alpha_0, \dots, \alpha_{n-1} \in C$, and all $\alpha \in C$,

$$\alpha \in \mathcal{U}^* \Leftrightarrow F(\alpha_0, \dots, \alpha_{n-1}, \alpha) = 1,$$

for any winning strategy F for player II in G_π. Thus the complexity of $\mathcal{U}^*$ depends (beyond that of C, π) on the complexity of $F \restriction \mathbb{R}^m, m = 1, 2, \dots$ (each of which is essentially a set of reals), for any winning strategy F for player II.

There are certain real-integer games which can be proved determined in AD only. In those instances we get reasonably good estimates for the winning strategies, which give us corresponding estimates for ultrafilters. Here is a relevant result.

THEOREM 4.1 (Woodin, unpublished). Assume ZF+DC+AD. If $A \subseteq {}^\omega\mathbb{R} \times {}^\omega\omega$ is co-Suslin, then the (real-integer) game corresponding to A is determined.

Here $A \subseteq {}^\omega\mathbb{R} \times {}^\omega\omega$ is **co-Suslin** iff $A' \subseteq {}^\omega\omega \times {}^\omega\omega$ given by $A'(\alpha, \beta) \Leftrightarrow A(\{(\alpha)_n\}, \beta)$ is co-Suslin, where $\alpha \mapsto \{(\alpha)_n\}$ is a recursive 1-1 correspondence between ${}^\omega\omega$ and ${}^\omega\mathbb{R}$.

We give below an alternative version and proof (motivated by the ideas of [Kec88A]) of that result, which also gives the definability estimates we want.

THEOREM 4.2. Assume ZF+DC+AD. Let $A \subseteq {}^\omega\mathbb{R} \times {}^\omega\omega$ and suppose $\neg A$ (viewed as a subset of ${}^\omega\omega \times {}^\omega\omega$ as above) carries a scale $\langle\varphi_n\rangle$ such that each corresponding relation $\leq^*_{\varphi_n}, <^*_{\varphi_n}$ belongs to a pointclass Γ_n, where $\Gamma_0 \subseteq \Gamma_1 \subseteq \dots \subseteq \Gamma_n \subseteq \dots$ and Γ_n is closed under $\wedge, \vee, \exists m \leq k, \forall m \leq k$ and recursive substitutions. Consider the game G

$$\begin{array}{lccccccl} \mathrm{I} & \alpha_0 & & \alpha_1 & & \dots & \alpha_j \in \mathbb{R} \\ \mathrm{II} & & i_0 & & i_1 & \dots & i_j \in \omega \end{array}$$

Player II wins iff $(\vec{\alpha}, \vec{i}) \notin A$.

If player I has no winning strategy in G, then player II has a winning strategy $F : {}^{<\omega}\mathbb{R} \to \omega$ such that for each $n, F \restriction \mathbb{R}^n \in \Game^2\Gamma_n$. (Here $\Game$ is the game quantifier, see [Mos80], and $\Game^2 = \Game\Game$).

PROOF. Fix a Turing degree $\boldsymbol{d}$. Then clearly player I does not have a winning strategy in the game where he plays reals $\alpha_i \leq_\mathrm{T} \boldsymbol{d}$. So, by the Third Periodicity

Theorem of [Mos80, 6E.1], player II has a winning strategy F_d in this restricted game, such that for each n the relation

$$\alpha_i \leq_T \boldsymbol{d} \wedge F_{\boldsymbol{d}}(\alpha_0, \ldots, \alpha_{n-1}) = i$$

is in $\Game\Gamma_n$, uniformly on $\boldsymbol{d}$, i.e., the relation

$$R_n(x, \alpha_0 \ldots \alpha_{n-1}, i) \Leftrightarrow \alpha_i \leq_T x \wedge F_{[x]_T}(\alpha_0, \ldots, \alpha_{n-1}) = i$$

is in $\Game\Gamma_n$. Define now $F(\alpha_0, \ldots, \alpha_{n-1}) = i \Leftrightarrow$ for a cone of $\boldsymbol{d}$'s, $F_{\boldsymbol{d}}(\alpha_0, \ldots, \alpha_{n-1}) = i$. Clearly $F \restriction \mathbb{R}^n \in \Game(\Game\Gamma_n) = \Game^2\Gamma_n$ and F is a winning strategy for player II in the game G. ⊣

Let us now mention some specific applications. We assume ZF+DC+AD below.

First let $\lambda < \boldsymbol{\delta}^1_\omega$ be a projective ordinal, let $\varphi\colon \mathbb{R} \to \lambda$ be a projective norm and, using the Moschovakis' Coding Lemma (see [Mos80, 7D.5]), let in the notation of §2, $C \subseteq \mathbb{R}$, $\pi\colon C \to \wp(\lambda)$ be also projective. (For π this means that the relation "$\varphi(x) \in \pi(y)$" is projective.) Then for any ultrafilter $\mathcal{U}$ on λ, $\mathcal{U}^*$ is projective (actually in some fixed level of the projective hierarchy) and $\beta(\lambda) < \boldsymbol{\delta}^1_\omega$.

One can of course state finer level-by-level versions of this result. However the recent work of Steve Jackson (see for example [Jac88]) provides a fairly complete analysis of ultrafilters on projective ordinals, which provides much more accurate estimates for the definability of ultrafilters.

Similarly, if $\lambda < \kappa^{\mathrm{KL}}$, every ultrafilter on λ is Kleene recursive in ${}^3E$ and a real (in the codes), and $\beta(\lambda) < \kappa^{\mathrm{KL}}$. If $\lambda < \kappa^{\mathbb{R}}$, then every ultrafilter on λ is hyperprojective in the codes and $\beta(\lambda) < \kappa^{\mathbb{R}}$. For $\lambda = \kappa^{\mathbb{R}}$ itself, we have that each ultrafilter on $\kappa^{\mathbb{R}}$ is Σ^*_m for some m (depending on $\mathcal{U}$). For the definition of the classes Σ^*_m see [Mos83]. So $\beta(\kappa^{\mathbb{R}}) \leq \boldsymbol{\delta}^*_\omega$. Martin has showed that also $\beta(\kappa^{\mathbb{R}}) \geq \boldsymbol{\delta}^*_\omega$, so in fact $\beta(\kappa^{\mathbb{R}}) = \boldsymbol{\delta}^*_\omega$.

Finally, if $\lambda < (\boldsymbol{\delta}^2_1)^{\mathrm{L}(\mathbb{R})}$, then every ultrafilter on λ is $(\boldsymbol{\Delta}^2_1)^{\mathrm{L}(\mathbb{R})}$ in the codes, and $\beta(\lambda) < (\boldsymbol{\delta}^2_1)^{\mathrm{L}(\mathbb{R})}$.

§5. Absoluteness of ultrafilters. The following fact is immediate from the analysis in §2.

COROLLARY 5.1. Assume ZF+DC+AD. Let M be an inner model of ZF+DC +AD containing $\mathbb{R}$, let $\lambda < \Theta^M$, and let C, π (as in §2) be in M. If F is a winning strategy for player II in G_π such that $F \restriction \mathbb{R}^n \in M$ for all $n \in \omega$, then every ultrafilter on λ is in M.

In particular we have

COROLLARY 5.2. Assume ZF+DC+AD. Let M be an inner model of ZF+DC +$\mathrm{AD}^{1/2}_{\mathbb{R}}$ containing $\mathbb{R}$, and let $\lambda < \Theta^M$. Then every ultrafilter on λ is in M, and thus, if λ is measurable, $M \models$ "λ is measurable".

Finally we can obtain the result for $\mathbf{L}(\mathbb{R})$ itself.

COROLLARY 5.3. Assume ZF+DC+AD. Let $\lambda < \Theta^{\mathbf{L}(\mathbb{R})}$. Then every ultrafilter on λ is in $\mathbf{L}(\mathbb{R})$, and if λ is measurable, $\mathbf{L}(\mathbb{R}) \models$ "λ is measurable".

PROOF. If $\mathbf{V} = \mathbf{L}(\mathbb{R})$, there is nothing to prove. If $\mathbf{V} \neq \mathbf{L}(\mathbb{R})$ then by [SVW82], $\mathbb{R}^{\#}$ exists. Then by a result of Solovay (unpublished—see however [MS83, p. 93]), every $A \in \mathbf{L}(\mathbb{R})$ admits a scale $\langle \varphi_n \rangle$ with $\leq^*_{\varphi_n}$, $<^*_{\varphi_n}$ in $\mathbf{L}(\mathbb{R})$, thus, by Theorem 4.2, the hypothesis of Corollary 5.1 is satisfied and we are done. ⊣

REFERENCES

STEPHEN JACKSON
[Jac88] AD *and the projective ordinals*, In Kechris et al. [CABAL iv], pp. 117–220.

ALEXANDER S. KECHRIS
[Kec85] *Determinacy and the structure of* $\mathbf{L}(\mathbb{R})$, ***Proceedings of symposia in pure mathematics***, vol. 42, American Mathematical Society, pp. 271–283.
[Kec88A] AD + Unif *is equivalent to* $\mathsf{AD}_{\mathbb{R}}^{1/2}$, In Kechris et al. [CABAL iv], pp. 98–102.

ALEXANDER S. KECHRIS, DONALD A. MARTIN, AND YIANNIS N. MOSCHOVAKIS
[CABAL iii] *Cabal seminar* 79–81, Lecture Notes in Mathematics, no. 1019, Berlin, Springer.

ALEXANDER S. KECHRIS, DONALD A. MARTIN, AND JOHN R. STEEL
[CABAL iv] *Cabal seminar* 81–85, Lecture Notes in Mathematics, no. 1333, Berlin, Springer.

ALEXANDER S. KECHRIS AND YIANNIS N. MOSCHOVAKIS
[CABAL i] *Cabal seminar* 76–77, Lecture Notes in Mathematics, no. 689, Berlin, Springer.

DONALD A. MARTIN AND JOHN R. STEEL
[MS83] *The extent of scales in* $\mathbf{L}(\mathbb{R})$, this volume, originally published in Kechris et al. [CABAL iii], pp. 86–96.

YIANNIS N. MOSCHOVAKIS
[Mos80] ***Descriptive set theory***, Studies in Logic and the Foundations of Mathematics, no. 100, North-Holland, Amsterdam.
[Mos83] *Scales on coinductive sets*, this volume, originally published in Kechris et al. [CABAL iii], pp. 77–85.

ROBERT M. SOLOVAY
[Sol78A] *A* $\underset{\sim}{\Delta}^1_3$ *coding of the subsets of* $^{\omega}\omega$, In Kechris and Moschovakis [CABAL i], pp. 133–150.

JOHN R. STEEL AND ROBERT VAN WESEP
[SVW82] *Two consequences of determinacy consistent with choice*, ***Transactions of the American Mathematical Society***, no. 272, pp. 67–85.

DEPARTMENT OF MATHEMATICS
CALIFORNIA INSTITUTE OF TECHNOLOGY
PASADENA, CA 91125, USA
E-mail: kechris@caltech.edu

THE TREE OF A MOSCHOVAKIS SCALE IS HOMOGENEOUS

DONALD A. MARTIN AND JOHN R. STEEL

§0. We work in the theory ZF+AD+$\mathsf{DC}_{\mathbb{R}}$ throughout this paper.

Let T be a tree on $\omega \times \kappa$, for some ordinal κ. The **projection of** T is the set of reals

$$\mathrm{p}[T] = \{x \in {}^{\omega}\omega : \exists f \in {}^{\omega}\kappa \forall n (x \restriction n, f \restriction n) \in T\}.$$

We say that $\mathrm{p}[T]$ is **Suslin** via T. A **homogeneity system for** T is a system $\langle \mu_s : s \in {}^{<\omega}\omega \rangle$ of countably additive, 0-1 valued measures[1] such that for all $s, t \in {}^{<\omega}\omega$ and $x \in {}^{\omega}\omega$

$$\mu_s(T_s) = 1,$$

and

$$s \subseteq t \Rightarrow \mu_t \text{ projects to } \mu_s,$$

and

$$x \in \mathrm{p}[T] \Rightarrow \langle \mu_{x \restriction n} : n < \omega \rangle \text{ is countably complete.}$$

Here, $T_s = \{u : (s, u) \in T\}$, and μ_t **projects to** μ_s iff whenever $\mu_s(A) = 1$, then $\mu_t(\{u : u \restriction \mathrm{dom}(s) \in A\}) = 1$, and the tower of measures $\langle \mu_{x \restriction n} : n < \omega \rangle$ is called **countably complete** just in case whenever $\langle A_n : n < \omega \rangle$ is a sequence of sets such that $\mu_{x \restriction n}(A_n) = 1$ for all n, then $\exists f \in {}^{\omega}\kappa \forall n (f \restriction n \in A_n)$.[2] If $\vec{\mu}$ is a homogeneity system for T, then we shall say $\mathrm{p}[T]$ is **homogeneously Suslin** via T and $\vec{\mu}$.

Which sets of reals are homogeneously Suslin? One can use partition cardinals to propagate homogeneity, and thereby show that all projective sets are homogeneously Suslin. (This is a result of Kunen and Martin; cf. [Kec81A, §6].) This method bogs down a bit past the projective sets, however, and leaves open, for example, whether all hyperprojective sets of reals are homogeneously Suslin. In the other direction, every homogeneously Suslin set is Suslin, and

[1] That is, countably complete ultrafilters.

[2] Taking $A_n = T_{x \restriction n}$, we see that the countable completeness of the tower $\langle \mu_{x \restriction n} : n < \omega \rangle$ implies $x \in \mathrm{p}[T]$.

The Cabal Seminar. Volume I: Games, Scales and Suslin Cardinals
Edited by A. S. Kechris, B. Löwe, J. R. Steel
Lecture Notes in Logic, 31

by [MS69], its complement in $\mathbb{R}$ is Suslin as well. Thus if $\mathbf{V} = \mathbf{L}(\mathbb{R})$, every homogeneously Suslin set is $\underset{\sim}{\mathbf{\Delta}}^2_1$.

In this paper we shall show that the homogeneously Suslin sets are precisely those Suslin sets whose complements are also Suslin. Thus if $\mathbf{V} = \mathbf{L}(\mathbb{R})$, a set of reals is homogeneously Suslin if and only if it is $\underset{\sim}{\mathbf{\Delta}}^2_1$. The homogeneous trees in question are not generated by the partition cardinal construction; rather they are trees associated to scales produced by Moschovakis' second periodicity construction.

This result is due to the first author. We also include a result of the second author concerning the additivity of the homogeneity measures produced by this construction.

§1. Our main result is

THEOREM 1.1. Assume ZF+AD+DC$_{\mathbb{R}}$; then for any $A \subseteq \mathbb{R}$, the following are equivalent:

(1) A is homogeneously Suslin,
(2) A and $\mathbb{R} \setminus A$ are Suslin.

PROOF. (1) ⇒ (2) follows at once from [MS69]. Now let A be Suslin and co-Suslin. Since $\varnothing$ is homogeneously Suslin, we may assume A is nonempty.

Let $\langle u_i : i < \omega \rangle$ be an enumeration without repetitions of ${}^{<\omega}\omega$. We write $n(u)$ for the unique $n \in \omega$ such that $u = u_n$. Let us choose our enumeration so that

$$(\text{dom}(u) \subseteq \text{dom}(v) \wedge \forall i \in \text{dom}(u)(u(i) \leq v(i))) \Rightarrow n(u) \leq n(v)).$$

This implies that proper initial segments of u are enumerated before u.[3] This enumeration lets us identify Lipschitz continuous functions on $\mathbb{R}$ (i.e. strategies for player II in games on ω) with reals. Namely, for any $\rho \in {}^{<\omega}\omega$, let $\rho^* \colon {}^{<\omega}\omega \to {}^{<\omega}\omega$ with domain $\{u : n(u) \in \text{dom}(\rho)\}$ be given by

$$\rho^*(u)(k) = \rho(n(u \restriction (k+1)),$$

whenever $k \in \text{dom}(u)$. Let $\rho^*(\varnothing) = \varnothing$. For any $\sigma \in {}^{\omega}\omega$, we let

$$\sigma^* = \bigcup_{n<\omega} (\sigma \restriction n)^*,$$

and we extend σ^* to a Lipschitz continuous function on $\mathbb{R}$, which we also call σ^*, by setting

$$\sigma^*(x) = \bigcup_{i<\omega} \sigma^*(x \restriction i).$$

[3]For example, we can let $n(u) \leq n(v)$ iff $\prod_{i \in \text{dom}(u)} p_i^{u(i)+1} \leq \prod_{i \in \text{dom}(v)} p_i^{v(i)+1}$, where p_i is the ith prime.

So ρ^* is a finite neighborhood condition in the space of all Lipschitz continuous functions on $\mathbb{R}$, and σ^* is the set of all neighborhood conditions coded into σ, as well as the Lipschitz continuous function meeting all these neighborhood conditions. We shall tend to use Greek letters $\rho, \sigma, \ldots$ for finite or infinite sequences of natural numbers when we are thinking of them as codes in this fashion, and Latin letters $s, x, \ldots$ otherwise.

Let

$$A^* = \{\sigma \in {}^{\omega}\omega : \forall x \in {}^{\omega}\omega(\sigma^*(x) \in A)\}$$

be the set of codes for Lipschitz functions mapping into A. It is easy to see that A is Lipschitz-reducible to A^*; simply send x to the code for the constantly x function. It will therefore suffice to show that A^* is homogeneously Suslin.

Let $\langle \varphi_n : n < \omega \rangle$ be a very good scale on A. Moschovakis' second periodicity construction [Mos80] produces a very good scale on A^*. It is most natural to index the norms in the Moschovakis scale with finite sequences, so we shall do that. Let $u \in {}^{<\omega}\omega$ and let $\sigma_1, \sigma_2 \in A^*$. Consider the game $G(u, \sigma_1, \sigma_2)$: Player I plays out $x_1 \in \mathbb{R}$, and player II plays out $x_2 \in \mathbb{R}$, the players alternating moves, with player I going first, as usual. Let $u^\frown x \in \mathbb{R}$ be the result of appending x to u, that is, $(u^\frown x)(n) = u(n)$ for $n \in \mathrm{dom}(u)$, and $(u^\frown x)(n) = x(n - \mathrm{dom}(u))$ otherwise. Since $\sigma_i \in A^*$, we have $\sigma_i^*(u^\frown x_i) \in A$ for $i = 1, 2$. We say

$$(x_1, x_2) \text{ wins } G(u, \sigma_1, \sigma_2) \text{ for player II} \Leftrightarrow \varphi_{\mathrm{dom}(u)}(\sigma_1^*(u^\frown x_1)) \leq \varphi_{\mathrm{dom}(u)}(\sigma_2^*(u^\frown x_2)).$$

Putting

$$\sigma_1 \leq_u \sigma_2 \Leftrightarrow \text{player II has a winning strategy in } G(u, \sigma_1, \sigma_2),$$

Moschovakis shows that each $\leq_u$ is a prewellorder of A^*. For $\sigma \in A^*$ and $u \in {}^{<\omega}\omega$, we let

$$\vartheta_u(\sigma) = \text{ rank of } \sigma \text{ in } \leq_u,$$

and we have by Moschovakis' arguments that $\langle \vartheta_u : u \in {}^{<\omega}\omega \rangle$ is a scale on A^*.[4] It is clear that $\vartheta_u(\sigma)$ only depends on $\sigma \restriction \{i : u \subseteq u_i \vee u_i \subseteq u\}$. It is also clear that

$$\vartheta_u(\sigma) \leq \vartheta_u(\tau) \Leftrightarrow \forall i \exists j (\vartheta_{u^\frown\langle i \rangle}(\sigma) \leq \vartheta_{u^\frown\langle j \rangle}(\tau).$$

The tree of the scale $\vec{\vartheta}$ is given by:

$$T = \{(\sigma \restriction n, \langle \vartheta_{u_0}(\sigma), \ldots, \vartheta_{u_{n-1}}(\sigma) \rangle) : \sigma \in A^*\}.$$

[4] To see the limit property: let $\sigma_i \to \sigma \pmod{\vec{\vartheta}}$. Let $x \in \mathbb{R}$; we must show $\sigma^*(x) \in A$. We can thin out the σ_i's so that $\sigma_{n+1} \leq_{x \restriction n} \sigma_n$ for all n; let Σ_n be a winning strategy for player II in $G(x \restriction n, \sigma_{n+1}, \sigma_n)$ witnessing this. Playing the Σ_n against each other simultaneously, we obtain z_n so that $z_n \to x$ and $\sigma_n^*(z_n) \to \sigma^*(x) \pmod{\vec{\varphi}}$. (Here z_n is defined by: $x \restriction n \subseteq z_n$, and for $k \geq n$, $z_n(k) = \Sigma_n(\langle x(n), z_{n+1}(n+1), \ldots, z_{n+1}(k) \rangle)$.) Since $\vec{\varphi}$ is a scale on A, $\sigma^*(x) \in A$.

By the limit property of scales, $\mathrm{p}[T] = A^*$. We shall show that T is homogeneous. The homogeneity measures μ_p on T_p come from a game in which the two players interact to produce a potential member of T_p.

We need a Suslin representation of the complement of A in order to define our game, so fix a tree S on some $\omega \times \kappa$ such that

$$\mathrm{p}[S] = \mathbb{R} \setminus A.$$

From S we get a Suslin representation of $\mathbb{R} \setminus A^*$: for $\sigma \in \mathbb{R} \setminus A^*$, we define the **leftmost witness that** $\sigma \notin A^*$ to be the pair (x, f), where $x \in \mathbb{R}$ and $f \in {}^\omega\kappa$ are defined inductively by

$$x(n) = \min\{k : \exists y \exists g((x \restriction n)^\frown\langle k\rangle \subseteq y \wedge f \restriction n \subseteq g \wedge (\sigma^*(y), g) \in [S])\},$$

and

$$f(n) = \min\{\alpha : \exists y \exists g(x \restriction (n+1) \subseteq y \wedge (f \restriction n)^\frown\langle\alpha\rangle \subseteq g \wedge (\sigma^*(y), g) \in [S])\}.$$

Now let $p \in {}^{<\omega}\omega$ and $X \subseteq T_p$. We shall say that $\mu_p(X) = 1$ if and only if player II has a winning strategy in the game $\mathrm{G}_p(X)$, which is played as follows. Players I and II alternate playing natural numbers, with player I moving first, as usual. At the end of a run player I has produced $\tau_1 \in \mathbb{R}$ and player II has produced τ_2. Set $\sigma_i = p^\frown\tau_i$. Player II wins $\mathrm{G}_p(X)$ if and only if either

(1) $\sigma_1 \notin A^*$ and $\sigma_2 \in A^*$, or

(2) $\sigma_1 \notin A^*$ and $\sigma_2 \notin A^*$, and letting (x_1, f_1) and (x_2, f_2) be the leftmost witnesses respectively, and k_i least such that $x_i \restriction k_i \notin \mathrm{dom}(p^*)$, we have $(x_1 \restriction k_1, f_1 \restriction k_1) \leq_{\mathrm{lex}} (x_2 \restriction k_2, f_2 \restriction k_2)$, or

(3) $\sigma_1 \in A^*$ and $\sigma_2 \in A^*$, and for all $u \in \mathrm{dom}(p^*)$, $\vartheta_u(\sigma_1) \leq \vartheta_u(\sigma_2)$, and $\langle \vartheta_{u_i}(\sigma_2) : i \in \mathrm{dom}(p)\rangle \in X$.

Here the lexicographic order $\leq_{\mathrm{lex}}$ is defined in a way which parallels our definition of the leftmost witnesses: $(s, t) \leq_{\mathrm{lex}} (u, v)$ iff $(s, t) = (u, v)$, or there is an $i \in \mathrm{dom}(s)$ such that $s \restriction i = u \restriction i, t \restriction i = v \restriction i$, and either $s(i) < u(i)$, or $s(i) = u(i)$ and $t(i) < v(i)$. So $\leq_{\mathrm{lex}}$ is a partial, not total, order on ${}^{<\omega}\omega \times {}^{<\omega}\kappa$. We also define $\leq_{\mathrm{lex}}$ on ${}^\omega\omega \times {}^\omega\kappa$ in the obvious way: $(x, f) \leq_{\mathrm{lex}} (y, g) \Leftrightarrow \forall k(x \restriction k, f \restriction k) \leq_{\mathrm{lex}} (y \restriction k, g \restriction k)$. Here the order is total.

It is clear that $\mu_p(T_p) = 1$, since player II can win $\mathrm{G}_p(T_p)$ by simply copying player I. It is equally clear that $\mu_p(\varnothing) = 0$, and that $X \subseteq Y \wedge \mu_p(X) = 1 \Rightarrow \mu_p(Y) = 1$. For the rest of the proof that the μ_p constitute a homogeneity system, we need some terminology concerning embeddings of one Lipschitz continuous function in another.

Let $\sigma, \tau, \pi \in {}^\omega\omega$. We say π **embeds** σ **in** τ iff $\sigma^* = \tau^* \circ \pi^*$. We say that π is an embedding **over** p, for $p \in {}^{<\omega}\omega$, iff $\pi^*(u) = u$ for all $u \in \mathrm{dom}(p^*)$. For any $\pi \in {}^\omega\omega$, we say

$$\pi \text{ is II-safe} \Leftrightarrow \forall u \in {}^{<\omega}\omega(u \leq_{\mathrm{lex}} \pi^*(u)),$$

and if π is an embedding over ρ,

$$\pi \text{ is I-safe over } \rho \Leftrightarrow \forall u \in {}^{<\omega}\omega \setminus \mathrm{dom}(\rho^*)(u <_{\mathrm{lex}} \pi^*(u)).$$

We say $\tau \in {}^{\omega}\omega$ is the **amalgamation of** $\langle \sigma_i : i \in X\rangle$ **via** $\langle \pi_i : i \in X\rangle$ iff each π_i embeds σ_i into τ for all i, and $\{\mathrm{ran}(\pi_i) : i \in X\}$ is a partition of ${}^{\omega}\omega$. We say the amalgamation is II-safe if all the π_i are II-safe, and over ρ if all the π_i are embeddings over ρ.

Remark. The following two facts help explain our interest in safe amalgamations:

(a) If τ is an amalgamation of $\langle \sigma_i : i \in X\rangle$ and each $\sigma_i \in A^*$, then $\tau \in A^*$. If the amalgamation is over ρ, then $\vartheta_u(\sigma_i) \le \vartheta_u(\tau)$ for all i and all $u \in \mathrm{dom}(\rho^*)$.

(b) If the amalgamation is via II-safe embeddings π_i, and $\tau \notin A^*$, then letting (x, f) be the leftmost witness that $\tau \notin A^*$, and $x = \pi_i(y)$, then $\sigma_i^*(y) = \tau^*(x)$, so (y, f) is a witness that $\sigma_i \notin A^*$ such that $(y, f) \le_{\mathrm{lex}} (x, f)$.

Claim 1.2. For any $\rho \in {}^{<\omega}\omega$ and $X \subseteq T_\rho$, either $\mu_\rho(X) = 1$ or $\mu_\rho(T_\rho \setminus X) = 1$.

Proof. Suppose $\mu_\rho(X) \ne 1$, and fix a winning strategy Σ for player I in $\mathrm{G}_\rho(X)$. Player II cannot quite afford to simply use Σ as his own strategy in $\mathrm{G}_\rho(T_\rho \setminus X)$, because of the demand that $\vartheta_u(\sigma_1) \le \vartheta_u(\sigma_2)$ for all $u \in \mathrm{dom}(\rho^*)$ in player II's payoff condition (3). So player II will also embed player I's play into his own as he proceeds. In order not to lose because of payoff condition (2), he will use a II-safe embedding. More precisely, player II's winning strategy in $\mathrm{G}_\rho(T_\rho \setminus X)$ is to insure that if player I plays out $x \in \mathbb{R}$, then player II's response y is such that $\tau = \rho^\frown y$ is a II-safe amalgamation over ρ of $\sigma = \rho^\frown x$ and $\psi = \rho^\frown\Sigma(y)$.

We show first that such a play (x, y) is a win for player II in $\mathrm{G}_\rho(T_\rho \setminus X)$. Suppose first that $\tau \in A^*$. Then both σ and ψ are in A^*, since they embed in τ. Since the embedding is over ρ,

$$\vartheta_u(\sigma) \le \vartheta_u(\tau) \text{ and } \vartheta_u(\psi) \le \vartheta_u(\tau)$$

for all $u \in \mathrm{dom}(\rho^*)$. Now (ψ, τ) comes from a play of $\mathrm{G}_\rho(X)$ according to player I's winning strategy Σ, and the second set of inequalities displayed therefore imply

$$\langle \vartheta_{u_0}(\tau), \dots, \vartheta_{u_{\mathrm{dom}(\rho)-1}}(\tau)\rangle \in T_\rho \setminus X.$$

But then the first set of inequalities show that (x, y), which gives rise to (σ, τ), is a win for player II under payoff clause (3).

Suppose next that $\tau \notin A^*$, and let (z, f) be the leftmost witness of this fact. Since the amalgamation yielding τ is II-safe, we can find $(w, g) \le_{\mathrm{lex}} (z, f)$ such that either (w, g) is the leftmost witness that $\sigma \notin A^*$ or (w, g) is the leftmost witness that $\psi \notin A^*$. (See Remark 1.) The latter is impossible, since it implies that $(\Sigma(y), y)$ is a defeat for Σ under payoff condition (2) of $G_\rho(X)$. The former implies that (x, y) is a win for player II in $G_\rho(T_\rho \setminus X)$ under payoff condition (2), as desired.

Finally, we must see that player II can play his y in response to x so that $\rho^\frown y$ is the desired amalgamation. For each $u \in \text{dom}(\rho^*)$, let

$$I_u^\rho = \{i : u^\frown\langle i\rangle \notin \text{dom}(\rho^*)\},$$

and

$$h_u^1, h_u^2 \colon I_u^\rho \to I_u^\rho$$

be order-preserving functions such that $\text{ran}(h_u^1)$ and $\text{ran}(h_u^2)$ partition I_u^ρ, and

$$k \leq h_u^i(k)$$

for all $k \in I_u^\rho$ and all $i \in \{0, 1\}$. We define our embeddings π_1 and π_2 by

$$\pi_i^*(v)(k) = h_{v \restriction k}^i(v(k))$$

if $v \restriction k \in \text{dom}(\rho^*)$ but $v \restriction (k+1) \notin \text{dom}(\rho^*)$, and

$$\pi_i^*(v)(k) = v(k)$$

otherwise. Notice that $\text{dom}(\pi_i^*(v)) = \text{dom}(\pi_i(v))$, and that $\pi_i^*(v)$ differs from v at at most one k, and for that k, $v(k) \leq \pi_i^*(v)(k)$. It follows that

$$\forall v(n(v) \leq n(\pi_i^*(v))).$$

Player II can now determine $y(k)$, granted $x \restriction (k+1)$, as follows: let u be such that

$$n(u) = \text{dom}(\rho) + k.$$

Then

$$u = \pi_i^*(v)$$

for exactly one pair i, v; moreover,

$$n(v) = \text{dom}(\rho) + j$$

for some $j \leq k$. We let

$$y(k) = x(j) \text{ if } i = 0$$

and

$$y(k) = \Sigma(y \restriction k)(j) \text{ if } i = 1.$$

We leave any further detail to the reader. ⊣ (Claim 1.2)

Claim 1.3. If $\mu_\rho(X_i) = 1$ for all $i < \omega$, then $\mu_\rho(\bigcap_i X_i) = 1$.

PROOF. Let Σ_i be a winning strategy for player II in $G_\rho(X_i)$. For each $u \in \mathrm{dom}(\rho^*)$, let

$$h_u^1, h_u^2 \colon I_u^\rho \to I_u^\rho$$

be as in the proof of Claim 1.2, but with

$$k < h_u^2(k)$$

for all $k \in I_u^\rho$ in addition. Let π_1 and π_2 be the II-safe embeddings over ρ determined from h_u^1 and h_u^2 as in Claim 1.2. Our additional property of h_u^2 guarantees that π_2 is in fact I-safe over ρ.

Our winning strategy Σ for player II in $\mathrm{G}_\rho(X)$ will behave as follows: suppose player I plays out $x \in {}^\omega\omega$, and let $\sigma = \rho^\frown x$. Let y be player II's response as dictated by Σ, and $\tau = \rho^\frown y$. Let φ be the amalgamation of σ and τ via π_1 and π_2. (Here π_2 is used to embed τ into φ.) Since σ and τ extend ρ and the amalgamation is over ρ, φ extends ρ. Let $\varphi = \rho^\frown z$, let $w_i = \Sigma_i(z)$, and let $\psi_i = \rho^\frown w_i$. Player II's strategy Σ will be to arrange that τ is a II-safe amalgamation over ρ of $\langle \psi_i : i < \omega \rangle$.

We first show that there is a strategy Σ such that plays (x, y) according to Σ have the property that there are z and w_i as above. Let $g_u^i \colon I_u^\rho \to I_u^\rho$ for $i \in \omega$ be nondecreasing functions whose ranges partition I_u^ρ, for each $u \in \mathrm{dom}(\rho^*)$. Let $\langle \eta_i : i \in \omega \rangle$ be the II-safe amalgamating functions these g_u^i determine (as in Claim 1.2). It is easy to see that there is a Lipschitz function which, given x, will produce y, z, and w_i so that the conditions above hold. Note first that $z \restriction (k+1)$, and hence $w_i \restriction (k+1)$, are determined by $x \restriction k$ and $y \restriction k$. For to determine $z(k)$, let $n(v) = \mathrm{dom}(\rho) + k$. If $v = \pi_1^*(r)$, then set $z(k) = x(n(r) - \mathrm{dom}(\rho))$, noting that $n(r) \le n(v)$, so that $n(r) - \mathrm{dom}(\rho) \le k$. If $v = \pi_2^*(r)$, then $n(r) < n(v)$ because of our additional requirement guaranteeing the I-safety of π_2. Thus it makes sense to set $z(k) = y(n(r) - \mathrm{dom}(\rho))$. It is enough, then, to determine $y(k)$ from $x \restriction (k+1)$ together with $z \restriction (k+1)$ and $w_i \restriction (k+1)$. Let $n(v) = \mathrm{dom}(\rho) + k$, and let $v = \eta_i^*(r)$, and note that $n(v) \le n(r)$. We then set $y(k) = w_i(n(r) - \mathrm{dom}(\rho))$.

We now verify that Σ wins $\mathrm{G}_\rho(\bigcap_{i<\omega} X_i)$ for player II. Let (x, y) be a play by Σ, with z and w_i for $i \in \omega$ associated as above. Let $\sigma = \rho^\frown x$, $\tau = \rho^\frown y$, $\varphi = \rho^\frown z$, and $\psi_i = \rho^\frown w_i$ for all i.

Suppose first that $\tau \notin A^*$, and let (p, f) be the leftmost witness to this fact. Let $p = \pi_i^*(q)$; then (q, f) is a witness that $\psi_i \notin A^*$, and $(q, f) \le_{\mathrm{lex}} (p, f)$ because π_i is II-safe. Now (z, w_i) is a play by Σ_i, which is winning for player II, and hence we have a witness (r, g) that $\varphi \notin A^*$ such that $(r \restriction k_1, g \restriction k_1) \le_{\mathrm{lex}} (q \restriction k_2, f \restriction k_2)$, where k_1, k_2 are least such that $r \restriction k_1 \notin \mathrm{dom}(\rho^*)$ and $q \restriction k_2 \notin \mathrm{dom}(\rho^*)$. We cannot have $r \in \mathrm{ran}(\pi_2^*)$, for if $r = \pi_2^*(s)$, then because π_2 is I-safe, we have $s \restriction k_1 <_{\mathrm{lex}} r \restriction k_1$, so (s, g) is a witness that $\tau \notin A^*$, and $(s \restriction k_1, g \restriction k_1) <_{\mathrm{lex}} (q \restriction k_2, f \restriction k_2)$, so $(s, g) <_{\mathrm{lex}} (q, f) \le_{\mathrm{lex}} (p, f)$, contrary to our choice of (p, f). Thus $r = \pi_1^*(s)$ for some s. But then (s, g) is a witness that $\sigma \notin A^*$, and

$$(s\restriction k_1, g\restriction k_1) \leq_{\text{lex}} (r\restriction k_1, g\restriction k_1) \leq_{\text{lex}} (p\restriction k_2, f\restriction k_2),$$

and since all embeddings above are over ρ, we have k_1 is least such that $s\restriction k_1 \notin \text{dom}(\rho^*)$ and k_2 is least such that $p\restriction k_2 \notin \text{dom}(\rho^*)$. It follows that (x, y) is a win for player II in $\text{G}_\rho(\bigcap X_i)$ under payoff condition (2).

Next, suppose $\tau \in A^*$. We may assume that $\sigma \in A^*$, as otherwise we are done in virtue of payoff condition (1). But then $\varphi \in A^*$, as it is an amalgamation of σ and τ, and $\psi_i \in A^*$ for all i, as τ is an amalgamation of $\langle \psi_i : i \in \omega \rangle$. Further, for $u \in \text{dom}(\rho^*)$ we have

$$\vartheta_u(\sigma) \leq \vartheta_u(\varphi) \leq \vartheta_u(\psi_i) \leq \vartheta_u(\tau)$$

for all i. (Here the first and last inequalities come from the properties of amalgamations, and the middle one holds because Σ_i wins for player II.) This shows player II has met his norm-inequality obligations. We also have, for $u \in \text{dom}(\rho^*)$,

$$\vartheta_u(\tau) \leq \vartheta_u(\varphi \leq \vartheta_u(\psi_i) \leq \vartheta_u(\tau),$$

for similar reasons. Thus

$$\langle \vartheta_{u_n}(\tau) : n \in \text{dom}(\rho) \rangle = \langle \vartheta_{u_n}(\psi_i) : n \in \text{dom}(\rho) \rangle \in X_i,$$

for all i. This implies that (x, y) is a win for player II in $\text{G}_\rho(\bigcap_{i<\omega} X_i)$ under payoff condition (3). ⊣ (Claim 1.3)

So μ_ρ is a countably complete ultrafilter concentrating on T_ρ. We now verify compatibility.

CLAIM 1.4. If $\rho, \varphi \in {}^{<\omega}\omega$, $\rho \subseteq \varphi$, $\mu_\rho(X) = 1$, and $Y = \{w : w\restriction \text{dom}(\rho) \in X\}$, then $\mu_\varphi(Y) = 1$.

PROOF. Fix a winning strategy Σ for player II in $\text{G}_\rho(X)$. Our strategy for player II in $\text{G}_\varphi(Y)$ is roughly the following. If player I plays σ extending φ, and ψ is the extension of ρ which Σ would play when σ is regarded as player I's play in $\text{G}_\rho(X)$, then player II will play a safe amalgamation τ of ψ and the parts of σ below u's in $\text{dom}(\varphi^*) \setminus \text{dom}(\rho^*)$.

More precisely, for $u \in \text{dom}(\rho^*)$, let

$$h_u : I_u^\rho \leftrightarrow I_u^\varphi$$

be an order-preserving bijection, where $I_u^\rho = \{i : u^\frown i \notin \text{dom}(\rho^*)\}$, and I_u^φ is defined similarly. Note that $i \leq h_u(i)$ for all i, as $I_u^\varphi \subseteq I_u^\rho$. We define a II-safe embedding over ρ by:

$$\pi^*(u^\frown\langle i\rangle^\frown v) = u^\frown\langle h_u(i)\rangle^\frown v$$

whenever $u \in \text{dom}(\rho^*)$ and $i \in I_u^\rho$, and of course $\pi^*(u) = u$ for all $u \in \text{dom}(\rho^*)$. We can now define τ from σ and ψ by letting

$$\tau^*(\pi^*(u)) = \psi^*(u),$$

and

$$\tau^*(v) = \sigma^*(v) \text{ for } v \notin \operatorname{ran}(\pi^*).$$

We leave it to the reader to check that $\tau(n)$ is determined by $\sigma\restriction n$ (via $\psi\restriction n$), for all n, so that we have legitimately described a strategy for player II in $G_\varphi(Y)$. (Notice here that τ does extend φ, because $\operatorname{ran}(\pi^*)$ is disjoint from $\operatorname{dom}(\varphi^*) \setminus \operatorname{dom}(\rho^*)$.) In order to see that this strategy wins for player II, suppose σ, τ, and ψ are as above.

Suppose first $\tau \notin A^*$, and let (x, f) be the leftmost witness to this fact. If $x \notin \operatorname{ran}(\pi^*)$, then (x, f) is a witness that $\sigma \notin A^*$, so player II has won under payoff condition (2). Now suppose $x = \pi^*(y)$, and let u be the longest initial segment of x, or equivalently y, such that $u \in \operatorname{dom}(\rho^*)$. Since (y, f) is a witness that $\psi \notin A^*$, and (σ, ψ) is a play by Σ, we have a leftmost witness (z, g) that $\sigma \notin A^*$ such that

$$(z\restriction(j+1), g\restriction(j+1)) \leq_{\text{lex}} (y\restriction(k+1), f\restriction(k+1)),$$

where j is largest such that $z\restriction j \in \operatorname{dom}(\rho^*)$ and k is largest such that $y\restriction k \in \operatorname{dom}(\rho^*)$. Note that $k = \operatorname{dom}(u)$ since $x\restriction(k+1) \notin \operatorname{dom}(\rho^*)$ and $\pi^*(y) = x$. If the inequality just displayed is strict, then we have by the II-safety of π^*

$$(z\restriction\ell, g\restriction\ell) <_{\text{lex}} (y\restriction i, f\restriction i) \leq_{\text{lex}} (x\restriction i, f\restriction i),$$

where $\ell \geq j+1$ is least such that $z\restriction\ell \notin \operatorname{dom}(\varphi^*)$, and $i \geq k+1$ is least such that $x\restriction i \notin \operatorname{dom}(\varphi^*)$. Thus player II has won $G_\varphi(Y)$ under clause (2). So suppose that

$$(z\restriction(j+1), g\restriction(j+1)) = (y\restriction(k+1), f\restriction(k+1)).$$

If $I_u^\rho = I_u^\varphi$, then we see from the definition of π^* that $x = y$, and also $j = k$ is largest so that $x\restriction k = z\restriction k$ is in $\operatorname{dom}(\varphi^*)$, so that again player II wins $G_\varphi(Y)$ under clause (2). Finally, if $I_u^\rho \neq I_u^\varphi$, then $y(k) < h_u(y(k)) = x(k)$, so $y\restriction(k+1) <_{\text{lex}} x\restriction(k+1)$, which leads to

$$(z\restriction\ell, g\restriction\ell) <_{\text{lex}} (x\restriction i, f\restriction i)$$

for the φ-critical ℓ and i, both of which are $\geq k+1$. So again, player II wins $G_\varphi(Y)$ under clause (2).

Next, suppose $\tau \in A^*$. If $\sigma \notin A^*$, we are done, so assume $\sigma \in A^*$. Thus (σ, ψ) is a win for player II in $G_\rho(X)$ under clause (3). We shall show that $\vartheta_u(\tau) = \vartheta_u(\psi)$ for all $u \in \operatorname{dom}(\rho^*)$ so that (σ, τ) is a win for player II in $G_\varphi(Y)$ under clause (3). We proceed by reverse induction on $\operatorname{dom}(u)$. Suppose that $\vartheta_v(\tau) = \vartheta_v(\psi)$ whenever $v \in \operatorname{dom}(\rho^*)$ and v is a proper extension of u. To see that $\vartheta_u(\psi) \leq \vartheta_u(\tau)$, it is enough to show that whenever v is any one-term extension of u, then there are one-term extensions r of u such that $\vartheta_v(\psi) \leq \vartheta_r(\tau)$. We can take $r = v$ if $v \in \operatorname{dom}(\rho^*)$, by induction. Otherwise, we take $r = \pi^*(v)$, which works because π^* embeds ψ into τ. To see that $\vartheta_u(\tau) \leq \vartheta_u(\psi)$, again let v be any one-term extension of u; we must find a one-term extension r of u such that $\vartheta_v(\tau) \leq \vartheta_r(\psi)$. Again, $r = v$ works if $v \in \mathrm{dom}(\rho^*)$. If $v \notin \mathrm{dom}(\varphi^*)$, then $r = (\pi^*)^{-1}(v)$ works, because π^* is an "isomorphism" between ψ^* below r and τ^* below v. Finally, if $v \in \mathrm{dom}(\varphi^*) \setminus \mathrm{dom}(\rho^*)$, then τ^* agrees with σ^* below v. But $\vartheta_u(\sigma) \leq \vartheta_u(\psi)$ because Σ is winning, so there is a one-term extension r of u such that $\vartheta_v(\tau) = \vartheta_v(\sigma) \leq \vartheta_r(\psi)$.

This completes the proof of Claim 1.4. ⊣ (Claim 1.4)

Finally, we must show the countable completeness of towers associated to $\sigma \in A^*$.

Claim 1.5. Let $\sigma \in A^*$, and suppose $\mu_{\sigma \restriction n}(X_n) = 1$ for all n; then there is an infinite sequence f such that $f \restriction n \in X_n$ for all n.

Proof. By Claim 1.4, we may as well assume that X_{n+1} projects into X_n, for all n. Let Σ_n be a winning strategy for player II in $\mathrm{G}_{\sigma \restriction n}(X_n)$. By playing all the Σ_n together in the usual way, we obtain τ_n for $n < \omega$ such that for each n,

$$\sigma \restriction n \subseteq \tau_n$$

and

$$(\tau_{n+1}, \tau_n) \text{ arises from a play by } \Sigma_n$$

in which player I is responsible for τ_{n+1} and player II is responsible for τ_n.[5]

We claim that $\tau_n \in A^*$ for all n. For if $\tau_n \notin A^*$, then since the Σ's win for player II, $\tau_k \notin A^*$ for all $k \geq n$, by induction on k. Letting (x_k, f_k) be the leftmost witness that $\tau_k \notin A^*$, payoff condition (2) implies that (x_k, f_k) converges to some (x, f). It is clear that (x, f) is a witness that $\sigma \notin A^*$, a contradiction.

Thus (τ_{n+1}, τ_n) falls under player II's payoff condition (3), for all n. But this implies that for each $u \in {}^{<\omega}\omega$, $\vartheta_u(\tau_k) \geq \vartheta_u(\tau_{k+1})$ for all $k > n(u)$, so that we can let

$$f(n) = \text{ eventual value of } \vartheta_{u_n}(\tau_k)$$

as $k \to \omega$. It is easy to see that $f \restriction n \in X_n$ for all n. ⊣ (Claim 1.5)

These claims complete the proof of Theorem 1.1. ⊣ (Theorem 1.1)

We conclude with two further observations.

First, the tree T constructed in the proof of Theorem 1.1 has the generic-codes property isolated by Kechris and Woodin [KW07].

Let T be the tree of a scale $\vec{\vartheta}$, and let $(r, v) \in T$. Consider the Banach-Mazur game $\mathrm{G}^{**}(r, v)$ on T in which at round n, player I plays $(r_{2n}, v_{2n}) \in T$, then player II plays $(r_{2n+1}, v_{2n+1}) \in T$, and we require $(r, v) \subseteq (r_0, v_0)$ and $(r_n, v_n) \subseteq (r_{n+1}, v_{n+1})$ for all n. Letting $x = \bigcup_n r_n$ and $f = \bigcup_n v_n$, we say player II wins $\mathrm{G}_{**}(r, v)$ iff for all $k \in \mathrm{dom}(r)$, $\vartheta_{u_k}(x) = v(k)$. (By the lower

[5]Formally, $\tau_n = (\sigma \restriction n)^\frown x_n$, where $x_n(k) = \Sigma_n(\langle \sigma(n) \rangle^\frown (x_{n+1} \restriction k))$.

semi-continuity of the scale $\vec{\vartheta}$, $\vartheta_{u_k}(x) \leq v(k)$.) Kechris and Woodin say that T has the generic codes property if for all $(r, v) \in T$, player II has a winning strategy in $G^{**}(r, v)$. For any Suslin set A, they construct a T with the generic codes property projecting to A.

It is quite easy to show that the T and $\vec{\vartheta}$ constructed in the proof of 1.1 have the generic codes property. To win $G^{**}(r, v)$, player II simply picks at the outset some $\sigma \in A^*$ such that $r \subseteq \sigma$ and $\vartheta_{u_k}(\sigma) = v(k)$ for all $k \in \mathrm{dom}(r)$. He then plays so that if (τ, f) is the output of the game, then σ is embedded into τ. Since player II can only make finite extensions, he can never prevent player II from further extending the embedding he is building.

Our second observation is that in certain natural situations, the homogeneity measures we have constructed are more than countably complete.

THEOREM 1.6. Let Γ be an ω-parametrized pointclass which is closed under recursive substitution, number quantification, and universal real quantification, and suppose that every set in Γ is Suslin and co-Suslin. Let $\underset{\sim}{\delta}$ be the supremum of the order types of prewellorders in $\Gamma \cap \check{\Gamma}$; then every set in Γ is $\underset{\sim}{\delta}$-homogeneously Suslin.

PROOF. We illustrate the proof in the special case $\Gamma = \Pi^1_3$. Let $\delta = \underset{\sim}{\delta}^1_3$. We shall define a complete Π^1_3 set A^* such that the homogeneity measures we get from the proof of 1.1 are δ-complete. In this connection, it will help if we are more careful in our choice of the ingredients A (which will be Σ^1_2), the scale $\vec{\varphi}$ on A, and the Suslin representation S of $\mathbb{R} \setminus A$.

Let $U \subseteq \omega \times \mathbb{R}$ be a universal Σ^1_2 set. Fix a map

$$H : \omega \twoheadrightarrow \mathrm{HF},$$

where HF is the set of all hereditarily finite sets. For $x \in {}^\omega\omega$, put $x \in \bar{A}$ iff

(a) for all n,

$$H(x(n)) = \langle r_n, v_n \rangle \in ({}^{<\omega}\omega \times {}^{<\omega}\omega),$$

where $\mathrm{dom}(r_n) = \mathrm{dom}(v_n) > 0$,

(b) $r_n \subseteq r_{n+1}$, and $r_n \neq r_{n+1}$, and

(c) letting e^x_i be the common value of $r_n(i)$ for all $i > n$, and defining $(x)_i \in {}^\omega\omega$ by

$$(x)_i(n) = v_{p+n}(i), \text{ where } p = \min\{k : i \in \mathrm{dom}(v_k)\},$$

we have

$$\exists i U(e^x_i, (x)_i).$$

Note that the set of $x \in {}^\omega\omega$ satisfying (a) and (b) in the definition of A is closed; let W_0 be the tree of this closed set. Put $x \in A$ iff $x \in \bar{A}$, or $x \notin [W_0]$.

Let $\langle \varphi_i^* : i < \omega \rangle$ be a very good Σ_2^1-scale on U. We get a Σ_2^1-scale on A in a routine way.[6] Let $x \in A$. If $x \notin [W_0]$, then set

$$\varphi_n(x) = \langle 0, k \rangle, \text{ where } k \text{ is least s.t. } x \restriction k \notin W_0.$$

Otherwise, $x \in \bar{A}$, and we have $U(e_i^x, (x)_i)$ for some i. Let i_0 be such that $\varphi_0^*(e_{i_0}^x, (x)_{i_0}) = \alpha$ is as small as possible, and i_0 is the least i such that $\varphi_0^*(e_i^x, (x)_i) = \alpha$, and set

$$\varphi_n(x) = \langle 1, \varphi_0^*(i_0, (x)_{i_0}), i_0, \varphi_1^*(i_0, (x)_{i_0}), \ldots, \varphi_n^*(i_0, (x)_{i_0}) \rangle$$

for all n. Here we use the lexicographic order to identify the range of φ_n with an ordinal. It is clear that $\vec{\varphi}$ is a Σ_2^1-scale on A. For a technical reason, we have not tried to make $\vec{\varphi}$ very good; it could happen that $\varphi_n(x_i)$ is eventually constant for each n, but the x_i do not converge. However, it *is* true that $n \leq m$ and $\varphi_n(x) < \varphi_n(y)$ imply $\varphi_m(x) < \varphi_m(y)$. This refinement property is enough to guarantee that the Moschovakis norms $\vec{\vartheta}$ on A^*, defined as in the proof of Theorem 1.1, constitute a scale on A^*. It is easy to check that if $\sigma \in A^*$ and $\sigma^*(u) \notin W_0$, then since $\varphi_{\mathrm{dom}(u)}$ is constant on the neighborhood determined by $\sigma^*(u)$, $\vartheta_u(\sigma^*) = 0$. This property, which might have been lost if we had tried to make $\vec{\varphi}$ very good, will be of use later.

Let S_0 be a tree on $\omega \times \kappa$ such that $\mathrm{p}[S_0] = (\omega \times \mathbb{R}) \setminus U$. We construct a tree S projecting to $\mathbb{R} \setminus A$. For $s \in W_0$, let $\langle r_n^s, v_n^s \rangle = H(s(n))$, and letting n be largest in $\mathrm{dom}(s)$, set

$$k(s) = \mathrm{dom}(r_n^s) \text{ and } e_i^s = r_n^s(i) \text{ for all } i < k(s),$$

and

$$(s)_i(j) = v_{p+j}^s(i), \text{ where } p = \text{ least } \ell \text{ s.t. } i \in \mathrm{dom}(v_\ell^s).$$

(The domain of $(s)_i$ is $(n - p) + 1$.) That is, e_i^s and $(s)_i$, for $i < k(s)$, are the common initial segments of all e_i^x and $(x)_i$ for $s \subseteq x$ determined by s. Now fix a bijection

$$c : \lambda \to {}^{<\omega}\kappa,$$

for some λ, with the property that $\mathrm{dom}(s) < \mathrm{dom}(t) \Rightarrow c^{-1}(s) < c^{-1}(t)$. For $s \in {}^{<\omega}\omega$ and $t \in {}^{<\omega}\lambda$, put (s, t) in S iff

(i) $s \in W_0$, $\mathrm{dom}(s) = \mathrm{dom}(t)$, and $\forall n \in \mathrm{dom}(s)$, $\mathrm{dom}(c(t(n))) = \mathrm{dom}(v_n^s)$, and

(ii) letting

$$(t)_i(j) = c(t(p + j))(i), \text{ where } p = \text{ least } \ell \text{ s.t. } i \in \mathrm{dom}(v_\ell^s),$$

(where j is such that $p + j < \mathrm{dom}(s)$, we have

$$(e_i^s, (s)_i, (t)_i) \in S_0$$

for all $i < k(s)$.

[6]The definability of the scale $\vec{\varphi}$, and of the Moschovakis scale $\vec{\vartheta}$ on A^* it induces, are actually not relevant to Theorems 1.1 or 1.6.

That is, S is the tree of attempts to build an (x, f) such that x satisfies (a) and (b) of the definition of $\bar{A}$, and for each $i < \omega$, $(f)_i$ (defined as above) witnesses that $\neg U(e_i^x, (x)_i)$.

Having defined the ingredients A, $\vec{\varphi}$, and S in the proof of Theorem 1.1, we let $\vec{\vartheta}$ be the Moschovakis scale on A^*, T be the tree of $\vec{\vartheta}$, and $\langle \mu_p : p \in {}^{<\omega}\omega \rangle$ be the homogeneity system for T defined there. We also carry over all the notation regarding Lipschitz maps from the proof of 1.1. Fix $\rho \in {}^{<\omega}\omega$, and suppose that $\mu_\rho(X_\alpha) = 1$ for all $\alpha < \gamma_0$, where $\gamma_0 < \delta$. We must show that $\mu_\rho(\bigcap_{\alpha<\gamma_0} X_\alpha) = 1$, that is, we must describe a winning strategy for player II in $\mathrm{G}_\rho(\bigcap_{\alpha<\gamma_0} X_\alpha)$. This strategy will be, roughly, to amalgamate plays according to strategies for player II in the games $\mathrm{G}_\rho(X_\alpha)$, as was done in the proof that μ_ρ is countably complete. The new problem, that there are uncountably many such strategies, makes the amalgamation somewhat more delicate, but we can use a rather direct transcription of an argument of Kunen.[7] That argument rests on the Coding Lemma.

Strategies for player II in games of the form $\mathrm{G}_\rho(Y)$ are Lipschitz functions on ${}^\omega\omega$, and so can be coded by reals $\Sigma \in {}^\omega\omega$ as we have done above. By the Coding Lemma, there is a Σ_3^1 set B such that

$$\forall \Sigma (B(\Sigma) \Rightarrow \exists \alpha < \gamma_0 (\Sigma \text{ is a winning strategy in } \mathrm{G}_\rho(X_\alpha))),$$

and

$$\forall \alpha < \gamma_0 \exists \Sigma (B(\Sigma) \wedge \Sigma \text{ is a winning strategy in } \mathrm{G}_\rho(X_\alpha)).$$

Since B is $\underset{\sim}{\Sigma}_3^1$, we can fix a real z_0 and an $a \in \omega$ such that for all Σ

$$B(\Sigma) \Leftrightarrow \exists y \neg U(a, [\Sigma, y, z_0]).$$

For use below, we spell out our coding of tuples here: if $s_0, \ldots, s_{k-1} \in {}^{<\omega}\omega$ are sequences of the same length n, then $[s_0, \ldots, s_{k-1}]$ is the sequence of length n such that for $i < n$, $[s_0, \ldots, s_{k-1}](i) = n(\langle s_0(i), \ldots, s_{k-1}(i) \rangle)$.) If $x_0, \ldots, x_{k-1} \in {}^\omega\omega$, then $[x_0, \ldots, x_{k-1}] = \bigcup_n [x_0 \restriction n, \ldots, x_{k-1} \restriction n]$. Finally, if $t = [s_0, \ldots, s_{k-1}]$, then we call t a k-code, and write $(t)^i = s_i$. (The more common $(t)_i$ notation has already been dedicated to something slightly different.)

Let π_1, π_2 be the II-safe embeddings over ρ defined in the proof of Claim 1.2 (induced by picking for each $u \in \mathrm{dom}(\rho^*)$ increasing maps h_u^1, h_u^2 from I_u^ρ to itself, whose ranges partition I_u^ρ). If $\sigma, \tau \in {}^\omega\omega$ extend ρ, then we write $\sigma \oplus \tau$ for their amalgamation via (π_1, π_2). Our strategy for player II in $\mathrm{G}_\rho(\bigcap_\alpha X_\alpha)$ will behave as follows: as player I produces σ extending ρ, we produce a τ extending ρ whose membership in A^* would code the truth of the statement

$$\forall \Sigma (B(\Sigma) \Rightarrow \Sigma(\sigma \oplus \tau) \in A^*).$$

[7]From his proof that all $\underset{\sim}{\delta}_n^1$ are measurable; see [Kec78A].

Notice that this statement is Π^1_3, and A^* is complete Π^1_3, so it is reasonable to try to construct such a τ. Of course, the statement coded by τ is self-referential. It is possible to bring in the recursion theorem at this point, but instead we shall simply give a direct construction of the desired τ.

More precisely, for each $u \in \mathrm{dom}(\rho^*)$, we want the restriction of τ^* to sequences extending u to code the truth of the statement

$$\forall\Sigma\forall y\forall x \supseteq u(U(a, [\Sigma, y, z_0]) \vee \Sigma(\sigma \oplus \tau)(x) \in A).$$

This is achieved by having τ^* regard the s in $u^\frown s$ as a 3-code for an initial segment of a possible Σ, y, and x, then compute an appropriate fragment of $\Sigma(\sigma \oplus \tau)(x)$, then give an output whose membership in A codes up the disjunction in the matrix of the formula just displayed. To make this work, we have to let $(s)^0$ give information about Σ quickly enough, so we make the convention: for any $s \in {}^{<\omega}\omega$, $\widetilde{s} = u_{s(0)}{}^\frown \dots {}^\frown u_{s(\mathrm{dom}(s)-1)}$, and for $x \in {}^{\omega}\omega$, $\widetilde{x} = \bigcup_n \widetilde{x\restriction n}$. It will be $\widetilde{s^0}$ which is considered by τ^* as a possible initial segment of some Σ in B.

We turn to the details of the construction of τ from σ. Suppose we are given $\tau\restriction n$ and $\sigma\restriction(n+1)$; we must define $\tau(n)$. We may assume $\mathrm{dom}(\rho) \leq n$, as otherwise $\tau(n) = \rho(n)$. Our amalgamation process yields $(\sigma \oplus \tau)\restriction(n+1)$. Letting $n = n(v)$, our job is to define $\tau^*(v)$. Let $\bar{v} = v\restriction(\mathrm{dom}(v) - 1)$. If $\tau^*(\bar{v}) \notin W_0$, then we just let $\tau^*(v)$ be any one-term extension of $\tau^*(\bar{v})$. Assume now that $\tau^*(\bar{v}) \in W_0$. Let us write

$$v = u^\frown w,$$

where u is the longest initial segment of v in $\mathrm{dom}(\rho^*)$. If w is not a 3-code, then again we just let $\tau^*(v)$ be any one-term extension of $\tau^*(\bar{v})$ which is not in W_0; note we can always leave W_0 in one step. So assume w is a 3-code, and let

$$\Gamma = \widetilde{(w)^0}.$$

This is the approximation to Σ determined by w, the approximations to y and x being $(w)^1$ and $u^\frown(w)^2$ respectively . If $(\sigma \oplus \tau)\restriction(n+1) \notin \mathrm{dom}(\Gamma^*)$ or $\rho \not\subseteq \Gamma^*((\sigma\oplus\tau)\restriction(n+1))$, then once again we let $\tau^*(v)$ be any one-term extension of $\tau^*(\bar{v})$ which is not in W_0. Assume then that $(\sigma \oplus \tau)\restriction(n+1) \in \mathrm{dom}(\Gamma^*)$, and that $\Gamma^*((\sigma\oplus\tau)\restriction(n+1))$ extends ρ. Clearly, $n(u^\frown(w)^2) \leq n(v)$, so we can set

$$s = (\Gamma^*((\sigma \oplus \tau)\restriction(n+1)))^*(u^\frown(w)^2),$$

and we have $\rho^*(u) \subset s$ and $\mathrm{dom}(s) = \mathrm{dom}(v)$. If $s \notin W_0$, then again we can let $\tau^*(v)$ be any one-term extension of $\tau^*(\bar{v})$ which is not in W_0. We are finally in what we shall call the **interesting case** in the definition of $\tau^*(v)$.

Let

$$k = \mathrm{dom}(u), \ell = \mathrm{dom}(\bar{v}), \text{ and } m = \mathrm{dom}(r^s_{k-1}).$$

We must define $\tau^*(v)(\ell)$, the earlier entries having been given by $\tau^*(\bar{v})$. We do this by setting

$$H(\tau^*(v)(\ell)) = \langle r, v \rangle,$$

where

$$r = \langle r^s_\ell(0), \dots, r^s_\ell(m-1), a, r^s_\ell(m), \dots, r^s_\ell(\mathrm{dom}(r^s_\ell - 1)) \rangle,$$

and

$$v = \langle v^s_\ell(0), \dots, v^s_\ell(m-1), [\Gamma^*, (w)^1, z_0](\ell - k), v^s_\ell(m), \dots, v^s_\ell(\mathrm{dom}(v^s_\ell - 1)) \rangle.$$

To be pedantic, we should have written $[\Gamma^* \restriction (\ell+1-k), (w)^1 \restriction (\ell+1-k), z_0 \restriction (\ell + 1 - k)]$ above. We leave it to the reader to check that $\tau^*(v) \in W_0$. This completes the definition of $\tau^*(v)$, and thereby our description of player II's putative winning strategy in $\mathrm{G}_p(\bigcap_\alpha X_\alpha)$.

Let (σ, τ) be a run of $\mathrm{G}_p(\bigcap_\alpha X_\alpha)$ according to this strategy. We must see player II wins.

Suppose first that $\tau \notin A^*$, and let (p, f) be the leftmost witness to this fact. Note that $\tau^*(p \restriction i)$ is defined using the interesting case, for all i. Let k be largest such that $p \restriction k \in \mathrm{dom}(\rho^*)$, and

$$p = u^\frown z,$$

where $u = p \restriction k$. Then

$$z = [\Sigma, y, \bar{x}]$$

must be a 3-code. Inspecting the tree S and our definition of $\tau^*(p)$, we see that

$$(a, [\Sigma, y, z_0], (f)_m) \in [S_0], \text{ where } m = \mathrm{dom}(r^{\tau^*(p)}_{k-1}),$$

which implies that $B(\Sigma)$ holds. Let $g \in {}^\omega\lambda$ be given as follows: if

$$c(f(j)) = \langle \alpha_0, \dots, \alpha_n \rangle,$$

then $g(j) = f(j)$ if $n < m$, and otherwise

$$c(g(j)) = \langle \alpha_0, \dots, \alpha_{m-1}, \alpha_{m+1}, \dots, \alpha_n \rangle.$$

Letting $x = u^\frown \bar{x}$ it is easy to see that our construction guarantees that (x, g) is a witness that $\Sigma^*(\sigma \oplus \tau) \notin A^*$, and moreover, $(x, g) \leq_{\mathrm{lex}} (p, f)$. (Note here that $g(j) = f(j)$ for $j < k$, and $g(j) < f(j)$ for $j \geq k$ because $c(g(j))$ is a sequence of length one less than that of $c(f(j))$.) The remainder of the argument goes as in the countable additivity proof: since Σ is winning, there is a witness (q, h) that $\sigma \oplus \tau \notin A^*$ such that $(q \restriction k_1, h \restriction k_1) \leq_{\mathrm{lex}} (x \restriction k_2, g \restriction k_2)$ for the appropriate k_1, k_2. The nature of the amalgamation and the fact that (p, f) is leftmost implies that (q, h) is a witness that $\sigma \notin A^*$, and that therefore (σ, τ) is a win for player II under payoff clause (2).

Next, suppose $\tau \in A^*$. If $\sigma \notin A^*$, we are done, so suppose $\sigma \in A^*$. Notice that if $u \in \text{dom}(\rho^*)$ and $B(\Sigma)$, then

$$\sup(\vartheta_u(\sigma), \vartheta_u(\tau)) \leq \vartheta_u(\sigma \oplus \tau) \leq \vartheta_u(\Sigma^*(\sigma \oplus \tau)),$$

because Σ is winning for player II.

Claim. For $u \in \text{dom}(\rho^*)$ and Σ such that $B(\Sigma)$, $\vartheta_u(\Sigma^*(\sigma \oplus \tau)) \leq \vartheta_u(\tau)$.

Proof of Claim. We describe a winning strategy Γ for player II in the Moschovakis comparison game $\text{G}(u, \Sigma^*(\sigma \oplus \tau), \tau)$. Fix a witness y to $B(\Sigma)$, so that

$$\neg U(a, [\Sigma, y, z_0]).$$

Suppose that player I plays out x extending u in this game (his first move being $x(\text{dom}(u))$). Let $\hat{u}$ be the longest initial segment of x which is in $\text{dom}(\rho^*)$, and

$$x = \hat{u}^\frown \bar{x}.$$

Then player II will respond with

$$\Gamma(x) = \hat{u}^\frown[w, y, z_0], \text{ where } \Sigma = \widetilde{w},$$

and w codes Σ sufficiently rapidly.[8] It is clear that this defines a strategy Γ for player II. To see that it wins, suppose player I has played x extending u. If $\Sigma^*(\sigma \oplus \tau)(x)$ is not a branch of W_0, then by our construction, neither is $\tau^*(\Gamma(x))$, and moreover both reals leave W_0 at the same time, so that $\varphi_n(\Sigma^*(\sigma \oplus \tau)(x)) = \varphi_n(\tau^*(\Gamma(x))) = \langle 0, k \rangle$ for all n, where k is this common departure time. Thus player II has won the Moschovakis comparison game. If $p = \Sigma^*(\sigma \oplus \tau)(x)$ is a branch of W_0, then by construction so is $q = \tau^*(\Gamma(x))$. By construction, setting $k = \text{dom}(\hat{u})$ and $m = \text{dom}(r^p_{k-1})$, we have

$$e^q_i = \begin{cases} e^p_i & \text{when } i < m \\ a & \text{when } i = m \\ e^p_{i-1} & \text{when } i > m \end{cases}$$

and

$$(q)_i = \begin{cases} (p)_i & \text{when } i < m \\ [\Sigma, y, z_0] & \text{when } i = m \\ (p)_{i-1} & \text{when } i > m. \end{cases}$$

Using these facts, and inspecting the definition of the φ_n, it is easy to see that $\varphi_n(p) \leq \varphi_n(q)$ for all n. The key point here is just that $U(a, (q)_m)$ is false, so it cannot lower the norm values. ⊣ (Claim)

[8]Namely, $\Sigma{\restriction}n(\sigma \oplus \tau{\restriction}(k+5)) \subseteq w{\restriction}(\widetilde{k - \text{dom}(\hat{u})})$ for all $k \geq \text{dom}(\hat{u})$.

We can now complete the proof of Theorem 1.6. We have that whenever $B(\Sigma)$ and $u \in \mathrm{dom}(\rho^*)$, then $\vartheta_u(\tau) = \vartheta_u(\Sigma^*(\sigma \oplus \tau))$. Fixing $\alpha < \gamma_0$ and Σ a winning strategy for player II in $G_\rho(X_\alpha)$ such that $B(\Sigma)$, this implies

$$\langle \vartheta_{u_0}(\tau), \dots, \vartheta_{u_k}(\tau) \rangle \in X_\alpha,$$

where $k = \mathrm{dom}(\rho) - 1$. Since this is true for all $\alpha < \gamma_0$, (σ, τ) is a win for player II under clause (3) in $\mathrm{G}_\rho(\bigcap_{\alpha<\gamma_0} X_\alpha)$, as desired. ⊣

REFERENCES

ALEXANDER S. KECHRIS

[Kec78A] AD *and projective ordinals*, In Kechris and Moschovakis [CABAL i], pp. 91–132.

[Kec81A] *Homogeneous trees and projective scales*, In Kechris et al. [CABAL ii], pp. 33–74.

ALEXANDER S. KECHRIS, DONALD A. MARTIN, AND YIANNIS N. MOSCHOVAKIS

[CABAL ii] *Cabal seminar* 77–79, Lecture Notes in Mathematics, no. 839, Berlin, Springer, 1981.

ALEXANDER S. KECHRIS AND YIANNIS N. MOSCHOVAKIS

[CABAL i] *Cabal seminar* 76–77, Lecture Notes in Mathematics, no. 689, Berlin, Springer, 1978.

ALEXANDER S. KECHRIS AND W. HUGH WOODIN

[KW07] *Generic codes for uncountable ordinals*, this volume, originally circulated manuscript, 2007.

DONALD A. MARTIN AND ROBERT M. SOLOVAY

[MS69] *A basis theorem for Σ^1_3 sets of reals*, ***Annals of Mathematics***, vol. 89 (1969), pp. 138–160.

YIANNIS N. MOSCHOVAKIS

[Mos80] ***Descriptive set theory***, Studies in Logic and the Foundations of Mathematics, no. 100, North-Holland, Amsterdam, 1980.

DEPARTMENT OF MATHEMATICS
UNIVERSITY OF CALIFORNIA
LOS ANGELES, CA 90095, USA
E-mail: dam@math.ucla.edu

DEPARTMENT OF MATHEMATICS
UNIVERSITY OF CALIFORNIA
BERKELEY, CA 94720, USA
E-mail: steel@math.berkeley.edu

WEAKLY HOMOGENEOUS TREES

DONALD A. MARTIN AND W. HUGH WOODIN

§1. Introduction and background. We work in ZF set theory, i.e., in ZFC without the axiom of Choice. We will give some sufficient conditions for the truth of "For all ordinals λ, every tree on $\omega \times \lambda$ is weakly homogeneous," and we will discuss the extent to which the projection of a weakly homogeneous tree must be the projection of a homogeneous tree.[1]

If λ is an ordinal number, then a **tree on** $\omega \times \lambda$ is a set T with the following properties.

(1) The members of T are pairs (s, t) with $s \in {}^{<\omega}\omega$, $t \in {}^{<\omega}\lambda$, and $\mathrm{lh}(s) = \mathrm{lh}(t)$.

(2) If $(s, t) \in T$ and $n < \mathrm{lh}(s)$, then $(s{\restriction}n, t{\restriction}n) \in T$.

In this paper, we shall mean by a **tree** a tree on $\omega \times \lambda$ for some ordinal number λ.

We shall treat the word "measure" as synonymous with "ultrafilter." We shall use "measure" almost exclusively for ultrafilters on sets ${}^n\lambda$ where $n < \omega$ and λ is a non-zero ordinal number. This includes the trivial ultrafilters on one-point sets ${}^0\lambda$.

By a **cardinal number**, we mean an initial ordinal, i.e., an ordinal number that cannot be injected into any smaller ordinal number.

If $n < \omega$, κ is an infinite cardinal number and λ is a non-zero ordinal number, let $\mathrm{MEAS}_n^{\kappa,\lambda}$ be the set of all κ-complete measures on ${}^n\lambda$.

If $m < n < \omega$ and $X \subseteq {}^m\lambda$, then define $\mathrm{ext}_n(X) = \{t \in {}^n\lambda : x{\restriction}m \in X\}$. If μ is a measure on ${}^n\lambda$, define

$$\mathrm{proj}_m(\mu) = \{X \subseteq {}^m\lambda : \mathrm{ext}_n(X) \in \mu\}.$$

It is easy to see that $\mathrm{proj}_m(\mu)$ is a measure on ${}^m\lambda$ and that it is as complete (i.e., as additive) as μ is.

For non-zero ordinal numbers λ, a **λ-tower of measures** is a sequence $\langle \mu_n : n < \omega \rangle$ such that

[1]Theorem 3.1 is due to Martin. Theorem 3.2 is a variation due to Woodin as are the results of the last section, except for Theorem 4.8 which is due to Steel. The proof given here of Theorem 3.1 is based on the proof of Theorem 3.2. Martin's original proof used partition cardinals. Theorem 3.1 and Theorem 3.2 were proved in the early 1980's.

The Cabal Seminar. Volume I: Games, Scales and Suslin Cardinals
Edited by A. S. Kechris, B. Löwe, J. R. Steel
Lecture Notes in Logic, 31

(i) for each n, $\mu_n \in \mathrm{MEAS}_n^{\aleph_1,\lambda}$;
(ii) for $m < n$, $\mu_m = \mathrm{proj}_m(\mu_n)$.

We say that a tower $\langle \mu_n : n < \omega \rangle$ is **countably complete** if, for every sequence $\langle X_n : n < \omega \rangle$ with $X_n \in \mu_n$ for all n, there is an $f : \omega \to \lambda$ such that $f(n) \in X_n$ for all n.

Let T be a tree on some $\omega \times \lambda$. Define

$$[T] = \{(x, f) : (\forall n < \omega)(x \restriction n, f \restriction n) \in T\};$$
$$\mathrm{p}[T] = \{x \in {}^{\omega}\omega : (\exists f)(x, f) \in [T]\};$$
$$T_s = \{t \in {}^{\mathrm{lh}(s)}\lambda : (s, t) \in T\} \text{ (for } s \in {}^{<\omega}\omega).$$

The tree T is **homogeneous** if there is a system $\langle \mu_s : s \in {}^{<\omega}\omega \rangle$ such that

(i) for each s, $\mu_s \in \mathrm{MEAS}_{\mathrm{lh}(s)}^{\aleph_1,\lambda}$ and $\mu_s(T_s) = 1$.
(ii) if $s_1 \subseteq s_2$ then $\mu_{s_1} = \mathrm{proj}_{\mathrm{lh}(s_1)}(\mu_{s_2})$.
(iii) for all $x \in \mathrm{p}[T]$ the λ-tower of measures $\langle \mu_{x \restriction n} : n < \omega \rangle$ is countably complete.

The tree T is **weakly homogeneous** if there is a system $\langle M_s : s \in {}^{<\omega}\omega \rangle$ such that

(i) for each s, M_s is a countable subset of $\mathrm{MEAS}_{\mathrm{lh}(s)}^{\aleph_1,\lambda}$ and T_s belongs to every member of M_s;
(ii) for all $x \in \mathrm{p}[T]$ there is a countably complete λ-tower of measures $\langle \mu_n : n < \omega \rangle$ such that each μ_n belongs to $M_{x \restriction n}$.

In this paper we show that every tree is weakly homogeneous if either of the following holds:

(1) $\mathrm{AD}_{\mathbb{R}}$, the Axiom of Determinacy for real games (i.e., for infinitely long games whose moves are real numbers);
(2) the universe comes from an inner model M of ZFC by taking the $M(\mathbb{R})$ of the Lévy collapse to ω of all ordinals less than some supercompact cardinal.

The proofs of these two theorems have a common core, to which §2 will be devoted. The two theorems will be deduced in §3. In §4 results related to the second theorem will be proved, including a weakening of its hypothesis. These results will be applied to the question of when a projection of a weakly homogeneous tree must be homogeneous.

If κ is a cardinal number and $\mathbb{A}$ is a set, then $\wp_\kappa(\mathbb{A})$ is the set of all subsets of $\mathbb{A}$ of size $< \kappa$.

Let κ and δ be cardinal numbers, let $\mathbb{A}$ be a nonempty set, and let $\mathcal{U}$ be an ultrafilter on $\wp_\kappa(\mathbb{A})$. $\mathcal{U}$ is:

(a) **fine** if, for every $x \in \mathbb{A}$, $\{a : x \in a\} \in \mathcal{U}$;
(b) **normal** if, for every $f : \wp_\kappa(\mathbb{A}) \to \wp(\mathbb{A})$, if $\{a : a \cap f(a) \neq \varnothing\} \in \mathcal{U}$ then there is an $x \in \mathbb{A}$ such that $\{a : x \in f(a)\} \in \mathcal{U}$.

It is easy to prove that $\mathcal{U}$ is normal just in case it is closed under diagonal intersections, i.e., just in case, for any $\langle Y_x : x \in \mathbb{A}\rangle$,

$$\text{if } Y_x \in \mathcal{U} \text{ for each } x \in \mathbb{A}, \text{ then } \{a : (\forall x \in a)\, a \in Y_x\} \in \mathcal{U}.$$

In the presence of the Axiom of Choice, it is usual to define $\mathcal{U}$ to be normal just in case, for every $f : \wp_\kappa(\mathbb{A}) \to \mathbb{A}$, if $\{a : f(a) \in a\} \in \mathcal{U}$, then f is constant on a set $\in \mathcal{U}$. Clearly $\mathcal{U}$ is normal in this sense if it is normal in our sense. Moreover, it is easy to prove that the converse holds if $\mathbb{A}$ can be wellordered.

If κ and $\lambda \geq \kappa$ are infinite cardinal numbers, then κ is **λ-supercompact** if there is a κ-complete, fine, normal ultrafilter on $\wp_\kappa(\lambda)$.

A different sort of example of a fine, normal ultrafilter exists under the hypothesis $\mathsf{AD}_\mathbb{R}$. We define a set $\mathcal{U}$ of subsets of $\wp_{\aleph_1}(\mathbb{R})$ as follows. For $X \subseteq \wp_{\aleph_1}(\mathbb{R})$, consider the real game G_X a play $\langle x_i : i < \omega\rangle$ of which is a win for player II just in case $\{x_i : i < \omega\} \in X$. We define

$$X \in \mathcal{U} \leftrightarrow \text{II has a winning strategy for } G_X.$$

Lemma 1.1 (Solovay, [Sol78B]). Assume $\mathsf{AD}_\mathbb{R}$. Then $\mathcal{U}$ is a countably complete, fine, normal ultrafilter on $\wp_{\aleph_1}(\mathbb{R})$.

Proof. Obviously $\wp_{\aleph_1}(\mathbb{R}) \in \mathcal{U}$, and obviously if $X \subseteq Y$ and $X \in \mathcal{U}$ then $Y \in \mathcal{U}$.

Suppose that strategies τ_1 and τ_2 witness $X_1 \in \mathcal{U}$ and $X_2 \in \mathcal{U}$ respectively. We get a strategy witnessing $X_1 \cap X_2 \in \mathcal{U}$ as follows. Let $E_1 = \{4n : n < \omega\}$ and let $E_2 = \{4n + 2 : n < \omega\}$. Let II's moves on E_i be gotten by playing τ_i delayed against an opponent whose moves are the x_j for $j < \omega \setminus E_i$.

An argument similar that just given for closure under intersections could be used to show directly that $\mathcal{U}$ is countably complete. But this is unnecessary, since AD implies—and so $\mathsf{AD}_\mathbb{R}$ implies—that every ultrafilter is countably complete. (See Proposition 28.1 of [Kan94].)

We thus can finish the proof that $\mathcal{U}$ is a countably complete ultrafilter by demonstrating that

$$(\forall X)(X \in \mathcal{U} \leftrightarrow (\wp_{\aleph_1}(\mathbb{R}) \setminus X) \notin \mathcal{U}).$$

If σ is a strategy for I witnessing that $(\wp_{\aleph_1}(\mathbb{R}) \setminus X) \notin \mathcal{U}$, then a strategy for II witnessing that $X \in \mathcal{U}$ is σ itself, delayed by a move. If τ is a strategy for II witnessing that $X \in \mathcal{U}$, then a strategy for I witnessing that $(\wp_{\aleph_1}(\mathbb{R}) \setminus X) \notin \mathcal{U}$ is gotten by choosing an arbitrary first move for I and then letting I get the x_{2i+1} by playing τ against an opponent who makes the moves $x_0, x_1, x_3, x_5, \ldots$.

Fineness is trivial, since player II can choose any given real as, for example, his first move.

For normality, let $f : \wp_{\aleph_1}(\mathbb{R}) \to \wp(\mathbb{R})$. Assume that

$$(\forall x \in \mathbb{R})\{a : x \notin f(a)\} \in \mathcal{U}.$$

Let
$$X = \{a : a \cap f(a) = \varnothing\}.$$
We must show that II has a winning strategy for G_X. Consider the following real game $\bar{G}$. A play $\langle x_i : i < \omega \rangle$ of $\bar{G}$ is a win for II just in case
$$x_0 \notin f(\{x_i : i < \omega\}).$$
Our assumption implies that, for any fixed x, II has a winning strategy for $\bar{G}$ when I is constrained to play $x_0 = x$. Thus I does not have a winning strategy in $\bar{G}$, and therefore II has a winning strategy τ. We can use τ to get a winning strategy for II in G_X. To do this, break the even natural numbers into infinite pieces E_i, $i < \omega$, in such a way that the least element of E_i is greater than i. Let II's moves on E_i be gotten by playing τ against an opponent who plays x_i followed by, in order, all the remaining x_j for $j < \omega \setminus E_i$. ⊣

Let Θ be the least non-zero ordinal number that is not a surjective image of $\mathbb{R}$. Clearly Θ is a cardinal number.

Let $0 < \lambda < \Theta$. By [Mos70A], AD implies that $\wp(\lambda)$ is a surjective image of $\mathbb{R}$, and hence that $\wp({}^n\lambda)$ is such a surjective image for each $n < \omega$. By [Kec88B], $\mathrm{AD}_{\mathbb{R}}$ implies that MEAS_n^λ is a surjective image of $\mathbb{R}$ for each $n < \omega$. It follows that $\mathrm{AD}_{\mathbb{R}}$ implies that $\bigcup_{n<\omega}(\wp({}^n\lambda) \cup \mathrm{MEAS}_n^\lambda)$ is a surjective image of $\mathbb{R}$.

Lemma 1.2. Assume $\mathrm{AD}_{\mathbb{R}}$ and let $0 < \lambda < \Theta$. There is a countably complete, fine, normal ultrafilter on
$$\wp_{\aleph_1}\Big(\bigcup_{n<\omega}(\wp({}^n\lambda) \cup \mathrm{MEAS}_n^\lambda)\Big).$$

Proof. Let $f : \mathbb{R} \to \bigcup_{n<\omega}(\wp({}^n\lambda) \cup \mathrm{MEAS}_n^\lambda)$ be a surjection. Let $\mathcal{U}$ be the ultrafilter given by Lemma 1.1. Define an ultrafilter $\mathcal{V}$ on $\wp_{\aleph_1}(\bigcup_{n<\omega}(\wp({}^n\lambda) \cup \mathrm{MEAS}_n^\lambda))$ by setting
$$X \in \mathcal{V} \leftrightarrow f^{-1}[X] \in \mathcal{U}.$$
It is easy to check that $\mathcal{V}$ inherits the properties of countable completeness, fineness, and normality from $\mathcal{U}$. ⊣

§2. The main lemmas. Throughout this section,

(A) let κ be an uncountable cardinal number;

(B) let λ be an ordinal number with $\kappa \leq \lambda$;

(C) let $\mathcal{V}$ be a κ-complete, fine, normal ultrafilter on $\wp_\kappa(\wp(\lambda) \cup \bigcup_{n<\omega} \mathrm{MEAS}_n^{\kappa,\lambda})$.

(D) Assume that the Axiom of Dependent Choice holds for relations on $\wp(\lambda)$.

(E) assume that $\bigcup_{n<\omega} \mathrm{MEAS}_n^{\kappa,\lambda}$ can be wellordered;

Remarks. (a) Under $\mathrm{AD}_{\mathbb{R}}$, Lemma 1.2 gives examples satisfying our assumptions with $\kappa = \aleph_1$. As long as $\lambda < \Theta$, Condition (D) follows from Uniformization, a consequence of $\mathrm{AD}_{\mathbb{R}}$. Condition (E) follows from (D) and Corollary 28.21 of [Kan94], a result of Kenneth Kunen.

(b) Under the Axiom of Choice, there are examples for all κ and λ such that $\kappa \leq \lambda$ and κ is 2^{2^λ}-supercompact.

We use "a" as a variable ranging over $\wp_\kappa(\bigcup_{n<\omega}(\wp({}^n\lambda) \cup \mathrm{MEAS}_n^{\kappa,\lambda}))$. When we say "for almost all a" or "for almost every a," we mean "for a set of a that belongs to $\mathcal{V}$." For $n < \omega$, let

$$\mathrm{meas}_n(a) = \mathrm{MEAS}_n^{\kappa,\lambda} \cap a.$$

For $n < \omega$ and $\mu \in \mathrm{MEAS}_n^{\kappa,\lambda}$, let

$$Y_\mu(a) = \bigcap_{X \in \mu \cap a} X.$$

Lemma 2.1. Let $n < \omega$. For almost all a,

(i) if μ_1 and μ_2 are distinct members of $\mathrm{meas}_n(a)$, then Y_{μ_1} and Y_{μ_2} are disjoint;

(ii) $\bigcup_{\mu \in \mathrm{meas}_n(a)} Y_\mu(a) = {}^n\lambda$.

Proof. Assume that (i) fails for almost all a. Since MEAS_n^λ is wellordered, let $a \mapsto \mu_1(a), \mu_2(a)$ be such that, for almost all a:

(a) $\mu_1(a) \in \mathrm{meas}_n(a) \wedge \mu_2(a) \in \mathrm{meas}_n(a)$;

(b) $\mu_1 \neq \mu_2$;

(c) $Y_{\mu_1(a)}(a) \cap Y_{\mu_2(a)}(a) \neq \varnothing$.

By normality there are μ_1 and μ_2 such that, for almost all a, $\mu_1(a) = \mu_1$ and $\mu_2(a) = \mu_2$. Let $X \subseteq {}^n\lambda$ be such that $X \in \mu_1$ and $X \notin \mu_2$. Fineness implies that, for almost all a, both X and ${}^n\lambda \setminus X$ belong to a, and so $Y_{\mu_1}(a) \subseteq X$ and $Y_{\mu_2}(a) \cap X = \varnothing$. This is a contradiction.

Now assume that (ii) fails for almost all a. Define

$$Z(a) = {}^n\lambda \setminus \bigcup_{\mu \in \mathrm{meas}_n(a)} Y_\mu(a).$$

Since ${}^n\lambda$ is wellordered, there is an f such that $f(a) \in Z(a)$ for almost all a. Define an element μ of $\mathrm{MEAS}_n^{\kappa,\lambda}$ by

$$X \in \mu \leftrightarrow f(a) \in X \text{ for almost all } a.$$

Fineness implies that μ belongs to almost all a. Since $Z(a)$ and $Y_\mu(a)$ are disjoint whenever $\mu \in a$, it follows that $f(a) \notin Y_\mu(a)$ for almost all a. Thus if

$$\mathcal{X}(a) = \{X \in \mu : f(a) \notin X\},$$

then, for almost all a, $a \cap \mathcal{X}(a) \neq \varnothing$. By normality, there is an X such that $X \in \mathcal{X}(a)$ for almost all a. By the definition of $\mathcal{X}$, $X \in \mu$, and $f(a) \notin X$ for almost all a. This contradicts the definition of μ. ⊣

If $\boldsymbol{\mu} = \langle \mu_i : i < \omega \rangle$ is a λ-tower of κ-complete measures, then for each a we define

$$\boldsymbol{Y_\mu}(a) = \{ f \in {}^\omega\lambda : (\forall i < \omega)\, f \restriction i \in Y_{\mu_i}(a) \}.$$

LEMMA 2.2. For almost all a,

if $\boldsymbol{\mu} = \langle \mu_i : i < \omega \rangle$ is a λ-tower of κ-complete measures and each $\mu_i \in a$, then $\boldsymbol{\mu}$ is countably complete if and only if $\boldsymbol{Y_\mu}(a) \neq \varnothing$. (†)

PROOF. Observe first that if we replace "if and only if" by "only if" then (†) holds for every a, by the definition of countable completeness.

To prove the lemma we shall first show that there is a function assigning to each a for which (†) fails a tower witnessing that failure. Then we shall apply normality to show that if the lemma is false then there is a single tower that witnesses failure for almost all a, and there can be no such tower.

Fix an a such that (†) fails for a.

Suppose that $\boldsymbol{\mu} = \langle \mu_i : i < \omega \rangle$ witnesses this failure. Then there is an $f : \omega \to \lambda$ such that

(1) $\boldsymbol{\mu}$ is a λ-tower of κ-complete measures and each $\mu_i \in a$;
(2) $f \in \boldsymbol{Y_\mu}(a)$;
(3) $\boldsymbol{\mu}$ is not countably complete.

By (3), there are sets X_i, $i < \omega$, such that $X_i \in \mu_i$ for each i but there is no $g : \omega \to \lambda$ such that $g \restriction i \in X_i$ for all i. For any such $\langle X_i : i < \omega \rangle$, there are functions $\bar{h}_i : X_i \to \lambda^+$ such that

(4) $(\forall i)(\forall t \in X_{i+1}) \Big((\forall j \leq i)\, t \restriction j \in X_j \to \bar{h}_{i+1}(t) < \bar{h}_i(t \restriction i) \Big)$.

The functions $\bar{h}_i$ can be extended to functions $h_i : {}^i\lambda \to \lambda^+$. By the definition of a tower, the h_i satisfy

(5) $(\forall i)(\{ t \in {}^{i+1}\lambda : h_{i+1}(t) < h_i(t \restriction i) \} \in \mu_{i+1})$.

Moreover any sequence $\langle h_i : i < \omega \rangle$ satisfying (5) determines sequences $\langle \bar{h}_i : i < \omega \rangle$ and $\langle X_i : i < \omega \rangle$ satisfying (4). Just set

$$\begin{aligned} X_0 &= {}^0\lambda; \\ X_{i+1} &= \{ t \in {}^{i+1}\lambda : h_{i+1}(t) < h_i(t \restriction i) \}; \\ \bar{h}_i &= h_i \restriction X_i. \end{aligned}$$

Note also that whether or not (5) is satisfied depends only on $\langle [h_i]_{\mu_i} : i < \omega \rangle$ and not on the full $\langle h_i : i < \omega \rangle$. (Our assumption (D) implies that the ordinals $[h_i]_{\mu_i}$ exist.)

Now forget our particular $\boldsymbol{\mu}$. We define a triple $(\boldsymbol{\mu}, f, g)$ as follows. Suppose that $\langle \mu_i : i < n \rangle$, $f \restriction n$, and $g \restriction n$ have been defined. For $i < n$, let h_i be such that $[h_i]_{\mu_i} = g(i)$. Assume inductively that $(\langle \mu_i : i < n \rangle, f \restriction n, \langle h_i : i < n \rangle)$ can be extended to at least one triple $(\boldsymbol{\mu}, f, \langle h_i : i < \omega \rangle)$ satisfying (1), (2), and (5). Note that whether this assumption holds depends only on $g \restriction n$ and not on the representative functions h_i. Let $(\mu_n, f(n), g(n))$ be lexicographically least preserving the extendability. Here the ordering of measures is one given by (E), and the orderings of λ and the ordinals are the standard ones.

Using (D) to pick functions h_i representating the ordinals $g(i)$, we see that our triple extends to a triple satisfying (1), (2), and (5). Thus $\boldsymbol{\mu}$ witnesses the failure of (†) for a. Our inductive definition depends on a, so let us call its components $\boldsymbol{\mu}(a)$, $f(a)$, and $g(a)$. These objects are defined for each a that does not satisfy (†), and so $\boldsymbol{\mu}(a)$ is a witness to the failure of (†) for every a for which it fails.

Assume that the lemma is false. By normality and countable completeness of $\mathcal{V}$, there is a tower $\boldsymbol{\mu}$ such that $\boldsymbol{\mu} = \boldsymbol{\mu}(a)$ for almost all a. Thus $\boldsymbol{\mu}$ is not countably complete. Let $\langle X_i : i < \omega \rangle$ witness this. By fineness, there is an a for which (†) fails such that each $X_i \in a$. For this a, $Y_{\mu_i}(a) \subseteq X_i$ for each i. But this is a contradiction. ⊣

For the rest of this section, let T be a tree on $\omega \times \lambda$.

LEMMA 2.3. For almost all a, the following five statements are true.
(1) For every $n < \omega$, a satisfies (i) and (ii) of Lemma 2.1.
(2) a satisfies (†) of Lemma 2.2.
(3) Both T_s and ${}^{\mathrm{lh}(s)}\lambda \setminus T_s$ belong to a for every $s \in {}^{<\omega}\omega$.
(4) If $m < n < \omega$, $X \subseteq {}^m\lambda$, and $X \in a$, then $\mathrm{ext}_n(X) \in a$.
(5) If $m < n < \omega$ and $\mu \in \mathrm{meas}_n(a)$, then $\mathrm{proj}_m(\mu) \in a$.

PROOF. The lemma follows easily from Lemma 2.1, Lemma 2.2, and the fact that $\mathcal{V}$ is countably complete, fine, and normal. ⊣

For the rest of this section, fix an a for which (1)–(5) are true. We shall write meas_n for $\mathrm{meas}_n(a)$, Y_μ for $Y_\mu(a)$, and $Y_{\boldsymbol{\mu}}$ for $Y_{\boldsymbol{\mu}}(a)$.

It follows immediately from (1) that for all $n < \omega$ and $t \in {}^n\lambda$ there is a unique $\mu \in \mathrm{meas}_n$ such that $t \in Y_\mu$. Let this unique μ be μ_t.

LEMMA 2.4. Let $n < \omega$ and let $t \in {}^n\lambda$. Let $m < n$ and let $s \in {}^m\lambda$ with $s \subseteq t$. Then

$$\mu_s = \mathrm{proj}_m(\mu_t).$$

PROOF. By (5), it is enough to show that $s \in Y_{\mathrm{proj}_m(\mu_t)}$. Let $X \in a \cap \mathrm{proj}_m(\mu_t)$. We must show that $s \in X$. By (4) and the definition of $\mathrm{proj}_m(\mu_t)$, $\mathrm{ext}_n(X) \in a \cap \mu_t$. Hence $Y_{\mu_t} \subseteq \mathrm{ext}_n(X)$. But this implies that $t \in \mathrm{ext}_n(X)$, which in turn implies that $s \in X$. ⊣

For each $f : \omega \to \lambda$, let
$$\boldsymbol{\mu}_f = \langle \mu_{f\restriction n} : n < \omega \rangle.$$
By Lemma 2.4, $\boldsymbol{\mu}_f$ is a tower of measures.

LEMMA 2.5. If $(x, f) \in [T]$, then $T_{x\restriction n} \in \mu_{f\restriction n}$ for each n.

PROOF. Since a satisfies (3), both $T_{x\restriction n}$ and ${}^n\lambda \setminus T_{x\restriction n}$ belong to a. By (1) it follows that either $Y_{\mu_{f\restriction n}} \subseteq T_{x\restriction n}$ or $Y_{\mu_{f\restriction n}} \cap T_{x\restriction n} = \varnothing$. Since $f_{x\restriction n}$ belongs to $Y_{\mu_{f\restriction n}} \cap T_{x\restriction n}$, the latter possibility is ruled out. ⊣

LEMMA 2.6. Let $(x, f) \in [T]$. Then
(a) $\boldsymbol{\mu}_f$ is a tower of measures;
(b) $\mu_{f\restriction n} \in a$ for all $n < \omega$;
(c) $T_{x\restriction n} \in \mu_{f\restriction n}$ for all $n < \omega$;
(d) $\boldsymbol{\mu}_f$ is countably complete.

PROOF. (a) follows from Lemma 2.4. (b) holds because $\mu_t \in a$ for all $t \in {}^{<\omega}\lambda$. (c) is an instance of Lemma 2.5. Assume that (d) does not hold. By Lemma 2.2, $\boldsymbol{Y_\mu}$ is empty. But that is impossible, since $f \in \boldsymbol{Y_\mu}$. ⊣

For $s \in {}^{<\omega}\omega$, let $M_s = \{\mu \in a : T_s \in \mu\}$.

LEMMA 2.7. If $\kappa = \aleph_1$, then the system $\langle M_s : s \in {}^{<\omega}\omega \rangle$ witnesses that T is weakly homogeneous.

PROOF. Lemma 2.6 implies that $\langle M_s : s \in {}^{<\omega}\omega \rangle$ meets all requirements for witnessing weak homogeneity of T except perhaps the requirement that the sets M_s be countable. If $\kappa = \aleph_1$, then a is countable, and the M_s are countable. ⊣

§3. The main theorems.

THEOREM 3.1. Assume $\mathrm{AD}_{\mathbb{R}}$. Every tree is weakly homogeneous.

PROOF. Let T be a tree. We first show that we may assume that T is on $\omega \times \lambda$ for some $\lambda < \Theta$. For each $x \in \mathrm{p}[T]$, let f_x be lexicographically least such that $(x, f) \in [T]$. The set
$$\{f_x(n) : x \in \mathrm{p}[T] \wedge n < \omega\}$$
is a surjective image of $\mathbb{R}$. Let π be a bijection from this set to an ordinal number $\lambda < \Theta$. Define a tree $\bar{T}$ on $\omega \times \lambda$ by
$$(s, t) \in \bar{T} \leftrightarrow (s, \pi^{-1}(t)) \in T.$$
Clearly $\mathrm{p}[\bar{T}] = \mathrm{p}[T]$, and any system witnessing that $\bar{T}$ is weakly homogeneous is essentially a system witnessing that T is weakly homogeneous. We may then assume $\bar{T} = T$ without loss of generality.

Let $\mathcal{V}$ be the ultrafilter given by Lemma 1.2. Let a satisfy (1)–(5) of Lemma 2.3. By Lemma 2.7, the system $\langle M_s : s \in {}^{<\omega}\omega\rangle$ defined as for Lemma 2.7 witnesses that T is weakly homogeneous. ⊣

THEOREM 3.2. Assume the Axiom of Choice. Let κ be supercompact. Let $\mathbf{V}^{\boldsymbol{B}}$ come by the Lévy collapse of all ordinals $< \kappa$ to ω. Let $N = \mathbf{V}(\mathbb{R}^{\mathbf{V}^{\boldsymbol{B}}})$. In N, every tree is weakly homogeneous.

PROOF. It suffices to prove the weakened version of the theorem in which "every tree" is replaced by "every tree in $\mathbf{V}$." For suppose that we have proved this weakened version. Let κ, $\boldsymbol{B}$, and N be as in the statement of the theorem. Let $T \in N$. For some $\alpha < \kappa$, T belongs to an inner model of N of the form $\mathbf{V}^{\boldsymbol{B}_\alpha}$, where $\mathbf{V}^{\boldsymbol{B}_\alpha}$ comes from $\mathbf{V}$ by the collapse of α to ω. Now κ is supercompact in $\mathbf{V}^{\boldsymbol{B}_\alpha}$, and $\mathbf{V}^{\boldsymbol{B}}$ comes from $\mathbf{V}^{\boldsymbol{B}_\alpha}$ by a Lévy collapse of all ordinals $< \kappa$ to ω. Moreover $N = \mathbf{V}^{\boldsymbol{B}_\alpha}(\mathbb{R}^{\mathbf{V}^{\boldsymbol{B}}})$. By our supposition, the theorem holds in $\mathbf{V}^{\boldsymbol{B}_\alpha}$. Hence T is weakly homogeneous in N.

Let κ, $\boldsymbol{B}$, and N be as in the statement of the theorem and let T be a tree in $\mathbf{V}$. Let λ be such that T is on $\omega \times \lambda$. If $\lambda < \kappa$, then T is countable in N and so is trivially weakly homogeneous. Assume then that $\kappa \leq \lambda$.

For the moment, let us work in $\mathbf{V}$. Let $\mathcal{V}$ be a κ-complete, fine, normal, ultrafilter on $\wp_\kappa(\wp(\lambda) \cup \bigcup_{n<\omega} \mathrm{MEAS}_n^{\kappa,\lambda})$. The assumptions (A)–(E) hold.

- Let a satisfy (1)–(5) of Lemma 2.3;
- For $t \in {}^{<\omega}\omega$, let μ_t be defined as for Lemma 2.4;
- Let $\langle M_s : s \in {}^{<\omega}\omega\rangle$ be as defined as for Lemma 2.7.

Since a satisfies (2) of Lemma 2.3, it satisfies the condition (†) of Lemma 2.2. By the proof of Lemma 2.2 and the definition of a λ-complete tower of measures, it follows that the following statement is true of a.

(∗) There is no triple $(\langle \mu_i : i < \omega\rangle, f, g)$ such that μ and f satisfy the conditions (1) and (2) of the proof of Lemma 2.2 and such that condition (5) from the same proof holds for sequences $\langle h_i : i < \omega\rangle$ for which $[h_i]_{\mu_i} = g(i)$ for every $i < \omega$.

For the rest of the proof of the theorem, we work in N.

For $\langle \mu_i : i < \omega\rangle$ such that each $\mu \in (\mathrm{MEAS}_n^{\kappa,\lambda})^{\mathbf{V}}$, we extend a previous definition by setting $\boldsymbol{Y}_{\boldsymbol{\mu}} = \{f : (\forall i \in \omega) f \restriction i \in Y_{\mu_i}\}$, where Y_{μ_i} is the $Y_{\mu_i}(a)$ defined in $\mathbf{V}$.

For $n < \omega$ and $\mu \in (\mathrm{MEAS}_n^{\kappa,\lambda})^{\mathbf{V}}$, define $\mu^+ \subseteq {}^n\lambda$ by

$$X \in \mu^+ \leftrightarrow (\exists Y \in \mu) Y \subseteq X.$$

Standard arguments show that

(1) μ^+ is a κ-complete (i.e., countably complete) measure on ${}^n\lambda$;

(2) for every $f : {}^n\lambda \to \mathbf{V}$ there is a $g : {}^n\lambda \to \mathbf{V}$ such that $g \in \mathbf{V}$ and $\{t : f(t) = g(t)\} \in \mu^+$.

It follows that

(3) $(\forall h)\big((h \in \mathbf{V} \wedge h\colon {}^n\lambda \to \mathrm{Ord}) \to [h]_{\mu^+} = ([h]_\mu)^{\mathbf{V}}\big)$.

Here is a version of the condition (†) of Lemma 2.2 that a satisfies with respect to measures μ^+.

Lemma 3.3. Let $\boldsymbol{\mu} = \langle \mu_i : i < \omega\rangle$, and assume that each $\mu_i \in (\mathrm{MEAS}_n^{\kappa,\lambda})^{\mathbf{V}}$. If $\boldsymbol{\mu}^+ = \langle \mu_i^+ : i < \omega\rangle$ is a λ-tower of κ-complete measures and each $\mu_i \in a$, then $\boldsymbol{\mu}^+$ is countably complete if and only if $\boldsymbol{Y}_{\boldsymbol{\mu}}(a) \neq \varnothing$. (†)$^+$

Proof. Since $Y_{\mu_i} \in \mu_i^+$ for each i, we need only prove the "if" part of (†)$^+$. By the definition of a λ-tower of measures, the fact that $(*)$ holds in $\mathbf{V}$ implies that there is in $\mathbf{V}$ no triple $(\langle \mu_i : i < \omega\rangle, f, g)$ such that:

(i) for $i < \omega$, $\mu_i \in (\mathrm{MEAS}_n^{\kappa,\lambda})^{\mathbf{V}}$;
(ii) for $j < i < \omega$, $\big(\mu_j = \mathrm{proj}_j(\mu_i)\big)^{\mathbf{V}}$;
(iii) $f\colon \omega \to \lambda$;
(iv) for $i < \omega$, $f(i) \in \boldsymbol{Y}_{\mu_i}$;
(v) $g\colon \omega \to \mathrm{Ord}$;
(vi) for $i < \omega$, for $h \in ({}^{{}^i\lambda}\lambda^+)^{\mathbf{V}}$, and for $h' \in ({}^{{}^{i+1}\lambda}\lambda^+)^{\mathbf{V}}$,
$$\big(([h]_{\mu_i})^{\mathbf{V}} = g(i) \wedge ([h']_{\mu_{i+1}})^{\mathbf{V}} = g(i+1)\big) \to \{t \in {}^{i+1}\lambda : h'(t) < h(t\restriction i)\} \in \mu_{i+1}.$$

By the absoluteness of wellfoundedness, there is no triple in N satisfying (i)–(vi).

Assume that $\boldsymbol{Y}_{\boldsymbol{\mu}} \neq \varnothing$, and let f witness this. Assume, to derive a contradiction, that $\boldsymbol{\mu}_f^+$ is not countably complete. Let $\langle h_i : i < \omega\rangle$ witness this. By property (2) of $\mu \mapsto \mu^+$, we may assume that each $h_i \in \mathbf{V}$. Let $g\colon \omega \to \mathrm{Ord}$ be defined by $g(i) = [h_i]_{\mu_i^+}$. By property (3) of $\mu \mapsto \mu^+$, $g(i) = ([h_i]_{\mu_i})^{\mathbf{V}}$ for each i. But then $(\langle \mu_i : i < \omega\rangle, f, g)$ has properties (i)–(vi), and we know that no triple can have these properties. ⊣

For $t \in {}^{<\omega}\lambda$, let $\mu_t^+ = (\mu_t)^+$. For $f\colon \omega \to \lambda$, let $\boldsymbol{\mu}_f = \langle \mu_{f\restriction n} : n < \omega\rangle$. The following Lemma does the work of Lemma 2.6.

Lemma 3.4. Let $(x, f) \in [T]$. Then

(a) $\boldsymbol{\mu}_f^+$ is a tower of measures;
(b) $\mu_{f\restriction n} \in a$ for all $n < \omega$;
(c) $T_{x\restriction n} \in \mu_{f\restriction n}$ for all $n < \omega$;
(d) $\boldsymbol{\mu}_f^+$ is countably complete.

Proof. Since clause (a) of Lemma 2.6 holds in $\mathbf{V}$, $\mu_{f\restriction i} \in (\mathrm{MEAS}_i^{\kappa,\lambda})^{\mathbf{V}}$ for $i < \omega$, and $\big(\mu_{f\restriction j} = \mathrm{proj}_j(\mu_{f\restriction i})\big)^{\mathbf{V}}$ for $j < i < \omega$.

From property (1) of $\mu \mapsto \mu^+$, it follows that $\mu_{f\restriction i}^+ \in \mathrm{MEAS}_i^{\kappa,\lambda}$.

From the definition of $\mu \mapsto \mu^+$, it follows that $\left(\mu^+_{f\restriction j} = \mathrm{proj}_j(\mu^+_{f\restriction i})\right)^{\mathbf{V}}$ for $j < i < \omega$. Hence clause (a) of the lemma holds.

Clauses (b) and (c) are the same as the corresponding clauses of Lemma 2.6.

Clause (d) follows from Lemma 3.3 in the same way as clause (d) of Lemma 2.6 followed from Lemma 2.2. ⊣

For $s \in {}^{<\omega}\omega$, define

$$M_s^+ = \{\mu^+ : (\mu \in \mathrm{meas}_n(a) \wedge \mu \in a \wedge T_s \in \mu\}.$$

The theorem follows from Lemma 3.4 and the fact that $\kappa = \aleph_1$ in N. ⊣

§4. **Further results and open problems.** The basic open problems concern to what extent it is true that the projection $\mathrm{p}[T]$ of a weakly homogeneous tree must be representable as the projection of a homogeneous tree. To avoid fairly simple counterexamples one must (and we do) assume that the collection

$$\{\mathrm{p}[T] : T \text{ is a weakly homogeneous tree}\}$$

is reasonably closed, for example that the collection is closed under complements. This question in turn involves variations on Theorem 3.2 where the large cardinal hypothesis on κ is reduced below that of the hypothesis that κ be supercompact.

A weakly homogeneous tree, T, is **κ-weakly homogeneous** if there is a witness to the weak homogeneity of T such that all the measures are κ-complete. Similarly, a homogeneous tree, T, is **κ-homogeneous** if there is a witness to the homogeneity of T such that all the measures are κ-complete.

A cardinal κ is **$\mathbf{V}_\alpha$-strong** if there exists an elementary embedding

$$j : \mathbf{V} \to M$$

with critical point κ such that $\mathbf{V}_\alpha \subseteq j(\mathbf{V}_\kappa)$. A cardinal κ is a **strong cardinal** if κ is $\mathbf{V}_\alpha$-strong for all ordinals α.

A cardinal κ is **$\mathbf{V}_\alpha$-1-strong** if there exists an elementary embedding

$$j : \mathbf{V} \to M$$

with critical point κ such that $\mathbf{V}_\alpha \subseteq j(\mathbf{V}_\kappa)$ and such that for all $\delta < \alpha$, δ is a strong cardinal in $\mathbf{V}$ if and only if δ is a strong cardinal in M. κ is a **1-strong** cardinal if κ is $\mathbf{V}_\alpha$-1-strong for all ordinals α.

THEOREM 4.1 (ZFC). *Suppose that κ is a regular cardinal which is a limit of 1-strong cardinals and that there is a strong cardinal above κ. Let $\mathbf{V}^B$ come by the Lévy collapse of all ordinals $< \kappa$ to ω. Let $N = \mathbf{V}(\mathbb{R}^{\mathbf{V}^B})$. In N, for every tree T the following are equivalent.*

1. *There is a weakly homogeneous tree, $\bar{T}$, such that $\mathrm{p}[T] = \mathrm{p}[\bar{T}]$.*
2. *There is a tree S such that $\mathrm{p}[T] = {}^{\omega}\omega \setminus \mathrm{p}[S]$.*

PROOF. By standard arguments (1) implies (2), and so we have only to prove that in N, (2) implies (1).

As in the proof of Theorem 3.2, it suffices to prove the weakened version of the theorem in which the trees S and T belong to $\mathbf{V}$. Clearly it suffices to produce the tree, $\bar{T}$, such that $\bar{T}$ is weakly homogeneous in $\mathbf{V}^{\boldsymbol{B}}$.

Clearly we may suppose that T is a tree on κ and by reshaping the tree T if necessary we may suppose that that for all $\gamma < \kappa$, if γ is strongly inaccessible then for all complete Boolean algebras, $\boldsymbol{B}_0 \in \mathbf{V}_\gamma$,

$$\mathbf{V}^{\boldsymbol{B}_0} \models \text{“p}[T] = \text{p}[T \restriction \gamma]\text{”}.$$

Fix $\kappa_0 < \kappa < \kappa_1$ such that κ_0 is a 1-strong cardinal and such that κ_1 is a strong cardinal. Let

$$j_0 \colon \mathbf{V} \to M_0$$

be an elementary embedding with critical point κ_0 such that $\kappa_1 < j_0(\kappa_0)$, $\mathbf{V}_{\kappa_1+\omega} \subseteq M_0$, and such that κ_1 is a strong cardinal in M_0.

Let

$$j_1 \colon M_0 \to M_1$$

be an elementary embedding in M_0 with critical point κ_1 such that $j_0(\kappa_0) < j_1(\kappa_1)$ and such that $M_0 \cap \mathbf{V}_{j_0(\kappa_0)+\omega} \subseteq M_1$.

For each $n < \omega$, let

$$\pi_n \colon \text{MEAS}_n^{\kappa_0,\kappa_0} \to \text{MEAS}_n^{\kappa_1,\kappa_1}$$

be the (unique) map such that for all $\mu \in \text{MEAS}_n^{\kappa_0,\kappa_0}$, for all $\nu \in \text{MEAS}_n^{\kappa_1,\kappa_1}$,

$$\pi_n(\mu) = \nu$$

if for all $A \subseteq {}^n\kappa_1$, $A \in \nu$ if and only if $j_1(A) \cap {}^n j_0(\kappa_0) \in j_0(\mu)$.

For each $t \in {}^n j_0(\kappa_0)$, let $\mu_t \in \text{MEAS}_n^{\kappa_0,\kappa_0}$ be such that for all $A \subseteq {}^n\kappa_0$, $A \in \mu_t$ if and only if $t \in j_0(A)$.

For each $s \in {}^{<\omega}\omega$, let

$$M_s = \{\pi_n(\mu_t) : t \in {}^n j_0(\kappa_0), n = \text{lh}(s), (s,t) \in j_0(T_0)\}.$$

A key point is that for each $s \in {}^{<\omega}\omega$, $|M_s| \leq |\mathbf{V}_{\kappa_0+2}| < \kappa_1$.

Thus there is a tree $\bar{T}$ on $\omega \times \kappa_1$ (with the tree $\bar{T}$ in $\mathbf{V}$) such that the sequence

$$\langle M_s : s \in {}^{<\omega}\omega \rangle$$

witnesses that $\bar{T}$ is weakly homogeneous in $\mathbf{V}^{\boldsymbol{B}}$ where $\mathbf{V}^{\boldsymbol{B}}$ comes by the Lévy collapse of all ordinals $< \kappa$ to ω.

We next show that

$$\mathbf{V}^{\boldsymbol{B}} \models \text{“p}[T] = \text{p}[\bar{T}]\text{”}.$$

This we do in two steps. First we show that

$$\mathbf{V}^{\boldsymbol{B}} \models \text{“p}[\bar{T}] \subseteq \text{p}[T]\text{”}.$$

If not then by absoluteness there exists

$$x \in \mathrm{p}[S] \cap \mathrm{p}[\bar{T}]$$

such that $x \in \mathbf{V}$. Let $\langle \nu_n : n < \omega \rangle$ be a countably complete tower of measures such that for each $n < \omega$,

$$\mu_n \in M_{x \restriction n}.$$

Let $\langle \mu_n : n < \omega \rangle$ be such that for each $n < \omega$, $\nu_n = \pi_n(\mu_n)$. It follows that

$$\langle \mu_n : n < \omega \rangle$$

is a countably complete tower at κ_0 and moreover that for each $n < \omega$,

$$\{t \in {}^n\kappa_0 \;:\; (x \restriction n, t) \in T_0\} \in \mu_n.$$

This implies that $x \in \mathrm{p}[T]$ which contradicts that $\mathrm{p}[T] \cap \mathrm{p}[S] = \varnothing$.

Next we show that

$$\mathbf{V}^{\boldsymbol{B}} \models \text{“}\mathrm{p}[T] \subseteq \mathrm{p}[\bar{T}]\text{”}.$$

Suppose $x \in \mathbf{V}^{\boldsymbol{B}}$ and

$$\mathbf{V}^{\boldsymbol{B}} \models \text{“}x \in \mathrm{p}[T]\text{”}.$$

Then we have

$$\mathbf{V}^{\boldsymbol{B}} \models \text{“}x \in \mathrm{p}[j_0(T_0) \restriction \kappa_1]\text{”},$$

by the reshaping of T. Let $f \in \mathbf{V}^{\boldsymbol{B}}$ be such that $f \in {}^{\omega}\kappa_1$ and such that in $\mathbf{V}^{\boldsymbol{B}}$, (x, f) is an infinite branch of $j_0(T_0) \restriction \kappa_1$. For each $n < \omega$, let $\mu_n = \mu_t$ where $t = f \restriction n$ and let $\nu_n = \pi_n(\mu_n)$. Thus

$$\langle \nu_n : n < \omega \rangle$$

is a tower of measures at κ_1 and for each $n < \omega$,

$$\nu_n \in M_{x \restriction n}.$$

Therefore it suffices to show that in $\mathbf{V}^{\boldsymbol{B}}$, the tower, $\langle \nu_n : n < \omega \rangle$, is countably complete. If not then again by absoluteness it follows that we can suppose that $f \in \mathbf{V}$. But then $\langle \mu_n : n < \omega \rangle \in \mathbf{V}$ and in $\mathbf{V}$, $\langle \mu_n : n < \omega \rangle$ is a countably complete tower of measures at κ_0. This implies that

$$\langle j_0(\mu_n) : n < \omega \rangle \in M_0$$

and that in M_0, $\langle j_0(\mu_n) : n < \omega \rangle$ is a countably complete tower of measures at $j_0(\kappa_0)$. But this in turn implies

$$\langle \nu_n : n < \omega \rangle \in M_0$$

and that in M_0, $\langle \nu_n : n < \omega \rangle$ is a countably complete tower of measures at κ_1. Since

$$\mathbf{V}_{\kappa_1+\omega} \subseteq M_0,$$

this implies that in $\mathbf{V}$, $\langle \nu_n : n < \omega \rangle$ is a countably complete tower of measures at κ_1, which is a contradiction.

This proves the claim that

$$\mathbf{V}^{B} \models \text{“}\mathrm{p}[T] = \mathrm{p}[\bar{T}]\text{”}.$$

Since $\bar{T}$ is weakly homogeneous in $\mathbf{V}^{B}$, $\bar{T}$ is as required. ⊣

The proof of Theorem 4.1 is actually an adaptation of the proof of the following theorem, for a proof see (Theorem 1.5.12) of [Lar04].

THEOREM 4.2 (ZFC). Suppose that κ is a Woodin cardinal and T is a tree. There exists $\gamma < \kappa$ such that if $\mathbf{V}^{B}$ comes by the Lévy collapse of γ to ω, then in $\mathbf{V}^{B}$, T is α-weakly homogeneous for all $\alpha < \kappa$.

One corollary of Theorem 4.2 is that the large cardinal hypothesis of Theorem 3.2 can be reduced significantly.

THEOREM 4.3 (ZFC). Suppose that κ is a measurable Woodin cardinal. Let $\mathbf{V}^{B}$ come by the Lévy collapse of of all ordinals $< \kappa$ to ω. Let $N = \mathbf{V}(\mathbb{R}^{\mathbf{V}^{B}})$. In N, every tree is weakly homogeneous.

PROOF. It suffices to show that if κ is a measurable cardinal and if T is a tree which is α-weakly homogeneous for all $\alpha < \kappa$, then T is κ-weakly homogeneous. This is essentially immediate. Fix the tree T and fix γ such that T is a tree on $\omega \times \gamma$. Let ν be a κ-complete uniform measure on κ.

For each $\alpha < \kappa$, let

$$\langle M^{\alpha}_{s} : s \in {}^{<\omega}\omega\rangle$$

witness that T is α-weakly homogeneous. For each $\alpha < \kappa$ and for each $s \in {}^{<\omega}\omega$, let

$$f^{\alpha}_{s} : \omega \to M^{\alpha}_{s}$$

be a surjection.

For each pair $(i, s) \in \omega \times {}^{<\omega}\omega$, define a measure μ^{s}_{i} on ${}^{\mathrm{lh}(s)}\gamma$ by $A \in \mu^{s}_{i}$ if

$$\{\alpha : A \in f^{\alpha}_{s}(i)\} \in \nu.$$

Clearly μ^{s}_{i} is a κ-complete measure on ${}^{\mathrm{lh}(s)}\gamma$ such that $T_s \in \mu^{s}_{i}$.

For each $s \in {}^{<\omega}\omega$, let

$$M_s = \{\mu^{s}_{i} : i < \omega\}.$$

We finish by showing that

$$\langle M_s : s \in {}^{<\omega}\omega\rangle$$

witnesses that T is κ-weakly homogeneous. Suppose that $x \in \mathrm{p}[T]$. Then for each $\alpha < \kappa$ there exists a countably complete tower,

$$\langle \mu^{\alpha}_{i} : i < \omega\rangle,$$

such that for each $i < \omega$, $\mu^{\alpha}_{i} \in M^{\alpha}_{x\restriction i}$.

Since ν is countably complete, for each $i < \omega$ there exists $i^* < \omega$ such that

$$\{\alpha < \kappa : \mu^{\alpha}_{i} = f^{\alpha}_{x\restriction i}(i^*)\} \in \nu.$$

Therefore

$$\left\langle \mu_{i*}^{x\restriction i} : i < \omega \right\rangle$$

is a countably complete tower such that for all $i < \omega$, $\mu_{i*}^{x\restriction i} \in M_{x\restriction i}$. ⊣

A proof of the following theorem can also be found in [Lar04], the theorem appears there in a slightly different form as (Theorem 3.3.8).

THEOREM 4.4 (ZFC). Suppose that δ is a Woodin cardinal and suppose that S and T are trees such that if $\mathbf{V}^B$ comes by the Lévy collapse of δ to ω then in $\mathbf{V}^B$,

$$\mathrm{p}[S] = {}^\omega\omega \setminus \mathrm{p}[T].$$

Then in $\mathbf{V}$, for all $\alpha < \delta$, S and T are α-weakly homogeneous.

As a corollary to Theorem 4.2 and Theorem 4.4 one obtains the following theorem.

THEOREM 4.5 (ZFC). Suppose that $\kappa < \delta$ are Woodin cardinals and κ is a limit of strong cardinals. Let $\mathbf{V}^B$ come by the Lévy collapse of of all ordinals $< \kappa$ to ω. Let $N = \mathbf{V}(\mathbb{R}^{\mathbf{V}^B})$. In N, for every tree T there exists a tree $\bar{T}$ such that $\mathrm{p}[T] = \mathrm{p}[\bar{T}]$ and such that $\bar{T}$ is weakly homogeneous.

The large cardinal hypothesis of Theorem 4.1 is significantly weaker than the large cardinal hypothesis of Theorem 4.5. But the conclusions do not seem that different; Theorem 4.1 obtains a symmetric Lévy extension of $\mathbf{V}$ in which every Suslin, co-Suslin, set $A \subseteq {}^\omega\omega$ is the projection of a weakly homogeneous tree, and Theorem 4.5 obtains obtains a symmetric Lévy extension of $\mathbf{V}$ in which every Suslin set $A \subseteq {}^\omega\omega$ is the projection of a weakly homogeneous tree. Does the latter really require the additional large cardinal assumptions? We note that by our main theorem, Theorem 3.1, the consistency strength of the theory,

$$\text{ZF+DC+“Every tree is weakly homogeneous”},$$

is at most that of the theory, ZF+DC+$\mathrm{AD}_{\mathbb{R}}$, and the consistency of the latter theory is well below that of the large cardinal hypothesis of Theorem 4.5. Similar considerations apply to Theorem 4.3, and so one can reasonably ask whether the large cardinal hypothesis for that theorem can also be reduced.

For each cardinal κ, let $\Gamma^\kappa_{\mathrm{wHom}}$ be the set of all $A \subseteq {}^\omega\omega$ such that A is the projection of a κ-weakly homogeneous tree, an let $\Gamma^\kappa_{\mathrm{Hom}}$ be the set of all $A \subseteq {}^\omega\omega$ such that A is the projection of a κ-homogeneous tree.

It is straightforward to verify that $\Gamma^\kappa_{\mathrm{wHom}}$ is closed under countable unions, countable intersections, and closed under both images preimages by continuous functions,

$$f : {}^\omega\omega \to {}^\omega\omega,$$

and moreover every set $A \in \Gamma^{\kappa}_{\mathrm{wHom}}$ is the image of a set $B \in \Gamma^{\kappa}_{\mathrm{Hom}}$ by a continuous function,

$$f : {}^{\omega}\omega \to {}^{\omega}\omega.$$

Therefore by consideration of Wadge games, one can easily prove the following lemma.

LEMMA 4.6 (ZFC). Suppose that κ is a cardinal such that $\Gamma^{\kappa}_{\mathrm{wHom}}$ is closed under complements. Then the following are equivalent.

1. $\Gamma^{\kappa}_{\mathrm{Hom}} = \Gamma^{\kappa}_{\mathrm{wHom}}$.
2. Every set in $\Gamma^{\kappa}_{\mathrm{wHom}}$ is determined.

Theorem 4.1 can be reformulated quite easily as follows.

THEOREM 4.7 (ZFC). Suppose that κ is a regular cardinal which is a limit of 1-strong cardinals and that there is a strong cardinal above κ. Let $\mathbf{V}^{B}$ come by the Lévy collapse of all ordinals $< \kappa$ to ω. Then in $\mathbf{V}^{B}$, $\Gamma^{\kappa}_{\mathrm{wHom}}$ is closed under complements.

Theorem 4.7 shows that one cannot hope to prove that

$$\Gamma^{\kappa}_{\mathrm{Hom}} = \Gamma^{\kappa}_{\mathrm{wHom}}$$

if one assumes just $\Gamma^{\kappa}_{\mathrm{wHom}}$ is closed under complements (and so is a σ-algebra). In fact assuming $\Gamma^{\kappa}_{\mathrm{wHom}}$ is closed under complements one cannot hope to prove even that $\Gamma^{\kappa}_{\mathrm{Hom}}$ contains all Δ^1_2 sets.

On the other hand there is the following theorem of Steel.

THEOREM 4.8 (Steel). (ZFC) Suppose that

$$\wp({}^{\omega}\omega) \cap \mathbf{L}({}^{\omega}\omega) \subseteq \Gamma^{\kappa}_{\mathrm{wHom}}.$$

Then $\wp({}^{\omega}\omega) \cap \mathbf{L}({}^{\omega}\omega) \subseteq \Gamma^{\kappa}_{\mathrm{Hom}}$.

The proof of Theorem 4.8 makes essential use of the *smallness* of $\wp({}^{\omega}\omega) \cap \mathbf{L}({}^{\omega}\omega)$, in particular the proof exploits the relationship of the determinacy hypothesis that all sets in $\mathbf{L}({}^{\omega}\omega)$ are determined and the existence of (iterable) inner models with infinitely many Woodin cardinals. The following theorem can be proved by the methods of [Lar04], details will appear in [WooC].

THEOREM 4.9 (ZFC). Suppose that κ is strongly compact. Then $\Gamma^{\kappa}_{\mathrm{wHom}}$ has the scale property and for each $A \in \Gamma^{\kappa}_{\mathrm{wHom}}$,

$$\wp({}^{\omega}\omega) \cap \mathbf{L}(A, {}^{\omega}\omega) \subseteq \Gamma_{\mathrm{wHom}}.$$

Building on this, and adapting the core model methods used to prove Theorem 4.8, one can very likely obtain the following theorem, [WooC].

THEOREM 4.10 (ZFC). Suppose that κ is strongly compact. Then $\Gamma^{\kappa}_{\mathrm{Hom}} = \Gamma^{\kappa}_{\mathrm{wHom}}$.

This suggests the following question. Fix a cardinal κ and suppose $A \subseteq {}^{\omega}\omega$ is such that

$$\wp({}^{\omega}\omega) \cap \mathbf{L}(A, {}^{\omega}\omega) \subseteq \Gamma^{\kappa}_{\mathrm{wHom}}.$$

Must $\wp({}^{\omega}\omega) \cap \mathbf{L}(A, {}^{\omega}\omega) \subseteq \Gamma^{\kappa}_{\mathrm{Hom}}$?

If the answer is yes then this would provide a different, and possibly more direct, proof of Theorem 4.10.

Finally, as noted above, the assumption that $\Gamma^{\kappa}_{\mathrm{wHom}}$ is closed under complements cannot imply (modulo inconsistency) that

$$\wp({}^{\omega}\omega) \cap \mathbf{L}_{\omega_1}({}^{\omega}\omega) \subseteq \Gamma^{\kappa}_{\mathrm{Hom}}.$$

Of course if $\Gamma^{\kappa}_{\mathrm{wHom}}$ is closed under complements then necessarily

$$\wp({}^{\omega}\omega) \cap \mathbf{L}_{\omega_1}({}^{\omega}\omega) \subseteq \Gamma^{\kappa}_{\mathrm{wHom}}.$$

Suppose that

$$\wp({}^{\omega}\omega) \cap \mathbf{L}_{\omega_1}({}^{\omega}\omega) \subseteq \Gamma^{\kappa}_{\mathrm{wHom}}.$$

Does this imply that $\wp({}^{\omega}\omega) \cap \mathbf{L}_{\omega_1}({}^{\omega}\omega)$ has the scale property? This may seem unlikely, but there is no known counterexample. For the natural models in which

$$\wp({}^{\omega}\omega) \cap \mathbf{L}_{\omega_1}({}^{\omega}\omega) \subseteq \Gamma^{\kappa}_{\mathrm{wHom}},$$

these are the Lévy collapse extensions given by Theorem 4.7 where the ground model, $\mathbf{V}$, is a reasonably closed extender model; it is true that $\wp({}^{\omega}\omega) \cap \mathbf{L}_{\omega_1}({}^{\omega}\omega)$ has the scale property. The proof of this remarkable fact divides into two cases. If

$$\mathbf{L}_{\omega_1}({}^{\omega}\omega) \models \mathsf{AD}$$

then the scale property holds by the Moschovakis periodicity theorems. If

$$\mathbf{L}_{\omega_1}({}^{\omega}\omega) \not\models \mathsf{AD}$$

then again $\wp({}^{\omega}\omega) \cap \mathbf{L}_{\omega_1}({}^{\omega}\omega)$ has the scale property by an analysis due to Steel. It is for this case that the assumption that the ground model be an extender model seems essential. If one could eliminate this assumption then very likely the answer to the question above is yes.

REFERENCES

Akihiro Kanamori

[Kan94] ***The higher infinite***, Springer-Verlag, Berlin, 1994.

Alexander S. Kechris

[Kec88B] *A coding theorem for measures*, this volume, originally published in Kechris et al. [Cabal iv], pp. 103–109.

Alexander S. Kechris, Donald A. Martin, and John R. Steel

[Cabal iv] *Cabal seminar* 81–85, Lecture Notes in Mathematics, no. 1333, Berlin, Springer, 1988.

Alexander S. Kechris and Yiannis N. Moschovakis

[Cabal i] *Cabal seminar* 76–77, Lecture Notes in Mathematics, no. 689, Berlin, Springer, 1978.

PAUL B. LARSON

[Lar04] ***The stationary tower: Notes on a course by W. Hugh Woodin***, University Lecture Series (AMS), vol. 32, Oxford University Press, Providence, RI, 2004.

YIANNIS N. MOSCHOVAKIS

[Mos70A] *Determinacy and prewellorderings of the continuum*, ***Mathematical logic and foundations of set theory. Proceedings of an international colloquium held under the auspices of the Israel Academy of Sciences and Humanities, Jerusalem, 11–14 November 1968*** (Y. Bar-Hillel, editor), Studies in Logic and the Foundations of Mathematics, North-Holland, Amsterdam-London, 1970, pp. 24–62.

ROBERT M. SOLOVAY

[Sol78B] *The independence of* DC *from* AD, In Kechris and Moschovakis [CABAL i], pp. 171–184.

W. HUGH WOODIN

[WooC] *Category, scales and determinacy*, In preparation.

DEPARTMENT OF MATHEMATICS
UNIVERSITY OF CALIFORNIA
LOS ANGELES, CA 90095, USA
E-mail: dam@math.ucla.edu

DEPARTMENT OF MATHEMATICS
UNIVERSITY OF CALIFORNIA
BERKELEY, CA 94720, USA
E-mail: woodin@math.berkeley.edu

BIBLIOGRAPHY

JOHN W. ADDISON

[Add59] *Some consequences of the axiom of constructibility*, ***Fundamenta Mathematicae***, vol. 46 (1959), pp. 123–135.

JOHN W. ADDISON AND YIANNIS N. MOSCHOVAKIS

[AM68] *Some consequences of the axiom of definable determinateness*, ***Proceedings of the National Academy of Sciences of the United States of America***, no. 59, 1968, pp. 708–712.

JAMES BAUMGARTNER, DONALD A. MARTIN, AND SAHARON SHELAH

[BMS84] ***Axiomatic set theory***. *Proceedings of the AMS-IMS-SIAM joint summer research conference held in Boulder, Colo., June* 19–25, 1983, Contemporary Mathematics, vol. 31, Amer. Math. Soc., Providence, RI, 1984.

HOWARD S. BECKER

[Bec78] *Partially playful universes*, In Kechris and Moschovakis [CABAL i], pp. 55–90.

[Bec79] ***Some applications of ordinal games***, Ph.D. thesis, UCLA, 1979.

[Bec80] *Thin collections of sets of projective ordinals and analogs of* **L**, ***Annals of Mathematical Logic***, vol. 19 (1980), pp. 205–241.

[Bec81] *Determinacy implies that $\aleph_2$ is supercompact*, ***Israel Journal of Mathematics***, vol. 40 (1981), no. 3–4, pp. 229–234.

[Bec85] *A property equivalent to the existence of scales*, ***Transactions of the AMS***, vol. 287 (1985), pp. 591–612.

HOWARD S. BECKER AND ALEXANDER S. KECHRIS

[BK84] *Sets of ordinals constructible from trees and the third Victoria Delfino problem*, In Baumgartner et al. [BMS84], pp. 13–29.

ANDREAS BLASS

[Bla75] *Equivalence of two strong forms of determinacy*, ***Proceedings of the American Mathematical Society***, vol. 52 (1975), pp. 373–376.

DOUGLAS R. BUSCH

[Bus76] *λ-Scales, κ-Souslin sets and a new definition of analytic sets*, ***The Journal of Symbolic Logic***, vol. 41 (1976), p. 373.

CHEN-LIAN CHUANG

[Chu82] ***The propagation of scales by game quantifiers***, Ph.D. thesis, UCLA, 1982.

DANIEL CUNNINGHAM

[Cun90] ***The real core model***, Ph.D. thesis, UCLA, 1990.

MORTON DAVIS

[Dav64] *Infinite games of perfect information*, ***Advances in game theory*** (Melvin Dresher, Lloyd S. Shapley, and Alan W. Tucker, editors), Annals of Mathematical Studies, vol. 52, 1964, pp. 85–101.

QI FENG, MENACHEM MAGIDOR, AND W. HUGH WOODIN

[FMW92] *Universally Baire sets of reals*, In Judah et al. [JJW92], pp. 203–242.

The Cabal Seminar. Volume I: Games, Scales and Suslin Cardinals
Edited by A. S. Kechris, B. Löwe, J. R. Steel
Lecture Notes in Logic, 31

LEO A. HARRINGTON

[Har73] ***Contributions to recursion theory in higher types***, Ph.D. thesis, MIT, 1973.

[Har78] *Analytic determinacy and* $0^{\#}$, ***The Journal of Symbolic Logic***, vol. 43 (1978), pp. 685–693.

LEO A. HARRINGTON AND ALEXANDER S. KECHRIS

[HK77] *Ordinal quantification and the models* $\mathrm{L}[T^{2n+1}]$, Mimeographed note, January 1977.

[HK81] *On the determinacy of games on ordinals*, ***Annals of Mathematical Logic***, vol. 20 (1981), pp. 109–154.

GREGORY HJORTH

[Hjo96] *Two applications of inner model theory to the study of* $\undertilde{\Sigma}^1_2$ *sets*, ***The Bulletin of Symbolic Logic***, vol. 2 (1996), no. 1, pp. 94–107.

[Hjo97] *Some applications of coarse inner model theory*, ***The Journal of Symbolic Logic***, vol. 62 (1997), no. 2, pp. 337–365.

[Hjo01] *A boundedness lemma for iterations*, ***The Journal of Symbolic Logic***, vol. 66 (2001), no. 3, pp. 1058–1072.

STEPHEN JACKSON

[Jac] *Non-partition results in the projective hierarchy*, to appear.

[Jac88] AD *and the projective ordinals*, In Kechris et al. [CABAL iv], pp. 117–220.

[Jac90] *A new proof of the strong partition relation on* ω_1, ***Transactions of the American Mathematical Society***, vol. 320 (1990), no. 2, pp. 737–745.

[Jac91] *Admissible Suslin cardinals in* $\mathrm{L}(\mathbb{R})$, ***The Journal of Symbolic Logic***, vol. 56 (1991), no. 1, pp. 260–275.

[Jac99] ***A computation of*** $\undertilde{\delta}^1_5$, vol. 140, Memoirs of the AMS, no. 670, American Mathematical Society, July 1999.

[Jac07] *Suslin cardinals, partition properties, homogeneity. Introduction to Part II*, this volume, 2007.

[Jac08] *Structural consequences of* AD, In Kanamori and Foreman [KF08].

STEPHEN JACKSON AND DONALD A. MARTIN

[JM83] *Pointclasses and wellordered unions*, In Kechris et al. [CABAL iii], pp. 55–66.

STEPHEN JACKSON AND RUSSELL MAY

[JM04] *The strong partition relation on* ω_1 *revisited*, ***Mathematical Logic Quarterly***, vol. 50 (2004), no. 1, pp. 33–40.

THOMAS JECH

[Jec02] ***Set theory***, second ed., Springer Monographs in Mathematics, Springer, 2002.

RONALD B. JENSEN

[Jen72] *The fine structure of the constructible hierarchy*, ***Annals of Mathematical Logic***, vol. 4 (1972), pp. 229–308.

H. JUDAH, W. JUST, AND W. HUGH WOODIN

[JJW92] ***Set theory of the continuum***, MSRI publications, vol. 26, Springer-Verlag, 1992.

AKIHIRO KANAMORI

[Kan94] ***The higher infinite***, Springer-Verlag, Berlin, 1994.

AKIHIRO KANAMORI AND MATTHEW FOREMAN

[KF08] ***Handbook of set theory into the 21st century***, Springer, 2008.

ALEXANDER S. KECHRIS

[Kec75] *The theory of countable analytical sets*, ***Transactions of the American Mathematical Society***, vol. 202 (1975), pp. 259–297.

[Kec77] AD *and infinite exponent partition relations*, Circulated manuscript, 1977.

[Kec78A] AD *and projective ordinals*, In Kechris and Moschovakis [Cabal i], pp. 91–132.

[Kec78B] *Forcing in analysis*, ***Higher set theory. Proceedings of a conference held at the Mathematisches Forschungsinstitut, Oberwolfach, April 13–23, 1977*** (Gert H. Müller and Dana Scott, editors), Lecture Notes in Mathematics, vol. 669, Springer, 1978, pp. 277–302.

[Kec78C] *On transfinite sequences of projective sets with an application to* $\underset{\sim}{\Sigma}^1_2$ *equivalence relations*, ***Logic colloquium '77*** (A. Macintyre, L. Pacholski, and J. Paris, editors), North-Holland, 1978, pp. 155–160.

[Kec81A] *Homogeneous trees and projective scales*, In Kechris et al. [Cabal ii], pp. 33–74.

[Kec81B] *Suslin cardinals, κ-Suslin sets, and the scale property in the hyperprojective hierarchy*, this volume, originally published in Kechris et al. [Cabal ii], pp. 127–146.

[Kec85] *Determinacy and the structure of* $\mathbf{L}(\mathbb{R})$, ***Proceedings of symposia in pure mathematics***, vol. 42, American Mathematical Society, 1985, pp. 271–283.

[Kec88A] AD + Unif *is equivalent to* $\mathsf{AD}_{\mathbb{R}}^{1/2}$, In Kechris et al. [Cabal iv], pp. 98–102.

[Kec88B] *A coding theorem for measures*, this volume, originally published in Kechris et al. [Cabal iv], pp. 103–109.

Alexander S. Kechris, Eugene M. Kleinberg, Yiannis N. Moschovakis, and W. Hugh Woodin

[KKMW81] *The axiom of determinacy, strong partition properties and nonsingular measures*, this volume, originally published in Kechris et al. [Cabal ii], pp. 75–100.

Alexander S. Kechris, Donald A. Martin, and Yiannis N. Moschovakis

[Cabal ii] *Cabal seminar* 77–79, Lecture Notes in Mathematics, no. 839, Berlin, Springer, 1981.

[Cabal iii] *Cabal seminar* 79–81, Lecture Notes in Mathematics, no. 1019, Berlin, Springer, 1983.

Alexander S. Kechris, Donald A. Martin, and Robert M. Solovay

[KMS83] *Introduction to Q-theory*, In Kechris et al. [Cabal iii], pp. 199–282.

Alexander S. Kechris, Donald A. Martin, and John R. Steel

[Cabal iv] *Cabal seminar* 81–85, Lecture Notes in Mathematics, no. 1333, Berlin, Springer, 1988.

Alexander S. Kechris and Yiannis N. Moschovakis

[KM77] *Recursion in higher types*, ***Handbook of mathematical logic*** (K. J. Barwise, editor), North-Holland, 1977, pp. 681–737.

[Cabal i] *Cabal seminar* 76–77, Lecture Notes in Mathematics, no. 689, Berlin, Springer, 1978.

[KM78B] *Notes on the theory of scales*, this volume, originally published in *Cabal Seminar* 76–77 [Cabal i], pp. 1–53.

Alexander S. Kechris and Robert M. Solovay

[KS85] *On the relative consistency strength of determinacy hypotheses*, ***Transactions of the American Mathematical Society***, vol. 290 (1985), no. 1, pp. 179–211.

Alexander S. Kechris, Robert M. Solovay, and John R. Steel

[KSS81] *The axiom of determinacy and the prewellordering property*, In Kechris et al. [Cabal ii], pp. 101–125.

Alexander S. Kechris and W. Hugh Woodin

[KW83] *Equivalence of determinacy and partition properties*, ***Proceedings of the National Academy of Sciences of the United States of America***, vol. 80 (1983), no. 6 i., pp. 1783–1786.

[KW07] *Generic codes for uncountable ordinals*, this volume, originally circulated manuscript, 2007.

JOHN KELLEY

[Kel55] ***General topology***, The University series in higher mathematics, Van Nostrand, Princeton, NJ, 1955.

RICHARD O. KETCHERSID

[Ket00] ***Toward*** $\mathsf{AD}_{\mathbb{R}}$ ***from the continuum hypothesis and an*** ω_1***-dense ideal***, Ph.D. thesis, Berkeley, 2000.

STEPHEN C. KLEENE

[Kle55] *Arithmetical predicates and function quantifiers*, ***Transactions of the American Mathematical Society***, vol. 79 (1955), pp. 312–340.

EUGENE M. KLEINBERG

[Kle77] ***Infinitary combinatorics and the axiom of determinacy***, Lecture Notes in Mathematics, vol. 612, Springer-Verlag, 1977.

[Kle82] *A measure representation theorem for strong partition cardinals*, ***The Journal of Symbolic Logic***, vol. 47 (1982), no. 1, pp. 161–168.

MOTOKITI KONDÔ

[Kon38] *Sur l'uniformization des complementaires analytiques et les ensembles projectifs de la seconde classe*, ***Japanese Journal of Mathematics***, vol. 15 (1938), pp. 197–230.

KAZIMIERZ KURATOWSKI

[Kur66] ***Topology***, vol. 1, Academic Press, New York and London, 1966.

PAUL B. LARSON

[Lar04] ***The stationary tower: Notes on a course by W. Hugh Woodin***, University Lecture Series (AMS), vol. 32, American Mathematical Society, Providence, RI, 2004.

AZRIEL LÉVY

[Lév66] *Definability in axiomatic set theory*, ***Logic, methodology and philosophy of science. Proceedings of the 1964 international congress.*** (Amsterdam) (Yehoshua Bar-Hillel, editor), Studies in Logic and the Foundations of Mathematics, North-Holland, 1966, pp. 127–151.

N. N. LUZIN AND P. S. NOVIKOV

[LN35] *Choix effectif d'un point dans un complemetaire analytique arbitraire, donne par un crible*, ***Fundamenta Mathematicae***, vol. 25 (1935), pp. 559–560.

RICHARD MANSFIELD

[Man69] ***The theory of*** Σ^1_2 ***sets***, Ph.D. thesis, Stanford University, 1969.

[Man70] *Perfect subsets of definable sets of real numbers*, ***Pacific Journal of Mathematics***, vol. 35 (1970), no. 2, pp. 451–457.

[Man71] *A Souslin operation on* Π^1_2, ***Israel Journal of Mathematics***, vol. 9 (1971), no. 3, pp. 367–379.

DONALD A. MARTIN

[MarA] *Borel and projective games*, to appear.

[MarB] *Games of countable length*, to appear.

[MarC] *On Victoria Delfino problem number* 1, notes.

[MarD] *Weakly homogeneous trees*, Circulated manuscript.

[Mar68] *The axiom of determinateness and reduction principles in the analytical hierarchy*, ***Bulletin of the American Mathematical Society***, vol. 74 (1968), pp. 687–689.

[Mar70A] *Measurable cardinals and analytic games*, ***Fundamenta Mathematicae***, (1970), no. LXVI, pp. 287–291.

[Mar70B] *Pleasant and unpleasant consequences of determinateness*, March 1970, unpublished manuscript.

[Mar73] *Countable Σ^1_{2n+1} sets*, 1973, circulated note.

[Mar75] *Borel determinacy*, ***Annals of Mathematics***, vol. 102 (1975), no. 2, pp. 363–371.

[Mar78] *Infinite games*, ***Proceedings of the international congress of mathematicatians, Helsinki 1978*** (Olli Lehto, editor), Finnish Academy of Sciences, 1978, pp. 269–273.

[Mar83A] *The largest countable this, that, and the other*, this volume, originally published in Kechris et al. [CABAL iii], pp. 97–106.

[Mar83B] *The real game quantifier propagates scales*, this volume, originally published in Kechris et al. [CABAL iii], pp. 157–171.

DONALD A. MARTIN, YIANNIS N. MOSCHOVAKIS, AND JOHN R. STEEL

[MMS82] *The extent of definable scales*, ***Bulletin of the American Mathematical Society***, vol. 6 (1982), pp. 435–440.

DONALD A. MARTIN AND ROBERT M. SOLOVAY

[MS] *Basis theorems for Π^1_{2k} sets of reals*, unpublished.

[MS69] *A basis theorem for Σ^1_3 sets of reals*, ***Annals of Mathematics***, vol. 89 (1969), pp. 138–160.

DONALD A. MARTIN AND JOHN R. STEEL

[MS83] *The extent of scales in* $\mathbf{L}(\mathbb{R})$, this volume, originally published in Kechris et al. [CABAL iii], pp. 86–96.

[MS89] *A proof of projective determinacy*, ***Journal of the American Mathematical Society***, vol. 2 (1989), pp. 71–125.

[MS94] *Iteration trees*, ***Journal of the American Mathematical Society***, vol. 7 (1994), pp. 1–73.

[MS07] *The tree of a Moschovakis scale is homogeneous*, this volume, 2007.

DONALD A. MARTIN AND W. HUGH WOODIN

[MW07] *Weakly homogeneous trees*, this volume, 2007.

WILLIAM J. MITCHELL AND JOHN R. STEEL

[MS94] ***Fine structure and iteration trees***, Lecture Notes in Logic, vol. 3, Springer-Verlag, Berlin, 1994.

YIANNIS N. MOSCHOVAKIS

[Mos67] *Hyperanalytic predicates*, ***Transactions of the American Mathematical Society***, vol. 129 (1967), pp. 249–282.

[Mos70A] *Determinacy and prewellorderings of the continuum*, ***Mathematical logic and foundations of set theory. Proceedings of an international colloquium held under the auspices of the Israel Academy of Sciences and Humanities, Jerusalem, 11–14 November 1968*** (Y. Bar-Hillel, editor), Studies in Logic and the Foundations of Mathematics, North-Holland, Amsterdam-London, 1970, pp. 24–62.

[Mos70B] *The Suslin-Kleene theorem for countable structures*, ***Duke Mathematical Journal***, vol. 37 (1970), no. 2, pp. 341–352.

[Mos71A] *Uniformization in a playful universe*, ***Bulletin of the American Mathematical Society***, vol. 77 (1971), pp. 731–736.

[Mos71B] *Uniformization in a playful universe*, ***Bulletin of the American Mathematical Society***, vol. 77 (1971), pp. 731–736.

[Mos73] *Analytical definability in a playful universe*, ***Logic, methodology, and philosophy of science IV*** (Patrick Suppes, Leon Henkin, Athanase Joja, and Gr. C. Moisil, editors), North-Holland, 1973, pp. 77–83.

[Mos74A] ***Elementary induction on abstract structures***, North-Holland, 1974.

[Mos74B] *Structural characterizations of classes of relations*, ***Generalized recursion theory. Proceedings of the 1972 Oslo symposium*** (Jens Erik Fenstad and Peter G. Hinman, editors), Studies in Logic and the Foundations of Mathematics, North-Holland, 1974, pp. 53–79.

[Mos78] *Inductive scales on inductive sets*, this volume, originally published in Kechris and Moschovakis [CABAL i], pp. 185–192.

[Mos80] ***Descriptive set theory***, Studies in Logic and the Foundations of Mathematics, no. 100, North-Holland, Amsterdam, 1980.

[Mos81] *Ordinal games and playful models*, In Kechris et al. [CABAL ii], pp. 169–201.

[Mos83] *Scales on coinductive sets*, this volume, originally published in Kechris et al. [CABAL iii], pp. 77–85.

JAN MYCIELSKI

[Myc64] *On the axiom of determinateness*, ***Fundamenta Mathematicae***, vol. 53 (1964), pp. 205–224.

[Myc66] *On the axiom of determinateness II*, ***Fundamenta Mathematicae***, vol. 59 (1966), pp. 203–212.

JAN MYCIELSKI AND STANISLAW SWIERCZKOWSKI

[MS64] *On the Lebesgue measurability and the axiom of determinateness*, ***Fundamenta Mathematicae***, vol. 54 (1964), pp. 67–71.

ITAY NEEMAN

[Nee] *Games of length* ω_1, ***Journal of Mathematical Logic***, to appear.

[Nee95] *Optimal proofs of determinacy*, ***The Bulletin of Symbolic Logic***, vol. 1 (1995), pp. 327–339.

[Nee02] *Optimal proofs of determinacy II*, ***Journal of Mathematical Logic***, vol. 2 (2002), pp. 227–258.

[Nee04] ***The determinacy of long games***, de Gruyter Series in Logic and its Applications, vol. 7, Walter de Gruyter, Berlin, 2004.

[Nee06] *Determinacy for games ending at the first admissible relative to the play*, ***The Journal of Symbolic Logic***, vol. 71 (2006), no. 2, pp. 425–459.

[Nee07] *Propagation of the scale property using games*, this volume, 2007.

MITCHELL RUDOMINER

[Rud99] *The largest countable inductive set is a mouse set*, ***The Journal of Symbolic Logic***, vol. 64 (1999), pp. 443–459.

JOSEPH R. SCHOENFIELD

[Sch61] *The problem of predicativity*, ***Essays on the foundations of mathematics*** (Y. Bar-Hillel et al., editors), Magnes Press, Jerusalem, 1961, pp. 132–139.

JACK H. SILVER

[Sil71] *Measurable cardinals and* Δ^1_3 *wellorderings*, ***Annals of Mathematics***, vol. 94 (1971), no. 2, pp. 141–446.

STEPHEN G. SIMPSON

[Sim78] ***A short course in admissible recursion theory***, ***Generalized recursion theory II***, Studies in Logic, vol. 94, North Holland, Amsterdam, 1978.

ROBERT M. SOLOVAY

[Sol66] *On the cardinality of* Σ^1_2 *set of reals*, ***Foundations of Mathematics: Symposium papers commemorating the 60th birthday of Kurt Gödel*** (Jack J. Bulloff, Thomas C. Holyoke, and S. W. Hahn, editors), Springer-Verlag, 1966, pp. 58–73.

[Sol67] *Measurable cardinals and the axiom of determinateness*, Lecture notes prepared in connection with the Summer Institute of Axiomatic Set Theory held at UCLA, Summer 1967.

[Sol70] *A model of set theory in which every set is Lebesgue measurable*, ***Annals of Mathematics***, vol. 92 (1970), pp. 1–56.

[Sol78A] *A* $\utilde{\Delta}^1_3$ *coding of the subsets of* ${}^{\omega}\omega$, In Kechris and Moschovakis [CABAL i], pp. 133–150.

[Sol78B] *The independence of* DC *from* AD, In Kechris and Moschovakis [CABAL i], pp. 171–184.

JOHN R. STEEL

[SteA] *The derived model theorem*, Available at `http://www.math.berkeley.edu/~steel`.

[SteB] *Scales in $K(\mathbb{R})$ at the end of a weak gap*, Preprint, available at `http://www.math.berkeley.edu/~steel`.

[Ste80] *More measures from* AD, mimeographed notes, 1980.

[Ste81A] *Closure properties of pointclasses*, In Kechris et al. [CABAL ii], pp. 147–163.

[Ste81B] *Determinateness and the separation property*, ***The Journal of Symbolic Logic***, vol. 46 (1981), no. 1, pp. 41–44.

[Ste83A] *Scales in* $\mathbf{L}(\mathbb{R})$, this volume, originally published in Kechris et al. [CABAL iii], pp. 107–156.

[Ste83B] *Scales on* Σ^1_1*-sets*, this volume, originally published in Kechris et al. [CABAL iii], pp. 72–76.

[Ste88] *Long games*, this volume, originally published in Kechris et al. [CABAL iv], pp. 56–97.

[Ste95A] $\mathbf{HOD}^{\mathbf{L}(\mathbb{R})}$ *is a core model below* Θ, ***The Bulletin of Symbolic Logic***, vol. 1 (1995), pp. 75–84.

[Ste95B] *Projectively wellordered inner models*, ***Annals of Pure and Applied Logic***, vol. 74 (1995), pp. 77–104.

[Ste96] ***The core model iterability problem***, Lecture Notes in Logic, no. 8, Springer-Verlag, Berlin, 1996.

[Ste05] PFA *implies* $\mathsf{AD}^{\mathbf{L}(\mathbb{R})}$, ***The Journal of Symbolic Logic***, vol. 70 (2005), no. 4, pp. 1255–1296.

[Ste07A] *The length-ω_1 open game quantifier propagates scales*, this volume, 2007.

[Ste07B] *Scales in* $\mathrm{K}(\mathbb{R})$, this volume, 2007.

[Ste08] *An outline of inner model theory*, In Kanamori and Foreman [KF08].

JOHN R. STEEL AND ROBERT VAN WESEP

[SVW82] *Two consequences of determinacy consistent with choice*, ***Transactions of the American Mathematical Society***, (1982), no. 272, pp. 67–85.

M. Y. SUSLIN

[Sus17] *Sur une définition des ensembles mesurables* B *sans nombres transfinis*, ***Comptes Rendus Hebdomadaires des Séances de l'Académie des Sciences***, vol. 164 (1917), pp. 88–91.

ROBERT VAN WESEP

[Van78A] *Separation principles and the axiom of determinateness*, ***The Journal of Symbolic Logic***, vol. 43 (1978), pp. 77–81.

[Van78B] *Wadge degrees and descriptive set theory*, In Kechris and Moschovakis [CABAL i], pp. 151–170.

JOHN VON NEUMANN

[vN49] *On rings of operators, reduction theory*, ***Annals of Mathematics***, vol. 50 (1949), pp. 448–451.

W. HUGH WOODIN

[WooA] unpublished.

[WooB] $\aleph_1$*-dense ideals*, to appear.

[WooC] *Category, scales and determinacy*, In preparation.

[Woo99] ***The axiom of determinacy, forcing axioms, and the nonstationary ideal***, De Gruyter Series in Logic and its Applications, Walter de Gruyter, Berlin, 1999.

A. STUART ZOBLE

[Zob00] ***Stationary reflection and the determinacy of inductive games***, Ph.D. thesis, U.C. Berkeley, 2000.

Lecture Notes in Logic

1. *Recursion Theory.* J. R. Shoenfield. (1993, reprinted 2001; 84 pp.)

2. *Logic Colloquium '90; Proceedings of the Annual European Summer Meeting of the Association for Symbolic Logic, Held in Helsinki, Finland, July 15–22, 1990.* Eds. J. Oikkonen and J. Väänänen. (1993, reprinted 2001; 305 pp.)

3. *Fine Structure and Iteration Trees.* W. Mitchell and J. Steel. (1994; 130 pp.)

4. *Descriptive Set Theory and Forcing: How to Prove Theorems about Borel Sets the Hard Way.* A. W. Miller. (1995; 130 pp.)

5. *Model Theory of Fields.* D. Marker, M. Messmer, and A. Pillay. (First edition, 1996; 154 pp. Second edition, 2006; 155 pp.)

6. *Gödel '96; Logical Foundations of Mathematics, Computer Science and Physics; Kurt Gödel's Legacy. Brno, Czech Republic, August 1996, Proceedings.* Ed. P. Hajek. (1996, reprinted 2001; 322 pp.)

7. *A General Algebraic Semantics for Sentential Objects.* J. M. Font and R. Jansana. (1996; 135 pp.)

8. *The Core Model Iterability Problem.* J. Steel. (1997; 112 pp.)

9. *Bounded Variable Logics and Counting.* M. Otto. (1997; 183 pp.)

10. *Aspects of Incompleteness.* P. Lindstrom. (First edition, 1997; 133 pp. Second edition, 2003; 163 pp.)

11. *Logic Colloquium '95; Proceedings of the Annual European Summer Meeting of the Association for Symbolic Logic, Held in Haifa, Israel, August 9–18, 1995.* Eds. J. A. Makowsky and E. V. Ravve. (1998; 364 pp.)

12. *Logic Colloquium '96; Proceedings of the Colloquium held in San Sebastian, Spain, July 9–15, 1996.* Eds. J. M. Larrazabal, D. Lascar, and G. Mints. (1998; 268 pp.)

13. *Logic Colloquium '98; Proceedings of the Annual European Summer Meeting of the Association for Symbolic Logic, Held in Prague, Czech Republic, August 9–15, 1998.* Eds. S. R. Buss, P. Hájek, and P. Pudlák. (2000; 541 pp.)

14. *Model Theory of Stochastic Processes.* S. Fajardo and H. J. Keisler. (2002; 136 pp.)

15. *Reflections on the Foundations of Mathematics; Essays in Honor of Solomon Feferman.* Eds. W. Seig, R. Sommer, and C. Talcott. (2002; 444 pp.)

16. *Inexhaustibility; A Non-Exhaustive Treatment.* T. Franzén. (2004; 255 pp.)

17. *Logic Colloquium '99; Proceedings of the Annual European Summer Meeting of the Association for Symbolic Logic, Held in Utrecht, Netherlands, August 1–6, 1999.* Eds. J. van Eijck, V. van Oostrom, and A. Visser. (2004; 208 pp.)

18. *The Notre Dame Lectures.* Ed. P. Cholak. (2005; 185 pp.)

19. *Logic Colloquium 2000; Proceedings of the Annual European Summer Meeting of the Association for Symbolic Logic, Held in Paris, France, July 23–31, 2000.* Eds. R. Cori, A. Razborov, S. Todorčević, and C. Wood. (2005; 408 pp.)

20. *Logic Colloquium '01; Proceedings of the Annual European Summer Meeting of the Association for Symbolic Logic, Held in Vienna, Austria, August 1–6, 2001.* Eds. M. Baaz, S. Friedman, and J. Krajíček. (2005; 486 pp.)

21. *Reverse Mathematics 2001.* Ed. S. Simpson. (2005; 401 pp.)

22. *Intensionality.* Ed. R. Kahle. (2005; 265 pp.)

23. *Logicism Renewed: Logical Foundations for Mathematics and Computer Science.* P. Gilmore. (2005; 230 pp.)

24. *Logic Colloquium '03: Proceedings of the Annual European Summer Meeting of the Association for Symbolic Logic, Held in Helsinki, Finland, August 14–20, 2003.* Eds. V. Stoltenberg-Hansen and J. Väänänen. (2006; 407 pp.)

25. *Nonstandard Methods and Applications in Mathematics.* Eds. N. J. Cutland, M. Di Nasso, and D. Ross. (2006; 248 pp.)

26. *Logic in Tehran: Proceedings of the Workshop and Conference on Logic, Algebra, and Arithmetic, Held October 18–22, 2003.* Eds. A. Enayat, I. Kalantari, and M. Moniri. (2006; 341 pp.)

27. *Logic Colloquium '02: Proceedings of the Annual European Summer Meeting of the Association for Symbolic Logic and the Colloquium Logicum, Held in Münster, Germany, August 3–11, 2002.* Eds. Z. Chatzidakis, P. Koepke, and W. Pohlers. (2006; 359 pp.)

28. *Logic Colloquium '05: Proceedings of the Annual European Summer Meeting of the Association for Symbolic Logic, Held in Athens, Greece, July 28–August 3, 2005.* Eds. C. Dimitracopoulos, L. Newelski, D. Normann, and J. Steel. (2007; 272 pp.)

29. *Logic Colloquium '04: Proceedings of the Annual European Summer Meeting of the Association for Symbolic Logic, Held in Torino, Italy, July 25–31, 2004.* Eds. A. Andretta, K. Kearnes, and D. Zambella. (2007; 220 pp.)

30. *Stable Domination and Independence in Algebraically Closed Valued Fields.* D. Haskell, E. Hrushovski, and D. Macpherson. (2007; 190 pp.)

31. *Games, Scales, and Suslin Cardinals: The Cabal Seminar, Volume I.* Eds. A. S. Kechris, B. Löwe, and J. R. Steel. (2008; 445 pp.)

Printed in the United States
By Bookmasters